An Invitation to Fractal Geometry

Fractal Dimensions, Self-Similarity, and Fractal Curves

GRADUATE STUDIES IN MATHEMATICS 247

An Invitation to Fractal Geometry

Fractal Dimensions, Self-Similarity, and Fractal Curves

Michel L. Lapidus
Goran Radunović

AMERICAN MATHEMATICAL SOCIETY
Providence, Rhode Island

2020 *Mathematics Subject Classification.* Primary 11M41, 28A12, 28A75, 28A80, 28B15, 30D10, 35P20, 42B20, 44A05, 58J32; Secondary 11M06, 30D30, 37C30, 37C45, 40A10, 42B35, 42B37, 44A10, 45Q05, 81Q20.

For additional information and updates on this book, visit
www.ams.org/bookpages/gsm-247

Library of Congress Cataloging-in-Publication Data

Names: Lapidus, Michel L. (Michel Laurent), 1956– author. | Radunović, Goran, author.
Title: An invitation to fractal geometry : fractal dimensions, self-similarity and fractal curves / Michel L. Lapidus, Goran Radunović.
Description: Providence, Rhode Island : American Mathematical Society, [2024] | Series: Graduate studies in mathematics, 1065-7339 ; volume 247 | Includes bibliographical references and index.
Identifiers: LCCN 2024036229 | ISBN 9781470476236 (hardcover) | ISBN 9781470478957 (paperback) | ISBN 9781470478964 (ebook)
Subjects: LCSH: Fractals. | AMS: Number theory – Zeta and L-functions: analytic theory – Other Dirichlet series and zeta functions. | Measure and integration – Classical measure theory – Contents, measures, outer measures, capacities. | Measure and integration – Classical measure theory – Length, area, volume, other geometric measure theory. | Measure and integration – Classical measure theory – Fractals. | Measure and integration – Set functions, measures and integrals with values in abstract spaces – Set functions, measures and integrals with values in ordered spaces. | Functions of a complex variable – Entire and meromorphic functions, and related topics – Representations of entire functions by series and integrals. | Partial differential equations – Spectral theory and eigenvalue problems – Asymptotic distribution of eigenvalues and eigenfunctions. | Harmonic analysis on Euclidean spaces – Harmonic analysis in several variables – Singular and oscillatory integrals (Calderón-Zygmund, etc.). | Integral transforms, operational calculus – Integral transforms, operational calculus – General transforms. | Global analysis, analysis on manifolds – Partial differential equations on manifolds; differential operators – Boundary value problems on manifolds. | Number theory – Zeta and L-functions: analytic theory – $\zeta(s)$ and $L(s,\chi)$. | Functions of a complex variable – Entire and meromorphic functions, and related topics – Meromorphic functions, general theory. | Dynamical systems and ergodic theory – Smooth dynamical systems: general theory – Zeta functions, (Ruelle-Frobenius) transfer operators, and other functional analytic techniques in dynamical systems | Dynamical systems and ergodic theory – Smooth dynamical systems: general theory – Dimension theory of dynamical systems. | Sequences, series, summability – Convergence and divergence of infinite limiting processes – Convergence and divergence of integrals. | Harmonic analysis on Euclidean spaces – Harmonic analysis in several variables – Function spaces arising in harmonic analysis. | Harmonic analysis on Euclidean spaces – Harmonic analysis in several variables – Harmonic analysis and PDE. | Integral transforms, operational calculus – Integral transforms, operational calculus – Laplace transform. | Integral equations – Inverse problems – Inverse problems. | Quantum theory – General mathematical topics and methods in quantum theory – Semiclassical techniques, including WKB and Maslov methods.
Classification: LCC QA614.86 .L3644 2024 | DDC 514/.742–dc23/eng/20240905
LC record available at https://lccn.loc.gov/2024036229

Graduate Studies in Mathematics ISSN: 1065-7339 (print)

DOI: https://doi.org/10.1090/gsm/247

10 9 8 7 6 5 4 3 2 1 29 28 27 26 25 24

In memory of my mentor and Ph.D. advisor

Gustave Choquet

and in gratitude to my mentor and Ph.D. advisor

Haïm Brezis.

In gratitude to my mathematics high school teachers

Jacqueline and Pierre Bret.

In memory of my mathematics professor at the Lycée Louis Le Grand

in Paris,

Jean-Daniel Bloch.

In loving memory of my mother and father,

Myriam and Serge Lapidus.

To my love, Odile,

on the eve of our 44th anniversary,

and at the dawn of a new beginning.

Michel L. Lapidus

To my brilliant wife, Marta, with all the love.

Goran Radunović

Contents

Preface

This book—entitled *An Invitation to Fractal Geometry* and subtitled *Fractal Dimensions, Self-Similarity, and Fractal Curves*—is aimed at bright undergraduate students, graduate students, and postdocs, as well as at professional mathematicians and other scientists (among them, physicists, computer scientists, and engineers) who are not necessarily experts in the field of fractal geometry but are curious to learn about its mathematical foundations and its connections with the rest of mathematics, as well as about its relevance to other fields.

In particular, we are assuming very few prerequisites, besides the mathematical maturity and knowledge acquired after having taken several standard mathematics undergraduate courses, such as basic and advanced calculus, differential equations, and (if possible) an elementary introduction to analysis. In fact, most of the necessary background in real analysis (including metric spaces and measure theory) is provided in the main text or in the appendices.

Even for the student or experienced mathematician with a solid background in analysis, some of the topics covered in the book, such as the Hausdorff metric, Hausdorff measure, and the main notions of fractal dimensions (Hausdorff dimension and Minkowski or box dimension), may not be that familiar and are typically not taught at the graduate level, let alone at the undergraduate level. Therefore, they may be of interest to a broad spectrum of readers, especially because unlike in most (or perhaps all) introductory textbooks on fractal geometry, they are treated in detail here. In particular, the construction and the main properties of Hausdorff measure are established and discussed at some length in a rigorous and essentially self-contained manner. A look at the detailed table of contents will also provide the interested reader with significant information about the main topics covered in this book.

The goal of this book is to provide an introduction to the subject of fractal geometry, by laying out its mathematical foundations and illustrating the developments of the general theory via a variety of concrete examples. We focus here on the most basic and emblematic examples, including the ternary Cantor set (and several of its variants), the Cantor curve (or Devil's staircase), the Sierpinski gasket and carpet, the Koch snowflake curve, the Weierstrass curve, as well as (with much less emphasis here) the Peano and Hilbert curves, among other plane-filling curves. From the mathematical point of view, all (but one) of these 'fractals' are examples of self-similar sets—or, more generally, of self-affine sets—defined via suitable iterated function systems (IFSs) comprised of contraction mappings. The Weierstrass curve, a celebrated example of a continuous but nowhere differentiable curve, is the single exception in this list; indeed, it is the attractor of a nonlinear and noncontractive IFS. It still exhibits, however, some kind of extended self-similarity.

As a result, a significant portion of the book is devoted to explaining how such self-similar (or, more generally, 'self-alike') sets can be precisely defined and studied from various points of view. In the process, we try to gently introduce the necessary concepts—such as Hausdorff metric and, later, the notions of Hausdorff measure and dimension. The related, but distinct, notions of Minkowski dimension and Minkowski content are also introduced and studied, because despite their less desirable mathematical properties, they are quite important in practice—as well as in the general theory—and play a key role in many parts of the subject, sometimes without the practitioners being even fully aware of it. In particular, along with various forms of self-similarity, they will be key to the sequel to this book, [**119**]—tentatively called *A Continuation of Fractal Geometry* and subtitled *Complex Dimensions and Vibrations of Fractal Drums, with Applications*—which is devoted to the study of complex fractal dimensions and their many applications in fractal geometry, dynamical systems, spectral geometry, harmonic analysis, and number theory.

More specifically, this second book focuses on the geometry and vibrational spectra of fractal strings and of fractal drums, as well as on the associated fractal zeta functions and their poles (the underlying complex dimensions), which provide detailed information about the intrinsic oscillations of fractal geometries. Fractal tube formulas and other explicit formulas are obtained which illustrate the direct connection between oscillations and complex dimensions, via fractal generalizations of Fourier and power series. The Minkowski dimension, which is a nonnegative real number, turns out to be the only complex dimension located on the real axis and having the largest real part. Fractality is then characterized (or, rather, defined) by the presence of nonreal complex dimensions.

The present book provides much of the purely fractal geometric background needed to comfortably read the second book. However, it is primarily aimed at being read independently of the second book. It shows, in particular, that the birth of fractal geometry is intimately connected with the history of mathematics, especially in the 19th and 20th centuries, via the developments of truly rigorous real analysis and its multiple ramifications, from the discovery of set theory by Cantor to the foundations of geometry as well as of analytic number theory and of Fourier analysis laid out by Fourier, Gauss, Riemann, Dirichlet, and Weierstrass, among many others.

This book grew, in part, out of lectures, Honors, undergraduate, and graduate courses, research and Honors seminars, given by the first author at the University of California, Riverside (as well as, in the early 1990s, at Yale University) over many years, and at research conferences and summer schools throughout the world.

During that period, notes of these lectures or courses were taken by a number of (then) undergraduate, graduate, and Ph.D. students of the first author, including Cheryl A. Griffith, Christina Q. He, Erin P. J. Pearse, John A. Rock, Hung (Tim) Lu, Britta K. Daudert, Scot P. Childress, Vicente Alvarez, Jonathan J. Sarhad, Hafedh Herichi, Nishu Lal, Robert G. Niemeyer, Michael A. Maroun, John Quinn, Timothy L. Cobler, Roby A. Scott, Sean Watson, Andrea Arauza Rivera, Frank Klostler, Edward K. Voskanian, Alexander M. Henderson, Adam Yassine, Therese-Marie Landry, Adam D. Richardson, Matthew D. Overduin, Will Hoffer, Shanna Dobson, Marco Munoz, Russell Ryan, and Gregory P. Leathrum, among others.

We are grateful for their input or feedback, especially that of Robert G. Niemeyer and John A. Rock, who were, at different times, collaborators of the first author on this project—but, unfortunately, had to leave it, due to other professional or personal commitments. We want to acknowledge here their significant initial contributions.

A Guide for the Instructor

We close this preface with a few words of advice for an instructor who would like to use this book as the textbook for an upper-division undergraduate course or a graduate course on fractal geometry (and related topics).

For an upper-division undergraduate course,[1] unless the students are already familiar with the basic notions of topology, metric spaces, and measure theory (including the Lebesgue theory of integration), we suggest beginning with a relatively quick overview of Chapters 2 and 3, followed by Chapters 4

[1]For a graduate course, our recommendation will be similar, except concerning Chapters 2 and 3, as well as Chapters 6 and 7, along with Appendix B.

and 5, with a selection of topics in Chapter 7 (especially in Sections 7.3 and 7.4), and then to proceed with Chapters 8 and 9.

In Chapter 8, as well as elsewhere in the book, the instructor may wish to omit at first the more technical proofs of the theorems. More specifically, here in Chapter 8, the instructor may wish to state some of the main results and properties, as well as to select those he or she wants to actually prove in class. (The proofs in Sections 8.1 and 8.3 are more accessible than later ones in Chapter 8.) The material in Section 8.2 (except for the key examples of the a-string and the Cantor string), as well as in Subsection 8.6.1, Sections 8.7 and 8.8, may be omitted upon a first reading (or in an introductory course).[2]

Chapter 9 alone might require a full course, although it is mostly written in this book as an introduction to the fascinating and open-ended subject of fractal curves—since, after all, most continuous curves are nowhere differentiable and "fractal". We note that the second sequel to this book, [**116**], will discuss many aspects of this topic, along with several closely related ones, in much greater depth.

After having briefly discussed the basic notions of a continuous parametrized curve and of its polygonal approximations (in Section 9.1), the instructor may wish to delve into his or her own selection of topics in Sections 9.4–9.7, regarding the Cantor curve (or Devil's staircase), in Section 9.4, the Koch (or snowflake) curve, in Section 9.5, the Sierpinski gasket and carpet (and their higher-dimensional analogs), in Section 9.6, as well as, finally, the Weierstrass continuous but nowhere differentiable curve,[3] in Section 9.7.[4]

As was alluded to earlier, for a graduate course, our recommendation is much the same as for an undergraduate course, except for the following key differences. We would suggest skipping most (but not all) of the material in Chapters 2 and 3 and referring to it if and when necessary, depending on the students' background in real analysis and measure theory. Two possible exceptions are the material in Section 2.3 on the Contraction Mapping Principle (also known as Picard's or Banach's Fixed Point Theorem), which will be key in Chapter 4 in order to rigorously define the attractor of an iterated function

[2]For an undergraduate course, it may be useful to "review" the twin notions of lower and upper limits (of a sequence or of a function), which are used, especially throughout much of Chapter 8 and are discussed in Appendix A. (Of course, Appendix A can be omitted for a graduate course.) On the other hand, still for an undergraduate course, we advise skipping Appendix B, upon a first reading, and referring the more advanced or brightest students to it, whenever necessary.

[3]The instructor may wish to refer to Section 8.2 (on the interchange of limits and lengths) and to Section 9.3, providing an introduction to the higher-dimensional theory of complex fractal dimensions—which is also the main subject of the latter part of the first sequel to this book, [**119**]—only if and when necessary.

[4]In addition, the Peano and Hilbert curves (among other plane- or space-filling curves) are discussed much more briefly in a few places in this book, including in Subsections 1.4.3 and 9.4.7, but will be addressed in more detail in its sequels, especially [**119**] and [**116**].

system (IFS)—and, in particular, of a self-similar system—as well as, more abstractly, the notion of a self-similar measure. The instructor may also wish to briefly go over parts of Chapter 3 (on Lebesgue integration) and of Appendix B (on Carathédory's approach to measure theory), which will provide the mathematical backbone of much of Chapters 5, 6, and 7 (on Hausdorff measure and its δ-approximations).

Contrary to the case of an undergraduate course, if the instructor is so inclined, he or she may cover not just the introductory Chapter 5 (on Hausdorff measure and dimension) but also give the complete construction of Hausdorff measure provided in this book in Chapters 6 and 7 (along with Appendix B), as well as discuss a selection of its basic properties (covered especially in Sections 7.3 and 7.4).

Finally, when going over parts of Chapters 8 and 9, and if the students are somewhat comfortable with basic complex analysis (of which a "lightning speed review" is given in Subsection 9.3.1), the instructor may want to also discuss the more basic aspects of the theory of complex dimensions for fractal strings (in Section 8.2) and for higher-dimensional fractal drums—introduced in Section 9.3 and used throughout Sections 9.4–9.7. Recall that the theory of complex dimensions will be presented in much greater detail in the sequel to this book, [**119**], as well as (from a different perspective) in parts of [**116**].

Overall—whether it is intended for upper-division undergraduate or graduate students, along with other potential readers—we hope that the material presented in this book will also serve as a learning vehicle for a number of topics in set theory, analysis, measure theory, topology, geometry, number theory, and the history of mathematics, as well as provide the reader with a glimpse of the deep unity, harmony, and beauty of mathematics.

Michel L. Lapidus, Riverside, California, USA
and Goran Radunović, Zagreb, Croatia
April 2024

List of Figures

Chapter 1

Chapter 4

Chapter 5

Chapter 7

Chapter 8

Chapter 9

Conventions

$f(x) = O(g(x))$, as $x \to +\infty$, if $f(x)/g(x)$ is bounded, as $x \to +\infty$,[1] 56

$f(x) = o(g(x))$, as $x \to +\infty$, if $f(x)/g(x)$ tends to zero, as $x \to +\infty$, 56

$f(x) \sim g(x)$, as $x \to +\infty$, if $f(x)/g(x)$ tends to 1, as $x \to +\infty$, 56

$f(x) \asymp g(x)$, as $x \to a$, with $f, g > 0$, if $f(x)/g(x)$ is bounded away from zero and infinity, as $x \to a$, 56

$f(x) \approx g(x)$, if f is approximated by g, 56

$x = [x] + \{x\}$ ($[x]$ and $\{x\}$, integer and fractional parts of $x \in \mathbb{R}$), 329

$A^c = X \setminus A$, complement of the set $A \subseteq X$ in X, 55

$A \setminus B = A - B$, the set of points in A that do not lie in B, 55

$\#A$, cardinality of the finite set A, 81

2^X, the power set (or class of all subsets) of the set X, 74

$\bigsqcup$, disjoint union, 60

$\overline{A} = \mathrm{cl}(A)$, the closure of A (the set of limit points of A), with $A \subseteq \mathbb{R}^m$ or $A \subseteq \mathbb{C}$ (or else $A \subseteq X$, an arbitrary metric space), 55

A^o (or $\mathring{A}$), the interior of A (the set of interior points of A), with $A \subseteq \mathbb{R}^m$ or $A \subseteq \mathbb{C}$ (or else $A \subseteq X$, an arbitrary metric space), 56

$\mathbb{N}$, the set of positive integers, $1, 2, 3, \ldots$, 32

$\mathbb{N}_0 = \mathbb{N} \cup \{0\}$, the set of nonnegative integers, $0, 1, 2, 3, \ldots$, 9

[1] Further notation is provided in the index of symbols at the end of the book.

$\mathbb{Z}$, the ring of integers, $\dots, -3, -2, -1, 0, 1, 2, 3, \dots$, 25

$\mathbb{Q}$, the field of rational numbers, 54

$\mathbb{R}$ and $\mathbb{C}$, the fields of real and complex numbers, respectively, 9 and 39

$(0, +\infty)$, the (multiplicative) half-line, 148

$\overline{\mathbb{R}} = [-\infty, +\infty]$, the extended real line $\mathbb{R} \cup \{\pm\infty\}$, 395

$[0, +\infty] = \overline{\mathbb{R}_+}$, the extended nonnegative half-line, 238

$i = \sqrt{-1}$, the imaginary unit (square root of -1); $e^{i\pi/2} = 1$, $e^{i\pi} = -1$, 25

$s = \sigma + it$, s is a complex number with real part $\sigma = \operatorname{Re}(s)$ and imaginary part $t = \operatorname{Im}(s)$, 25

$\bar{s} = \sigma - it$, the complex conjugate of $s = \sigma + it$, 64

$\log x$, the natural logarithm of $x > 0$, 21

$\log_a x = \log x / \log a$, the logarithm of $x > 0$, with base a, 21

$\mathbb{R}^m$,[2] the m-dimensional Euclidean space, with $m \in \mathbb{N}$, 7

$\{\operatorname{Re}(s) < \alpha\} := \{s \in \mathbb{C} : \operatorname{Re}(s) > \alpha\}$, with $\alpha \in [-\infty, +\infty]$, the open right half-plane $\{s \in \mathbb{C} : \operatorname{Re}(s) > \alpha\}$; if $\alpha = \pm\infty$, it is $\emptyset$ (resp., $\mathbb{C}$), 403

$\liminf_{n\to\infty} x_n = \underline{\lim}_{n\to\infty} x_n$, lower limit of the sequence $\{x_n\}_{n=1}^\infty$, 511

$\limsup_{n\to\infty} x_n = \overline{\lim}_{n\to\infty} x_n$, upper limit of the sequence $\{x_n\}_{n=1}^\infty$, 511

$\liminf_{\varepsilon\to 0^+} f(\varepsilon) = \underline{\lim}_{\varepsilon\to 0^+} f(\varepsilon)$, lower limit of the function $f = f(\varepsilon)$, defined on $(0, +\infty)$, say, 514

$\limsup_{\varepsilon\to 0^+} f(\varepsilon) = \overline{\lim}_{\varepsilon\to 0^+} f(\varepsilon)$, upper limit of the function $f = f(\varepsilon)$, defined on $(0, +\infty)$, say, 514

d_m, Euclidean distance on $\mathbb{R}^m$, 50

$|\boldsymbol{x} - \boldsymbol{y}| = \|\boldsymbol{x} - \boldsymbol{y}\|_m = d_m(\boldsymbol{x}, \boldsymbol{y})$, Euclidean distance from $\boldsymbol{x}$ to $\boldsymbol{y}$, 51

$|\boldsymbol{x}| = \|\boldsymbol{x}\| = |\boldsymbol{x}|_m (= \|\boldsymbol{x}\|_m = d_m(\boldsymbol{x}, \boldsymbol{0}))$, Euclidean norm (or magnitude) of $\boldsymbol{x} \in \mathbb{R}^m$, 51

---

[2]Occasionally, $\mathbb{R}^m$ is equipped with a metric (or norm) other than the Euclidean metric (or norm), in which case it is always specified.

Let no one be allowed
to drive us from the paradise
that Cantor created for us
David Hilbert, 1925

[...] *big whirls have little whirls*
that feed on their velocity
and little whirls have lesser whirls
and so on to viscosity
Rhyme by Lewis F. Richardson
(about the self-similarity of turbulence), 1922.
Cited in [**146**, p. 45]; see also [**512**].

Bottomless wonders spring from simple rules,
which are repeated without end
Benoît B. Mandelbrot

I see it, but I don't believe it!
Georg Cantor, in a letter to Richard Dedekind
[**466**, p. 3], 1877

Part 1

Preliminary Material

Chapter 1

Introduction to Concepts in Fractal Geometry

It is iconic that fractals, many of which were invented as examples of pathological behavior, turn out not to be pathological at all. In fact, they are the rule in the universe. Shapes which are not fractals are the exception.

I love Euclidean geometry but it is quite clear that it does not give a reasonable [re]presentation of the world. Mountains are not cones, clouds are not spheres, trees are not cylinders, neither does lightning travel in a straight line. Almost everything around us is non-Euclidean.

Benoît B. Mandelbrot

I am so in favor of the actual infinite that instead of admitting that Nature abhors it, as is commonly said, I hold that Nature makes frequent use of it everywhere, in order to show more effectively the perfections of its Author. Thus I believe that there is no part of matter which is not—I do not say divisible—but actually divisible; and consequently the least particle ought to be considered as a world full of an infinity of different creatures.

Georg Cantor, as quoted in [**538**]

1.1. Monsters

The word *fractal* was first coined by Benoit Mandelbrot in 1975 to describe mathematical 'monsters' that exhibit highly irregular structure. These objects may have been called monsters because they reveal just how strange, unintuitive, and even ugly the truth can be.

The use of the word monster to describe fractal objects turns out to be rather ironic. To be monstrous is to be unnatural, and yet Mandelbrot's seminal work *The Fractal Geometry of Nature* [**419**] reveals that nature itself exhibits fractal structure in a seemingly endless variety of ways. Even well-disseminated mathematical notions such as limit, continuity, differentiability, and the fundamental theorem of calculus have been shown to describe monstrosities which, historically, have made mathematicians remarkably uncomfortable. These mathematical monsters, studied by the likes of Weierstrass, Cantor, von Koch, Sierpinski, Julia, and Hausdorff, were largely thought to be aberrations not worth further investigation. Nothing could be further from the truth.

Fractal geometry provides an approach to the study of highly irregular natural and mathematical phenomena that defy conventional analytical techniques. The tools of the fractal trade allow for a more complete understanding of not only such mathematical monsters but also the strengths and shortcomings of the conventional ideas. As will be seen in this text, conventional ideas cannot adequately describe the structure of fractals. Moreover, fractal geometry provides a fertile foundation for new ideas regarding the mathematical structure of natural objects and deep relationships between seemingly disparate fields of mathematical study.

1.1.1. A Brief History of Fractal Geometry. A definition for *fractal* first proposed by Benoit Mandelbrot is as follows:

> *A set S is a fractal set if its Hausdorff dimension*
> *is strictly greater than its topological dimension.*

His definition was an attempt to clarify what is meant by the "highly irregular" structure exhibited by mathematical monsters. Unfortunately, and as we will see later on, this definition leaves out examples of objects which many would consider to be fractal, as Mandelbrot knew. In fact, he almost immediately regretted having given such a definition; see [**419**] and its introduction.

Much like the word 'life,' there is no completely agreed upon definition of 'fractal.' Nonetheless, there are various properties such that the more of these a set has, the more likely it is considered to be fractal. Roughly speaking, a set S may be considered fractal if it satisfies most of the following properties:

(i) S and its local structure cannot be readily described with classical analytical tools.

Figure 1.1. A prefractal approximation of the Cantor set C, as described in Example 1.1. As the figure suggests, it is *very* sparse. In fact, it is the complement in $[0, 1]$ of an open set whose total length is 1, and yet it is very richly structured.

(ii) S exhibits some form of self-similarity at all scales.

(iii) S contains intricate structure at all scales.

(iv) The construction of S is straightforward, often (possibly) recursive.

The well-known Cantor set, which we denote by C, exhibits each of the above properties. See Figures 1.1, 1.2, and 1.3, along with Example 1.1. In particular, the Cantor set C contains exact copies of itself at scales of the form 3^{-k} for all nonnegative integers k, contains no nontrivial intervals, has length zero, and is constructed with a simple iterative procedure. Even so, C is so richly structured that it can be put into bijective correspondence with the unit interval $[0, 1]$. The Cantor set plays a central role throughout the text because it allows us to demonstrate many classical results in fractal geometry.

Georg Cantor [**69**] (see also [**68**], [**70**]) introduced his famous set in 1883 in an attempt to find a set whose cardinality is between that of the integers and the real numbers.[1] He was not the only mathematician to directly confront mathematical monsters. For instance, in 1875, Henry Smith [**560**] gave the initial construction of the Cantor set and, in 1872, Karl Weierstrass [**607**] (see also [**608–610**]) presented the construction and proof of a family of continuous but nowhere differentiable functions. History would go on to show that continuous functions on compact intervals are, in large part, nowhere differentiable.

In 1890, Guiseppe Peano [**479**] (see also [**478**]) showed that the unit square $[0, 1]^2$ is the continuous image of the unit interval $[0, 1]$, which is one of the most astonishing (and historically unsettling) mathematical truths to ever be revealed.

In 1904, Helge von Koch [**603**] described what is now known as the Koch curve, a nonrectifiable and nowhere differentiable curve; i.e., it has infinite length and does not have a tangent line at any point. Furthermore, it is contained in a bounded subset of the plane, is the limit of a sequence of natural polygonal approximations, and exhibits exact self-similarity.

In 1915, Waclaw Sierpinski [**551**] shared the construction of what came to be known as the Sierpinski gasket (or Sierpinski triangle), a construction which is very straightforward and yet yields remarkably rich and intricate structure. At about the same time, he also introduced the so-called Sierpinski carpet (or

[1]This is the celebrated (Cantor) Continuum Hypothesis, which—as we now know—cannot be solved within the standard (i.e., Zermelo–Fraenkel) set theory; it is said to be undecidable; see, e.g., [**90**].

Sierpinski square), which—in contrast to the Sierpinski gasket—has striking universality properties.

In 1918, Gaston Julia [**277**] (see also [**244**]) showed that the iteration of certain rational functions produces full families of incredibly ornate fractals, now called Julia sets. Almost simultaneously (in fact, earlier), Pierre Fatou (see, e.g., [**178, 179**]) introduced and studied the same remarkable fractals, including Julia sets and Fatou sets. These beautiful mathematical objects have become the staple of modern complex dynamics.

Benoit Mandelbrot published his seminal book *The Fractal Geometry of Nature* [**419**] in 1983, in which he revealed, among many other things, the elegant connections between Julia sets and what may now be considered as the emblem of fractal geometry: the Mandelbrot set; see Figure 1.13 in Section 1.6.

We will elaborate upon and rigorously discuss several of these famous fractals—along with many others—in due course throughout the text, including in Chapter 9. The latter chapter is dedicated to the study of well-known fractal curves such as the Devil's staircase (or Cantor–Lebesgue curve), in Section 9.4, the Koch and snowflake curves, in Section 9.5, the Sierpinski gasket and carpet, along with their higher-dimensional analogs, in Section 9.6, the Weierstrass nowhere differentiable curve, in Section 9.7—and, finally (but much more briefly), in Subsection 1.4.3 and Remark 9.63 towards the end of Subsection 9.4.7, the Peano and Hilbert curves, along with other plane and space-filling curves.

To begin our mathematical discussion, we consider the seemingly innocuous question discussed in Chapter 5 of [**419**]: "How long is the coast of Britain?"

1.1.2. The Coastline of Britain. What is the perimeter of the state of Colorado? It is about 1,320 miles, and this answer is easy to derive. Colorado is a rectangle 280 miles long and 380 miles wide and its perimeter is, of course, readily computed as twice the length plus twice the width.

Now, determining the length of the coastline of Britain is a far more delicate problem. In particular, the coast of Britain has many twists and sharp turns and it is not smooth along any stretch and at any reasonable scale. As such, attempts to compute its length using classical tools like a simple measuring stick of a fixed length or the more sophisticated use of integration to determine arc length lead to very unsatisfactory results: Regardless of the implemented technique, the length of the coast of Britain seems to be *unbounded*.

Consider a family of maps of Britain which have finer and finer detail. A map at the largest scale may show the larger bays and peninsulas, but not the smaller ones. As the scale decreases, the finer and finer maps reveal smaller and smaller bays, peninsulas, and fluctuations, each adding to the total approximate length of the coast. The result is a sequence of values for the total approximate length of the coast which grows without bound.

If the reader were asked to determine the number of points or the volume of the coast of Britain, the answers would likely be both immediate and uninformative. There are infinitely many points contained in sets of arbitrarily small volume in many mathematical approximations of the coast of Britain. Counting points and measuring area each instinctively seem to be the wrong way to go about measuring the *size* of the coast of Britain. However, given the discussion above, so do attempts to measure its length.

A classic paper with a title similar to that of this subsection is the 1967 *Science* article by Benoît B. Mandelbrot, [**415**]—entitled *How long is the coast of Britain? Statistical self-similarity and fractional dimension*—which (along with the three later books by the same author, [**418**], [**419**] and [**420**]; see also [**422**], [**426**]) has played a key historic role in introducing and stressing the importance of fractality to a broad scientific public.[2]

We have now arrived at a fundamental question which underlies the study of fractal geometry as a whole: What is the proper way to measure a given set?

1.2. A Preliminary Fractal Toolbox

The trouble with determining the length of the coastline of Britain is indicative of a behavior which is a common feature among fractal sets: They behave badly with respect to classical notions of measurement like length, area, volume, or cardinality (the number of points in a set). Such bad behavior may well have led to the nickname of "mathematical monsters" for fractals. However, this moniker belies the fundamental truth that, from a mathematical perspective, the study of fractal geometry was in part born out of the desire to understand the depth and variety of ways in which one may understand the size and structure of a set, function, and other mathematical objects.

1.2.1. Measuring Sets. In the classical setting, a subset S of Euclidean space $\mathbb{R}^m$, where m is a positive integer, is implicitly associated with a mathematical tool which is to be used to provide a measurement of the set in some appropriate way. In particular, sets of the same dimension are to be measured using the same tool: A finite set is measured by counting the number of points contained in the set (generalized with *cardinality*), intervals and other 1-dimensional objects have their lengths measured, surfaces and 2-dimensional objects have area, 3-dimensional objects have volume... but like the coast of Britain, these

[2]One should also stress here the seminal scientific contributions of the British polymath Lewis Fry Richardson regarding the importance of scaling laws and self-similarity, the length of borders and coastlines, turbulence and weather prediction, as well as many other relevant topics; see, e.g., L. F. Richardson [**509–512**] and P. G. Drazin [**146**], along with the appropriate references therein.

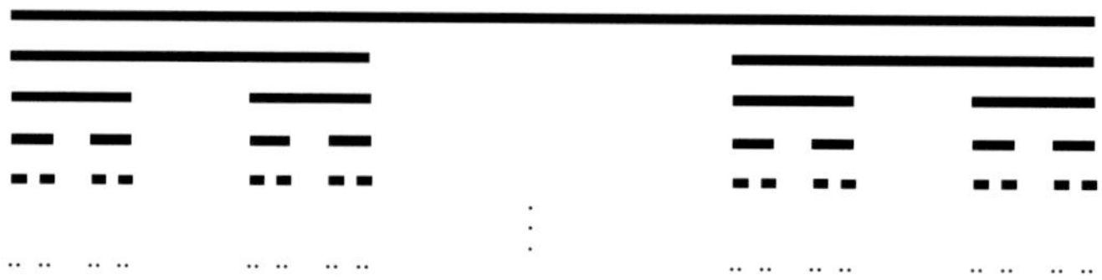

Figure 1.2. The construction of the Cantor set C by repeated "middle-third removal", as described in Example 1.1. The intervals at each step of the construction are enlarged for clarity and the approximation of C is given at the bottom of the figure.

familiar forms of measurement fall short of providing meaningful information when it comes to fractal sets.

For a concrete mathematical example, consider the Cantor set C (see Figures 1.1, 1.2, and 1.3). It is perhaps the simplest example of a fractal set and yet it may also be the most important one because it exhibits all of the key properties that make it indisputably fractal. Among these properties are its defiance in the face of classical notions of measurement like length and cardinality: It has the length of a single point (C has length zero) and yet can it be put into a bijective correspondence with the continuum of real numbers (C is uncountable).

Example 1.1 (*"Middle-Third Removal" Construction of the Cantor Set* C)**.** There are a number of ways to construct C, and perhaps the most well-known is the heuristic middle-third removal construction. This particular construction readily reveals that C has no length.

(i) Consider the unit interval $I = [0,1]$. Remove the open middle-third interval $(1/3, 2/3)$ from I. (ii) Remove the open middle third of the resulting components $[0, 1/3]$ and $[2/3, 1]$. (iii) Continue removing open middle thirds from the remaining components, ad infinitum; see Figure 1.2.

To see that C has length zero, gather the lengths of the open intervals removed in the construction into a sequence $\mathcal{L}_C$:

$$\mathcal{L}_C = (1/3, 1/9, 1/9, 1/27, 1/27, 1/27, 1/27, \ldots). \tag{1.1}$$

By a mild abuse of notation, it will be convenient to identify $\mathcal{L}_C$ as a *multiset* where each of the distinct lengths are counted according to multiplicity as follows:

$$\mathcal{L}_C = \{3^{-n} : 3^{-n} \text{ occurs with multiplicity } 2^{n-1}, n \in \mathbb{N}\}. \tag{1.2}$$

It turns out that the total length of the removed intervals is 1 and the remaining fractal C has length zero; see Example 1.5 below.

Remark 1.2. The sequence of lengths $\mathcal{L}_C$ encodes much more information with regard to the Cantor set C and its complement in the unit interval $[0,1]\backslash C$

than just length. For instance, we will use $\mathcal{L}_C$ to determine various notions of dimension associated with C, including Hausdorff dimension, Minkowski dimension, and complex dimensions. Moreover, $\mathcal{L}_C$ is an example of a *fractal string*; see Subsection 1.3.7 and Definition 1.24. Fractal strings play a key role (i.e., they are important tools in our fractal toolbox) in the approach to fractal geometry described in this text; see Section 8.2 for a brief introduction and, especially, the first part of the sequel to this book, [**119**].

Remark 1.3. Another useful way to understand the structure of the Cantor set C is to consider the ternary (base 3) representation of elements of $[0,1]$. That is, for each $x \in [0,1]$ there is a sequence $\{a_j\}_{j=1}^{\infty}$ of digits, where $a_j \in \{0,1,2\}$ for all $j \in \mathbb{N}$, such that $x = 0.a_1a_2\ldots_3 = \sum_{j=0}^{\infty} a_j/3^j$. Such (ternary or base 3) representations are not unique, in general, but representations with nonterminating expansions (i.e., representations for which only finitely many of the digits a_j are zero) are unique. In this setting, the Cantor set C is the subset of $[0,1]$ whose elements have a nonterminating ternary representation given entirely in terms of 0's and 2's. That is,

$$C = \{x \in [0,1] \,:\, x = 0.a_1a_2\ldots_3\,, a_j \in \{0,2\}, \text{ for all } j \in \mathbb{N}\}. \tag{1.3}$$

The Cantor set C can immediately be put into a bijective correspondence with $\{0,2\}^{\mathbb{N}}$, the set of all one-sided sequences of 0's and 2's. Indeed, removing each element of $[0,1]$ which necessarily has a 1 in its ternary expansion leaves us with the Cantor set C. Employing Cantor's diagonalization allows us to prove that C is uncountable. (This technique is used to show, for instance, that the unit interval $[0,1]$ is uncountable by associating each $x \in [0,1]$ with a binary representation (base 2) in $\{0,1\}^{\mathbb{N}}$.)

In general, one may replace the base 3 in (1.3) with a positive integer $N \geq 2$ and allow $a_j \in \{0,1,\ldots,N-1\}$, for all $j \in \mathbb{N}_0 := \mathbb{N} \cup \{0\}$. This allows for a different encoding of the points in the unit interval $[0,1]$ and emulates the identification of points in the *attractor* of an *iterated function system*; see Chapter 4.

It was claimed above that the length of the Cantor set C is zero, but a precise definition of "length" is needed before this claim can be proved.

1.2.2. Lebesgue Outer Measure. A simple notion of length defined for an interval $I \subseteq \mathbb{R}$ with endpoints a and b is given by the distance $|b-a|$. Although this notion is defined only for intervals, we can use it to create the much more robust *Lebesgue outer measure* which defines a precise notion of length for an arbitrary subset of the real line.

For a closed interval $I = [a,b]$, the nonnegative real number $|b-a|$ denotes the *length* of I. The idea is to cover $A \subseteq \mathbb{R}$ by collections of countably many intervals and define the Lebesgue outer measure of A, denoted $\mathscr{L}^*$, to be the greatest lower bound of the sum of the lengths over all such collections.

Definition 1.4 (Lebesgue Outer Measure on $\mathbb{R}$). The *Lebesgue outer measure*[3] of $A \subseteq \mathbb{R}$, denoted $\mathscr{L}^*(A)$, is given by

$$\mathscr{L}^*(A) = \inf\left\{ \sum_{j=1}^{\infty} |b_j - a_j| \,:\, A \subseteq \bigcup_{j=1}^{\infty} [a_j, b_j], j \in \mathbb{N} \right\}.$$

Example 1.5. To determine $\mathscr{L}^*(C)$, note that the Cantor set C is contained in 2^n intervals of length $1/3^n$, for each $n \in \mathbb{N}_0$; see Figure 1.2. Hence, $0 \leq \mathscr{L}^*(C) \leq (2/3)^n$, for each $n \geq 0$, and so, $\mathscr{L}^*(C) = 0$.

Moreover, recall from Example 1.1 that the Cantor set C is the complement in $[0, 1]$ of a disjoint union of a collection of open intervals with lengths given by

$$\mathcal{L}_C = \{3^{-n} \,:\, 3^{-n} \text{ occurs with multiplicity } 2^{n-1}, n \in \mathbb{N}\}. \tag{1.4}$$

Hence, the set $\Omega_C = [0, 1]\backslash C$, which is called the *Cantor string*, has Lebesgue outer measure given by the sum of the series whose terms are the distinct lengths of $\mathcal{L}_C$ counted with their multiplicities. It turns out that $\mathscr{L}^*([0, 1]\backslash C)$ is given by the sum of a geometric series:

$$\mathscr{L}^*(\Omega_C) = \mathscr{L}^*([0, 1]\backslash C) = \sum_{n=1}^{\infty} \frac{2^{n-1}}{3^n} = \frac{1/3}{1 - 2/3} = 1. \tag{1.5}$$

As we shall see throughout the text, series like this one encode a lot of information (well beyond length) for sets like the Cantor string Ω_C.

Lebesgue outer measure $\mathscr{L}^*$ is a precursor to a family of tools, namely *measures*, which are a significant portion of our fractal toolbox. A bit more specifically, Lebesgue measures can be defined in a completely analogous fashion for subsets of $\mathbb{R}^m$; see Section 3.2.

For the time being, we consider other tools which will prove to be very helpful in our endeavor to understand the structure of fractals.

1.2.3. Self-Similarity. In this subsection, we discuss how a subset $A \subseteq \mathbb{R}^m$ exhibits *self-similarity* when it contains copies of itself at arbitrarily small scales. To make a scaled copy of A, we can take a real-valued scalar $\lambda > 0$ and a vector $\boldsymbol{x} \in \mathbb{R}^m$ and define $\lambda A + \boldsymbol{x} = \{\lambda \mathbf{a} + \boldsymbol{x} \,:\, \mathbf{a} \in A\}$. Then $\lambda A + \boldsymbol{x}$ is a copy of A scaled by the factor λ and translated by $\boldsymbol{x}$. Each of the examples we consider in this subsection contain copies of themselves scaled by arbitrarily small values of λ.

[3]The definition of Lebesgue outer measure on $\mathbb{R}$ requires a restriction of the domain to a σ-algebra of subsets of $\mathbb{R}$ known as the (Lebesgue) measurable sets; see Section 3.2.

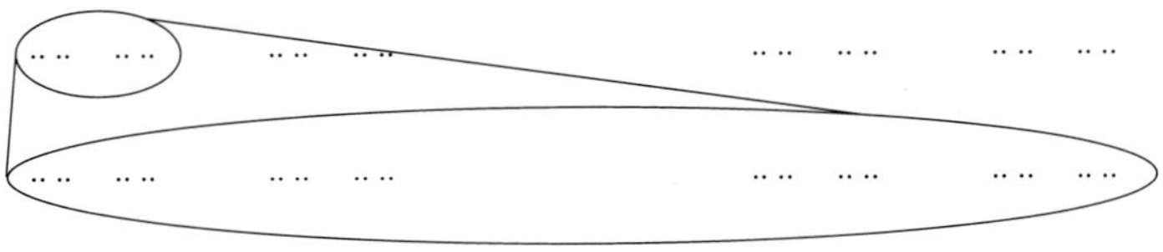

Figure 1.3. The Cantor set is self-similar, as described in Example 1.6. In this picture, we have that $9([0,1/9] \cap C) = C$.

Figure 1.4. The set F_1 is self-similar, as explained in Example 1.7. Here, we have that $4([0,1/4]^2 \cap F_1) = F_1$.

Example 1.6 (*Cantor Set*; [**69**]). The Cantor set C contains *exact* copies of itself at all scales of the form $\lambda_n = 1/3^n$ for each positive integer n; see Figure 1.3. Actually, and more importantly in terms of the precise notion of self-similarity given in Chapter 4, C is exactly the union of 2 disjoint copies of itself, each scaled by a factor of $\lambda = 1/3$. We have that

$$C = \frac{1}{3}C \cup \left(\frac{1}{3}C + \frac{2}{3}\right). \tag{1.6}$$

As we will see, if a set A satisfies a *Similarity Equation* in the way C satisfies (1.6), the number of copies and the corresponding scaling factors provide a wealth of information on the structure of A. In fact, such a Similarity Equation *defines* the corresponding nonempty, compact set A uniquely.

Example 1.7 (*A 1-Dimensional Fractal*). The set denoted F_1 also contains exact copies of itself at all scales of the form $\lambda_n = 1/4^n$ for each positive integer n: see Figure 1.4. In particular, F_1 is exactly the union of 4 disjoint copies of itself, each scaled by a factor of $\lambda = 1/4$. We have that

$$F_1 = \bigcup_{j=1}^{4} \left(\frac{1}{4}F_1 + \mathbf{t}_j\right), \tag{1.7}$$

where $\mathbf{t}_1 = (0,0), \mathbf{t}_2 = (3/4,0), \mathbf{t}_3 = (3/4,3/4)$, and $\mathbf{t}_4 = (0,3/4)$. In fact, as will be seen in Chapter 4 (Theorem 4.17), F_1 is the unique nonempty compact subset of $\mathbb{R}^2$ satisfying the above Similarity Equation (1.7). It turns out that F_1 is a 1-dimensional, totally disconnected, and nowhere dense self-similar set in $\mathbb{R}^2$, of area 0.

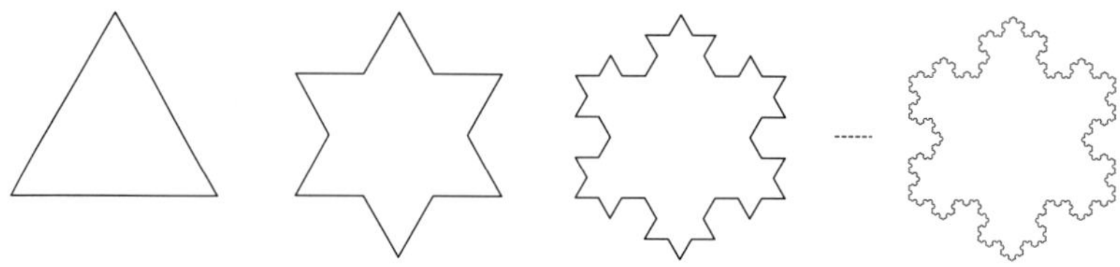

Figure 1.5. A heuristic construction of the Koch snowflake curve $K_S = \partial\Omega_{KS}$, as described in Example 1.8. The first step—namely, replacing a line segment with 4 copies scaled by a factor of 1/3 arranged end to end, as depicted in the figure—indirectly encodes all the information necessary to construct K_S.

Heuristically, the fact that F_1 has fractal dimension 1 can be seen as follows, in the language of Chapters 4 and 5 below.[4] It is a self-similar set and is defined by four similarity transformations, each of scaling ratio 1/4. Hence, its similarity dimension D_S is given by

$$D_S = \frac{\log 4}{\log(1/4)^{-1}} = \frac{\log 4}{\log 4} = 1. \tag{1.8}$$

Example 1.8 (*Koch Snowflake Domain and Curve*; [**603**]). Consider an equilateral triangle (of side lengths 1). Remove the middle third of each side and replace each middle segment with two line segments of length 1/3 that of the original side connected at a 60° angle pointed outward and attached to the endpoints of the remaining segments of the original triangle. Iterate the procedure by replacing each of the new sides with two line segments of length 1/3 pointed outward, repeated ad infinitum. The resulting object and the area it surrounds is called the *Koch snowflake domain*, which we denote by Ω_{KS}. The Koch snowflake domain is a simply connected and bounded open subset of $\mathbb{R}^2$ with finite area but with a boundary, denoted $K_S = \partial\Omega_{KS}$ and called the *Koch snowflake curve*, of infinite length; see Figure 1.5. Indeed, each step of the construction increases the length of the boundary by a factor of 4/3, indicating that the length of $K_S = \partial\Omega_{KS}$ is infinite. Note that Ω_{KS} is not self-similar, in the strict sense, since it does not contains exact scaled copies of itself. However, its boundary $K_S = \partial\Omega_{KS}$ is self-similar. We refer the reader to Example 4.31 where a closely related example, the Koch curve K_C, is defined and investigated.

The Koch snowflake domain Ω_{KS}, its boundary $K_S = \partial\Omega_{KS}$, the Koch snowflake curve, as well as the Koch curve (or von Koch curve) K_C, will all be studied in some detail in Section 9.5, where they will be denoted $\Omega_{KS}, \mathbf{\Gamma}_{KS} = \partial\Omega_{KS}$, and $\mathbf{\Gamma}_{KC}$, respectively; see also Example 4.31.

Example 1.9 (*Sierpinski Gasket*; [**551**], [**554**]). Consider an equilateral triangle with sides of length 1. Remove the open middle inverted equilateral triangle

[4]The full justification of this statement is provided by Moran's Theorem (Theorem 5.40) in Section 5.4.

Figure 1.6. A heuristic construction of the Sierpinski gasket **SG**, as described in Example 1.9. The first step, removing the open middle equilateral triangle from the unit equilateral triangle, indirectly encodes all the information necessary to construct **SG**.

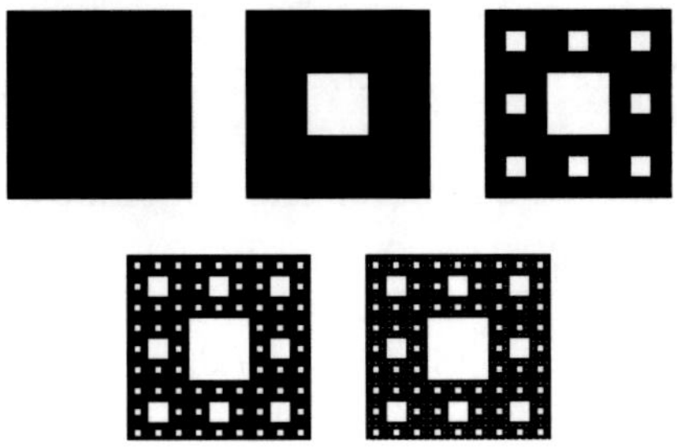

Figure 1.7. A heuristic construction of the Sierpinski carpet **SC**, as described in Example 1.10. The first step, removing the open middle square $(1/3, 2/3)^2$ from the unit square $[0,1]^2$, indirectly encodes all the information necessary to construct **SC**.

whose corners are the midpoints of the sides of the original triangle. This leaves three closed upright equilateral triangles that meet at the midpoints of the sides of the original triangle. Iterate the procedure of removing the middle inverted triangle ad infinitum; see Figure 1.6. The resulting object is the *Sierpinski gasket* **SG**, an intricate self-similar subset of $\mathbb{R}^2$ that has no area (i.e., zero area). Indeed, each step of the construction reduces the total area of the remaining upright triangles by a factor of $3/4$.

The Sierpinski gasket will be further studied in Section 9.6; see also Example 4.33.

Example 1.10 (*Sierpinski Carpet*; [**551**], [**552**]). Consider the unit square $[0,1]^2$. Remove the open middle square with sides of length $1/3$. Then, out of each of the eight remaining squares, remove the open middle square with sides of length $1/9$. Iterate the procedure ad infinitum. The resulting object is called the *Sierpinski carpet*, which we denote by **SC**; see Figure 1.7. The Sierpinski carpet seems to be a richer set, somehow, than the boundary ∂K_S of the Koch snowflake domain K_S. Yet, the area of the Sierpinski carpet is zero as well. Much like the middle-third removal construction of the Cantor set C discussed in Example 1.1 above, all of the girth of the initial set (an interval for C, a square for **SC**) is removed, leaving a sparse set.

The Sierpinski carpet—along with its higher-dimensional counterparts—will be further discussed in Section 9.6 from a geometric and topological point of view; see also Example 4.34 in Section 4.3.

Remark 1.11. In order to analyze self-similar sets and other fractals constructed with relatively simple iterated processes, it is often convenient to use finite (or countable) approximations, much as one would use the sum of the area of a finite number of rectangles to compute (and in fact, define) the integral of a nonnegative real-valued function over some interval. As it turns out, the "skeletons" one obtains by iterating such a process a finite number of times, perhaps starting with a simple object like a singleton or a box, provide very nice approximations to the otherwise intractable fractal objects defined by iterating the process indefinitely. For example, one may take the set of endpoints of the removed intervals at any stage of the construction of the Cantor set C in Examples 1.1 and 1.6 as a nice *prefractal* approximation of C. Moreover, as developed in Chapter 4, such prefractals provide us with a specific notion of *limit* that allows us to concretely define self-similar sets and other fractals.

The Cantor set C, the boundary ∂K_S of the Koch snowflake domain K_S, and the Sierpinski carpet SC are all examples of self-similar sets. Each one of them is poorly understood by using classical notions of measurement, such as length and area. These difficulties give rise to the development of mathematically robust notions of *dimension*.

1.3. Notions of Dimension

How does one know which tool should be used to measure a given set? The sets we have considered so far indicate that the answer to this question may not be easy to find, or even worse, that we do not yet have the appropriate set of tools. Before we address that question, consider another: Given a particular tool (length, area, volume, etc.), to which sets will it assign a meaningful measure? To answer this question, consider some simple objects and the classical notion of volume.

Consider the following subsets of $\mathbb{R}^3$: a singleton $\{x\}$, the unit interval $[0,1]\times\{0\}\times\{0\}$, the unit square $[0,1]\times[0,1]\times\{0\}$, and the unit cube $[0,1]\times[0,1]\times[0,1]$. What is the volume of each of these objects? Of course, the singleton, the line segment, and the square each have no volume (i.e., they have volume zero), and the cube has volume 1. Did that not seem like a strange question? Instinctively, the appropriate tool for measuring the cube is known to be the volume. However, the fact that three very different sets each have volume zero indicates that volume is somehow the *wrong* tool for the purpose of measuring sets which are not like the cube. Indeed, volume is a tool best reserved for measuring sets which are 3-*dimensional*. In fact, this notion of volume is really 3-dimensional Lebesgue outer measure, as specified in Section 3.2.

So, what do we mean by *dimension* and how does the dimension of a set relate to the way in which one could (or should) measure the set? Throughout this text, we provide a variety of answers to these questions, but first we lay out some properties that a notion of dimension should satisfy.

1.3.1. Typical Properties of a Dimension.

Suitable notions of dimension are expected to satisfy most of the following properties. Let X denote a metric space (say, $X = \mathbb{R}^m$) and let dim denote an unspecified notion of dimension on X, taking nonnegative values on all (suitable) nonempty subsets of X.

(i) *Normalization*: $\dim(\emptyset) = -\infty$.[5]

(ii) *Monotonicity*: If $A \subseteq B$, then $\dim(A) \leq \dim(B)$.

(iii) *Stability*:

(a) *Finite Stability*: If $A, B \subseteq X$, then

$$\dim(A \cup B) = \max\{\dim(A), \dim(B)\}.$$

(b) *Countable Stability*: If $A_n \subseteq X$ for all $n \in \mathbb{N}$, then

$$\dim\left(\bigcup_{n=1}^{\infty} A_n\right) = \sup\{\dim(A_n) : n \geq 1\}.$$

Countable stability is sometimes referred to as *σ-stability*. Furthermore, it is easy to check that countable stability implies finite stability, whereas the converse statement is clearly not true, in general.

(iv) *Bi-Lipschitz Invariance*: If $A \subseteq X$ and $f : X \to X$ is bi-Lipschitz,[6] then $\dim(A) = \dim(f(A))$.

(v) *Range of Values*: If $X = \mathbb{R}^m$ and $A \subseteq X$, then $0 \leq \dim(A) \leq m$.

Also, if X is an m-dimensional space, in the sense of Lebesgue (for example, if $X = \mathbb{R}^m$), and if $\dim_T$ denotes the *topological dimension* associated with the metric space X, then $0 \leq \dim_T(A) \leq \dim(A) \leq m$.[7]

(vi) *Open Sets*: If Ω is a nonempty open subset of $X = \mathbb{R}^m$, then we have that $\dim(\Omega) = m$. In particular, $\dim(\mathbb{R}^m) = m$.

Notions of dimension that satisfy the properties (i), (ii), (iii)(a), or (iii)(b), are said to be *normal, monotonic, finitely stable,* or *countably stable*, respectively.

Remark 1.12. The three primary (real-valued) notions of dimension studied in this text are the Hausdorff, box-counting, and Minkowski dimensions. These are denoted $\dim_H$, $\dim_B$, and $\dim_M$, respectively.

---

[5] See Subsection 9.3.8 (and [**119**]) for an explanation of our choice of convention; we note that the usual convention adopted in the literature is $\dim(\emptyset) = -1$.

[6] See Definition 2.30 in Section 2.2 for the notion of a bi-Lipschitz map.

[7] In part (v), we assume that $A \neq \emptyset$. Indeed, by convention, $\dim_T(\emptyset) = \dim(\emptyset) := -\infty$.

Hausdorff dimension satisfies all of the above properties (i)–(vi), whereas Minkowski dimension—and hence also box-counting dimension, by the last equality in (1.9) below—satisfies all of these properties, except for the countable stability (i.e., (iii)(b)); indeed, it is only finitely stable, in the sense of (iii)(a).

Hausdorff dimension is not easily defined, but it is a very powerful mathematical tool. The definition and a thorough discussion of Hausdorff dimension can be found in Chapter 5; see, in particular, Definition 5.13 in Section 5.2. Preliminary versions of the box-counting (or box) and Minkowski dimensions are provided in the present chapter in Definitions 1.21 and 1.22 below, respectively, but a more thorough analysis of these notions begins in Chapter 8; see, in particular, Section 8.1 (including Definition 8.3), Section 8.3 (including Definition 8.22), and Sections 8.4–8.5.

It is well known that if $X = \mathbb{R}^m$, then $\dim_B = \dim_M$, as proved in Section 8.4 (Theorem 8.29). Moreover, for any bounded and nonempty set $A \subseteq \mathbb{R}^m$, we have that

$$0 \leq \dim_T(A) \leq \dim_H(A) \leq \dim_B(A) = \dim_M(A) \leq m. \tag{1.9}$$

Note that, still for $A \neq \emptyset$, $\dim_T(A)$ takes its values in $\mathbb{N}_0$ (or, really, in $\{0, \ldots, m\}$), whereas $\dim_H(A)$, $\dim_B(A)$, and $\dim_M(A)$ take their values in $[0, +\infty)$ (or, really, in $[0, m]$). An example for which the above inequalities in (1.9) are strict is provided in Example 1.16 below. Also, it turns out that the Hausdorff, box-counting, and Minkowski dimensions are all normal, monotonic, bi-Lipschitz invariant, and *finitely* stable, but that the box-counting and (equivalently) Minkowski dimensions are not *countably* stable. These and related facts are discussed in Chapters 5 and 8, accordingly.

In some texts, the requirement that $\dim(\Omega) = m$ in part (vi) above is replaced with $0 \leq \dim(\Omega) \leq m$. This is done, for instance, in the context of the Assouad dimension. For the latter notion of dimension, see the original paper by P. Assouad [**12**] and the book by J. M. Fraser [**199**].

In anticipation of Chapter 5 and Chapter 8, let us consider the following simple example, which shows that—in contrast to the Hausdorff dimension—the Minkowski (and the box) dimensions are not countably stable.

Example 1.13. Let $X = \mathbb{R}$ and

$$A_n := \left\{\frac{1}{n}\right\}, \text{ for each } n \in \mathbb{N}, \text{ and } A := \bigcup_{n=1}^{\infty} A_n = \left\{\frac{1}{n} : n \geq 1\right\}.$$

Then, $\dim_M(A_n) = \dim_B(A_n) = \dim_H(A_n) = 0$, for all $n \in \mathbb{N}$, but $\dim_M(A) = \dim_B(A) = 1/2$, which is strictly greater than

$$\sup\{\dim_M(A_n) = \dim_B(A_n) : n \geq 1\} = 0.$$

By contrast, we have $\dim_H(A) = \sup\{\dim_H(A_n) : n \geq 1\} = 0$, since A is countable and in accord with the countable stability of the Hausdorff dimension.

Example 1.14.

(i) Verify the assertions made in Example 1.13.

(ii) Extend Example 1.13 to the case when $A := \{1/n^a : n \in \mathbb{N}\}$, with $a > 0$ arbitrary, as well as to the case when $X := \mathbb{R}^m$, for any given $m \in \mathbb{N}$.

[*Hint*: In the latter situation, consider, for instance, $B_n := A_n \times [0,1]^{n-1}$.]

Exercise 1.15. Let dim be a notion of dimension which is monotonic. Then show that if $A_n \subseteq X$, for all $n \in \mathbb{N}$, then we always have the following inequality: $\sup\{\dim(A_n) : n \geq 1\} \leq \dim\left(\bigcup_{n=1}^{\infty} A_n\right)$.

Example 1.16. Let $S_1 := \{-1/n : n \geq 1\}$ and S_2 be the (unique) self-similar subset of the unit interval $[0,1]$ that satisfies $S_2 = \frac{1}{8}S_2 \cup \left(\frac{1}{8}S_2 + \frac{7}{8}\right)$. Now, let $S \subseteq [-1,1]$ be given by $S := S_1 \cup S_2$. Then, as shown below, we have that $\dim_T(S) < \dim_H(S) < \dim_B(S)$. Also, it can be proved that $\dim_T(S_1) = \dim_T(S_2) = 0$—and thus, by the finite additivity of topological dimension, we have $\dim_T(S) = 0$.

Furthermore, since S_1 is the union of countably many singletons $\{-1/n\}$, each of which satisfies $\dim_H(\{-1/n\}) = 0$, and Hausdorff dimension is countably stable, we have that $\dim_H(S_1) = 0$. Also, by Moran's Theorem (see Theorem 5.40), we have that $\dim_H(S_2) = 1/3$ since $t := 1/3$ is the unique positive real number that satisfies the equation $2/8^t = 1$. So, by the finite additivity of Hausdorff dimension, we have $\dim_H(S) = 1/3$.

Next, and again by Moran's Theorem (Theorem 5.40), we have $\dim_B(S_2) = 1/3$. Alternately, this can be shown via a direct computation (using Definition 1.21 and a calculation similar to that in (1.13) below) since S_2 can be covered by 2^n closed intervals of length $1/8^n$ for each $n \in \mathbb{N}$. Also, as stated in Example 1.13 and shown in Chapter 8, Section 8.2, we have that $\dim_B(S_1) = 1/2$, despite the fact that singletons have box-counting dimension equal to zero. (Recall from Example 1.13 that S_1 provides an example of a set for which countable stability is not satisfied by the box-counting dimension.) So, by the finite stability of box-counting dimension, $\dim_B(S) = 1/2$—and hence, as claimed above, we conclude that $\dim_T(S) = 0 < \dim_H(S) = 1/3 < \dim_B(S) = 1/2$.

1.3.2. Topological Dimension. The first specific notion of dimension of a set we consider is tied to the topology of the space in which the set is embedded. We need several definitions.

Definition 1.17 (*Open Cover and Refinement*). Let A be a subset of a topological space X. A collection $\mathcal{U}$ of open subsets of X is an *open cover* of A if

$A \subseteq \bigcup_{U \in \mathcal{U}} U$. An open cover $\mathcal{U}'$ of A is a *refinement* of $\mathcal{U}$ if every $U' \in \mathcal{U}'$ is a subset of some $U \in \mathcal{U}$. In this case, $\mathcal{U}'$ is said to *refine* $\mathcal{U}$.

Definition 1.18 (*Topological Dimension*). Let A be a nonempty subset of a topological space X and $n \in \mathbb{N}_0$. The *topological dimension* of A, denoted $\dim_T(A)$, is equal to n if n is the smallest nonnegative integer such that every finite open cover $\mathcal{U}$ of A admits a finite refinement $\mathcal{U}$ of A in which no point of A is included in more than $n+1$ elements; in this case, we let $\dim_T(A) := n$. If no such minimal $n \in \mathbb{N}_0$ exists, A is said to have *infinite topological dimension* and we then write $\dim_T(A) = +\infty$. Otherwise, A is said to be *n-dimensional*. Furthermore, by convention, we let $\dim_T(\emptyset) := -\infty$.

Topological dimension aligns with intuition in that it identifies finite sets as 0-dimensional, line segments as 1-dimensional, squares as 2-dimensional, and cubes as 3-dimensional. Moreover, sets which are *homeomorphic* to (roughly, smooth and nontrivial images of) a finite set, a line segment, a square, or a cube, are 0-, 1-, 2-, or 3-dimensional, respectively. Hence, topological dimension can be used to determine which measuring tool (cardinality, length, area, volume, etc.) is appropriate for a given set.

In particular, also in agreement with one's intuition, a point has topological dimension 0, a line has topological dimension 1, while a plane—and, more generally, a k-dimensional linear subspace of $\mathbb{R}^m$, for any integer k in $\{0, 1, \ldots, m\}$—has topological dimension 2 and k, respectively.

Now, what does topological dimension have to say about the fractals we have seen thus far? As a *totally disconnected* subset of $[0, 1]$, the Cantor set C contains no (nonempty) intervals and is, in fact, 0-dimensional. Also, since the boundary ∂K_S of the Koch snowflake domain K_S and the Sierpinski carpet **SC** have no area but contain line-like segments, ∂K_S and **SC** are both 1-dimensional.

So, combined with the fact that $\mathcal{L}^*(C) = 0$,[8] it seems that an appropriate tool for measuring the size of C is cardinality. However, we will show that C can be put into a bijective correspondence with the unit interval $[0, 1]$; hence, C is uncountably infinite. Similarly, ∂K_S is infinitely long as well as **SC**; so despite being 1-dimensional, length does not seem to be the appropriate tool for measuring the size of ∂K_S and **SC**.

The monstrous structure of the Cantor set C is being revealed. Topological dimension tells us that cardinality is the tool that should be used to measure C. However, cardinality sees C as having the same size as the unit interval $[0, 1]$: they are both uncountable; more specifically, they both have the cardinality of the continuum, that is, the cardinality of the real line. On the other hand,

[8] See Definition 1.4 and Example 1.5 above.

Lebesgue outer measure $\mathcal{L}^*$ (see Definition 1.4 above)[9] sees C as having the same length as the empty set $\emptyset$ ($\mathcal{L}^*(C) = \mathcal{L}^*(\emptyset) = 0$), or even, as any finite or countable set. Similar unsatisfactory results hold for many of the self-similar sets defined above and in Chapter 4 below.

The notion of topological dimension will be further studied in Section 8.6, where it will also be compared with several key notions of fractal dimensions.

We next seek refined notions of measurements in order to assign meaningful values for the size of these sets and refined notions of dimensions to tell us which notion of measurement is appropriate for a given set.

1.3.3. Counting Boxes. For certain sets, there is a relationship between the number of boxes of a specified size which are required to cover a given set and the topological dimension of the set. Specifically, for a given set $A \subseteq \mathbb{R}^3$, let $\varepsilon > 0$ be the common side lengths of the cubes, let N_ε be the minimum number of cubes with sides of length $\varepsilon > 0$ which are required to cover A, and let D_T denote the topological dimension of A. Consider Table 1.1.

Table 1.1. Notions of dimensions and counting boxes.

set	ε	N_ε	D	D_T
singleton	1/3	1	0	0
unit interval	1/3	3	1	1
unit square	1/3	9	2	2
unit cube	1/3	27	3	3
Cantor set C	1/3	2	$\log_3 2$	0

With the exception of the Cantor set C, the topological dimension of each set in the table is given by the real solution D of the following equation:

$$N_\varepsilon = \varepsilon^D. \tag{1.10}$$

Furthermore, the value $\varepsilon = 1/3$ is not special. Whenever $\varepsilon = 1/n$ for some positive integer n, (1.10) holds when A is either a singleton, the unit interval $[0,1]$ (embedded in $\mathbb{R}^3$ as $[0,1] \times \{0\} \times \{0\}$), the unit square $[0,1]^2$ (embedded in $\mathbb{R}^3$ as $[0,1] \times [0,1] \times \{0\}$), or the unit cube $[0,1]^3$.

Formula (1.10) itself does not provide a useful definition of dimension. Indeed, the line segment $[0,2] \times \{0\} \times \{0\}$ (which has topological dimension 1) is covered by a minimum number $N_{1/n} = 2n$ of cubes of side lengths $1/n$. Thus, each $n \in \mathbb{N}$ yields a different value for D as a (real) solution to (1.10), given by $D = \frac{\log N_\varepsilon}{-\log \varepsilon} = \frac{\log 2n}{\log n} = \frac{\log 2}{\log n} + 1$, where log denotes the natural logarithm.

[9]See also Chapter 3, especially Section 3.4.

Nonetheless, with various refinements, (1.10) leads to a couple of very useful notions of dimension, namely the *similarity* and *box-counting* dimensions.

Note one last bit of information regarding (1.10) before discussing more precise notions of dimension. We have seen that the Cantor set C seems to be neither 0-dimensional nor 1-dimensional. To elaborate, note that in the case of the Cantor set C, setting $\varepsilon = 3^{-n}$ yields $N_{3^{-n}} = 2^n$ and so, $0 < D = \log_3 2 = \log 2/\log 3 < 1$ for each $n \in \mathbb{N}$. We have come upon a key concept in fractal geometry: *The dimension of a set need not be an integer.*

1.3.4. Similarity Dimension. Consider the repeated copy, reduce, and paste procedure for the construction of the Koch snowflake curve ∂K_S sketched out in Example 1.8 above. Analogous procedures work for the Cantor set C and the Sierpinski carpet **SC** as well. As discussed immediately below in the case of C, each step consists of replacing a given interval with 2 copies scaled by a factor of 1/3. For **SC**, each step consists of replacing a given square with 8 copies with side lengths scaled by 1/3; see Figure 1.7 in Subsection 1.2.3 above.

The number of copies, along with the amount by which the original image is reduced, can be used to define a notion of dimension for the resulting set. Moreover, this notion of dimension agrees with the topological dimension in many cases and provides more meaningful values in others.

For each $n \in \mathbb{N}$, the unit interval $[0,1]$ is identical to n $(1/n)$-scaled copies of itself where the copies are pasted together end to end. Similarly, the interval $[0,2]$ is identical to n copies of itself, where each copy is $(1/n)$-th the size of the original. The unit square $[0,1] \times [0,1]$ is identical to n^2 copies of itself, where each copy is scaled (linearly) by $1/n$ and the copies are pasted together accordingly.

The Cantor set C is identical to two copies of itself scaled down by a factor of 1/3 and placed a length of 1/3 away from each other; see Figures 1.2 and 1.3. Analogously, each 'side' of the Koch snowflake curve ∂K_S is identical to four scaled copies of itself, each scaled (linearly) by a factor of 1/3, pasted together accordingly; see Figures 1.2 and 1.5 in Subsections 1.2.1 and 1.2.3 above, respectively.

The discussion above leads to the notion of *similarity dimension* for self-similar sets. We refrain from giving the technical definition of similarity dimension until Chapter 4, but loosely speaking, if a set A is equal to the union of $N \in \mathbb{N}\backslash\{1\}$ nearly disjoint copies of itself with scales given by r_j, for $j = 1, \ldots, N$, then the similarity dimension of A is the unique real (and necessarily positive) solution D_S of the *Moran equation*

$$\sum_{j=1}^{N} r_j^t = 1, \quad t \in \mathbb{R}. \tag{1.11}$$

Example 1.19. Table 1.1 yields the following with respect to (1.11):

(i) For $[0,1]$, $3(1/3)^t = 1$ implies $D_S = D = 1$.

(ii) For $[0,1]^2$, $9(1/3)^t = 1$ implies $D_S = D = 2$.

(iii) For $[0,1]^3$, $27(1/3)^t = 1$ implies $D_S = D = 3$.

(iv) For the Cantor set C, in Examples 1.1 and 1.5, we have $2(1/3)^t = 1$; hence, we have that $0 = D_T < D_S = D = \log_3 2 < 1$.

(v) For the set F_1, in Example 1.7, we have $4(1/4)^t = 1$; thus, we have that $0 = D_T < D_S = D = 1$.

(vi) For the Koch snowflake curve (or boundary of the Koch snowflake domain) ∂K_S, in Example 1.8, we have $4(1/3)^t = 1$; hence, $1 = D_T < D_S = D = \log_3 4 < 2$.

(vii) For the Sierpinski gasket **SG**, in Example 1.9, we have $3(1/2)^t = 1$; and so, we have that $1 = D_T < D_S = D = \log_2 3 < 2$.

(viii) For the Sierpinski carpet **SC**, in Example 1.10, we have $8(1/3)^t = 1$; thus, we have that $1 = D_T < D_S = D = \log_3 8 < 2$.

Note that in the cases of C, F_1, ∂K_S, **SG**, and **SC**, the similarity dimension exceeds the topological dimension. This agrees, at least in spirit, with the definition of fractality first posited by Mandelbrot; see Subsection 1.1.1.

A significant limitation of similarity dimension is that it only applies to self-similar sets. The next notion of dimension we discuss, the *box-counting dimension*, is defined for arbitrary bounded sets. One useful property of the box-counting dimension is that it agrees with the similarity dimension for any given (sufficiently nice) self-similar set. Moreover, it is one of the most important notions of dimension encountered in this text—and, arguably, the most important one in numerical, physical, engineering, geological, and biological applications.

1.3.5. Box-Counting Dimension.

The preliminary tools needed to define *box-counting dimension* are essentially the same as those used in (1.10). For a set $A \subseteq \mathbb{R}^m$ and $\varepsilon > 0$, let $N_\varepsilon(A)$ denote the smallest number of cubes (in dimension m) with sides of length ε whose union contains A.

Remark 1.20. Throughout the text, the notation $N_\varepsilon(A)$ is used to denote various types of counting functions, depending on the context. For instance, on occasion $N_\varepsilon(A)$ denotes the maximum number of disjoint closed balls with radii ε and centers in A.

Definition 1.21 (*Box-Counting Dimension*)**.** For a nonempty bounded set $A \subseteq \mathbb{R}^m$, the *box-counting dimension* (or *box dimension*, in short) of A, denoted

$\dim_B(A)$, is given by the following limit (when it exists):[10]

$$\dim_B(A) = \lim_{\varepsilon\to 0^+} \frac{\log N_\varepsilon(A)}{-\log \varepsilon}. \tag{1.12}$$

Furthermore, by convention, we let $\dim_B(\emptyset) := -\infty$.

The box-counting dimension is central to our approach to fractal geometry and it has many interesting properties, but we refrain from delving too deeply into these topics until Chapter 8. For now, we shall simply determine the box-counting dimension for some of the sets we have considered up to this point.

Consider, once again, the Cantor set C. For each $n \in \mathbb{N}$, C is covered by 2^n closed intervals of length 3^{-n}. Therefore,

$$\dim_B(C) = \lim_{n\to\infty} \frac{\log N_{3^{-n}}(C)}{-\log 3^{-n}} = \lim_{n\to\infty} \frac{\log 2^n}{\log 3^n} = \log_3 2. \tag{1.13}$$

Note that the similarity dimension and the box-counting dimension of the Cantor set are equal: $D_S = \log_3 2 = \dim_B(C)$.

Analogously, $\dim_B(\partial K_S) = \log_3 4$, $\dim_B([0,1]) = 1$, $\dim_B([0,1]^3) = 3$, etc. In fact, we will see that, for (nonempty) open sets, box-counting dimension agrees with topological dimension and for suitable self-similar sets, a result of Moran's Theorem (Theorem 5.40 in Chapter 5), box-counting dimension agrees with similarity dimension.

In the next section, we discuss another important tool which appears in various places and in various forms throughout the text.

1.3.6. Minkowski Dimension. Another very useful notion of fractal dimension — that will turn out to coincide with the box-counting dimension (see Section 8.4)— is the Minkowski dimension, which will be the central topic of Chapter 8.[11]

Definition 1.22 (*Minkowski Dimension*)**.** For a nonempty bounded set $A \subseteq \mathbb{R}^m$, the *Minkowski dimension* of A, denoted $\dim_M(A)$, is given by the following limit (when it exists):[12]

$$\dim_M(A) = m - \lim_{\varepsilon\to 0^+} \frac{\log \mathrm{vol}_m(A_\varepsilon)}{\log \varepsilon}, \tag{1.14}$$

where A_ε is the ε-neighborhood of A comprising all of the points in $\mathbb{R}^m$ within ε of A. Furthermore, by convention, we let $\dim_M(\emptyset) := -\infty$.

[10] Note that for $0 < \varepsilon < 1$, we have that $\log \varepsilon < 0$ and hence, $-\log \varepsilon = \log(1/\varepsilon) > 0$.

[11] See, especially, Section 8.1 (including Theorem 8.10), along with Sections 8.4–8.6.

[12] It is also sometimes called the Minkowski–Bouligand dimension of A, due to the significant work on this subject by Georges Bouligand in **[56]**.

Remark 1.23.

(a) In Chapter 8, we shall see alternate definitions of the Minkowski dimension (or Minkowski–Bouligand dimension), $\dim_M(A)$, along with refined notions of upper and lower Minkowski dimensions, $\overline{\dim}_M(A)$ and $\underline{\dim}_M(A)$, expressed in terms of upper and lower limits (as $\varepsilon \to 0^+$) of suitable quantities involving $\mathrm{vol}^m(A_\varepsilon)/\varepsilon^{m-d}$, with $d \geq 0$. Then, $\dim_M(A)$ is said to exist when $\overline{\dim}_M(A) = \underline{\dim}_M(A)$. Furthermore, in this case, $\dim_M(A)$ is defined as being equal to this common value.

(b) Heuristically, we clearly have that, for all sufficiently small $\varepsilon > 0, N_\varepsilon(A) \approx \frac{\mathrm{vol}^m(A_\varepsilon)}{\varepsilon^{m-D}}$, where $D := \dim_M(A)$. In fact, we will see in Theorem 8.29 of Chapter 8 that $\dim_M(A) = \dim_B(A)$, provided one of these fractal dimensions exists and, more generally, that $\overline{\dim}_M(A) = \overline{\dim}_B(A)$ and $\underline{\dim}_M(A) = \underline{\dim}_B(A)$.

1.3.7. Fractal Strings and Zeta Functions. A fractal string is a rather simple mathematical object that contains a wealth of information. In this section, we introduce fractal strings along with a few related objects and discuss how they fit into our fractal toolbox.

Definition 1.24 (*Fractal String*; **[353]**). A *fractal string* $\mathcal{L} = \{\ell_j\}_{j=1}^\infty$ is a nonincreasing sequence of positive real numbers which tends to zero. That is, $\ell_j > 0$ for each $j \in \mathbb{N}$ and $\lim_{j\to\infty} \ell_j = 0$.[13]

Remark 1.25. The values ℓ_j which constitute a fractal string $\mathcal{L}$ are referred to as the *lengths* or *scales* of the fractal string. It will often be convenient to denote a fractal string $\mathcal{L}$ as a set of *distinct* lengths l_n counted according to multiplicity m_n for each positive integer n. That is, the countable set

$$\{l_n : l_n \text{ occurs with multiplicity } m_n, n \in \mathbb{N}\} \tag{1.15}$$

provides an equivalent form of a fractal string $\mathcal{L}$.

Throughout this text, zeta functions of the following form are considered.

Definition 1.26 (*Geometric Zeta Function*; **[370]**, **[375]**). Let $\mathcal{L}$ be a fractal string. Then, the *geometric zeta function* of $\mathcal{L}$, denoted $\zeta_\mathcal{L}$, is given by[14]

$$\zeta_\mathcal{L}(t) := \sum_{j=1}^\infty \ell_j^t \tag{1.16}$$

[13] In this definition, we can also allow the trivial case when $\mathcal{L}$ is a finite sequence of positive numbers. Furthermore, one usually requires that $\mathcal{L}$ has finite total length: $\sum_{j=1}^\infty \ell_j < \infty$; in this case, $\mathcal{L}$ is called an *ordinary fractal string*.

[14] It is also given by (1.16), for all $t \in \mathbb{C}$ with $\mathrm{Re}(t) > D_\mathcal{L}$.

for all $t > D_{\mathcal{L}}$, where $D_{\mathcal{L}}$ is the *dimension* of $\mathcal{L}$ given by

$$D_{\mathcal{L}} := \inf\left\{\alpha \in \mathbb{R} : \zeta_{\mathcal{L}}(\alpha) = \sum_{j=1}^{\infty} \ell_j^{\alpha} < \infty\right\}. \tag{1.17}$$

The dimension $D_{\mathcal{L}}$ is also known as the *abscissa of convergence* of $\zeta_{\mathcal{L}}$ and it is analogous to the radius of convergence of a power series. That is, each value acts as a cut-off point beyond which the corresponding series diverges.

If $\mathcal{L}$ is an ordinary fractal string (i.e., if $\sum_{j=1}^{\infty} \ell_j < \infty$; see footnote 13), then we have that $0 \leq D_{\mathcal{L}} \leq 1$.[15]

Classically, the notion of a *fractal string* is tied to a relatively simple geometric object. Specifically, the classic definition of an *ordinary fractal string*, from ([**352, 353**], [**318–320**], [**339**], [**340**], [**370**], [**374**], [**375**], [**329**], [**330**]), is as follows.

Definition 1.27 (*Ordinary Fractal String*; [**353**]). An *ordinary fractal string* Ω is a bounded open subset of the real line (or, more generally, an open subset of $\mathbb{R}$ with finite length).

A given ordinary fractal string Ω can be expressed uniquely as the union of an at most countable collection of pairwise disjoint (and bounded) open intervals $\{I_j\}_{j\geq 1}$; see [**189**, Proposition 9] or Theorem 2.43 at the end of Section 2.2 below. As such, when Ω comprises an infinite number of distinct intervals, the lengths ℓ_j of these (bounded) disjoint intervals are taken to be a fractal string $\mathcal{L}$.[16] Then (as, e.g., in [**329**]), Ω is called a *geometric realization* of the fractal string $\mathcal{L} = \{\ell_j\}_{j\geq 1}$.

Example 1.28 (*The Cantor String*; [**353**], [**370**], [**375**]). We saw an ordinary fractal string earlier in Example 1.6 of Subsection 1.2.3. Namely, the complement of the Cantor set in the unit interval is the Cantor string $\Omega_C = [0,1]\backslash C$.[17] The corresponding set of lengths is the fractal string

$$\mathcal{L}_C = \{3^{-n} : 3^{-n} \text{ occurs with multiplicity } 2^{n-1}, n \in \mathbb{N}\}. \tag{1.18}$$

The geometric zeta function of the Cantor string $\mathcal{L}_C$, denoted ζ_C, is given, for all sufficiently large (real) values of t, by

$$\zeta_C(t) = \zeta_{\mathcal{L}_C}(t) = \sum_{n=1}^{\infty} \frac{2^{n-1}}{3^{nt}}. \tag{1.19}$$

[15] The only exception is when $\mathcal{L}$ is a finite string, in which case the right-hand side of (1.17) is equal to $-\infty$, so that $D_{\mathcal{L}} = \max(-\infty, 0) = 0$.

[16] Again, we can also allow here the trivial case when Ω consists of a disjoint union of finitely many (bounded) open intervals.

[17] This ordinary fractal string was originally introduced in [**319, 320**], [**352, 353**], while its geometric zeta function was obtained in [**369–371, 374, 375**].

It turns out that $\zeta_C(t)$ is a geometric series with initial term 3^{-t} and common ratio $2\cdot 3^{-t}$; hence, the dimension of the Cantor string, denoted $D_C = D_{\mathcal{L}_C}$, is found readily. We have

$$\zeta_C(t) = \sum_{n=1}^{\infty} \frac{2^{n-1}}{3^{nt}} = \frac{3^{-t}}{1-2\cdot 3^{-t}}. \tag{1.20}$$

Note that ζ_C converges for all $t > \log_3 2$ and diverges at $t = \log_3 2$, as evidenced by the fact that $D_C = D_{\mathcal{L}_C} = \log_3 2$ is the unique real solution of the equation $1 - 2\cdot 3^{-t} = 0$. Hence, D_C coincides with the similarity dimension $D_S = \log_3 2$ and the box-counting dimension $\dim_B C$ of the Cantor set C. As we will see in Section 8.2, and as noted earlier, this is far from a coincidence.

The use of zeta functions of the forms given in (1.16) and (1.20) allows for the definition of *complex dimensions* ([**370**, **375**], [**356**], [**329**]), dimensions which take on *complex* values. In the context of ordinary fractal strings Ω, complex dimensions encode geometric and spectral information about the corresponding bounded open subset of the real line that cannot be obtained from the other real-valued notions of dimension addressed in this text.

Roughly speaking, the set of complex dimensions $\mathcal{D}_{\mathcal{L}}$ of a fractal string $\mathcal{L}$ is the set of points at which an appropriately extended version of the geometric zeta function $\zeta_{\mathcal{L}}$ (its domain is carefully extended to a suitable connected open subset of the complex plane $\mathbb{C}$) has a singularity—for example, most often, a pole.

Example 1.29 (*Complex Dimensions of the Cantor String*; [**370**], [**375**]). By the Principle of Analytic Continuation (to be briefly reviewed in Subsection 9.3.1), the closed form for the geometric zeta function of the Cantor string ζ_C given by the rightmost expression in (1.20) allows us to extend the domain to the whole complex plane $\mathbb{C}$ and determine the set of complex dimensions of the Cantor string, denoted $\mathcal{D}_C$. That is, we have

$$\zeta_C(s) = \frac{3^{-s}}{1-2\cdot 3^{-s}}, \quad \text{for all } s \in \mathbb{C}. \tag{1.21}$$

We then immediately find the complex dimensions to be given by

$$\mathcal{D}_C = \left\{\omega = \log_3 2 + \frac{2iz\pi}{\log 3} \, : \, z \in \mathbb{Z}\right\}, \tag{1.22}$$

where each complex number ω in (1.22) satisfies the equation $1 - 2\cdot 3^{-\omega} = 0$.

Remark 1.30. Note that the set of complex dimensions of the Cantor string, $\mathcal{D}_C$, is *exactly* the set of complex roots of the corresponding Moran equation $\sum_{j=1}^{N} r_j^s = 2\cdot 3^{-s} = 1$, where the real number t in (1.11) is replaced with the complex number s. As will be seen in the sequel to this book, [**119**], and as developed in [**375**, Chapters 2 and 3], the complex solutions of the Moran equation corresponding to a self-similar set are, in a general setting, the complex dimensions of the ordinary fractal string (called a *self-similar string*) that is given by the complement of the self-similar set.

Fractal strings stem from more than just bounded open subsets of the real line. As developed in the aforementioned sequel to this book, [**119**] (motivated by [**375**] and [**356**]), zeta functions and complex dimensions contain lots of information on objects used to define corresponding fractal strings. In particular, fractal complex dimensions provide very detailed information about the *geometric oscillations* that are intrinsic to fractal geometries. For fractal strings, this will be briefly discussed in Section 8.2 and, in much greater detail, in the first part of the sequel to this book, [**119**]. For higher-dimensional fractals, this is illustrated throughout Chapter 9 (on fractal curves)—and, especially, discussed in much greater detail in the second part of the aforementioned sequel to this book, [**119**], based on the higher-dimensional theory of complex dimensions developed in [**356**] (as well as in [**357–367**]).

Fractality as the Presence of Nonreal Complex Dimensions. Whether for fractal strings (the 1-dimensional case) or for higher-dimensional geometric objects, the theory of complex dimensions has led to a significantly more accurate definition of fractality than the one introduced by Mandelbrot in the early 1980s and discussed earlier in this introduction; see the very beginning of Subsection 1.1.1.[18] Accordingly (see [**370**], [**374**], [**375**], [**356**], [**329**], [**330**]), a geometric object is said to be *fractal* if it has at least one *nonreal* complex dimension.[19,20]

All known "fractals" for which the complex dimensions can be calculated either exactly or numerically satisfy this definition. Many examples will be discussed in Section 8.2 and, especially, in Chapter 9, while (in addition to loc. cit.), an accessible account of the theory of complex dimensions for fractal strings, as well as for higher-dimensional fractals and fractal drums, will be provided with increasing sophistication (and complexity) in the sequels to this book, [**119**], [**116**], and [**117**].

In the next section, we take a break from developing our fractal toolbox in order to discuss other ways in which fractal structures may arise.

[18]For example, because in each case, the Hausdorff and topological dimensions coincide (with values 1 and 2, respectively), neither the Cantor curve (or Devil's staircase)—discussed in Subsection 1.4.1 —nor the Peano curve (discussed in Subsection 1.4.3) is "fractal", according to Mandelbrot's definition; see the very beginning of Subsection 1.1.1 above. However, both the Cantor curve (also called the Cantor graph in Subsection 1.4.1 below) and the Peano curve are fractal according to the definition discussed here, requiring the presence of *nonreal* complex dimensions; see Subsection 9.4.7 and Remark 9.63 (towards the end of Subsections 9.4.7), respectively, along with [**119**] and [**116**].

[19]It then has at least two nonreal complex dimensions, since nonreal complex dimensions come in complex conjugate pairs.

[20]In fact, most classic fractals have infinitely many complex dimensions (necessarily, countably many).

1.4. Fractal Functions and Mathematical Monsters

In this section, we briefly explore several classic mathematical monsters that arise as functions—along with their graphs, the associated fractal curves—and investigate some of their properties.

1.4.1. A Surprising Pair of Functions: The Devil's Staircase. There is a surjective (i.e., onto) mapping between the Cantor set C and the closed unit interval $[0,1]$ that has a relatively straightforward construction. The idea is to exploit the similarities between the ternary expansions of the points in the Cantor set C and the binary expansions of points in $[0,1]$. Additionally, this mapping extends to a classical mathematical monster known as the Devil's staircase, the graph of the Cantor function f_{DS}, and denoted by $\mathbf{\Gamma}_{DS}$.[21]

Example 1.31. Let $x = .x_1x_2\ldots_3 \in C$ and $y = .y_1y_2\ldots_2 \in [0,1]$ and consider only nonterminating expansions (i.e., the expansion contains only finite strings of 0 digits) in both sets.[22] Define $f : C \to [0,1]$ using the rule $y_n = x_n/2$ for each $n \in \mathbb{N}$. For example, this map sends 1/4 to 1/3:

$$\frac{1}{4} = .0202\ldots_3 = \sum_{n=1}^{\infty} \frac{2}{3^{2n}} \mapsto \sum_{n=1}^{\infty} \frac{1}{2^{2n}} = .0101\ldots_2 = \frac{1}{3}.$$

Thus, the function f is an example of the counterintuitive notion that there is a set with Lebesgue outer measure 0 which can nevertheless be mapped *onto* a set of Lebesgue outer measure 1.[23]

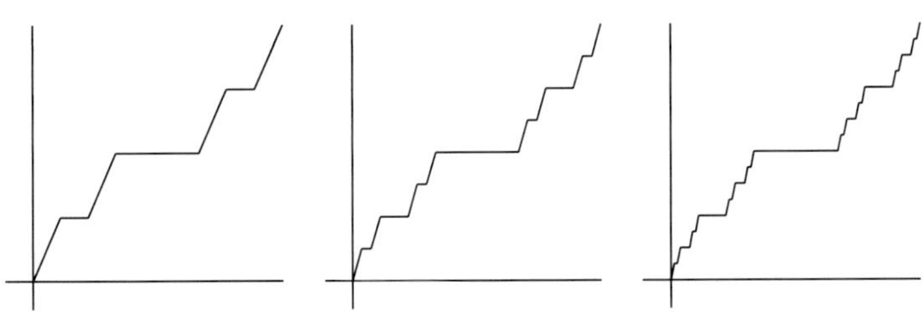

Figure 1.8. There are various ways to construct the Devil's staircase $\mathbf{\Gamma}_{DS}$ (the graph of the Cantor function f_{DS}); see Example 1.32, along with Exercises 1.34 and 1.35. This figure depicts a construction via a uniformly convergent sequence of piecewise linear functions.

[21] We refer the interested reader to the paper by J. F. Fleron, **[188]**, for a detailed history of the Cantor function f_{DS} and of the Cantor set C; see also, e.g., **[57]**, **[94]**, **[101]**, **[102]**, **[145]**, and **[216]**, along with the relevant references therein, for further information and various generalizations, some of which will be discussed later in this book.

[22] Here and henceforth, $.x_1x_2\ldots_3$ (resp., $.y_1y_2\ldots_2$) denotes "the" base 3 expansion of x (resp., "the" base 2 expansion of y). More generally, if $x \in [0,1]$ and $b \in \mathbb{N}\backslash\{1\}$, then "the" base b expansion of x is given by $x = .x_1x_2\ldots_b := \sum_{n=1}^{\infty} x_n/b^n$, where $x_n \in \{0,1,\ldots,b-1\}$, for all $n \in \mathbb{N}$.

[23] Actually, since each of these sets is Lebesgue measurable, C (resp., $[0,1]$) has 1-dimensional Lebesgue measure 0 (resp., 1).

Example 1.32 (*Devil's Staircase or Cantor Graph*). The *Devil's staircase* is the graph of a function $f_{DS} : [0,1] \to [0,1]$, called the *Cantor function*, that has derivative identically equal to zero on the complement of the Cantor set and yet maps the unit interval onto itself continuously; see Figure 1.8. In order to construct the Cantor function f_{DS}, keep in mind that each element in $[0,1]\backslash C$ *must* contain a digit 1 in its ternary expansion, we proceed as follows:

(i) Express x in the domain $[0,1]$ in base 3 (ternary).

(ii) If x contains a 1 and the first such digit occurs at index $k \in \mathbb{N}$, set $x_n = 0$ for all $n \geq k$.

(iii) Replace each remaining digit 2 with a 1.

(iv) Interpret the result in base 2 (binary). This defines the value of $f_{DS}(x)$.

Note that this procedure extends the function f from Example 1.31, initially defined on the Cantor set C, to the continuous function f_{DS} on $[0,1]$. That is, $f_{DS}|_C = f$, where $f_{DS}|_C$ is the restriction of f_{DS} to C.

Intriguingly, the Devil's staircase is flat on the Cantor string $\Omega_C = [0,1]\backslash C$. That is, $f_{DS}|_{\Omega_C}$ is differentiable and $(f_{DS}|_{\Omega_C})'$ is identically zero. To see this, note that $\Omega_C = \bigcup_{j=1}^{\infty} I_j$, where $\{I_j\}_{j=1}^{\infty}$ is a countable collection of pairwise disjoint open intervals. For each $j \in \mathbb{N}$, there is a unique index $n_j \in \mathbb{N}$ and a unique collection of digits $z_1, \ldots, z_{n_j-1}$ such that $z_n \in \{0,2\}$ for each $n = 1, \ldots, n_j - 1$ and the ternary expansion of each $x = .x_1x_2\ldots_3 \in I_j$ satisfies

$$x_n := \begin{cases} z_n, & n = 1, \ldots, n_j - 1, \\ 1, & n = n_j, \end{cases} \tag{1.23}$$

and there is at least one index $k > n_j$ such that $x_k \neq 2$. Then, for each $x \in I_j$, $f_{DS}(x) = .y_1y_2\ldots_2$, where

$$y_n := \begin{cases} z_n/2, & n = 1, \ldots, n_j - 1, \\ 1, & n = n_j, \\ 0, & n > n_j. \end{cases} \tag{1.24}$$

Hence, $f_{DS}|_{I_j}$ is constant for each $j \in \mathbb{N}$. The fact that f_{DS} is continuous is the object of Exercise 1.33. Alternate proofs of the continuity of f_{DS} are suggested in Exercises 1.34 and 1.35.

Exercise 1.33.

(i) Show that the Cantor–Lebesgue function f_{DS}, as defined in Example 1.32, is continuous.

(ii) Deduce from (i) that f_{DS} maps $[0,1]$ *onto* $[0,1]$.

[*Hint*: Note that $f_{DS}(0) = 0$ and $f_{DS}(1) = 1$; then use the Intermediate Value Theorem.]

The following two exercises provide two alternative (but related) definitions of the Cantor–Lebesgue function f_{DS}, as well as alternative proofs of the continuity of f_{DS}.

Exercise 1.34 (*First Alternative Construction of f_{DS}*). Let g_{DS} be the piecewise constant function from $[0,1] \setminus C$ to itself defined as follows: $g_{DS} = 1/2$ on $(1/3, 2/3)$, $g_{DS} = 1/4$ on $(1/9, 2/9)$, $g_{DS} = 3/4$ on $(7/9, 8/9)$, and so on. In other words, g_{DS} is constant on each of the open deleted intervals of the ternary Cantor set C.

(i) Show that g_{DS} is uniformly continuous on $[0,1] \setminus C$.

(ii) Deduce from (i) that g_{DS} admits a unique continuous extension on $[0,1]$, still denoted by g_{DS} (for notational simplicity). Also, deduce that g_{DS} is uniformly continuous on all of $[0,1]$.

(iii) Show that g_{DS} is nondecreasing on $[0,1] \setminus C$ and that the unique continuous extension of g_{DS} to all of $[0,1]$ is also nondecreasing and can be given for any $x \in [0,1]$ by

$$g_{DS}(x) = \sup\{g_{DS}(y) : y \in [0,1] \setminus C,\ y < x\}. \tag{1.25}$$

In particular, show that $g_{DS}(0) = 0$, $g_{DS}(1) = 1$, and determine the value of g_{DS} at each of the ternary points of $[0,1]$ (i.e., at each point x in $[0,1]$ of the form $x = m/3^n$, for some $m \in \mathbb{N}_0$ not divisible by 3, and for some $n \in \mathbb{N}$).

(iv) Alternatively, assuming that g_{DS} is only initially defined on $[0,1]\setminus C$ in the manner described just before part (i), show that if $g_{DS}(x)$ is given by (1.25) for any $x \in [0,1] \setminus C$, then g_{DS} is well defined and continuous on $[0,1]$.

(v) Prove that $g_{DS} = f_{DS}$, where f_{DS} is the Cantor–Lebesgue function defined in Example 1.32.

Exercise 1.35 (*Second Alternative Construction of f_{DS}*). For each integer $n \geq 1$, define the piecewise linear (or step) function $h_{DS,n}$, in a way similar to g_{DS} in Exercise 1.34, except that one must stop at the n-th step of the construction of the Cantor set. More specifically, $h_{DS,1}(0) = 0$, $h_{DS,1}(1) = 1$, $h_{DS,1} = 1/2$ on $[1/3, 2/3]$, and $h_{DS,1}$ is then (uniquely) extended (by linear interpolation) to all of $[0,1]$ in order to be a continuous, piecewise linear (really, piecewise affine) function on $[0,1]$.

(i) Show that the sequence of continuous functions $\{h_{DS,n}\}_{n=1}^{\infty}$ converges uniformly on $[0,1]$ to a continuous function on $[0,1]$, denoted by h_{DS}.

[*Hint*: Show that $\{h_{DS,n}\}_{n=1}^{\infty}$ satisfies the uniform Cauchy criterion on $[0,1]$ and then apply a well-known theorem about uniform limits of continuous functions.]

(ii) Prove that $h_{DS} = f_{DS}$. Deduce from part (v) of Exercise 1.34 that we have $h_{DS} = g_{DS} = f_{DS}$, the Cantor–Lebesgue function.

Remark 1.36. The reader familiar with aspects of Lebesgue's theory of integration (see, e.g., Chapter 3 for an introduction) will recognize that the Cantor–Lebesgue function f_{DS} (which also coincides with the functions g_{DS} and h_{DS} defined in Exercise 1.34 and Exercise 1.35, respectively) is yet another of these mathematical monsters from classical analysis. Indeed, for example in light of Exercise 1.34, $f_{DS} = g_{DS}$ is differentiable and $f'_{DS} = 0$ almost everywhere. More specifically, f_{DS} has derivative zero at every point of $[0, 1]$ except on a set of Lebesgue measure zero, namely, except on the Cantor set C. However, f_{DS} is certainly not constant on $[0, 1]$ since $f_{DS}(0) = 0$ and $f_{DS}(1) = 1$. In fact, according to part (ii) of Exercise 1.33, $f_{DS}([0, 1]) = [0, 1]$.

At first, this seems to contradict the Fundamental Theorem of Calculus from Riemann's theory of integration but that is clearly not the case because f_{DS} is not continuously differentiable, and not even differentiable, on all of $[0, 1]$.[24] The reader familiar with some of the more sophisticated aspects of Lebesgue's theory of integration (see, e.g., [**5**], [**57**], [**193**], [**521**], [**522**], or [**523**]) will also realize that the hypotheses of Lebesgue's version of the Fundamental Theorem of Calculus are also not satisfied here. Indeed, f_{DS} is continuous but not absolutely continuous on $[0, 1]$; see loc. cit.[25]

Once again, we see that historically, one of the key origins of 'monstrous fractals' lies in the discoveries of important and counterintuitive counterexamples to classic theorems of mathematical analysis. Another well-known illustration of that is provided by Cantor's independent discovery of the so-called Cantor set driving his investigation of the problem (studied by Riemann in particular) of sets of uniqueness for Fourier series—along with his remarkable development of key parts of modern set theory, including the theory of cardinals and ordinals, in particular.[26] (See, e.g., [**321**].)

Exercise 1.37. Show that f_{DS} is not differentiable at any point of the ternary Cantor set C.

[*Hint*: It may be easier to work with g_{DS}, as defined in Exercise 1.34, and to use the fact that $g_{DS} = f_{DS}$, as stated in part (v) of Exercise 1.34.]

Remark 1.38. The Devil's staircase (the graph of f_{DS} or, equivalently, according to Exercise 1.34 and Exercise 1.35, respectively, the graph of g_{DS} and h_{DS}), along with some of its variants, has played a very important role as a key

[24] Actually, one can also show that f_{DS} is not differentiable at any point of the ternary Cantor set C; see Exercise 1.37.

[25] In fact, a classic theorem (see, e.g., G. B. Folland's book, [**193**, Theorem 3.35]) characterizes the absolutely continuous functions on a compact interval (such as $[0, 1]$ here) as those functions which are almost everywhere differentiable functions and coincide with the antiderivative of their (almost everywhere defined) derivative.

[26] Roughly speaking, a subset $A \subseteq [0, 1]$ is said to be a *set of uniqueness* if for any (sufficiently nice) periodic function φ on $\mathbb{R}$, $\varphi(x) = 0$, for all $x \in A$, implies that all of the Fourier coefficients of φ are equal to zero.

motivation for the development (by the first author and his collaborators) of the mathematical theory of complex dimensions. Indeed, as will be further explained in Part 3 of this book (Chapter 9), it appeared from the beginning to the first author that the paradoxes associated with Mandelbrot's definition of 'fractality' (in terms of the nonequality of the Hausdorff and topological dimensions) could be resolved by appealing to a much finer notion of fractality, based on the intrinsic oscillations inherent to complicated (or rough) geometries.

As we will see in Chapter 9 (Subsection 9.4.7), the Devil's staircase (or one of its simpler variants) is 'fractal' if one uses the presence of nonreal complex dimensions for defining fractality, whereas paradoxically, and counter to any geometer's intuition, it is not fractal according to Mandelbrot's definition of fractality.

We note that most of the time in this book, f_{DS} will simply be called the Cantor function rather than the Cantor–Lebesgue function. Also, along with its graph, the Devil's staircase, f_{DS} will be studied in more detail in Section 9.4 of Chapter 9 on fractal curves. Furthermore, in Example 4.93 of Section 4.10, the Devil's staircase is described as an inhomogeneous self-affine set. Moreover, we shall see in Section 9.4 that the two descriptions of the Devil's staircase—either as the Cantor graph or as a self-affine set—are equivalent.

1.4.2. Weierstrass's Nowhere Differentiable Function. One of the most startling results in the history of mathematics was the discovery by Karl Weierstrass of a class of functions which are continuous everywhere but nowhere differentiable. In 1872, [**607**], and in later years (see [**608–610**]), he published the following theorem.

Theorem 1.39 (Nowhere Differentiability of the Weierstrass Function; K. Weierstrass, [**607–610**], 1872–1886). *Consider the function $\mathcal{W} = \mathcal{W}_{\lambda,b}$ on the real line given by*

$$\mathcal{W}(x) = \sum_{n=0}^{\infty} \lambda^n \cdot \sin(b^n \pi x), \tag{1.26}$$

where $0 < \lambda < 1$, b is a positive odd integer, and $\lambda b > 1 + (3/2)\pi$. Then, $\mathcal{W}$ is continuous on all of $\mathbb{R}$ but is not differentiable at any $x \in \mathbb{R}$. Accordingly, the function $\mathcal{W}$ is said to be nowhere differentiable *on $\mathbb{R}$.*

In other words, a function $\mathcal{W} = \mathcal{W}_{\lambda,b}$ that satisfies the hypotheses of Weierstrass's Theorem, which we call a *Weierstrass function*, cannot be approximated by a linear function at any point. No matter how much one zooms into the graph of a Weierstrass function at a point $(x, \mathcal{W}(x))$, the graph is *not* monotone and, as such, no derivative exists there; in other words, its graph, called the *Weierstrass curve*—associated with $\mathcal{W}$ and denoted $\mathbf{\Gamma}_{\mathcal{W}}$—does not have a

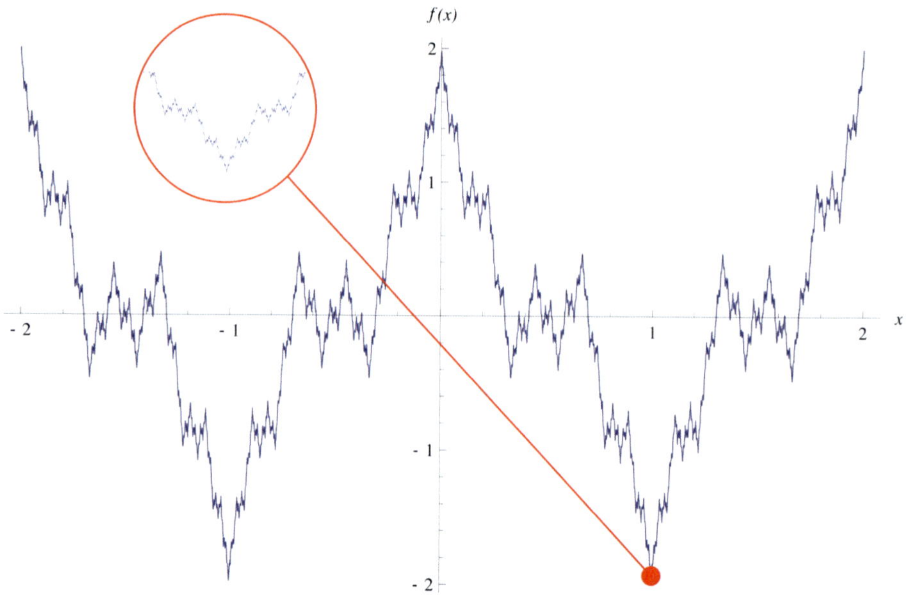

Figure 1.9. The Weierstrass curve, an example of a continuous but nowhere differentiable fractal curve, as discussed in Subsection 1.4.2 (and Section 9.7).

tangent line anywhere. See Figure 1.9 in this chapter for a picture of $\mathbf{\Gamma}_W$, along with Figure 9.30 in Subsection 9.7.2 for a depiction of its prefractal polygonal approximations.

We will return to this fascinating topic in Section 9.7 below, as well as in much greater detail, in the second sequel to this book, [**116**].

1.4.3. Peano's Space-Filling Curve. In 1877–1878, Cantor proved that there exists a surjective (or onto) mapping from the unit interval $[0, 1]$ onto the unit square $[0, 1]^2$; see [**68**] and [**466**].[27] Despite the elegant simplicity of his proof (sketched below), Cantor wrote in his 1877 letter to Richard Dedekind, "I see it, but I don't believe it!"; see [**466**, p. 3].[28] The idea is to find a preimage z in $[0, 1]$ for each point (x, y) in $[0, 1]^2$. Take the decimal expansion for $x = 0.x_1x_2x_3\ldots$ and $y = 0.y_1y_2y_3\ldots$ and define the preimage $z = 0.x_1y_1x_2y_2x_3y_3\ldots$.

Later, in 1890, Giuseppe Peano [**479**] pushed the issue further and endeavored to prove the existence of a curve that 'fills' a square (namely, $[0, 1]^2$); see

[27] Earlier, in 1874, Cantor [**67**] proved—via the so-called Diagonalization Method (or Diagonal Argument) that there could not exist a bijection between $\mathbb{N}$, the set of positive integers, and $[0, 1]$, the unit interval—thereby setting the stage for modern set theory. We note that Cantor's Diagonal Argument was later published as a systematic method in his 1891 paper [**71**].

[28] In that same 1877 letter to Dedekind, where his proof was first sketched (using continued fractions, in particular), he also asked the following more general question (which, in modern terms, is tantamount to asking about the invariance property of the topological dimension; see Subsection 8.6.1 below): "Can a surface (e.g., the unit square) be put into relation with a curve (e.g, the unit square)?"; see loc. cit. and, e.g., [**148**], [**251**], and [**462**].

Figure 1.10. The Peano curve, a plane- (or, rather, square)-filling curve, as described in Subsection 1.4.3.

[**479**]. He created a procedure that builds a *continuous* map from a line segment onto a square (namely, $[0,1]^2$). This procedure is very similar to the construction of the Koch snowflake curve discussed in Example 1.8 of Subsection 1.2.3 above:

(i) Start with a line segment called the initiator.

(ii) Make 9 copies of the line segment scaled by a factor of $1/3$.

(iii) Arrange the 9 segments end to end at right angles as indicated in Figure 1.10. This is called the generator.

(iv) Replace each of the 9 line segments in the generator with a copy of the generator scaled by a factor of $1/3$.

(v) Repeat this process ad infinitum, scaling by $1/3$ at each step.

The outcome of this procedure, when it is imbued with an appropriate parametrization, is a continuous function whose domain is a line segment and whose image is a square; see Figure 1.10.

This disturbing result bothered mathematicians who were under the impression that line segments and squares were topologically distinct: There does not exist a homeomorphism (a continuous bijection with continuous inverse) between a 1-dimensional object like a line segment and a 2-dimensional object like a square. It turns out that Peano's space-filling construction is not injective (i.e., not one-to-one) and that homeomorphisms preserve topological dimension (see Subsection 8.6.1); the latter result was not obtained in full until after years of careful study.

Another celebrated (and simpler) example of a plane-filling curve is the Hilbert curve, introduced in 1891 by David Hilbert in [**247**]; see Figure 9.12 near Remark 9.63 in Section 9.4 for a picture of this fractal curve.

In Chapter 9, we will briefly return in several places to the intriguing subject of plane-filling curves, and we will further discuss it in the sequels to this book, especially [**119**] and [**116**].[29]

1.5. Randomly Constructed Fractals

The notion of fractality is not limited to the types of sets and functions explored above and which are defined in a deterministic fashion. In fact, replacing some of the steps in such constructions with some random process or procedure allows us to extend the analytical tools defined in this text to a much broader collection of sets and functions. And, of course, many examples of randomly constructed fractals are found in nature.

Consequently, even though they will not be the focus of this book, random fractals play an important role in the applications of fractal geometry to physics, chemistry, astronomy, engineering, high technology and computer science, image processing, biology, medicine, geology, economics and finance, among other fields. See, e.g., [**10**], [**27**], [**28,29**], [**50**], [**62**], [**65**], [**72**], [**149–151**], [**172,175**], [**194**], [**209**], [**217**], [**218**], [**225,226**], [**227**], [**230**], [**236**], [**274**], [**308**], [**313–315**], [**320**], [**356**, Section 6.2], [**372**, Part 2], [**375**, Section 13.4], [**382**], [**383–385**], [**386–388**], [**418–420**], [**422–426**], [**454**], [**475**], [**491**], [**494**], [**506**], [**534**], [**542**], [**558**], [**559**], [**568**], [**570**], [**611**], [**621**], [**624**], and [**628,629**], along with the relevant references therein.

We point out, in particular, the books by C. J. Bishop and Y. Peres, [**50**], entitled *Fractals in Probability and Analysis*, by G. F. Lawler, [**382**], entitled *Conformally Invariant Processes in the Plane*, by P. Mörters and Y. Peres, [**454**], entitled *Brownian Motion*, by D. Revuz and M. Yor, [**506**], entitled *Continuous Martingales and Brownian Motion*, by (or about) M. Yor, [**624**], entitled (in English) *Marc Yor: The Passion of Brownian Motion*, along with the book by G. W. Johnson and the first author, [**274**], entitled *The Feynman Integral and Feynman's Operational Calculus*.

We also mention the survey articles (or lecture notes) by M. T. Barlow, [**27**], on *Diffusions on fractals*, by B. Duplantier, [**149**], on *Conformal quantum gravity and boundary quantum gravity*, by B. Duplantier, R. Rhodes, S. Sheffield, and V. Vargas, [**151**], on *Log-correlated Gaussian fields: An overview*, by J.-F. Le Gall, [**386, 387**], on *Random geometry on the sphere* and on *Brownian geometry*, respectively, as well as by W. Werner, [**611**], on *Random planar curves*

[29]References on the Peano curve and other plane or space-filling curves (such as the Hilbert curve) include [**533**], [**541**], [**531**], and [**37**], along with the books by H. Sagan [**532**] and M. Bader [**15**], both with the same main title, *Space-Filling Curves*.

and Schramm–Loewner evolutions. Finally, we note that in their 2001 paper, [**383**]—based, in particular, on their joint work in [**384**]—G. F. Lawler, O. Schramm, and W. Werner proved Mandelbrot's Conjecture stated in [**419**], according to which the Hausdorff dimension of the outer boundary of a planar Brownian path is 4/3; they established, in the process, several physically motivated conjectures made by B. Duplantier (see, e.g., [**149**] and the relevant references therein). A closely related and very fruitful research direction—at the intersection of mathematical physics, probability theory, and fractal geometry—is the study of critical percolation and of the associated critical exponents—conducted by S. Smirnov in [**558**], as well as by that same author and W. Werner in [**559**].

1.5.1. Brownian Motion. In 1827, while looking through a microscope at particles found in pollen grains immersed in water, the botanist Robert Brown noted that the particles moved through the water in a highly irregular fashion but he was unable to determine the mechanisms that caused this strange motion. Unknowingly, Brown effectively provided the first description of the random movement of particles caused by a bombardment of molecules. Later in 1905, Albert Einstein [**159**] published a physical study that explained in precise detail how the motion Brown observed was a result of the particles being moved by individual water molecules.[30] In turn, this led to the work of Jean Perrin [**488**] on the discontinuous structure of matter and the experimental determination of Avogadro's number for which he was awarded the Nobel Prize in Physics in 1926. We refer, e.g., to [**274**, Chapter 2] and the relevant references therein for a detailed history of Brownian motion, as well as of its physical and mathematical investigations.

In 1923–1924, Norbert Wiener defined in [**615–617**] (see also [**618**]) a rigorous mathematical model of random paths that behaved similarly to the motion observed by Brown and Perrin, among many others; see, e.g., [**274**, Chapters 2–4]. Roughly speaking, this model generated functions with values in $\mathbb{R}$, called *Brownian motions*, whose graphs—or *Brownian paths*, the paths traced out by such functions—were so irregular as to be (almost surely) (3/2)-dimensional; see Figure 1.11 for a typical Brownian path.[31] Hence, with probability 1, the graph of a Brownian path has Hausdorff (or Minkowski) dimension 3/2, which

[30] At about the same time, the Polish physicist Marian Schmoluchowski had developed a similar theory of Brownian motion and of diffusions. Earlier, in his 1900 Ph.D. thesis *Théorie de la spéculation* ("The Theory of Speculation"), the French mathematician Louis Bachelier had provided the first mathematical theory of Brownian motion, motivated by his study of the stock market. (See, e.g., Chapter 2 in [**274**].)

[31] See footnote 32 for the case of an m-dimensional Brownian motion.

Figure 1.11. A typical Brownian path.

means intuitively that it tends to wander around a lot, throughout the half-plane $\{(t,u) : t \geq 0,\ u \in \mathbb{R}\}$, but *not* to "fill" this half-plane; see, e.g., [**454**, Theorem 4.29].[32] By contrast, the topological dimension of the graph of a Brownian path is almost surely (i.e., with probability 1) equal to 1; indeed, such a path is almost surely the graph of a continuous function.

An even more striking property of Brownian paths is that, almost surely, such curves are *nowhere differentiable*. In other words, they are random realizations of Weierstrass-type curves. Much more is known on this subject; for example, with probability 1, a Brownian path, viewed as the graph of a continuous function $f : [0,+\infty) \to \mathbb{R}$, is Hölder continuous of any order $\alpha < 1/2$, but it is not Hölder continuous of order $\alpha > 1/2$.[33] Recall that f is said to be *(locally) Hölder continuous of order* $\alpha \in (0,1]$ at $u \geq 0$ if there exists $\delta > 0$ (with $\delta < u$ if $u > 0$) and $m \in (0,+\infty)$ such that $|f(u) - f(v)| \leq M \cdot |u-v|^{\alpha}$, for all $v \in (u-\delta, u+\delta)$ (or for all $v \in (0,\delta)$ if $u = 0$). Also, f is said to be *(locally) Hölder continuous of order* $\alpha \in (0,1]$ if it satisfies this property at any $u \geq 0$.

[32]More generally, for Brownian motion with values in $\mathbb{R}^m$, almost surely, a Brownian path has Hausdorff dimension 3/2 if $m = 1$ and 2 for $m \geq 2$. Hence, in the case of $\mathbb{R}^2$, it tends to wander throughout the entire plane.

[33]More specifically, it fails to be locally Hölder continuous (of order $\alpha > 1/2$) at any point.

We refer, e.g., to [**506**], [**454**, Chapter 4], and [**274**, Chapters 2–4], along with the relevant references therein, for more information about the basic properties of Brownian motion and of Brownian paths.

Finally, we mention that for a "typical" Brownian path—viewed as a (random) continuous parametrized curve f from $[0, +\infty) \to \mathbb{R}$—if one fixes an output $f(x_0)$, the set of preimages of that point is a type of (random) Cantor-like set. In particular, the zero-set of Brownian motion, that is, $\{t \geq 0 : f(t) = 0\}$, the set of times at which a Brownian path crosses the t-axis, is a random Cantor set which is (1/2)-dimensional. More precisely, with probability 1, the Minkowski and Hausdorff dimensions of the zero-set of a Brownian path coincide and are equal to 1/2, whereas the topological dimension of such a path is equal to 0. More generally, for a class of stochastic processes called α-processes, where $\alpha \in (0,1)$ is arbitrary, the zero-set has (almost surely) Minkowski dimension equal to α. (The special value $\alpha = 1/2$ corresponds to the usual Brownian motion.)[34]

1.5.2. Statistically Self-Similar Sets and Other Random Fractals. Other types of classic stochastic processes and their fractal generalizations arise naturally in this context. These are examples of statistically self-similar or self-affine—or else conformally invariant—random fractals. Often (but not always), random fractals are natural extensions of their deterministic counterparts, for instance, random Cantor sets, random Koch curves—which, in practice, are better models of coastlines than the classic (deterministic) Koch curves—as well as random Sierpinski or Bedford–McMullen carpets. See, e.g., the sample list of relevant references provided in the introduction to this section.

1.6. Complex Dynamics and Fractals: Julia Sets and the Mandelbrot Set

Complex dynamics—the study of the iteration of functions of one (or several) complex variable(s)—has a long and rich history, which intersects in multiple ways with the history and the development of fractal geometry, especially via the introduction of Julia and Fatou sets, towards the beginning of the 20th century, as well as more than half a century later, of the Mandelbrot set.

Complex dynamics is not the main focus of this book but many of the concepts and techniques presented in this book (and its sequels, [**119**], [**116**], [**117**]) can be applied in this context as well. We will limit ourselves to a few definitions, remarks, and references, which the interested reader can consult for

[34]See [**227**], as described in [**375**, Section 13.4], where these properties are further studied and refined from the point of view of the theory of complex dimensions of random fractal strings, along with [**454**] and the relevant references therein (for the standard theory of fractal dimensions in this probabilistic context).

further information. We will also restrict ourselves to the traditional case of a single complex variable and keep the exposition as simple and brief as possible.

For two informative books on the history of complex dynamics, see D. S. Alexander [**2**] and D. S. Alexander, F. Iavernaro, and A. Rosa [**3**]. For an interesting survey article, mainly focused on the more recent history of complex dynamics, see the paper by M. Rees [**504**].

A sample list of (undergraduate or graduate) textbooks on (or related to) complex dynamics is as follows: P. Montel [**447**], A. F. Beardon [**33**], L. Carleson and T. W. Gamelin [**73**], R. L. Devaney [**138**], J. W. Milnor [**444**], along with H.-O. Peitgen, H. Jürgens, and D. Saupe [**483**]. A research monograph on this subject is that of C. T. McMullen [**437**],[35] while a list of relevant collective volumes includes A. Bonifant, M. Lyubich, and S. Sutherland [**53**], D. Cerveau [**77**], R. L. Devaney and L. Keen [**139**], as well as Tan Lei [**394**].[36]

Furthermore, several survey articles on complex dynamics are those of P. Blanchard [**51**], A. Douady [**142**], and C. T. McMullen [**436**].

Moreover, two very original and accessible books, which are not traditional textbooks, are *The Beauty of Fractals* [**484**], by H.-O. Peitgen and P. Richter—which, in particular, discusses Julia sets and the Mandelbrot set, along with the Julia-type fractal sets associated with Cayley's complexified Newton Method [**75**] for finding *complex* roots of equations—and *Indira's Pearls*: *The Vision of Felix Klein* [**455**], by D. Mumford, C. Series, and D. Wright, which focuses, in particular, on Kleinian groups (acting in the hyperbolic plane) and their limit sets, along with their analogy with the complex dynamics of rational maps (acting in the extended complex plane or Riemann sphere, $\widetilde{\mathbb{C}} := \mathbb{C} \cup \{\infty\}$).[37,38]

In addition, a *sample* of research articles on the subject includes, beginning with the classic work of P. Fatou [**178, 179**] and G. Julia [**277**] (where the celebrated Fatou and Julia sets were introduced) and continuing with the work of C. L. Siegel [**548**],[39] A. Avila, J. Kahn, M. Lyubich, and W. Shen [**13**], A. Douady and J. H. Hubbard [**143,144**], J. H. Hubbard [**255**], M. Lyubich [**412**], C. T. McMullen [**435**], D. Sullivan [**579,580**], Tan Lei [**393**], J.-C. Yoccoz [**623**],

---

[35]The research monograph [**437**], entitled *Complex Dynamics and Renormalization*, addresses and develops, in particular, Sullivan's conjectures (and Program), proposed in [**580**] and motivated, in part, by M. J. Feigenbaum's 1980 work on "Universality" in nonlinear chaotic dynamical systems [**185**].

[36]The latter reference is dedicated to the study of the Mandelbrot set.

[37]This beautiful analogy between Kleinian groups and complex dynamics was originally proposed and developed by D. Sullivan in 1985 in [**579**]; it is now known as the *Sullivan Dictionary*.

[38]See, e.g., the relevant parts of the excellent collective volume by T. Bedford, M. Keane, and C. Series [**35**], for an introduction to the topic of limit sets of Kleinian groups and its relationship with hyperbolic geometry, dynamical systems, and fractal geometry; see also, e.g., [**438**].

[39]See also the book by C. L. Siegel and J. K. Moser [**549**].

and B. B. Mandelbrot [**417**][40] (where the celebrated Mandelbrot set was introduced, in 1980).[41]

Finally, we point out that complex dynamics (and, in particular, the topic of Julia sets) plays an important role in the study of analysis on fractals and the spectral theory of Laplacians on certain self-similar sets. It gave rise, especially, to the so-called *Spectral Decimation Method* (or *Spectral Decimation*, in short), introduced in this context in physics in the mid-1990s by R. Rammal and G. Toulouse [**501**] as well as by R. Rammal [**500**].[42] Then, in the early 1990s, Spectral Decimation (in the initial case of the Sierpinski gasket) was made mathematically rigorous by M. Fukushima and T. Shima [**207**] and later, it was further extended, in particular, by L. Malozemov and A. Teplyaev [**414**]. Moreover, the Method of Spectral Decimation has been applied in many contexts and papers, including by J. Kigami and M. L. Lapidus in [**294**] and [**295**], J. Kigami (see, e.g., [**293**]), A. Teplyaev (see, e.g., [**586, 587**] and the relevant references therein), G. Derfel, P. Grabner, and F. Vogl [**131**], C. Sabot [**528–530**], R. S. Strichartz (see, e.g., [**574**]), N. Lal and M. L. Lapidus ([**311**] and [**312**]). The latter paper, [**312**], is a survey article on the Spectral Decimation Method where the interested reader can find many other references on this subject and its direct connections with spectral dynamics—including suitable real Julia sets which can be viewed as the spectra of Laplacians on certain self-similar fractals or graphs.

1.6.1. Julia Sets and the Mandelbrot Set. We close this section by giving a brief introduction to Julia sets and the Mandelbrot set, by mostly limiting ourselves (in order to facilitate and shorten the discussion) to the classic quadratic family $\{f_c\}_{c\in\mathbb{C}}$, where

$$f_c(z) := z^2 + c, \quad \text{for all } z \in \mathbb{C}. \tag{1.27}$$

Here, the complex parameter will play a key role in the definition of the Mandelbrot set. We further note that the definition of the Fatou and Julia sets can also be extended to rational functions (see, e.g., [**33**])—or even to entire functions and more general meromorphic functions (see, e.g., [**138, 139**], [**73**], and the appropriate references therein); see Remark 1.40 below.

Given $n \in \mathbb{N}_0$, let us denote by $f_c^n = (f_c)^n$ the n-fold iterate of f_c, obtained by composing f_c with itself n times.[43] Given an initial point $z_0 \in \mathbb{C}$, we consider

[40] See also B. B. Mandelbrot's books [**419**] and [**425**].

[41] Independently and at about the same time, but with somewhat different motivations (coming both from dynamical systems and Kleinian group theory), R. Brooks and J. P. Matelski also discovered a slightly different form of the Mandelbort set, in [**58**].

[42] It seems that, much earlier and for different reasons, the French probabilist Paul Lévy had already introduced an analogous method in his study of Markov chains.

[43] We adopt the usual convention according to which $f_c^0 := I$, the identity map of $\mathbb{C}$.

the sequence $\{z_n\}_{n=0}^{\infty}$ defined inductively by

$$z_{n+1} = f_c(z_n), \quad \text{for all } n \in \mathbb{N}; \tag{1.28}$$

i.e., $z_n = f_c^n(z_0)$, for all $z_0 \in \mathbb{C}$ and $n \in \mathbb{N}_0$.

Then, the *filled-in Julia set* of f_c, denoted $K_c = K(f_c)$, is given by

$$K_c = K(f_c) := \{z_0 \in \mathbb{C} : \{z_n\}_{n=0}^{\infty} \text{ is bounded (in } \mathbb{C})\}, \tag{1.29}$$

while the *Julia set* of f_c, denoted $J_c = J(f_c)$, is defined as the (topological) boundary of K_c (in $\mathbb{C}$):

$$J_c := \partial K_c \; (:= \partial(K_c)). \tag{1.30}$$

The *Fatou set* F_c of f_c is, by definition, the complement of the Julia set J_c in the Riemann sphere $\widetilde{\mathbb{C}} := \mathbb{C} \cup \{\infty\}$:

$$F_c = \widetilde{\mathbb{C}} \setminus J_c \quad (\text{or vice versa, } J_c = \widetilde{\mathbb{C}} \setminus F_c). \tag{1.31}$$

Intuitively, from a dynamical point of view, the Fatou set is the subset of $\widetilde{\mathbb{C}}$ on which the dynamics is ordinary (i.e., regular or "tame"), whereas the Julia set (its complement in $\widetilde{\mathbb{C}}$) is the subset of $\widetilde{\mathbb{C}}$ on which the dynamics is chaotic; i.e., for example, the underlying dynamical system exhibits sensitive dependence on the initial conditions.[44]

A key theorem—due to Fatou and Julia (see [**178**] and [**277**])[45]—is that given any value of the parameter $c \in \mathbb{C}$, the corresponding Julia set $J_c = J(f_c)$ is either totally disconnected (i.e., roughly, is a "Cantor dust" in the complex plane)—and hence of topological dimension zero—or else connected (i.e., roughly, made of a single piece); see, e.g., the left-hand side and the right-hand side of Figure 1.12, respectively.

Accordingly, it is now simple to define the *Mandelbrot set* M, as follows:

$$M := \{c \in \mathbb{C} : J_c \text{ is connected}\}. \tag{1.32}$$

Another (equivalent) characterization of the Mandelbrot set is given by

$$M = \{c \in \mathbb{C} : \{f_c^n(0)\}_{n=0}^{\infty} \text{ is bounded (in } \mathbb{C})\}. \tag{1.33}$$

The simplest example of a Julia set is obtained by choosing $c := 0$ (i.e., $f_0(z) = z^2$, for all $z \in \mathbb{C}$). Then, in light of (1.28)–(1.30), and since $f_0^n(z) = z^{2^n}$, for all $z \in \mathbb{C}$, we see that the filled-in Julia set K_0 is the closed unit disk of $\mathbb{C}$—and hence that the Julia set $J_0 = \partial K_0$ is the unit circle:

$$K_0 = \{z \in \mathbb{C} : |z| \leq 1\} \quad \text{and} \quad J_0 = \{z \in \mathbb{C} : |z| = 1\}. \tag{1.34}$$

Clearly, in this exceptional case, the Julia set J_0 is not a fractal, whatever definition of fractality one wishes to adopt.

[44] See, e.g., [**138**], for a more precise mathematical definition of a chaotic dynamical system—or, equivalently, of "chaos", although (just as is the case of "fractality") there is no universal agreement on what this definition should entail.

[45] See, e.g., [**444**] for a modern exposition.

Figure 1.12. The left-hand side (resp., the right-hand side) depicts an example of a totally disconnected (resp., connected) and fractal Julia set J_c, associated with a quadratic polynomial $f_c(z) = z^2 + c$, for all $z \in \mathbb{C}$; see Subsection 1.6.1. Here, on the left-hand side (resp., on the right-hand side), $c := 0.427 - 0.376i$ (resp., $c := -0.473 - 0.554i$).

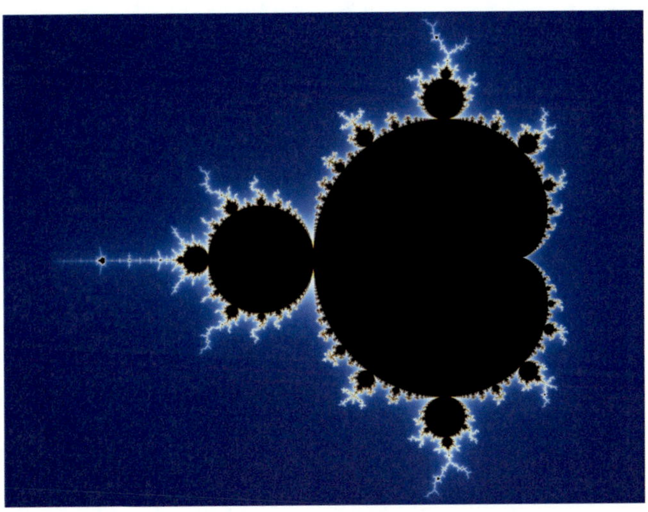

Figure 1.13. The Mandelbrot set M, as defined in Subsection 1.6.1. It is represented in the complex plane, viewed as the *parameter plane*; i.e., M is the set of parameters $c \in \mathbb{C}$ satisfying (1.32) or, equivalently, (1.33).

Typically, however, when $c \neq 0$, the Julia set J_c is a "fractal". For example, if $c \in M$ is sufficiently small—and hence lies in the main "cardioid" of the Mandelbrot set—D. Sullivan [**579**] showed in 1985 that the corresponding Julia set J_c is a quasicircle, the boundary of a quasidisk.[46] Furthermore, according to Sullivan's Theorem in [**579**], J_c has "fractal dimension" between 1 and 2 and, in fact, can take any value between 1 and 2. In the present case, it is known that the Hausdorff and Minkowski (or box-counting) dimensions of J_c coincide; hence, "fractal dimension" can be interpreted unambiguously in this context.

Another highly nontrivial result connecting in a deep manner complex dynamics and fractal geometry is the following celebrated result of M. Shishikura [**547**], published in 1998 and according to which the boundary ∂M of the Mandelbrot set M has "fractal dimension" 2. More precisely, it has both Minkowski and Hausdorff dimensions 2:

$$\dim_M(\partial M) = \dim_H(\partial M) = 2, \tag{1.35}$$

which is the maximal possible dimension since $\partial M \subseteq \mathbb{C}$ and $\mathbb{C}$ is identified with $\mathbb{R}^2$.[47]

Shishikura's results in [**547**], the proof of which makes use of sophisticated techniques from complex dynamics, implies that there are Julia sets J_c (with $c \in \partial M$) for which the fractal dimension (i.e., either the Hausdorff or Minkowski dimension) is equal to 2, the maximum possible fractal dimension:

$$\dim_M(J_c) = \dim_H(J_c) = 2. \tag{1.36}$$

Accordingly, such Julia sets are of an extreme complexity for these values of the parameter $c \in \partial M$.[48] In light of (1.35), the same statement holds true for the boundary ∂M of the Mandelbrot set, which is perhaps one of the most—if not the most—complicated "fractal" geometric object in mathematics.

The reader unfamiliar with complex analysis (and dynamical systems) may wish to omit the following remark upon a first reading.

Remark 1.40 (*Fatou and Julia Sets, Revisited*). In the modern mathematical theory of complex dynamics (see, e.g., [**51**], [**33**]), a rational function—or, more generally, a meromorphic function f—is viewed as a map from the Riemann sphere $\widetilde{\mathbb{C}} = \mathbb{C} \cup \{\infty\}$ to itself, where $\widetilde{\mathbb{C}}$ (viewed as a compact Riemann surface) is equipped with the so-called chordal metric (see, e.g., [**33**], [**375**], [**356**]).

---

[46] A *quasidisk* is the image of the unit disk (in $\mathbb{C}$) under a quasiconformal map of $\mathbb{C}$. Roughly speaking, a *quasiconformal map* is a homeomorphism which sends an infinitesimally small disk (of $\mathbb{C}$) onto the interior of an infinitesimally small ellipse, with a uniformly bounded eccentricity; see, e.g., Ch. Pommerenke [**495**], V. G. Maz'ja [**432**], P. W. Jones [**275**], M. L. Lapidus [**318**], as well as M. L. Lapidus and M. M. H. Pang [**346**], along with the relevant references therein. For example, the Koch snowflake domain (see Section 9.5 below) is a quasidisk while its boundary, the Koch snowflake curve (ibid) is a quasicircle.

[47] It suffices to show (as is done in [**547**]) that $\dim_H(\partial M) = 2$. Indeed, as is the case for any bounded subset of $\mathbb{R}^2$ (see Section 8.5), $\dim_H(\partial M) \leq \dim_M(\partial M)$, and, clearly, $\dim_M(\partial M) \leq 2$ (since $\partial M \subseteq \mathbb{R}^2$).

[48] Note, however, that—unlike the Peano curve—they are not plane-filling curves, even though they have dimension 2.

Then, one usually defines the *Fatou set* $F = F(f)$ of f first, as the "set of normality" of f (or, rather, of the sequence of its forward iterates $\{f^n\}_{n\in\mathbb{N}_0}$), and one then defines the *Julia set* $J = J(f)$ as the complement (in $\widetilde{\mathbb{C}}$) of the Fatou set $F = F(f)$. By construction, F is open (in $\widetilde{\mathbb{C}}$) and hence, J is closed (in $\widetilde{\mathbb{C}}$). There exist examples (of non-quadratic maps) for which J is not compact in $\mathbb{C}$ (i.e., J is not bounded), and even for which $J = \mathbb{C}$; see, e.g., [**33**], [**51**], or [**444**].

From a dynamical point of view, an alternative description of the Fatou set F and the Julia set J is as follows (see, e.g., loc. cit. or [**138**], for the appropriate definitions): the Riemann sphere $\widetilde{\mathbb{C}}$ splits into two disjoint sets, $F = F(f)$ and $J = J(f)$, which are, respectively, the set F of "normal" (essentially, stable) points F and the closure J of the set of repelling periodic points. From this characterization, it is also clear that F is open while J is closed.

For completeness, we mention that, by definition, the *set of normality* of f is the set of points $z \in \widetilde{\mathbb{C}}$ for which there exists an open neighborhood U of z such that the restriction to U of the sequence of iterates of f is *normal* (in the sense of Montel [**447**]), i.e., is relatively compact in the topology of uniform convergence on every compact subset of U—or, equivalently (by the Arzela–Ascoli Theorem, see, e.g., [**193**] or [**522**]), if the sequence of functions $\{f^n\}_{n=0}^{\infty}$ is equicontinuous on every compact subset of U. Further recall from "elementary analysis" (cf. loc. cit.) that a family of functions $\mathcal{F}$ is normal, in the above sense, iff one can extract a convergent subsequence (here, in the topology of the local uniform convergence on U) from every given sequence of functions contained in $\mathcal{F}$.

1.7. Chaotic Dynamics and Fractals: Strange Attractors

There is, of course, a vast mathematical and scientific literature on the subject of chaotic dynamical systems and on its many connections to fractals. Our object here is not to give an extensive discussion of this subject but merely to mention a few relevant references.

Many such references are given in Section 1.6 above and in Section 1.8 below, including the undergraduate and/or graduate textbooks [**483**] (appropriately called *Chaos and Fractals*), by H.-O. Peitgen, H. Jürgens, and D. Saupe, and [**138**] (entitled *An Introduction to Chaotic Dynamical Systems*), by R. L. Devaney.

We would also like to point out [**97**], which is a very interesting collection of reprints of classic papers on chaotic dynamical systems and their application to physics, biology, and other scientific fields; it is artfully edited by P. Cvitanović and entitled *Universality in Chaos*.

We also mention the following books by two masters of chaos theory, Edward N. Lorenz[49] and David Ruelle. Namely, the book [**411**], by E. N. Lorenz and entitled *The Essence of Chaos*, as well as the book [**526**], by D. Ruelle and entitled *Chaotic Evolution and Strange Attractors*. As is stated by Lorenz in the preface to his book [**411**], "I have come across a phenomenon that later came to be called 'chaos'—seemingly random and unpredictable behavior that nevertheless proceeds according to precise and often easily expressed rules."

Various mathematical definitions of chaos have been provided in the literature; they are not all equivalent and often capture different and complementary aspects of chaotic dynamics; see, e.g., [**138**]. In essence, all of these definitions require that the underlying dynamical system exhibit "sensitivity to initial conditions"; in other words, a very tiny change in the initial condition of the system may give rise to a wildly different time evolution (or orbit) of the system. Additional requirements may include (as, e.g., in [**138**]) that the set of periodic points of the underlying (discrete) dynamical system be dense (in the state or phase space) or that the so-called Lyapunov exponent of the system be strictly positive (which implies, roughly, that two initially nearby orbits will eventually diverge exponentially fast from each other), among other technical conditions—such as "topological transitivity"; see, e.g., [**138**].

Many dynamical systems are not chaotic on their entire phase space but are only chaotic in some invariant subset of phase space—which, in the most interesting cases, is also an *attractor* (in an intuitively obvious sense) for the orbits of the system. Typically, such a chaotic attractor exhibits a fractal and very complicated structure (in many different senses). Following the original suggestion of D. Ruelle and F. Takens in 1971 in [**527**], it is then called a *strange attractor*. Famous examples of strange attractors include the *Lorenz attractor*—which arose in Edward N. Lorenz's very simplified and insightful model of the "chaotic" time evolution of a weather system—and the Hénon attractor—which emerged out of the French mathematician and astronomer Michel Hénon's attempt in the 1970s to model the chaotic orbits of celestial bodies, such as stars, evolving under the influence of the gravitational force; see [**237**]. We refer to Figures 1.14 and 1.15, respectively, for a depiction of the Lorenz attractor and of the Hénon attractor.[50]

Another well-known example of a strange attractor is the Rössler attractor, which was discovered in 1976 by the German biochemist Otto Eberhard Rössler

[49]E. N. Lorenz is an American meteorologist, also trained as an applied mathematician; he is the discoverer (in 1969) of the celebrated "Butterfly Effect" and of the "Lorenz attractor"; see [**410**]. His work made him the modern father of chaos theory. As a result, he is widely regarded as the contemporary discoverer of chaos, many years after Henri Poincaré saw the first glimpse of chaos in his work on the three body problem in celestial mechanics; see, e.g., [**32**].

[50]See, e.g., [**38**] for a rigorous study of the Hénon map giving rise to the Hénon attractor.

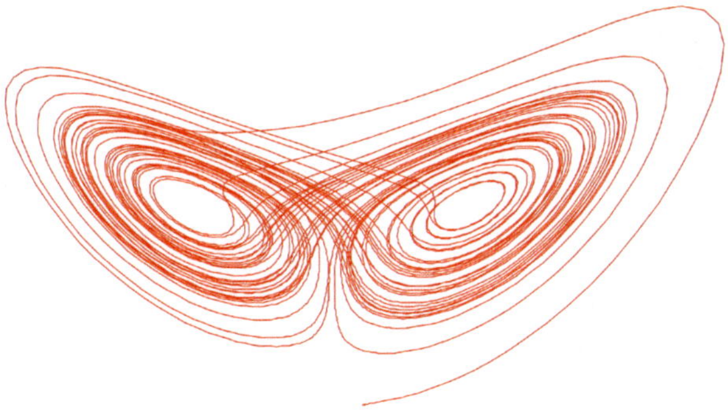

Figure 1.14. The Lorenz attractor: a celebrated example of strange attractor for a nonlinear chaotic dynamical system providing an extremely simplified model of the weather.

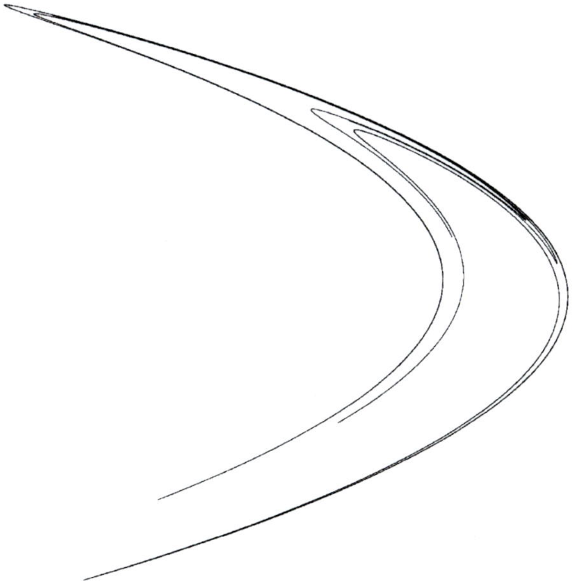

Figure 1.15. The Hénon attractor: another classic example of strange attractor for a nonlinear chaotic dynamical system describing (in a very simplified manner) the chaotic orbits of celestial bodies.

[**519**]; see Figure 1.16. It is the (chaotic) attractor for a system of three nonlinear differential equations (thus, corresponding to a continuous time chaotic dynamical system). Interestingly, Rössler interpreted it as being a model for "taffy pulling" (associated with the repeated folding of hot candy).

We point out that in his joint book [**482**], entitled *Encounter with Chaos: Self-Organized Hierarchical Complexity in Semiconductor Experiments*, Rössler

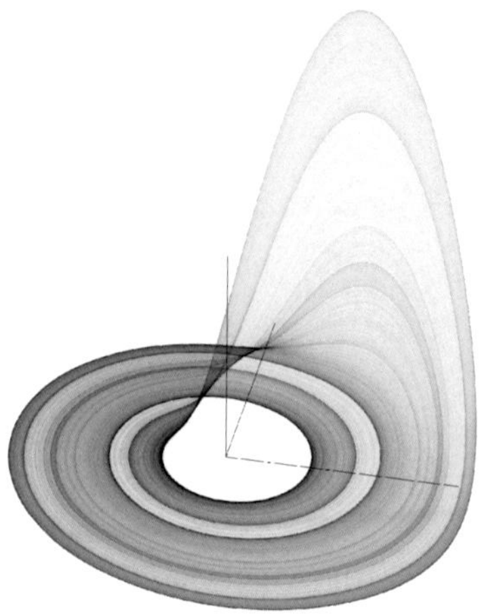

Figure 1.16. The Rössler attractor, the amazing strange attractor of a nonlinear continuous time, chaotic dynamical system modeling "taffy pulling".

provides an interesting account of work on chaotic dynamics, from a practical and physical perspective.

Strange attractors are pervasive in the study of (typically nonlinear) chaotic dynamical systems and their prevalence and complexity are one of the key reasons why the interdisciplinary fields of chaos and of fractals are intimately connected. We refer, e.g., the interested reader to [**32**], [**138, 139**], [**410**], [**411**], [**525, 526**], [**527**], [**237**], [**519**], [**482**], [**567**], [**568**], [**96, 97**], [**493, 494**], [**419**], [**425**], [**475**], [**483**] for further information about this fascinating topic and its multifaceted relationship with fractal geometry.

1.8. A Sample of Books on Fractal Geometry

Textbooks (at the undergraduate or graduate level), along with accessible research monographs or memoirs and collective edited volumes, on various aspects of fractal geometry—in relation to, for example, analysis, measure theory, harmonic analysis, spectral theory, probability theory, dynamical systems, number theory, geometric measure theory, differential equations, quasicrystals and aperiodic tilings, mathematical and theoretical physics (including statistical mechanics, condensed matter physics and astronomy or cosmology), as well as other scientific applications (to, e.g., engineering, computer science, chemistry, geology, economics and finance, along with the biological and medical sciences), without forgetting the arts (e.g., computer graphics, painting, architecture, and music)—include the texts by the following authors:

P. Alonso Ruiz, J. P. Chen, L. G. Rogers, R. S. Strichartz and A. Teplyaev [**7**], M. Baake and U. Grimm [**14**], M. Bader [**15**], C. Bandt, K. J. Falconer, and M. Zähle [**19**], C. Bandt, J. Flachsmeyer, and H. Haase [**20**], M. F. Barnsley

([**28**] and [**29**]), M. F. Barnsley, R. L. Devaney, B. B. Mandelbrot, H.-O. Peitgen, D. Saupe and R. F. Voss [**31**], C. J. Bishop and Y. Peres [**50**], A. Bunde and S. Havlin [**62**], L. Calvet and A. J. Fisher [**65**], D. Carfi, M. L. Lapidus, E. P. J. Pearse and M. van Frankenhuijsen ([**72**], Parts I & II), C. David and M. L. Lapidus ([**115–117**]), C. David, M. L. Lapidus, and G. Radunović [**119**],[51] G. David and S. Semmes ([**121**], [**543**]), R. L. Devaney [**138**], R. L. Devaney and L. Keen [**139**], G. A. Edgar ([**152**] and [**153**]), C. J. G. Evertsz, H.-O. Peitgen, and R. F. Voss [**170**], K. J. Falconer ([**171**], [**172**], [**175**], and [**176**]), F. Family and T. Vicsek [**177**], J. Feder [**180**], H. Federer [**182**], M. Fleischmann, D. J. Tildesley, and R. C. Ball [**192**], M. L. Frame and N. Cohen [**194**], M. L. Frame and B. B. Mandelbrot [**195**], J. M. Fraser [**199**], J. Fröhlich [**204**], H. Furstenberg [**208**], J. Gleick [**215**], J.-F. Gouyet [**217**], A. Harrison [**230**], C. Q. He and M. L. Lapidus [**235**], A. Heck and J. M. Perdang [**236**], H. Herichi and M. L. Lapidus [**238**], A. M. Hinz, S. Klavžar, and C. Petr [**250**], M. Jarnicki and P. Pflug [**271**], J.-P. Kahane and R. Salem [**281**], J. Kigami [**293**], M. L. Lapidus ([**326**] and [**330**]), M. L. Lapidus, G. Radunović, and D. Žubrinić [**356**], M. L. Lapidus and M. van Frankenhuijsen ([**370**], [**374**], and [**375**], along with the "Mandelbrot Jubilee", [**372**, Parts 1 & 2]), H. Lauwerier [**381**], G. F. Lawler [**382**], A. Le Méhauté [**395**, **396**], S. L. Lipscomb [**408**], B. B. Mandelbrot ([**418–420**] and [**422–425**]), B. B. Mandelbrot and R. L. Hudson [**426**], P. Mattila [**430**], D. Mumford, C. Series, and D. Wright [**455**], T. Nakayama and K. Yakubo [**460**], R. G. Niemeyer, E. P. J. Pearse, J. A. Rock, and T. Samuel [**463**], T. F. Nonnenmacher, G. A. Losa, and E. R. Weibel [**467**], L. Nottale [**468**], T. N. Palmer [**475**], H.-O. Peitgen, H. Jürgens, and D. Saupe [**483**], H.-O. Peitgen and P. Richter [**484**], Ya. B. Pesin [**491**], Ya. B. Pesin and V. Climenhaga [**492**], I. Peterson [**493**, **494**], J. Rataj and M. Zähle [**503**], H. Sagan [**532**], B. Sapoval [**534**], M. R. Schroeder [**542**], M. Senechal [**544**], S. Sternberg [**567**], I. Stewart [**568**], I. Stewart and A. C. Clarke [**570**], R. S. Strichartz [**574**], Tan Lei [**394**], and C. Tricot [**593**].

This is, by far, not an exhaustive list and is geared towards textbooks, research monographs, and collective volumes published in the English language.

Two classic books on the key notions of Hausdorff dimension (or measure) [**518**] and of topological dimension [**260**] are, respectively, authored by C. A. Rogers and by W. Hurewicz and H. Wallman.

The book edited by G. A. Edgar, [**152**], and entitled *Classics on Fractals*, consists of a very interesting collection of reprints (in English translation, whenever needed) of many of the original and classic papers which eventually gave rise to fractal geometry, as we know it nowadays. This collection includes papers by Georg Cantor, Karl Weierstrass, Helge von Koch, Abram S.

[51] References [**119**], [**116**], and [**117**] are the forthcoming sequels to the present book.

Besicovitch (and his collaborators), Felix Hausdorff, Georges Bouligand, Constantin Carathéodory, Paul Lévy, Georges de Rham, Lev Semenovich Pontryagin and Lev Genrikovich Schnirelmann, Andrey Nikolaievich Kolmogorov and Vladimir Mikhailovich Tihomirov, Patrick A. P. Moran, John M. Marstrand, and Benoît B. Mandelbrot.

Finally, we mention that the survey article [**321**] by the first author (M. L. Lapidus) of the present book—entitled *Fractal geometry and applications —An introduction to this volume*, and which serves as an introduction to the two parts of the Mandelbrot Jubilee, [**372**]—provides a personal perspective on some of the history and the mathematical origins of the subject of fractal geometry.

Exercises

Exercise 1.41. Prove that if $A \subseteq \mathbb{R}^m$ and $\dim_B(A)$ exists, then for each $\lambda > 0$, we have that $\dim_B(\lambda A)$ exists and $\dim_B(\lambda A) = \dim_B(A)$.

Exercise 1.42.

(a) Show that if $A \subseteq \mathbb{R}^m$ and $\dim_B(A)$ exists, then, for any map $\varphi : \mathbb{R}^m \to \mathbb{R}^m$ which is either a translation, a rotation, or a reflection of $\mathbb{R}^m$, we have that $\dim_B(\varphi(A))$ exists and $\dim_B(\varphi(A)) = \dim_B(A)$.

(b) Deduce from part (a) and Exercise 1.41 that if the map φ is now assumed to be a similarity transformation of $\mathbb{R}^m$,[52] then the conclusion of part (a) still holds.

[52]That is, $\varphi : \mathbb{R}^m \to \mathbb{R}^m$ is the composite of a translation, a rotation (and possibly, also, a reflection) of $\mathbb{R}^m$.

Chapter 2

Metric Spaces and Fixed Point Theorem

Mathematics is the most beautiful and most powerful creation of the human spirit. Mathematics is as old as Man.

Stefan Banach

The basic ideas and simplest facts of set-theoretic topology are needed in the most diverse areas of mathematics: the concepts of topological and metric spaces, of compactness, the properties of continuous functions and the like are often indispensable.

Pavel S. Alexandrov and Heinz Hopf, 1935, [**4**].

In reality, the origin of the notion of derivatives is in the vague feeling of the mobility of things, and of the greater or lesser speed with which phenomena take place; this is well expressed by the terms fluent *and* fluxion, *which were used by Newton and which we may believe were borrowed from the ancient mathematician Heraclitus.*

Émile Picard

A mathematician is a person who can find analogies between theorems; a better mathematician is one who can see analogies between proofs, and the best mathematician can notice analogies between theories. One can imagine that the ultimate mathematician is one who can see analogies between analogies.

Stefan Banach

Many of the fundamental results on the real line $\mathbb{R}$ covered in an undergraduate course on real analysis follow from properties of the absolute value function. For instance, notions of limit and topological concepts such as continuity, open sets, and closed sets can all be expressed in terms of the distance given by $|x-y|$, for any $x, y \in \mathbb{R}$. It turns out that these concepts extend readily to more abstract settings when one considers an appropriately defined *metric* on a given set. Loosely speaking, a metric is a function that defines a notion of distance between two elements of a set. Later, in Chapter 4 below, we will use the properties of a special metric known as the *Hausdorff metric* to construct fractal subsets in Euclidean space, including many of the examples discussed in Chapter 1.

Metrics can also provide a backbone for the definition of *measures*. A measure is a set-valued function that assigns a meaningful notion of size to sets. Classical notions of length, area, and volume are all examples of measures. In Chapter 3 below, we discuss some of the basic properties of measures. However, a full, technical discussion of measure theory is beyond the scope of this text, so we focus on the definitions and properties that suit our needs; see, however, Appendix B.

2.1. Definition and Properties

The following definition allows one to concretely define a notion of distance between points in a given set.

Definition 2.1 (*Metric*). A *metric* on a set X is a $[0,+\infty)$-valued map $d\colon X \times X \to [0,+\infty)$ satisfying the following properties, for all $\boldsymbol{x}, \boldsymbol{y}, \boldsymbol{z} \in X$:

(i) $d(\boldsymbol{x}, \boldsymbol{y}) = 0$ iff $\boldsymbol{x} = \boldsymbol{y}$ (*Definiteness*);

(ii) $d(\boldsymbol{x}, \boldsymbol{y}) = d(\boldsymbol{y}, \boldsymbol{x})$ (*Symmetry*);

(iii) $d(\boldsymbol{x}, \boldsymbol{y}) \leq d(\boldsymbol{x}, \boldsymbol{z}) + d(\boldsymbol{z}, \boldsymbol{y})$ (*Triangle Inequality*).

Moreover, a *metric space* (X, d) is a set X equipped with a metric d.

Remark 2.2. Given a metric space (X, d) and a subset $Y \subseteq X$, the *subspace* $(Y, d|_Y)$ is the metric space on the set Y with metric $d|_Y$ defined as the restriction of the metric d to the domain $Y \times Y$. That is,

$$d|_Y(\boldsymbol{x}, \boldsymbol{y}) := d(\boldsymbol{x}, \boldsymbol{y}), \text{ for all } \boldsymbol{x}, \boldsymbol{y} \in Y. \tag{2.1}$$

Example 2.3 (*Classic Metric on* $\mathbb{R}^m$). The Pythagorean Theorem allows us to define the classic notion of Euclidean distance on $\mathbb{R}^m$, called the *Euclidean distance* and denoted by $d_m = d_{m,2}$, for each $m \in \mathbb{N}$. Accordingly, the distance between two points $\mathbf{x} = (x_1, \ldots, x_m)$ and $\mathbf{y} = (y_1, \ldots, y_m)$ in the Euclidean space $\mathbb{R}^m$ is given by[1,2]

$$d_{m,2}(\mathbf{x},\mathbf{y}) := \sqrt{\sum_{j=1}^{m}(x_j - y_j)^2}. \tag{2.2}$$

Remark 2.4. Unless otherwise specified, throughout the text a Euclidean space $\mathbb{R}^m$ is assumed to be equipped with its classic metric d_m and *Euclidean norm* $\|\cdot\|$, given by $\|\mathbf{x}\| := \sqrt{\sum_{j=1}^{m} x_j^2}$. Note that $d_m(\mathbf{x},\mathbf{y}) = \|\mathbf{x}-\mathbf{y}\|$ and $\|\mathbf{x}\| = d_m(\mathbf{x},\mathbf{0})$, where $\mathbf{0} := (0,\ldots,0)$.

Example 2.5 (*Discrete Metric*). On a given nonempty set X, one can define a wide variety of metrics. Perhaps the simplest metric on X is the *discrete metric*, denoted d_D and given for all $\boldsymbol{x},\boldsymbol{y} \in X$ by

$$d_D(\boldsymbol{x},\boldsymbol{y}) := \begin{cases} 1, & \boldsymbol{x} \neq \boldsymbol{y}, \\ 0, & \boldsymbol{x} = \boldsymbol{y}. \end{cases} \tag{2.3}$$

Example 2.6 (*Taxicab Metric*). Another metric that can be defined on $\mathbb{R}^m$ for any $m \in \mathbb{N}$ is the *taxicab metric*, denoted d_T. It is defined by

$$d_T(\mathbf{x},\mathbf{y}) := \sum_{j=1}^{m} |x_j - y_j|, \text{ for all } \mathbf{x},\mathbf{y} \in \mathbb{R}^m.$$

In the case where $m = 2$, this metric determines the distance between $\mathbf{x},\mathbf{y}$ in $\mathbb{R}^2$ by moving along horizontal and vertical lines, but not along any other line. The metric d_T has properties very similar to those of d_m, but as discussed below, they 'see' the underlying space $\mathbb{R}^m$ rather differently.

Example 2.7 (*Metrics for Continuous Functions*). Metrics can be defined on spaces (or sets) which are much more abstract and elaborate than a Euclidean space like $\mathbb{R}^m$. Consider $C[0,1]$, the space of continuous functions on the closed unit interval $[0,1]$.

One metric on $C[0,1]$, denoted δ_1, determines the 'distance', $\delta_1(f,g)$, between two functions $f,g \in C[0,1]$ as the area between the graphs of the functions over the interval $[0,1]$. That is, δ_1 is defined by

$$\delta_1(f,g) := \int_0^1 |f(x) - g(x)|dx, \text{ for all } f,g \in C[0,1]. \tag{2.4}$$

Another metric on $C[0,1]$, denoted δ_∞, determines the 'distance' between two functions $f,g \in C[0,1]$ as the supremum of the distance between heights

[1] For clarity, we adopt the following convention throughout this book: We will usually write, for example, $\mathbf{x} = (x_1,\ldots,x_m) \in \mathbb{R}^m$, while we will write $\boldsymbol{x} \in \mathbb{R}^m$ (or, more generally, $\boldsymbol{x} \in X$) when no underlying components of $\boldsymbol{x}$ are explicitly involved. In the latter case, when $m = 1$, we simply write x instead of $\boldsymbol{x}$.

[2] For simplicity's sake, we often write d_m instead of $d_{m,2}$, when no ambiguity may arise.

of the functions evaluated at the same value in $[0,1]$. More specifically, δ_∞ is defined by

$$\delta_\infty(f,g) := \sup\{|f(x)-g(x)| : x \in [0,1]\}. \tag{2.5}$$

The example of a metric space which is of interest throughout this text is the Hausdorff metric d_H on the space $\mathbb{K}(\mathbb{R}^m)$ of nonempty compact subsets of $\mathbb{R}^m$. Further discussion of the general properties of metrics is in order before investigating the structure of the metric space $(\mathbb{K}(\mathbb{R}^m), d_H)$ and, in particular, before discussing the mathematical construction of fractals in this setting; see Section 4.1 below.[3]

The notion of *diameter* defined below assigns a useful value for the size, in a certain respect, of a subset of a metric space.

Definition 2.8 (*Diameter*). Let (X,d) be a metric space and let $A \subseteq X$. The *diameter* of A, denoted $\operatorname{diam}(A)$, is given by

$$\operatorname{diam}(A) := \sup\{d(\boldsymbol{x},\boldsymbol{y}) : \boldsymbol{x},\boldsymbol{y} \in A\}. \tag{2.6}$$

By convention, we let $\operatorname{diam}(\emptyset) := +\infty$ since the diameter is given by a supremum. It follows that $\operatorname{diam}(A) < \infty$ if and only if (iff, in short) A is bounded (i.e., $A \subseteq B$, for some ball B) and nonempty.

Proposition 2.9. *Let (X,d) be a metric space. Then, the following statements clearly hold*:

(i) *For $A \subseteq X$,* $\operatorname{diam}(A) = 0$ *iff A is a singleton.*

(ii) *If $A \subseteq B \subseteq X$, then* $\operatorname{diam}(A) \le \operatorname{diam}(B)$.

2.2. Analysis and Topology

In this section, we briefly review several basic notions from topology, the theory of metric spaces, and analysis. The reader who is familiar with these elementary notions may take note of the terminology and the notation and then move directly to Section 2.3.

Metrics allow for notions of distance, and hence notions of limits, to be concretely defined in a variety of abstract settings.

Definition 2.10 (*Convergent Sequence in a Metric Space*). Let (X,d) be a metric space. A sequence $\{\boldsymbol{x}_n\}_{n=1}^\infty \subseteq X$ *converges* to $\boldsymbol{x} \in X$ if for every $\varepsilon > 0$, there is $N \in \mathbb{N}$ such that for all $n \ge N$, we have $d(\boldsymbol{x}_n,\boldsymbol{x}) < \varepsilon$. In this case, we say that $\{\boldsymbol{x}_n\}_{n=1}^\infty$ is *convergent* and $\boldsymbol{x}$ is its (necessarily unique) *limit*. Also, we write $\boldsymbol{x}_n \to \boldsymbol{x}$, as $n \to \infty$, or $\lim_{n\to\infty} \boldsymbol{x}_n = \boldsymbol{x}$.

[3] We also consider the case when $\mathbb{R}^m$ is replaced by any (complete) metric space X.

Exercise 2.11. Show that the function $d : X \times X \to [0, +\infty)$ is continuous.

[*Hint*: Use (and prove) the fact that in a metric space, continuity can be expressed in terms of sequences.]

Topological concepts are also readily defined in the context of metric spaces.

Definition 2.12 (*Balls and Neighborhoods*). Let (X, d) be a metric space, $\boldsymbol{x}_0 \in X$, and $r > 0$. The *open ball centered at* $\boldsymbol{x}_0$ *of radius* r is the set $B_{\boldsymbol{x}_0}(r)$ given by $B_{\boldsymbol{x}_0}(r) := \{\boldsymbol{x} \in X : d(\boldsymbol{x}_0, \boldsymbol{x}) < r\}$. The *closed ball centered at* $\boldsymbol{x}_0$ *of radius* r is the set $\overline{B_{\boldsymbol{x}_0}}(r)$ given by $\overline{B_{\boldsymbol{x}_0}}(r) := \{\boldsymbol{x} \in X : d(\boldsymbol{x}_0, \boldsymbol{x}) \le r\}$. Also, a set U is a *neighborhood* of $\boldsymbol{x}_0$ if it contains an open ball centered at $\boldsymbol{x}_0$, i.e., if for some $r > 0$, we have $B_{\boldsymbol{x}_0}(r) \subseteq U$.

Example 2.13. For a given $r > 0$, the set $B_{\boldsymbol{x}_0}(r)$ in $(\mathbb{R}^3, d_3)$ is reasonably called an *open ball* since its boundary is indeed a sphere. In $(\mathbb{R}, d_1)$, a 'ball' $B_{\boldsymbol{x}_0}(r)$ is an open interval of length $2r$ with midpoint $\boldsymbol{x}_0$, whereas in $(\mathbb{R}^2, d_2)$, a 'ball' $B_{\boldsymbol{x}_0}(r)$ is an open disk of radius r with center $\boldsymbol{x}_0$.

Now, in the case of the taxicab metric d_T on $\mathbb{R}^3$, the word "ball" may not seem as appropriate. Indeed, in this setting, a ball $B_{\boldsymbol{x}_0}(r)$ is shaped like a "diamond" with triangular faces. More specifically, it is a regular octahedron; as an exercise, the interested reader may wish to check that the counterpart of this latter statement is also true in $(\mathbb{R}^m, d_T)$, where $m \ge 3$.

The reader is encouraged to describe, in his or her own words, what balls look like in the metric spaces given in Examples 2.5 and 2.7.

Exercise 2.14. Determine the balls of $(\mathbb{R}^m, d_{m,p})$, where[4]

$$d_{m,p}(\mathbf{x}, \mathbf{y}) := \left(\sum_{j=1}^{m} |x_j - y_j|^p \right)^{1/p}, \text{ for all } \mathbf{x}, \mathbf{y} \in \mathbb{R}^m, \tag{2.7}$$

where $p \ge 1$ is a real number. (Observe that when $p = 2$, we have $d_{m,2} = d_m$, the Euclidean metric, in our previous notation.) When $m = 2$, sketch these balls for various values of p and determine their limiting shapes when $p \to 1^+$, $p \to +\infty$, and $p \to 2^+$, respectively.

A more difficult, but classic, exercise consists in showing that $d_{m,p}$ is actually a metric on $\mathbb{R}^m$, for any $p \ge 1$. You may also address this question.

In addition to balls, *rectangles* play an important role in the analysis of fractals and, in particular, the construction of the Lebesgue measures.

Definition 2.15 (*Rectangles in Euclidean Spaces*). Consider a Euclidean space $(\mathbb{R}^m, d_m)$, where $d_m = d_{m,2}$. A *rectangle* in $\mathbb{R}^m$ is a Cartesian product of m

[4]Here, as before, $\mathbf{x} = (x_1, \dots, x_m) \in \mathbb{R}^m$ and $\mathbf{y} = (y_1, \dots, y_m) \in \mathbb{R}^m$.

intervals. An *open rectangle* is a rectangle R of the form[5] $R = \prod_{j=1}^{m}(a_j, b_j) := (a_1, b_1) \times (a_2, b_2) \times \cdots \times (a_m, b_m)$. A *closed rectangle* R is of the form $R = \prod_{j=1}^{m}[a_j, b_j] := [a_1, b_1] \times [a_2, b_2] \times \cdots \times [a_m, b_m]$.

Remark 2.16. Definition 2.12 allows for an immediate alternative to Definition 2.10 for convergence in a metric space in terms of neighborhoods as follows: Let (X, d) be a metric space. A sequence $\{\boldsymbol{x}_n\}_{n=1}^{\infty} \subseteq X$ *converges* to $\boldsymbol{x} \in X$ if for every $\varepsilon > 0$, there is a positive integer $N \in \mathbb{N}$ such that for all $n \geq N$, we have $\boldsymbol{x}_n \in B_{\boldsymbol{x}_0}(\varepsilon)$; i.e., $d(\boldsymbol{x}_0, \boldsymbol{x}_n) < \varepsilon$.

Definition 2.17 (*Cauchy Sequence*)**.** Let (X, d) be a metric space. A sequence $\{\boldsymbol{x}_n\}_{n=1}^{\infty} \subseteq X$ is *Cauchy* if for every $\varepsilon > 0$, there is $N \in \mathbb{N}$ such that for all $m, n \geq N$, we have $d(\boldsymbol{x}_m, \boldsymbol{x}_n) < \varepsilon$.

It is well known that, in $(\mathbb{R}, d_1)$—or, more generally, in the Euclidean space $(\mathbb{R}^m, d_m)$, for any integer $m \geq 1$—a sequence converges iff it is Cauchy. However, this fact relies on a key property of the real line that does not hold for metric spaces in general: When a sequence $\{\boldsymbol{x}_n\}_{n=1}^{\infty} \subseteq \mathbb{R}$ is Cauchy, there exists a unique $\boldsymbol{x} \in \mathbb{R}$ such that $\lim_{n\to\infty} \boldsymbol{x}_n = \boldsymbol{x}$. The existence of such a limit is by no means guaranteed, as is discussed in Example 2.19 just below. First, we need another definition.

Definition 2.18 (*Complete Metric Space*)**.** A metric space (X, d) is *complete* if every Cauchy sequence in X converges to an element of X.

Example 2.19. As is well known, the metric space $(\mathbb{R}, d_1)$ is complete, but the subspace $(\mathbb{Q}, d_1|_{\mathbb{Q}})$ is not.[6] Indeed, the real line $\mathbb{R}$ satisfies the *least upper bound property*: Every subset $S \subseteq \mathbb{R}$ that is bounded above has a least upper bound, called the supremum of S and denoted by $\sup S$. Accordingly, the Bolzano–Weierstrass Theorem (see, e.g., **[522]** or **[571]**) can be used to prove that $(\mathbb{R}, d_1)$ is complete.

On the other hand, the set of rational numbers $\mathbb{Q}$, does *not* satisfy the least upper bound property with respect to the metric $d_1|_{\mathbb{Q}}$. Consider a sequence $\{q_n\}_{n=1}^{\infty} \subseteq \mathbb{Q} \subseteq \mathbb{R}$ that converges to $\sqrt{2} \in \mathbb{R}\backslash\mathbb{Q}$, such as the increasing sequence of rational numbers with finite decimal expansions that approximate $\sqrt{2}$ given by $\{q_n\}_{n=1}^{\infty} = (1, 1.4, 1.41, 1.414, 1.4142, \ldots)$. Such a sequence is definitely Cauchy. However, in spite of the fact that each of its terms is rational, the sequence approaches an irrational value, namely, $\sqrt{2}$.

Exercise 2.20. Show that the metric space $(\mathbb{R}^m, d_{m,p})$ is complete, for any real number $p \geq 1$.[7] Generalize this property to $(\mathbb{R}^m, d_{m,\infty})$, where the metric $d_{m,\infty}$

[5] Here and in the sequel, the symbol $\times$ denotes the Cartesian product of sets.

[6] Here, $d_1|_{\mathbb{Q}}$ denotes the restriction of the metric d_1 to the subset $\mathbb{Q}$ of $\mathbb{R}$.

[7] The metric $d_{m,p}$ for $1 \leq p \leq \infty$ is defined in formula (2.7) of Exercise 2.14.

is given by $d_{m,\infty}(\mathbf{x},\mathbf{y}) := \max\{|x_j - y_j| : j = 1,\dots,m\}$, with $\mathbf{x} = (x_1,\dots,x_m) \in \mathbb{R}^m$ and $\mathbf{y} = (y_1,\dots,y_m) \in \mathbb{R}^m$.

The following definition provides a number of topological concepts that will appear throughout the text.

Definition 2.21. Let (X,d) be a metric space. The *complement* of $S \subseteq X$, denoted by S^c, is the set of all points in X that are not in S; i.e., $S^c := X\backslash S$. A subset G of X is *open* if, for each $\boldsymbol{x} \in G$, there is a neighborhood U of $\boldsymbol{x}$ such that $U \subseteq G$. A subset F of X is *closed* if its complement $F^c = X\backslash F$ is open. Moreover, a subset A of X is *disconnected* if it can be written as the disjoint union of two nonempty open sets. Otherwise, A is said to be *connected.* The largest connected subset of A containing a point $\boldsymbol{x} \in A$ is called the *connected component* of $\boldsymbol{x}$. Finally, a set A is *totally disconnected* if, for any $\boldsymbol{x},\boldsymbol{y} \in A$ with $\boldsymbol{x} \neq \boldsymbol{y}$, there exist disjoint open sets U and V such that $\boldsymbol{x} \in U$, $\boldsymbol{y} \in V$, and $A \subseteq U \cup V$.

Remark 2.22. The statement "A is open (resp., closed) in X" means that "A is an open (resp., closed) subset of X". For a given metric space (X,d), the following properties are readily verified: X and the empty set $\emptyset$ are both open and closed; arbitrary unions of open sets are open; finite intersections of open sets are open; arbitrary intersections of closed sets are closed; and finite unions of closed sets are closed.

Example 2.23. In the metric space $(\mathbb{R}, d_1)$, and for real numbers $a < b$, we have: Intervals of the form (a,b), $(a,+\infty)$, or $(-\infty,b)$ are open; intervals of the form $[a,b]$, $[a,+\infty)$, or $(-\infty,b]$ are closed; and intervals of the form $(a,b]$ and $[a,b)$ are neither open nor closed. In $(\mathbb{R}^m, d_m)$, open rectangles are open and closed rectangles are closed; see Definition 2.15 above.

In a metric space (X,d), we have that for each $\boldsymbol{x}_0 \in X$ and any $\varepsilon > 0$, the open ball $B_{\boldsymbol{x}_0}(\varepsilon)$ is open and, hence, any neighborhood of $\boldsymbol{x}_0$ contains an open neighborhood of $\boldsymbol{x}_0$. Similarly, the closed ball $\overline{B_{\boldsymbol{x}_0}}(r)$ is closed.

Definition 2.24 (*Limit Point*). Let (X,d) be a metric space and let $A \subseteq X$. A *limit point* of A is any point $\boldsymbol{x}_0$ of X such that, for all $\varepsilon > 0$, $B_{\boldsymbol{x}_0}(\varepsilon) \cap A \neq \emptyset$.[8] The set of limit points of A is denoted by $\overline{A}$ (or $cl(A)$). Clearly, $A \subseteq \overline{A}$.

In other words, $\boldsymbol{x}_0$ is a limit point of A (i.e., $\boldsymbol{x}_0 \in \overline{A}$) if every neighborhood of $\boldsymbol{x}_0$ contains an element of A. Equivalently, $\boldsymbol{x}_0$ is a limit point of A if there exists a sequence $\{\boldsymbol{x}_n\}_{n=1}^{\infty}$ of points of A such that $\boldsymbol{x}_n \to \boldsymbol{x}_0$, as $n \to \infty$. Note that $\overline{A}$ coincides with the *closure* of A, the smallest closed subset of X containing A.

[8] By contrast, $\boldsymbol{x}_0 \in X$ is an *accumulation point* of A if it is a limit point of A such that, in Definition 2.24, the nonempty set $B_{\boldsymbol{x}_0}(\varepsilon) \cap A$ is not reduced to the singleton $\{\boldsymbol{x}_0\}$. Equivalently, $\boldsymbol{x}_0 \in X$ is an accumulation point of A if there exists a sequence $\{\boldsymbol{x}_n\}_{n=1}^{\infty}$ of *distinct* points of A such that $\boldsymbol{x}_n \to \boldsymbol{x}$, as $n \to \infty$.

Exercise 2.25. Write down the precise definitions of the limit and of the continuity of a function from a metric space to another one. (Reason by analogy with the standard case encountered in elementary mathematics.)

Notation 2.26.

(i) For two functions $f, g : \mathbb{R} \to \mathbb{R}$, with g nonnegative, we write $f(x) = O(g(x))$, as $x \to +\infty$, if there exists a positive real number c such that for all sufficiently large x, $|f(x)| \leq cg(x)$.

More generally, if there exists C such that $|f(x)| \leq Cg(x)$, for all x sufficiently close to some value $a \in \mathbb{R} \cup \{\pm\infty\}$, then we write $f(x) = O(g(x))$, as $x \to a$. The same comment applies to parts (ii) and (iii) just below.

(ii) Let $f, g : \mathbb{R} \to \mathbb{R}$ be two functions such that g does not vanish in a neighborhood of a; see Remark 2.27 for the general case. If $\lim_{x\to a} f(x)/g(x) = 0$, then we write $f(x) = o(g(x))$, as $x \to a$. If $\lim_{x\to a} f(x)/g(x) = 1$, then we write $f(x) \sim g(x)$, as $x \to a$. Of course, f and g only need to be defined in a neighborhood of a.

(iii) If there exist positive constants c_1 and c_2 such that $c_1|f(x)| \leq g(x) \leq c_2|f(x)|$, for all x sufficiently close to some value $a \in \mathbb{R} \cup \{\pm\infty\}$, then we write $f(x) \asymp g(x)$, as $x \to a$. Also, if f is approximated by g, in some sense, then we write $f(x) \approx g(x)$.[9]

Remark 2.27. In part (ii) of Notation 2.26, one can allow g to vanish (even infinitely often) near a by stating instead that $f(x) \sim g(x)$, as $x \to a$, means that $g(x) = f(x)(1 + o(1))$, as $x \to a$, where $o(1)$ denotes a function tending to zero, as $x \to a$. The advantage of this more general and symmetric definition is that then, $\sim$ becomes an equivalence relation on the set of functions defined in a neighborhood of a.

In the sequel, for sequences (instead of for functions of a real variable), we will adopt entirely similar definitions to those introduced in Notation 2.26 and Remark 2.27.

Definition 2.28 (*Closure and Interior. Boundary, Density, and Perfectness*). Let (X, d) be a metric space and let $S \subseteq X$. Recall that the *closure* of S, denoted $\overline{S}$, was introduced in Definition 2.24. The *interior* of S, denoted S^o (or $\mathring{S}$), is the union of all open sets contained in S; i.e., S^o is the largest open set contained in S. Moreover, the *boundary* of S, denoted ∂S, is given by $\partial S := \overline{S} \backslash S^o$. Also, S is said to be *perfect* if it is closed and does not have any isolated point, i.e., if S coincides with its set of accumulation points; see footnote 8. Finally, a set $B \subseteq X$ is *dense* in $A \subseteq X$ if $\overline{B} = A$.

[9] Entirely analogous definitions as in (i)–(iii) can be given, for example, if $x \to a^+$ (provided this makes sense).

Remark 2.29. For a given metric space (X, d) and a set $S \subseteq X$, the following statements are readily verified (proofs are left as exercises): S is closed iff $S = \overline{S}$; furthermore, $\overline{S}$ and ∂S are closed; S^o is open; $\partial S = \partial(S^c)$; for each point $x \in \partial S$, every neighborhood of x contains a point from S and a point from S^c, the complement of S in X; and $B \subseteq A$ is dense in A (i.e., $\overline{B} = A$) iff for every $\boldsymbol{a} \in A$ and $\varepsilon > 0$, there exists $\boldsymbol{b} \in B$ such that $\boldsymbol{b} \in B_{\boldsymbol{a}}(\varepsilon)$.

As with the definition of convergence of sequences in a metric space in Definition 2.10, the definition of continuity for functions from one metric space to another is very similar to the classical ε-δ definition of continuity for functions from $\mathbb{R}$ to $\mathbb{R}$.[10] Indeed, one simply replaces the instances of absolute value with the appropriate metrics. By necessity of concision, we omit the corresponding formal definition.

The following definition provides a few additional useful concepts that are analogous to continuity.

Definition 2.30 (*Hölder and Lipschitz Continuous Functions*). Let (X_1, d_1) and (X_2, d_2) be metric spaces and let $f : X_1 \to X_2$ be a function. Then, given $\alpha \in (0, 1]$, f is called a *Hölder continuous function of Hölder exponent* α if

$$d_2(f(x), f(y)) \leq c \cdot (d_1(x, y))^{\alpha}, \text{ for all } \boldsymbol{x}, \boldsymbol{y} \in X_1, \tag{2.8}$$

for some positive constant $c \in \mathbb{R}$. In the case when $\alpha = 1$, the function f is called *Lipschitz*. If there are $c_1, c_2 \in \mathbb{R}$ such that $0 < c_1 \leq c_2 < \infty$ and

$$c_1 \cdot (d_1(x, y)) \leq d_2(f(x), f(y)) \leq c_2 \cdot (d_1(x, y)), \text{ for all } x, y \in X_1, \tag{2.9}$$

then f is said to be *bi-Lipschitz*.

Many of the sets discussed in this text are *compact*.

Definition 2.31 (*Compact Space*). Let (X, d) be a metric space and let $A \subseteq X$. An *open cover* of A is a collection of open subsets of X whose union contains A. Given an open cover of A, a *finite subcover* is a finite subcollection of the open cover whose (finite) union contains A. A subset K of X is *compact* if every open cover of K has a finite subcover. In this case, we say that 'K is compact in X'.

Remark 2.32. Using mathematical notation, compactness is defined as follows: Let (X, d) be a metric space and let $K \subseteq X$. The set K is *compact* if for any open cover $\{G_\lambda\}_{\lambda \in \Lambda}$ (where Λ is some indexing set and each G_λ is open in X), there exists a finite subcollection $\{G_{\lambda_1}, \ldots, G_{\lambda_q}\} \subseteq \{G_\lambda\}_{\lambda \in \Lambda}$ such that $K \subseteq \bigcup_{j=1}^{q} G_{\lambda_j}$.

[10]See also Exercises 2.11 and 2.25 above.

The notion of compactness is very useful but notoriously difficult to understand when encountered for the first time. Thankfully, in the case of the Euclidean spaces $(\mathbb{R}^m, d_m)$, compactness can be characterized rather concretely in terms of closed and *bounded* sets.

Definition 2.33 (*Bounded Set*). Let (X, d) be a metric space. A subset $A \subseteq X$ is *bounded* if there is a point $x_0 \in X$ and a real number $M > 0$ such that $A \subseteq \{\boldsymbol{x} \in X : d(\boldsymbol{x}, \boldsymbol{x}_0) \leq M\}$. In other words, A is bounded if it is contained in a ball.

In the setting of metric spaces, compactness is characterized in terms of sequences as follows, in Theorem 2.35.

Definition 2.34 (*Sequentially Compact Space*). Let (X, d) be a metric space. A set $K \subseteq X$ is *sequentially compact* if each sequence in K has a subsequence that converges to a point in K.

Theorem 2.35 (See, e.g., [**456**, Theorem 28.2]). *Let (X, d) be a metric space and let $K \subseteq X$. Then, K is compact iff K is sequentially compact.*

Theorem 2.36 (Heine–Borel Theorem; see, e.g., [**456**, Theorem 27.3]). *In a Euclidean space $(\mathbb{R}^m, d_m)$, a set $S \subseteq \mathbb{R}^m$ is compact iff S is closed and bounded.*

Example 2.37. By Theorem 2.36, closed rectangles of the form $[a_1, b_1] \times [a_2, b_2] \times \cdots \times [a_m, b_m]$ are compact. Also, we immediately have that a finite union of compact subsets of a metric space is compact.

Remark 2.38. We caution the reader, however, that in general metric spaces, a compact subset is always closed and bounded but that the converse is not usually true.

The following theorem (along with the exercise following it) provides a useful, and purely topological, way of rephrasing the compactness property.

Theorem 2.39 (Finite Intersection Property). *Let $\{K_\lambda\}_{\lambda \in \Lambda}$ be a collection of* compact *subsets in the metric space (X, d) such that the intersection of every finite subcollection of $\{K_\lambda\}_{\lambda \in \Lambda}$ is nonempty. Then, $\bigcap_{\lambda \in \Lambda} K_\lambda$ is nonempty.*

Proof. (When $X = \mathbb{R}^m$). Fix $K_{\lambda_0} \in \{K_\lambda\}_{\lambda \in \Lambda}$ and assume that no point of K_{λ_0} belongs to every K_λ. Note that by the Heine–Borel Theorem (Theorem 2.36), each K_λ is compact. Hence, each K_λ^c is open and the collection $\{K_\lambda^c\}_{\lambda \in \Lambda}$ is an open cover of K_{λ_0}. Since K_{λ_0} is compact, there is a finite subcollection $\{K_{\lambda_1}^c, \ldots, K_{\lambda_q}^c\} \subseteq \{K_\lambda^c\}_{\lambda \in \Lambda}$ such that $K_{\lambda_0} \subseteq \bigcup_{j=1}^q K_{\lambda_j}^c$. However, this means that $\bigcap_{j=0}^q K_{\lambda_j} = \emptyset$, which contradicts the hypothesis. □

Exercise 2.40 (*Characterization of Compactness*). Show that in any metric space X, $K \subseteq X$ is compact iff it satisfies the Finite Intersection Property—as stated in Theorem 2.39, except for the fact that $\{K_\lambda\}_{\lambda\in\Lambda}$ is now assumed to consist of *closed* (rather than compact) subsets of K.

The next theorem, which complements and strengthens Theorem 2.39, under suitable hypotheses, provides us with a useful technique for constructing fractal sets. It is often used, in practice, to construct accumulation points (or limit points) of fractals viewed as 'limits' of prefractal approximations of self-similar sets, such as the Cantor set, the Sierpinski gasket, the Sierpinski carpet, or the Koch (snowflake) curve, and to associate an (infinite) address to such points; see Chapter 4.

Theorem 2.41. *Let $\{K_n\}_{n=1}^\infty$ be a sequence of nonempty compact subsets of the* complete *metric space (X, d) such that $K_{n+1} \subseteq K_n$, for all $n \geq 1$, and* $\operatorname{diam}(K_n) \to 0$, *as $n \to \infty$. Then, $\bigcap_{n=1}^\infty K_n = \{\boldsymbol{x}_0\}$, for some (necessarily unique) point $\boldsymbol{x}_0 \in \mathbb{R}^m$.*

Proof. Since each K_n is nonempty, we can obtain a sequence $\{\boldsymbol{x}_n\}_{n=1}^\infty$ such that $\boldsymbol{x}_n \in K_n$, for every $n \geq 1$. Then, by the hypothesis about the diameters, $\{\boldsymbol{x}_n\}_{n=1}^\infty$ is a Cauchy sequence in (X, d). Indeed, given n_1 and n_2 in $\mathbb{N}$, we let $n := \max(n_1, n_2)$ and we then have that $\boldsymbol{x}_{n_1}, \boldsymbol{x}_{n_2} \in K_n$. Hence, $|\boldsymbol{x}_{n_1} - \boldsymbol{x}_{n_2}| \leq \operatorname{diam}(K_n) \to 0$, as $n_1, n_2 \to \infty$ (which clearly implies that $n \to \infty$). Thus, given $\varepsilon > 0$, there exists $N \geq 1$ such that $m_1, m_2 \geq N$ implies that $|\boldsymbol{x}_{m_1} - \boldsymbol{x}_{m_2}| < \varepsilon$, as claimed.

Now, since $\{\boldsymbol{x}_n\}_{n=1}^\infty$ is a Cauchy sequence in the complete metric space (X, d), it is convergent in X. Thus, there exists $\boldsymbol{x}_0 \in X$ such that $\boldsymbol{x}_n \to \boldsymbol{x}_0$ in (X, d). Next, fix $n \in \mathbb{N}$. We have that $\boldsymbol{x}_q \in K_n$, for any integer $q \geq n$. Hence, since K_n is closed, $\boldsymbol{x}_0 := \lim_{q\to\infty} \boldsymbol{x}_q$ belongs to K_n. Since $n \in \mathbb{N}$ was arbitrary, we conclude that $\boldsymbol{x}_0 \in \bigcap_{n=1}^\infty K_n$, as desired.

Finally, assume that $\boldsymbol{x}_0, \boldsymbol{y}_0 \in \bigcap_{n=1}^\infty K_n$. Then, $|\boldsymbol{x}_0 - \boldsymbol{y}_0| \leq \operatorname{diam}(K_n) \to 0$, as $n \to \infty$. It follows that $|\boldsymbol{x}_0 - \boldsymbol{y}_0| = 0$ and so, $\boldsymbol{x}_0 = \boldsymbol{y}_0$. We have thus shown that $\bigcap_{n=1}^\infty K_n = \{\boldsymbol{x}_0\}$, as required. □

A useful fact regarding continuous functions is that they preserve compactness. The proof of the following theorem is left as an exercise.

Theorem 2.42 (*Continuous Image of a Compact Set*). *Let (X_1, d_1) and (X_2, d_2) be metric spaces and let $f : X_1 \to X_2$ be a continuous function. We have that if S is compact in X_1, then $f(S)$ is compact in X_2.*

Given a complete metric space (X, d), the space of nonempty compact subsets of X is denoted by $\mathbb{K}(X)$. Much of the remainder of the text will concern the case where X is a Euclidean space $\mathbb{R}^m$, for some $m \in \mathbb{N}$. Most of the fractal sets investigated in this book are elements of $\mathbb{K}(\mathbb{R}^m)$.

We close this discussion by recalling and proving a well-known key topological property of open subsets of the real line (see, e.g., [**81**, **522**, **571**]), which will play an important role in several places in this book, for example, when studying the Cantor set and its natural topological generalizations (in Section 4.9) or, briefly, the notion of a fractal string (in Section 8.2) and, in much greater detail, in the sequel to this book, [**119**].

Theorem 2.43 (Disjoint Decomposition of Open Subsets of $\mathbb{R}$). *Let Ω be a (nonempty) open subset of $\mathbb{R}$. Then, there exists a finite or countable collection of pairwise disjoint, nonempty open intervals $\{I_j\}_{j\geq 1}$ of $\mathbb{R}$ contained in Ω and such that Ω is the disjoint union of this family:*

$$\Omega = \bigsqcup_{j\geq 1} I_j. \tag{2.10}$$

Up to reordering, this family $\{I_j\}_{j\geq 1}$ is unique. In fact, it consists of the connected components *of Ω, i.e., here, the largest subintervals (necessarily open, since Ω is open) contained in Ω.*

If, in addition, Ω is bounded—or, more generally, has finite length (i.e., $\mathrm{vol}_1(\Omega) < \infty$, where $\mathrm{vol}_1(\cdot)$ denotes the 1-dimensional Lebesgue measure of $\mathbb{R}$; see Chapter 3)—then each interval I_j has finite length (denoted ℓ_j) and $\mathrm{vol}_1(\Omega) = \sum_{j\geq 1} \ell_j$, which is clearly independent of the ordering of the sequence of intervals $\{I_j\}_{j\geq 1}$, as should be the case.

Proof. We focus on the existence and uniqueness (up to permutation) of the disjoint decomposition $\{I_j\}_{j\geq 1}$ of Ω into open intervals. Actually, the uniqueness part follows at once from the proof below, which will show that the I_j's are the *connected components* of Ω, as is stated in the theorem. The end of the statement of the theorem concerning the case when Ω is bounded (or, more generally, of finite length) is an immediate consequence of the disjoint decomposition (2.10), which implies the stated formula for $\mathrm{vol}_1(\Omega)$, by the (finite or) countable additivity of (Lebesgue measure on $\mathbb{R}$) $\mathrm{vol}_1 = \mathscr{L}^1$ and since $\mathscr{L}^1(I_j) = \mathrm{vol}_1(I_j)$ is equal to ℓ_j, the length of the interval I_j, for each j; see Chapter 3 below.

We must establish the existence of the disjoint decomposition (2.10). For clarity, the proof is divided into several steps.

Step 1 (Equivalence Relation on Ω). *Let $\Omega \subseteq \mathbb{R}$ be open, and given $x, y \in \Omega$, write $x \sim y$ iff $[x, y] = \overline{xy}$, the segment connecting x and y, is contained in Ω : $[x, y] \subseteq \Omega$.*[11] *Then, $\sim$ is an equivalence relation on Ω.*

[11] Here, without loss of generality, we may assume that $x \leq y$; otherwise, interchange the roles of x and y. Then, $[x, y] = \overline{xy} := \{\alpha y + (1-\alpha)x : \alpha \in [0, 1]\} \subseteq \mathbb{R}$.

Indeed, the reflexivity and the symmetry of $\sim$ are immediate to verify since if $x = y \in \Omega$, then $[x, y] = \{x\} \subseteq \Omega$, which implies that $x \sim x$. Also, note that, in general, $[x, y] = [\min\{x, y\}, \max\{x, y\}]$, which implies that $x \sim y \Leftrightarrow y \sim x$, as required.

Next, the transitivity of $\sim$ can be checked as follows. Let $x, y, z \in \Omega$ be such that $x \sim y$ and $y \sim z$. A moment's reflection (or drawing a picture) shows that, without loss of generality, we may assume that $x \leq y \leq z$. Since $x \sim y$ and $y \sim z$, we have that $[x, y] \subseteq \Omega$ and $[y, z] \subseteq \Omega$. Hence, clearly, we have that $[x, z] = [x, y] \cup [y, z] \subseteq \Omega$, which implies that $x \sim z$, as desired. This completes the proof of Step 1.

In the remainder of the proof, given $t \in \Omega$, we denote by $\mathring{t}$ the equivalence class of t (with respect to the equivalence relation $\sim$ defined in Step 1). We also denote by A the quotient space of Ω by $\sim$, so that $A = \Omega/\sim$, identified (as a set, via the Axiom of Choice) with a family of representatives (one representative per equivalence class). Hence, we have the following disjoint union:

$$\Omega = \bigsqcup_{t \in A} \mathring{t}. \tag{2.11}$$

Step 2 (Connectivity of the Equivalence Classes). *Each equivalence class $\mathring{t}$ (for $t \in A$) is an interval of $\mathbb{R}$ (necessarily contained in Ω). Hence, it is a* connected *subset of $\mathbb{R}$ (and thus also of Ω).*

Indeed, fix $t \in A$; then, by definition, $\mathring{t} \subseteq \Omega$. Furthermore, by a well-known characterization of the intervals of $\mathbb{R}$ (see, e.g., [**522**]), it suffices to check that $\mathring{t}$ is a *convex* subset of $\mathbb{R}$; i.e., for any $x, y \in \mathring{t}$, with $x \leq y$ (say), we have that $[x, y] \subseteq \Omega$. But, this is almost obvious since, by definition, $x \sim t$ and $y \sim t$, hence also $t \sim y$. Thus, by the transitivity of $\sim$, $x \sim y$ and so $[x, y] \subseteq \Omega$, as required. This concludes the proof of Step 2.

Step 3 (Openness of the Equivalence Classes). *Each equivalence class $\mathring{t}$ (for $t \in A$) is an* open *subset of $\mathbb{R}$ (or, equivalently, of Ω since $\mathring{t} \subseteq \Omega$ and Ω is open in $\mathbb{R}$)—and hence, by Step 2, it is a nonempty and* open interval *contained in Ω.*

Indeed, let $x \in \mathring{t}$. Thus $x \in \Omega$, by construction. Now, since Ω is open, there exists $\delta > 0$ such that the closed interval $J := [x - \delta, x + \delta]$ is contained in Ω. We now claim that $J \subseteq \mathring{t}$. Indeed, let $y \in J$; i.e., $x - \delta \leq y \leq x + \delta$. Then, since J is an interval, either $[x, y]$ or $[y, x]$ is contained in J (according to whether $x \leq y$ or $y \leq x$)—and hence also is contained in Ω. This precisely means that $x \sim y$ (i.e., $y \sim x$). But since $x \in \mathring{t}$ (i.e., $x \sim t$), we must also have that $y \in \mathring{t}$ (i.e., $y \sim t$), by the symmetry and transitivity of $\sim$. Hence, $J \subseteq \mathring{t}$, as claimed. Finally, $\mathring{t}$ is nonempty since $t \in \mathring{t}$. This completes the proof of Step 3.

Step 4 (Countability of the Quotient Space). *The quotient space $A := \Omega/\sim$ is at most countable; i.e., A is finite or countable. Consequently, the disjoint decomposition* (2.10) *holds.*

Indeed, in order to prove this statement, it suffices to exhibit an injection from A into $\mathbb{N}$, the set of positive integers. In light of Steps 1–3—and, in particular, of (2.11)—we have the following disjoint union:

$$\Omega = \bigsqcup_{a\in A} I_a, \tag{2.12}$$

where each I_a is a nonempty open interval of $\mathbb{R}$. Note that, given any $\alpha \in A$, by the density of $\mathbb{Q}$ in $\mathbb{R}$, we can choose a rational number $r_a \in I_a$ (again, implicitly, by using the Axiom of Choice). This provides an injection from A into $\mathbb{Q}$, defined by the map $a \mapsto r_a$ from A to $\mathbb{Q}$. (Indeed, if $\alpha, \beta \in A$ with $\alpha \neq \beta$, then $r_\alpha \neq r_\beta$ because $r_\alpha \in I_\alpha, r_\beta \in I_\beta$, and $I_\alpha \cap I_\beta = \emptyset$.) Since $\mathbb{Q}$ is countably infinite, it follows that A is at most countable, as claimed—and hence, in light of (2.12) (and after relabeling accordingly the elements of A), we deduce that the (finite or countable) disjoint decomposition (2.10) holds, as desired. This concludes the proof of Step 4—and thus also of the entire theorem, in light of our preliminary discussion at the beginning of the proof. □

The following key section is dedicated to setting up and proving a very well-known and important theorem, known as the *Contraction Mapping Principle.*[12]

2.3. Contraction Mapping Principle

This section covers the statement and proof of the very useful Contraction Mapping Principle. Among many other things, this theorem guarantees the existence of fractals and other sets constructed via iterated function systems, the main topic of Chapter 4.

Definition 2.44 (*Lipschitz Constant*). Let $f : X \to Y$, where (X, d_X) and (Y, d_Y) are metric spaces. The *Lipschitz constant* of f, denoted by $\mathrm{Lip}(f)$, is given by

$$\mathrm{Lip}(f) := \sup_{x\neq y} \frac{d_Y(f(\boldsymbol{x}), f(\boldsymbol{y}))}{d_X(\boldsymbol{x}, \boldsymbol{y})}, \text{ where } \boldsymbol{x}, \boldsymbol{y} \in X. \tag{2.13}$$

Furthermore, if $d_Y(f(\boldsymbol{x}), f(\boldsymbol{y})) = d_X(\boldsymbol{x}, \boldsymbol{y})$, for all $\boldsymbol{x}, \boldsymbol{y} \in X$, then f is called an *isometry* and then, clearly, $\mathrm{Lip}(f) = 1$.

[12] It is also known as the Banach (or the Picard) Fixed Point Theorem, among other names; see Stefan Banach's 1922 paper, [**17**]. Within the context of the theory of differential equations, and integral equations, a precursor of this theorem is the Picard–Lindelöf Theorem; see Ernst Lindelöf's 1894 article, [**403**].

Remark 2.45. A useful, alternative definition of Lip(f) is given as follows:

$$\text{Lip}(f) = \inf\{r \geq 0 : d_Y(\psi(\boldsymbol{x}), \psi(\boldsymbol{y})) \leq r \cdot d_X(\boldsymbol{x}, \boldsymbol{y}), \boldsymbol{x}, \boldsymbol{y} \in X\}. \tag{2.14}$$

In general, Lip$(f) \in [0, +\infty]$. However, if Lip$(f) < \infty$, then the function f is said to be *Lipschitz* (or *Lipschitz continuous*). This is the case iff there exists $M > 0$ such that $d_Y(f(\boldsymbol{x}), f(\boldsymbol{y})) \leq M \cdot d_X(\boldsymbol{x}, \boldsymbol{y})$, for all $\boldsymbol{x}, \boldsymbol{y} \in X$.

Exercise 2.46. Show the following properties concerning Lipschitz mappings between metric spaces:

(i) Show that a Lipschitz function, as defined in Remark 2.45, is continuous and even uniformly continuous.

(ii) Show that a Lipschitz function $f : X \to Y$ sends a Cauchy sequence of (X, d_X) onto a Cauchy sequence of (Y, d_Y).

(iii) Prove that the composition of two Lipschitz functions is also Lipschitz and give an estimate for the resulting Lipschitz constant, as given in Definition 2.44 or Remark 2.45.

(iv) A bijective (i.e., one-to-one and onto) map $f : X \to Y$ is said to be *bi-Lipschitz* if both f and its inverse f^{-1} are Lipschitz.[13] If possible, give an estimate for Lip(f^{-1}), expressed in terms of Lip(f). Moreover, show that all of the metric notions (such as Cauchy sequences, completeness, etc.) are preserved under bi-Lipschitz equivalence.

Exercise 2.47. Show the following two properties concerning isometries between metric spaces:

(i) Show that an isometry $f : X \to Y$ is always one-to-one (i.e., injective) but not necessarily onto (i.e., surjective), even if $X = Y$.

(ii) Show that if an isometry $f : X \to Y$ is bijective (i.e., according to part (i), if it is onto), then its inverse map $f^{-1} : Y \to X$ is also an isometry. More generally, if f is not onto, then show that $f^{-1} : f(X) \subseteq Y \to X$ is also an isometry.

Exercise 2.48. Show the following properties regarding isometries in Euclidean spaces and their matrix representations:

(i) Show that the isometries $f : \mathbb{R}^m \to \mathbb{R}^m$ from Euclidean space $(\mathbb{R}^m, d_m)$ to itself are necessarily linear, provided $f(\mathbf{0}) = \mathbf{0}$—or, more generally, are affine, if the origin $\mathbf{0}$ of $\mathbb{R}^m$ is not necessarily a fixed point of f; that is, there exists an $m \times m$ matrix A with real entries such that $f(\boldsymbol{x}) = A\boldsymbol{x}$, for all $\boldsymbol{x} \in \mathbb{R}^m$, in the first case, and there also exists $\boldsymbol{u} \in \mathbb{R}^m$, such that $f(\boldsymbol{x}) = A\boldsymbol{x} + \boldsymbol{u}$, for all $\boldsymbol{x} \in \mathbb{R}^m$, in the second case.

[*Hint*: Use the Polarization Identity.]

---

[13]Two metric spaces are said to be *bi-Lipschitz equivalent* if there is a bi-Lipschitz map between them.

(ii) Furthermore, deduce from part (i) of Exercise 2.47 (and from an elementary result from linear algebra) that f is bijective and hence that the matrix A is invertible; i.e., $\det A \neq 0$.

(iii) Characterize the matrices A arising in parts (i) and (ii). More specifically, show that they are orthogonal matrices; i.e., they satisfy the identity $A \cdot A^t \, (= A^t \cdot A) = I$, where A^t denotes the transpose of A, and hence have determinant ± 1. [The case when $\det A = 1$ (resp., $\det A = -1$) corresponds to the case when f and the rotation associated with A are orientation preserving (resp., orientation reversing).]

(iv) Finally, use part (iii) in order to determine (both, geometrically and algebraically) all of the isometries of $\mathbb{R}^2$ (resp., of $\mathbb{R}^3$). Generalize your answer to $\mathbb{R}^m$, with $m \geq 1$.

Definition 2.49 (*Contraction*). Let (X, d) be a metric space. A function $\psi : X \to X$ is a *contraction* on X if there exists a real number r such that $0 < r < 1$ and

$$d(\psi(\boldsymbol{x}), \psi(\boldsymbol{y})) \leq r \cdot d(\boldsymbol{x}, \boldsymbol{y}), \text{ for all } \boldsymbol{x}, \boldsymbol{y} \in X. \tag{2.15}$$

If equality holds for all $\boldsymbol{x}, \boldsymbol{y} \in X$ in (2.15), then ψ is said to be a *contracting similarity*:

$$d(\psi(\boldsymbol{x}), \psi(\boldsymbol{y})) = r \cdot d(\boldsymbol{x}, \boldsymbol{y}), \text{ for all } \boldsymbol{x}, \boldsymbol{y} \in X. \tag{2.16}$$

Contracting similarities are also referred to as *contracting similitudes*.

Definition 2.50 (*Scaling Ratio*). The *scaling ratio* of a contraction ψ, denoted by r_ψ, is the Lipschitz constant of ψ; that is, $r_\psi := \mathrm{Lip}(\psi)$.

Remark 2.51. Assume that ψ satisfies (2.16), as at the end of Definition 2.49; furthermore, let $r = r_\psi$. If $r > 0$ (instead of $0 < r < 1$), then ψ is called a *similarity* (or *similitude*), or else a *similarity transformation*.

Exercise 2.52. Show that if $X = \mathbb{R}^m$, the m-dimensional Euclidean space $(\mathbb{R}^m, d_m)$, then the similarities (resp., contractive similarities) of $\mathbb{R}^m$ are obtained by composing (direct or orientation-preserving) rotations, reflections, translations, and homotheties of scaling ratio $r > 0$ (resp., $r \in (0,1)$).

[*Hint*: Use the results of Exercise 2.48.]

Exercise 2.53. Identify $\mathbb{R}^2$ and $\mathbb{C}$, viewed as the complex plane. Then, determine all the similarities (also called *"similitudes"*) of $\mathbb{C}$. More specifically, show that they are exactly of the form (for some uniquely determined constants $a, b \in \mathbb{C}$) $\psi(z) = az + b$, for all $z \in \mathbb{C}$, if ψ is orientation preserving, or else $\psi(z) = a\overline{z} + b$, for all $z \in \mathbb{C}$, if ψ is orientation reversing. (Here, $\overline{z}$ denotes the complex conjugate of $z \in \mathbb{C}$.) Deduce that ψ is a contractive similarity iff $|a| < 1$.

Contracting similarities on $\mathbb{R}^m$ are completely characterized as follows. (See Exercise 2.48 and Exercise 2.52 just above.)

Proposition 2.54. *A function* $\varphi : \mathbb{R}^m \to \mathbb{R}^m$ *is a contracting similarity iff, for all* $\boldsymbol{x} \in \mathbb{R}^m$, $\varphi(\boldsymbol{x}) = r \cdot O(\boldsymbol{x}) + \mathbf{t}_0$, *where* $0 < r < 1$, O *is an orthogonal matrix (necessarily of determinant* ± 1*), and* $\mathbf{t}_0 \in \mathbb{R}^m$ *is fixed. In this case,* r *is equal to the scaling ratio of* φ*; that is,* $r = r_\varphi$.[14]

We refer to [**261**, Section 2.3, Proposition 1] for the statement and proof of a more general version of Proposition 2.54 in which the similarities are not necessarily contracting (r is allowed to be positive).

Exercise 2.55. Consider the (direct) rotations of $\mathbb{R}^m$ or, equivalently, the (real $m \times m$) orthogonal matrices of determinant 1, that is, the set of "direct" or orientation-preserving isometries of the Euclidean space $(\mathbb{R}^m, d_m)$. Show that they form a group, called the *direct orthogonal group* and often denoted by $SO(m)$. Prove that this group is commutative (or abelian) if $m = 1$ or $m = 2$, but that it is noncommutative (or nonabelian) if $m \geq 3$.

An immediate consequence of Definition 2.49 is that every contraction is a continuous function; see part (i) of Exercise 2.46 above.

The Contraction Mapping Principle, Theorem 2.58 below, is a tremendously useful theorem. In Chapter 4, we use it to construct fractal sets, in particular *self-similar sets*. As shown below in Theorem 2.61, Newton's Method for determining (or approximating) the zeros of a certain type of function is a consequence of the Contraction Mapping Principle. Also, in the study of ordinary differential equations, the Contraction Mapping Principle is used to guarantee the existence of unique solutions to initial-value problems. (See, e.g., [**49**, Chapter 6], [**522**], or [**571**, Chapter 11], along with Exercise 2.68 below.)

In order to prove the Contraction Mapping Principle, Theorem 2.58, we use a well-known result regarding geometric series as stated in the following lemma (or, rather, we use an analog of the proof of this lemma). Recall that, by definition, a *series* of real numbers $\sum_{n=0}^{\infty} x_n$ converges if the sequence of its partial sums $s_n := \sum_{j=0}^{n} x_j$ converges as a sequence in $(\mathbb{R}, d_1)$ and the *sum* of the series is the limit of the partial sums.

Remark 2.56. An important class of series consists of the p-series. Consider series of the form $\sum_{n=1}^{\infty}(1/n^p)$, where p is a given positive real number. A classic result in real analysis states that such a p-series converges iff $p > 1$.

[14]The same result holds, of course, if φ is not assumed to be contracting, except for the fact that the scaling ratio r is no longer assumed to be less than 1.

Lemma 2.57 (*Sum of a Geometric Series*). *Let $z \in \mathbb{C}$ with $|z| < 1$. Then, the geometric series $\sum_{n=0}^{\infty} z^n$ converges and its sum is given by*

$$\sum_{n=0}^{\infty} z^n = \frac{1}{1-z}. \tag{2.17}$$

Proof. The geometric series $\sum_{n=0}^{\infty} z^n$ has partial sums s_n given by

$$s_n := \sum_{j=0}^{n} z^j = 1 + z + z^2 + \cdots + z^n,$$

for each $n \in \mathbb{N}_0$. Hence, $(1-z)s_n = 1 - z^{n+1}$, which yields $s_n = \frac{1-z^{n+1}}{1-z}$. Then, since $|z| < 1$, taking the limit as n tends to infinity yields

$$\lim_{n\to\infty} s_n = \lim_{n\to\infty} \frac{1-z^{n+1}}{1-z} = \frac{1}{1-z};$$

i.e., the series $\sum_{n=0}^{\infty} z^n$ converges to the sum $1/(1-z)$. □

The above well-known elementary formula (2.17) in Lemma 2.57 will be useful in various parts of this book (especially in Section 8.2 and in Chapter 9, when dealing with complex dimensions and fractal zeta functions), as well as in its sequels, [**119**], [**116**], and [**117**].

We are now ready to state and prove the Contraction Mapping Principle, also called the (Picard–Banach) Fixed Point Theorem.

Theorem 2.58 (Contraction Mapping Principle). *Let (X, d) be a complete metric space. If $\psi : X \to X$ is a contraction on X with scaling ratio $r_\psi = r$, then there is a unique $\boldsymbol{x}^* \in X$, called* the fixed point *of ψ, such that*

$$\psi(\boldsymbol{x}^*) = \boldsymbol{x}^*. \tag{2.18}$$

Moreover, given any initial point $\boldsymbol{x}_0 \in X$, the sequence $\{\boldsymbol{x}_n\}_{n=0}^{\infty}$ given by

$$\boldsymbol{x}_n := \psi(\boldsymbol{x}_{n-1}) = \psi^n(\boldsymbol{x}_0), \text{ for every } n \in \mathbb{N}, \tag{2.19}$$

converges to $\boldsymbol{x}^*$.

Proof. Let $\boldsymbol{x}_0 \in X$. Let $m, n \in \mathbb{N}_0$ and, without loss of generality, assume $m < n$. Note that by the Triangle Inequality, we have

$$d(\boldsymbol{x}_m, \boldsymbol{x}_n) \leq d(\boldsymbol{x}_m, \boldsymbol{x}_{m+1}) + d(\boldsymbol{x}_{m+1}, \boldsymbol{x}_{m+2}) + \cdots + d(\boldsymbol{x}_{n-1}, \boldsymbol{x}_n). \tag{2.20}$$

Since ψ is a contraction with scaling ratio $0 < r < 1$, we have

$$d(\boldsymbol{x}_k, \boldsymbol{x}_{k+1}) = d(\psi(\boldsymbol{x}_{k-1}), \psi(\boldsymbol{x}_k)) \leq r \cdot d(\boldsymbol{x}_{k-1}, \boldsymbol{x}_k),$$

for any $k \in \mathbb{N}$. By induction, we obtain

$$d(\boldsymbol{x}_k, \boldsymbol{x}_{k+1}) \leq r^k \cdot d(\boldsymbol{x}_0, \psi(\boldsymbol{x}_0)), \text{ for all } k \in \mathbb{N}_0. \tag{2.21}$$

Hence, (2.20) and (2.21) imply that for all $m \in \mathbb{N}_0$ and $n \in \mathbb{N}$ such that $m < n$,

$$d(\boldsymbol{x}_m, \boldsymbol{x}_n) \leq (r^m + r^{m+1} + \cdots + r^{n-1}) \cdot d(\boldsymbol{x}_0, \psi(\boldsymbol{x}_0)); \tag{2.22}$$

furthermore, since $0 < r < 1$, we have

$$r^m + r^{m+1} + \cdots + r^{n-1} = r^m \cdot \left(\frac{1 - r^{n-m}}{1 - r}\right) \leq \frac{r^m}{1 - r}. \tag{2.23}$$

Thus, (2.22) and (2.23) imply that for all integers $m \geq 0$ and $n \geq 1$ such that $m < n$,

$$d(\boldsymbol{x}_m, \boldsymbol{x}_n) \leq \left(\frac{r^m}{1 - r}\right) \cdot d(\boldsymbol{x}_0, \psi(\boldsymbol{x}_0)). \tag{2.24}$$

Therefore, $\{\boldsymbol{x}_n\}_{n=0}^{\infty}$ is a Cauchy sequence in X.[15] Since X is complete, $\{\boldsymbol{x}_n\}_{n=0}^{\infty}$ has a limit, say, $\boldsymbol{x}^* \in X$. It follows that

$$\boldsymbol{x}^* = \lim_{n\to\infty} \boldsymbol{x}_n = \lim_{n\to\infty} \psi(\boldsymbol{x}_{n-1}).$$

Since ψ is a contraction, it is clearly continuous; see part (i) of Exercise 2.46 above. Thus, we have $\lim_{n\to\infty} \psi(\boldsymbol{x}_{n-1}) = \psi(\lim_{n\to\infty} \boldsymbol{x}_{n-1}) = \psi(\boldsymbol{x}^*)$. It follows that $\boldsymbol{x}^* = \psi(\boldsymbol{x}^*)$ and, hence, ψ has a fixed point in X (namely, $\boldsymbol{x}^*$), obtained as the limit of the sequence $\{\boldsymbol{x}_n\}_{n=0}^{\infty}$.

Now, suppose that $\boldsymbol{x} = \psi(\boldsymbol{x})$ and $\boldsymbol{y} = \psi(\boldsymbol{y})$ are fixed points of ψ. Then, $d(\boldsymbol{x}, \boldsymbol{y}) = d(\psi(\boldsymbol{x}), \psi(\boldsymbol{y})) \leq r \cdot d(\boldsymbol{x}, \boldsymbol{y})$. Since $0 < r < 1$, the above inequality holds only when $d(\boldsymbol{x}, \boldsymbol{y}) = 0$. Thus, by part (i) of Definition 2.1, we must have $\boldsymbol{x} = \boldsymbol{y}$ and, therefore, the fixed point $\boldsymbol{x}^*$ is unique, as required. This completes the proof of the theorem. □

The following useful proposition shows how well a given point $\boldsymbol{x}_0$ and its iterates under a contraction ψ (i.e., $\psi^k(\boldsymbol{x}_0)$, for all nonnegative integers k) approximate the unique fixed point $\boldsymbol{x}^*$.

Proposition 2.59. *Let (X, d) be a complete metric space and let ψ be a contraction on X with scaling ratio $r_\psi \in [0, 1)$ and (necessarily unique) fixed point $\boldsymbol{x}^* \in X$. Let $\boldsymbol{x}_0$ be any point in X and let $\{\boldsymbol{x}_n\}_{n=0}^{\infty} = \{\psi^n(\boldsymbol{x}_0)\}_{n=0}^{\infty}$, as defined in (2.19) of Theorem 2.58. Then,*

$$d(\boldsymbol{x}_0, \boldsymbol{x}^*) \leq \left(\frac{1}{1 - r_\psi}\right) \cdot d(\boldsymbol{x}_0, \psi(\boldsymbol{x}_0)), \text{ and} \tag{2.25}$$

$$d(\psi^k(\boldsymbol{x}_0), \boldsymbol{x}^*) \leq r_\psi^k \cdot d(\boldsymbol{x}_0, \boldsymbol{x}^*), \text{ for all } k \in \mathbb{N}_0. \tag{2.26}$$

---

[15] Indeed, if we let $n := m + p$, with $p \in \mathbb{N}_0$ (and including the trivial case when $m = n$), then (2.24) implies that $d(\boldsymbol{x}_m, \boldsymbol{x}_{m+p}) \to 0$, as $m \to \infty$, uniformly in $p \in \mathbb{N}_0$.

Proof. The Triangle Inequality for d and the fact that $\boldsymbol{x}^*$ is a fixed point of ψ imply that

$$\begin{aligned} d(\boldsymbol{x}_0, \boldsymbol{x}^*) &\leq d(\boldsymbol{x}_0, \psi(\boldsymbol{x}_0)) + d(\psi(\boldsymbol{x}_0), \boldsymbol{x}^*) = d(\boldsymbol{x}_0, \psi(\boldsymbol{x}_0)) + d(\psi(\boldsymbol{x}_0), \psi(\boldsymbol{x}^*)) \\ &\leq d(\boldsymbol{x}_0, \psi(\boldsymbol{x}_0)) + r_\psi \cdot d(\boldsymbol{x}_0, \boldsymbol{x}^*). \end{aligned}$$

Since $r_\psi \in [0, 1)$ and hence $(1 - r_\psi) > 0$, inequality (2.25) now follows immediately.

Next, by (2.15), $d(\psi(\boldsymbol{x}_0), \boldsymbol{x}^*) = d(\psi(\boldsymbol{x}_0), \psi(\boldsymbol{x}^*)) \leq r_\psi \cdot d(\boldsymbol{x}_0, x^*)$. So, by induction, $d(\psi^k(\boldsymbol{x}_0), \boldsymbol{x}^*) = d(\psi^k(\boldsymbol{x}_0), \psi^k(\boldsymbol{x}^*)) \leq r_\psi^k \cdot d(\boldsymbol{x}_0, \boldsymbol{x}^*)$, for all $k \in \mathbb{N}_0$ (since the case $k = 0$ is obvious). This yields (2.26) and concludes the proof of the proposition. □

Remark 2.60. The proof of the Contraction Mapping Principle given above is called a *constructive proof* because it not only guarantees the existence of a unique fixed point for a given contraction but it also provides an explicit (algorithmic) method for constructing the fixed point. Namely, the fixed point $\boldsymbol{x}^*$ is the limit of the iterates of the given contraction applied to a chosen initial point $\boldsymbol{x}_0 \in X$. Moreover, the construction of the fixed point is completely independent of the choice of initial point $\boldsymbol{x}_0$, a fact that is very important in the construction of many fractals and, in particular, of self-similar sets; see Chapter 4.

Newton's Method is a well-known technique that allows one to find a zero (or root) of a certain type of function through an iterative process. The fact that this process produces a zero follows from the Contraction Mapping Principle (Theorem 2.58).

Theorem 2.61 (Newton's Method for Finding Roots). *Assume that f is a twice continuously differentiable real-valued function on the compact interval $[a, b]$ (with first and second derivatives denoted f' and f''). Furthermore, assume that we have $f(x^*) = 0$ and $f'(x^*) \neq 0$, with $x^* \in (a, b)$. Then, there is a nonempty open interval $\tilde{I} \subseteq (a, b)$ such that $x^* \in \tilde{I}$ and, for any $x_0 \in \tilde{I}$, the sequence $\{x_n\}_{n=0}^\infty$ converges to x^*, where $\{x_n\}_{n=0}^\infty$ starts at x_0 and is defined recursively by*

$$x_n = x_{n-1} - \frac{f(x_{n-1})}{f'(x_{n-1})}, \quad \text{for all } n \in \mathbb{N}. \tag{2.27}$$

Proof. Since f' and f'' are continuous on $[a, b]$, by Theorem 2.42 and the Heine–Borel Theorem (Theorem 2.36), they are bounded on $[a, b]$ by some $M_1 > 0$ and $M_2 > 0$, respectively. Also, there exists a closed interval $J \subseteq [a, b]$ containing x^* in its interior and such that for all $x \in J$, we have $f'(x) \neq 0$. By the Mean Value Theorem, for each $x \in J$ there exists $c_1 \in J$ such that

$$|f(x) - f(x^*)| = |f(x)| = |f'(c_1)||x - x^*| \leq M_1 \cdot |x - x^*|. \tag{2.28}$$

Since J is closed and bounded in $\mathbb{R}$, it is compact (by Theorem 2.36). Hence, the continuous and positive function $|f'|^2$ achieves its minimum on J; so, its minimum value on J is strictly positive and can be denoted M_3^{-1}, with $M_3 > 0$.

Now, the function $g : J \to \mathbb{R}$ given by $g(x) := x - (f(x)/f'(x))$ is differentiable on J. Furthermore,

$$g'(x) = 1 - \frac{(f'(x))^2 - f(x)f''(x)}{(f'(x))^2} = \frac{f(x)f''(x)}{(f'(x))^2}.$$

Since $f(x^*) = 0$, we have $g'(x^*) = 0$. Thus, in light of (2.28) and the paragraph following it, we have that (with $M := M_1M_2M_3$), for all $x \in J$,

$$|g'(x)| = \frac{|f(x)f''(x)|}{|(f'(x))^2|} \leq M \cdot |x - x^*| \leq C \cdot |x - x^*|, \tag{2.29}$$

where we have chosen the positive constant $C \geq M$ so large that

$$I := \overline{B_{x^*}}(1/2C)(\subseteq B_{x^*}(1/2M)) \subseteq \mathring{J}(\subseteq J), \tag{2.30}$$

which is possible since $x^* \in J$. We note for later purposes that, by construction, $I := \overline{B_{x^*}}(1/2C)$ is a closed and bounded (i.e., compact) subinterval of J. In particular, it is a *complete* metric space.

Also, since $f(x^*) = 0$, we have $g(x^*) = x^* - (f(x^*)/f'(x^*)) = x^*$; that is, x^* is a fixed point of g.

So, by the inequality in (2.29), if $x \in J$ and $|x - x^*| \leq 1/2C$ (i.e., if $x \in I$), then $|g'(x)| \leq 1/2$. By the Mean Value Theorem applied once again, we deduce that, for all $x, y \in I$, there exists $c_2 \in \mathring{I}$ such that

$$|g(x) - g(y)| = |g'(c_2)| \cdot |x - y| \leq (1/2) \cdot |x - y|. \tag{2.31}$$

In particular, with $y := x^*$, we deduce from (2.31) that $g(x) \in I$, for all $x \in I = \overline{B_{x^*}}(1/2C)$. Indeed,

$$|g(x) - x^*| \leq (1/2) \cdot |x - x^*| \leq 1/4C \leq 1/2C, \tag{2.32}$$

so that $g : J \to J$; i.e., g is a self-map of the complete metric space J. Furthermore, it also follows from (2.31) that g is a contraction on J. Therefore, by the Contraction Mapping Principle (Theorem 2.58), for any $x_0 \in I$, the sequence $\{x_n\}_{n=0}^{\infty}$ defined by (2.27) converges to the fixed point x^*. It now suffices to let $\tilde{I} := \mathring{I}$, which is an open subinterval of $[a, b]$ with the desired property. This completes the proof of the theorem. □

Exercises

Exercise 2.62. Verify that, for each $m \in \mathbb{N}$, the function d_m defined in Example 2.3 is a metric on $\mathbb{R}^m$.

Exercise 2.63. Verify that the functions δ_1 and δ_∞ defined in Example 2.7 are metrics on $C([0, 1])$.

Exercise 2.64 (*Limit of a Cauchy Sequence*). Show that a Cauchy sequence $\{\boldsymbol{x}_n\}_{n=1}^{\infty}$ in a metric space X converges in X iff it has a limit point in X (or, equivalently, iff it has a convergent subsequence in X). In this case, the limit point is unique and necessarily coincides with the limit of $\{\boldsymbol{x}_n\}_{n=1}^{\infty}$ (or, equivalently, with the limit of a convergent subsequence in X).

Exercise 2.65 (*Continued Fractions*). Let $x \in \mathbb{R}$ and let $\{x_n\}_{n=1}^{\infty} := \{p_n/q_n\}_{n=1}^{\infty}$ be the sequence of reduced fractions (or "convergents") of x obtained by using the well-known continued fraction approximation procedure (or algorithm). Show that $\{x_n\}_{n=1}^{\infty}$ is a Cauchy sequence in $\mathbb{Q} \subseteq \mathbb{R}$ and that it converges to x in $\mathbb{R}$ (but not necessarily in $\mathbb{Q}$). Consequently, it is convergent in $\mathbb{Q}$ iff $x \in \mathbb{Q}$.

[*Advice*: Recall that, in standard notation, $x = [a_0, a_1, \ldots, a_n]$, with $a_0 := [x] \in \mathbb{Z}$ (the integer part of $x \in \mathbb{R}$) and $a_j \in \mathbb{N}_0$ (for all $j \geq 1$). Also, for every $n \in \mathbb{N}_0$, we have that

$$\frac{p_n}{q_n} = a_0 + \cfrac{1}{a_1 + \frac{1}{a_2 + \cdots + \frac{1}{a_n}}},$$

stopped at the n-th step, while x is given by a similar expression ("continued fraction"), but (possibly) no longer limited to a finite step (unless $x \in \mathbb{Q}$). For an elementary exposition, see [**291**], for example.]

Exercise 2.66. Prove all of the statements made in Remarks 2.22 and 2.29.

Exercise 2.67. Let $r > 0$ and let X be a nonempty set. For each $\boldsymbol{x} \in X$, determine the structure of the open ball $B_{\boldsymbol{x}}(r)$ when X is equipped with the discrete metric d_D defined by (2.3) in Example 2.5.

Exercise 2.68 (*Existence and Uniqueness of Solutions of Initial-Value Problems: Picard's Theorem*). Let $x_0 \in \mathbb{R}$ and $\boldsymbol{y}_0 \in \mathbb{R}^m$, and let $[a, b]$ and B be compact neighborhoods of x_0 and $\boldsymbol{y}_0$ in $\mathbb{R}$ and $\mathbb{R}^m$, respectively, such that the function $\varphi\colon [a, b] \times B \to \mathbb{R}^m$ is jointly continuous and Lipschitz with respect to the second variable, in the following sense: There exists $M > 0$ such that

$$|\varphi(x, \boldsymbol{y}) - \varphi(x, \boldsymbol{z})| \leq M \cdot |\boldsymbol{y} - \boldsymbol{z}|, \text{ for all } x \in [a, b] \text{ and } \boldsymbol{y}, \boldsymbol{z} \in B. \tag{2.33}$$

The primary purpose of this problem is to show that there exists a nonempty open interval I centered at x_0 such that the initial-value problem (IVP, in short) $\boldsymbol{y}' = \varphi(x, \boldsymbol{y})$, $\boldsymbol{y}(x_0) = \boldsymbol{y}_0$, has a unique solution on I, denoted $\boldsymbol{y} = \boldsymbol{y}(x)$. That is, $\boldsymbol{y}$ is of class C^1 on I and its derivative $\boldsymbol{y}'(x)$ equals $\varphi(x, \boldsymbol{y}(x))$, for all $x \in I$; also, $\boldsymbol{y}(x_0) = \boldsymbol{y}_0$.

(i) Show that the above IVP is equivalent to the integral equation

$$\boldsymbol{y}(x) = \boldsymbol{y}_0 + \int_{x_0}^{x} \varphi(t, \boldsymbol{y}(t))dt, \text{ for all } x \in I. \tag{2.34}$$

(ii) Show that for a small enough $\delta > 0$, and if we take the interval J_δ to be of the form $J_\delta := [x_0 - \delta, x_0 + \delta] \subseteq (a, b)$, the (typically nonlinear) integral operator $L : X \to X$, given by

$$L(\boldsymbol{y})(x) := \boldsymbol{y}_0 + \int_{x_0}^{x} \varphi(t, \boldsymbol{y}(t))dt, \text{ for all } x \in J_\delta, \tag{2.35}$$

is well defined and is a contraction.[16] (Use the Lipschitz condition (2.33).)

(iii) Deduce from the Contraction Mapping Principle (Theorem 2.58) and from parts (i) and (ii) of the present exercise that there exists $\delta > 0$ small enough such that the above IVP has a unique solution on the open interval $I := (x_0 - \delta, x_0 + \delta)$ (and even on the closed interval $J_\delta = [x_0 - \delta, x_0 + \delta]$).

[*Hint*: Observe that the integral equation (2.34) is equivalent to the fixed point equation $L(\boldsymbol{y}) = \boldsymbol{y}$, with $\boldsymbol{y} \in X$.]

(iv) (*Linear Case*: *Global Solution*). Assume that the above mapping, $\varphi : I \times \mathbb{R}^m \to \mathbb{R}^m$, is given by $\varphi(x, \boldsymbol{y}) := A(x)\boldsymbol{y} + \boldsymbol{\rho}$, for all $x \in I$ and $\boldsymbol{y} \in \mathbb{R}^m$, where $\boldsymbol{\rho} \in \mathbb{R}^m$, I is a given nonempty open interval of $\mathbb{R}$ containing x_0, and the map $x \mapsto A(x)$ is a continuous map with values in $M_m(\mathbb{R})$, the ring of $m \times m$ matrices with real entries. (The statement according to which $A(\cdot)$ is continuous means that each of the entries of $A(x)$ depends continuously on x.)[17] Then, show that the IVP $\boldsymbol{y}' = A\boldsymbol{y} + \boldsymbol{\rho}$, $\boldsymbol{y}(x_0) = \boldsymbol{y}_0$, with $x_0 \in I$ and $\boldsymbol{y}_0 \in \mathbb{R}^m$ given (but arbitrary), has a unique solution defined on all of I. Such a solution is said to be a *global solution* to the IVP, as opposed to the one obtained in part (iii) in the nonlinear case, which is referred to as a *local solution* to the first IVP.

Also, show that this result (in part (iv)) can be applied to prove the existence and uniqueness of global solutions to IVPs associated to linear systems of differential equations as well as, in particular, via a well-known procedure, to higher-order linear differential equations (with leading coefficient 1).

---

[16] Here, $X := C(J_\delta)$ denotes the space of continuous functions on J_δ, equipped with the metric δ_∞ of uniform convergence on J_δ: Accordingly, for every $\boldsymbol{y}_1, \boldsymbol{y}_2 \in X$, we have that $\delta_\infty(\boldsymbol{y}_1, \boldsymbol{y}_2) := \sup_{x \in J_\delta} |\boldsymbol{y}_1(x) - \boldsymbol{y}_2(x)|$. The space (X, δ_∞) is well known to be complete; see, e.g., [**522**] or [**571**], along with Theorem 9.9 in Section 9.2 below.

[17] Equivalently, one can view the space of $m \times m$ real matrices as being identified with $\mathbb{R}^{m^2}$, equipped with the Euclidean metric $d_{m^2} = d_{m^2,2}$ on $\mathbb{R}^{m^2}$, defined in Example 2.3.

Chapter 3

Measure Theory and Integrals

If only I had the Theorems! Then I should find the proofs easily enough.

Bernhard Riemann

Any argument where one supposes an arbitrary choice to be made an uncountably infinite number of times ... [is] outside the domain of mathematics.

Émile Borel

This new integral of Lebesgue is proving itself a wonderful tool. I might compare it to a modern Krupp gun, so easily does it penetrate barriers which were impregnable.

Edward Burr van Vleck

The only teaching that a professor can give, in my opinion, is that of thinking in front of his students.

Henri Lebesgue

The role of a metric on a set X is to define distance between points in X. A *measure* on X, on the other hand, provides a notion of size for subsets of X. In this section, measures are defined and examples are discussed, but we refrain from providing a full introduction to measure theory. Specifically, the technical details regarding σ-*algebras* and *measurable sets* are mostly left out of this chapter but can be found in Appendix B, as well as in Chapters 6 and 7.

3.1. Elementary Families and Applications

In this section, we consider the key notions of algebras of sets and of elementary families of sets; they will be used throughout this chapter (as well as in Appendix B and in Chapters 6 and 7).

Notation 3.1. In the sequel, it is convenient to use the notation 2^X in order to denote the *power set* of X, i.e., the collection of subsets of a given set X.

Definition 3.2 (*Algebra*). Let X be a set. A nonempty collection $\mathcal{A} \subseteq 2^X$ is called an *algebra* (of subsets of X)—or, in short, an *algebra* of X—if (with $A^c := X \setminus A$ denoting the *complement* of A in X)

(i) $A, B \in \mathcal{A} \Rightarrow A \cup B \in \mathcal{A}$ and

(ii) $A \in \mathcal{A} \Rightarrow A^c \in \mathcal{A}$.

By De Morgan's Laws, we also have that if $\mathcal{A}$ is an algebra, then $A, B \in \mathcal{A} \Rightarrow A \cap B \in \mathcal{A}$. By induction, finite unions and finite intersections of sets in $\mathcal{A}$ are also in $\mathcal{A}$. Also, given $A, B \in \mathcal{A}$, then the *difference set* $A \setminus B := A \cap B^c \in \mathcal{A}$.

Furthermore, we must have that $\emptyset \in \mathcal{A}$ and $X \in \mathcal{A}$. Indeed, since $\mathcal{A}$ is nonempty, let $A \in \mathcal{A}$; then, $\emptyset = A \cap A^c \in \mathcal{A}$ and $X = A \cup A^c \in \mathcal{A}$ (or, alternatively, $X = \emptyset^c \in \mathcal{A}$).

Moreover, as stated in the following proposition, given a collection $\mathcal{C}$ of subsets of X, there is a smallest algebra (necessarily unique) that contains $\mathcal{C}$; it is called *the algebra generated by* $\mathcal{C}$. For the proof, see Exercise 3.121 at the end of this chapter.

Proposition 3.3. *Let $\mathcal{C}$ be a collection of subsets of X. Then, there is a smallest algebra $\mathcal{A}$ of X containing $\mathcal{C}$; i.e., there is an algebra $\mathcal{A}$ of X such that*

(i) *$\mathcal{C} \subseteq \mathcal{A}$ and*

(ii) *if $\mathcal{B}$ is an algebra of X such that $\mathcal{C} \subseteq \mathcal{B}$, then $\mathcal{A} \subseteq \mathcal{B}$.*

Sketch of the Proof. Define $\mathcal{A}$ to be the intersection of all algebras of X that contain $\mathcal{C}$; see Exercise 3.121 for the details. □

Definition 3.4 (*Algebra Generated by a Collection*). Given a collection $\mathcal{C}$ of subsets of X, the (necessarily unique) smallest algebra $\mathcal{A}$ of X containing $\mathcal{C}$ given by Proposition 3.3 is called the *algebra generated by* $\mathcal{C}$.

Definition 3.5 (*Elementary Family*). Let X be a set. A collection $\mathcal{E} \subseteq 2^X$ is called an *elementary family* of X if

(i) $\emptyset \in \mathcal{E}$,

(ii) $A, B \in \mathcal{E} \Rightarrow A \cap B \in \mathcal{E}$, and

(iii) given $A \in \mathcal{E}$, its complement A^c is a finite disjoint union of elements of $\mathcal{E}$. That is, $A^c = \bigsqcup_{k=1}^{K} C_k$, where $K \in \mathbb{N}$, $C_k \in \mathcal{E}$ for all $k = 1, \ldots, K$, and $\bigsqcup_{k=1}^{K} C_k$ denotes the *disjoint union* of the sets C_k.

Proposition 3.6 (Characterization of the Algebra Generated by an Elementary Family). *If $\mathcal{E}$ is an elementary family of X, then $\mathcal{R} \subseteq 2^X$, given by $\mathcal{R} :=$ {finite disjoint unions of elements of $\mathcal{E}$}, is the algebra of X generated by $\mathcal{E}$; i.e., it is the smallest algebra of X containing $\mathcal{E}$.*

Proof. If $A, B \in \mathcal{E}$, write $B^c = \bigsqcup_{k=1}^{K} C_k$, where $C_k \in \mathcal{E}$ for each $k = 1, \ldots, K$. Then, since by Definition 3.5, $\mathcal{E}$ is closed under finite intersections and, by part (iii) of that definition, we have that $A \cup B = (A \backslash B) \sqcup B \in \mathcal{R}$, and $A \backslash B = A \cap B^c = \bigsqcup_{k=1}^{K} (A \cap C_k) \in \mathcal{R}$. Hence, $A \cup B$ and $A \backslash B$ are finite disjoint unions of members of $\mathcal{E}$ and so, $A \backslash B \in \mathcal{R}$ and $A \cup B \in \mathcal{R}$. By induction, it follows that if $A_1, \ldots, A_n \in \mathcal{E}$, then $\bigcup_{j=1}^{n} A_j \in \mathcal{R}$.

Now, let us show that $\mathcal{R}$ is an algebra. We immediately have that $\emptyset \in \mathcal{R}$ and, by the definition of $\mathcal{R}$ along with the above argument, $\mathcal{R}$ is closed under finite unions. In order to see that $\mathcal{R}$ is closed under complements, let $A \in \mathcal{R}$ and write $A = \bigsqcup_{m=1}^{n} A_m$, where $\{A_m\}_{m=1}^{n}$ is a finite disjoint family of elements of $\mathcal{E}$. Since $A_m \in \mathcal{E}$, we have that $A_m^c = \bigsqcup_{j=1}^{J_m} B_{m,j}$, where $\{B_{m,j}\}_{j=1}^{J_m}$ is a finite disjoint family of elements of $\mathcal{E}$, for $m = 1, \ldots, n$. Then,

$$A^c = \left(\bigcup_{m=1}^{n} A_m \right)^c = \bigcap_{m=1}^{n} A_m^c = \bigcap_{m=1}^{n} \left(\bigsqcup_{j=1}^{J_m} B_{m,j} \right) = \bigcap_{m=1}^{n} \left(\bigsqcup_{j_m=1}^{J_m} B_{m,j_m} \right),$$

and so, $A^c = \bigsqcup \{ B_{1,j_1} \cap \cdots \cap B_{n,j_n} : 1 \le j_m \le J_m, 1 \le m \le n \}$; i.e., A^c is a finite disjoint union of elements of $\mathcal{E}$. Thus, $A^c \in \mathcal{R}$ and $\mathcal{R}$ is an algebra.

Lastly, since $\mathcal{E} \subseteq \mathcal{R}$ and $\mathcal{R}$ is an algebra, we have that the algebra generated by $\mathcal{E}$ is a subset of $\mathcal{R}$. Also, $\mathcal{R}$ is a subset of the algebra generated by $\mathcal{E}$ since the latter contains all finite unions of elements of $\mathcal{E}$. □

Example 3.7. Consider the collection of open, closed, or half-open intervals of $\mathbb{R}$ given by[1]

$$\mathcal{E} := \{(a, b), [a, b], [a, b), (a, b] : a \le b, -\infty \le a, b \le +\infty\}. \tag{3.1}$$

We then have that $\mathcal{E}$ is an elementary family of $\mathbb{R}$. Indeed, the case of an open interval with $a = b$ implies that $\emptyset \in \mathcal{E}$. Also, given $(a, b), (c, d) \in \mathcal{E}$, we

[1] Here, for simplicity's sake, we call *half-open interval* any interval of the form $[a, b)$ or $(a, b]$, with a, b as in (3.1). Furthermore, in (3.1), as well as throughout this example and related discussions in this chapter and in Appendix B, we use the following implicit convention: If I is any of the intervals appearing in (3.1) and if at least one of its endpoints belongs to I and is equal to $\pm\infty$ (which implies that I is unbounded), then we replace I by the interval $I \cap \mathbb{R} \subseteq \mathbb{R}$. For example, according to this prescription, $I := [-\infty, b)$ becomes $I \cap \mathbb{R} = (-\infty, b)$, irrespective of whether $b \in \mathbb{R}$ or $b = +\infty$.

have that $(a,b) \cap (c,d)$ is either empty, (a,c), (a,d), or (c,b). Furthermore,[2] $(a,b)^c = (-\infty, a] \sqcup [b, +\infty)$. One reasons entirely similarly with other choices of elements of $\mathcal{E}$ in order to show that $\mathcal{E}$ is closed under finite intersections and complements. Therefore, $\mathcal{E}$ is an elementary family of $\mathbb{R}$.

Now, let $\mathcal{A}$ denote the algebra of $\mathbb{R}$ generated by $\mathcal{E}$, where $\mathcal{E}$ is the elementary family defined by (3.1). That is, according to Proposition 3.6,

$$\begin{aligned} \mathcal{A} &= \{\text{finite disjoint unions of open, closed, or half-open intervals}\} \\ &= \{\text{finite disjoint unions of elements of } \mathcal{E}\}. \end{aligned} \tag{3.2}$$

The *length function*, denoted by λ (or by ℓ), is defined as follows. Given an open interval (a,b), with $-\infty \leq a \leq b \leq +\infty$,

$$\lambda((a,b)) = \ell((a,b)) := \begin{cases} +\infty, & \text{if } (a,b) \text{ is unbounded,} \\ b-a, & \text{if } (a,b) \text{ is bounded (i.e., if } a,b \in \mathbb{R}\text{),} \end{cases}$$

and entirely similarly for closed and half-open intervals. Namely,

$$\lambda([a,b]) = \lambda([a,b)) = \lambda((a,b]) := b - a \ (= \lambda((a,b))).$$

The length function λ extends to $\mathcal{A}$ as follows. For $A \in \mathcal{A}$, we have $A = \bigsqcup_{j=1}^{J} I_j$ where $\{I_j := (a_j, b_j)\}_{j=1}^{J}$ is a finite collection of pairwise disjoint open intervals. Let $\lambda(A)$ be given by

$$\lambda(A) := \sum_{j=1}^{J} \lambda((a_j,b_j)) = \sum_{j=1}^{J} (b_j - a_j), \tag{3.3}$$

and entirely similarly for other finite disjoint unions of elements of $\mathcal{E}$ (on which λ can easily be shown to be well defined).

The length function readily extends to open subsets of $\mathbb{R}$ since any such set is the disjoint union of an at most countable (and necessarily unique, up to reordering) collection of open intervals, $\{I_j\}_{j=1}^{\infty}$. Suppose Ω is an open subset of $\mathbb{R}$ such that $\Omega = \bigsqcup_{j=1}^{\infty} I_j$; see Theorem 2.43 in Section 2.2 above. Then, we have that $\lambda(\Omega) := \sum_{j=1}^{\infty} \lambda((a_j,b_j)) = \sum_{j=1}^{\infty} (b_j - a_j)$.

These results (and their extensions to higher dimensions) will be used to construct Lebesgue measure on $\mathbb{R}$ (as well as on $\mathbb{R}^m$) in Section 3.4.

Remark 3.8. Note that if $A := \bigsqcup_{j=1}^{J} L_j$ is a union of pairwise disjoint intervals of $\mathbb{R}$, where each interval L_j (for $j = 1, \ldots, J$) is either open, closed, or half-open (in the sense of Example 3.7 just above), then $\lambda(A)$ is *uniquely defined* by $\lambda(A) := \sum_{j=1}^{J} \lambda(L_j)$, with $\lambda(L_j) = \ell(L_j)$, the length of the interval L_j, given as in Example 3.7, for $1 \leq j \leq J$.

---

[2]Note that if $a = -\infty$ (resp., if $b = +\infty$), then $(-\infty, a] = \emptyset$ (resp., then $[b, +\infty) = \emptyset$).

Indeed, if $\{M_k\}_{k=1}^K$ is another family of pairwise disjoint intervals of $\mathbb{R}$ such that $A = \bigsqcup_{k=1}^K M_k$, then $A = A \cap A = \bigsqcup_{j,k} (L_j \cap M_k)$, where, here and thereafter, the indices j and k range in $\{1, \ldots, J\}$ and $\{1, \ldots, K\}$, respectively. Observe that each set $L_j \cap M_k$ is also an interval. Furthermore, since the length of intervals is clearly a finitely additive set function, we must have that $(\lambda(A) =) \sum_j \lambda(L_j) = \sum_{j,k} \lambda(L_j \cap M_k) = \sum_k \lambda(M_k)$, as desired. It follows that $\lambda(A)$ *is independent of the choice of the representation of* A *as a finite disjoint union of intervals*, i.e., of elements of the elementary family $\mathcal{E}$ defined in Example 3.7.

Exercise 3.9. Extend the results of Example 3.7 and of Remark 3.8 from $\mathbb{R}$ to $\mathbb{R}^m$, where the integer $m \geq 1$ is arbitrary. The resulting elementary family is called the *elementary family of (m-dimensional) rectangles of* $\mathbb{R}^m$. It consists of all open, closed, or half-open (bounded or unbounded) rectangles of $\mathbb{R}^m$—or, simply, of *rectangles* (of $\mathbb{R}^m$), when no ambiguity may arise.

[*Hint*: Work with the collection $\mathcal{S} = \mathcal{S}_{\mathcal{R}}$ of all open, closed, or half-open rectangles $R = \prod_{j=1}^m I_j$ of $\mathbb{R}^m$ defined in Remark 3.63 of Section 3.4 below (with each interval I_j of $\mathbb{R}$ being either open, closed, or half-open), and show, much as in Example 3.7, that it is an elementary family of $\mathbb{R}^m$. You may wish to begin by considering the case when $m = 2$.]

3.2. Measures

A *measure* μ on a set X is, loosely speaking, a function defined for suitable subsets of X that satisfies certain properties, including the following one:

For any countable collection $\{A_j\}_{j=1}^\infty$ of *pairwise disjoint*[3] subsets of X, we have that

$$\mu\left(\bigcup_{j=1}^{\infty} A_j\right) = \sum_{j=1}^{\infty} \mu(A_j) \quad \textit{(Countable Additivity)}. \tag{3.4}$$

In general, there are many such functions (i.e., measures) defined for *all* subsets of X. However, if $X = \mathbb{R}^m$, the m-dimensional Euclidean space, then—provided we require countable additivity (as above), along with certain natural geometric properties (such as invariance under displacements)—a restriction of the domain of μ to certain subsets of X is required. In other words, there is no good notion of "volume" defined for all subsets of $\mathbb{R}^m$; see, e.g., the book [**589**] for a thorough discussion of the corresponding celebrated "Banach–Tarski Paradox".

[3]A collection of sets is *pairwise disjoint* if any two distinct sets in the collection have an empty intersection: Here, $A_j \cap A_k \neq \emptyset$, for $j \neq k$.

Moreover, for instance, the Lebesgue outer measure $\mathscr{L}^*$ (see Definition 3.10 below for the notion of an outer measure), defined on all subsets of $\mathbb{R}$ is not countably additive. Even worse, $\mathscr{L}^*$ is not even *finitely* additive. That is, there exist disjoint sets $A, B \subseteq \mathbb{R}$ such that $\mathscr{L}^*(A \cup B) < \mathscr{L}^*(A) + \mathscr{L}^*(B)$. Also, the right side of (3.4) implicitly requires that a countable union of sets in the domain of μ is in the domain of μ. Constantine Carathéodory realized that a solution to these problems amounts to restricting the domain of the Lebesgue outer measure $\mathscr{L}^*$ to a *σ-algebra* of *measurable sets*.

In the sequel, we will briefly outline Carathéodory's use of outer measures in order to construct important classes of measures. This entails, in particular, the very subtle and clever definition of measurable sets relative to an outer measure (also called measurable sets in the sense of Carathéodory). These Carathéodory measurable sets form what is called a σ-algebra (a central notion in measure theory; see Definition 3.12). In particular, the collection of such sets is closed under (finite or) countable unions; see Theorem 3.19. Then, as was alluded to above, given an outer measure (necessarily defined on all subsets of X), one can restrict it to the σ-algebra of measurable sets (in the sense of Carathéodory) in order to obtain a measure; see Theorem 3.26. These results will be completed in Section 3.3, beginning with the construction of Carathéodory's measure induced by a nonnegative set function (see Theorem 3.48 and Definition 3.50) and of which the central result is Theorem 3.57.

The above brief outline of the remainder of this section and of Section 3.3 describes some of the key results of Carathéodory's theory. These results constitute the technical backbone of basic measure theory and provide a powerful way for constructing many of the most important measures, such as Lebesgue measure on $\mathbb{R}^m$ (see Section 3.4) and Hausdorff measure (see Section 5.1 and Chapters 6–7), which play a key role in fractal geometry as well as in the theory of dynamical systems (and well beyond). The rather technical proofs of the main results of Carathéodory's theory are provided in full detail in Appendix B. Furthermore, the applications of the theory to the construction and the study of the main properties of Hausdorff measure (and its approximations) are discussed in Chapter 5 and, especially in Chapters 6 and 7.

The next definition is key to Carathéodory's approach to measure theory (see Appendix B).

Definition 3.10 (*Outer Measure*). Let X be a set. An *outer measure* on X is a set function $\mu^* : 2^X \to [0, +\infty]$ which satisfies the following properties:

(i) *Normalization*: $\mu^*(\emptyset) = 0$.

(ii) *Monotonicity*: For $A, B \subseteq X$, $A \subseteq B \Rightarrow \mu^*(A) \leq \mu^*(B)$.

(iii) *Countable Subadditivity*: For any countable collection $\{A_j\}_{j=1}^{\infty}$ of subsets of X, we have that[4]

$$\mu^*\left(\bigcup_{j=1}^{\infty} A_j\right) \leq \sum_{j=1}^{\infty} \mu^*(A_j). \tag{3.5}$$

For any $A \subseteq X$, the value of $\mu^*(A)$ is called the *outer measure* of A.

We employ the method of Constantine Carathéodory in order to define the measurability of a set as follows.

Definition 3.11 (*Carathéodory Measurable Set*). Let X be a set and let μ^* be an outer measure on X. A set $E \subseteq X$ is said to be *measurable with respect to* μ^* (or *measurable in the sense of Carathéodory*, or else μ^**-measurable*) if for every subset $A \subseteq X$, we have

$$\mu^*(A) = \mu^*(A \cap E) + \mu^*(A \cap E^c). \tag{3.6}$$

The collection of sets which are measurable with respect to μ^* is denoted $\mathcal{M} := \mathcal{M}_{\mu^*}$.

Definition 3.12 (σ*-Algebra*). Let X be a set. A collection of subsets of X, denoted $\mathcal{A}$, is called a σ*-algebra* of X (or on the set X) provided

(i) $X \in \mathcal{A}$,

(ii) if $A \in \mathcal{A}$, then $A^c \in \mathcal{A}$, and

(iii) if $\{A_j\}_{j=1}^{\infty}$ is any countable collection of sets in $\mathcal{A}$, then we have that $\bigcup_{j=1}^{\infty} A_j \in \mathcal{A}$.

Exercise 3.13. Let $\mathcal{A}$ be a σ-algebra on a set X. Then:

(i) Show that $\emptyset \in \mathcal{A}$ and show that if $\{A_j\}_{j=1}^{\infty}$ is any countable collection of sets in $\mathcal{A}$, then $\bigcap_{j=1}^{\infty} A_j \in \mathcal{A}$.

[*Hint*: Use De Morgan's Law, according to which we have that $(\bigcap_{j=1}^{\infty} A_j)^c = \bigcup_{j=1}^{\infty} A_j^c$.]

(ii) Show that if $\{A_j\}_{j=1}^{k}$ is any finite collection of sets in $\mathcal{A}$ (with $k \in \mathbb{N}$), then $\bigcup_{j=1}^{k} A_j \in \mathcal{A}$ and $\bigcap_{j=1}^{k} A_j \in \mathcal{A}$. Deduce, in particular, that a σ-algebra is an algebra (in the sense of Definition 3.2).

[*Hint*: Recall from part (i) of the exercise that $\emptyset \in \mathcal{A}$.]

(iii) Show that if $A, B \in \mathcal{A}$, then $A \setminus B$, $B \setminus A$, and $A \triangle B$ must all belong to $\mathcal{A}$.[5]

[4] In light of (i), it easily follows from (iii) that μ^* is also finitely subadditive.

[5] Recall that $A \setminus B := \{x \in A : x \notin B\}$ and $A \triangle B := (A \setminus B) \sqcup (B \setminus A)$; they are called, respectively, the *difference set* and the *symmetric difference* set of A and B.

Example 3.14. Given a set X, then $\mathcal{A} := \{\emptyset, X\} \subseteq 2^X$ is a σ-algebra of X, called the *trivial* σ-algebra on X. It is the smallest σ-algebra on X.

Example 3.15. Given a set X, then $\mathcal{A} := 2^X$, the power set of X, is a σ-algebra on X. It is the largest σ-algebra on X.

Example 3.16. Let $X := \mathbb{R}$, the set of real numbers. Prove that $\mathcal{A} := \{A \subseteq X : A \text{ or } A^c \text{ is countable}\}$ is a σ-algebra on X. In fact, the same result holds if X is assumed to be any uncountable set (instead of just $\mathbb{R}$).

Exercise 3.17. Prove all of the assertions made in Examples 3.14–3.16.

The following definition is also a simple proposition; it parallels Definition 3.4, justified by Proposition 3.3, in the case of algebras. The reader is asked to justify it in Exercise 3.122, which is the exact counterpart for σ-algebras of Exercise 3.121 formulated for algebras.

Definition 3.18 (*σ-Algebra Generated by a Collection*). Given a collection $\mathcal{C}$ of subsets of a set X, then there exists a (necessarily unique) smallest σ-algebra $\mathcal{S}$ of X containing $\mathcal{C}$; it is called the *σ-algebra generated by $\mathcal{C}$*.

The following theorem is a central result in measure theory, but its proof is beyond the scope of the main part of this text. (However, its full proof is provided in Appendix B; see the first part of Theorem B.3 and its proof.) Basically, it states that for a given outer measure, the collection of measurable sets is a σ-algebra.

Theorem 3.19 (Carathéodory's Measurability Theorem). *Let X be a set and let μ^* be an outer measure on X. Then, the collection $\mathcal{M} = \mathcal{M}_{\mu^*}$ of sets that are measurable with respect to μ^* is a σ-algebra.*

There are many different σ-algebras one could choose to provide the foundation of a measure on X. However, the types of sets that are primarily of interest for this text are elements of a specific σ-algebra (usually associated with a given Euclidean space $\mathbb{R}^m$), known as the *Borel sets*. In most contexts, as throughout this text, every Borel set is a measurable set.

Definition 3.20 (*Borel Sets and Borel σ-Algebra*). Let (X, d) be a metric space. The collection of *Borel sets* of X or *Borel σ-algebra* of X, denoted $\mathcal{B} = \mathcal{B}(X)$, is the intersection of all σ-algebras of subsets of X that contain all of the open sets. It is the smallest σ-algebra containing all of the open sets. Hence, it is the σ-algebra generated by the open sets or, equivalently, by the closed sets; see Definition 3.18 above.

Remark 3.21. For the measures examined throughout this text and a given metric space (X, d), where X is equipped with an outer measure μ^*, under a mild suitable condition on μ^* (namely, that μ^* is a *metric* outer measure, in the sense recalled in Section 6.2 below), we have that the open sets are measurable, in the sense of Carathéodory. (The proof of this key fact will be provided in Section 6.2 below.) Hence, every Borel set is measurable (i.e., $\mathcal{B} \subseteq \mathcal{M}$) and so, any set which is a countable union or intersection of open sets (or closed sets) is a Borel set and is therefore measurable. Moreover, all of the *attractors* of *iterated function systems* defined in Chapter 4 are Borel sets because, being compact, they are closed subsets of X.

Now that we know what a σ-algebra is, we can truly define the general notion of a measure on a σ-algebra (or on the associated measurable space), as well as the corresponding notion of a measure space.

Definition 3.22 (*Measurable Space*). A *measurable space* is a pair $(X, \mathcal{M})$ comprising a set X and σ-algebra $\mathcal{M}$ of subsets of X. A set $A \subseteq X$ is *measurable* (or $\mathcal{M}$-measurable, if any ambiguity may arise) if $A \in \mathcal{M}$.

Definition 3.23 (*Measure on a σ-Algebra; Measure Space*). Let $(X, \mathcal{M})$ be a measurable space. A *measure* is a set function $\mu : \mathcal{M} \to [0, +\infty]$ for which

(i) $\mu(\emptyset) = 0$ and

(ii) for any countable collection $\{A_j\}_{j=1}^{\infty} \subseteq \mathcal{M}$ of pairwise disjoint measurable subsets of X, we have that

$$\mu\left(\bigcup_{j=1}^{\infty} A_j\right) = \sum_{j=1}^{\infty} \mu(A_j) \quad (\textit{Countable Additivity}). \tag{3.7}$$

A *measure space* $(X, \mathcal{M}, \mu)$ is a measurable space $(X, \mathcal{M})$ together with a measure μ defined on $\mathcal{M}$.

Example 3.24 (*Counting Measure*). Perhaps the simplest example of a measure on a given set X is the *counting measure*, denoted $\#(\cdot)$. The measure space $(X, 2^X, \#)$ is defined as follows: If $A \subseteq X$ is finite, then $\#(A)$ is defined to be the number of elements in A; if $A \subseteq X$ is infinite, then $\#(A) := +\infty$.

Example 3.25 (*Dirac Measure*). Another simple measure associated with a set X is the *Dirac measure* at $\boldsymbol{x}$, which is denoted $\delta_{\boldsymbol{x}}$, for each $\boldsymbol{x} \in X$. (It is also known as a *point mass* at $\boldsymbol{x}$). The measurable space $(X, 2^X, \delta_{\boldsymbol{x}})$, known as the *Dirac measure space*, is defined as follows: For a fixed $\boldsymbol{x} \in X$ and each $A \subseteq X$, we have

$$\delta_{\boldsymbol{x}}(A) := \begin{cases} 1, & \text{if} \quad \boldsymbol{x} \in A, \\ 0, & \text{if} \quad \boldsymbol{x} \notin A. \end{cases} \tag{3.8}$$

In general, the restriction of an outer measure μ^* to the corresponding σ-algebra $\mathcal{M}$ of measurable sets is a measure. This extremely important result is due to Carathéodory. Its rather technical proof is provided in Appendix B, to which we refer the interested reader. (See the second part of Theorem B.3 and its proof.)

Theorem 3.26 (Carathéodory's Restriction Theorem). *Let X be a set and let μ^* be an outer measure on X. If μ is the restriction of μ^* to the σ-algebra $\mathcal{M}$ of sets that are measurable with respect to μ^*, then $(X, \mathcal{M}, \mu)$ is a measure space. Hence, μ is a measure on $(X, \mathcal{M})$ and is, in particular, countably additive.*

Definition 3.27. The restriction of Lebesgue outer measure $\mathscr{L}^*$ from Definition 3.10 (or, more specifically, from Definition 1.4) to its collection of measurable sets is the *Lebesgue measure on* $\mathbb{R}$, denoted $\mathscr{L}^1$.

Lemma 3.28. *If $A \subseteq \mathbb{R}$ is countable, then $\mathscr{L}^*(A) = \mathscr{L}^1(A) = 0$.*[6]

Proof. Let $\varepsilon > 0$. Since A is countable, the collection of its elements can be written as a sequence $\{a_k\}_{k=1}^{\infty}$. Consider the countable cover of A given by $\left\{\overline{B_{a_k}}\left(\varepsilon/2^{k+2}\right) := \left[a_k - \left(\varepsilon/2^{k+2}\right), a_k + \left(\varepsilon/2^{k+2}\right)\right] : k \in \mathbb{N}\right\}$. Note that for each $k \in \mathbb{N}$, we have $\mathscr{L}^*\left(\overline{B}_{a_k}\left(\varepsilon/2^{k+2}\right)\right) = \varepsilon/2^{k+1}$. Thus, by Lemma 2.57, we have $0 \leq \mathscr{L}^*(A) \leq \sum_{k=1}^{\infty}\left(\varepsilon/2^{k+1}\right) = \varepsilon/2 < \varepsilon$. Since $\varepsilon > 0$ was arbitrary, we deduce that $\mathscr{L}^*(A) = 0$, as claimed. □

Measure spaces have a number of useful properties, several of which are stated in the next proposition.

Proposition 3.29 (Key Properties of Measures). *Let $(X, \mathcal{M}, \mu)$ be a measure space. Then:*

(i) *For any finite disjoint collection $\{A_k\}_{k=1}^{n}$ of measurable sets,*

$$\mu\left(\bigcup_{k=1}^{n} A_k\right) = \sum_{k=1}^{n} \mu(A_k) \quad \text{(Finite Additivity).} \tag{3.9}$$

(ii) *If A and B are measurable sets and $A \subseteq B$, then we have that $\mu(A) \leq \mu(B)$* (Monotonicity).

(iii) *If A and B are measurable sets such that $A \subseteq B$ and $\mu(A) < \infty$, then $\mu(B \backslash A) = \mu(B) - \mu(A)$* (Excision).
Moreover, if, in addition, $\mu(A) = 0$, then $\mu(B \backslash A) = \mu(B)$.

(iv) *For any countable collection $\{A_k\}_{k=1}^{\infty}$ of measurable sets and any measurable set A such that $A \subseteq \bigcup_{k=1}^{\infty} A_k$, we must have that*

$$\mu(A) \leq \sum_{k=1}^{\infty} \mu(A_k) \quad \text{(Countable Subadditivity).} \tag{3.10}$$

[6]Here, $\mathscr{L}^*(A)$ is defined by (3.22), as in Definition 3.65 below, when $m = 1$.

(v) *If* $\{A_k\}_{k=1}^{\infty}$ *is any countable collection of measurable sets such that* $A_k \nearrow A$, *as* $k \to \infty$ *(i.e.,* $A_k \subseteq A_{k+1}$, *for all* $k \geq 1$ *and* $A := \bigcup_{k=1}^{\infty} A_k$*), then* A *is measurable and* $\mu(A_k) \nearrow \mu(A)$, as $k \to \infty$ (First Monotonicity Property or Left Continuity). *In particular, we have* $\lim_{k\to\infty} \mu(A_k) = \mu(A)$.

(vi) *If* $\{A_k\}_{k=1}^{\infty}$ *is any countable collection of measurable sets such that* $A_k \searrow A$, *as* $k \to \infty$ *(i.e.,* $A_{k+1} \subseteq A_k$, *for all* $k \geq 1$ *and* $A := \bigcap_{k=1}^{\infty} A_k$*) and* $\mu(A_{k_0}) < \infty$, *for some integer* $k_0 \geq 1$, *then* $\mu(A_k) \searrow \mu(A)$, as $k \to \infty$ (Second Monotonicity Property or Right Continuity). *In particular, we have* $\lim_{k\to\infty} \mu(A_k) = \mu(A)$.

Proof. The proof of statements (i)–(iii) is left as an easy exercise for the reader; see Exercise 3.30 below.

(iv) Construct a sequence of sets $\{B_k\}_{k=1}^{\infty}$, as follows: $B_1 := A_1$, $B_2 := A_2 \setminus A_1, \ldots, B_k := A_k \setminus \left(\bigcup_{j=1}^{k-1} A_j\right)$, for each integer $k \geq 1$ (with the convention according to which $A_0 := \emptyset$). Then, clearly, the B_k's are *pairwise disjoint*,

$$\bigcup_{j=1}^{k} A_j = \bigsqcup_{j=1}^{k} B_j, \text{ for each } k \geq 1, \text{ and also, } A := \bigcup_{j=1}^{\infty} A_j = \bigsqcup_{j=1}^{\infty} B_j. \tag{3.11}$$

Now, in light of the first equality of (3.11) and since the B_j's are pairwise disjoint, we deduce from part (i) that

$$\mu\Big(\bigcup_{j=1}^{k} A_j\Big) = \mu\Big(\bigcup_{j=1}^{k} B_j\Big) = \sum_{j=1}^{k} \mu(B_j) \leq \sum_{j=1}^{\infty} \mu(B_j)$$
$$= \mu\Big(\bigcup_{j=1}^{\infty} B_j\Big) = \mu\Big(\bigcup_{j=1}^{\infty} A_j\Big) = \mu(A),$$

where the last three equalities follow from the countable additivity of μ (since the B_j's are pairwise disjoint) and from the latter part of (3.11). This establishes part (iv). Note that in this proof, we could have $\mu(\bigcup_{j=1}^{\infty} A_j) = +\infty$. Also, in light of part (i), the collection $\{A_k\}$ could be finite; see Exercise 3.31.

(v) First, assume that $\mu(A_k) < \infty$, for all $k \geq 0$. Let $A_0 := \emptyset$ and $C_k := A_k \setminus A_{k-1}$, for each $k \geq 1$. Then, by part (iii), $\mu(C_k) := \mu(A_k) - \mu(A_{k-1})$, for each $k \geq 1$, and so, by using twice the latter identity and recalling that $A = \bigcup_{j=1}^{\infty} A_j$, we deduce that $\mu(A_k) = \sum_{j=1}^{k} \mu(C_j) \nearrow \sum_{j=1}^{\infty} \mu(C_j) = \mu(A)$, as $k \to \infty$. Note that in the first and second equalities of this last equation, we have used the fact that (in light of the above definition of the C_j's), for each integer $k \geq 1$, $\bigcup_{j=1}^{k} C_j = A_k$ and $\bigcup_{j=1}^{\infty} C_j = A$.

Finally, we observe that if $\mu(A_{k_0}) = +\infty$, for some $k_0 \geq 1$, then $\mu(A_k) = +\infty$, for each $k \geq k_0$ and $\mu(A) \geq \mu(A_{k_0})$. Hence, $\mu(A) = +\infty$ and so, clearly, $\mu(A_k) \nearrow \mu(A)$, as $k \to \infty$. This concludes the proof of part (v).

(vi) This statement easily follows from part (v) (and part (iii)) by passing to the complements of the sets involved. Indeed, for example, according to De Morgan's law, $A^c = \bigcup_{k=1}^{\infty} A_k^c$, with $A := \bigcap_{k=1}^{\infty} A_k$. Supplying the details is left as an exercise for the interested reader. □

Exercise 3.30. Prove parts (i)–(iii) in Proposition 3.29. Also find an example showing that in part (iii) (resp., in part (vi)), the assumption according to which $\mu(A) < \infty$ (resp., $\mu(A_{k_0}) < \infty$, for some $k_0 \geq 1$) is necessary.

Exercise 3.31. Formulate and prove the finite subadditivity of a measure, i.e., the finite analog of property (iv) in Proposition 3.29.

Definition 3.32 (*Sets of Measure Zero*)**.** Let $(X, \mathcal{M}, \mu)$ be a measure space. Then, $A \in \mathcal{M}$ is said to be a set of *measure zero* (or, more precisely, of *μ-measure zero*) or else a *μ-null set* if $\mu(A) = 0$.

The next proposition and the result stated in Exercise 3.34 following it provide some of the main properties of the very important class of the sets of measure zero.

Proposition 3.33. *Let $(X, \mathcal{M}, \mu)$ be a measure space. Then, the collection of subsets of μ-measure zero (or μ-null sets) is closed under finite or countable unions.*

Proof. If $\{A_j\}_{jh\geq 1}$ is an at most countable family of μ-null sets, then (with $A := \bigcup_{j\geq 1} A_j$) the finite or countable additivity of μ shows that the following inequalities hold: $0 \leq \mu(A) \leq \sum_{j\geq 1} \mu(A_j) = 0$. Hence, $\mu(A) = 0$. Note that $A \in \mathcal{M}$ since $\mathcal{M}$ is a σ-algebra. □

Exercise 3.34. Show that a *measurable* subset of a set of measure zero is also of measure zero; i.e., if $A, B \in \mathcal{M}$ are such that $A \subseteq B$ and $\mu(B) = 0$, then $\mu(A) = 0$.

Remark 3.35. The above exercise motivates the definition of a *complete* measure, given in Definition B.5 of Appendix B, that is, a measure μ such that any subset of a μ-null set is *measurable* and hence, by Exercise 3.34, is also μ-null. This technical assumption is often useful to simplify various results in measure theory. In fact, as is discussed in Section B.2 of Appendix B, any measure μ can be uniquely extended to become complete, on a possibly larger σ-algebra; see Theorem B.6 and Exercise B.8 of Appendix B. Moreover, the Carathéodory Restriction Theorem (Theorem 3.26 above or part of Theorem B.3) always provides a complete measure μ (out of a given outer measure μ^*) on the σ-algebra of μ^*-measurable sets; see Theorem B.6, Corollary B.7, and Exercise B.8 in Appendix B.

The following example discusses perhaps the most famous set of measure zero, namely, the classic ternary Cantor set, introduced in Example 1.1. This topic was already briefly discussed in Example 1.5, but with much less detail since at that early stage in the book, we did not have at our disposal the main definitions and tools of measure theory.

Example 3.36 (*Ternary Cantor Set*). If C is the standard classic ternary Cantor set (as defined in Example 1.1) and if $\mu := \mathcal{L}^1$ is ordinary Lebesgue measure on $\mathbb{R}$ (or on $[0,1]$), as constructed in Section 3.4 below, then C has measure zero. Therefore, C is an uncountable set (and even a set which is in bijection with $[0,1]$ or, equivalently, with $\mathbb{R}$) and hence is quite "large" from the point of view of set theory (it has the cardinality of the continuum), but it is "very small" from the point of view of measure theory because it has Lebesgue measure zero.

To see why the classic Cantor set C has Lebesgue measure zero (i.e., has "length" zero), first observe that C is Lebesgue measurable because its complement C^c in $[0,1]$ is open and hence is measurable.[7] (See the sketch of the construction of $\mathcal{L}^1$ provided at the end of Section 3.1; see also Section 3.4 below.) Next, recall that, by construction of the Cantor set C, its complement C^c (in $[0,1]$) is the countable disjoint union of one interval of length $1/3$ (namely, the open interval $(1/3, 2/3)$), two intervals of length $1/9$ (namely, $(1/9, 2/9)$ and $(7/9, 8/9)$), four intervals of length $1/27, \ldots$, and more generally, 2^{j-1} intervals of length 3^{-j}, for each integer $j \geq 1$; see Example 1.1. It then follows from this discussion that, by the countable additivity of Lebesgue measure $\mathcal{L}^1$,

$$\mathcal{L}^1(C^c) = \sum_{j=1}^{\infty} 2^{j-1} \cdot 3^{-j} = \frac{1}{2} \cdot \frac{2}{3} \sum_{j=0}^{\infty} \left(\frac{2}{3}\right)^j = \frac{1}{3} \cdot \frac{1}{1 - \frac{2}{3}} = \frac{1}{3} \cdot 3 = 1.$$

We conclude that since $C^c = [0,1] \setminus C$ has full measure 1 in $[0,1]$, its complement C in $[0,1]$ has measure zero. Indeed, we deduce from part (iii) of Proposition 3.29, that $\mathcal{L}^1(C) = \mathcal{L}^1([0,1]) - \mathcal{L}^1(C^c) = 1 - 1 = 0$, as claimed.

Exercise 3.37 (*Existence of Fat Cantor Sets*). Modify the construction of the classic ternary Cantor set given in Example 1.1 in order to obtain a (topological) Cantor set contained in $[0,1]$ (that is, a compact subset of $[0,1]$ which is totally disconnected and without any isolated point) but which has (strictly) positive 1-dimensional Lebesgue measure. Such a Cantor-like set is often called a "fat Cantor set" in the scientific literature.

Exercise 3.38 (*Generalized Cantor Sets*). Generalize the construction of a fat Cantor set in Exercise 3.37 just above, in order to show that for any $\alpha \in [0,1)$, there exists a (topological) Cantor set contained in $[0,1]$ and which has Lebesgue measure α.[8] More specifically, construct a one-parameter family

[7] Every open subset of $\mathbb{R}$ is a finite or countable disjoint union of intervals; see Theorem 2.43. Hence, it is measurable because it is Borel measurable, as a finite or countable union of Borel sets.

[8] Here, a *topological Cantor set* is understood as being a metric space that is homeomorphic to the ternary Cantor set C. Hence, in particular, it is perfect, totally disconnected, and has an empty interior; it is also countably infinite and has, in fact, the cardinality of the continuum.

$(C_\alpha)_{\alpha\in[0,1)}$ of Cantor-like sets or "generalized Cantor sets" such that $\mathcal{L}^1(C_\alpha) = \alpha$, for every $\alpha \in [0,1)$, and such that the case when $\alpha = 0$ corresponds to the classic ternary Cantor set $C_0 := C$, while the case when $0 < \alpha < 1$ corresponds to "the" fat Cantor set from Exercise 3.37.

Definition 3.39 (*Property Holding Almost Everywhere*). Let $(X, \mathcal{M}, \mu)$ be a measure space. A property is said to hold *almost everywhere with respect to* μ (written μ-a.e. or simply a.e.) if the set $A \subseteq X$ for which it fails belongs to $\mathcal{M}$ and satisfies $\mu(A) = 0$.

Examples of properties that hold almost everywhere are equalities and inequalities between two (measurable)[9] real-valued (or $[0, +\infty]$-valued) functions f and g on X, which play a key role in Lebesgue's integration theory (as outlined in Section 3.6). Indeed, sets of measure zero are the "negligible" sets in measure theory and hence, one cannot distinguish, for example, between two functions f, g which are such that $f = g$ μ-a.e. Accordingly, in Lebesgue's integration theory, one identifies two such functions and therefore works with equivalence classes $[f]$ of almost everywhere equal functions; see Exercise 3.40.

Exercise 3.40. Show that the relation $f = g$ μ-a.e., as described just above, is an equivalence relation on the set of real-valued or $[0, +\infty]$-valued (measurable) functions on the measurable space $(X, \mathcal{M}, \mu)$. Note that the main significance of the measurability assumption of the functions f and g is that it implies that the set $\{x \in X : f(x) \neq g(x)\}$ is measurable (i.e., it belongs to $\mathcal{M}$), as required by Definition 3.39.

Definition 3.41 (*Finite, σ-Finite, and Probability Measures*). Let $(X, \mathcal{M}, \mu)$ be a measure space. The measure μ is called *finite* if $\mu(X) < \infty$ and it is called *σ-finite* if X is the union of a finite or countable collection of measurable sets, each of which has finite measure.

Finally, a (necessarily finite) measure μ is said to be a *probability measure* if $\mu(X) = 1$; i.e., μ has "total mass" 1. Then, μ takes its values in $[0, 1]$.

The *support* of a measure is, loosely speaking, the set on which the mass of the measure is concentrated.

Definition 3.42 (*Support of a Measure*). Let $(X, \mathcal{M}, \mu)$ be a measure space equipped with a metric d_X such that every Borel set is measurable. The *support* of μ, denoted $\operatorname{supp}(\mu)$, is the intersection of all closed sets $E \subseteq X$ such that $\mu(E^c) = \mu(X \backslash E) = 0$. Hence, $S := \operatorname{supp}(\mu)$ is a closed subset of X and $\mu(S^c) = 0$; furthermore, it is the smallest closed subset of X having this property.

The properties of a measure laid out in the next proposition follow readily from Definitions 3.23 and 3.42 as well as properties of sequences and series. As such, the proof is omitted.

[9] See Definition 3.101 below.

Proposition 3.43. *Let $(X, \mathcal{M}, \mu)$ be a measure space equipped with a metric d_X such that every Borel set is measurable. Then, we have that*

(i) $\mathrm{supp}(\mu)$ *is closed and*

(ii) *for $A \subseteq X$, if $A \cap \mathrm{supp}(\mu) = \emptyset$, then $\mu(A) = 0$.*

Remark 3.44. We shall see in Section 3.4 that Lebesgue measure on $\mathbb{R}^m$ is such that every Borel set is measurable, as is assumed in Definition 3.42;[10] indeed, every open set is measurable. More generally, as will be verified in Chapter 6, Section 6.2, the same property holds (for any $t \geq 0$) for the (t-dimensional) Hausdorff measure $\mathcal{H}^t$ on a metric space (X, d_X). This follows from the fact that $\mathcal{L}^m$ and $\mathcal{H}^t$ come from metric outer measures (in the sense of Definition 3.58) as well as from the results discussed in Section 3.3 below—especially Theorems 3.48 and 3.61, along with their counterparts provided and established in Appendix B.

Remark 3.45. Many notions of measures will be investigated in this book. However, most of the sets under consideration in this book, whether fractal or not, are measurable in an appropriate sense. Hence, unless otherwise stated, we assume that all sets in the remainder of the text are measurable with respect to the measure under consideration. Additionally, the notation for measurable sets is suppressed and we say that μ is a measure on X or we refer to a pair (X, μ) (or, more precisely, a triple $(X, \mathcal{M}, \mu)$, where $\mathcal{M}$ is the underlying σ-algebra) as a measure space.

3.3. Extension of a Premeasure to a Measure

In this section, we consider a very useful procedure for constructing an outer measure μ^*, starting with a suitable set function μ (in practice, the length function or the volume function if the underlying set X is $\mathbb{R}$ or $\mathbb{R}^m$, respectively, or else a power of the diameter function if X is a metric space). Then, under suitable hypotheses, and applying the key Carathéodory Measurability and Restriction Theorems (Theorems 3.19 and 3.26, respectively; see also Sections B.1 and B.3 in Appendix B), the central result of this section (Theorem 3.48) implies that the measure $\overline{\mu}$ obtained by restricting the outer measure μ^* yields the Lebesgue outer measure $\overline{\mu} = \mathcal{L}^1$ on $\mathbb{R}$ (or, more generally, $\overline{\mu} = \mathcal{L}^m$ on $\mathbb{R}^m$, for any integer $m \geq 1$); see Section 3.4. Furthermore, $\overline{\mu}$ extends the original set function μ and (under one additional mild hypothesis) this measure extension is unique. Moreover, in the case of a metric space (X, d_X), when applied to the set function $\mu(E) := (\mathrm{diam}(E))^t$, for some fixed real number $t \geq 0$, this construction yields (after having taken a suitable limit) the t-dimensional

[10] Here, the σ-algebra of Borel sets (or *Borel σ-algebra*) is the σ-algebra of $\mathbb{R}^m$ generated by the open sets (or, equivalently, the closed sets) in $\mathbb{R}^m$.

Hausdorff outer measure $\overline{\mu} = \mathcal{H}^t$ on (X, d_X), which plays a central role in fractal geometry; see Chapter 5, especially Section 5.1, along with Chapters 6 and 7. We note that the fact that, for any $t \geq 0$, $\mathcal{H}^t$, the t-dimensional Hausdorff measure, exists and is a (countably additive) measure is crucial in order to define the pervasive notion of Hausdorff dimension and for establishing several of its key properties.

As is usual in this chapter, the more technical proofs of the main results are provided in Appendix B (or else, in one notable case, in Section 6.2).

Definition 3.46 (*Cover or $\mathcal{S}$-Cover*). If $\mathcal{S} \subseteq 2^X$, $\mathcal{S} \neq \emptyset$, where X is a set, and given $E \subseteq X$, we say that the sequence $\{A_j\}_{j\geq 1}$ of elements of $\mathcal{S}$ is a *cover* of E (or, more specifically, an $\mathcal{S}$-*cover* of E if some ambiguity may arise) if $E \subseteq \bigcup_{j\geq 1} A_j$.

Remark 3.47. In Definition 3.46 above as well as in Theorem 3.48 below, we could also consider finite covers (that is, finite sequences $\{A_j\}_{j=1}^{n_0}$ of elements of $\mathcal{S}$ such that $E \subseteq \bigcup_{j=1}^{n_0} A_j$). However, since (in the notation of Theorem 3.48 below) $\emptyset \in \mathcal{S}$ and $\mu(\emptyset) = 0$, we can then associate to this finite cover the countably infinite cover $\{A_j\}_{j=1}^{\infty}$ defined by $A_j := \emptyset$ for all integers $j \geq n_0 + 1$. Hence, the value of $\mu^*(E)$ in (3.12) below would not change since, by construction, $\sum_{j=1}^{\infty} \mu(A_j) = \sum_{j=1}^{n_0} \mu(A_j)$. Thus, in the sequel (and, in particular, in the statement and the proof of Theorem 3.48), unless specifically mentioned otherwise, we will only consider countably infinite covers (or $\mathcal{S}$-covers).

The following important theorem (Theorem 3.48) is due to Carathéodory and yields the standard procedure for constructing outer measures in the applications. Its elegant and beautifully simple proof can be found in many textbooks on real analysis and measure theory (e.g., [**91**, **193**, **521**, **522**]). We provide it here rather than in Appendix B because, unlike the proofs of the Carathéodory Measurability, Restriction, and Extension Theorems (Theorems 3.19, 3.26, and 3.57, respectively) it is not too technical and its overall structure (in the key step of the proof of the countable additivity) can be easily apprehended. At this point, the reader may wish to review the definition of an outer measure given earlier in Definition 3.10.

Theorem 3.48 (Carathéodory's Construction of Outer Measures). *Let $\mathcal{S}$ be a collection of subsets of a set X such that $\emptyset \in \mathcal{S}$, and let $\mu : \mathcal{S} \to [0, +\infty]$ be a set function such that $\mu(\emptyset) = 0$. Define $\mu^* : 2^X \to [0, +\infty]$ by setting $\mu^*(\emptyset) = 0$ and, for any nonempty $E \subseteq X$, by letting*

$$\mu^*(E) := \inf\left\{ \sum_{j=1}^{\infty} \mu(A_j) : \{A_j\}_{j=1}^{\infty} \text{ is a cover (i.e., an } \mathcal{S}\text{-cover) of } E \right\}. \tag{3.12}$$

Then, μ^* is an outer measure on X.

Proof. The fact that μ^* satisfies the normalization and monotonicity properties of an outer measure (see Definition 3.10) follows immediately from the definition of μ^*.

In short, the reason why $\mu^*(\emptyset) = 0$ is that $\{\emptyset\}$ (or, rather, the sequence $\{A_j := \emptyset\}_{j=1}^{\infty}$) is a cover of $\emptyset$ and hence that $0 \leq \mu^*(\emptyset) \leq \mu(\emptyset) = 0$; this establishes the normalization property of μ^*.

Moreover, if E and F are subsets of $\mathcal{S}$ such that $E \subseteq F$, then a moment's reflection shows that any cover (i.e., $\mathcal{S}$-cover) of F is also a cover (i.e., an $\mathcal{S}$-cover) of E and hence, in light of (3.12) defining $\mu^*(E)$ (and $\mu^*(F)$) as an infimum, we must have that $\mu^*(E) \leq \mu^*(F)$, as desired. (Observe that if $S \subseteq T \subseteq [0, +\infty]$, then $\inf T \leq \inf S$.) This completes the proof of the monotonicity property of μ^*.

Next, in order to show that μ^* is countably subadditive, we proceed as follows. Let $\{E_j\}_{j=1}^{\infty}$ be a sequence of subsets of X, and let $E := \bigcup_{j=1}^{\infty} E_j$. We must show that

$$\mu^*(E) \leq \sum_{j=1}^{\infty} \mu^*(E_j). \tag{3.13}$$

Equivalently, we will assume from now on that $\sum_{j=1}^{\infty} \mu^*(E_j) < \infty$ (since otherwise, the inequality (3.13) is obvious) and show that for every $\varepsilon > 0$,

$$\mu^*(E) \leq \sum_{j=1}^{\infty} \mu^*(E_j) + \varepsilon. \tag{3.14}$$

Clearly, this last inequality will then imply (3.13), by letting $\varepsilon \to 0^+$.

Fix $\varepsilon > 0$. Now, for each fixed integer $j \geq 1$, and in light of the counterpart of (3.12) defining $\mu^*(E_j)$ as an infimum,[11] there exists a countable $\mathcal{S}$-cover $\{A_{j,k}\}_{k=1}^{\infty}$ of E_j such that

$$\sum_{k=1}^{\infty} \mu(A_{j,k}) \leq \mu^*(E_j) + \frac{\varepsilon}{2^j}. \tag{3.15}$$

Then, clearly, the countable collection $\{A_{j,k}\}_{j,k=1}^{\infty}$ of elements of $\mathcal{S}$ is an $\mathcal{S}$-cover of $E := \bigcup_{j=1}^{\infty} E_j$, as we next explain in more detail for the reader's convenience.

Indeed, since for each fixed $j \geq 1$, the sequence $\{A_{j,k}\}_{k=1}^{\infty}$ is an $\mathcal{S}$-cover of E_j, it follows from Definition 3.46 that $A_{j,k} \in \mathcal{S}$ for all $k \geq 1$ and also that $E_j \subseteq \bigcup_{k=1}^{\infty} A_{j,k}$. Hence, the family of elements of $\mathcal{S}$, $\{A_{j,k}\}_{j,k=1}^{\infty} = \{A_{j,k}\}_{(j,k)\in\mathbb{N}\times\mathbb{N}}$, is countable (since the Cartesian product $\mathbb{N} \times \mathbb{N}$ is countable) and after having taken the union (over all integers $j \geq 1$) of the latter family of elements of $\mathcal{S}$, we deduce that $E = \bigcup_{j=1}^{\infty} E_j \subseteq \bigcup_{j=1}^{\infty} \bigcup_{k=1}^{\infty} A_{j,k} = \bigcup_{j,k=1}^{\infty} A_{j,k}$.

[11] Recall that without loss of generality, we may assume that $\sum_{l=1}^{\infty} \mu^*(E_l) < \infty$; hence, $\mu^*(E_j) < \infty$, for each $j \geq 1$.

In light of Definition 3.46, this last equation shows that $\{A_{j,k}\}_{j,k=1}^{\infty}$ is an $\mathcal{S}$-cover of E, as claimed—and hence, we deduce from (3.12) defining $\mu^*(E)$ that

$$\mu^*(E) \le \sum_{j,k=1}^{\infty} \mu(A_{j,k}) = \sum_{j=1}^{\infty}\sum_{k=1}^{\infty} \mu(A_{j,k}), \tag{3.16}$$

where the last equality in (3.16) follows from a discrete version of Fubini's Theorem; see Exercise 3.49 below.

Also, by summing (over all $j \ge 1$) the inequalities (3.15), we obtain that

$$\sum_{j=1}^{\infty}\sum_{k=1}^{\infty} \mu(A_{j,k}) \le \sum_{j=1}^{\infty} \mu^*(E_j) + \varepsilon, \tag{3.17}$$

where on the right-hand side of (3.17), we have used the easily verified fact that $\sum_{j=1}^{\infty} \varepsilon \cdot 2^{-j} = \varepsilon \cdot 1 = \varepsilon$; simply use (2.17) in Lemma 2.57, with $z := 1/2$.

We combine the inequalities (3.16) and (3.17) to deduce that for all $\varepsilon > 0$, the desired inequality (3.14) holds. This implies that (3.13) holds and hence that μ^* is countably subadditive, as claimed. We conclude that μ^* is an outer measure, thereby completing the proof of the theorem. □

In the course of the proof of Theorem 3.48 (see (3.16) and the comment following it), we have used part (i) of Exercise 3.49 below. It is a discrete version of Fubini's Theorem (as well as of the Fubini–Tonelli Theorem) and is, in fact, a prerequisite for proving the general version of Fubini's Theorem in Lebesgue's integration theory.

Exercise 3.49.

(i) Let $\{a_{j,k}\}_{j,k=1}^{\infty}$ be a double sequence of elements of $[0, +\infty]$. Then, show that $\sum_{j,k=1}^{\infty} a_{j,k} = \sum_{j=1}^{\infty}\sum_{k=1}^{\infty} a_{j,k} = \sum_{k=1}^{\infty}\sum_{j=1}^{\infty} a_{j,k}$, where the equalities hold in $[0, +\infty]$.

(ii) Extend the result of part (i) to the case where $\{a_{j,k}\}_{j,k=1}^{\infty}$ is a *summable* (and hence also absolutely convergent) double sequence of elements of $\mathbb{R}$ (or of $\mathbb{C}$); i.e., $\sum_{j,k=1}^{\infty} |a_{j,k}| < \infty$.[12]

(iii) Show that the result of part (ii) fails, in general, if $\{a_{j,k}\}_{j,k=1}^{\infty}$ is assumed to be conditionally convergent and, hence, is not summable.

Definition 3.50 (*Carathéodory Measure Associated with a Set Function*). Given a set function μ, the outer measure μ^* defined as in Theorem 3.48 (via (3.12)) is called *the outer measure induced by* μ. The measure $\overline{\mu}$ obtained by restricting μ^* to the σ-algebra $\mathcal{M} = \mathcal{M}_{\mu^*}$ of μ^*-measurable sets is called *the Carathéodory measure induced by* μ.

[12]The conclusion of part (i) remains the same under this assumption, except that the corresponding equality holds in $\mathbb{R}$ or $\mathbb{C}$, instead of in $[0, +\infty]$.

Remark 3.51. In Definition 3.50, the fact that $(X, \mathcal{M}, \mu)$ is a measurable space, and hence that $\mathcal{M} = \mathcal{M}_{\mu^*}$ is a σ-algebra on X and $\overline{\mu}$ is a measure on $(X, \mathcal{M})$, is justified by an application of Carathéodory's Measurability and Restriction Theorems (Theorem 3.19 and Theorem 3.26, respectively, or else Theorem B.3 in Appendix B, where a complete proof of this key result is provided).

Remark 3.52. It is important to note that without additional assumptions on $\mathcal{S} \subseteq 2^X$ and on the set function $\mu : \mathcal{S} \to [0, +\infty]$, it is not necessarily true that the outer measure μ^* (and hence also the Carathéodory measure $\overline{\mu}$) extends μ, that is, coincides with μ on $\mathcal{S}$. All we can say, in general, is that the following equality and inequality hold:[13]

$$\mu^*(E) = \overline{\mu}(E) \leq \mu(E), \text{ for every } E \in \mathcal{S}. \tag{3.18}$$

The goal of the various forms of the Carathéodory Extension Theorem (see Theorem 3.57 below, along with Theorem B.12 in Appendix B) is to find sufficient conditions on the collection $\mathcal{S}$ and on the set function μ guaranteeing that μ^* and hence also $\overline{\mu}$ (since, by construction, $\mu^* = \overline{\mu}$ on $\mathcal{M} \supseteq \mathcal{S}$) do extend μ from $\mathcal{S}$ to 2^X (for the case of μ^*) or to $\mathcal{M}$ (for the case of $\overline{\mu}$), as well as an additional hypothesis on μ guaranteeing the uniqueness of the measure extension from $\mathcal{S}$ to the σ-algebra $\mathcal{M}$.

Definition 3.53 (*Measure Extension, Premeasure, and Semialgebra*). Let $\mathcal{S}$ be a collection of subsets of a set X and let $\mu : \mathcal{S} \to [0, +\infty]$ be a set function. Then:

(i) (*Measure Extension*). A measure ν is an *extension* of μ if the elements of $\mathcal{S}$ are measurable with respect to ν and $\nu(A) = \mu(A)$, for all $A \in \mathcal{S}$.

(ii) (*Premeasure*). The set function μ is a *premeasure* if it is finitely additive (as in (3.9), but with the added condition that $\bigcup_{k=1}^n A_k \in \mathcal{S}$, which is automatically satisfied if $\mathcal{S}$ is a semialgebra, in the sense of (iii) just below) and countably subadditive (as in (3.5), but with the added condition that $\bigcup_{j=1}^\infty A_j \in \mathcal{S}$) and, if $\emptyset \in \mathcal{S}$, then $\mu(\emptyset) = 0$.

(iii) (*Semialgebra*). If $\mathcal{S}$ is nonempty with the property that whenever $A, B \in \mathcal{S}$, we have $A \cap B \in \mathcal{S}$ and there is a finite disjoint collection $\{E_j\}_{j=1}^n$ of sets in $\mathcal{S}$ for which $A \backslash B = \bigcup_{j=1}^n E_j$, then $\mathcal{S}$ is called a *semialgebra* (or an *elementary family*, as in Definition 3.5; see Remarks 3.54 and 3.55).

Remark 3.54. The definition of a semialgebra given in part (iii) of Definition 3.53 is easily seen to be precisely equivalent to that of an elementary family, as

[13] To see why the inequality in (3.18) is true, it suffices to consider the following trivial countable $\mathcal{S}$-cover of a given $E \in \mathcal{S}$: $\{A_j\}_{j=1}^\infty$, where $A_1 := E$ and $A_j := \emptyset$ for any $j \geq 2$. Then, clearly, by (3.12) and since $\mu(\emptyset) = 0$, we have $\mu^*(E) \leq \mu(E)$. Since $\mu^*(E) = \overline{\mu}(E)$ (because $\mathcal{S} \subseteq \mathcal{M}$ and $\mu^* = \overline{\mu}$ on $\mathcal{M}$), we deduce that (3.18) holds.

given in Definition 3.5. We leave the verification of this statement as an exercise for the interested reader.

Remark 3.55. In the mathematical literature, an algebra (of sets) is also called a ring, while a semialgebra is also called a semiring (or, according to Remark 3.54, an elementary family, in the sense of Definition 3.5).

Remark 3.56. When $\mathcal{S}$ is a semialgebra (i.e., an elementary family), as is assumed in Theorem 3.57 below, in order to verify the countable subadditivity of a premeasure μ on $\mathcal{S}$ (see part (ii) of Definition 3.53), it suffices to verify that (3.10) holds for any sequence $\{A_j\}_{j=1}^{\infty}$ of *pairwise disjoint* elements of $\mathcal{S}$. This follows from the finite additivity of a premeasure and by considering the sequence of disjoint elements of $\mathcal{S}$ defined by $B_1 = A_1$ and $B_j = A_j \backslash \bigcup_{k=1}^{j-1} A_k$, for all $j \geq 2$, so that $A = B$, where $B := \bigsqcup_{j=1}^{\infty} B_j$. Then, if the A_j's belong to $\mathcal{S}$ (but are not necessarily disjoint) and $A \in \mathcal{S}$, we have that $\mu(A) = \mu(B) \leq \sum_{j=1}^{\infty} \mu(B_j) \leq \sum_{j=1}^{\infty} \mu(A_j)$.[14] Note that for each $j \geq 1$, $\mu(B_j) \leq \mu(A_j)$ since $B_j \subseteq A_j$ and μ is monotonic because it is finitely additive; see part (ii) of Proposition 3.29 above.

In practice, to check the countable subadditivity of μ on an elementary family $\mathcal{S}$ (such as the half-open, half-closed intervals in the case of the construction of Lebesgue measure or, more generally, of Lebesgue–Stieltjes measures on $\mathbb{R}$), one uses the simpler criterion given in Remark 3.56.

The next fundamental theorem provides a practical way to construct many measures, such as Lebesgue measure on $\mathbb{R}$ (or on $\mathbb{R}^m$)—and, more generally (at least in one dimension), Lebesgue–Stieltjes measures on $\mathbb{R}$. It is usually known as the Carathéodory Extension Theorem in the literature, although in its present general form, it is attributed to both Carathéodory and Hahn. Actually, by one of these unfortunate twists of mathematical history, it is neither due to Carathéodory nor to Hahn, originally, but was first stated and proved by Fréchet in [**201**]. Note, however, that both its statement and its proof rely on all essential aspects of Carathéodory's theory discussed in Appendix B and in this chapter (in Sections 3.1–3.3), including Theorem 3.19, Theorem 3.26, and Theorem 3.48, namely, Carathéodory's Measurability and Restriction Theorems as well as the Construction of Outer Measures Theorem (Theorem 3.48). The proof of the Carathéodory Extension Theorem is provided in Appendix B, where we emphasize the original form of the theorem (in which the semialgebra $\mathcal{S}$ is replaced by an algebra $\mathcal{A}$ and the premeasure μ is assumed to be a measure on $\mathcal{A}$, in the same sense as in Definition 3.23 except that in order to obtain the countable additivity stated in (3.7), we require that $\bigcup_{j=1}^{\infty} A_j \in \mathcal{A}$).

[14] Here and henceforth, for simplicity, we frequently write "disjoint" instead of "pairwise disjoint".

See Theorem B.12 and its proof, Corollary B.13, and the ensuing (essentially complete) proof of Theorem 3.57 provided in Appendix B.

Theorem 3.57 (Carathéodory's Extension Theorem: Extension and Uniqueness). *Let $\mu : \mathcal{S} \to [0, +\infty]$ be a premeasure on a semialgebra (or an elementary family; see Remark* 3.54) *$\mathcal{S}$ of X. Then, it follows that the Carathéodory measure $\overline{\mu}$ induced by μ (as in Definition* 3.50) *is an extension of μ.*

Additionally, if μ is σ-finite,[15] *then so is $\overline{\mu}$ and $\overline{\mu}$ is the* unique *measure on the σ-algebra of μ^*-measurable sets that extends μ.*

When applying Theorem 3.57 to concrete situations, in order to know that many subsets (in fact, all Borel subsets) of a given metric space are measurable, we need to introduce the following notion of a metric outer measure

Definition 3.58 (*Metric Outer Measure*). Let (X, d) be a metric space and let $\mu*$ be an outer measure on X. Then, μ^* is said to be a *metric outer measure* if

$$\mu^*(A \cup B) = \mu^*(A) + \mu^*(B) \tag{3.19}$$

provided $\text{dist}(A, B) > 0$, where

$$\text{dist}(A, B) := \inf\{d(\boldsymbol{a}, \boldsymbol{b}) : \boldsymbol{a} \in A, \boldsymbol{b} \in B\}. \tag{3.20}$$

Note that, in light of (3.20), if $\text{dist}(A, B) > 0$, then A and B must be disjoint.[16] Clearly, the converse is not true, in general.[17] However, the following exercise points out an important case when the converse of this statement is also true.

Exercise 3.59. Show that if A, B are two *disjoint compact* subsets of Euclidean space $\mathbb{R}^m = (\mathbb{R}^m, d_m)$ (or, more generally, of any metric space (X, d)), then $\text{dist}(A, B) > 0$.

[*Hint*: Without loss of generality, you may assume that A and B are nonempty since, otherwise, $\text{dist}(A, B) = +\infty > 0$. Then, consider the real-valued function $f(\boldsymbol{a}, \boldsymbol{b}) := d(\boldsymbol{a}, \boldsymbol{b}) = |\boldsymbol{a} - \boldsymbol{b}|$, for $(\boldsymbol{a}, \boldsymbol{b}) \in A \times B$, and show that it has a (strictly) positive minimum value on the nonempty compact subset $A \times B$ of $\mathbb{R}^2$.]

---

[15] More specifically, this means that there exists $\{E_n\}_{n=1}^{\infty} \subseteq \mathcal{S}$ such that $\mu(E_n) < \infty$, for all integers $n \geq 1$, and $X = \bigcup_{n \geq 1} E_n$. Note that since $\emptyset \in \mathcal{S}$, we may also allow $\{E_n\}_{n \geq 1}$ to be a finite family.

[16] The only exception is when A or B is empty, in which case $\text{dist}(A, B) = +\infty$, by convention (since $\inf(\emptyset) := +\infty$).

[17] Indeed, consider $A := [0, 1)$ and $B := [1, 2]$, viewed as subsets of $\mathbb{R}$. Then, A and B are disjoint but $\text{dist}(A, B) = 0$.

Example 3.60. In Exercise B.18 in Appendix B, the reader is asked to show that m-dimensional Lebesgue measure $\mathcal{L}^m$ is an outer measure on $\mathbb{R}^m = (\mathbb{R}^m, d_m)$, for any integer $m \geq 1$.

Much more generally, it is shown in Chapter 7 (Section 7.1), using key results from Chapter 6 (Section 6.3) and Appendix B, that on any metric space (X, d) and for any real number $t \geq 0$, the t-dimensional Hausdorff outer measure $\mathcal{H}^t$ (as defined in Section 5.1 below, following the general construction given in Theorem 3.48) is a metric outer measure on (X, d). In light of Theorem 3.61 below, it follows that every Borel subset of (X, d) is measurable with respect to Hausdorff measure $\mathcal{H}^t$ (as defined in Section 5.1, following the procedure described in Theorem 3.57). This fact, when applied to any compact subset of Euclidean space $(\mathbb{R}^m, d_m)$, is especially important because it shows that any (Euclidean) "fractal" is measurable (in the sense of Hausdorff outer measure).

Theorem 3.61 (Borel Measure Induced by a Metric Outer Measure). *Let μ^* be a metric outer measure (in the sense of Definition* 3.58) *on a metric space (X, d). Then, all the Borel subsets of X (and, in particular, all compact subsets of X) are measurable with respect to μ^*. Hence, $\mathcal{B}(X) \subseteq \mathcal{M}$, where $\mathcal{M}$ is the σ-algebra of measurable subsets of X (as in Theorem* 3.57) *and $\mathcal{B}(X)$ is the σ-algebra of Borel subsets of X (as in Definition* 3.20).

Consequently, the Carathéodory measure $\overline{\mu}$ induced by μ^* on $\mathcal{M}$ (via Theorem 3.48 *and Definition* 3.50), *when restricted to $\mathcal{B}(X)$, induces a Borel measure—i.e., a measure on $(X, \mathcal{B}(X))$—which is denoted $\widetilde{\mu}$ and called* the Borel measure associated with the metric outer measure μ^*.

The proof of Theorem 3.61 is provided in Section 6.2 below; see Theorem 6.14 and its proof.

3.4. Lebesgue Measure

The Lebesgue measures on Euclidean spaces play a central role in this book and in mathematical analysis in general. These measures provide the concrete notions of length, area, and volume (as well as generalizations to higher dimensions) for subsets of Euclidean space. For instance, the Lebesgue measure $\mathcal{L}^1$ defined in Definition 3.27 *is* the definition of length for (measurable) subsets of $\mathbb{R}$. As done in that section, before defining Lebesgue measure on $\mathbb{R}^m$, we first need to define a preliminary notion of volume for rectangles.

Definition 3.62 (*Volume of Rectangles in $\mathbb{R}^m$*). The *m-dimensional volume* of a closed rectangle R in $\mathbb{R}^m$, denoted $\text{vol}^m(R)$, is the product of the lengths of its component intervals. That is, for each $R = \prod_{j=1}^{m} [a_j, b_j] \subseteq \mathbb{R}^m$, we have

$$\operatorname{vol}^m(R) := \prod_{j=1}^{m} |b_j - a_j|. \tag{3.21}$$

Similarly, we define as follows the volume $\operatorname{vol}^m(R)$ of any rectangle R of $\mathbb{R}^m$. Namely, if $R = \prod_{j=1}^{m} I_j$, where, for each $j = 1, \ldots, m$, I_j is a closed, open, or half-open interval of $\mathbb{R}$, then[18] $\operatorname{vol}^m(R) := \prod_{j=1}^{m} \ell(I_j)$, if R is bounded, and $\operatorname{vol}^m(R) := +\infty$, otherwise (i.e., if at least one of the component intervals I_j is unbounded).

Note that the 1-dimensional volume of an interval in $\mathbb{R}$ is its length. That is, for $m = 1$ and an interval I with endpoints a and b, we have $\operatorname{vol}^1(I) = |b - a|$ (and $\operatorname{vol}^1(I) = +\infty$, if I is unbounded).

Remark 3.63. In keeping with Example 3.7 of Section 3.1 above, in the definition of a rectangle $R = \prod_{j=1}^{m} I_j$ of $\mathbb{R}^m$, we could allow each interval I_j to be either open, closed, or half-open (i.e., of the form $(a_j, b_j), [a_j, b_j]$, or else $[a_j, b_j)$ or $(a_j, b_j]$, with $-\infty \le a_j \le b_j \le +\infty$) and still define $\operatorname{vol}^m(R)$ by (3.21), provided the rectangle R is bounded (i.e., provided each interval I_j is bounded). Also, if any of the intervals I_j is unbounded (i.e., if the rectangle R is unbounded), then $\operatorname{vol}^m(R) := +\infty$.[19]

Remark 3.64. The idea behind the definition of Lebesgue measure on $\mathbb{R}^m$, denoted $\mathcal{L}^m$, is to cover $A \subseteq \mathbb{R}^m$ by collections of countably many closed rectangles and define the measure of A to be the greatest lower bound (i.e., the infimum) of the sum of the volumes over all such collections. Technically, such a procedure only defines an *outer* measure on subsets of $\mathbb{R}^m$. The restriction of such an outer measure to the corresponding measurable sets yields the true Lebesgue measure on $\mathbb{R}^m$. However, this amounts to an application of Theorem 3.26, as is explained in Appendix B; see, especially, Section B.4. As such, the details are omitted here and the Lebesgue measure on $\mathbb{R}^m$ is defined below as the corresponding outer measure. Additionally, all of the subsets of $\mathbb{R}^m$ considered in this text are indeed measurable, so we implicitly assume subsets of $\mathbb{R}^m$ are measurable.

Also, by 'Lebesgue measure', we mean Lebesgue measure on $\mathbb{R}^m$, for some $m \in \mathbb{N}$.

---

[18] Recall that $\ell(I_j) = \operatorname{vol}^1(I_j)$ denotes the length of the interval I_j.

[19] Recall that we use here the same implicit convention as in footnote 1 of Example 3.7. Accordingly, each component interval I_j containing $\pm\infty$ is replaced by the interval $I_j \cap \mathbb{R} \subseteq \mathbb{R}$.

Definition 3.65 (Lebesgue Measure on $\mathbb{R}^m$). The *m-dimensional outer Lebesgue measure* of a set $A \subseteq \mathbb{R}^m$, denoted $\mathscr{L}^m(A)$, is given by[20]

$$\inf\left\{\sum_{k=1}^{\infty} \operatorname{vol}^m(R_k) : A \subseteq \bigcup_{k=1}^{\infty} R_k, R_k = \prod_{j_k=1}^{m} [a_{j_k}, b_{j_k}], k \in \mathbb{N}\right\} \tag{3.22}$$

$$= \inf\left\{\sum_{k=1}^{\infty} \operatorname{vol}^m(R_k) : A \subseteq \bigcup_{k=1}^{\infty} R_k, R_k \text{ is a rectangle of } \mathbb{R}^m, k \in \mathbb{N}\right\}.$$

Note that when $m = 1$, Definition 3.65 agrees with the definition of Lebesgue outer measure in Definition 3.10 and in Theorem 3.48 (where $\mu := \operatorname{vol}^m$ and $\mathcal{S}$ is either the collection of all closed rectangles of $\mathbb{R}^m$ or the elementary family $\mathcal{S} = \mathcal{S}_{\mathcal{R}}$ of all open, closed, or half-open rectangles of $\mathbb{R}^m$, respectively)—and, after restricting to measurable sets, with the definition of Lebesgue measure $\mathscr{L}^1$ on $\mathbb{R}$ given in Definition 3.27. The fact that the Lebesgue measure $\mathscr{L}^m$ is indeed a measure follows from an application of the Carathéodory–Hahn Theorem 3.57. This issue will be discussed in a lot more detail in Remark 3.67 below and in Appendix B.

Remark 3.66 (*Length, Area, and Volume*). The classical notions of length, area, and volume are special cases of Lebesgue measure when $m = 1, 2,$ or 3, respectively. As mentioned above, $\mathscr{L}^1$ defines length for subsets of $\mathbb{R}$. But, also, $\mathscr{L}^2$ defines area for subsets of $\mathbb{R}^2$ and $\mathscr{L}^3$ defines volume for subsets of $\mathbb{R}^3$. Of course, the m-dimensional volume of a rectangle is its m-dimensional Lebesgue measure. That is, for any rectangle $R \subseteq \mathbb{R}^m$, we have $\mathscr{L}^m(R) = \operatorname{vol}^m(R)$. Thus, in order to aid intuition, it is common to denote $\mathscr{L}^m(A)$ for any (Lebesgue measurable)[21] set $A \subseteq \mathbb{R}^m$ by $\operatorname{vol}^m(A)$.

In the next remark, we deduce from the Carathéodory theory developed in Appendix B and further completed in Section 6.2 below some of the main properties of Lebesgue outer measure $\mu^* = \mathscr{L}^m$, Lebesgue measure $\overline{\mu} = \mathscr{L}^m$, and, finally, of the associated Borel measure $\widetilde{\mu} = \mathscr{L}^m$.

Remark 3.67 (*Application of Carathéodory's Theory: Lebesgue Outer Measure vs. Lebesgue Measure*). For any $A \subseteq \mathbb{R}^m$, $\mathscr{L}^m(A)$, as given by (3.22), stands for the *Lebesgue outer measure* of A. More specifically, in (3.22), $\mu^* =: \mathscr{L}^m$ is the outer measure associated via Carathéodory's construction to the set function $\mu := \operatorname{vol}^m(\cdot)$ defined by (3.21) on the semialgebra (or elementary family) $\mathcal{S} := \mathcal{S}_{\mathcal{R}}$ of (open, closed, or half-open) rectangles of $\mathbb{R}^m$; see Theorem 3.48, along with Theorem 3.57 and parts (ii) and (iii) of Definition 3.53. It can be

[20] A moment's reflection shows that the equality between the two infima holds in (3.22). Furthermore, observe that in (3.22), we can clearly restrict our attention to the case when all of the rectangles R_k involved are bounded; indeed, if R_k is unbounded, then $\operatorname{vol}^m(R_k) = +\infty$.

[21] A set $A \subseteq \mathbb{R}^m$ is *Lebesgue measurable* if it belongs to the σ-algebra of measurable sets associated with Lebesgue measure $\mathscr{L}^m$; see Appendix B for a thorough discussion.

easily checked that $\mathcal{S}_{\mathcal{R}}$ is a semialgebra on $\mathbb{R}^m$ and, with a little more effort, that $\mu := \mathrm{vol}^m(\cdot)$ is a premeasure (in the sense of part (ii) of Definition 3.53) on this semialgebra (or elementary family), as required by the hypotheses of Theorem 3.57 (the Carathéodory–Hahn Theorem, which is fully proved in Appendix B, Theorem B.12). See also Exercise 3.75 or, in the broader context of Lebesgue–Stieltjes measures (but when $m = 1$), Example B.20 and (when $m \geq 1$) Exercises B.17 and B.18 in Appendix B, where the reader is asked, in particular, to show that $\mathcal{S}_{\mathcal{R}}$ is a semialgebra and that $\mu = \mathrm{vol}^m(\cdot)$ is indeed a premeasure on $\mathcal{S}_{\mathcal{R}}$.

Recall that since $\mathcal{L}^m$ (as defined by (3.22)) is an outer measure on $\mathbb{R}^m$, it follows from the Carathéodory Measurability and Restriction Theorems (Theorems 3.19 and 3.26, restated in Theorem B.3 and fully proved in Section B.1 of Appendix B) that $\mathcal{L}(\mathbb{R}^m)$, the class of Lebesgue measurable sets of $\mathbb{R}^m$ (i.e., of subsets of $\mathbb{R}^m$ which are measurable in the sense of Carathéodory with respect to the Lebesgue outer measure $\mu^* := \mathcal{L}^m$; see Definition 3.11) is a σ-algebra and that the restriction $\overline{\mu} = \mathcal{L}^m$ of $\mu^* = \mathcal{L}^m$ to this σ-algebra is a measure on the measurable space $(\mathbb{R}^m, \mathcal{L}(\mathbb{R}^m))$. The measure $\overline{\mu} := \mathcal{L}^m$, defined on the measure space $(\mathbb{R}^m, \mathcal{L}(\mathbb{R}^m))$, is called the *Lebesgue measure* on $\mathbb{R}^m$ or, sometimes, the *m-dimensional Lebesgue measure* (on $\mathbb{R}^m$).

Here, for clarity, we adopt the notation (used in Section 3.3 and in Appendix B) μ^* and $\overline{\mu}$, respectively, for the Lebesgue outer measure and the Lebesgue measure on $\mathbb{R}^m$. On the other hand, for notational simplicity, and unless there can be any ambiguity in a given context, we will also denote them indifferently by $\mathcal{L}^m$; that is, most of the time, we will not distinguish (in our notation, $\mathcal{L}^m$) between the Lebesgue outer measure and the Lebesgue measure on $\mathbb{R}^m$; which of these two notions is being referred to should be clear from the context. The same comment can be applied to $\widetilde{\mu} = \mathcal{L}^m$, the Borel measure associated with either μ^* or $\overline{\mu}$,[22] and which we discuss next.

One can check directly that $\mathcal{B}(\mathbb{R}^m)$, the Borel σ-algebra of $\mathbb{R}^m$, is contained in $\mathcal{M}_{\mu^*} := \mathcal{L}(\mathbb{R}^m)$, the σ-algebra of Lebesgue measurable sets. For this purpose, it suffices to show that the (closed) rectangles in $\mathbb{R}^m$ generate $\mathcal{B}(\mathbb{R}^m)$; see Exercise 3.76 below.

Alternatively, one can show (see Exercise 3.78) that the Lebesgue outer measure $\mu^* = \mathcal{L}^m$ is a metric outer measure (in the sense of Definition 3.58 or of Definition 6.11). It then follows at once from Theorem 3.61 or Theorem 6.14 (which is fully proved in Section 6.2) that $\mathcal{B}(\mathbb{R}^m) \subseteq \mathcal{M}_{\mu^*} = \mathcal{L}(\mathbb{R}^m)$.

Accordingly, one can restrict the Lebesgue measure $\overline{\mu} = \mathcal{L}^m$ to $\mathcal{B}(\mathbb{R}^m)$ in order to obtain a Borel measure on $\mathbb{R}^m$, denoted by $\widetilde{\mu} = \mathcal{L}^m$ and still often

[22] For example, in Proposition 3.70 below, $\mathcal{L}^m$ can be interpreted indifferently as $\mu^*, \overline{\mu}$, or $\widetilde{\mu}$.

referred to as Lebesgue measure on $\mathbb{R}^m$ or else called the Lebesgue–Borel measure on $\mathbb{R}^m$; see Corollary 6.16 in Section 6.2.

On the other hand, as was mentioned earlier, one can show that $\mathcal{S} = \mathcal{S}_{\mathcal{R}}$, the collection of closed (or, alternately, of half-closed, half-open) rectangles in $\mathbb{R}^m$, is a semialgebra on $\mathbb{R}^m$ (in the sense of part (ii) of Definition 3.53) and that $\mu = \mathrm{vol}^m(\cdot)$ is a premeasure on this semialgebra (also in the sense of part (ii) of Definition 3.53). Furthermore, it is easy to check that μ is σ-finite on $\mathbb{R}^m$; see Exercise 3.79. It then follows from Theorem 3.57 (the Carathéodory Extension Theorem for semialgebras, fully proved in Subsection B.3.2) that $\overline{\mu} = \mathscr{L}^m$ (and hence also $\widetilde{\mu} = \mathscr{L}^m$) *uniquely extends* $\mu = \mathrm{vol}^m(\cdot)$ from $\mathcal{S} = \mathcal{S}_{\mathcal{R}}$ to all of $\mathcal{M}_{\mu^*} = \mathscr{L}(\mathbb{R}^m)$ and thus also to $\mathcal{B}(\mathbb{R}^m)$.[23]

We will see in Theorem 6.14 and Corollary 6.15 that $\overline{\mu} = \mathscr{L}^m$ is a *complete* measure, in the sense of Definition B.5 of Section B.2; that is, any subset of a Lebesgue measurable set of measure zero (also called a Lebesgue null set) is also Lebesgue measurable and hence also of measure zero (or is said to be 'Lebesgue null'):

$$\begin{aligned}&\mathscr{L}^m(A) = 0, \text{ with } A \in \mathscr{L}^m(\mathbb{R}^m) \text{ and } B \subseteq A\\ &\qquad \Rightarrow B \in \mathscr{L}^m(\mathbb{R}^m) \text{ and thus } \mathscr{L}^m(A) = 0.\end{aligned}$$

This would be true for $\mu^* = \mathscr{L}^m$ without the assumption that $A \in \mathscr{L}^m(\mathbb{R}^m)$ and the conclusion that $B \in \mathscr{L}^m(\mathbb{R}^m)$, but the statement made just above is significantly more precise and technically useful in the development and the applications of the theory of the Lebesgue integral (as is briefly discussed in Section 3.6 below).

Moreover, since $\mu = \mathrm{vol}^m(\cdot)$ is σ-finite, as was noted just above, it follows from the (strong) uniqueness part of Theorem 3.57 (established in Subsection B.3.2 of Appendix B) that $\overline{\mu} = \mathscr{L}^m$ is the unique measure extension of $\widetilde{\mu}$ from the Borel σ-algebra $\mathcal{B}(\mathbb{R}^m)$ to the σ-algebra $\mathcal{M}_{\mu^*} := \mathscr{L}(\mathbb{R}^m)$ of Lebesgue measurable sets in $\mathbb{R}^m$. Consequently, the Lebesgue measure $\overline{\mu} = \mathscr{L}^m$ is the completion (in the sense of Definition B.5 and Theorem B.6 of Section B.2) of the Lebesgue–Borel measure $\widetilde{\mu} = \mathscr{L}^m$.

In summary, and continuing to use the notation introduced in the general Carathéodory theory developed in Appendix B and in Section 6.2, we have at our disposal three basic tools, namely, (i) a metric outer measure μ^* acting on $(\mathbb{R}^m, 2^{\mathbb{R}^m})$; (ii) a measure $\overline{\mu}$ associated with μ^* via the Carathéodory construction and acting on the measure space $(\mathbb{R}^m, \mathscr{L}(\mathbb{R}^m))$; and (iii) a Borel measure $\widetilde{\mu}$ associated with $\overline{\mu}$, which acts on the measure space $(\mathbb{R}^m, \mathcal{B}(\mathbb{R}^m))$ and is obtained by restricting $\overline{\mu}$ to be Borel σ-algebra $\mathcal{B}(\mathbb{R}^m)$.

[23]This statement consists of two parts: (i) $\mathscr{L}^m$ extends $\mathrm{vol}^m(\cdot)$ and (ii) this extension is unique. Neither of these facts is easy to establish directly, as the reader can verify. (See also Exercise 3.77, for statement (i), along with [**57**, Chapters 5 and 6].)

Just as in the case of Lemma 3.28, it turns out that arbitrary (finite or) countable sets, such as $\mathbb{N}, \mathbb{Z}, \mathbb{N} \times \mathbb{N}, \mathbb{Z}^3$, and $\mathbb{Q}^k$ (for any $k \in \mathbb{N}$), have Lebesgue measure zero.

Lemma 3.68. *If A is a* (finite or) *countable subset of $\mathbb{R}^m$, then $\mathcal{L}^m(A) = 0$.*

Proof. It follows immediately from Definition 3.65 that $\mathcal{L}^m(\{\boldsymbol{x}\}) = 0$, for each $\boldsymbol{x} \in \mathbb{R}^m$, since $\boldsymbol{x}$ is contained in arbitrarily small rectangles. Suppose $A = \{\mathbf{a}_j\}_{j=1}^{\infty} \subseteq \mathbb{R}^m$. Then, by the countable additivity of $\mathcal{L}^m$, we have $\mathcal{L}^m(A) = \sum_{j=1}^{\infty} \mathcal{L}^m(\{\mathbf{a}_j\}) = 0$. □

Lebesgue measure satisfies the following properties with respect to scaling, translation, and rotation. These are nice tools that are (probably) familiar properties of classical notions of length, area, and volume. First, we recall and expand upon some notation from Subsection 1.2.3.

Notation 3.69. Let $A \subseteq \mathbb{R}^m, \boldsymbol{x} \in \mathbb{R}^m, \lambda > 0$, and let O be an $m \times m$ orthogonal matrix. Then, $\lambda \cdot OA + \boldsymbol{x} := \{\lambda \cdot O\mathbf{a} + \boldsymbol{x} \ : \ \mathbf{a} \in A\}$. In particular, we denote the *translate* of A by $A + \boldsymbol{x} := \{\mathbf{a} + \boldsymbol{x} : \mathbf{a} \in A\}$, while we denote the *scaled* (or *λ-scaled*) version of A by $\lambda A := \{\lambda \mathbf{a} : \mathbf{a} \in A\}$.

The proof of the following proposition follows readily from the definitions; see Exercise 3.72.

Proposition 3.70 (Geometric Properties of Lebesgue Measure $\mathcal{L}^m$). *Let $A \subseteq \mathbb{R}^m$. Then:*

(i) (Translation Invariance). $\mathcal{L}^m(A + \boldsymbol{x}) = \mathcal{L}^m(A)$, *for all $\boldsymbol{x} \in \mathbb{R}^m$.*

(ii) (Scale Covariance). $\mathcal{L}^m(\lambda A) = \lambda^m \cdot \mathcal{L}^m(A)$, *for all $\lambda > 0$.*

(iii) (Rotation Invariance). $\mathcal{L}^m(OA) = \mathcal{L}^m(A)$, *for any $m \times m$ orthogonal matrix O.*

Property (i) above says that $\mathcal{L}^m$ is translation invariant. That is, translating a set does not change the value of $\mathcal{L}^m$. Likewise, property (iii) says that $\mathcal{L}^m$ is invariant under rotation.

Property (ii) of Proposition 3.70 says that when a set $A \subseteq \mathbb{R}^m$ has the components of its elements scaled by a factor of λ, the Lebesgue measure of the resulting set λA is obtained by scaling the Lebesgue measure of A by a factor of λ^m. In particular, as is well known to every school child, at least intuitively, a square that has its side lengths doubled has its area increased by a factor of $2^2 = 4$, a cube that has its side lengths doubled has its volume increased by a factor of $2^3 = 8$, etc.

Example 3.71. To aid the understanding of the scaling property (ii) in Proposition 3.70, consider the following sets: $I = [0, 1] \subseteq \mathbb{R}$, $I^2 = [0, 1]^2 \subseteq \mathbb{R}^2$, and

$I^3 = [0,1]^3 \subseteq \mathbb{R}^3$. By Proposition 3.70 (and using the notation $\mathscr{L}^m = \mathrm{vol}^m$), we have

$$\mathrm{vol}^1(2I) = \mathrm{vol}^1([0,2]) = 2 = 2^1 = 2 \cdot \mathrm{vol}^1(I),$$
$$\mathrm{vol}^2(2I^2) = \mathrm{vol}^2([0,2]^2) = 4 = 2^2 = 2^2 \cdot \mathrm{vol}^2(I^2), \text{ and}$$
$$\mathrm{vol}^3(2I^3) = \mathrm{vol}^2([0,2]^3) = 8 = 2^3 = 2^3 \cdot \mathrm{vol}^3(I^3).$$

Exercise 3.72. Prove Proposition 3.70.

[*Hint*: To start with, use Definition 3.65 and Notation 3.69.]

Example 3.73. Recall from Example 1.5 that the length of the Cantor set C is 0. That is, $\mathscr{L}^1(C) = \mathrm{vol}^1(C) = 0$. However, C is an infinite set, so the counting measure defined in Example 3.24 also fails to give us meaningful information about the structure of C. Even worse, C is *uncountable*: There is no bijection f such that $f : \mathbb{N} \to C$. The Cantor set is too big to be the image of a surjective function whose domain is a countable set. Is there a way to measure C that will assign to it a meaningful nonzero and finite value? Yes, but we need more tools in order to be able to answer the question.[24]

Consider the following subsets of $\mathbb{R}^3$ (which are discussed in Section 1.3): a singleton $U_0 = \{\boldsymbol{x}\}$, the unit interval $A_1 = [0,1] \times \{0\} \times \{0\}$, the unit square $A_2 = [0,1] \times [0,1] \times \{0\}$, and the unit cube I^3. We have $\mathrm{vol}^3(A_0) = \mathrm{vol}^3(A_1) = \mathrm{vol}^3(A_2) = 0$, and, of course, $\mathrm{vol}^3(I^3) = 1$. Note that the 3-dimensional Lebesgue measure $\mathscr{L}^3 = \mathrm{vol}^3$ cannot see a difference between A_0, A_1, and A_2, but it does assign a meaningful value for the volume of the unit cube I^3. Indeed, the unit cube I^3 is intuitively understood to be "3-dimensional" and, moreover, A_0, A_1, and A_2 are *not* "3-dimensional." It seems reasonable to conclude that the notion of volume provided by $\mathscr{L}^3 = \mathrm{vol}^3$ should not be used to measure anything but "3-dimensional" sets...but what exactly is a "3-dimensional" set?

In light of the above discussion and Examples 1.5, 3.71, and 3.73, we have come across a key issue in the analysis of fractals: How should a given set be measured and what is its size? One way to address this two-part problem is to define the notions of dimension and measure in tandem that, together, use dimension to identify which measure, if any, will assign a meaningful value (perhaps nonzero and finite) as the size of the set.

The combination of box-counting dimension (loosely defined in Definition 1.21 and examined thoroughly in Section 8.3) and m-dimensional Lebesgue measure (see Definition 3.65) provides an example of a partial solution.

Example 3.74. Let U be a nonempty open subset of $\mathbb{R}^m$. Then, we have that $\dim_B(U) = m$. This fact is proved in Chapter 8. Now, suppose $n \in \mathbb{N}$ is such

[24] It will turn out that the Hausdorff measure of C (see Chapters 5–7, along with Example B.25 in Appendix B) provides an answer to this question.

that $n \geq m$ and $f : \mathbb{R}^m \to \mathbb{R}^n$ is an isometry, but not necessarily onto. Then,

$$\mathscr{L}^n(f(U)) = \begin{cases} \mathscr{L}^m(U), & \text{if } n = m, \\ 0, & \text{if } n > m. \end{cases} \tag{3.23}$$

So, given an open set $U \subseteq \mathbb{R}^m$, the box-counting (or Minkowski) dimension $\dim_B$ determines U to be m-dimensional and, in turn, m-dimensional Lebesgue measure $\mathscr{L}^m$ is the only measure for which it is even possible that $\mathscr{L}^m(U)$ be nonzero and finite.

Consider the ternary Cantor set C. As determined by (1.13), the box-counting or Minkowski dimension of C is $\log_3 2$, but there is no $\log_3 2$-dimensional Lebesgue measure that would give C a meaningful measure value. A generalization of Lebesgue measure is needed, a family of measures that can assign meaningful measure values to sets with noninteger box-counting dimension such as C, the Koch snowflake curve ∂K_S, the Sierpinski gasket **SG**, and the Sierpinski carpet **SC** (see Subsection 1.2.3). Ideally, such a measure should also satisfy the scaling and invariance properties that generalize the results of Proposition 3.70 (which hold for Lebesgue measure). That is, we seek to define a family of t-dimensional measures $\{\mu^t : t \geq 0\}$ parametrized by real numbers $t \geq 0$ such that for each t, any set $A \subseteq \mathbb{R}^m$, and for any $\lambda > 0$, we have $\mu^t(\lambda A) = \lambda^t \cdot \mu^t(A)$, and such that each μ^t is translation and rotation invariant.

A primary goal of Chapters 5, 6, 7, and 8 (along with Appendix B) is to construct such measures (or contents) and corresponding tools, namely, *Hausdorff measure* and *Hausdorff dimension* (Chapters 5–7 and Appendix B), box-counting content and dimension, as well as *Minkowski content* and *Minkowski dimension* (Chapter 8), respectively.

Exercise 3.75. Show that $\mathcal{S} = \mathcal{S}_{\mathcal{R}}$, the collection of closed rectangles in $\mathbb{R}^m$, is a semialgebra on $\mathbb{R}^m$ and that $\mu := \mathrm{vol}^m(\cdot)$ is a premeasure on $\mathcal{S}_{\mathcal{R}}$.

Exercise 3.76. Show directly that the collection of (m-dimensional) rectangles generates $\mathcal{B}(\mathbb{R}^m)$, the Borel σ-algebra of $\mathbb{R}^m$.

[*Hint*: It suffices to show that the σ-algebra generated by the rectangles contains every open subset of $\mathbb{R}^m$. This follows from Lemma 7.35 of Chapter 7, according to which any open set U in $\mathbb{R}^m$ is a countable union of closed dyadic cubes of $\mathbb{R}^m$, but you should also try to prove it directly.]

Exercise 3.77.

(i) Prove directly that if $R = \prod_{j=1}^m [a_j, b_j]$, with $-\infty < a_j \leq b_j < \infty$ for $j = 1, \ldots, m$, is a closed rectangle in $\mathbb{R}^m$ and if $\widetilde{\mu} = \mathscr{L}^m$ is the Borel measure associated with Lebesgue measure $\overline{\mu} = \mathscr{L}^m$, then[25] we have that $\mathscr{L}^m(R) = \widetilde{\mu}(R) = \overline{\mu}(R) = \mathrm{vol}^m(R) := \prod_{j=1}^{m}(b_j - a_j)$.

(ii) Find a direct proof of the fact that $\overline{\mu} = \mathscr{L}^m$ (or else that $\widetilde{\mu} = \mathscr{L}^m$), which is an extension of $\mu = \mathrm{vol}^m(\cdot)$ from $\mathcal{S} = \mathcal{S}_{\mathcal{R}}$ (the collection of all closed rectangles of $\mathbb{R}^m$) to $\mathscr{L}(\mathbb{R}^m)$ (or else to $\mathcal{B}(\mathbb{R}^m)$), according to part (i), is in fact, the *unique* extension of μ to the indicated σ-algebra.

[*Advice*: If you have difficulty solving question (i) or (ii), we suggest to the reader consulting Chapters 5 and 6 of the very interesting undergraduate textbook by D. Bressoud, [**57**], where an "elementary" (but far from simple) proof of these facts is provided.]

Exercise 3.78. Show that the Lebesgue outer measure on $\mathbb{R}^m$, $\mu^* = \mathscr{L}^m$, is a metric outer measure, in the sense of Definition 3.58. Then, deduce from this result that, in light of Theorem 3.61 or of Theorem 6.14, $\mathcal{B}(\mathbb{R}^m) \subseteq \mathcal{M}_{\mu^*} = \mathscr{L}(\mathbb{R}^m)$.

[*Hint*: You may reason by analogy with the proof of the corresponding result for the δ-approximate Hausdorff measure; see part (i) of Theorem 6.18 in Chapter 6.]

Exercise 3.79. Show that $\mu = \mathrm{vol}^m(\cdot)$ is σ-finite, and, hence, deduce from the uniqueness statement in Theorem 3.57 that $\overline{\mu} = \mathscr{L}^m$ (resp., $\widetilde{\mu} = \mathscr{L}^m$) is the unique extension of $\mu = \mathrm{vol}^m(\cdot)$ from $\mathcal{S} = \mathcal{S}_{\mathcal{R}}$ to all of $\mathscr{L}(\mathbb{R}^m)$ (resp., to $\mathcal{B}(\mathbb{R}^m)$).

Our next and final exercise for now is somewhat difficult, without further hints or background.

Exercise 3.80.

(i) Show that every subset A of $\mathbb{R}^m$ (or of $[0,1]^m$) with positive Lebesgue measure (i.e., $A \in \mathscr{L}(\mathbb{R}^m)$ and $\mathscr{L}^m(A) > 0$) contains a nonmeasurable set N (i.e., $N \subseteq A$ and $N \notin \mathscr{L}^m(A)$).

[*Hint*: First, assume for simplicity that $m = 1$. Then, use the Axiom of Choice and consider the equivalence relation, $x \sim y$ iff $x - y \in \mathbb{Q}$, on $[0,1]$ or on a suitable subset thereof.]

(ii) Explain why, in question (i), we had to assume that $A \in \mathscr{L}(\mathbb{R}^m)$ had positive Lebesgue measure.

[*Hint*: Reason by contradiction and use the completeness of Lebesgue measure $\mathscr{L}^m$; see Section B.2 and Section B.3 in Appendix B.]

Exercise 3.81. Let f_D be the Cantor function (also called the Cantor–Lebesgue function). Recall that $f_D : [0,1] \to [0,1]$ is continuous, nondecreasing, and surjective (i.e., $f_D([0,1]) = [0,1]$). See Example 1.32, along with Exercises 1.33,

[25]Throughout this exercise and the remaining exercises for Section 3.4, we use the notation $\mu^* = \mathscr{L}^m$, $\overline{\mu} = \mathscr{L}^m$, and $\widetilde{\mu} = \mathscr{L}^m$ introduced in Remark 3.67.

1.34 and 1.35. Then, let $\varphi_D : [0,1] \to [0,2]$ be given by $\varphi_D(x) := x + f_D(x)$, for all $x \in [0,1]$.

(i) Show that φ_D is continuous, nondecreasing, and surjective (i.e., here, $\varphi_D([0,1]) = [0,2]$). Furthermore, show that φ_D is also injective and hence is a bijection from $[0,1]$ onto $[0,2]$.

(ii) Deduce from part (i) that φ_D is a homeomorphism from $[0,1]$ onto $[0,2]$ (i.e., both φ_D and its inverse map $\Psi_D := \varphi_D^{-1}$ are continuous) and, therefore, that both φ_D and $\Psi_D = \varphi_D^{-1}$ send Borel sets onto Borel sets.

(iii) Let C be the ternary Cantor set. Show that $\varphi_D(C)$ is a Borel subset of $[0,2]$[26] and has positive Lebesgue measure.

(iv) In light of the results of part (iii) and Exercise 3.80, there exists a non-measurable subset N_D of $\varphi_D(C)$; i.e., $N_D \notin \mathscr{L}(\mathbb{R}^m)$. Next, consider $E_D \subseteq \mathbb{R}$ given by $E_D := \Psi_D(N_D) = \varphi_D^{-1}(N_D)$. Show that E_D is Lebesgue measurable but is *not* Borel measurable; i.e., $E_D \in \mathscr{L}(\mathbb{R})$ but $E_D \notin \mathcal{B}(\mathbb{R})$.

[*Hint*: By construction, $E_D \subseteq C$; hence, $E_D \in \mathscr{L}(\mathbb{R})$ since $\mathscr{L}^1$ is a complete measure and $\mathscr{L}^1(C) = 0$. However, $E_D \notin \mathcal{B}(\mathbb{R})$ because otherwise, in light of part (ii), $N_D = \varphi_D(E_D) \in \mathcal{B}(\mathbb{R})$, which is a contradiction.]

(v) Note that N_D is not Borel measurable and by part (iii), neither is its image under the continuous map $\Psi_D, E_D = \Psi_D(N_D)$. This raises the following natural, but difficult, question:[27]

Can you find a continuous function $f : [0,1] \to \mathbb{R}$ and a Borel subset B of $[0,1]$ such that $f(B)$ is *not* Borel measurable?

[*Hint*: The answer should be yes but concrete examples are not easy to come by. Also, clearly, $[0,1]$ does not play any particular role here and could be replaced by other Borel subsets of $\mathbb{R}$, including $\mathbb{R}$ itself.]

3.5. Content and the Riemann Integral

Historically, the notions of (Jordan) content and Riemann integral were developed before the introduction of Lebesgue measure (on $\mathbb{R}^m$) and Lebesgue integral, respectively. The main goal of this section is to provide several of the basic properties of the Jordan content and to give a simple characterization of the Jordan measurability of a bounded subset of $\mathbb{R}$ (see Proposition 3.84) as well as of the Riemann integrability of a bounded function on a compact interval $[a,b] \subseteq \mathbb{R}$ (see Theorem 3.95). The notions of Jordan content (or content, in short) and of Riemann integral, however, have serious shortcomings. For example, the Jordan content fails to be finitely additive, while the space of Riemann integrable functions is not complete (in the sense of metric spaces).

---

[26] Recall that in $\mathbb{R}^m$ (and, in particular, in $\mathbb{R}$), a Borel set is Lebesgue measurable, in light of Exercise 3.78.

[27] In fact, for some time, Henri Lebesgue himself thought erroneously that the answer to this question should be negative.

Hence, it is often very difficult, if not impossible, to justify the interchange of limits and Riemann integrals or else of Riemann integrals and (countably infinite) sums; see, respectively, Exercise 3.90 and the relevant discussion towards the end of Section 3.6.

We note that the notion of Jordan content—also called Jordan–Peano measure in the literature, even though it is not a measure in the modern sense used in the present chapter—was introduced by the Italian mathematician Giuseppe Peano ([**478**], 1887) and the French mathematician Camille Jordan ([**276**], 1892).

Definition 3.82 (*Outer and Inner Jordan Contents*). Let $A \subseteq \mathbb{R}^m$ be bounded. A *cover of* the set A is a collection of subsets of $\mathbb{R}^m$ whose union contains A. Let $\mathfrak{C}_A^e$ be the collection of all *finite* covers of A comprising m-dimensional rectangles. The *outer content of* A, denoted $\mathscr{C}^*(A)$, is given by

$$\mathscr{C}^*(A) := \inf\left\{\sum_{j=1}^{n} \operatorname{vol}^m(R_j) \,:\, R_1, \ldots, R_n \in \mathfrak{C}_A^e\right\}. \tag{3.24}$$

Let $\mathfrak{C}_A^i$ denote the collection of all pairwise disjoint *finite* collections of open m-dimensional rectangles which are subsets of A. The *inner content of* A, denoted $\mathscr{C}_*(A)$, is given by

$$\mathscr{C}_*(A) := \sup\left\{\sum_{j=1}^{n} \operatorname{vol}^m(R_j) \,:\, R_1, \ldots, R_n \in \mathfrak{C}_A^i\right\}. \tag{3.25}$$

The above notions of outer and inner contents are often called *outer Jordan content* and *inner Jordan content*, respectively.

Obviously, in light of (3.24) and (3.25), we have that for every bounded set $A \subseteq \mathbb{R}^m, 0 \leq \mathscr{C}_*(A) \leq \mathscr{C}^*(A) < \infty$. Also, it is easy to check that $\mathscr{C}^*$ is monotonic; i.e., $A \subseteq B$ implies that $\mathscr{C}^*(A) \leq \mathscr{C}^*(B)$. Indeed, this follows at once from (3.25) and the fact that $C_A^i \subseteq C_B^i$ if $A \subseteq B$.

Definition 3.83 (*Jordan Measurable Set*). A bounded set $A \subseteq \mathbb{R}^m$ is *Jordan measurable* if its outer and inner (Jordan) contents coincide: $\mathscr{C}^*(A) = \mathscr{C}_*(A)$. In this case, the common value is called the *Jordan measure* (or *Jordan content*) of A and is denoted $\mathscr{C}(A)$. The collection of Jordan measurable sets is denoted $\mathcal{M}_J$.

The next result provides a useful relationship between the inner and outer contents of a set, as well as a key characterization of the Jordan measurability of a set. We omit its proof, which can be deduced, for example, from the discussion in [**193**, pp. 71–73], where a seemingly different (but equivalent) definition of inner and outer Jordan content is adopted; see also Exercise 3.91 below.

Proposition 3.84 (Characterization of Jordan Measurable Sets). *Let S be a bounded subset of $\mathbb{R}^m$, with (topological) boundary ∂S. Then,*

$$\mathscr{C}^*(S) = \mathscr{C}_*(S) + \mathscr{C}^*(\partial S). \tag{3.26}$$

Hence, S is Jordan measurable iff $\mathscr{C}^*(\partial S) = 0$.

Like the Lebesgue outer measure on $\mathbb{R}$, the outer content on $\mathbb{R}$ is not finitely additive; see, respectively, Exercise 3.92 and part (i) of Exercise 3.90.

Example 3.85. The set $\mathbb{Q} \cap [0,1]$ is a Borel set, but it is *not* Jordan measurable. Since $\mathbb{Q} \cap [0,1]$ contains no nonempty open intervals, we immediately have $\mathscr{C}_*(\mathbb{Q}\cap[0,1]) = 0$. Also, since $\mathbb{Q}$ is dense in $\mathbb{R}$, we have that $\partial(\mathbb{Q}\cap[0,1]) = [0,1]$ and it can be shown that $\mathscr{C}^*([0,1]) = 1$; see Exercise 3.87 below. Therefore, by Proposition 3.84, we have that $\mathbb{Q} \cap [0,1]$ is *not* Jordan measurable.

Remark 3.86. Example 3.85 proves that $\mathcal{B} \not\subseteq \mathcal{M}_J$, at least in the context of subsets of $\mathbb{R}$. It can also be shown that $\mathcal{M}_J \not\subseteq \mathcal{B}$. However, it is true that $\mathcal{B} \subseteq \mathcal{M}$ *and* $\mathcal{M}_J \subseteq \mathcal{M}$, where $\mathcal{M} = \mathscr{L}(\mathbb{R})$ denotes the set of Lebesgue measurable subsets of $\mathbb{R}$.

Exercise 3.87. Show that $[0,1]$ is Jordan measurable and that its Jordan measure is equal to 1 : $\mathscr{C}([0,1]) = 1$.

Example 3.88. The Cantor set C is Jordan measurable and has Jordan measure (or content) zero. To see this, note that for each $n \in \mathbb{N}$, C is contained in the union of 2^n closed intervals of length $1/3^n$. Furthermore, for each $\varepsilon > 0$, there is $n_0 \in \mathbb{N}$ such that $\mathscr{C}^*(C) \leq (2/3)^{n_0} < \varepsilon$. Thus, $\mathscr{C}^*(C) = 0$. Hence, since $0 \leq \mathscr{C}_*(C) \leq \mathscr{C}^*(C)$, we also have that $\mathscr{C}_*(C) = 0$, from which we deduce that $\mathscr{C}(C)$ exists and is equal to zero; i.e., C is Jordan measurable and has Jordan measure (or content) zero, as desired.

Note that the fact that $\mathscr{C}^*(C) = 0$ is consistent with Proposition 3.84. Indeed, since C does not contain any nonempty open interval, $\mathscr{C}_*(C) = 0$. Furthermore, since C is perfect (i.e., C is a closed set without any isolated points or, equivalently, in which every point is an accumulation point), we also have that $\partial C = C$; thus, $\mathscr{C}^*(\partial C) = \mathscr{C}^*(C) = 0$. Hence, $0 = \mathscr{C}^*(C) = 0 + 0 = \mathscr{C}_*(C) + \mathscr{C}^*(\partial C)$, in agreement with (3.26) of Proposition 3.84.[28]

Exercise 3.89. Extend the definition of outer and inner contents in Definition 3.82 to any unbounded subset of $\mathbb{R}^m$. What are $\mathscr{C}^*(A)$ and $\mathscr{C}_*(A)$ equal to if A is unbounded? Does Proposition 3.84 hold, then?

[28]Of course, alternatively (and more simply), the fact that $\mathscr{C}_*(C) = 0$ and $\mathscr{C}^*(\partial C) = 0$ also follows, respectively, from the fact that $0 \leq \mathscr{C}_*(C) \leq \mathscr{C}^*(C)$ and the monotonicity of $\mathscr{C}^*$ (since $\partial C \subseteq C$).

Exercise 3.90.

(i) Show by means of a suitably chosen example that the (Jordan) outer content on $\mathbb{R}^m$ is *not* finitely additive (and hence, also, *not* countably additive).

(ii) Is it true that a finite or countable union of bounded subsets of $\mathbb{R}$ with outer content zero is still of outer content zero?

(iii) Is it true that a finite or countable union of Jordan measurable subsets of $\mathbb{R}^m$ is still Jordan measurable?

[*Hint*: It may be helpful to use question (iv) below, along with Proposition 3.84.]

(iv) If $\{S_j\}_{j\geq 1}$ is a finite or countable union of bounded sets $S_j \subseteq \mathbb{R}^m$ such that $\mathscr{C}^*(\partial S_j) = 0$, for each j, is it true that $\mathscr{C}^*(S) = 0$, where $S := \bigcup_{j\geq 1} S_j$?

Exercise 3.91.

(i) Give a proof of Proposition 3.84 when $m = 1$.

[*Hint*: Think of using dyadic intervals, that is, intervals of $\mathbb{R}$ of the form $[q/2^n, (q+1)/2^n]$, where $n \in \mathbb{N}_0$, $q \in \mathbb{Z}$, and q is odd.]

(ii) Try to Prove Proposition 3.84 in the general case, i.e., in $\mathbb{R}^m$.

[*Hint*: Replace the dyadic intervals of part (i) by dyadic cubes, as described towards the beginning of Subsection 7.4.1.]

Exercise 3.92. Show that the Lebesgue outer measure $\mu^* = \mathscr{L}^m$ on $\mathbb{R}^m$ is not finitely additive, although it is, of course, countably subadditive.

A precursor to the definition of the Lebesgue integral, and its generalization discussed in Section 3.6, is the following classical notion known as the Riemann integral which is defined for bounded functions on compact intervals. As given below, Definition 3.94 is not the original definition of the Riemann integral, but it is equivalent and it is more closely related to the notion of (Lebesgue) integral defined in Section 3.6.

In the sequel, $[a, b]$ is a compact interval of $\mathbb{R}$, so that $-\infty < a \leq b < +\infty$, with $a, b \in \mathbb{R}$.

Definition 3.93 (*Step Function*). A function $\varphi : [a, b] \to \mathbb{R}$ is a *step function* if there are a finite collection of points $a := x_0 < x_1 < \cdots < x_n =: b$ and a finite collection of real numbers $c_1, \dots, c_n$ such that, for $k = 1, \dots, n$,

$$\varphi(x) = c_k \text{ whenever } x_{k-1} < x < x_k. \tag{3.27}$$

The *Riemann integral of a step function* φ on a compact interval $[a, b]$, denoted $R\int_a^b \varphi$, is given by

$$R\int_a^b \varphi := \sum_{k=1}^{n} c_k(x_k - x_{k-1}). \tag{3.28}$$

Definition 3.94 (*Upper and Lower Riemann Integrals; Riemann Integrable Function*). Let $f : [a,b] \to \mathbb{R}$ be a bounded function. The *upper* (*resp., lower*) *Riemann integral of* f on $[a,b]$, denoted $\overline{R}\int_a^b f$ (resp., $\underline{R}\int_a^b f$), is given by

$$\overline{R}\int_a^b f := \inf\left\{R\int_a^b \varphi : \varphi \text{ is a step function}, \varphi \geq f\right\} \text{ and} \tag{3.29}$$

$$\underline{R}\int_a^b f := \sup\left\{R\int_a^b \varphi : \varphi \text{ is a step function}, \varphi \leq f\right\}. \tag{3.30}$$

A bounded function f on $[a,b]$ is *Riemann integrable on* $[a,b]$ if

$$\overline{R}\int_a^b f = \underline{R}\int_a^b f. \tag{3.31}$$

In this case, the common value in (3.31) is called the *Riemann integral of* f *over* $[a,b]$ and is denoted by $\int_a^b f$. The collection of functions that are Riemann integrable on $[a,b]$ is denoted $\mathcal{R}([a,b])$.

Theorem 3.95 provides two characterizations of Riemann integrability: one in terms of outer content, the other in terms of Lebesgue measure. The proof is omitted. We note that the proof of the equivalence between (i) and (iii), which is the most frequently encountered in the literature, can be found in many textbooks, including [**522**], [**521**], and [**193**].

We refer to Appendix A (Section A.2) for the definition of the upper and lower limits (denoted, respectively, by $\limsup_{x\to x_0} f$ and $\liminf_{x\to x_0} f$) of a function $f = f(x)$, as $x \to x_0$.

Theorem 3.95 (Characterization of Riemann Integrability). *Let* $f : [a,b] \to \mathbb{R}$ *be a bounded function. Then, the following statements are equivalent*:

(i) *The function* f *is Riemann integrable on* $[a,b]$.

(ii) *For each* $\tau > 0$, *we have* $\mathscr{C}^*(S_\tau) = 0$, *where*

$$S_\tau := \left\{x_0 \in [a,b] : \limsup_{x\to x_0} f - \liminf_{x\to x_0} f \geq \tau\right\}. \tag{3.32}$$

(iii) *The function* f *is continuous* $\mathscr{L}^1$-a.e. *on* $[a,b]$.[29]

We note that the quantity $Osc(f,x_0) := \limsup_{x\to x_0} f - \liminf_{x\to x_0} f$ appearing on the right-hand side of (3.32) in order to define $S_\tau = S_\tau(x_0)$ is called the *oscillation* of f at x_0 and that it can be alternatively defined as follows: $Osc(f,x_0) := \inf_J \operatorname{diam}(f(J))$, where J runs through all of the bounded open subintervals of $[a,b]$ centered at x_0 and contained in $[a,b]$. We leave it as an

[29] Recall that it means that the set E of points of $[a,b]$ at which f is *not* continuous is a set of Lebesgue measure zero: $\mathscr{L}^1(E) = 0$; see also Exercise 3.98.

exercise for the reader to check the equivalence between the two definitions of $Osc(f, x_0)$, as well as to check that for every x_0 in $[a, b]$, $Osc(f, x_0) = 0 \Leftrightarrow f$ is continuous at x_0—and hence that $Osc(f, x_0) > 0$ iff f is discontinuous at x_0.

Exercise 3.96. Try to prove Theorem 3.95 or at least the equivalence between (i) and (ii), sometimes known as the *Jordan Content Riemann Integrability Criterion*.

[*Hint* (for one of the implications): Assume that $\mathscr{C}^*(S_\tau) > 0$, for some $\tau > 0$, and find a suitable positive lower bound for $U(P, f) - L(P, f)$, where $U(P, f)$ and $L(P, f)$ denote, respectively, the upper and the lower Riemann (or, rather, Darboux) sums of f with respect to a partition P of $[a, b]$.]

Example 3.97. Consider *Dirichlet's function*, $\chi_{\mathbb{Q}\cap[0,1]}$, given by

$$\chi_{\mathbb{Q}\cap[0,1]}(x) := \begin{cases} 1, & x \in \mathbb{Q} \cap [0, 1], \\ 0, & x \in [0, 1]\backslash\mathbb{Q}, \end{cases} \tag{3.33}$$

where, much as in Definition 3.103 below, $\chi_{\mathbb{Q}\cap[0,1]}$ is the characteristic function of the rational numbers in the unit interval $[0, 1]$. Since $\mathbb{Q} \cap [0, 1]$ and $[0, 1]\backslash\mathbb{Q}$ are both dense in $[0, 1]$, Dirichlet's function is discontinuous at every point in $[0, 1]$. Since $\mathscr{L}^1([0, 1]) = 1 > 0$, Theorem 3.95 implies that Dirichlet's function is not Riemann integrable on $[0, 1]$.

Alternately, it can be shown that, for $0 < \tau \leq 1$, we have $S_\tau = [0, 1]$ and, therefore, $\mathscr{C}^*(S_\tau) > 0$. It is also easy to verify directly that

$$\overline{R}\int_0^1 \chi_{\mathbb{Q}\cap[0,1]} = 1 \neq 0 = \underline{R}\int_0^1 \chi_{\mathbb{Q}\cap[0,1]}.$$

Exercise 3.98.

(a) Show that if $f : [a, b] \to \mathbb{R}$ is Riemann integrable (as in part (i) of Theorem 3.95), then $E \subseteq [a, b]$ is a Borel set, where E is given by $E := \{x \in [a, b] : f \text{ is discontinuous at } x\}$.

(b) Deduce from part (a) and Theorem 3.95 that if condition (i) of Theorem 3.95 holds, then (according to part (iii) of Theorem 3.95, in particular), $\mathscr{L}^1(E) = 0$, where, in the notation of Remark 3.67 above, $\mathscr{L}^1$ denotes indifferently the Lebesgue measure on $\mathbb{R}$ (or its restriction to $[a, b]$, in the sense of Exercise 3.124 below), $\mathscr{L}^1 = \overline{\mu}$, or the associated Borel measure on $\mathbb{R}$ (or on $[a, b]$), $\mathscr{L}^1 = \widetilde{\mu}$.

Exercise 3.99. Use, in particular, the characterization of Riemann integrability given in part (iii) of Theorem 3.95 in order to show that every bounded and Riemann integrable function on the compact interval $[a, b]$ is Lebesgue integrable on $[a, b]$.

[*Hint*: Observe that by the aforementioned characterization, f is Borel measurable and that since f is bounded (except on a set of Lebesgue measure zero), it is Lebesgue integrable on $[a, b]$; see the relevant results in Section 3.6 below.]

We caution the reader that it is perhaps not so reasonable to expect to be able to solve the next exercise without further hints and, for question (iii), without having had prior exposure to the Lebesgue theory of integration (Section 3.6 below and well beyond).

Exercise 3.100 (*Lebesgue Completes Riemann*).

(i) Show that the space $\mathcal{E} := \mathscr{L}^1([a,b])$ of Riemann integrable functions on the compact interval $[a,b]$ is *not* complete, when equipped with the (pseudo) metric d given by $d(f,g) := \int_a^b |f(t) - g(t)|dt$, for all $f, g \in \mathcal{E}$.

(ii) Show that the space $\widetilde{\mathcal{E}} = C([a,b])$ of continuous (real-valued) functions is *not* complete (in the sense of metric spaces),[30] when it is equipped with the restriction of d (as introduced in part (i)) to $\widetilde{\mathcal{E}} \subseteq \mathcal{E}$.[31]

[*Advice*: You may wish to solve question (ii) first.]

(iii) Anticipating the next section (i.e., Section 3.6), show that the *completion* of the (pseudo) metric space $\mathcal{E}$ (in part (i)) as well as of the metric space $\widetilde{\mathcal{E}}$ (in part (ii)) is equal to (or, more precisely, can be identified with) $L^1([a,b])$, the space of (real-valued) *Lebesgue integrable* functions on the compact interval $[a,b]$, modulo the (almost everywhere) equivalence relation $\sim$ defined by $f \sim g$ iff $f = g$ $\mathscr{L}^1$-a.e., and equipped with the (actual) metric induced by $d(f,g) := \int_a^b |f(t) - g(t)|dt$.

In short, the corresponding slogan is that on a compact interval $[a,b]$, "*the space of Lebesgue integrable functions is the completion of the space of Riemann integrable functions*".

3.6. Lebesgue Integration

The definition of integration used throughout this book, Definition 3.111, is a generalization of the definition of integration on $(\mathbb{R}, \mathscr{L}^1)$ due to Henri Lebesgue. This definition subsumes Riemann's definition of integration (Definition 3.94) and it provides a sufficiently general setting for our purposes. However, most of the proofs of results stated in this section are beyond the scope of this book and are omitted. Indeed, they can be found in many books, a selection of which is given below.

We next give a brief outline of the construction of the Lebesgue integral. In the development of the Lebesgue integration theory, one typically proceeds

[30]The *completion* $\widehat{M}$ of a given metric space M is unique, up to isomorphism (i.e., up to a bijective isometry). It is "the" complete metric space in which M can be embedded isometrically as a dense subset of $\widehat{M}$. As a classic example, we have that $\mathbb{R}$ (the set of real numbers) is the completion of $\mathbb{Q}$ (the set of rational numbers). Such a metric completion is unique, up to an isometric isomorphism.

[31]Note that now, when restricted to $\widetilde{\mathcal{E}}$, d is a true metric on $\widetilde{\mathcal{E}}$; that is, for all $f, g \in \widetilde{\mathcal{E}}$, $d(f,g) = 0 \Rightarrow f = g$.

step by step:

Step 1: First, one defines the Lebesgue integral of a simple function, the abstract analog of a step function on $\mathbb{R}$ or on a compact interval $[a, b] \subseteq \mathbb{R}$.

Step 2: Second, one considers the case of a nonnegative (and measurable) function.

Step 3: Third, one deals with the case of a (measurable) real-valued function f by simply writing $f = f_+ - f_-$, where f_+ and f_- denote, respectively, the positive and negative part of f, and one applies to $f_\pm$ the definition of the Lebesgue integrals $\int f_\pm$ given in Step 2.

Step 4: Finally, if one wishes to also work with a (measurable) complex-valued function f (which may be very useful in many contexts), one just has to break up f into its real and imaginary parts, $\mathrm{Re}(f)$ and $\mathrm{Im}(f)$, and to apply Step 3 to each of these real-valued (and measurable) functions.

One special feature of the construction of the Lebesgue integral is worth pointing out, namely, the fact that if f is an arbitrary measurable and *nonnegative* function, then its Lebesgue integral, $\int f$, is always well defined, as the supremum of the integral of nonnegative *simple* functions approximating it, whether it is equal to a nonnegative real number or to $+\infty$. This is entirely analogous to the corresponding property of series, $\sum_{n=1}^{\infty} a_n$, of nonnegative terms (i.e., $a_n \geq 0$, for all $n \geq 1$), for which its sum $\sum_{n=1}^{\infty} a_n$ is always well defined, as the supremum of its partial sums $\sum_{n=1}^{N} a_n$ (with $N \geq 1$ arbitrary), but now viewed as an extended nonnegative real number, i.e., as an element of $[0, +\infty]$. Of course, just as such a series of nonnegative terms is said to be convergent if $\sum_{n=1}^{\infty} a_n < \infty$, the nonnegative (and measurable) function f is said to be Lebesgue integrable if $\int f < \infty$.

For standard references on the theory of the Lebesgue integral, with various levels of sophistication, see, for example, [**57**], [**522**], or [**571**] (at the undergraduate level) and [**5**], [**91**], [**193**], [**523**], and [**521**] or [**189**] (at the graduate level).

We begin the outline of the construction of the Lebesgue integral by extending the notion of measurability from subsets (of the same set X, equipped with a σ-algebra) to real-valued functions defined on the same set X.[32]

Definition 3.101 (*Measurable Function*). Let $(X, \mathcal{M})$ be a measurable space and suppose $f : X \to [-\infty, +\infty]$. Then, f is *measurable* with respect to $\mathcal{M}$ if for each $c \in \mathbb{R}$, the set $\{\boldsymbol{x} \in X : f(\boldsymbol{x}) > c\}$ is measurable.

[32]The extension of this notion to complex-valued functions f is immediate via the usual procedure, described in Step 4 above, namely, by separating the real and imaginary parts of f.

Measurable functions have numerous useful properties, as is described in the next result.

Theorem 3.102 (Properties of Measurable Functions). *Let $(X, \mathcal{M}, \mu)$ be a measure space. Then, the following properties hold:*

(i) *For any pair of measurable real-valued functions on X, f and g, and any pair of real numbers α and β, we have that $|f|$, $\alpha f + \beta g$, fg, $\max\{f, g\}$, and $\min\{f, g\}$ are measurable.*

(ii) *If $f : X \to \mathbb{R}$ is measurable and $\varphi : \mathbb{R} \to \mathbb{R}$ is continuous, then the composition $\varphi \circ f$ is measurable.*

(iii) *If $f = g$ μ-a.e. on X, then f is measurable iff g is measurable.*[33]

(iv) *If $\{f_n\}_{n=1}^{\infty}$ is a sequence of measurable functions on X, then the functions $\sup\{f_n\}$, $\inf\{f_n\}$, $\limsup\{f_n\}$, and $\liminf\{f_n\}$ are measurable. Furthermore, if $f_n \to f$ pointwise μ-a.e. on X, then f is measurable.*[34]

(v) (Egoroff's Theorem). *Suppose $\{f_n\}_{n=1}^{\infty}$ is a sequence of measurable functions that converges pointwise μ-a.e. on X to a function f that is finite μ-a.e. on X. Then, for each $\varepsilon > 0$, there is a measurable subset A_ε of X for which $f_n \to f$* uniformly on A_ε and $\mu(X \backslash A_\varepsilon) < \varepsilon$.

Definition 3.103 (*Characteristic and Step Functions*). Let $(X, \mathcal{M})$ be a measurable space. The *characteristic function* of a measurable set A, denoted χ_A, is given defined by

$$\chi_A(x) := \begin{cases} 1, & \boldsymbol{x} \in A, \\ 0, & \boldsymbol{x} \in A^c \ (:= X \backslash A). \end{cases} \tag{3.34}$$

A real-valued function ψ on X is *simple* if there is a finite collection of measurable sets $\{A_k\}_{k=1}^{n}$ and a finite set of real numbers $\{c_k\}_{k=1}^{n}$ such that

$$\psi(x) = \sum_{k=1}^{n} c_k \cdot \chi_{A_k}(x). \tag{3.35}$$

Remark 3.104. Every simple function is measurable. Also, a given simple function can always be written (necessarily uniquely) in such a way that, for each $k = 1, \ldots, n$, the measurable sets A_k are pairwise disjoint and the real numbers c_k are distinct. Indeed, in this case, we must have that $A_k = \{\boldsymbol{x} \in X : f(\boldsymbol{x}) = c_k\}$, for each $k = 1, \ldots, n$, where $\{c_k\}_{k=1}^{n}$ are the finitely many distinct values of f.

The following theorem is essential to the definition of integration used in this book.

[33]Given two real-valued measurable functions f and g on X, we write that $f = g$ a.e (resp., $f \leq g$ a.e.) if the measurable set $\{\boldsymbol{x} \in X : f(\boldsymbol{x}) \neq g(\boldsymbol{x})\}$ (resp., $\{\boldsymbol{x} \in X : f(\boldsymbol{x}) \leq g(\boldsymbol{x})\}$) is of μ-measure zero.

[34]Technically, we need to assume here that either $(X, \mathcal{M}, \mu)$ is complete (in the sense of Definition B.5 of Section B.2) or that the convergence of the sequence $\{f_n\}$ holds pointwise on X.

Theorem 3.105 (Approximation of Measurable Functions by Simple Functions). *Let $(X, \mathcal{M}, \mu)$ be a measure space and let f be a measurable function on X. Then, there is a sequence $\{\psi_n\}_{n=1}^{\infty}$ of simple functions on X converging pointwise to f and such that $|\psi_n| \leq |f|$ on X, for all $n \in \mathbb{N}$.*

If, in addition, X is a countable union of sets with finite measure (i.e., if μ is σ-finite), then the sequence $\{\psi_n\}_{n=1}^{\infty}$ can be chosen so that each ψ_n is zero outside a set of finite measure.

Moreover, if f is nonnegative, then the sequence $\{\psi_n\}_{n=1}^{\infty}$ can be chosen to be nondecreasing and such that each ψ_n is nonnegative.

As was explained earlier at the beginning of this section, Lebesgue integration is defined in four steps: first, for nonnegative simple functions, then, for nonnegative measurable functions, and lastly, for general real-valued (and then, complex-valued) measurable functions.

Definition 3.106 (*Lebesgue Integral of a Nonnegative Simple Function*). Let $(X, \mathcal{M}, \mu)$ be a measure space and let ψ be a nonnegative simple function on X given, without loss of generality, by $\psi = \sum_{k=1}^{n} c_k \cdot \chi_{A_k}$, where each c_k is nonnegative and $A_k = \{\boldsymbol{x} \in X : \psi(\boldsymbol{x}) = c_k\}$. The *(Lebesgue) integral of ψ over X*, denoted $\int_X \psi \, d\mu$ (or just $\int \psi$), is given by[35]

$$\int_X \psi \, d\mu := \sum_{k=1}^{n} c_k \cdot \mu(A_k), \tag{3.36}$$

with the convention that the right-hand side is equal to $+\infty$ if, for some $k \in \mathbb{N}$, $\mu(A_k) = +\infty$ and $c_k > 0$. For a measurable subset A of X, the *integral of ψ over A with respect to μ*, denoted $\int_A \psi \, d\mu$, is given by

$$\int_A \psi \, d\mu := \int_X \psi \cdot \chi_A \, d\mu. \tag{3.37}$$

Exercise 3.107. Show that under the assumption of Definition 3.106, the integral $\int_X \psi d\mu$ is independent of the way the function ψ is written as a simple function, as in Definition 3.106.

Definition 3.108 (*Lebesgue Integral of a Nonnegative Measurable Function*). Let $(X, \mathcal{M}, \mu)$ be a measure space and let f be a nonnegative measurable function on X. The *(Lebesgue) integral of f over X*, denoted $\int_X f \, d\mu$, is given by

$$\int_X f \, d\mu := \sup\left\{\int_X \psi \, d\mu \; : \; \psi \text{ simple}, 0 \leq \psi \leq f\right\}. \tag{3.38}$$

[35] In Lebesgue's integration (and measure) theory, besides the aforementioned rules concerning the manipulation of the symbols $\pm\infty$, the appropriate convention is to let $0 \cdot (\pm\infty) := 0$.

For a measurable subset A of X, the *integral of f over A with respect to μ*, denoted $\int_A f\, d\mu$, is defined by $\int_A f\, d\mu = \int_X f \cdot \chi_A\, d\mu$. A nonnegative measurable function f is said to be *(Lebesgue) integrable over X with respect to μ* if $\int_X f\, d\mu < \infty$.

In order to define the integral of a *real-valued* measurable function f, it suffices to identify f as a difference of nonnegative measurable functions as follows.

Definition 3.109 (*Positive and Negative Parts of a Real-Valued Function*). Let $(X, \mathcal{M})$ be a measurable space and let f be a real-valued measurable function on X. The *positive part* and *negative part* of f, denoted f^+ and f^-, respectively, are given on X by

$$f^+(\boldsymbol{x}) := \max\{f(\boldsymbol{x}), 0\} = \begin{cases} f(\boldsymbol{x}), & \text{if } f(\boldsymbol{x}) \geq 0, \\ 0, & \text{if } f(\boldsymbol{x}) < 0, \end{cases} \tag{3.39}$$

$$f^-(\boldsymbol{x}) := \max\{-f(\boldsymbol{x}), 0\} = \begin{cases} -f(\boldsymbol{x}), & \text{if } f(\boldsymbol{x}) \leq 0, \\ 0, & \text{if } f(\boldsymbol{x}) > 0. \end{cases} \tag{3.40}$$

Remark 3.110. We have that, on X, $f = f^+ - f^-$, $|f| = f^+ + f^-$, and $|f|, f^+$, and f^- are nonnegative and measurable.

Definition 3.111 (*Lebesgue Integrability and Integral of a Measurable Function*). Let $(X, \mathcal{M}, \mu)$ be a measure space. A measurable (*real-valued*) function f on X is *integrable over X with respect to μ* if, as nonnegative measurable functions, f^+ and f^- are integrable over X with respect to μ. In this case, the *(Lebesgue) integral of f over X with respect to μ*, denoted $\int_X f\, d\mu$, is given by

$$\int_X f\, d\mu := \int_X f^+\, d\mu - \int_X f^-\, d\mu. \tag{3.41}$$

For a measurable subset A of X, the *integral of f over A with respect to μ*, denoted $\int_A f\, d\mu$, is defined by $\int_A f\, d\mu = \int_X f \cdot \chi_A\, d\mu$.

Remark 3.112. If f is complex-valued, then by definition, f is *(Lebesgue) integrable* if both its real part $\mathrm{Re}(f)$ and its imaginary part $\mathrm{Im}(f)$ are integrable (in the sense of Definition 3.111). Also, in this case, the *(Lebesgue) integral of* f over X, denoted $\int_X f d\mu$ (or simply $\int f$), is given by

$$\int_X f d\mu = \int_X \mathrm{Re}(f) d\mu + i \int_X \mathrm{Im}(f) d\mu. \tag{3.42}$$

Exercise 3.113. Show that if f is measurable and complex-valued, then f is Lebesgue integrable iff its modulus $|f|$ satisfies the inequality $|f| \leq g$, μ-a.e. on X, for some (Lebesgue) integrable nonnegative function g.

In particular, it follows that a measurable function f is Lebesgue integrable iff $|f|$ is Lebesgue integrable.

There are many useful properties of integration. Next, we point out the basic properties (in Theorem 3.114) and then, some of the most fundamental properties (in Theorem 3.116). The latter properties demonstrate the great power and flexibility of the Lebesgue integral and do not have counterparts in the Riemann theory of integration.

Theorem 3.114 (Basic Properties of the Lebesgue Integral). *Let $(X, \mathcal{M}, \mu)$ be a measure space. Then:*

(i) (Linearity). *If f and g are integrable over X and $\alpha, \beta \in \mathbb{C}$, then $\alpha f + \beta g$ is integrable over X and $\int_X (\alpha f + \beta g)\, d\mu = \alpha \int_X f\, d\mu + \beta \int_X g\, d\mu$.*

(ii) (Monotonicity). *If f and g are $\mathbb{R}$-valued integrable functions over X and $f \leq g$ μ-a.e. on X, then $\int_X f\, d\mu \leq \int_X g\, d\mu$.*

(iii) (Additivity). *If A and B are disjoint measurable subsets of X and f is integrable over X, then $\int_{A\cup B} f\, d\mu = \int_A f\, d\mu + \int_B f\, d\mu$.*

(iv) (Chebyshev's Inequality). *If f is a nonnegative measurable function and $c > 0$, then $\mu(\{\boldsymbol{x} \in X : f(\boldsymbol{x}) \geq c\}) \leq (1/c) \int_X f\, d\mu$.*

(v) *If f is a bounded measurable function over X that vanishes outside of a set of finite measure, then f is integrable over X.*

Exercise 3.115. Prove Theorem 3.114. In particular, prove Chebyshev's Inequality (part (iv) of the theorem).

There are also many useful properties of integration when it comes to sequences of measurable sets or of measurable functions.

Theorem 3.116 (Interchange of Limits and of Lebesgue Integrals). *Let $(X, \mathcal{M}, \mu)$ be a measure space and let $\{f_n\}_{n=1}^{\infty}$ be a sequence of $\mathbb{C}$-valued measurable functions on X. Then:*

(i) (Continuity of Integration for Unions). *If f is integrable over X and $\{A_n\}_{n=1}^{\infty}$ is a nondecreasing sequence of measurable sets such that $\bigcup_{n=1}^{\infty} A_n = X$, then $\int_X f\, d\mu = \lim_{n\to\infty} \int_{A_n} f\, d\mu$.*

(ii) (Continuity of Integration for Intersections). *If f is integrable over X and $\{B_n\}_{n=1}^{\infty}$ is a sequence of measurable sets such that $B_n \supseteq B_{n+1}$, for all $n \in \mathbb{N}$, then $\int_{\bigcap_{n=1}^{\infty} B_n} f\, d\mu = \lim_{n\to\infty} \int_{B_n} f\, d\mu$.*

(iii) (Fatou's Lemma). *If each f_n is nonnegative, $f_n \to f$ pointwise μ-a.e. on X, then f is nonnegative and measurable and*

$$\int_X f\, d\mu \leq \liminf_{n\to\infty} \int_X f_n\, d\mu.$$

(iv) (Monotone Convergence Theorem). *If each f_n is nonnegative, the sequence $\{f_n\}_{n=1}^{\infty}$ is nondecreasing, and $f(\boldsymbol{x}) = \lim_{n\to\infty} f_n(\boldsymbol{x})$,* for μ-a.e. $\boldsymbol{x} \in X$, *then $\int_X f\, d\mu = \lim_{n\to\infty} \int_X f_n\, d\mu$.*

(v) (Lebesgue's Dominated Convergence Theorem). *If $f_n \to f$ pointwise* μ-a.e. *on X, f is measurable, and there exists a nonnegative function g which is integrable over X and such that $|f_n| \leq g$* μ-a.e. on X, for all $n \in \mathbb{N}$, *then f is integrable over X and $\int_X f\, d\mu = \lim_{n\to\infty} \int_X f_n\, d\mu$.*

Proposition 3.117 (Lebesgue Integral and Null Sets). *Let $(X, \mathcal{M}, \mu)$ be a measure space. Then:*

(i) *If A is a measurable subset of X of measure zero (i.e., $\mu(A) = 0$), then if $f : X \to \mathbb{C}$ is measurable, we have that $\int_A f d\mu = 0$.*

(ii) *Conversely, provided the measurable function f is assumed to be nonnegative, then $\int_X f d\mu = 0 \Rightarrow f = 0$,* μ-a.e. on X. *Hence, still for $f \geq 0$, then for any $A \subseteq X$ measurable, we have that $\int_A f d\mu = 0 \Leftrightarrow f = 0$,* μ-a.e. on A.

Exercise 3.118. Prove Proposition 3.117.

[*Hint*: For part (i), you may assume without loss of generality that $f \geq 0$. For part (ii), you may use a relevant part of Theorem 3.116. Alternatively, you may use Chebyshev's Inequality (part (iv) of Theorem 3.114).]

Exercises

Exercise 3.119. Replace (a, b) by $(a, b]$ in the discussion of $\mathscr{L}^1$, Lebesgue measure on $\mathbb{R}$.

Exercise 3.120. What are the natural elementary families on $\mathbb{R}^m$?

Exercise 3.121 (*Generated Algebra*). The purpose of this exercise is to show that given a collection $\mathcal{C}$ of subsets of a given set X, there is a smallest algebra $\mathcal{A}$ (of X) containing $\mathcal{C}$; it is called *the algebra generated by* $\mathcal{C}$.

(i) First, show that $\mathcal{A} := \bigcap_{\mathcal{C}\subseteq\mathcal{B}} \mathcal{B}$ is an algebra (of X) containing $\mathcal{C}$.

(ii) Prove that if $\mathcal{D}$ is an algebra (of X) containing $\mathcal{C}$, then $\mathcal{A}$, as defined in part (i), is a subalgebra of $\mathcal{D}$; i.e., $\mathcal{A} \subseteq \mathcal{D}$.

(iii) Conclude from parts (i) and (ii) that $\mathcal{A}$, as given in (i), is the smallest algebra containing $\mathcal{C}$ and that it is the unique such algebra.

Exercise 3.122 (*Generated σ-Algebra*). Let X be a set and let $\mathcal{C}$ be a collection of subsets of X. Then, show that there exists a (necessarily unique) σ-algebra $\mathcal{A}$ which is the smallest σ-algebra of X containing $\mathcal{C}$; it is called the *σ-algebra generated by* $\mathcal{C}$.

[*Hint*: Go through exactly the same steps as in parts (i)–(iii) of Exercise 3.121 above, but now with the word "algebra" replaced by "σ-algebra".]

Exercise 3.123. Show directly that for any integer $m \geq 1$, the m-dimensional Lebesgue outer measure $\mathscr{L}^m$ is a metric outer measure on Euclidean space $\mathbb{R}^m = (\mathbb{R}^m, d_m)$.

Exercise 3.124 (*Restriction of a Measure*). Let $(X, \mathcal{M}, \mu)$ be a measure space and let $A \in \mathcal{M}$. Show that $\mathcal{M}_A := \{A \cap E : E \in \mathcal{M}\}$ is a σ-algebra on A and that μ_A, given by $\mu_A(D) = \mu(D)$, for all $D \in \mathcal{M}_A$, is well defined and is a measure on $(A, \mathcal{M}_A)$, called the *restriction of μ to* (the measurable set) A. The resulting measure space $(A, \mathcal{M}_A, \mu_A)$ is called *the restriction of the measure space* $(X, \mathcal{M}, \mu)$ to A.

[*Hint*: Observe that $D \in \mathcal{M}_A \Leftrightarrow D \in \mathcal{M}$ and $D \subseteq A$.]

The following exercise is longer and perhaps more difficult than the other exercises in this chapter. It also makes use of several key results provided in Appendix B; see, especially, the hint given at the end of the exercise.

Exercise 3.125 (*Product Measure*). Let $(X, \mathcal{M}, \mu)$ and $(Y, \mathcal{N}, \nu)$ be two measure spaces, with μ and ν being assumed to be σ-finite, for simplicity. Let $\mathcal{M} \times \mathcal{N}$ be the σ-algebra on the Cartesian product $X \times Y$ generated by the *rectangles* $\{A \times B : A \in \mathcal{M}, B \in \mathcal{N}\}$, as defined in Exercise 3.122 above. Furthermore, let $\mu \times \nu$ (or $\mu \otimes \nu$) be given by

$$(\mu \times \nu)(A \times B) := \mu(A) \cdot \nu(B), \text{ for all } A \in \mathcal{M} \text{ and } B \in \mathcal{N}, \tag{3.43}$$

and then *uniquely* extended to $\mathcal{M} \times \mathcal{N}$, as explained in the hint just below.

(i) Show that $\mu \times \nu$ is well defined and is a measure on the measurable space $(X \times Y, \mathcal{M} \times \mathcal{N})$. It is called the *product measure* (or *tensor product*) of the measures μ and ν.

[*Hint*: Prove that the σ-algebra $\mathcal{M} \times \mathcal{N}$ is generated by the algebra $\mathcal{A}$ consisting of the finite disjoint unions of rectangles. Furthermore, show that $\mu \times \nu$ first extends in the obvious way, in light of (3.43), and uniquely defines a premeasure on the algebra $\mathcal{A}$, in the sense of Definition B.9 in Appendix B. Then, apply the Algebra Version of the Carathéodory Extension Theorem (Theorem B.12 in Appendix B) in order to show that $\mu \times \nu$ extends to the σ-algebra $\mathcal{M} \times \mathcal{N}$.]

(ii) Moreover, prove that if μ and ν are σ-finite, then so is $\mu \times \nu$. Hence, deduce the uniqueness part of Theorem B.12 from the uniqueness of the extension of $\mu \times \nu$, initially defined on rectangles by (3.43).

Exercise 3.126 (*Product of Lebesgue Measures*). Show that, with the notation (and results) of Exercise 3.125, if we let $X := \mathbb{R}^m$, $Y := \mathbb{R}^n$ (with $m, n \geq 1$ arbitrary integers), $\mu := \mathscr{L}^m$ and $\nu := \mathscr{L}^n$ (the Lebesgue measures on $\mathbb{R}^m$ and $\mathbb{R}^n$, respectively), then their product measure coincides with $\mathscr{L}^{m+n}$, the Lebesgue measure on $\mathbb{R}^{m+n}$: $\mathscr{L}^{m+n} = \mathscr{L}^m \times \mathscr{L}^n$, interpreted either as the usual Lebesgue measure or as a Borel measure on $\mathbb{R}^{m+n}$ (as in Exercise 3.127 below).

[*Hint*: Show, in particular, that $\mathcal{M}^{(m+n)} = \mathcal{M}^{(m)} \times \mathcal{M}^{(n)}$, where, for any integer $k \geq 1$, $\mathcal{M}^{(k)}$ denotes the σ-algebra of Lebesgue measurable sets on $\mathbb{R}^k$, i.e., the measurable sets with respect to $\mathscr{L}$, the Lebesgue outer measure on $\mathbb{R}^k$.]

Exercise 3.127 (*Product of Borel Measures*). Show that if X and Y are metric spaces, then $\mathcal{B}(X \times Y) = \mathcal{B}(X) \times \mathcal{B}(Y)$. Deduce that if μ and ν are Borel measures on X and Y, respectively, then their product measure $\mu \times \nu$ is a Borel measure on $X \times Y$.

Exercise 3.128. Show that, with the notation of Exercises 3.125 and 3.126, the following properties hold:

(a) The measurable space $(\mathbb{R}^m, \mathcal{M}^{(m)}) \times (\mathbb{R}^n \times \mathcal{M}^{(m)})$ can be identified with the measurable space $(\mathbb{R}^{m+n}, \mathcal{M}^{(m+n)})$.[36]

(b) The measurable space $(\mathbb{R}^m, \mathcal{B}(\mathbb{R}^m)) \times (\mathbb{R}^n, \mathcal{B}(\mathbb{R}^n))$ can be identified with the measurable space $(\mathbb{R}^{m+n}, \mathcal{B}(\mathbb{R}^m \times \mathbb{R}^n))$.

[*Hint*: Use, in particular, part (ii) of Exercise 3.125.]

[36]Here, for any $k \in \mathbb{N}$, $\mathcal{M}^{(k)}$ denotes the σ-algebra of Lebesgue measurable subsets of $\mathbb{R}^k$.

Part 2

Dimension Theory

Chapter 4

Iterated Function Systems and Self-Similarity

The early sense of self-similarity as an organizing principle came from the limitations on the human experience of scale. [...]

Self-similarity is symmetry across scale. It implies recursion pattern inside of pattern.

James Gleick, [**215**]

A cloud is made of billows upon billows upon billows that look like clouds. As you come closer to a cloud, you don't get something smooth, but irregularities at a smaller scale.

Benoît B. Mandelbrot

Iterated function systems provide a concise method for constructing many fractals and, more generally, compact sets. Indeed, the Cantor set C, the Koch curve K_C, and the Sierpinski carpet **SC** in Examples 1.1, 1.8, and 1.10, respectively, are all examples of fractals constructed using iterated function systems. In this chapter, a mathematical procedure for constructing compact sets through iterated function systems is developed.

First, the chapter begins with a discussion of a particular metric space that provides the context for the mathematical construction of many of the fractals investigated throughout this text.

4.1. Hausdorff Metric

In this section, the Hausdorff metric d_H and the metric space $(\mathbb{K}, d_H)$ are defined. Some definitions are in order.

Definition 4.1. Consider the Euclidean space $(\mathbb{R}^m, d_m)$. The *distance* between a point $\boldsymbol{x} \in \mathbb{R}^m$ and a set $A \subseteq \mathbb{R}^m$ is given by

$$\text{dist}(\boldsymbol{x}, A) := \inf\{d_m(\boldsymbol{x}, \boldsymbol{a}) : \boldsymbol{a} \in A\}. \tag{4.1}$$

Furthermore, for $A, B \subseteq \mathbb{R}^m$, let $\rho(A, B) \in [0, +\infty]$ be given by

$$\rho(A, B) := \sup\{\text{dist}(\boldsymbol{a}, B) : \boldsymbol{a} \in A\}. \tag{4.2}$$

Finally, in the sequel, we denote by $\mathbb{K}(\mathbb{R}^m)$ the collection of all nonempty compact subsets of $\mathbb{R}^m$.[1]

Remark 4.2. Note that ρ fails to be a metric when restricted to $\mathbb{K}(\mathbb{R}^m)$ since it is not symmetric. To see this, consider a pair of sets $A, B \in \mathbb{K}(\mathbb{R}^m)$ such that B is a proper subset of A (that is, $B \subseteq A$ and $B \neq A$). Then, $\rho(B, A) = 0$ but $\rho(A, B) > 0$ since there exists a point $\boldsymbol{a}_0 \in A$ such that $\boldsymbol{a}_0 \notin B$ and, since B is closed, $\text{dist}(\boldsymbol{a}_0, B) > 0$. Nonetheless, dist and ρ are helpful notions in that they underlie the structure of the *Hausdorff metric* d_H defined in (4.3) below.

Theorem 4.3. *Let* $\boldsymbol{x} \in \mathbb{R}^m$ *and* $A, B, C \in \mathbb{K}(\mathbb{R}^m)$. *Then, the following properties of* dist *and* ρ *hold*:

(i) $\text{dist}(\boldsymbol{x}, A) = 0$ *iff* $\boldsymbol{x} \in A$.

(ii) $\rho(A, B) = 0$ *iff* $A \subseteq B$.

(iii) *There exists* $\boldsymbol{a}_{\boldsymbol{x}} \in A$ *such that* $\text{dist}(\boldsymbol{x}, A) = d_m(\boldsymbol{x}, \boldsymbol{a}_{\boldsymbol{x}})$.

(iv) *There exist* $\boldsymbol{a}_0 \in A$ *and* $\mathbf{b}_0 \in B$ *such that* $\rho(A, B) = d_m(\boldsymbol{a}_0, \mathbf{b}_0)$. *Hence,* $\rho(A, B) < \infty$.

(v) $A \subseteq B \Rightarrow \text{dist}(\boldsymbol{x}, B) \leq \text{dist}(\boldsymbol{x}, A)$.

(vi) $A \subseteq B \Rightarrow \rho(C, B) \leq \rho(C, A)$.

(vii) $\rho(A \cup B, C) = \max\{\rho(A, C), \rho(B, C)\}$.

(viii) $\rho(A, B) \leq \rho(A, C) + \rho(C, B)$.

Proof. In order to prove property (i), let $\boldsymbol{x} \in A$; then, we immediately have $\text{dist}(\boldsymbol{x}, A) = 0$. Now, suppose $\text{dist}(\boldsymbol{x}, A) = 0$. Then, $\boldsymbol{x} \in \overline{A}$. Since A is compact, it is closed and hence, $\overline{A} = A$. Therefore, $\boldsymbol{x} \in A$.

To prove property (ii), suppose $A \subseteq B$ and let $\boldsymbol{a} \in A$. Since $\boldsymbol{a} \in B$, property (i) implies that $\text{dist}(\boldsymbol{a}, B) = 0$. So, $\rho(A, B) = 0$. Now, assume that $\rho(A, B) = 0$ and let $\boldsymbol{a} \in A$. Then, $0 \leq d(\boldsymbol{a}, B) \leq \rho(A, B) = 0$; so, property (i) yields $\boldsymbol{a} \in B$. Hence, $A \subseteq B$.

[1]Observe that if $A, B \in \mathbb{K}(\mathbb{R}^m)$, then $0 \leq \rho(A, B) < \infty$; the latter strict inequality follows from part (iv) of Theorem 4.3.

To prove (iii), let $\{\boldsymbol{a}_n\}_{n=1}^{\infty} \subseteq A$ be a sequence such that $d_m(\boldsymbol{x}, \boldsymbol{a}_n) \leq \text{dist}(\boldsymbol{x}, A) + (1/n)$. Since A is sequentially compact, there is a subsequence $\{\boldsymbol{a}_{n_k}\}_{k=1}^{\infty}$ of $\{\boldsymbol{a}_n\}_{n=1}^{\infty}$ which converges to a point $\boldsymbol{a}_{\boldsymbol{x}} \in A$. We have

$$\begin{aligned}\text{dist}(\boldsymbol{x}, A) \leq d_m(\boldsymbol{x}, \boldsymbol{a}_{\boldsymbol{x}}) &\leq d_m(\boldsymbol{x}, \boldsymbol{a}_{n_k}) + d_m(\boldsymbol{a}_{n_k}, \boldsymbol{a}_{\boldsymbol{x}}) \\ &\leq \text{dist}(\boldsymbol{x}, A) + \frac{1}{n_k} + d_m(\boldsymbol{a}_{n_k}, \boldsymbol{a}_{\boldsymbol{x}}).\end{aligned}$$

Letting $k \to \infty$ yields $\text{dist}(\boldsymbol{x}, A) = d_m(\boldsymbol{x}, \boldsymbol{a}_{\boldsymbol{x}})$.

To prove property (iv), let $\{\boldsymbol{a}_n\}_{n=1}^{\infty} \subseteq A$ be a sequence such that $\text{dist}(\boldsymbol{a}_n, B) \to \rho(A, B)$, as $n \to \infty$.[2] By property (iii), for each $n \in \mathbb{N}$ there is a $\boldsymbol{b}_n \in B$ such that $\text{dist}(\boldsymbol{a}_n, B) = d_m(\boldsymbol{a}_n, \boldsymbol{b}_n)$. Since A is sequentially compact, there is a subsequence $\{\boldsymbol{a}_{n_k}\}_{k=1}^{\infty}$ of $\{\boldsymbol{a}_n\}_{n=1}^{\infty}$ that converges to some $\boldsymbol{a}_0 \in A$. Also, since B is sequentially compact, there is a subsequence $\{\boldsymbol{b}_{n_{k_j}}\}_{j=1}^{\infty}$ of $\{\boldsymbol{b}_{n_k}\}_{k=1}^{\infty}$ which converges to some $\boldsymbol{b}_0 \in B$. Consequently, we must have that $d(\boldsymbol{a}_{n_{k_j}}, \boldsymbol{b}_{n_{k_j}}) \to d(\boldsymbol{a}_0, \boldsymbol{b}_0)$, as $j \to \infty$. Hence,

$$\rho(A, B) = \lim_{j \to \infty} \text{dist}(\boldsymbol{a}_{n_{k_j}}, B) = \lim_{j \to \infty} d_m(\boldsymbol{a}_{n_{k_j}}, \boldsymbol{b}_{n_{k_j}}) = d_m(\boldsymbol{a}_0, \boldsymbol{b}_0).$$

To prove property (v), suppose $A \subseteq B$. Then, obviously, given any $\boldsymbol{a} \in A$, we have that $\boldsymbol{a} \in B$. Therefore, for each $\boldsymbol{x} \in \mathbb{R}^m$, it follows that

$$\text{dist}(\boldsymbol{x}, A) = \inf_{\boldsymbol{a} \in A}\{d_m(\boldsymbol{x}, \boldsymbol{a})\} \geq \inf_{\boldsymbol{b} \in B}\{d_m(\boldsymbol{x}, \boldsymbol{b})\} = \text{dist}(\boldsymbol{x}, B).$$

To prove property (vi), suppose $A \subseteq B$. By property (v), we have $\text{dist}(\boldsymbol{c}, B) \leq \text{dist}(\boldsymbol{c}, A)$, for all $\boldsymbol{c} \in C$. Hence, $\rho(C, B) \leq \rho(C, A)$.

To prove property (vii), note that we immediately have

$$\begin{aligned}\rho(A \cup B, C) &= \max\left\{\sup_{\boldsymbol{a} \in A}\{\text{dist}(\boldsymbol{a}, C)\}, \sup_{\boldsymbol{b} \in B}\{\text{dist}(\boldsymbol{b}, C)\}\right\} \\ &= \max\{\rho(A, C), \rho(B, C)\}.\end{aligned}$$

Finally, to prove property (viii), note that by property (iii), we have that for each $\boldsymbol{a} \in A$, there is some $\boldsymbol{c}_{\boldsymbol{a}} \in C$ such that $\text{dist}(\boldsymbol{a}, C) = d_m(\boldsymbol{a}, \boldsymbol{c}_{\boldsymbol{a}})$. Hence,

$$\begin{aligned}\text{dist}(\boldsymbol{a}, B) &\leq \inf_{\boldsymbol{b} \in B}\{d_m(\boldsymbol{a}, \boldsymbol{c}_{\boldsymbol{a}}) + d_m(\boldsymbol{c}_{\boldsymbol{a}}, \boldsymbol{b})\} = d_m(\boldsymbol{a}, \boldsymbol{c}_{\boldsymbol{a}}) + \inf_{\boldsymbol{b} \in B}\{d_m(\boldsymbol{c}_{\boldsymbol{a}}, \boldsymbol{b})\} \\ &= \text{dist}(\boldsymbol{a}, C) + \text{dist}(\boldsymbol{c}_{\boldsymbol{a}}, B) \leq \rho(A, C) + \rho(C, B).\end{aligned}$$

Thus, $\rho(A, B) = \sup_{\boldsymbol{a} \in A}\{\text{dist}(\boldsymbol{a}, B)\} \leq \rho(A, C) + \rho(C, B)$, as desired. □

The Hausdorff metric d_H is defined in terms of the functions dist and ρ as follows. The fact that d_H is indeed a metric is verified in Proposition 4.9.

[2] Such a sequence exists because ρ is defined as a supremum; see formula (4.2).

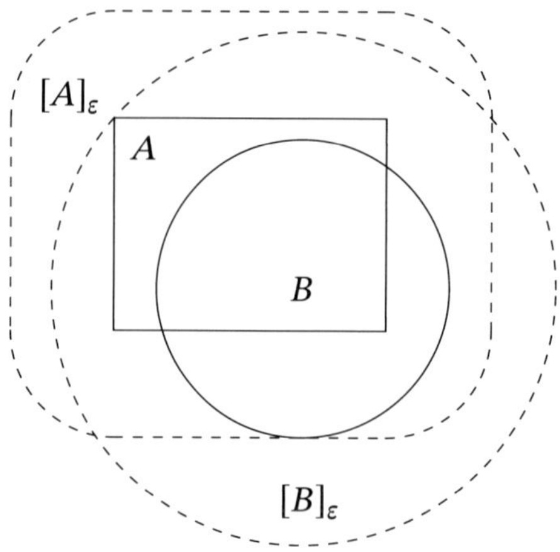

Figure 4.1. Two compact sets $A, B \in \mathbb{K}$ and their ε-neighborhoods $[A]_\varepsilon$ and $[B]_\varepsilon$, where $\varepsilon = d_H(A, B)$. Note that ε is just large enough so that $A \subseteq [B]_\varepsilon$ and $B \subseteq [A]_\varepsilon$.

Definition 4.4 (*Hausdorff Metric*)**.** Let $A, B \in \mathbb{K}(\mathbb{R}^m)$. Then, the *Hausdorff metric* d_H is given on $\mathbb{K}(\mathbb{R}^m)$ by

$$d_H(A,B) := \max\{\rho(A,B), \rho(B,A)\} \tag{4.3}$$

$$= \max\left\{\sup_{\boldsymbol{a}\in A}\{\mathrm{dist}(\boldsymbol{a},B)\}, \sup_{\boldsymbol{b}\in B}\{\mathrm{dist}(\boldsymbol{b},A)\}\right\}.$$

Corollary 4.5 follows immediately from property (iv) of Theorem 4.3 and Definition 4.4.

Corollary 4.5. *Let $A, B \in \mathbb{K}(\mathbb{R}^m)$. Then, there exist $\boldsymbol{a}_0 \in A$ and $\boldsymbol{b}_0 \in B$ such that $d_H(A, B) = d_m(\boldsymbol{a}_0, \boldsymbol{b}_0)$.*

An equivalent form of the Hausdorff metric is given in terms of the ε-neighborhoods of sets.

Definition 4.6 (*Closed ε-Neighborhood*)**.** Let $A \subseteq \mathbb{R}^m$. Then, given $\varepsilon > 0$, the *closed ε-neighborhood* of A, denoted $[A]_\varepsilon$ and also called the *ε-tubular neighborhood* of A, is defined by

$$[A]_\varepsilon := \{\boldsymbol{x} \in \mathbb{R}^m : d_m(\boldsymbol{x}, \boldsymbol{a}) \le \varepsilon, \text{ for some } \boldsymbol{a} \in A\} \tag{4.4}$$

$$= \{\boldsymbol{x} \in \mathbb{R}^m : \mathrm{dist}(\boldsymbol{x}, A) \le \varepsilon\}.$$

In other words, $[A]_\varepsilon$ is the set of points in $\mathbb{R}^m$ within distance ε from a point in A or, equivalently, from A itself.[3]

[3] Since the distance function $f(\boldsymbol{x}) := \mathrm{dist}(\boldsymbol{x}, A)$ is continuous, $[A]_\varepsilon = f^{-1}([0,\varepsilon])$ is closed, as the inverse image of a closed set. Also, since (for fixed $\boldsymbol{x} \in \mathbb{R}^m$) the function $d_m(\boldsymbol{x}, \cdot)$ is continuous, it achieves its minimum on the compact set A, at some point $\boldsymbol{a} \in A$.

Lemma 4.7. *Let $A, B \in \mathbb{K}(\mathbb{R}^m)$. Then,*

$$d_H(A,B) = \inf\{\varepsilon > 0 : A \subseteq [B]_\varepsilon \text{ and } B \subseteq [A]_\varepsilon\}. \tag{4.5}$$

Rephrased: Given any $\varepsilon > 0$, we have that

$$d_H(A,B) \le \varepsilon \Leftrightarrow A \subseteq [B]_\varepsilon \text{ and } B \subseteq [A]_\varepsilon. \tag{4.6}$$

Proof. Suppose $\varepsilon \ge d_H(A,B) \ge \rho(A,B)$. If $\boldsymbol{x} \in A$, then $\operatorname{dist}(\boldsymbol{x}, B) \le \rho(A,B) \le \varepsilon$. Hence, $A \subseteq [B]_\varepsilon$. A similar argument yields $B \subseteq [A]_\varepsilon$.

Now, suppose $\varepsilon < d_H(A,B)$. Without loss of generality, we may assume that $0 < \varepsilon < \rho(A,B)$. Let $\boldsymbol{a} \in A$ and $\boldsymbol{b} \in B$ such that $d_m(\boldsymbol{a},\boldsymbol{b}) = d_H(A,B)$. Then $\boldsymbol{a} \notin [B]_\varepsilon$, which implies $A \not\subseteq [B]_\varepsilon$.

In order to see why (4.6) holds, we proceed as follows. First, note that we have already proved just above the implication $\Rightarrow$ in (4.6). Then, we assume, for example, that $A \subseteq [B]_\varepsilon$. It follows from the definition of $[B]_\varepsilon$ in (4.4) that every $\boldsymbol{a} \in A$ satisfies the inequality $\operatorname{dist}(\boldsymbol{a}, B) \le \varepsilon$. Hence, in light of (4.2), we have that $\rho(A,B) \le \varepsilon$. Since, by interchanging the roles of A and B, we see that $B \subseteq [A]_\varepsilon \Rightarrow \rho(B,A) \le \varepsilon$, we conclude, in light of (4.3), that (4.6) holds, as desired. □

Remark 4.8. It follows from the above proof of Lemma 4.7 that for any $A, B \in \mathbb{K}(\mathbb{R}^m)$,

$$\rho(A,B) \le \varepsilon \Leftrightarrow A \subseteq [B]_\varepsilon. \tag{4.7}$$

In light of the first equality in (4.3), we thus also deduce that (4.6) holds.

Proposition 4.9. *The Hausdorff metric d_H is a metric on $\mathbb{K}(\mathbb{R}^m)$, the space of nonempty compact subsets of $\mathbb{R}^m$.*

Proof. Let $A, B, C \in \mathbb{K}(\mathbb{R}^m)$. We immediately have that $d_H(A,B) \ge 0$ since $\rho(A,B) \ge 0$ and $\rho(B,A) \ge 0$. Similarly, $d_H(A,B) < \infty$. Now, we have $d_H(A,B) = 0$ iff $\rho(A,B) = \rho(B,A) = 0$. So, by part (ii) of Theorem 4.3, we have $d_H(A,B) = 0$ iff $A \subseteq B$ and $B \subseteq A$, i.e., iff $A = B$.

Next, d_H is symmetric since

$$d_H(A,B) = \max\{\rho(A,B), \rho(B,A)\} = \max\{\rho(B,A), \rho(A,B)\} = d_H(B,A).$$

Lastly, in order to verify that d_H satisfies the Triangle Inequality, note that by property (viii) of Theorem 4.3 and by Definition 4.4, we have

$$\begin{aligned} \rho(A,B) &\le \rho(A,C) + \rho(C,B) \le d_H(A,C) + d_H(C,B) \quad \text{and} \\ \rho(B,A) &\le \rho(B,C) + \rho(C,A) \le d_H(B,C) + d_H(C,A), \end{aligned}$$

as desired. Thus, $d_H(A,B) \le d_H(A,C) + d_H(C,B)$, since d_H is symmetric. □

Theorem 4.10. *$(\mathbb{K}(\mathbb{R}^m), d_H)$ is a complete metric space.*

Proof. Let $\{A_n\}_{n=1}^{\infty}$ be a subsequence of a given Cauchy sequence in the metric space $(\mathbb{K}(\mathbb{R}^m), d_H)$ such that $d_H(A_n, A_{n+1}) < 2^{-n}$, for each $n \in \mathbb{N}$.[4] For any $N \in \mathbb{N}$ and $\boldsymbol{x} \in A_N$, there is a sequence $\{\boldsymbol{a}_n\}_{n=1}^{\infty}$ such that $\boldsymbol{a}_N = \boldsymbol{x}$, $\boldsymbol{a}_n \in A_n$ and $d_m(\boldsymbol{a}_n, \boldsymbol{a}_{n+1}) \leq 2^{-n}$, for each $n \geq N$. Namely, for $n < N$ we can let $\boldsymbol{a}_n \in A_n$ be arbitrary while for $n \geq N$ we find an appropriate $\boldsymbol{a}_{n+1} \in A_{n+1}$ by using the fact that $d_H(A_n, A_{n+1}) < 2^{-n}$. Such a sequence $\{\boldsymbol{a}_n\}_{n=1}^{\infty}$ is Cauchy in $(\mathbb{R}^m, d_m)$ since, by the Triangle Inequality, $d_m(\boldsymbol{a}_n, \boldsymbol{a}_j) \leq 2^{-N+1}$, for all $j, n \geq N$. Since $(\mathbb{R}^m, d_m)$ is complete, there exists some $\boldsymbol{a} \in \mathbb{R}^m$ such that $\boldsymbol{a}_n \to \boldsymbol{a}$, as $n \to \infty$. In addition, note that again by the Triangle Inequality, $d_m(\boldsymbol{a}_n, \boldsymbol{a}_k) \leq 2^{-n+1}$, for all $k \geq n \geq N$. Hence, by taking the limit as k tends to infinity, we obtain that $d_m(\boldsymbol{a}_n, \boldsymbol{a}) \leq 2^{-n+1}$, for all $n \geq N$.

Define A to be the set of all $\boldsymbol{a} \in \mathbb{R}^m$ such that there exists a subsequence $\{A_n\}_{n=1}^{\infty}$ of the given Cauchy sequence with a sequence of points $\boldsymbol{a}_n \in A_n$ constructed as above such that $\boldsymbol{a}_n \to \boldsymbol{a}$, as $n \to \infty$, in $(\mathbb{R}^m, d_m)$. Then, A is nonempty since each set in the given Cauchy sequence is nonempty and we have shown a moment ago that there exists at least one $\boldsymbol{a} \in A$. Furthermore, for any $\boldsymbol{a} \in A$ and any $n \in \mathbb{N}$, there is some $\boldsymbol{a}_n \in A_n$ such that $d_m(\boldsymbol{a}_n, \boldsymbol{a}) < 2^{-n+1}$. Hence, we have that $A \subseteq [A_n]_{2^{-n+1}}$, for any $n \in \mathbb{N}$, and thus, $\overline{A}$ is nonempty, closed, and bounded. By the Heine–Borel Theorem (Theorem 2.36) and the properties of closed sets, it follows that $\overline{A} \in \mathbb{K}(\mathbb{R}^m)$.

Now, let $\varepsilon > 0$. Then, there exists some $N \in \mathbb{N}$ such that $0 < 2^{-N} < \varepsilon$. Let $n \geq N$. By the above argument, for any $\boldsymbol{x} \in A_n$ there is some Cauchy and hence, convergent sequence $\{\boldsymbol{a}_n\}_{n=1}^{\infty}$ converging to $\boldsymbol{a} \in \overline{A}$ such that $\boldsymbol{a}_n = \boldsymbol{x}$ and $d_m(\boldsymbol{x}, \boldsymbol{a}) \leq 2^{-n+1} \leq 2^{-N}$. Hence, $A_n \subseteq [\overline{A}]_{2^{-N}} \subseteq [\overline{A}]_{\varepsilon}$. Moreover, by Lemma 4.7 and the containment $A \subseteq [A_n]_{2^{-n+1}}$ established a moment ago, we have $d_H(A_n, \overline{A}) < \varepsilon$, for all $n \geq N$. That is, $A_n \to \overline{A}$ as $n \to \infty$ in $(\mathbb{K}(\mathbb{R}^m), d_H)$, as desired. In light of Exercise 2.64 above. This completes the proof. □

Propositions 4.9 and Theorem 4.10 along with the Contraction Mapping Principle (Theorem 2.58) combine to guarantee the existence of fractals constructed using an *iterated function system* (IFS). An IFS, as defined in Section 4.2, is a collection of contractions on $\mathbb{R}^m$ which, when taken together in a specific way to define a function from $\mathbb{K}(\mathbb{R}^m)$ to itself, is a contraction. Hence, by the Contraction Mapping Principle (Theorem 2.58), one may construct compact sets, including fractals, as fixed points of contractions on the space $(\mathbb{K}(\mathbb{R}^m), d_H)$ of nonempty compact sets equipped with the Hausdorff metric.

[4] As can be easily checked by induction on $n \in \mathbb{N}$, such a sequence always exists, for any given Cauchy sequence.

Remark 4.11. The definition of the Hausdorff metric d_H can be extended in the obvious way if $\mathbb{R}^m$ is replaced by any metric space E, equipped with the metric d. Then, all of the results of this section (i.e., Section 4.1) extend without change, except that for the counterpart of Theorem 4.10 to hold, (E, d) must be assumed to be a complete metric space, as is the case of Euclidean space $(\mathbb{R}^n, d_m)$. More specifically, under this latter assumption, and if we denote by $\mathbb{K}(E)$ the space of nonempty compact subsets of E, then $(\mathbb{K}(E), d_H)$ is a complete metric space. Similarly, the general definition of an IFS and that of its attractor given in Section 4.2 just below extends at once to any complete metric space.

4.2. Iterated Function Systems

An *iterated function system* (IFS) on $(\mathbb{R}^m, d_m)$ generates a unique fixed point (in fact, a nonempty compact set) when it is interpreted as a contraction on $(\mathbb{K}(\mathbb{R}^m), d_H)$, the space of nonempty compact subsets of $\mathbb{R}^m$ equipped with the Hausdorff metric, as defined in Section 4.1.

Definition 4.12.

(i) (*Iterated Function System: IFS*). An *iterated function system* (IFS) $\mathbf{\Phi}$ on $(\mathbb{R}^m, d_m)$ is a finite collection of two or more contractions on $(\mathbb{R}^m, d_m)$. That is, for some $N \in \mathbb{N}$ where $N \geq 2$, we have that $\mathbf{\Phi} = \{\varphi_j\}_{j=1}^N$, where each φ_j is a contraction on $(\mathbb{R}^m, d_m)$, for $j = 1, \dots, N$.

(ii) (*Self-Similar System*). If, in addition, for each $j = 1, \dots, N$, φ_j is a contracting similarity on $(\mathbb{R}^m, d_m)$ with scaling ratio r_j, then $\mathbf{\Phi}$ is said to be a *self-similar system* with *scaling vector* $\mathbf{r} = (r_1, \dots, r_N)$.

Remark 4.13. By a mild abuse of notation and terminology, an iterated function system $\mathbf{\Phi} = \{\varphi_j\}_{j=1}^N$ on $(\mathbb{R}^m, d_m)$ also denotes the map $\mathbf{\Phi} : \mathbb{K}(\mathbb{R}^m) \to \mathbb{K}(\mathbb{R}^m)$ given by

$$\mathbf{\Phi}(A) := \bigcup_{j=1}^N \varphi_j(A), \quad \text{for all } A \in \mathbb{K}(\mathbb{R}^m). \tag{4.8}$$

In this context, iterated function systems are contractions on the metric space $(\mathbb{K}(\mathbb{R}^m), d_H)$, a fact proved in Proposition 4.15. Another definition is in order, first.

Definition 4.14 (*Prefractal*). Let $\mathbf{\Phi}$ be an iterated function system on $\mathbb{R}^m$ with attractor $F_{\mathbf{\Phi}}$ (as defined in Theorem 4.17 and Definition 4.18 below) and let $A \in \mathbb{K}(\mathbb{R}^m)$. For $E \in \mathbb{K}(\mathbb{R}^m)$, let $\mathbf{\Phi}^0(E) := E$ and for each $n \in \mathbb{N}$, let $\mathbf{\Phi}^n(E) := \mathbf{\Phi}(\mathbf{\Phi}^{n-1}(E))$. Then, for each $n \in \mathbb{N}_0$, the *n-th prefractal approximation of* $F_{\mathbf{\Phi}}$ *with respect to* A, or simply the *n-th prefractal*, is the n-th iterate of the function $\mathbf{\Phi}$ applied to A given by $\mathbf{\Phi}^n(A)$.

Proposition 4.15. *Let* $\boldsymbol{\Phi} = \{\varphi_j\}_{j=1}^N$ *be an iterated function system on* $(\mathbb{R}^m, d_m)$ *with scaling vector* $\mathbf{r} = (r_1, \dots, r_N)$.[5] *Then, the map* $\boldsymbol{\Phi} : \mathbb{K}(\mathbb{R}^m) \to \mathbb{K}(\mathbb{R}^m)$, *as defined in* (4.8), *is a contraction with respect to the Hausdorff metric* d_H.

Proof. Since each contraction φ_j is continuous, for $j = 1, \dots, N$, it follows from Theorem 2.42 that each of the images $\varphi_j(A)$ is compact. Furthermore, by a straightforward induction argument, $\boldsymbol{\Phi}^n(A)$ is compact for each $n \in \mathbb{N}_0$.

Now, let $A, B \in \mathbb{K}(\mathbb{R}^m)$. For any $\varepsilon > 0$ such that $\varphi_j(B) \subseteq [\varphi_j(A)]_\varepsilon$, and for $j = 1, \dots, N$, we have

$$\boldsymbol{\Phi}(B) = \bigcup_{j=1}^N \varphi_j(B) \subseteq \left[\bigcup_{j=1}^N \varphi_j(A)\right]_\varepsilon = [\boldsymbol{\Phi}(A)]_\varepsilon, \quad \text{and so,}$$

$$\begin{aligned} d_H(\boldsymbol{\Phi}(A), \boldsymbol{\Phi}(B)) &\leq \max_{j=1,\dots,N} \{d_H(\varphi_j(A), \varphi_j(B))\} \\ &\leq \left(\max_{j=1,\dots,N} r_j\right) \cdot d_H(A, B). \end{aligned}$$

Also, we have $0 < r_j < 1$, for each $j = 1, \dots, N$. Thus, $\boldsymbol{\Phi}$ is a contraction on $(\mathbb{K}(\mathbb{R}^m), d_H)$ with scaling ratio $r_{\boldsymbol{\Phi}} \leq \max_{j=1,\dots,N} r_j < 1$, as desired. □

Remark 4.16. A consequence of the above proof is that the contraction ratio $r_{\boldsymbol{\Phi}}$ of an IFS $\boldsymbol{\Phi} := \{\varphi_j\}_{j=1}^N$ (i.e., of the map $\boldsymbol{\Phi}$ associated with this IFS) is bounded above by the maximum of the scaling ratios r_j of the corresponding contractions φ_j.

Proposition 4.15, along with the Contraction Mapping Principle (see Theorems 2.58 and 4.10), immediately yield the following important theorem as a corollary.

The following key result was obtained by John E. Hutchinson in his influential 1981 paper, entitled *Fractals and self-similarity*, [**261**]. The result itself (or rather, the techniques and tools leading to it) was essentially a special case of a more general statement—already known, in particular, to researchers in the area of optimal transport as well as in probability and statistics—establishing, in particular, the existence and uniqueness of certain (self-similar) probability measures (with given probability vectors), themselves now very useful in fractal geometry; see Section 4.8 below for the relevant references and the main result, proved by means of the Monge–Kantorovich (instead of the Hausdorff) metric [**283–285**].

[5] More specifically, for each $j = 1, \dots, N$, $r_j = \mathrm{Lip}(\varphi_j) \in (0, 1)$ is the Lipschitz constant (also called scaling ratio) of the contraction φ_j.

Theorem 4.17 (Existence and Uniqueness of an Invariant Set Under an IFS; J. E. Hutchinson, [**261**], 1981). *Let $\mathbf{\Phi}$ be an iterated function system on $(\mathbb{R}^m, d_m)$. Then, $\mathbf{\Phi} : \mathbb{K}(\mathbb{R}^m) \to \mathbb{K}(\mathbb{R}^m)$, as defined in (4.8), has a unique fixed point $F_{\mathbf{\Phi}} \in \mathbb{K}(\mathbb{R}^m)$. That is, there is a unique nonempty compact set $F_{\mathbf{\Phi}}$ such that $F_{\mathbf{\Phi}} = \mathbf{\Phi}(F_{\mathbf{\Phi}}) = \bigcup_{j=1}^{N} \varphi_j(F_{\mathbf{\Phi}})$.*

Definition 4.18 (*Attractor of an IFS*). Let $\mathbf{\Phi}$ be an iterated function system on $(\mathbb{R}^m, d_m)$. Then, the unique fixed point $F_{\mathbf{\Phi}} \in \mathbb{K}(\mathbb{R}^m)$ which is invariant with respect to $\mathbf{\Phi}$ is called the *attractor* of $\mathbf{\Phi}$ (or, indifferently, the *invariant set* of $\mathbf{\Phi}$). Note that it is a nonempty compact subset of $\mathbb{R}^n$.

The statement of the Contraction Mapping Principle (Theorem 2.58) guarantees the existence of a unique fixed point $\boldsymbol{x}^*$ for a given contraction ψ. Also, in the proof, the construction of $\boldsymbol{x}^*$ is shown to be *independent of the choice of the initial point* x_0. In the setting of iterated function systems, this means that the attractor $F_{\mathbf{\Phi}}$ can be constructed by choosing *an arbitrary nonempty compact set* $A \in \mathbb{K}(\mathbb{R}^m)$ and iterating the function $\mathbf{\Phi}$. The following proposition—sometimes inappropriately called the Collage Theorem (although, in the literature, this name is in part justified by the applications to image processing; see, e.g., [**28, 29**])—expands upon this point and provides an estimate as to how well a given prefractal approximates the attractor. Note that it is an immediate (and well-known) consequence of the abstract Contraction Mapping Principle (Theorem 2.58) and its proof (see Proposition 2.59), which has applications in the theory of differential and partial differential equations (see Exercise 2.68 above), as well as in numerical analysis (see, for example, Newton's Method, Theorem 2.61) and scientific computation, among many other fields.

Proposition 4.19. *Let $\mathbf{\Phi} = \{\varphi_j\}_{j=1}^N$ be an iterated function system on $(\mathbb{R}^m, d_m)$ with scaling vector $\mathbf{r} = (r_1, \ldots, r_N)$ and attractor $F_{\mathbf{\Phi}}$. Also, let $r = \max\{r_j : j = 1, \ldots, N\}$ and $A \in \mathbb{K}(\mathbb{R}^m)$. Then,*

$$d_H(A, F_{\mathbf{\Phi}}) \le \left(\frac{1}{1-r}\right) \cdot d_H(A, \mathbf{\Phi}(A)), \text{ and} \tag{4.9}$$

$$d_H(\mathbf{\Phi}^n(A), F_{\mathbf{\Phi}}) \le r^n \cdot d_H(A, F_{\mathbf{\Phi}}), \text{ for all } n \in \mathbb{N}_0. \tag{4.10}$$

In particular, for any $A \in \mathbb{K}(\mathbb{R}^m)$, the sequence $\{\mathbf{\Phi}^n(A)\}_{n=0}^{\infty}$ converges to the attractor $F_{\mathbf{\Phi}}$, in the Hausdorff metric d_H on $\mathbb{R}^m$.

Proof. The results follow at once from Proposition 2.59. □

The following corollary of Proposition 4.19 says that a nonempty compact set is arbitrarily close to a self-similar set.

Corollary 4.20. *Let $A \in \mathbb{K}(\mathbb{R}^m)$ and $\varepsilon > 0$. Then, there exists a self-similar system $\mathbf{\Phi} = \{\varphi_j\}_{j=1}^N$ with attractor $F_{\mathbf{\Phi}}$ and such that $d_H(A, F_{\mathbf{\Phi}}) < \varepsilon$.*

Exercise 4.21. Prove Corollary 4.20.

[*Hint*: By the compactness of A, given $\varepsilon > 0$, we can find a finite cover $\{B_j\}_{j=1}^N$ by balls B_j of radii at most $\varepsilon/4$. Also, for each $j = 1, \ldots, N$, we can find a similitude φ_j (of $\mathbb{R}^m$) of scaling ratio $r_j < 1/2$ and such that $\varphi_j(A) \subseteq B_j$. Then, use Lemma 4.7 to estimate $d_H(A, \mathbf{\Phi}(A))$.]

Remark 4.22. To our knowledge, the term *iterated function system* (IFS) was introduced by M. F. Barnsley and S. Demko in their 1985 paper, [**30**], entitled *Iterated function systems and the global construction of fractals*, providing interesting linear and nonlinear examples of IFSs, several of which are directly relevant to (or motivated by) the theory of dynamical systems. The notion of an IFS (part (i) of Definition 4.12 above) was precisely the one used (without a specific name) by J. E. Hutchinson in his aforementioned 1981 paper [**261**], as the main hypothesis of Theorem 4.17 above. The term "iterated function system" was popularized by M. F. Barnsley, in particular, in his 1993 book *Fractals Everywhere* (and its earlier edition), [**28**], along with his later 2006 book *Superfractals*, [**29**], where many interesting applications are discussed in a very accessible manner (amenable to undergraduate students, in particular).

4.3. Self-Similar Sets

In this section, the iterated function systems (IFSs) that generate the Cantor set C, the Koch curve K_C, the Sierpinski gasket **SG**, and the Sierpinski carpet **SC** are given explicitly. Each of these sets is self-similar; that is, in the sense of Definition 4.24 below, the underlying IFS is a self-similar system, also in the sense of part (ii) of Definition 4.12. Indeed, each of them is the attractor (i.e. *invariant set*) of a self-similar system and, moreover, each one contains exact copies of itself at arbitrarily small scales.

We begin our discussion with a prototypical example—namely, the ternary Cantor set—which will then be abstracted in Definition 4.24 to obtain the general notion of a self-similar set.

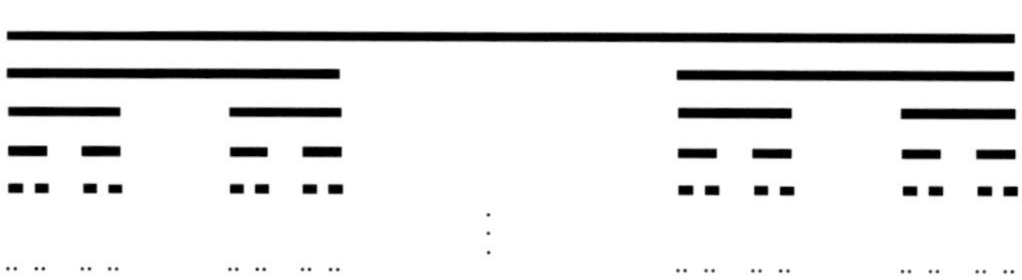

Figure 4.2. An illustration of the construction and the self-similarity of the classic ternary Cantor set C, as described in Examples 1.1 and 4.23. As the attractor of the self-similar system $\mathbf{\Phi}_C$, the Cantor set C is the unique nonempty compact set such that $\mathbf{\Phi}_C(C) = \varphi_1(C) \cup \varphi_2(C) = C$.

Example 4.23 (*Self-Similarity of the Cantor Set*). Consider the following pair of contracting similarities defined on $\mathbb{R}$:

$$\varphi_1(x) = \frac{1}{3}x, \quad \varphi_2(x) = \frac{1}{3}x + \frac{2}{3}. \tag{4.11}$$

The corresponding self-similar system $\mathbf{\Phi}_C = \{\varphi_1, \varphi_2\}$ has scaling vector $\mathbf{r} = (1/3, 1/3)$ and, moreover, the Cantor set C is *defined to be* its attractor; see Figure 4.2. Expanding upon Example 1.6, since the Cantor set is the attractor of $\mathbf{\Phi}_C$, it satisfies the following *Self-Similarity Equation*:

$$C = \mathbf{\Phi}_C(C) = \varphi_1(C) \cup \varphi_2(C) = \frac{1}{3}C \cup \left(\frac{1}{3}C + \frac{2}{3}\right). \tag{4.12}$$

The notion of self-similarity discussed in Figure 4.2 can be generalized as follows: For every $n \in \mathbb{N}$, $C = 3^n \cdot ([0, 1/3^n] \cap C)$. Hence, for any $\varepsilon > 0$, there is $N \in \mathbb{N}$ such that $1/3^N < \varepsilon$ and yet, $C = 3^N \cdot ([0, 1/3^N] \cap C)$. That is, C contains exact (scaled) copies of itself at arbitrarily small scales.

In conclusion, of course, the Cantor set is a self-similar set, in the sense of Definition 4.24 just below.

Definition 4.24 (*Self-Similar Set*). Let $\mathbf{\Phi} := \{\varphi\}_{j=1}^N$ be an arbitrary self-similar system (i.e., consisting of $N \geq 2$ contractive similitudes φ_j of $\mathbb{R}^m$ with scaling ratio $r_j \in (0, 1)$, for $j = 1, \ldots, N$), as in part (ii) of Definition 4.12 above (or as in Definition 4.26 below). Then, F, the attractor (or invariant set) of $\mathbf{\Phi}$, in the sense of Definition 4.18 above and as uniquely provided by Theorem 4.17, is called the *self-similar set associated with* $\mathbf{\Phi}$ or simply—when no ambiguity may arise or when the knowledge of a suitable choice of the self-similar system $\mathbf{\Phi}$ is unimportant—a *self-similar set*. By construction, F belongs to $\mathbb{K}(\mathbb{R}^m)$.

Remark 4.25. Conversely, given $K \in \mathbb{K}(\mathbb{R}^m)$, we say that K is *self-similar*, or that it is a *self-similar set*, if there exists a self-similar system $\mathbf{\Phi} = \{\varphi_j\}_{j=1}^N$, as in Definition 4.24 above, such that K is the (necessarily unique) self-similar set (or attractor) associated with $\mathbf{\Phi}$. We caution the reader that, for a given $K \in \mathbb{K}(\mathbb{R}^m)$, such a self-similar system $\mathbf{\Phi}$ is far from unique; see, e.g., Remark 4.28 below combined with Example 4.23 above for a concrete illustration of this statement.

The following important definition was partially contained in the statement of Definition 4.24 just above.

Definition 4.26 (*Self-Similar System*). Let $\mathbf{\Phi} := \{\varphi_j\}_{j=1}^N$ be an IFS (in the sense of part (i) of Definition 4.12), with an integer $N \geq 2$. Then, $\mathbf{\Phi}$ is said to be a *self-similar system* if it consists of contractive similitudes of $\mathbb{R}^m$, i.e., if each map φ_j (for $j = 1, \ldots, N$) is a contractive similitude of $\mathbb{R}^m$ (necessarily with scaling ratio $r_j \in (0, 1)$). Furthermore, as in part (ii) of Definition 4.12, $\boldsymbol{r} := (r_1, \ldots, r_N) \in (0, 1)^N$ is called the *scaling vector* of $\mathbf{\Phi}$.

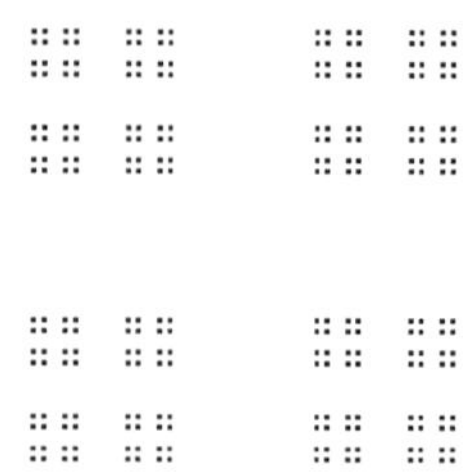

Figure 4.3. The self-similarity of the Cantor dust $C \times C$, where C is the classic ternary Cantor set; see Exercise 4.27

Exercise 4.27 (*Self-Similarity of the Cantor Dust*). Show that the *Cantor dust* $F := C \times C \subseteq \mathbb{R}^2$, as described in Figure 4.3 and where C is the classic ternary Cantor set, is a self-similar set of $\mathbb{R}^2$, and exhibit an associated self-similar system $\boldsymbol{\Phi}_F$ (along with its scaling vector $\boldsymbol{r}$); see Figure 4.3.

Remark 4.28. Although a self-similar system $\boldsymbol{\Phi}$ uniquely determines its attractor (i.e., the self-similar set associated with $\boldsymbol{\Phi}$, in the sense of Definition 4.24), a given attractor is in fact the attractor of an infinite collection of self-similar systems. For instance, consider the self-similar system $\boldsymbol{\Phi}_0 = \{\varphi_{0,j}\}_{j=1}^4$ on $\mathbb{R}$ given (for all $\boldsymbol{x} \in \mathbb{R}$) by

$$\varphi_{0,1}(x) = \frac{1}{9}x, \quad \varphi_{0,2}(x) = \frac{1}{9}x + \frac{2}{9},$$
$$\varphi_{0,3}(x) = \frac{1}{9}x + \frac{6}{9}, \quad \varphi_{0,4}(x) = \frac{1}{9}x + \frac{8}{9}.$$

Then, the Cantor set C is the attractor of both $\boldsymbol{\Phi}_C$ and $\boldsymbol{\Phi}_0$; see Figure 4.2.

Exercise 4.29. Try to precisely define the notion of a *minimal* IFS—and, in particular, of a *minimal* self-similar system. Furthermore, show how to use it and determine the minimal self-similar systems of several of the examples of self-similar sets discussed in this section, including the Cantor set, the Sierpinski gasket and carpet, the Koch curve, along with the Menger sponge, in Examples 4.23, 4.30, 4.31, and 4.33–4.35.

Example 4.30 (*The* 1*-Dimensional Fractal* F_1*, as a Self-Similar Set in* $\mathbb{R}^2$). Consider the following contracting similarities defined on $\mathbb{R}^2$:

$$\varphi_j(\boldsymbol{x}) = \frac{1}{4}\boldsymbol{x} + \mathbf{t}_j, \; j = 1, 2, 3, 4, \tag{4.13}$$

where $\mathbf{t}_1 = (0,0), \mathbf{t}_2 = (3/4, 0), \mathbf{t}_3 = (3/4, 3/4)$, and $\mathbf{t}_4 = (0, 3/4)$. The corresponding self-similar system $\boldsymbol{\Phi}_{F_1} = \{\varphi_j\}_{j=1}^4$ has scaling vector $\mathbf{r} = (1/4, 1/4, 1/4, 1/4)$ and attractor F_1; see Figure 1.4 in Section 1.2 and Example 1.7. As the attractor of $\boldsymbol{\Phi}_{F_1}$, it satisfies the Self-Similarity Equation (1.7) and, moreover, for every $n \in \mathbb{N}$, $F_1 = 4^n \cdot ([0, 1/4^n]^2 \cap F_1)$. As with the Cantor set C, F_1 contains exact (scaled) copies of itself at arbitrarily small scales. Thus, F_1 is a self-similar set in $\mathbb{R}^2$; nevertheless, as is explained in Example 1.7 and in Subsection 1.3.4, it is 1-dimensional; see also Moran's Theorem (Theorem 5.40).

Example 4.31 (*The Koch Curve as a Self-Similar Set*). The Koch curve K_C is also defined to be the attractor of a self-similar system. Consider the following contracting similarities of $\mathbb{R}^2$ given as follows, for all $\boldsymbol{x} \in \mathbb{R}^2$:

$$\begin{aligned}
\varphi_1(\boldsymbol{x}) &= \frac{1}{3}\begin{pmatrix} 1 & 0 \\ 0 & 1 \end{pmatrix}\boldsymbol{x} = \frac{1}{3}\boldsymbol{x}, \\
\varphi_2(\boldsymbol{x}) &= \frac{1}{3}\begin{pmatrix} 1/2 & -\sqrt{3}/2 \\ \sqrt{3}/2 & 1/2 \end{pmatrix}\boldsymbol{x} + \begin{pmatrix} 1/3 \\ 0 \end{pmatrix}, \\
\varphi_3(\boldsymbol{x}) &= \frac{1}{3}\begin{pmatrix} 1/2 & \sqrt{3}/2 \\ -\sqrt{3}/2 & 1/2 \end{pmatrix}\boldsymbol{x} + \begin{pmatrix} 1/2 \\ \sqrt{3}/2 \end{pmatrix}, \\
\varphi_4(\boldsymbol{x}) &= \frac{1}{3}\begin{pmatrix} 1 & 0 \\ 0 & 1 \end{pmatrix}\boldsymbol{x} + \begin{pmatrix} 2/3 \\ 0 \end{pmatrix} = \frac{1}{3}\boldsymbol{x} + \begin{pmatrix} 2/3 \\ 0 \end{pmatrix}.
\end{aligned} \tag{4.14}$$

Let E be the compact set with boundary given by the triangle with vertices $(0, 0)$, $(1, 0)$, and $(1/2, \sqrt{3}/6)$. The contracting similarities $\varphi_1, \varphi_2, \varphi_3$, and φ_4 each contract the triangle E by a factor of $1/3$. That is, $\mathbf{r} = (1/3, 1/3, 1/3, 1/3)$. Also, φ_2 rotates E by $60° = \pi/3$ rad and translates E by $(1/3, 0)$, φ_3 rotates E by $-60° = -\pi/3$ rad and translates E by $(1/2, \sqrt{3}/6)$, and φ_4 translates E by $(2/3, 0)$; see Figure 4.4.

The self-similar system $\boldsymbol{\Phi}_{KC} = \{\varphi_1, \varphi_2, \varphi_3, \varphi_4\}$ has the Koch curve K_C as its attractor; namely, K_C is the self-similar set associated with $\boldsymbol{\Phi}_{KC}$, in the sense of Definition 4.24. Like the Cantor set C, the Koch curve K_C contains exact (scaled) copies of itself at all scales. Indeed, for each $n \in \mathbb{N}$, we have that $K_C = 3^n \cdot ([0, 1/3^n]^2 \cap K_C)$.

The Koch curve K_C is very closely related to the Koch snowflake curve $K_S = \partial\Omega_{KS}$ discussed in Example 1.8. In fact, three appropriately rotated and translated copies of K_C create the Koch snowflake curve $K_S = \partial\Omega_{KS}$; see Section 9.5 below.

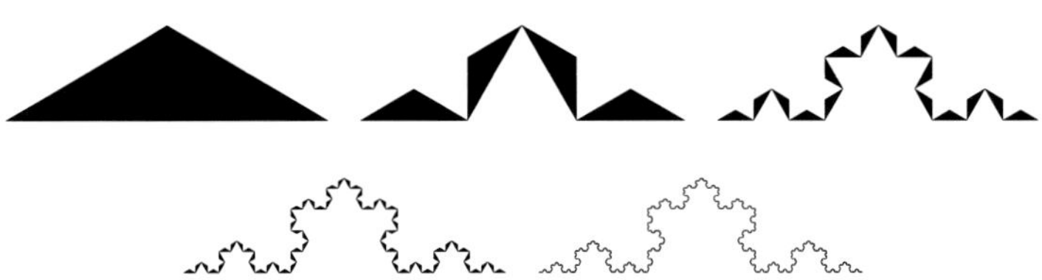

Figure 4.4. The first five prefractal approximations (i.e., prefractals) to the Koch curve K_C, constructed via the self-similar system $\boldsymbol{\Phi}_{KC}$ described in Example 4.31.

Remark 4.32 (*The Attractor K_C, Viewed as the Koch Curve*). We will show in Section 9.5 that the attractor K_C—viewed here as a self-similar set—coincides with the classic Koch curve $\mathbf{\Gamma}_{KC}$; see Theorem 9.70 in Subsection 9.5.2 below. Here, $\mathbf{\Gamma}_{KC}$ is viewed as a continuous curve obtained as a suitable limit (uniformly and hence also in $(\mathbb{K}(\mathbb{R}^2), d_H)$) of its natural polygonal approximations. This continuous and self-similar curve is studied in detail in Section 9.5 and, more precisely, in [**116**], the second sequel to this book.

Example 4.33 (*The Sierpinski Gasket, Viewed as a Self-Similar Set: Standard IFS*). The Sierpinski gasket **SG** is also defined to be the attractor of a self-similar system. Consider the following contracting similarities defined on $\mathbb{R}^2$:

$$\begin{aligned}\varphi_1(\boldsymbol{x}) &= \frac{1}{2}\begin{pmatrix}1 & 0\\ 0 & 1\end{pmatrix}\boldsymbol{x} = \frac{1}{2}\boldsymbol{x},\\ \varphi_2(\boldsymbol{x}) &= \frac{1}{2}\begin{pmatrix}1 & 0\\ 0 & 1\end{pmatrix}\boldsymbol{x} + \begin{pmatrix}1/2\\ 0\end{pmatrix},\\ \varphi_3(\boldsymbol{x}) &= \frac{1}{2}\begin{pmatrix}1 & 0\\ 0 & 1\end{pmatrix}\boldsymbol{x} + \begin{pmatrix}1/4\\ \sqrt{3}/4\end{pmatrix}.\end{aligned} \tag{4.15}$$

Let $E \subseteq \mathbb{R}^2$ be the compact set with boundary given by the equilateral triangle with vertices $(0,0)$, $(1,0)$, and $(1/2, \sqrt{3}/2)$. The contracting similarities φ_1, φ_2, and φ_3 each contract the triangle E by a factor of $1/2$. That is, $\mathbf{r} = (1/2, 1/2, 1/2)$; see Figures 4.6 and 4.5.

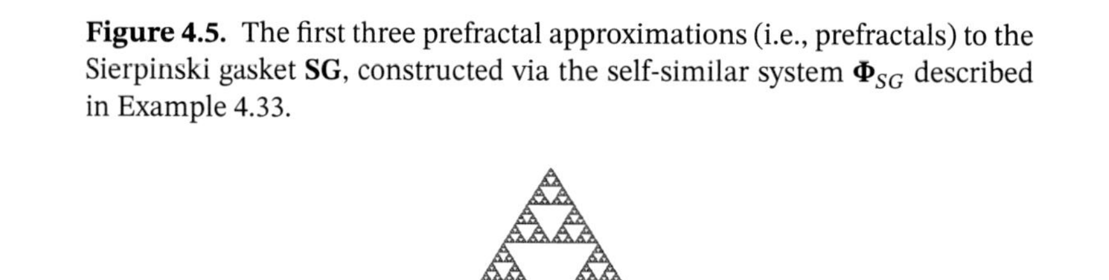

Figure 4.5. The first three prefractal approximations (i.e., prefractals) to the Sierpinski gasket **SG**, constructed via the self-similar system $\mathbf{\Phi}_{SG}$ described in Example 4.33.

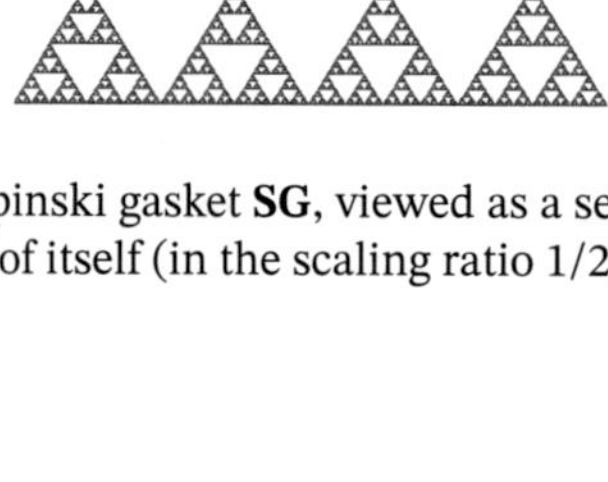

Figure 4.6. The Sierpinski gasket **SG**, viewed as a self-similar set: **SG** is the union of three copies of itself (in the scaling ratio 1/2).

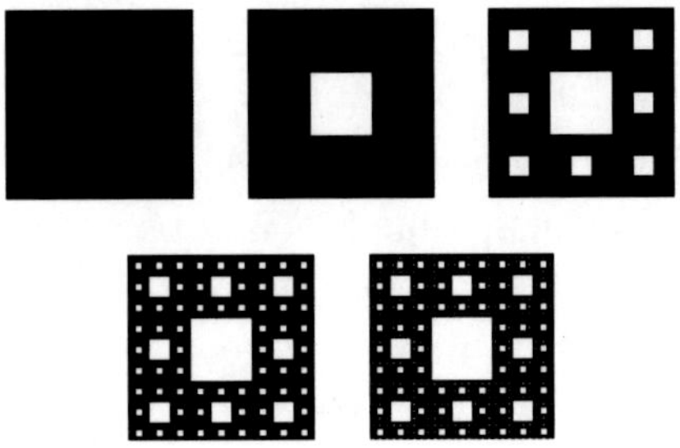

Figure 4.7. The first five prefractal approximations (i.e., prefractals) to the Sierpinski carpet **SC**, constructed via the self-similar system to $\mathbf{\Phi}_{SC}$ described in Example 4.34.

The self-similar system $\mathbf{\Phi}_{SG} = \{\varphi_1, \varphi_2, \varphi_3\}$ has the Sierpinski gasket **SG** as its attractor. Like the Cantor set C and the Koch curve K_C, the Sierpinski gasket **SG** contains exact (scaled) copies of itself at all scales. Indeed, for each $n \in \mathbb{N}$, we have $\mathbf{SG} = 2^n \cdot ([0, 1/2^n]^2 \cap \mathbf{SG})$. The Sierpinski gasket **SG** is discussed in more detail in Section 9.6.

Example 4.34 (*The Sierpinski Carpet as a Self-Similar Set*). Consider the unit square $E = [0,1]^2 \subseteq \mathbb{R}^2$ and the following contracting similarities of $\mathbb{R}^2$ defined as follows. Given $j_1, j_2 \in \{0,1,2\}$, let

$$\varphi_{j_1,j_2}(\boldsymbol{x}) := \frac{1}{3}\boldsymbol{x} + \begin{pmatrix} j_1/3 \\ j_2/3 \end{pmatrix}, \text{ for all } \boldsymbol{x} \in \mathbb{R}^2. \tag{4.16}$$

The self-similar system $\mathbf{\Phi}_{SC} = \{\varphi_{j_1,j_2} : j_1, j_2 \in \{0,1,2\}\} \backslash \{\varphi_{1,1}\}$ has the Sierpinski carpet **SC** as its attractor. Note that each of the 8 contracting similarities φ_{j_1,j_2} (of which $\mathbf{\Phi}_{SC}$ is comprised) contracts the unit square $E = [0,1]^2$ by a factor of 1/3 and then translates it by a suitable vector. Thus, the scaling ratios are given by $r_{\varphi_{j_1,j_2}} = 1/3$, for all allowed pairs of j_1 and j_2.

Like the Cantor set C, Koch curve K_C, and Sierpinski gasket **SG**, the Sierpinski carpet **SC** contains exact (scaled) copies of itself at all scales. Indeed, for each $n \in \mathbb{N}$, we have $\mathbf{SC} = 3^n \cdot ([0, 1/3^n]^2 \cap \mathbf{SC})$. The Sierpinski carpet **SC** is discussed in further detail in Section 9.6.

Example 4.35 (*The Menger Sponge as a Self-Similar Set*). An analog of the Cantor set C and the Sierpinski carpet **SC** embedded in $\mathbb{R}^3$ is the Menger sponge **MS**; see Figure 9.26 in Subsection 9.6.2. Before we construct **MS**, first consider the unit cube $I^3 := [0,1]^3$ as the attractor of the following self-similar system. Given $j_1, j_2, j_3 \in \{0,1,2\}$, let

$$\varphi_{j_1,j_2,j_3}(\boldsymbol{x}) := \frac{1}{3}\boldsymbol{x} + \begin{pmatrix} j_1/3 \\ j_2/3 \\ j_3/3 \end{pmatrix}, \text{ for all } \boldsymbol{x} \in \mathbb{R}^2. \tag{4.17}$$

The self-similar system $\mathbf{\Phi}_{I^3} = \{\varphi_{j_1,j_2,j_3} : j_1, j_2, j_3 \in \{0,1,2\}\}$ has the unit cube I^3 as its attractor. Note that each of the φ_{j_1,j_2,j_3} maps I^3 to 27 cubes of side lengths 1/3 (and volume 1/27) whose union is I^3.

Heuristically, the Menger sponge **MS** is constructed by removing the following seven cubes: the middle cube from each face of the unit cube and the cube in the very center of the unit cube. More specifically, the Menger sponge is the attractor of the self-similar system $\mathbf{\Phi}_{MS}$ comprising the 20 contracting similarities of $\mathbf{\Phi}_{I^3}$ whose indices have at most one 1. In other words, $\mathbf{\Phi}_{MS} = \mathbf{\Phi}_{I^3} \backslash \{\varphi_{1,1,1}, \varphi_{1,1,0}, \varphi_{1,1,2}, \varphi_{1,0,1}, \varphi_{1,2,1}, \varphi_{0,1,1}, \varphi_{2,1,1}\}$. See Figure 9.26 in Subsection 9.6.2 for a picture of the Menger sponge **MS**.

Like the other examples of self-similar sets discussed above, the Menger sponge **MS** contains exact (scaled) copies of itself at all scales. Indeed, for each $n \in \mathbb{N}$, we have $\mathbf{MS} = 3^n([0, 1/3^n]^3 \cap \mathbf{MS})$. Additionally, it can be shown that the volume of the Menger sponge is 0.

The following proposition is essentially Remark 3.4 of [**261**]. It gives an interesting characterization of line segments in terms of self-similar systems. Note that line segments are indeed self-similar sets (i.e., they are attractors of self-similar systems that contain copies of themselves at all scales), though they are not fractal.

Proposition 4.36 (Characterization of Line Segments). *A compact connected set $A \subseteq \mathbb{R}^m$ is a line segment iff A is invariant with respect to a self-similar system $\mathbf{\Phi}$ comprising $N \geq 2$ contracting similarities such that $\sum_{j=1}^{N} r_j = 1$.*

A proof of Proposition 4.36 can be found in Chapter 5 of [**518**]; see also [**261**] for the proof.

Example 4.37 (*Compact Intervals as Self-Similar Sets: The Unit Interval*). The closed unit interval $I = [0,1]$ can be defined as the attractor of a variety of self-similar systems. Consider the following contracting similarities defined on $\mathbb{R}$:

$$\varphi_j(x) := \frac{1}{10}x + \frac{j}{10}, \text{ where } j = 0, \ldots, 9. \tag{4.18}$$

Each of the contracting similarities φ_j contract the unit interval I by a factor of $1/10$. The self-similar system $\mathbf{\Phi}_I = \{\varphi_j\}_{j=0}^{9}$ has the unit interval I as its attractor.

Another self-similar system that has I as it attractor is the following:

$$\varphi_{0,1}(x) = \frac{1}{6}x, \quad \varphi_{0,2}(x) = \frac{1}{3}x + \frac{1}{6}, \quad \varphi_{0,3}(x) = \frac{1}{2}x + \frac{1}{2}.$$

Let $\mathbf{\Phi}_{I,0} = \{\varphi_{0,j}\}_{j=1}^{3}$ denote the corresponding self-similar system.

Yet another self-similar system with attractor I is given as follows:

$$\varphi_{1,1}(x) = \varphi_{1,2}(x) = \frac{1}{2}x, \qquad \varphi_{2,1}(x) = \varphi_{2,2}(x) = \frac{1}{2}x + \frac{1}{2}.$$

Let $\mathbf{\Phi}_{I,1} := \{\varphi_{j_1,j_2}\}_{j_1,j_2=1}^{2}$ denote the corresponding self-similar system.

It is easy to check that $I = \mathbf{\Phi}_I(I) = \mathbf{\Phi}_{I,0}(I) = \mathbf{\Phi}_{I,1}(I)$. Hence, the closed unit interval I is indeed the attractor of each of the self-similar systems $\mathbf{\Phi}_I$, $\mathbf{\Phi}_{I,0}$, and $\mathbf{\Phi}_{I,1}$. Of course, these self-similar systems are rather different from one another. For instance, the images of I under φ_j and φ_k intersect in at most one point when $j \neq k$ (and a similar statement holds for the images of under the contracting similarities in $\mathbf{\Phi}_{I,0}$), whereas under $\varphi_{1,1} = \varphi_{1,2}$, the images of I overlap completely.

Remark 4.38. The self-similar systems $\mathbf{\Phi}_I$ and $\mathbf{\Phi}_{I,0}$ are said to satisfy the *Open Set Condition* (OSC), an indication that the images of the attractor under the contracting similarities do not overlap much; see Definition 4.65 below. On the other hand, $\mathbf{\Phi}_{I,1}$ does *not* satisfy the Open Set Condition, indicating that a significant amount of redundancy occurs when the unit interval I is constructed as the attractor of $\mathbf{\Phi}_{I,1}$. We refer to Definition 4.65 and the surrounding material below for a more detailed analysis of the consequences of the Open Set Condition.

The form of the scaling ratios also plays a significant role in the analysis of self-similar sets. Note that each of the contracting similarities in $\mathbf{\Phi}_I$ has the same scaling ratio (1/10), whereas there are three distinct scaling ratios (1/6, 1/3, and 1/2) associated with the contracting similarities in $\mathbf{\Phi}_{I,0}$. The self-similar system $\mathbf{\Phi}_I$ is said to be *lattice* and $\mathbf{\Phi}_{I,0}$ is *nonlattice*.[6] The terms lattice and nonlattice are defined in Section 4.6 below, and a preliminary investigation of the differences between such systems is also given therein.

Of course, not every iterated function system is a self-similar system, as discussed in the following section.

4.4. Self-Affine Sets

The examples considered in Section 4.3 above exhibit *strict* self-similarity: They contain *exact* scaled copies of themselves at all scales. However, fractal objects may exhibit less restrictive self-similarity properties. For instance, if an iterated function system comprises contractions that are given by *affine* maps, the resulting attractor is said to be *self-affine* and features a type of skewed self-similarity at all scales.

[6] However, the nonlattice self-similar system $\mathbf{\Phi}_{I,0}$ is *nongeneric* (in the sense of [**371**] and [**375**, Chapter 3]), because $\log 6 = \log 3 + \log 2$; i.e., $\log 2$ and $\log 3$ (the logarithms of the distinct scaling ratios) are not rationally related but satisfy (at least one) nontrivial, linear relation over the field $\mathbb{Q}$ of rational numbers.

Definition 4.39 (*Self-Affine System and Self-Affine Set*). A map $f : \mathbb{R}^m \to \mathbb{R}^m$ is *affine* if it is given by $f(\boldsymbol{x}) := T\boldsymbol{x} + \mathbf{t}_0$, where $\mathbf{t}_0 \in \mathbb{R}^m$ is fixed and T is a nonsingular $m \times m$ matrix.[7] If an iterated function system $\boldsymbol{\Psi}$ comprises only affine contractions, then $\boldsymbol{\Psi}$ and its attractor are said to be *self-affine*. More specifically, the IFS is said to be a *self-affine system* and its attractor $F_{\boldsymbol{\Phi}}$ is said to be a *self-affine set* (in $\mathbb{R}^m$). Note that, clearly, all self-similar systems and sets are self-affine.

Remark 4.40. Note that, in general, an affine map (or affine transformation) f is a combination of a rotation, reflection, translation, and dilation (or homothety) which contract with (typically) different ratios in various directions. In $\mathbb{R}^2$, a square is transformed into a parallelogram and, in any $\mathbb{R}^m$, a sphere is transformed into an ellipsoid. Also, it follows immediately from Definitions 2.49 and 4.39 that a contracting similarity is affine. Hence, as was mentioned just above, self-similar sets are always self-affine.

In the special case when $m = 1$, an affine map $f : \mathbb{R} \to \mathbb{R}$ is simply given by $f(x) = tx + t_0$, where $t, t_0 \in \mathbb{R}$ are fixed. It immediately follows that affine contractions on $\mathbb{R}$ are in fact contracting similarities. In other words, there is no difference between 1-dimensional self-affine and self-similar systems or sets (i.e., in $\mathbb{R}$).

Example 4.41 (*The Barnsley Fern as a Self-Affine Set*). The Barnsley fern $\mathbf{B_F}$ (see [**28**]) is the unique attractor of the iterated function system (IFS) $\boldsymbol{\Phi}_B := \{\varphi_j\}_{j=1}^4$ on $\mathbb{R}^2$ given by the following four contractions:

$$\varphi_1(\boldsymbol{x}) = \begin{pmatrix} .00 & .00 \\ .00 & .16 \end{pmatrix} \boldsymbol{x}, \quad \varphi_2(\boldsymbol{x}) = \begin{pmatrix} .85 & .04 \\ -.04 & .85 \end{pmatrix} \boldsymbol{x} + \begin{pmatrix} .00 \\ 1.6 \end{pmatrix},$$

$$\varphi_3(\boldsymbol{x}) = \begin{pmatrix} .20 & -.26 \\ .23 & .22 \end{pmatrix} \boldsymbol{x} + \begin{pmatrix} .00 \\ 1.6 \end{pmatrix}, \quad \varphi_4(\boldsymbol{x}) = \begin{pmatrix} -.15 & .28 \\ .26 & .24 \end{pmatrix} \boldsymbol{x} + \begin{pmatrix} .00 \\ .44 \end{pmatrix}.$$

Clearly, $\mathbf{B_F}$ is an example of a self-affine set which is *not* self-similar.[8] An image of the *Barnsley fern* $\mathbf{B_F}$ is provided in Figure 4.8. Despite the fact that each of the leaflets in the image of $\mathbf{B_F}$ is a scaled copy of the whole, $\mathbf{B_F}$ is *not* technically a self-similar set, since the scalings are different in the horizontal and vertical directions.

Remark 4.42. In order to obtain the picture depicted in Figure 4.8, one must necessarily use a computer. In fact, more is needed: $\boldsymbol{B_F}$ should really be suitably interpreted as a random fractal, viewed as the attractor of a randomized

[7]Recall that a matrix is nonsingular if it is an invertible square matrix. Hence, it is a square matrix with nonzero determinant.

[8]Strictly speaking, $\boldsymbol{B_F}$ is not self-affine, in the sense of Definition 4.39; indeed, φ_1 is singular because its linear part has determinant 0. Nevertheless, it is applied very infrequently since its associated probability is $p_1 = 0.01$. However, it is needed in order to fill in the stem of the fern in Figure 4.8; see Remark 4.42. We will ignore this fine point here.

Figure 4.8. The Barnsley fern $\mathbf{B_F}$ discussed in Example 4.41 and viewed as a self-affine (but *not* self-similar) set.

IFS, $\mathbf{\Phi}_B = \{\varphi_j\}_{j=1}^4$, in which, roughly speaking, $\varphi_1, \varphi_2, \varphi_3$, and φ_4 are, respectively, given probabilities $p_1 = 0.01, p_2 = 0.85$, and $p_3 = p_4 = 0.07$. For more detail, see [**28**] (second revised edition, Academic Press, 2000); see also the apparently unpublished manuscript entitled "*V-variable fractals and superfractals*", by M. F. Barnsley, J. E. Hutchinson, and Ö. Stenflo (e-print: `arXiv:math/0312314 [math.PR]`, 2003).

Exercise 4.43. Can you find a closed form formula for the various fractal (e.g., Hausdorff, Minkowski, or box-counting) dimensions of the Barnsley fern $\mathbf{B_F}$? If not, obtain numerical approximations of $\dim_B(\mathbf{B_F}) = \dim_M(\mathbf{B_F})$ via, for example, the box-counting method described in Subsections 1.3.3 and 1.3.5 above as well as in Chapter 8 below; see, especially, Sections 8.3 and 8.4 for more details. Obtain a family of variants of the Barnsley fern and answer similar questions for the resulting ferns.

Example 4.44 (*Bedford–McMullen Carpets*). Consider the unit square $E = [0,1]^2$ and the self-affine system $\mathbf{\Psi_{BM}} := \{\psi_j\}_{j=1}^3$ determined by the following contractions, for all $\boldsymbol{x} \in \mathbb{R}^2$:

$$\begin{aligned}\psi_1(\boldsymbol{x}) &= \begin{pmatrix} 1/3 & 0 \\ 0 & 1/2 \end{pmatrix} \boldsymbol{x}, \\ \psi_2(\boldsymbol{x}) &= \begin{pmatrix} 1/3 & 0 \\ 0 & 1/2 \end{pmatrix} \boldsymbol{x} + \begin{pmatrix} 1/3 \\ 1/2 \end{pmatrix}, \\ \psi_3(\boldsymbol{x}) &= \begin{pmatrix} 1/3 & 0 \\ 0 & 1/2 \end{pmatrix} \boldsymbol{x} + \begin{pmatrix} 2/3 \\ 0 \end{pmatrix}.\end{aligned} \tag{4.19}$$

Each ψ_j (for $j = 1, 2, 3$) is an affine contraction but *not* a contracting similarity. The unique attractor **BM** of the IFS $\mathbf{\Psi_{BM}}$ is called the *Bedford–McMullen carpet* associated with $\mathbf{\Psi_{BM}}$. The Bedford–McMullen carpet **BM**—along with

Figure 4.9. An example of a Bedford–McMullen carpet **BM**, as described in Example 4.44. It is a key example of a self-affine (but *not* self-similar) set, such that $\dim_H(\mathbf{BM}) \neq \dim_M(\mathbf{BM}) = \dim_B(\mathbf{BM})$, so that $\dim_H(\mathbf{BM}) < \dim_M(\mathbf{BM})$.

its generalizations (the $p \times q$ **BM** carpets, still denoted **BM**, or else $\mathbf{BM}_{p\times q}$) described below, when $p \neq q$—is a key example of a self-affine but *not* self-similar set, for which, in addition, the Hausdorff and Minkowski dimensions can be calculated "in closed form" but do *not* coincide, as we next discuss: $\dim_H(\mathbf{BM}) \neq \dim_M(\mathbf{BM}) = \dim_B(\mathbf{BM})$. See Figure 4.9 for a picture of a Bedford–McMullen self-affine carpet.

Let us now discuss the aforementioned well-known generalization of the above construction; see, e.g., [**173**, Chapter 4] or [**198**]. Fix $(p,q) \in \mathbb{N}^2$ such that $p > q > 1$. Divide the unit square $E := [0,1]^2$ into a $p \times q$ grid; hence, the grid consists of pq congruent rectangles of side lengths $1/p$ and $1/q$.[9] Select a subset $\mathcal{R}$ of the resulting collection of rectangles, and consider the IFS $\mathbf{\Psi}_{\mathbf{BM}} = \mathbf{\Psi}_{\mathbf{BM},p\times q,\mathcal{R}} = \{\psi_j\}_{j=1}^{\#\mathcal{R}}$ composed of the contractive affine maps ψ_j (for $j = 1, \dots, \#\mathcal{R}$) which map E into each given rectangle R_j in $\mathcal{R}$, while preserving the orientation.[10] Thus, the linear part of the matrix representing each ψ_j depends only on (p,q): It is just the diagonal 2×2 matrix with diagonal entries $1/p$ and $1/q$, so that, for $j = 1, \dots, \#\mathcal{R} =: P$,

$$\psi_j(\boldsymbol{x}) = \begin{pmatrix} 1/p & 0 \\ 0 & 1/q \end{pmatrix} \boldsymbol{x} + \boldsymbol{u}_j, \tag{4.20}$$

for all $\boldsymbol{x} \in \mathbb{R}^2$ and for some suitable choice of $\boldsymbol{u}_j \in \mathbb{R}^2$.

[9] Clearly, all of these rectangles are squares iff $p = q$.

[10] In the sequel, we denote by $\mathcal{R}_{\max}$ the set of all pq admissible rectangles.

The unique attractor of the self-affine IFS $\mathbf{\Psi_{BM}} = \mathbf{\Psi}_{\mathbf{BM},p\times q,\mathcal{R}}$ is denoted, for simplicity, **BM** (or else $\mathbf{BM}_{p\times q,\mathcal{R}}$) and is called the *$(p \times q, \mathcal{R})$-Bedford–McMullen carpet.* It is self-affine but, in general, is *not self-similar*. Such self-affine carpets—which naturally generalize both the Sierpinski carpet **SC** and the first Bedford–McMullen carpet discussed at the beginning of this example, as the reader can easily verify—were introduced in 1984 independently by T. Bedford in [**34**] and by C. T. McMullen in [**434**].[11] In addition, it can be shown (see [**34, 434**] and loc. cit.) that

$$\dim_M(\mathbf{BM}) = \dim_B(\mathbf{BM}) = \frac{\log Q}{\log q} + \frac{\log(P/Q)}{\log p}, \tag{4.21}$$

$$\dim_H(\mathbf{BM}) = \frac{\log\left(\sum_{k=1}^{Q} N_k^{\log q/\log p}\right)}{\log q}, \tag{4.22}$$

where $P := \#\mathcal{R}$ (as before) is the number of rectangles in $\mathcal{R}$ (or, equivalently, of maps in the underlying IFS $\mathbf{\Psi_{BM}}$), Q is the number of (nonempty) columns (of the corresponding $p\times q$ matrix) containing at least one rectangle in $\mathcal{R}$, and, for each $k = 1, \ldots, Q$, $N_k \in \mathbb{N}$ is the number of rectangles in $\mathcal{R}$ corresponding to the k-th nonempty column (of the same matrix).[12]

Remark 4.45. The so-called packing dimension (see Subsection 8.9.3 below), $\dim_P(\mathbf{BM})$, is also given (see [**34, 434**]) by the last equality of (4.21)—and hence coincides with $\dim_M(\mathbf{BM}) = \dim_B(\mathbf{BM})$—whereas the Assouad dimension (see [**12, 199**]) of **BM**, denoted by $\dim_A(\mathbf{BM})$, is given (see J. M. Mackay [**413**]) as follows:

$$\dim_A(\mathbf{BM}) = \frac{\log Q}{\log q} + \frac{\max_{k=1,\ldots,Q}(\log N_k)}{\log p}. \tag{4.23}$$

In general, one has (in Euclidean space $\mathbb{R}^m$, see Sections 8.3, 8.4, and 8.9 below) $\dim_T \leq \dim_H \leq \dim_P \leq \dim_M = \dim_B \leq m$, while here, we have that $\dim_T(\mathbf{BM}) \in \{0, 1\}$ and

$$\begin{aligned} 0 \leq \dim_T(\mathbf{BM}) &\leq \dim_H(\mathbf{BM}) \leq \dim_P(\mathbf{BM}) \\ &= \dim_M(\mathbf{BM}) = \dim_B(\mathbf{BM}) \leq \dim_A(\mathbf{BM}) < 2, \end{aligned} \tag{4.24}$$

with, typically, strict inequalities between $\dim_H(\mathbf{BM})$ and $\dim_M(\mathbf{BM}) = \dim_P(\mathbf{BM})$; see Exercise 4.46.

Exercise 4.46. Find different choices of p, q, and $\mathcal{R}$ in the above construction of self-affine Bedford–McMullen carpets $\mathbf{BM} = \mathbf{BM}_{p\times q,\mathcal{R}}$ for $p \neq q$ (so that **BM** is *not* self-similar, at least when $\mathcal{R} \neq \mathcal{R}_{\max}$) such that we have strict inequalities

[11] The Bedford–McMullen self-affine carpets, along with various types of generalizations, have been further studied from different points of view, e.g., in [**23, 198–200, 316, 485**].

[12] The computation of $\dim_M(\mathbf{BM}) = \dim_B(\mathbf{BM})$ is due independently to T. Bedford in [**34**] and to C. T. McMullen in [**434**].

in (4.24) between $\dim_T(\mathbf{BM})$, $\dim_H(\mathbf{BM})$, and $\dim_M(\mathbf{BM})$; namely, with the notation of Remark 4.45,

$$\begin{aligned} 0 &\le \dim_T(\mathbf{BM}) < \dim_H(\mathbf{BM}) < \dim_P(\mathbf{BM}) \\ &= \dim_M(\mathbf{BM}) = \dim_B(\mathbf{BM}) < \dim_A(\mathbf{BM}) < 2. \end{aligned} \tag{4.25}$$

Can you find examples for which $\dim_T(\mathbf{BM})$ equals 0 (resp., 1) or for which $\dim_T(\mathbf{BM}) = \dim_H(\mathbf{BM})$?

In particular, for the Bedford–McMullen carpet $\mathbf{BM} = \mathbf{BM}_{3\times 2,\mathcal{R}_{\max}}$ first discussed at the beginning of Example 4.44 (and for which $p = 3$, $q = 2$, and $\mathcal{R} = \mathcal{R}_{\max}$ consists of all 6 admissible rectangles), show that (4.25) holds and calculate the explicit values of the corresponding fractal dimensions (especially $\dim_M(\mathbf{BM})$ and $\dim_H(\mathbf{BM})$).

We refer the reader to Example 4.93 in Section 4.10 below for a discussion of the very important example of the Devil's staircase $\mathbf{\Gamma}_{DS}$ (or graph of the Cantor function f_{DS}), which is a key example of an inhomogeneous self-affine set. We point out that in Section 9.4 of Chapter 9 on fractal curves and their complex dimensions, we will provide a lot more information about the Devil's staircase, viewed as a self-affine and rectifiable fractal curve.

4.5. Symbolic Dynamics: Words and Addresses

The central topic of this section can be viewed as being a part of *symbolic dynamics* (as specialized to IFSs), within the broader theory of dynamical systems; see, e.g., [**35, 138, 172, 293, 375, 491**].

For a given iterated function system (IFS) $\mathbf{\Phi} = \{\varphi_j\}_{j=1}^N$ on $\mathbb{R}^m$, there is a method for assigning an address to each of the points in the corresponding attractor $F_{\mathbf{\Phi}}$ which, in fact, is a generalization of the ternary expansions discussed in Remark 1.3, the ubiquitous notion of decimal expansions of points in the unit interval $[0, 1]$.

Some additional definitions and notation are in order.

Notation 4.47. Let $\mathbf{\Phi} = \{\varphi_j\}_{j=1}^N$ be an iterated function system on $\mathbb{R}^m$. For each $k \in \mathbb{N}_0$, $\mathcal{J}_k := \{1, \dots, N\}^k = S^k$ denotes the set of all finite sequences of length k in the symbols (or letters) $\{1, \dots, N\}$ (i.e., *words of length k*). The set $S := \{1, \dots, N\}$ is called the *alphabet*. For $k = 0$, $\mathcal{J}_0$ denotes the set comprising the empty word. The set of all *finite words* in the symbols $\{1, \dots, N\}$ is denoted by $\mathcal{J} := \bigcup_{k=0}^{\infty} \mathcal{J}_k$, where $\mathcal{J}_k$ is the set of finite words of length $k \in \mathbb{N}_0$, and $\mathcal{J}_\infty$ denotes the set of all one-sided infinite sequences (or *infinite words*), indexed by $\mathbb{N}$, in the symbols $\{1, \dots, N\}$. The set $\mathcal{J}_\infty$ is often called the *code space* (or *shift space*) in the literature on symbolic dynamics or on fractal geometry.

For $J \in \mathcal{J}$, $|J|$ denotes the number of components of (i.e., the *length of the word*) J and set $|J| = \infty$ iff $J \in \mathcal{J}_\infty$. For a word $J \in \mathcal{J} \cup \mathcal{J}_\infty$ and each $n \in \mathbb{N}$

where $n \leq |J|$, $J|n$ denotes the truncation of J at its n-th component and $J|0$ denotes the empty word. A word J is also written by concatenating and we write $J|n = j_1 j_2 \dots j_n$ if J begins with $j_1, j_2, \dots, j_n$. For each $n \in \mathbb{N}$, all $J \in \mathfrak{J}_n$, and $j \in \{1, \dots, N\}$, $Jj := j_1 j_2 \dots j_n j \in \mathfrak{J}_{n+1}$.

For $J \in \mathfrak{J}$, define the contracting map φ_J by[13]

$$\varphi_J := \varphi_{j_1} \circ \varphi_{j_2} \circ \cdots \circ \varphi_{j_{|J|}}. \tag{4.26}$$

The scaling ratio of φ_J is given by $r_J = \prod_{q=1}^{|J|} r_{\pi_q(J)}$, where $\pi_q(\cdot)$ is the projection of a word onto its q-th component (i.e., its q-th letter).

For the empty word $J|0$, let $\varphi_{J|0}$ denote the identity map and define $r_{J|0} := 1$. For any set $A \subseteq \mathbb{R}^m$ and a finite word $J \in \mathfrak{J}$, let $A_J := \varphi_J(A)$.

Theorem 4.48. *Let $\mathbf{\Phi} = \{\varphi_j\}_{j=1}^N$ be an iterated function system on $\mathbb{R}^m$ with scaling vector $\mathbf{r} = (r_1, \dots, r_N)$ and attractor $F_{\mathbf{\Phi}} = F$. Then, the following statements hold:*

(i) *For each $A \in \mathbb{K}(\mathbb{R}^m)$ and $J \in \mathfrak{J}_\infty$, we have*

$$\lim_{n\to\infty} \operatorname{diam}(A_{J|n}) = 0.$$

(ii) *For each $J \in \mathfrak{J}$, we have $F_J = \bigcup_{j=1}^N F_{Jj}$.*

(iii) *For each $J \in \mathfrak{J}_\infty$ and $n \in \mathbb{N}$, we have $F_{J|n-1} \supseteq F_{J|n}$. Also, $\bigcap_{n=0}^\infty F_{J|n}$ is a singleton whose single element is denoted $\boldsymbol{x}_J$.*[14] *Furthermore, we have that $\boldsymbol{x}_J \in F$.*

(iv) *For each $J, I \in \mathfrak{J}$ and $H \in \mathfrak{J}_\infty$, we have $\varphi_J(F_I) = F_{JI}$ and $\varphi_J(\boldsymbol{x}_H) = \boldsymbol{x}_{JH}$.*[15]

(v) *For each $A \in \mathbb{K}(\mathbb{R}^m)$ and $J \in \mathfrak{J}_\infty$, we have*

$$\lim_{n\to\infty} \operatorname{dist}(\boldsymbol{x}_J, A_{J|n}) = 0,$$

and similarly with $d_H(\{\boldsymbol{x}_J\}, A_{J|n})$ instead of $\operatorname{dist}(\boldsymbol{x}_J, A_{J|n})$, *for each $n \geq 1$; i.e., $A_{J|n} \to \{\boldsymbol{x}_J\}$, in $(\mathbb{K}(\mathbb{R}^m), d_H)$.*

(vi) *Let $A \in \mathbb{K}(\mathbb{R}^m)$ be such that $\mathbf{\Phi}(A) := \bigcup_{j=1}^N \varphi_j(A) \subseteq A$. Then,*

$$F_{\mathbf{\Phi}} = F = \bigcap_{n=0}^{\infty} \mathbf{\Phi}^n(A). \tag{4.27}$$

[13] Observe the order in which the maps are composed, in the definition of φ_J given in (4.26). Furthermore, note that in the special case when $\mathbf{\Phi} = \{\varphi_J\}_{j=1}^N$ is a self-similar system, and if J is not the empty word, then φ_J is a contracting *similarity* because the composition of (contracting) similarities is also a (contracting) similarity.

[14] That is, $\bigcap_{n=0}^\infty F_{J|n} = \{\boldsymbol{x}_J\}$.

[15] Here, for example, JI denotes the concatenation of J and I, in that order.

(vii) *For each $\boldsymbol{x} \in F$, there exists some $J \in \mathcal{J}_\infty$ such that $\boldsymbol{x} = \boldsymbol{x}_J$; this infinite word J is called an* infinite address *of $\boldsymbol{x}$. Furthermore, $F = \bigcup_{J\in\mathcal{J}_\infty}\{\boldsymbol{x}_J\}$.*

(viii) *If $\varphi_j(F) \cap \varphi_k(F) = \emptyset$ whenever $j \neq k$, for $j, k \in \{1, \ldots, N\}$, then F is totally disconnected.*[16]

Proof. In order to prove property (i), note that by the definitions of contraction and diameter, we have $0 \leq \operatorname{diam}(\varphi_j(A)) \leq r_j \cdot \operatorname{diam}(A)$, for each $j = 1, \ldots, N$. Let $r := \max\{r_j : j = 1, \ldots, N\}$. Then, for each $J \in \mathcal{J}_\infty$ and $n \in \mathbb{N}$, we have[17]

$$0 \leq \operatorname{diam}(A_{J|n}) = \operatorname{diam}(\varphi_{J|n}(A)) \leq r_{J|n} \cdot \operatorname{diam}(A) \leq r^n \cdot \operatorname{diam}(A).$$

Since $0 < r < 1$, we have $\operatorname{diam}(A_{J|n}) \to 0$, as $n \to \infty$.

In order to prove property (ii), note that for each $n \in \mathbb{N}$, we have

$$F = \bigcup_{j=1}^{N} \varphi_j(F) = \bigcup_{i,j=1}^{N} \varphi_i(\varphi_j(F)) = \bigcup_{i,j=1}^{N} F_{ij} = \cdots = \bigcup_{J\in\mathcal{J}_n} F_J.$$

So, for a fixed $J \in \mathcal{J}$, we have (since $\varphi_J\varphi_j = \varphi_{Jj}$)

$$F_J = \varphi_J(F) = \varphi_J\Big(\bigcup_{j=1}^{N} \varphi_j(F)\Big) = \bigcup_{j=1}^{N} \varphi_{Jj}(F) = \bigcup_{j=1}^{N} F_{Jj}.$$

This establishes property (ii) and the following part of property (iii): For each $J \in \mathcal{J}_\infty$ and $n \in \mathbb{N}$, we have $F_{J|n-1} \supseteq F_{J|n}$.

In order to complete the proof of property (iii), note that the sequence $\{F_{J|n}\}_{n=0}^{\infty}$ is a decreasing sequence of nonempty compact sets.[18] Hence, its intersection is nonempty and compact, by Theorem 2.39.[19] Also, we have $\bigcap_{m=0}^{\infty} F_{J|m}^{\infty} \subseteq F_{J|n}$, for each $n \in \mathbb{N}_0$. So, by part (ii) of Proposition 2.9, for each $n \in \mathbb{N}$, we have $0 \leq \operatorname{diam}\big(\bigcap_{m=0}^{\infty} F_{J|m}\big) \leq \operatorname{diam}(F_{J|n})$. By property (i) (and upon letting $n \to \infty$), $\operatorname{diam}\big(\bigcap_{m=0}^{\infty} F_{J|m}\big) = 0$ and thus by part (i) of Proposition

[16]An important example for which the hypotheses of part (viii) are satisfied is the classic Cantor set C, viewed as the self-similar set $F_{\boldsymbol{\Phi}}$ which is the attractor of the self-similar system (or IFS) $\boldsymbol{\Phi} = \{\varphi_1, \varphi_2\}$ given in Example 4.23, formula (4.11). Indeed, if $C_j = \varphi_j(C)$, for $j = 1, 2$; then, clearly, $C_1 = C \cap [0, 1/3]$ and $C_2 = C \cap [2/3, 1]$, with $C_1 \cap C_2 = \emptyset$.

[17]Observe that, given $n \in \mathbb{N}$ and $H \in \mathcal{J}_n$, with $H := j_1, \ldots, j_n$, we have that $\operatorname{Lip}(\varphi_H) \leq r_H := r_{j_1} \cdots r_{j_n} = \prod_{k=1}^{n} \operatorname{Lip}(\varphi_{j_k}) \leq r^n$, where $r := \max_{1\leq k\leq n} r_{j_k}$. We clearly have the following inequalities (see Definition 2.44 and Remark 2.45):

$$\operatorname{diam}(A_H) = \operatorname{diam}(\varphi_H(A)) \leq r_H \cdot \operatorname{diam}(A) \leq r^n \cdot \operatorname{diam}(A).$$

Also note that in the important special case when $\boldsymbol{\Phi} = \{\varphi_j\}_{j=1}^{N}$ is a self-similar system, then $\operatorname{Lip}(\varphi_H) = r_H = r_{j_1\ldots j_n} = r_{j_1} \cdots r_{j_n}$, whereas for a general IFS, we only have the inequality $\operatorname{Lip}(\varphi_H) \leq r_H := r_{j_1} \cdots r_{j_n}$.

[18]Note that for any $K \in \mathcal{J}$ and any $A \in \mathbb{K}(\mathbb{R}^m)$, we have that $A_K = \varphi_K(A) \in \mathbb{K}(\mathbb{R}^m)$, as the continuous image of a nonempty and compact subset of $\mathbb{R}^m$. In particular, for any $J \in \mathcal{J}_\infty$ and $n \in \mathbb{N}_0$, $F_{J|n} \in \mathbb{K}(\mathbb{R}^m)$—as well as, more generally, $A_{J|n} \in \mathbb{K}(\mathbb{R}^m)$, for all $A \in \mathbb{K}(\mathbb{R}^m)$.

[19]Indeed, due to the fact that the sequence is nonincreasing, it clearly satisfies the Finite Intersection Property; see Theorem 2.39.

2.9, $\bigcap_{m=0}^{\infty} F_{J|m}$ is a singleton. (Its unique element is denoted $\boldsymbol{x}_J$.) To conclude the proof of part (iii), observe that since $F = \bigcup_{k=1}^{N} \varphi_k(F)$, we have that $\varphi_k(F) \subseteq F$, for $k = 1, \ldots, N$. Hence, by finite induction, $\varphi_H(F) \subseteq F$, for any finite word H. In particular, $F_{J|n} \subseteq F$, for any $n \in \mathbb{N}_0$, and so $\{\boldsymbol{x}_J\} = \bigcap_{n=0}^{\infty} F_{J|n} \subseteq F$; i.e., $\boldsymbol{x}_J \in F$.

In order to prove property (iv), note that for $J, I \in \mathcal{J}$ we clearly have that $\varphi_J(F_I) = \varphi_J(\varphi_I(F)) = \varphi_{JI}(F) = F_{JI}$. Also, for $H \in \mathcal{J}_\infty$, we have[20] $\varphi_J(\boldsymbol{x}_H) \in \varphi_J\left(\bigcap_{n=0}^{\infty} F_{H|n}\right)$, but $\varphi_J\left(\bigcap_{n=0}^{\infty} F_{H|n}\right) = \bigcap_{n=0}^{\infty} \varphi_J(F_{H|n}) = \{\boldsymbol{x}_{JH}\}$.

In order to prove property (v), observe that for each $J \in \mathcal{J}_\infty$ with $J = j_1 j_2 \ldots$ and for each $n \in \mathbb{N}_0$, we have that

$$\begin{aligned} \operatorname{dist}(\boldsymbol{x}_J, A_{J|n}) &= \operatorname{dist}(\varphi_{J|n}(\boldsymbol{x}_{j_{n+1}j_{n+2}\ldots}), \varphi_{J|n}(A)) \le r_{J|n} \cdot \operatorname{dist}(\boldsymbol{x}_{j_{n+1}j_{n+2}\ldots}, A) \\ &\le r_{J|n} \cdot \sup\{d_m(\boldsymbol{x}, \boldsymbol{a}) : \boldsymbol{x} \in F, \boldsymbol{a} \in A\} \le r^n \cdot \sup\{d_m(\boldsymbol{x}, \boldsymbol{a}) : \boldsymbol{x} \in F, \boldsymbol{a} \in A\}, \end{aligned}$$

where $r := \max\{r_j : j = 1, \ldots, N\}$.[21] Since $0 < r < 1$, we deduce, as desired, that $\lim_{n\to\infty} \operatorname{dist}(\boldsymbol{x}_J, A_{J|n}) = 0$.[22] We leave the proof of the second statement in part (v) as an exercise for the interested reader.

In order to prove property (vi), if $A \in \mathbb{K}(\mathbb{R}^m)$ such that $\boldsymbol{\Phi}(A) \subseteq A$, then $\boldsymbol{\Phi}^n(A) \subseteq \boldsymbol{\Phi}^{n-1}(A)$, for each $n \in \mathbb{N}$. Hence, the sequence $\{\boldsymbol{\Phi}^n(A)\}_{n=0}^{\infty}$ is a nonincreasing sequence of nonempty compact sets whose intersection $\bigcap_{n=0}^{\infty} \boldsymbol{\Phi}^n(A)$ is nonempty and compact, by Theorem 2.39. Moreover, since $\{\boldsymbol{\Phi}^n(A)\}_{n=0}^{\infty}$ is nonincreasing, $\boldsymbol{\Phi}(\bigcap_{n=0}^{\infty} \boldsymbol{\Phi}^n(A)) = \bigcap_{n=0}^{\infty} \boldsymbol{\Phi}^n(A)$. That is, the set $\bigcap_{n=0}^{\infty} \boldsymbol{\Phi}^n(A)$ is invariant with respect to $\boldsymbol{\Phi}$ and $\bigcap_{n=0}^{\infty} \boldsymbol{\Phi}^n(A) \in \mathbb{K}(\mathbb{R}^m)$. By the uniqueness of the attractor of the IFS $\boldsymbol{\Phi}$ (see Theorem 4.17 and Definition 4.18), we conclude that $F_{\boldsymbol{\Phi}} = F = \bigcap_{n=0}^{n} \boldsymbol{\Phi}^n(A)$, as desired.

In order to prove property (vii), suppose $\boldsymbol{x} \in F_{\boldsymbol{\Phi}} = F$ and $A \in \mathbb{K}(\mathbb{R}^m)$ is such that $\boldsymbol{\Phi}(A) \subseteq A$. Then, by property (vi), there is some (not necessarily unique) $J \in \mathcal{J}_\infty$ such that $\boldsymbol{x} \in \varphi_{J|n}(A)$, for each $n \in \mathbb{N}_0$. Therefore, $\boldsymbol{x} = \boldsymbol{x}_J$ and $F = \bigcup_{K \in \mathcal{J}_\infty} \{\boldsymbol{x}_K\}$.[23] Note that the word J obtained just above is independent of the set A, as long as $\boldsymbol{\Phi}(A) \subseteq A$.

Finally, in order to prove property (viii), suppose that $\varphi_j(F) \cap \varphi_k(F) = \emptyset$ whenever $j \neq k$, and let $J, J' \in \mathcal{J}_\infty$ such that $\boldsymbol{x}_J \neq \boldsymbol{x}_{J'}$. Then, for some $n \in \mathbb{N}$, we have $J|n \neq J'|n$. Therefore, $F_{J|n}$ and $F_{J'|n}$ are disjoint, and $\boldsymbol{x}_J \in F_{J|n}$ and $\boldsymbol{x}_{J'} \in F_{J'|n}$. Since $F_{J|n}$ and $F_{J'|n}$ are disjoint and closed, there exists some $\varepsilon > 0$

[20] A moment's reflection shows that $J(H|n) = (JH)_{|n}$, for any $n \in \mathbb{N}_0$.

[21] Note that by property (iii), we have that $\boldsymbol{x}_L \in F$, where $L := j_{n+1}j_{n+2}\cdots$.

[22] Note that the supremum at the end of the last displayed equation is finite. Indeed, F and A are compact and nonempty subsets of $\mathbb{R}^m$ (because $F \times A$ itself is compact), so that the distance function d_m (which is a continuous function) achieves its maximum on the Cartesian product $F \times A \subseteq \mathbb{R}^m \times \mathbb{R}^m$.

[23] Recall from the end of property (iii) that $\boldsymbol{x}_K \in F$, for any $K \in \mathcal{J}_\infty$.

such that their interiors, $[F_{J|n}]_\varepsilon^o$ and $[F_{J'|n}]_\varepsilon^o$, are disjoint and open subsets of $\mathbb{R}^m$; furthermore, $\boldsymbol{x}_J \in [F_{J|n}]_\varepsilon^o$ and $\boldsymbol{x}_{J'} \in [F_{J'|n}]_\varepsilon^o$. Hence, F is totally disconnected. □

Property (vii) of Theorem 4.48 just above motivates the following definition.

Definition 4.49 (*Address*). Let $\boldsymbol{\Phi} = \{\varphi_j\}_{j=1}^N$ be an iterated function system on $\mathbb{R}^m$ with attractor $F_{\boldsymbol{\Phi}}$, and let $\boldsymbol{x} \in F_{\boldsymbol{\Phi}}$. An *address* of $\boldsymbol{x}$ is an infinite word $J \in \mathfrak{J}_\infty$ such that $\boldsymbol{x} = \boldsymbol{x}_J$. According to property (vii) of Theorem 4.48, such an address always exists.

Remark 4.50. Property (vii) of Theorem 4.48 asserts that for each $\boldsymbol{x} \in F$, there exists some infinite word $J \in \mathfrak{J}_\infty$ such that $\boldsymbol{x} = \boldsymbol{x}_J$, but such a J is not necessarily unique. For instance, consider the self-similar system $\boldsymbol{\Phi}_I$ from Example 4.37, which provides the decimal expansions for points in the closed unit interval $I = [0,1]$. Consider the point $x = 0.1 \in [0,1] = I$. Then, the infinite words $J_1 = 1000\ldots =: \overline{1}$ (a single one followed by zeros) and $J_2 = 0999\ldots = 0\overline{9}$ (a single zero followed by nines) satisfy $0.1 = x_{J_1} = x_{J_2}$.

Remark 4.51 (*Lack of Uniqueness of the Representation*). Elaborate on Remark 4.50 just above by discussing for a variety of examples (e.g., the classic Cantor set, the Koch curve, the Sierpinski gasket and carpet, the Menger sponge, the Bedford–McMullen carpets, and the Devil's staircase; see Sections 4.3, 4.4 above and 4.10 below) the issue of the uniqueness or lack thereof of the representation (via an address in symbolic dynamics) of the points of the attractor of a given IFS; see Theorem 4.48 above for the relevant context and notation. Furthermore, try to formulate and prove a general theorem, based on your own observations motivated by the study of some of these examples. In particular, which points of a (suitable) self-similar set have a finite (resp., infinite) representation? Is the latter representation unique? Do these points also have an infinite (resp., a finite) representation?

Example 4.52 (*Ultrametric on $\mathfrak{J}_\infty$*).

(a) Consider the *code space* (or *shift space*) $\mathfrak{J}_\infty = S^{\mathbb{N}}$, where $S := \{1,\ldots,N\}$. Show that $\mathfrak{J}_\infty$ is a compact metric space, when equipped with the distance

$$d_{SD}(\boldsymbol{x},\boldsymbol{y}) = 2^{-v_{SD}(\boldsymbol{x},\boldsymbol{y})}, \tag{4.28}$$

where $v_{SD}(\boldsymbol{x},\boldsymbol{y}) := \min\{n \in \mathbb{N} : j_n \neq k_n\}$ and $\boldsymbol{x} = j_1 j_2 \ldots$, $\boldsymbol{y} = k_1 k_2 \ldots$ are arbitrary nonempty infinite words. Furthermore, show that d_{SD} is an ultrametric[24] (with associated ultrametric—and, hence, nonarchimedean—valuation v_{SD}). Moreover, prove that $\mathcal{J}_\infty$ is homeomorphic to the (ternary) Cantor set C—and hence is a topological Cantor set which therefore has the same topological and set-theoretic properties as C (e.g., it is perfect, totally disconnected, uncountable and is, in fact, in bijection with $[0, 1]$ or, equivalently, with $\mathbb{R}$), Finally, extend the metric and the topology of $\mathcal{J}_\infty$ to $X := \mathcal{J}_\infty \cup \mathcal{J}$.

(b) (*Shift-Map*). Consider the *shift-map* (or *one-sided shift*) σ given (for all $\boldsymbol{x} = j_1 j_2 \ldots \in \mathcal{J}_\infty$) by $\sigma(\boldsymbol{x}) := j_2 j_3 \cdots$. Show that $\sigma : \mathcal{J}_\infty \to \mathcal{J}_\infty$ is continuous, where $\mathcal{J}_\infty$ is equipped with the topology induced by the metric introduced in part (a). It is clearly surjective but not injective; briefly explain why. What are the inverse branches of σ? What are the periodic points (of period $n \in \mathbb{N}$) of σ (i.e., the points $\boldsymbol{x} \in \mathcal{J}_\infty$ such that $\sigma^n(\boldsymbol{x}) = \boldsymbol{x}$)? Is the set of all periodic points of $\mathcal{J}_\infty$ (of *any* period $n \geq 1$) dense in $\mathcal{J}_\infty$? Is σ a contracting map or an expanding map (or neither) of the metric space $(\mathcal{J}_\infty, d_{SD})$?

4.6. Lattice and Nonlattice Self-Similar Systems

In the present section, we discuss a result extending to higher-dimensional self-similar sets some aspects of a method developed by M. L. Lapidus and M. van Frankenhuijsen (see, especially, [**370, 371, 374**] and [**375**, Section 3.4]) based on Diophantine approximation and enabling us to approximate nonlattice self-similar fractal strings by a suitable sequence of lattice self-similar fractal strings. This extension is obtained in [**137**] and is due to K. Dettmers, R. Giza, C. Knox, R. Morales, and J. A. Rock; see, especially, Theorem 4.63 below. The original method from [**370, 371, 374, 375**]—alongside with its consequences for the quasiperiodic distribution of the complex dimensions of nonlattice self-similar strings—will be discussed in the sequel to this book, [**119**].

Definition 4.53 (*Lattice vs. Nonlattice Self-Similar System*)**.** Let $\boldsymbol{\Phi}$ be a self-similar system with scaling vector $\mathbf{r} = (r_j)_{j=1}^N$. Then, $\boldsymbol{\Phi}$ is said to be *lattice* if there exist $r \in (0, 1)$ and integers k_j such that $r_j = r^{k_j}$, for $j = 1, 2, \ldots, N$. Otherwise, $\boldsymbol{\Phi}$ is said to be *nonlattice*.

It is helpful to consider the following slight variation of the notion of a scaling vector; see [**375**, Chapters 3 and 7], [**136**], and [**137**].

Definition 4.54. Let $\boldsymbol{\Phi}$ be a self-similar system with scaling vector $\mathbf{r} = (r_j)_{j=1}^N$. The *weight vector* of $\boldsymbol{\Phi}$ is given by $\boldsymbol{w} = (w_j)_{j=1}^N := (-\log r_j)_{j=1}^N$. The components w_j of a weight vector are called *weights*.

[24]This means that d_{SD} is a metric on $\mathcal{J}_\infty$ which satisfies the following strong form of the Triangle Inequality (also called the *Ultrametric Triangle Inequality*):

$$d_{SD}(\boldsymbol{x}, \boldsymbol{z}) \leq \max\{d_{SD}(\boldsymbol{x}, \boldsymbol{y}), d_{SD}(\boldsymbol{y}, \boldsymbol{z})\}, \text{ for all } \boldsymbol{x}, \boldsymbol{y}, \boldsymbol{z} \in \mathcal{J}_\infty. \tag{4.29}$$

Remark 4.55 (*Additive vs. Multiplicative Group Formulation*). Clearly, in Definition 4.53, it suffices to consider the *distinct* scaling ratios of $\mathbf{\Phi}$. Accordingly, following [**371, 374, 375**], we can rephrase this definition (Definition 4.53) in terms of additive/multiplicative groups. Namely, the self-similar system $\mathbf{\Phi}$ is *lattice* iff the multiplicative group generated by the *distinct* scaling ratios of $\mathbf{\Phi}$ (see Remark 4.56) is of rank one, as a subgroup of the multiplicative group $(0, +\infty)$; then, $r \in (0, 1)$, chosen as in Definition 4.53, is called a *multiplicative generator*. Equivalently, $\mathbf{\Phi}$ is lattice iff the additive group generated by (minus) the logarithms of the *distinct* scaling ratios of $\mathbf{\Phi}$ is of rank one, as a subgroup of $\mathbb{R}$; then, $-\log r = \log(r^{-1}) > 0$, with r given as in Definition 4.53, is called an *additive generator*. Finally, of course, the self-similar system $\mathbf{\Phi}$ is said to be *nonlattice* iff it is not lattice. Furthermore, it is called *generic nonlattice* iff the above (additive or multiplicative) group is of maximal rank $\widetilde{M}$, where $\widetilde{M}$ is the number of *distinct* scaling ratios of $\mathbf{\Phi}$ and $\widetilde{M} \geq 2$.

The reader may wish to keep the following comment in mind when reading the remainder of this section.

Remark 4.56. For the simplicity of exposition, we do not change the notation adopted here. However, as in done in [**370, 371, 374, 375**] in order to define the lattice/nonlattice dichotomy (as in Lemma 4.58 and Definition 4.53), and in order to state the corresponding results obtained in loc. cit. and in the present section, one should work here with the $\widetilde{M}$-tuples of scaling ratios $\mathbf{r}' = (r_1', \ldots, r_{\widetilde{M}}')$ and weights $\boldsymbol{w}' = (w_1', \ldots, w_{\widetilde{M}}')$ with $\widetilde{M} \leq N$, consisting of the *distinct* scaling ratios and hence also weights.

Accordingly, the Cantor set, with scaling vector $\mathbf{r} = (1/3, 1/3)$ and so with weight vector $\boldsymbol{w} = (\log 3, \log 3)$ and, hence, with $\mathbf{r}' = 1/3$ and so $\boldsymbol{w}' = \log 3$, is defined by a lattice self-similar system or is a *lattice self-similar set*, in short. An entirely similar comment could be made about the Sierpinski gasket, the Sierpinski carpet, the Koch curve, and the Menger sponge, which are all defined by lattice self-similar systems or are lattice self-similar sets, in short. By contrast, a self-similar system with (distinct) scaling ratios $\mathbf{r}' = (1/2, 1/3, 1/5)$ (and hence with distinct weights $\boldsymbol{w}' = (\log 2, \log 3, \log 5)$) is nonlattice since, for example, $\log 3/\log 2$ is irrational. For the same reason, a self-similar system with scaling vector $\mathbf{r}' = (1/2, 1/3)$—and so with weight vector $\boldsymbol{w}' = (\log 2, \log 3)$—is nonlattice.

Remark 4.57 (*Lattice/Nonlattice Dichotomy*). The *lattice/nonlattice dichotomy* is often called the *arithmetic/nonarithmetic dichotomy* in the literature—especially in probability theory and statistics (e.g., in renewal theory and in queuing theory); see, e.g., W. Feller's book [**186**]. In the present context of self-similar sets, it was first used by S. P. Lalley in [**313–315**] and then by R. S. Strichartz [**572, 573**], M. L. Lapidus [**320, 322–324, 326, 329–331**], M. L. Lapidus and M. van Frankenhuijsen [**369–371, 373, 374**] and [**375**], the first author (M. L. Lapidus) and his collaborators—including J. Kigami, B. Hambly, E. P. J. Pearse, S. Winter, H. Lu, N. Lal, R. de Santiago, S. A. Roby, J. A. Rock, K. E. Ellis, M. C. Mackenzie, G. Radunović, C. David, E. K. Voskanian, and D. Žubrinić (see, e.g., [**107–113, 116, 119, 136, 160, 227, 294, 295, 311, 312, 333–338, 347–351, 356–365, 376–378**]), as well as by many other authors (see, e.g., K. J. Falconer [**174, 175**], D. Gatzouras [**209**], J. Kigami [**293**], S. Kombrink et al. [**202, 304, 305**], K. Dettmers et al. [**137**], E. P. J. Pearse et al. [**480, 481**], and C. Sabot [**528–530**]).

We now implement in this context the method of approximation of nonlattice self-similar sets by a sequence of lattice self-similar sets developed by M. L. Lapidus and M. van Frankenhuijsen in [**370, 371, 374, 375**] and used, e.g., in several of the above references, including in [**376–378**].[25]

As defined above, each weight w_j is positive since $0 < r_j < 1$ for each $j = 1, \dots, N$. Also, the scaling ratios of a given self-similar system allow one to determine whether the self-similar system is lattice or nonlattice, as stated in the following lemma (see [**375**, Section 3.4]).

Lemma 4.58. *A self-similar system (or a scaling vector* $\mathbf{r}$*) is nonlattice iff at least one ratio of weights of the form* w_j/w_1 *is irrational. It is lattice, otherwise.*

Lemma 4.59 is a reformulation of the simultaneous Diophantine approximation spelled out in Lemma 3.16 of [**375**],[26] which says that if at least one of a finite set of real numbers $\alpha_1, \dots, \alpha_N$ is irrational, then these numbers can be approximated by rational numbers *with a common denominator*. Thus, one can find integers q such that for each $j = 1, \dots, N$, the product $q\alpha_j$ is within a small distance to the nearest integer.

Lemma 4.59. *Let* $w_1, w_2, \dots, w_N$ *be positive weights such that at least one ratio of weights* w_j/w_1 *is irrational, for some* $j = 2, \dots, N$*. Then, for every* $\delta > 1$*, there exist integers* q *and* $k_1, \dots, k_N$ *such that* $1 \le q < \delta^{N-1}$ *and*[27]

$$|qw_j - k_j \cdot w_1| \le w_1/\delta, \text{ for } j = 1, \dots, N. \tag{4.30}$$

Remark 4.60. Note that $|qw_j - k_j w_1| \neq 0$ when w_j/w_1 is irrational, so that $q \to \infty$, as $\delta \to \infty$.

[25] In these references, the method is developed and implemented for (generalized) self-similar strings—and, in particular, for self-similar sets in $\mathbb{R}$. However, because it relies on the Diophantine approximation of the underlying scaling ratios—and the scaling ratios involved are allowed to be arbitrary numbers in $(0, 1)$ satisfying a mild compatibility condition (see [**375**, Chapter 3])—it can easily be adapted and extended to the higher-dimensional case, namely, to self-similar sets in $\mathbb{R}^m$, for any $m \in \mathbb{N}$.

[26] See, e.g., Theorem 1A and the remark following Theorem 1E of [**539**].

[27] In keeping with Remark 4.56 above, in Lemma 4.59 and its later applications, N should really be replaced by $\widetilde{M}$ (necessarily with $\widetilde{M} \le N$), the number of *distinct* scaling ratios.

The following lemma is an immediate consequence of the approximation procedure of nonlattice vectors (or weights) by lattice vectors (or weights) developed by M. L. Lapidus and M. van Frankenhuijsen in [**370,371,374,375**]; see [**375**, Theorem 3.19] where it is stated significantly more precisely (by taking into account the associated multiplicities, in particular).

Lemma 4.61. *Let* $\mathbf{r} = (r_j)_{j=1}^N$ *be a nonlattice scaling vector. Then, there exists a sequence* $\{\mathbf{r}_M\}_{M=1}^\infty$ *of lattice scaling vectors, each of order* N*, such that* $\mathbf{r}_M \to \mathbf{r}$*, componentwise as* $M \to \infty$.

Proof. Let $\mathbf{r} = (r_j)_{j=1}^N$ be a nonlattice scaling vector and let $\boldsymbol{w} = (w_j)_{j=1}^N$ be the associated weight vector. By Lemma 4.58, at least one of the ratios w_j/w_1 is irrational and, thus, the simultaneous Diophantine approximation in Lemma 4.59 applies to the weights of $\boldsymbol{w}$. Let $M \in \mathbb{N}$. Since $M+1 > 1$, we have by Lemma 4.59 that there exist integers q and $k_1, \ldots, k_N$ such that $1 \leq q < (M+1)^{N-1}$ and $|qw_j - k_j \cdot w_1| \leq w_1/(M+1)$, for $j = 1, \ldots, N$. Dividing by q on both sides of the above inequality yields $|w_j - (k_j \cdot w_1)/q| \leq w_1/(q(M+1)) \leq w_1/(M+1)$, for $j = 1, \ldots, N$. Now, let $\widetilde{w} = w_1/q$ and $\mathbf{r}_M = (r_{M,j})_{j=1}^N$ such that $r_{M,j} = e^{-k_j \cdot \widetilde{w}}$. Observe that $r_{M,j} = e^{-k_j \cdot \widetilde{w}} = (e^{-w_1})^{k_j/q} = ({r_1}^{1/q})^{k_j}$, for $j = 1, \ldots, N$. Hence, $\mathbf{r}_M$ is indeed a scaling vector of order N since $0 < r_{M,j} < 1$. Furthermore, $\mathbf{r}_M$ is lattice since for each $j = 1, \ldots, N$, we have $r_{M,j} = r^{k_j}$, where $r := {r_1}^{1/q}$. Now, $r_{M,j} - r_j = \int_{k_j \cdot \widetilde{w}}^{w_j} e^{-x} dx$. Since w_j and $k_j \cdot \widetilde{w}$ are both positive, we deduce that $|r_{M,j} - r_j| = |\int_{k_j \cdot \widetilde{w}}^{w_j} e^{-x} dx| \leq |w_j - k_j \cdot \widetilde{w}| \leq w_1/(M+1)$, for $j = 1, \ldots, N$. Thus, we have shown that $|r_{M,j} - r_j| \leq w_1/(M+1)$, for $j = 1, \ldots, N$. Hence, for each $j = 1, \ldots, N$ and as $M \to \infty$, we have $|r_{M,j} - r_j| \to 0$. That is, we have constructed a sequence of order N lattice scaling vectors, $\{\mathbf{r}_M\}_{M=1}^\infty$, such that $\mathbf{r}_M \to \mathbf{r}$ componentwise, as desired. □

As developed throughout this section (and, especially, in the sequel to this book, [**119**]), there is an important dichotomy among self-similar systems: A given self-similar system is either *lattice* or *nonlattice*. A first step in understanding this dichotomy is Theorem 4.63 which ensures that a given self-similar set can be approximated as well as one likes by a lattice self-similar system, much as was established in [**371, 374, 375**], where the complex dimensions of nonlattice self-similar strings were shown to have a quasiperiodic structure because (as is shown in loc. cit.) self-similar strings can be approximated by sequences of lattice strings.[28]

One result that stems from the Diophantine approximation stated in Lemma 4.59 is that a given self-similar set can be approximated as well as desired by a *lattice* self-similar set—just as nonlattice self-similar strings can be approximated as well as possible by a sequence of lattice self-similar strings,

[28] At this stage, we advise the reader to refer to Remark 4.56 above.

as is shown in [**371**, **374**, **375**] and will be discussed in [**119**], the sequel to this book. In the higher-dimensional case of self-similar sets (or of self-similar systems) in $\mathbb{R}^m$, this is stated and formally proved in Theorem 4.63 below, with additional help from the following lemma. Both Lemma 4.62 and Theorem 4.63 can be found in [**137**].

Lemma 4.62. *Let $\mathbf{\Phi} = \{\varphi_j\}_{j=1}^N$ be a self-similar system on $\mathbb{R}^m$ and let $A \in \mathbb{K}(\mathbb{R}^m)$. Then, with $\mathcal{J}$ defined as in Section 4.5, the collection of functions $\{\varphi_J\}_{J\in\mathcal{J}}$ is uniformly bounded on A. That is, there exists $c > 0$ such that for all $J \in \mathcal{J}$ and $\boldsymbol{x} \in A$, we have that $|\varphi_J(\boldsymbol{x})| \leq c$, for all $\boldsymbol{x} \in \mathbb{R}^m$.*

Proof. Since $\mathbf{\Phi}$ is a contraction on $(\mathbb{K}(\mathbb{R}^m), d_H)$, by Proposition 4.15, it follows that there exists a unique compact set $F \in \mathbb{K}(\mathbb{R}^m)$ such that $\mathbf{\Phi}^n(A) \to F$, as $n \to \infty$, for any given $A \in \mathbb{K}(\mathbb{R}^m)$. Let $\varepsilon > 0$. Then, there exists $N_0 \in \mathbb{N}$ such that for all $n \geq N_0$, we have $\mathbf{\Phi}^n(A) \subseteq [F]_\varepsilon$. Thus, there are only finitely many integers $k \geq 0$ such that $\mathbf{\Phi}^k(A) \nsubseteq [F]_\varepsilon$. Let

$$c := \sup\{|\boldsymbol{x}| \; : \; \boldsymbol{x} \in \mathbf{\Phi}^k(A) \text{ where } \mathbf{\Phi}^k(A) \nsubseteq [F]_\varepsilon, \text{ or } \boldsymbol{x} \in [F]_\varepsilon\}.$$

Since the sets $\mathbf{\Phi}^k(A)$ and $[F]_\varepsilon$ are compact and are therefore bounded, we have $c < \infty$. This concludes the proof of the lemma. □

We can now state the following result, which extends to the higher-dimensional case (i.e., to self-similar systems in $\mathbb{R}^m$, with $m \geq 1$ arbitrary) and interprets in terms of the Hausdorff metric, aspects of an aforementioned 2003 result (in [**371**]) obtained by the first author and M. van Frankenhuijsen relying on Diophantine approximation and concerning the approximation of nonlattice self-similar strings by sequences of lattice self-similar strings. See [**369**–**371**, **374**] and [**375**, Section 3.4]; see also the further numerical and experimental work by the same authors and E. K. Voskanian in [**376**], alongside with their later work [**377**, **378**] on diffraction measures for the quasiperiodic patterns of the complex dimensions of self-similar strings, also to be discussed in some detail in [**119**].

Theorem 4.63 (Approximation of Nonlattice Self-Similar Sets by Lattice Self-Similar Sets; K. Dettmers et al., [**137**], 2017). *Let $\mathbf{\Phi} = \{\varphi_j\}_{j=1}^N$ be a nonlattice self-similar system on $\mathbb{R}^m$ with attractor $F_{\mathbf{\Phi}}$. Then, there exists a sequence of lattice self-similar systems $\{\mathbf{\Phi}_M\}_{M=1}^\infty$, where $\mathbf{\Phi}_M = \{\varphi_{M,j}\}_{j=1}^N$ with attractor F_M for each $M \in \mathbb{N}$, such that $F_M \to F_{\mathbf{\Phi}}$, in the Hausdorff metric d_H as $M \to \infty$.*

In light of the definition of Hausdorff metric provided in Section 4.1, the proof of Theorem 4.63 follows relatively easily from the various results stated in the present section; see, especially, Lemmas 4.59, 4.61, 4.58, and 4.62. We leave it as a good exercise for the interested reader to assemble a proof of this theorem or else to refer to [**137**].

4.7. Moran's Equation and Similarity Dimension

In this section, we introduce the Open Set Condition (OSC) and the similarity dimension of a self-similar set, both of which will play an important role in the sequel, especially when used in conjunction with Moran's Theorem (Theorem 5.40), to be established in the next chapter in Section 5.4. We begin our discussion with a simple preliminary result.

Theorem 4.64. *Let* $\{r_j\}_{j=1}^N$ *be a finite collection of real numbers such that* $0 < r_j < 1$, *for each* $j = 1, \ldots, N$. *For* $t \in \mathbb{R}$, *let* $f(t) := \sum_{j=1}^N r_j^t$. *Then, there is a unique nonnegative real number* t_0 *such that*[29]

$$f(t_0) = \sum_{j=1}^{N} r_j^{t_0} = 1. \tag{4.31}$$

Furthermore, $0 \leq t_0 \leq m$. *Also,* $t_0 = 0$ *iff* $N = 1$.[30]

Proof. For $t \in \mathbb{R}$ and each $j = 1, \ldots, N$, define $f_j(t) = r_j^t$. Since each function f_j is differentiable and $f = \sum_{j=1}^N f_j$, we have that f is continuous and differentiable. In fact, $f'(t) = \sum_{j=1}^N r_j^t \log(r_j) < 0$, for each $t \in \mathbb{R}$, where f' denotes the derivative of f. Also, we have $f(0) = N \geq 1$ and $\lim_{t \to +\infty} f(t) = 0$. Thus, since f is strictly decreasing, there exists a unique $t_0 \geq 0$ such that $f(t_0) = 1$, and we clearly have $f(0) = 1$ iff $N = 1$. □

In the context of a self-similar system $\boldsymbol{\Phi}$, with scaling vector given by $\mathbf{r} = (r_1, \ldots, r_N)$, and which satisfies the *Open Set Condition* (see Definition 4.65 just below), the equation on the right side of (4.31) is known as the *Moran equation* associated with $\boldsymbol{\Phi}$. The unique nonnegative real number t_0 satisfying the Moran equation (4.31) plays the role of the *similarity dimension*, D_S, of the attractor of $\boldsymbol{\Phi}$; see Subsection 1.3.4 above and Definition 4.67 below. Recall from Theorem 4.64 that $t_0 > 0$ since $N \geq 2$.

Definition 4.65 (*Open Set Condition and Strong Open Set Condition*). A self-similar system (or, more generally, an iterated function system) $\boldsymbol{\Phi} = \{\varphi_j\}_{j=1}^N$ on $\mathbb{R}^m$ is said to satisfy the *Open Set Condition* (OSC, in short) if there exists a nonempty open set $V \subseteq \mathbb{R}^m$ such that $\varphi_j(V) \subseteq V$ for each $j = 1, \ldots, N$, and the images $\varphi_j(V)$ are pairwise disjoint, also for $j = 1, \ldots, N$. Such a set V is called a *feasible* open set. If, in addition, $V \cap F \neq \emptyset$, where F is the attractor of $\boldsymbol{\Phi}$, then $\boldsymbol{\Phi}$ is said to satisfy the *Strong Open Set Condition* (SOSC).

[29] In fact, t_0 is the unique *real* solution of (4.31).

[30] When applied to a self-similar system of scaling vector $\mathbf{r} = (r_1, \ldots, r_N)$, we must have that $N \geq 2$ and, hence, $r_0 > 0$.

It turns out that, in the context of self-similar systems, the Open Set Condition implies the Strong Open Set Condition. This fact is proved in [**536**].

Theorem 4.66. *A self-similar system satisfies the Open Set Condition iff it satisfies the Strong Open Set Condition.*

Definition 4.67 (*Similarity Dimension*). Let $\mathbf{\Phi}$ be a self-similar system that satisfies the Open Set Condition with scaling ratios $\mathbf{r} = (r_1, \dots, r_N)$ and attractor $F_{\mathbf{\Phi}}$. Then, as in Subsection 1.3.4, the *similarity dimension* of $F_{\mathbf{\Phi}}$, denoted D_S, is the unique real solution of the Moran equation associated with $\mathbf{\Phi}$. That is, D_S is the unique nonnegative (in fact, positive, since $N \geq 2$) real number such that $\sum_{j=1}^{N} r_j^{D_S} = 1$.

Remark 4.68. Note that in Definition 4.67, the notion of a similarity dimension is *not* defined for attractors of iterated function systems, in general. Nonetheless, the unique nonnegative (actually, positive) real number t_0 satisfying Moran's equation (4.31) plays a role in understanding the values of various notions of fractal dimensions for attractors of iterated function systems.

Moreover, in the chapters dealing with the Hausdorff and Minkowski dimensions, specifically, Chapter 5 and Chapter 8, respectively, it will be shown that, in the context of a self-similar system that satisfies the Open Set Condition with attractor F, the similarity dimension D_S is actually equal to the Hausdorff, Minkowski, and box-counting dimensions of F; see, especially, Moran's Theorem (Theorem 5.40).

The next example illustrates the Open Set Condition (OSC), along with the closely related notion of a feasible open set, both introduced in Definition 4.65. It shows that all of the classic self-similar fractals (along with their associated self-similar systems), the Cantor set, the Koch curve, the Sierpinski gasket, the Sierpinski carpet, and the Menger sponge, along with the associated self-similar systems—as discussed in the relevant examples of Section 4.3—satisfy the Open Set Condition (OSC)—and hence also the Strong Open Set Condition (in the sense of Definition 4.65), according to Theorem 4.66. In each case, we will indicate an appropriate feasible open set V, also in the sense of Definition 4.65.

Example 4.69 (*Classic Self-Similar Sets and OSC*). The following classical self-similar sets (and the associated indicated self-similar systems) satisfy the Open Set Condition (OSC).

(a) (*Cantor Set*). Let $\mathbf{\Phi}_C = \{\varphi_1, \varphi_2\}$ be the self-similar system of $\mathbb{R}$ defined in formula (4.11) and of which the (ternary) Cantor set C is the attractor (or invariant set); see Example 4.23 of Section 4.3.

Let $V_j := \varphi_j(V)$, for $j = 1, 2$, where $V := (0, 1)$ is the open unit interval. Then, the open set $V \subseteq \mathbb{R}$ is clearly a feasible open set for $(C, \mathbf{\Phi}_C)$ (or for C, in short).[31] Indeed, $V_1 = (0, 1/3)$ and $V_2 = (2/3, 1)$, each of which is contained in $V = (0, 1)$; furthermore, $V_1 \cap V_2 = \emptyset$.[32] It follows that C—or, rather, $(C, \mathbf{\Phi}_C)$—satisfies the OSC.

(b) (*Koch Curve*). Consider the Koch curve, K_C, viewed as a self-similar set and with associated self-similar system $\mathbf{\Phi}_{KC} = \{\varphi_j\}_{j=1}^4$, as defined in (4.14); see Example 4.31 of Section 4.3.

At this stage, we invite the reader to review the construction of the Koch snowflake curve $K_S = \partial\Omega_{KS}$ described in Example 1.8 of Section 1.2, as well as of the Koch curve K_C given in Example 4.31 in Section 4.3. Let $V := T(A, B, D)$ be the interior of the (equilateral) triangle with vertices A, B, and D given by $A := (0, 0)$, $B := (1/2, \sqrt{3}/2)$, and $D := (1, 0)$. Then, with an analogous notation, we have that, for $j = 1, \ldots, 4$, $V_j := \varphi_j(V) = T(A_j, B_j, D_j)$, the open (equilateral) triangle with vertices respectively given by

$$
\begin{aligned}
&A_1 := (0,0),\ B_1 := \left(\frac{1}{6}, \frac{\sqrt{3}}{6}\right),\ D_1 := \left(\frac{1}{3}, 0\right),\\
&A_2 := \left(\frac{1}{3}, 0\right),\ B_2 := \left(\frac{1}{6}, \frac{\sqrt{3}}{6}\right),\ D_2 := \left(\frac{1}{2}, \frac{\sqrt{3}}{6}\right),\\
&A_3 := \left(\frac{1}{2}, \frac{\sqrt{3}}{6}\right),\ B_3 := \left(\frac{5}{6}, \frac{\sqrt{3}}{6}\right),\ D_3 := \left(\frac{2}{3}, 0\right),\\
&A_4 := \left(\frac{2}{3}, 0\right),\ B_4 := \left(\frac{5}{6}, \frac{\sqrt{3}}{6}\right),\ D_4 := (1, 0).
\end{aligned}
\tag{4.32}
$$

Clearly, in light of Figure 4.10, we have that $V_j \subseteq V$, for $j = 1, \ldots, 4$, and the elements of the family $\{V_j\}_{j=1}^4$ are pairwise disjoint. Indeed, the latter statement follows from the fact that any two of the closed triangles $\{\overline{V_j} = \overline{T}(A_j, B_j, D_j)\}_{j=1}^4$ are either disjoint or intersect at either one or two common vertices; see again Figure 4.10, which the reader may wish to complete by representing the points A_j, B_j, D_j, for $j = 1, \ldots, 4$. It follows that V is a feasible open set for $(K_C, \mathbf{\Phi}_{KC})$ and that the Koch curve (or, rather, $(K_C, \mathbf{\Phi}_{KC})$) satisfies the OSC.

(c) (*Sierpinski Gasket*). Let $\mathbf{\Phi}_{SG} = \{\varphi_j\}_{j=1}^3$ be the self-similar system of $\mathbb{R}^2$ given in formula (4.15) and of which the Sierpinski gasket **SG** is the attractor (when viewed as a self-similar set); see Example 4.33.

Let $V_j := \varphi_j(V)$, for $j = 1, 2, 3$, where $V := T(A, B, C)$ is the initial open (equilateral) triangle, with vertices A, B, D (given exactly as in part (b)), in the

[31] *Caution*: The fact that the attractor $F_{\mathbf{\Phi}}$ of a given self-similar system (or, more generally, IFS) satisfies the OSC depends on the choice of the associated self-similar system (or, more generally, of the associated IFS). An example illustrating this fact is provided towards the end of Example 4.71 below.

[32] Observe that V_1 and V_2 are nothing but the interiors of the two remaining closed intervals (of length 1/3) obtained in the first step of the construction of the Cantor set C by repeated "middle-third removals"; see Example 1.1, along with the accompanying Figure 1.2, in Subsection 1.2.1.

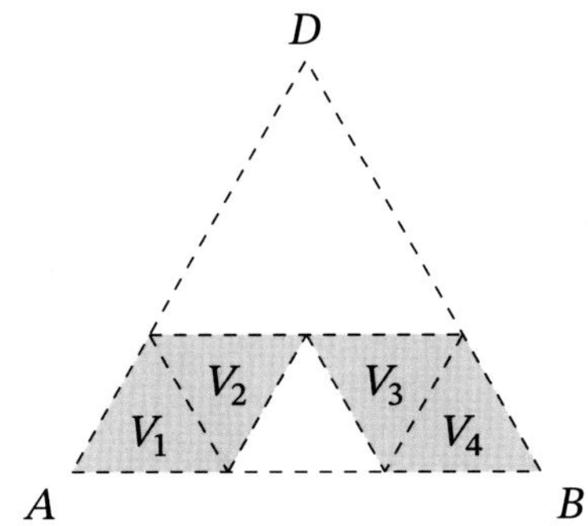

Figure 4.10. A feasible open set V (the larger open equilateral triangle $V := T(A, B, D)$) for the Koch curve K_C, along with its images $V_j := \varphi_j(V)$, for $j = 1, \ldots, 4$, under the elements φ_j of the self-similar system $\mathbf{\Phi}_{KC} = \{\varphi_j\}_{j=1}^4$.

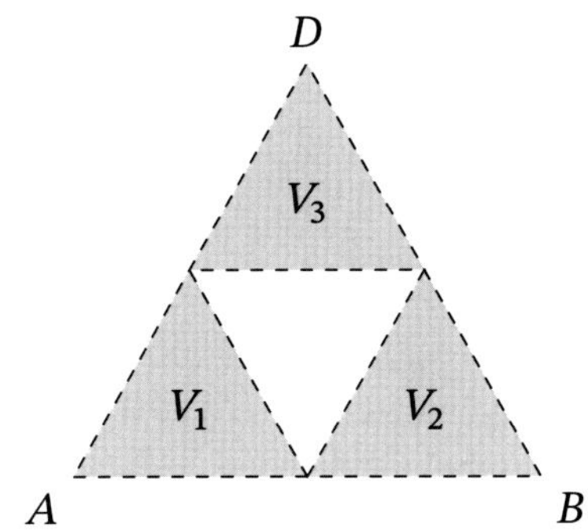

Figure 4.11. A feasible open set V (the initial and larger open equilateral triangle $V := T(A, B, D)$) for the Sierpinski gasket **SG**, along with its images $V_j := \varphi_j(V)$, for $j = 1, 2, 3$, under the elements φ_j of the self-similar system $\mathbf{\Phi}_{SG} = \{\varphi_j\}_{j=1}^3$.

construction of the Sierpinski gasket described in Example 1.9 of Section 1.2.[33] Then, each set V_j (for $j = 1, 2, 3$) is an open (equilateral) triangle. Also, clearly, $V_j \subseteq V$ (still for $j = 1, 2, 3$). Furthermore, the closures $\overline{V_j}$ are either pairwise disjoint or intersect at one (boundary) point, their common vertex; see Figure 4.11. It follows that V is a feasible open set for **SG**—or, rather, for $(\mathbf{SG}, \mathbf{\Phi}_{SG})$—and hence that the Sierpinski gasket **SG** satisfies the OSC.

[33]Note that V_1, V_2, V_3 are just the interiors of the three closed equilateral triangles (of side lengths $1/2$) obtained in the first step of the construction of the Sierpinski gasket by repeated removal of (open) equilateral triangles of smaller and smaller sizes; see Example 1.9, along with the accompanying Figure 1.6 in Subsection 1.2.3.

(d) (*Sierpinski Carpet*). Let

$$\mathbf{\Phi}_{SC} = \{\varphi_{j_1j_2}\}_{j_1,j_2\in\{0,1,2\},(j_1,j_2)\neq(1,1)} \tag{4.33}$$

denote the self-similar system of $\mathbb{R}^2$, of which the Sierpinski carpet **SC** is the attractor (when viewed as a self-similar set of $\mathbb{R}^2$); see Example 4.34 of Section 4.3.

Let $V_{j_1j_2} = \varphi_{j_1j_2}(V)$, for $j_1, j_2 \in \{0,1,2\}$, with $(j_1, j_2) \neq (1,1)$, where $V := (0,1)^2$ is the open unit square of $\mathbb{R}^2$.[34] Note that $V_{j_1j_2} \subseteq V$, for all j_1, j_2 as above, and that the open squares in the family $\{V_{j_1j_2}\}_{(j_1,j_2)\neq(1,1)}$ are obviously pairwise disjoint. It follows that $V = (0,1)^2$ is a feasible open set for **SC**—or, rather, for $(\mathbf{SC}, \mathbf{\Phi}_{SC})$—and hence that the Sierpinski gasket **SC** satisfies the OSC.

(e) (*Menger Sponge*). This example is entirely analogous to that of the Sierpinski carpet **SC** discussed in part (d) just above. It suffices to replace the self-similar system $\mathbf{\Phi}_{SC}$ for **SC** by its counterpart $\mathbf{\Phi}_{MS}$ for **MS**, i.e., having the self-similar set **MS** as its attractor; see Example 4.35 of Section 4.3 above, along with Figure 9.26 in Subsection 9.6.2 below. We must also replace the open unit square $(0,1)^2$ of $\mathbb{R}^2$ by the open unit cube $V := (0,1)^3$ of $\mathbb{R}^3$. Then, $\mathbf{\Phi}_{MS}(V) := \{\varphi_{j_1j_2}(V)\}$—where $j_1, j_2 \in \{0,1,2\}$ and $(j_1, j_2) \neq (1,1)$—comprises $27-7 = 20$ open and pairwise disjoint cubes (of side lengths 1/3); furthermore, each element of the family $\mathbf{\Phi}_{MS}(V)$ is contained in V. Therefore, $V := (0,1)^3$ is a feasible open set for **MS**—or, rather, for $(\mathbf{MS}, \mathbf{\Phi}_{MS})$—and hence the Menger sponge **MS** satisfies the Open Set Condition (OSC).

Example 4.70. Draw a figure depicting and labeling the disjoint open squares (resp., cubes) of $\mathbb{R}^2$ (resp., of $\mathbb{R}^3$) of which $\mathbf{\Phi}_{SC}(V)$ (resp., $\mathbf{\Phi}_{MS}(V)$) is comprised; see part (d) (resp., (e)) of Example 4.69 just above.

Example 4.71 (*OSC and Similarity Dimension*). Each of the self-similar systems discussed in the examples of Section 4.3, except for $\mathbf{\Phi}_{I,1}$ in Example 4.37, satisfies the Open Set Condition (OSC). We refer to Example 1.19 for the similarity dimensions of the unit interval $[0,1]$, the unit square $[0,1]^2$, the unit cube $[0,1]^3$, the Cantor set C, the Koch snowflake curve $K_S = \partial\Omega_{KS}$, the Sierpinski gasket **SG**, and the Sierpinski carpet **SC**; see, especially, Example 4.69 above.

In the case of the Menger sponge **MS**, the self-similar system $\mathbf{\Phi}_{\mathbf{MS}}$ satisfies the OSC with the open unit cube $V := (0,1)^3$ as a feasible open set, and each

[34]Observe that $\{V_{j_1j_2}\}_{(j_1,j_2)\neq(1,1)}$ consists of the interiors of all the eight closed squares (of side lengths 1/3) obtained after the first step of the construction of the Sierpinski carpet **SC** by repeated deletion of open squares; see Example 1.10, along with Figure 1.7, in Subsection 1.2.3.

of its 20 contracting similarities has scaling ratio 1/3; see part (e) of Example 4.69. Hence, $2 < D_S = \log_3 20 < 3$.

Now, the unit interval $I := [0,1]$ is the attractor of the three distinct self-similar systems found in Example 4.37 given by $\mathbf{\Phi}_I, \mathbf{\Phi}_{I,0}$, and $\mathbf{\Phi}_{I,1}$. We have that $\mathbf{\Phi}_I$ and $\mathbf{\Phi}_{I,0}$ both satisfy the OSC with the open interval $V := (0,1)$ as a feasible open set. Hence, they both yield a similarity dimension for $I = [0,1]$. Specifically, for $\mathbf{\Phi}_I$ the Moran equation is given by $\sum_{j=1}^{10}(1/10)^t = 10\,(1/10)^t = 1$, $t \in \mathbb{R}$; hence, $D_S = 1$. For $\mathbf{\Phi}_{I,0}$, the Moran equation is given by $\sum_{j=1}^{3} r_j^t = (1/6)^t + (1/3)^t + (1/2)^t = 1$, $t \in \mathbb{R}$; thus, $D_S = 1$ in this case as well. It turns out that every self-similar set is the attractor of an infinite number of distinct self-similar systems. However, for a given self-similar set and among such self-similar systems, the ones that satisfy the OSC each yield the same value as the corresponding similarity dimension. This fact will be proved as a consequence of Moran's Theorem (Theorem 5.40).

On the other hand, the attractor of the self-similar system $\mathbf{\Phi}_{I,1}$ does *not* satisfy the OSC. Note that $\mathbf{\Phi}_{I,1}$ comprises 4 contracting similarities, each with scaling ratio 1/2; thus, $t_0 = 2$ is the unique solution to the equation $\sum_{j=1}^{4}(1/2)^t = 4\,(1/2)^t = 1$, $t \in \mathbb{R}$. Hence, a condition such as the OSC is required in the definition of the similarity dimension if it is desired, for a self-similar set, that this value depend on the set but *not* on a particular choice of self-similar system.

Exercise 4.72.

(a) Show that the closed unit cube $I = Q_0 := [0,1]^m$ in $\mathbb{R}^m$—viewed as the self-similar set which is the attractor of the self-similar system $\mathbf{\Phi}_{Q_0}$ defined in Example 4.95 of Section 4.10 below—satisfies the OSC.

(b) Show that the unit interval $I := [0,1]$—viewed as the self-similar set which is the attractor of the self-similar system $\mathbf{\Phi}_{I,1}$ defined in Example 4.37—does *not* satisfy the OSC.

The following Strong Separation Condition in Definition 4.73 should be compared with the Open Set Condition (OSC) and the Strong Open Set Condition (SOSC) introduced in Definition 4.65 above. We leave it to the interested reader to compare these three conditions.

Definition 4.73 (*Strong Separation Condition*). Let $\mathbf{\Psi} = \{\psi_j\}_{j=1}^N$ denote an iterated function system on $\mathbb{R}^m$ with attractor F. Then, $\mathbf{\Psi}$ and F are said to be strongly separated if the images of F under the contracting similarities ψ_j are pairwise disjoint and, moreover, if $0 < \eta < \infty$, where

$$\eta := \sup\{\alpha : \|\boldsymbol{x} - \boldsymbol{y}\|_m > \alpha, \forall \boldsymbol{x} \in \psi_j(F), \boldsymbol{y} \in \psi_k(F), j \neq k, j,k = 1,\dots,N\}.$$

Remark 4.74. The value of η in Definition 4.73 is equivalently given by $\eta = \inf\{\|\boldsymbol{x} - \boldsymbol{y}\|_m : \boldsymbol{x} \in \psi_j(F), \boldsymbol{y} \in \psi_k(F), j \neq k, j, k = 1, \ldots, N\}$; i.e., η is the minimum distance between any pair of the (pairwise disjoint) compact sets $\psi_1(F)$, $\ldots, \psi_N(F)$. Also, by part (viii) of Theorem 4.48, F is totally disconnected when F is strongly separated. Additionally, if $\boldsymbol{\Psi}$ is strongly separated, then it satisfies the OSC.

Example 4.75. The Cantor set C and the 1-dimensional fractal F_1 are both strongly separated. However, none of the other self-similar sets discussed in Example 4.71 is strongly separated. For the Cantor set C, $\delta = 1/3$, and for F_1, $\delta = 1/2$.

4.8. Self-Similar Measures

In Section 4.2, the existence of a unique attractor for a given iterated function system is guaranteed by the Contraction Mapping Principle (Theorem 2.58). Moreover, attractors are in fact constructed by following the proof of this theorem, as established in Theorem 4.17. Another useful application of the Contraction Mapping Principle is the guarantee of the existence of *self-similar measures*. In order to construct these measures, the corresponding metric space and contractions need to be defined. Also, the contractions of interest build upon self-similar systems in a natural way.

Notation 4.76. Let $\mathfrak{M}$ denote the set of (regular) Borel measures μ on $\mathbb{R}^m$ with bounded support (in the sense of Definition 3.42) and finite mass (i.e., $\mu(\mathbb{R}^m) < \infty$). Let $\mathfrak{M}^1$ denote the subset of $\mathfrak{M}$ comprising measures μ with mass 1 (i.e., $\mu(\mathbb{R}^m) = 1$). Hence, $\mathfrak{M}^1$ consists of probability measures.

The following metric is called the *Monge–Kantorovich metric* (or the *Wasserstein metric* [**605**]). It was introduced by L. V. Kantorovich and G. Sh. Rubinstein in 1958 in [**285**] (see also Kantorovich's work in [**283, 284**]) and, much later, it was used by Hutchinson in [**261**, Section 4.3] and elsewhere in the literature in the present context of weighted IFSs.

Definition 4.77. The *L-metric* is the metric on $\mathfrak{M}^1$ given for any $\mu, \nu \in \mathfrak{M}^1$ by

$$L(\mu, \nu) = \sup\left\{\int_{\mathbb{R}^m} f d\mu - \int_{\mathbb{R}^m} f d\nu : f : \mathbb{R}^m \to \mathbb{R}, \mathrm{Lip}(f) \leq 1\right\}. \tag{4.34}$$

Exercise 4.78. Show that L is indeed a metric on $\mathfrak{M}^1$.

The next result is key to being able to apply the Contraction Mapping Principle.

Theorem 4.79. *The metric space $(\mathfrak{M}^1, L)$ is complete.*

Definition 4.80 (*Weighted Iterated Function System*). A *weighted iterated function system* (*weighted* IFS) is a pair $(\mathbf{\Phi}, \mathbf{p})$, where $\mathbf{\Phi}$ is an iterated function system that satisfies the Open Set Condition and $\mathbf{p} = (p_1, \dots, p_N)$ is a probability vector (i.e., $0 \le p_j \le 1$, for $j = 1, \dots, N$, and $\sum_{j=1}^{N} p_j = 1$).

As with iterated function systems, by a mild abuse of notation, a pair $(\mathbf{\Phi}, \mathbf{p})$ also denotes a function which happens to be a contraction; see Remark 4.83 below for the definition of the resulting map.

Proposition 4.81. $(\mathbf{\Phi}, \mathbf{p})$ *is a contraction on* $(\mathfrak{M}, L)$.

Theorem 4.82 (Existence and Uniqueness of a Self-Similar Measure Associated with a Weighted IFS). *Given an IFS* $\mathbf{\Phi}$ *and a probability vector* $\mathbf{p} = (p_1, \dots, p_N)$ *(or, equivalently, a weighted iterated function system* $(\mathbf{\Phi}, \mathbf{p})$ *as in Definition* 4.80*), there exists a unique fixed point* μ *of the contraction map* $(\mathbf{\Phi}, \mathbf{p})$ *associated with* $\mathbf{\Phi}$ *and* $\mathbf{p}$ *and acting on* $\mathfrak{M}^1$:

$$(\Phi, \mathbf{p})(\mu) = \mu; \tag{4.35}$$

i.e., μ *is the unique probability measure (i.e.,* $\mu \in \mathfrak{M}^1$*) which satisfies the* Self-Similarity Equation

$$\mu = \sum_{j=1}^{N} p_j \cdot \mu \circ \varphi_j^{-1}, \tag{4.36}$$

where $\mathbf{\Phi} := \{\varphi_j\}_{j=1}^{N}$ *and for each* $j = 1, \dots, N$, $\mu \circ \varphi_j^{-1} \in \mathfrak{M}^1$ *is the probability measure induced by the map* φ_j *and the probability measure* $\mu \in \mathfrak{M}^1$*:*[35]

$$(\mu \circ \varphi_j^{-1})(B) := \mu(\varphi_j^{-1}(B)), \tag{4.37}$$

for all $B \in \mathcal{B}(\mathbb{R}^m)$*, where* $\mathcal{B}(\mathbb{R}^m)$ *denotes the Borel* σ*-algebra of* $\mathbb{R}^m$.

Proof. In light of Theorem 4.79 and Proposition 4.81, this result follows at once from the Contraction Mapping Principle (Theorem 2.58). □

Remark 4.83. The (contraction) map $(\mathbf{\Phi}, \mathbf{p})$ is defined as follows, as a self-map of $\mathfrak{M}^1$:

$$(\mathbf{\Phi}, \mathbf{p})(\nu) = \sum_{j=1}^{N} p_j \cdot \nu \circ \varphi_j^{-1}, \text{ for all } \nu \in \mathfrak{M}^1, \tag{4.38}$$

where $\mathbf{\Phi} = \{\varphi_j\}_{j=1}^{N}$ and $\mathbf{p} = (p_1, \dots, p_N)$, as in Definition 4.80 and Theorem 4.82. Here, for each $j = 1, \dots, N$, $\nu \circ \varphi_j^{-1}$ is defined exactly as in (4.37).

[35]In measure theory, this map is called the *image measure* of μ under the map φ_j.

Definition 4.84 (*Self-Similar Measure*). The unique (Borel) probability measure defined in Theorem 4.82 by the fixed point equation (4.35)—or, equivalently, by the Self-Similarity Equation (4.36)—is called the *self-similar measure associated with the weighted iterated function system* $(\mathbf{\Phi}, \mathbf{p})$ (as given by Definition 4.80) or, equivalently, the *self-similar measure associated with the IFS* $\mathbf{\Phi} = \{\varphi_j\}_{j=1}^N$ *and with probability vector* $\mathbf{p} = (p_1, \dots, p_N)$.

In several of the following exercises, namely, in Exercises 4.85 and 4.86, the interested reader is asked to provide the omitted proofs of the main results needed in order to establish Theorem 4.82, which is the key result of the present section. The reader is invited to review Section 4.2 where the analogous results are fully proved in the case of the Hausdorff metric d_H acting on the space $\mathbb{K}(\mathbb{R}^m)$, the space of nonempty compact subsets of $\mathbb{R}^m$.

Exercise 4.85. Show that $(\mathfrak{M}^1, L)$ is a complete metric space; i.e, prove Theorem 4.79.

Exercise 4.86. Prove Proposition 4.81, according to which the self-map $(\mathbf{\Phi}, \mathbf{p})$ of $\mathfrak{M}^1$ is a contraction on the (complete) metric space $(\mathfrak{M}^1, L)$.

Exercise 4.87 (*Hausdorff vs. Monge–Kantorovich Metric*). Let $\mathcal{E}$ denote the subset of $\mathfrak{M}^1$ consisting of all Dirac measures $\delta_{\boldsymbol{x}}$, where $\boldsymbol{x} \in \mathbb{R}^m$ is arbitrary. Then, consider the restriction to $\mathcal{E}$, denoted by $L_{\mathcal{E}}$, of the Monge–Kantorovich metric L on $\mathfrak{M}^1$.

Compare $L_{\mathcal{E}}$ and the Hausdorff metric d_H on $\mathbb{R}^m$ (as defined in Section 4.1 above). More specifically, determine whether or not the metric spaces $(\mathbb{R}^m, d_H)$ and $(\mathcal{E}, L_{\mathcal{E}})$ are bijectively isometric.

Exercise 4.88 (*Self-Similar Measures and Sets*). Under the hypotheses and with the notation of Theorem 4.82, show that the support of the self-similar measure μ coincides with the attractor (or the self-similar set) F associated with the self-similar system $\mathbf{\Phi}$ defining μ: $\operatorname{supp}(\mu) = F$.

Exercise 4.89 (*Self-Affine Measures and Beyond*). Extend Definition 4.84 and Theorem 4.82 to self-affine measures (rather than self-similar measures), as well as, more generally to invariant measures associated with arbitrary IFSs.

Remark 4.90. A sample of references on (or related to) self-similar measures and their natural generalizations includes [**72, 109, 110, 136, 160, 172, 175, 261, 283–285, 292–295, 303, 304, 311, 312, 323, 324, 332, 366, 370, 372, 374, 375, 380, 486, 487, 528–530, 572–575, 586, 587**], along with the collective volume [**35**] in which, in particular, the role of Bernoulli measures (important examples of self-similar measures) in symbolic dynamics and ergodic theory is discussed in some detail.

4.9. Perfect, Nowhere Dense Sets

We refer to D. Bressoud's book, [**57**], for a characterization of perfect, nowhere dense subsets of $\mathbb{R}$ (see [**57**, Theorem 4.2]), as well as for motivations for the introduction and the study of Smith–Volterra–Cantor sets (SVC sets), to be discussed in Example 4.92 below.

Some readers may wish to review the statement of Theorem 2.43 in Section 2.2 about the well-known decomposition of the open subsets of $\mathbb{R}$ as (at most) countable disjoint unions of open intervals.

Theorem 4.91. *Let $S \subseteq \mathbb{R}$ be a nonempty, bounded, perfect, nowhere dense set. Let $a := \inf S$ and $b := \sup S$. Then, there is a countably infinite collection of open intervals $\{I_j\}_{j\in\mathbb{N}}$ such that $I_j \subseteq [a,b]$ for each $j \in \mathbb{N}$, S is uncountable, and S is the set of accumulation points of the set of endpoints of the intervals I_j.*

Proof. Since S is perfect, it contains its accumulation points and, in particular, $a, b \in S$. It is not difficult to check that $S^c \cap [a,b]$, the complement of S in $[a,b]$, is an open set comprising a countable union of disjoint open intervals; see Theorem 2.43. Let $\{I_j\}_{j\in\mathbb{N}}$ denote an ordering of this countable collection of open intervals. Since S is perfect, every point of S is an accumulation point. Thus, a and b are not endpoints of any of the intervals I_j and, in fact, a right endpoint of one interval I_j is never the left endpoint of another. So, if $S^c \cap [a,b]$ comprised only finitely many intervals, then S would have to be a closed interval of positive length. However, since S is nowhere dense, $S^c \cap [a,b]$ must be a union of countably many open intervals.

Now, since S is perfect, each endpoint of each I_j is an accumulation point of S. So, since S is nowhere dense, if $x \in S$, then every neighborhood of x contains at least one of the intervals I_j and, therefore, at least one of the endpoints of such an interval. Hence, x is an accumulation point of the set of endpoints of the intervals I_j.

To show that S is uncountable, it suffices to find an injective map from $2^{\mathbb{N}}$, the set of sequences of 0's and 1's, into S. Let r_j and l_j denote the right and left endpoints, respectively, of I_j for each $j \in \mathbb{N}$. Let $\{y_n\} \in 2^{\mathbb{N}}$ and let $[a_0, b_0] := [a,b]$. Given $[a_{n-1}, b_{n-1}]$, for $n \in \mathbb{N}$, let I_{j_n} denote the first open interval in the ordered collection $\{I_j\}_{j\in\mathbb{N}}$ which is contained in $[a_{n-1}, b_{n-1}]$. Let

$$[a_n, b_n] := \begin{cases} [a_{n-1}, l_{j_n}], & \text{if } y_n = 0, \\ [r_{j_n}, b_{n-1}], & \text{if } y_n = 1. \end{cases} \tag{4.39}$$

Then, for each $n \in \mathbb{N}$, either $a_n = a$ or $a_n = r_{j_n}$ and likewise either $b_n = b$ or $b_n = l_{j_n}$. Hence, $a_n \neq b_n$ and $I_{j_n} \subsetneq (a_n, b_n)$, for each $n \in \mathbb{N}$. So, the intervals $[a_n, b_n]$ are nested (i.e., $[a_n, b_n] \subseteq [a_{n-1}, b_{n-1}]$, for each $n \in \mathbb{N}$) and the right endpoints b_n form a bounded, nonincreasing sequence that converges to some $y \in S$. Define $f : 2^{\mathbb{N}} \to S$ by $f(\{y_n\}) = y$.

It remains to show that f is injective. Let $\{x_j\}, \{y_j\} \in S$ such that $\{x_j\} \neq \{y_j\}$ and let $k \in \mathbb{N}$ denote the first index for which $x_k \neq y_k$. If $y_k = 0$, then $f(\{y_j\}) < l_{k_n}$, and if $y_k = 1$, then $f(\{y_j\}) > l_{k_n}$. In either case, $f(\{y_j\}) \neq f(\{x_j\})$ and, therefore, f is injective. □

The classic (ternary) Cantor set C (see, e.g., Example 1.1 in Subsection 1.2.1 and Example 1.6 in Subsection 1.2.3) is an example of a nonempty, perfect, nowhere dense subset of $\mathbb{R}$. As noted in Example 4.23, the Cantor set C is also the attractor of a self-similar system. It turns out that iterated function systems provide a method for constructing a wide variety of nonempty, perfect, nowhere dense sets with fractal structure, but there are also other methods for doing so. The Smith–Volterra–Cantor sets described below provide such a method which essentially generalizes the "middle-third removal" construction of the Cantor set C described in Example 1.1.

Example 4.92 (*Smith–Volterra–Cantor Sets*). Let $n \in \mathbb{N}$ be such that $n \geq 3$. Mimic the construction of the Cantor set C from Example 1.1 as follows:

(i) Remove an open interval of length $1/n$ centered at $1/2$ from the unit interval $S_0 := [0, 1]$, leaving a compact set S_1 given by

$$S_1 = [0, 1/2 - 1/2n] \cup [1/2 + 1/2n, 1]. \tag{4.40}$$

(ii) From the center of each of the two connected components of S_1, remove the open interval of length $1/n^2$ (i.e., 2 intervals of length $1/n^2$ are removed). This leaves a compact set S_2 comprising $2^2 = 4$ connected components of equal length.

(iii) Repeating the process in a recursive fashion shows that for each positive integer $k \geq 1$, S_k comprises 2^k connected components. Then S_{k+1} is constructed by removing an open interval of length $1/n^{k+1}$ from the center of each of the connected components of S_k.

(iv) For this $n \in \mathbb{N}$, the *Smith–Volterra–Cantor set* $SVC(n)$ is the nonempty compact set defined by the intersection of all the nonempty compact sets S_k constructed above; i.e., $SVC(n) := \bigcap_{k \in \mathbb{N}} S_k$. Observe that for $n = 3$, $SVC(3) = C$, the classic (ternary) Cantor set.

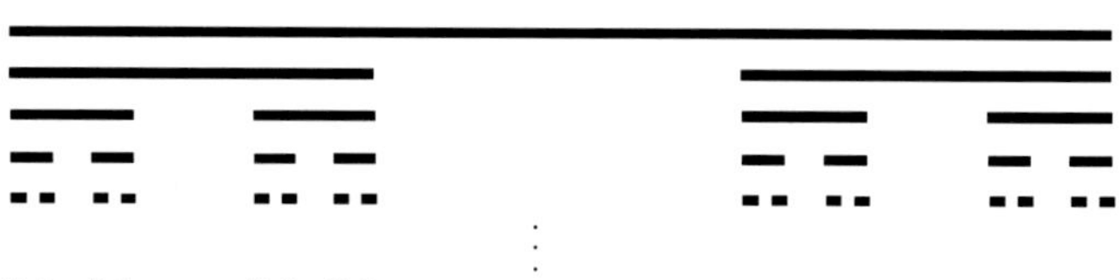

Figure 4.12. The construction of the Cantor set C by repeated "middle-third removal", just as in Figure 1.2 in Subsection 1.2.1.

(v) For each $n \geq 4$, and by the well-known Theorem on Nested Intervals (with decreasing lengths tending to zero), $SVC(n)$ is a nonempty, compact, perfect, nowhere dense subset of $[0,1]$ *with positive length*; see Figure 4.12 for the case of the classic (ternary) Cantor set C. In fact, the open set of $\mathbb{R}$ given by $\Omega_{(n)} := [0,1]\backslash SVC(n)$ is an ordinary fractal string with the following associated fractal string $\mathcal{L}_{(n)}$:

$$\mathcal{L}_{(n)} := \left\{ l_k := \frac{1}{n^k} : l_k \text{ occurs with multiplicity } 2^{k-1}, k \in \mathbb{N} \right\}.$$

As such, the length (more specifically, the 1-dimensional Lebesgue measure) of $SVC(n)$ is 1 minus the sum of the lengths of the removed intervals. This length is therefore given by

$$\mathscr{L}^1(SVC(n)) = 1 - \sum_{k=1}^{\infty} \frac{2^{k-1}}{n^k} = 1 - \frac{1/n}{1-(2/n)} = \frac{n-3}{n-2}. \tag{4.41}$$

Note that the series in (4.41) is, in fact, a geometric series; so, Lemma 2.57 applies and the equations above hold. Also, the 1-dimensional Lebesgue measure of the ordinary fractal string $[0,1]\backslash SVC(n)$ is given by evaluating its geometric zeta function $\zeta_{\mathcal{L}_{(n)}}(s)$ at $s = 1$ (see Definition 1.26). That is,

$$\mathscr{L}^1([0,1]\backslash SVC(n)) = \sum_{k=1}^{\infty} \frac{2^{k-1}}{n^k} = \frac{n-3}{n-2} = \zeta_{\mathcal{L}_{(n)}}(1). \tag{4.42}$$

4.10. Additional Examples

In this section, we discuss several additional examples of (possibly generalized) attractors of IFSs, including, especially, the Devil's staircase (or Cantor graph) which is neither self-similar nor self-affine, in the usual sense. It is, however, a self-affine set in a suitably extended sense, as we see next.

Example 4.93 (*Devil's Staircase: A Key Example of an Inhomogeneous Self-Affine Set*). The Devil's staircase $\mathbf{\Gamma}_{DS}$ (or graph of the Cantor function f_{DS})—introduced in Subsection 1.4.1 above (see, especially, Example 1.32 and Exercises 1.33–1.35) as well as to be studied in more detail in Section 9.4 below—is a very important example of a fractal set which is *not* self-similar (in the strict sense of the term; compare to Definition 4.24 in Section 4.3 above) but is instead self-affine—in a slightly more extended sense than in Definition 4.39 in Section 4.4 above. Namely, $\mathbf{\Gamma}_{DS}$ is an *inhomogeneous self-affine set*, as we now explain, following the discussion provided in the introduction of [**356**].

First of all, just like an ordinary self-affine set, the Devil's staircase exhibits a striking dissymmetry between the horizontal and the vertical axes.[36] More specifically, the scaling rate in the vertical axis is equal to 2^{-n}, at the n-th step

[36] Here and henceforth, we paraphrase the beginning of Remark 1.2.1, page 26, in [**356**].

of the construction of $\mathbf{\Gamma}_{DS}$ (or of f_{DS}), whereas the scaling rate in the horizontal direction is equal to 3^{-n} also, at the n-th step of the construction. "Since $3^{-n}/2^{-n} \to 0$, as $n \to \infty$, there is a dramatic change in the elongation in the vertical direction near the points of the graph of f_{DS} (i.e., the Devil's staircase $\mathbf{\Gamma}_{DS}$), corresponding to the points of the middle-third Cantor set C." (See loc. cit.) This is a reflection of the fact that $\mathbf{\Gamma}_{DS}$ is self-affine, albeit in a somewhat generalized sense.

Secondly, as was alluded to just above, the Devil's staircase (or graph of the Cantor function f_{DS}) is an *inhomogeneous self-affine set* (but *not* an inhomogeneous self-similar set), in the following precise sense (see Figure 1.8 in Subsection 1.4.1):[37,38]

$$\mathbf{\Gamma}_{DS} = \mathbf{\Phi}_{DS}(\mathbf{\Gamma}_{DS}) \cup \widetilde{\mathbf{\Gamma}}_{DS}, \quad \text{where} \tag{4.43}$$

$$\mathbf{\Phi}(\mathbf{\Gamma}_D S) = \boldsymbol{M}\mathbf{\Gamma}_{DS} \cup \left(\boldsymbol{M}\mathbf{\Gamma}_{DS} + \left(\frac{2}{3}, \frac{1}{2}\right)^T\right), \quad \text{with} \tag{4.44}$$

$$\boldsymbol{M} := \begin{pmatrix} 1/3 & 0 \\ 0 & 1/2 \end{pmatrix}, \quad \boldsymbol{M}\mathbf{\Gamma}_{DS} := \{\boldsymbol{M}\boldsymbol{x} : \boldsymbol{x} \in \mathbf{\Gamma}_{DS}\}, \tag{4.45}$$

and the line segment $\widetilde{\mathbf{\Gamma}}_{DS} := [1/3, 2/3] \times \{1/2\}$ is the *nonhomogeneous part* of the *Inhomogeneous Self-Affinity Equation* (4.43). Furthermore, for example, $\boldsymbol{M}\mathbf{\Gamma}_{DS}$ in (4.44) and (4.45) just above corresponds to the "left third" of the Devil's staircase $\mathbf{\Gamma}_{DS}$; i.e., it is the compact subset of $\mathbb{R}^2$ (or of $[0,1]^2$, here) given by[39]

$$\begin{aligned} \boldsymbol{M}\Gamma_{DS} &= \left\{\boldsymbol{x} = (x_1, x_2) \in \mathbf{\Gamma}_{DS} : 0 \le x_1 \le \frac{1}{3}\right\} \\ &= \left\{\boldsymbol{x} = (t, f_{DS}(t)) : t \in \left[0, \frac{1}{3}\right]\right\}, \end{aligned} \tag{4.46}$$

and similarly for $\boldsymbol{M}\mathbf{\Gamma}_{DS} + (2/3, 1/3)^T$ appearing on the right-hand side of (4.44). Namely, it corresponds to the "right third" of $\mathbf{\Gamma}_{DS}$:

$$\begin{aligned} \boldsymbol{M}\mathbf{\Gamma}_{DS} + (2/3, 1/3)^T &= \left\{\boldsymbol{x} = (x_1, x_2) \in \mathbf{\Gamma}_{DS} : \frac{2}{3} \le x_1 \le 1\right\} \\ &= \left\{\boldsymbol{x} = (t, f_{DS}(t)) : t \in \left[\frac{2}{3}, 1\right]\right\}. \end{aligned} \tag{4.47}$$

Here, as in elementary geometry and linear algebra, the 2×2 matrix $\boldsymbol{M}$ defined in the left-hand side of (4.45) and appearing in (4.44)–(4.47) above, is called the *affinity matrix*; it is the matrix representing the affinity transformation of $\mathbb{R}^2$ which helps one translate in geometric terms the aforementioned

[37] The notion of an inhomogeneous self-affine set was introduced by M. F. Barnsley and S. Demko in [**30**]; see, e.g., [**197, 198**] for later work on this subject.

[38] Here and thereafter, in (4.44), the notation $(2/3, 1/3)^T$ (the transpose of the row vector $(2/3, 1/3)$) stands for the column vector $\begin{pmatrix} 2/3 \\ 1/2 \end{pmatrix}$.

[39] Note that $\boldsymbol{M}\mathbf{\Gamma}_{DS}$ is obtained from the fractal curve $\mathbf{\Gamma}_{DS}$ by scaling $\mathbf{\Gamma}_{DS}$ by the factor 1/3 in the horizontal direction and then by the factor 1/2 in the vertical direction (or vice versa).

dissymmetric *scaling laws* (or scaling relations) underlying the Devil's staircase $\mathbf{\Gamma}_{DS}$. For more information concerning the scaling relations—along with the associated fundamental equations satisfied by the Cantor function f_{DS}—we refer the interested reader to Subsection 9.4 below, in Chapter 9 on fractal curves and their complex dimensions.

Moreover, for more information concerning inhomogeneous self-affine (as well as inhomogeneous self-similar) sets, we refer to [**356**], pages 26–27 and Remark 2.1.87 of Subsection 2.1.6 of that book, as well as to several other examples discussed throughout [**356**], along with the relevant references therein; see also Exercise 4.94 just below.

Exercise 4.94. Justify all of the statements made in Example 4.93 above. In particular, show by suitably adapting the statement and proof of Theorem 4.17 in Section 4.2 (about the existence and uniqueness of an attractor for a contractive IFS)—that the fixed point equation (4.43) has a unique solution, given by the inhomogeneous self-affine set $\mathbf{\Gamma}_{DS}$. Moreover, attempt to formalize, in the process, the notion of an *inhomogeneous self-affine set*—and, in particular, of an inhomogeneous self-similar set—as well as to justify their existence and uniqueness for a given inhomogeneous self-affine (or self-similar) fixed point equation.

The next two examples (Examples 4.95 and 4.96)—which generalize the trivial case of the unit interval $[0,1]$ to dimension 2 and dimension $m \geq 1$, respectively—show very clearly that not every self-similar set is "fractal", in any reasonable sense. Conventionally, we work with the unit cube and the m-dimensional cube, in Example 4.95 and Example 4.96, respectively, but, of course, entirely analogous results can be obtained for a square or an m-dimensional cube, of arbitrary side lengths $a > 0$.

Example 4.95 (*Unit Square as a Self-Similar Set of* $\mathbb{R}^2$). Let $S := [0,1]^2$ denote the unit square. Then, S is a self-similar set of $\mathbb{R}^2$. More specifically, it is the attractor (or invariant set) of the contractive self-similar system (or IFS) $\mathbf{\Phi}_S = \{\varphi_j\}_{j=0}^3$ consisting of the 4 contractive similarity transformations of $\mathbb{R}^2$ (of common scaling ratios 1/2) given by $\varphi_j(\boldsymbol{x}) = (1/2)\boldsymbol{x} + \boldsymbol{u}_j$, for $j = 0, \ldots,$ 3, where $\boldsymbol{u}_0, \ldots, \boldsymbol{u}_3$ are the following translation vectors of $\mathbb{R}^2$ (identified here with their coordinates in the standard basis of $\mathbb{R}^2$):

$$\boldsymbol{u}_0 := (0,0),\ \boldsymbol{u}_1 := \left(\frac{1}{2},0\right),\ \boldsymbol{u}_2 := \left(0,\frac{1}{2}\right),\ \text{and}\ \boldsymbol{u}_3 := \left(\frac{1}{2},\frac{1}{2}\right). \tag{4.48}$$

Note that $\varphi_j : S \to S$, for $j = 0, \ldots, 3$. Actually, it is immediate to check that $S = \bigcup_{j=0}^3 \varphi_j(S)$—which, since $S \in \mathbb{K}(\mathbb{R}^2)$, implies (by Theorem 4.17 above) that the unit square $S = [0,1]^2$ is, indeed, the unique attractor of $\mathbf{\Phi}_S$. Observe that, for $j = 0, \ldots, 3$, the subsets $\varphi_j(S)$ of S (i.e., the "scaled copies" of S) are

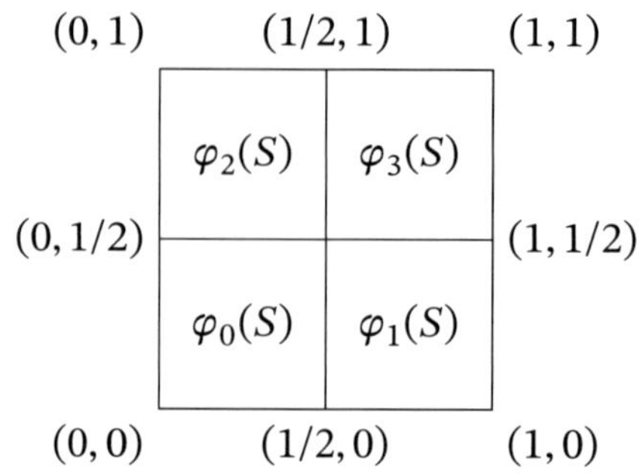

Figure 4.13. The unit square S viewed as a self-similar set of $\mathbb{R}^2$ (with scaling ratios $r_0 = \cdots = r_3 = 1/2$): $S = \bigcup_{j=0}^{3} \varphi_j(S)$; see Example 4.95.

themselves squares (of side lengths 1/2, now) and intersect only along their common boundaries (and so, along line segments of lengths 1/2); see Figure 4.13.

For pedagogical reasons, we started with the $m = 2$ case in Example 4.95 just above. However, it is now a simple matter to deal with the general case of the m-dimensional unit cube, with $m \geq 1$ arbitrary. It just requires a suitable formalization of the previous example.

Example 4.96 (*The m-Dimensional Unit Cube as a Self-Similar Set*). Let $m \geq 1$ be an arbitrary integer and let $Q_0 := [0,1]^m \subseteq \mathbb{R}^m$ denote the m-dimensional unit cube; so that $Q_0 = [0,1] \subseteq \mathbb{R}$, the unit interval, if $m = 1$, and $Q_0 = S = [0,1]^2 \subseteq \mathbb{R}^2$, the unit square, if $m = 2$ (as in Example 4.95). Then, clearly, Q_0 is a self-similar set in $\mathbb{R}^m$. In fact, it is the unique attractor (or invariant set) of the contractive self-similar system (or IFS) $\mathbf{\Phi}_{Q_0} := \{\varphi_w\}_{w\in\Sigma_0}$, where $\Sigma_0 := \{0, 1/2\}^m$. More specifically, here, this IFS consists of the $2^m = \#(\Sigma_0)$ contractive similarity transformations φ_w (with $w \in \Sigma_0$) of $\mathbb{R}^m$ (of common scaling ratios 1/2) given by $\varphi_w(\boldsymbol{x}) := (1/2)\boldsymbol{x} + \boldsymbol{u}_w$, for all $\boldsymbol{x} \in \mathbb{R}^m$ and $w \in \Sigma_0$), where $\boldsymbol{u}_w$ is the translation vector of $\mathbb{R}^m$ defined by its coordinates w (with respect to the standard orthonormal basis of $\mathbb{R}^m$). Note that $\varphi_w : Q_0 \to Q_0$ (i.e., $\varphi_w(Q_0) \subseteq Q_0$), for each $w \in \Sigma_0$. Actually, it is easy to check (and is also geometrically obvious) that $Q_0 = \bigcup_{w\in\Sigma_0} \varphi_w(Q_0)$. Hence, since $Q_0 \in \mathbb{K}(\mathbb{R}^m)$ verifies the above fixed point equation, it follows from the uniqueness part of Theorem 4.17 that Q_0 is the (unique) attractor of $\mathbf{\Phi}_{Q_0}$, as claimed.

Exercise 4.97. Can you think of other self-similar sets of $\mathbb{R}^m$—other than the unit cube (discussed in Example 4.96) Q_0 or its trivial variations—which are *not* "fractal", in any reasonable sense?

Exercise 4.98. Clearly, the Peano curve (discussed, in particular, in Subsection 1.4.3; see Figure 1.10 in Chapter 1) should be viewed as a fractal curve.

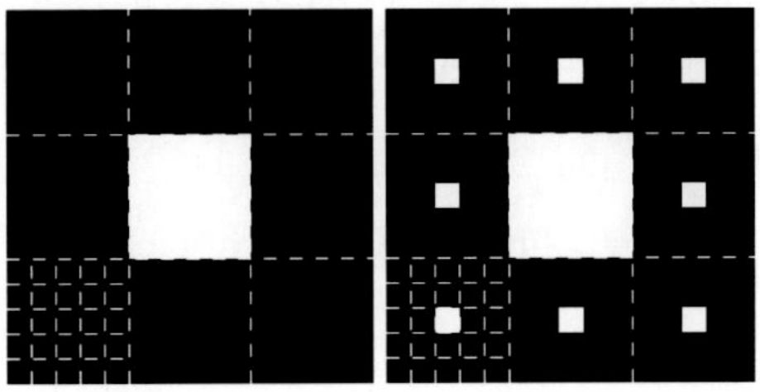

Figure 4.14. The first- and second-level approximations of the non-self-similar "1/3" Sierpinski carpet; see Example 4.99.

However, as a subset of $\mathbb{R}^2$, it coincides with the unit square $S = [0,1]^2$. Can you then explain in which sense the "plane-filling" Peano curve can be considered as a fractal? Furthermore, explain why the Peano curve can be viewed as a self-similar set and exhibit an associated self-similar system.

Example 4.99 (*Non-Self-Similar Carpet*). Consider a sequence of rational numbers of the form $\{1/q_k : q_k = 2k+1, k \geq 1\}$. Next consider the set $E := I \times I$, where $I := [0,1]$. Remove the open middle square with side lengths $1/q_1 = 1/3$. The resulting set, called a "1/3" Sierpinski carpet, is depicted in Figure 4.14.

Now, $E \setminus (1/3, 2/3) \times (1/3, 2/3)$ is made up of 8 closed squares with side lengths measuring $1/3$. Consider one of these squares: We may partition each of these squares into $q_2^2 = 25$ many squares, each with side lengths measuring $1/3 \cdot 1/5$. We continue this process. In general, one continues to remove the open middle square of side lengths $1/(q_1 \cdot q_2 \cdot \cdots \cdot q_n)$ from the remaining square of side lengths $1/(q_1 \cdot q_2 \cdot \cdots \cdot q_{k-1})$. Continue this process ad infinitum. The resulting limiting object is called a *"non-self-similar" Sierpinski carpet*; see Figure 4.14.

Exercise 4.100. Give a proof of Theorem 4.63 from Section 4.6, based on the results stated in that section.

Exercise 4.101. True or false: Every self-similar set in $\mathbb{R}$ is either connected or totally disconnected. Is there an analogous dichotomy in $\mathbb{R}^m$, for $m \geq 2$?

Chapter 5

Introduction to Hausdorff Measure and Dimension

Our goal in this chapter is to first give a straightforward introduction to the key notions of Hausdorff measure and Hausdorff dimension, in Sections 5.1 and 5.2, respectively. A lot more information, details, and complete justifications of the corresponding main theorems will be provided in Chapters 6 and 7, along with Appendix B. In Sections 5.3–5.5, we also discuss various helpful techniques for calculating—or, at least, estimating—the Hausdorff dimension of a fractal set. Finally, in Section 5.6, we establish several properties related to the Hausdorff density of fractal sets, as well as to the Hausdorff dimension of Cartesian products of fractal sets.

5.1. Hausdorff Measure

In this chapter, we focus on subsets of Euclidean space $\mathbb{R}^m$, equipped with the usual Euclidean metric d_m and Euclidean norm (or length) $||\cdot||$:

$$d_m(\boldsymbol{x}, \boldsymbol{y}) = ||\boldsymbol{x} - \boldsymbol{y}||, \text{ for all } \boldsymbol{x}, \boldsymbol{y} \in \mathbb{R}^m. \tag{5.1}$$

We stress that throughout this section, and unless explicitly stated otherwise, *t is an arbitrary real number such that $t \geq 0$.*

Definition 5.1 (*δ-Cover and Open δ-Cover*). A countable collection of sets $\{A_j\}_{j=1}^{\infty}$ is a *δ-cover* of a set $A \subseteq \mathbb{R}^m$ if $A \subseteq \bigcup_{j=1}^{\infty} A_j$ and $\operatorname{diam}(A_j) \leq \delta$. If, in addition, each A_j is open, then $\{A_j\}_{j=1}^{\infty}$ is called an *open δ-cover* of A. Suppose $t \geq 0$ and $\delta > 0$. Let the *t-dimensional, δ-approximate Hausdorff measure* of A be given by

$$\mathscr{H}_\delta^t(A) := \inf\left\{\sum_{j=1}^{\infty}(\operatorname{diam}(A_j))^t \;:\; \{A_j\}_{j=1}^{\infty} \text{ is a } \delta\text{-cover of } A\right\}. \tag{5.2}$$

Lemma 5.2. *Let $A \subseteq \mathbb{R}^m$, $t \geq 0$, and $\delta > 0$. Then,*

$$\mathscr{H}_\delta^t(A) = \inf\left\{\sum_{j=1}^{\infty}(\operatorname{diam}(U_j))^t \;:\; \{U_j\}_{j=1}^{\infty} \text{ is an open } \delta\text{-cover of } A\right\}.$$

Proof. The proof is left as an exercise for the interested reader. □

Note that the only difference between Definition 5.1 and Lemma 5.2 is that the δ-covers in Definition 5.1 comprise arbitrary sets of sufficiently small diameter whereas δ-covers in Lemma 5.2 comprise arbitrary *open sets* of sufficiently small diameter.

Moreover, observe that since, by convention, $\operatorname{diam}(\emptyset) := +\infty$—and so, $(\operatorname{diam}(\emptyset))^t := +\infty$ (even for $t = 0$)—we can work with either finite or countably infinite δ-covers or open δ-covers in Definition 5.1 or Lemma 5.2, respectively.

Lemma 5.3. *For each $t \geq 0$ and $A \subseteq \mathbb{R}^m$, then, as δ tends to zero from the right, the limit of $\mathscr{H}_\delta^t(A)$ exists in $[0, +\infty]$.*

Proof. As δ decreases, the collection of sets with diameters less than or equal to δ is reduced and, hence, so is the collection of permissible δ-covers of A. Since $\mathscr{H}_\delta^t(A)$ is an infimum over such δ-covers, we have that $\mathscr{H}_\delta^t(A)$ is a nonincreasing function of δ on the interval $(0, +\infty)$. Hence, $\mathscr{H}_\delta^t(A)$ is nondecreasing as $\delta > 0$ tends to zero—or, actually, $0 < \delta_1 \leq \delta_2$ implies that $\mathscr{H}_{\delta_1}^t(A) \geq \mathscr{H}_{\delta_2}^t(A)$.

So, if $\mathscr{H}_\delta^t(A)$ is bounded, it follows from elementary analysis that the limit $\lim_{\delta\to0^+}\mathscr{H}_\delta^t(A)$ exists in $[0, +\infty)$. Otherwise, $\lim_{\delta\to0^+}\mathscr{H}_\delta^t(A) = +\infty$. In addition, in either case, the limit is equal to the supremum of $\mathscr{H}_\delta^t(A)$ taken over all $\delta > 0$—or, equivalently, over all $\delta \in (0, 1]$. □

In light of Lemma 5.3 and its proof, we can state the following definition.

Definition 5.4 (*Hausdorff Outer Measure*). For $t \geq 0$ and a set $A \subseteq \mathbb{R}^m$, the *t-dimensional Hausdorff outer measure* of A, denoted $\mathscr{H}^t(A)$, is given by

$$\mathscr{H}^t(A) := \lim_{\delta\to0^+}\mathscr{H}_\delta^t(A) = \sup_{\delta>0}\mathscr{H}_\delta^t(A) = \sup_{0<\delta\leq1}\mathscr{H}_\delta^t(A). \tag{5.3}$$

Note that, in general, $\mathscr{H}^t(A) \in [0, +\infty]$.

The following key result is central to the theory of Hausdorff measure. It will be fully established in Chapters 6 and 7.

Theorem 5.5 (Hausdorff Outer Measure and Hausdorff Measure). *Fix $t \geq 0$. Then:*

(i) (Hausdorff Outer Measure). $\mathscr{H}^t$*, as given in Definition* 5.4*, is a metric outer measure on* $\mathbb{R}^m$ *(see Definition* 3.58 *in Section* 3.3*), called the* t-dimensional Hausdorff outer measure *on* $\mathbb{R}^m$.

(ii) (Hausdorff Measure). *Therefore, according to Theorem* 3.61 *(also in Section* 3.3*), the* metric *outer measure* $\mathscr{H}^t$*—when restricted to* $\mathcal{B}(\mathbb{R}^m)$*, the Borel* σ*-algebra of* $\mathbb{R}^m$ *(see Definition* 3.20 *in Section* 3.2*)—induces a Borel* measure *on* $\mathbb{R}^m$.

Remark 5.6. It is straightforward to show that $\mathscr{H}^t$ is an outer measure (see Definition 3.10 in Section 3.2) for each $t \geq 0$, but it is rather more difficult to show that if $\{A_j\}_{j=1}^{\infty}$ is a countable collection of pairwise disjoint Borel sets, then $\mathscr{H}^t\left(\bigcup_{j=1}^{\infty} A_j\right) = \sum_{j=1}^{\infty} \mathscr{H}^t(A_j)$, for each $t \geq 0$. As in the construction of the Lebesgue measure $\mathscr{L}^m$, this fact follows from an application of the Carathéodory–Hahn Theorem (Theorem 3.57, along with Theorem 3.61). For any $\delta > 0$, we first prove the counterpart of Theorem 5.5 for the δ-approximate outer measure $\mathscr{H}^t_\delta$ (by using Theorem 3.61, in particular) and then deduce from it Theorem 5.5 via Lemma 5.3 and Definition 5.4.

As is mentioned in Remark 5.7 just below, a completely self-contained proof of Theorem 5.5 (as well as of its δ-approximate counterpart) is provided in Chapters 6 and 7, along with Appendix B.

Remark 5.7 (*δ-Approximate Hausdorff Outer Measure and Measure*). In fact, as is alluded to in Remark 5.6, and still for any given fixed $t \geq 0$, an entirely analogous theorem (to Theorem 5.5) is first proved for $\mathscr{H}^t_\delta$ (for any $\delta > 0$). In other words, $\mathscr{H}^t_\delta$, as given in Definition 5.1 and Lemma 5.2, is a metric outer measure on $\mathbb{R}^m$, called the *(t-dimensional) δ-approximate Hausdorff outer measure* on $\mathbb{R}^m$. Therefore, in light of Theorem 3.61, it induces on $(\mathbb{R}^m, \mathcal{B}(\mathbb{R}^m))$ a Borel measure on $\mathbb{R}^m$, still denoted by $\mathscr{H}^t_\delta$ and now called the *(t-dimensional) δ-approximate Hausdorff measure* on $\mathbb{R}^m$.

Theorem 5.5, as stated above, then follows easily from this key result. The detailed proof of this (δ-approximate) counterpart of Theorem 5.5 is provided in Chapter 6 (Section 6.3), while Theorem 5.5 itself is deduced from this result in Chapter 7 (Section 7.1).

Remark 5.8 (*Proportionality of $\mathscr{H}^m$ and $\mathscr{L}^m$, for any $m \in \mathbb{N}$*). It can be shown (see Subsections 7.4.2 and 7.4.6) that for any given $m \in \mathbb{N}$, the m-dimensional Hausdorff measure is a constant multiple of m-dimensional Lebesgue measure; see Theorems 7.53 and 7.66. That is, if ω_m denotes the volume of an

m-dimensional ball of radius 1, given by

$$\omega_m = \begin{cases} \pi^{m/2}/(2^m(m/2)!\,), & \text{if } m \text{ is even,} \\ \pi^{(m-1)/2}((m-1)/2)!\,/m!\,, & \text{if } m \text{ is odd,} \end{cases} \tag{5.4}$$

then[1]

$$\mathscr{H}^m(A) = 2^m \cdot \omega_m^{-1} \cdot \mathscr{L}^m(A), \tag{5.5}$$

for any Borel subset A of $\mathbb{R}^m$. In particular, $\mathscr{H}^1(A) = \mathscr{L}^1(A)$, for all $A \in \mathscr{B}(\mathbb{R})$. Furthermore, the following statements hold:

(i) If A is finite, then $\mathscr{H}^0(A)$ is the number of points in A.

(ii) If A is a smooth curve, then $\mathscr{H}^1(A)$ is its length.[2]

(iii) If A is a smooth surface, then $\mathscr{H}^2(A) = (4/\pi) \cdot \mathrm{vol}^2(A)$.

(iv) In general, if A is a smooth n-dimensional submanifold of $\mathbb{R}^m$, then $\mathscr{H}^n(A) = 2^n \cdot \omega_n^{-1} \cdot \mathscr{L}^n(A) = 2^n \cdot \omega_n^{-1} \cdot \mathrm{vol}^n(A)$, where ω_n is given in (5.4).

Proposition 5.9. *Let $A \subseteq \mathbb{R}^m$ and let $f : A \to \mathbb{R}^m$ be a Hölder continuous function (see Definition 2.30) such that*

$$\|f(\boldsymbol{x}) - f(\boldsymbol{y})\| \le c \cdot \|\boldsymbol{x} - \boldsymbol{y}\|^\alpha, \text{ for all } \boldsymbol{x}, \boldsymbol{y} \in \mathbb{R}^m \tag{5.6}$$

and for some positive constants α and c. Then, for each $t \ge 0$,

$$\mathscr{H}^{t/\alpha}(f(A)) \le c^{t/\alpha} \cdot \mathscr{H}^t(A). \tag{5.7}$$

Proof. Let $\{A_j\}_{j=1}^\infty$ be a δ-cover of A. Since

$$\mathrm{diam}(A \cap A_j) \le c \cdot (\mathrm{diam}(A \cap A_j))^\alpha \le c \cdot (\mathrm{diam}(A_j))^\alpha,$$

for all $j \in \mathbb{N}$, we have that $\{f(A \cap A_j)\}_{j=1}^\infty$ is a $c \cdot \delta^\alpha$-cover of $f(A)$. Hence, $\sum_{j=1}^\infty (\mathrm{diam}\, f(A \cap A_j))^{t/\alpha} \le c^{t/\alpha} \cdot (\sum_{j=1}^\infty (\mathrm{diam}(A_j))^t)$, which implies that $\mathscr{H}^{t/\alpha}_{c\delta^\alpha}(f(A)) \le c^{t/\alpha} \cdot \mathscr{H}^t(A)$. Thus, in light of Definition 5.4, letting δ tend to zero yields (5.7). □

Remark 5.10. If $\alpha = 1$ in Proposition 5.9, then f is Lipschitz and $\mathscr{H}^t(f(A)) \le c^t \cdot \mathscr{H}^t(A)$, for each $t \ge 0$.

Compare the following result with Proposition 3.70 for Lebesgue measure $\mathscr{L}^m$.

Proposition 5.11 (Action Under Scaling, Rotations, and Translations). *Let $\lambda > 0$, let $\mathbf{v} \in \mathbb{R}^m$ be a fixed vector, and let O be an $m \times m$ orthogonal matrix, necessarily representing a rotation of $\mathbb{R}^m$ (in the standard orthonormal basis of*

[1] We use here the normalization introduced in Chapter 7, especially in Subsections 7.4.2 and 7.4.6.

[2] See also Proposition 5.61 in Section 5.5 below.

$\mathbb{R}^m$). *If* $A \subseteq \mathbb{R}^m$ *and* $0 \leq t \leq m$, *then*

(i) $\mathcal{H}^t(A + \mathbf{v}) = \mathcal{H}^t(A)$ (Translation Invariance),

(ii) $\mathcal{H}^t(\lambda A) = \lambda^t \cdot \mathcal{H}^t(A)$ (Scale Covariance), *and*

(iii) $\mathcal{H}^t(OA) = \mathcal{H}^t(A)$ (Rotation Invariance).

Consequently, if $\varphi : \mathbb{R}^m \to \mathbb{R}^m$ *is a similarity with scaling ratio* (*Lipschitz constant*) $\mathrm{Lip}(\varphi) = \lambda > 0$, *then* (*still for all* $t \geq 0$),[3]

(iv) $\mathcal{H}^t(\varphi(A)) = \lambda^t \cdot \mathcal{H}^t(A)$ (Similarity Covariance).

Proof. Parts (i) and (iii) follow immediately from the fact that for any fixed vector $\mathbf{v}$, orthogonal matrix O, and $\boldsymbol{x}, \boldsymbol{y} \in A$, we have

$$\|\boldsymbol{x} - \boldsymbol{y}\| = \|O\boldsymbol{x} - O\boldsymbol{y}\| = \|(\boldsymbol{x} + \mathbf{v}) - (\boldsymbol{x} + \mathbf{v})\|.$$

For part (ii), consider the map $f : \mathbb{R}^m \to \mathbb{R}^m$, given by $f(\boldsymbol{x}) := \lambda \cdot \boldsymbol{x}$, for all $\boldsymbol{x} \in \mathbb{R}^m$. Then, the fact that f is invertible implies that we have both

$$\|f(\boldsymbol{x}) - f(\boldsymbol{y})\| = \lambda \cdot \|\boldsymbol{x} - \boldsymbol{y}\| \quad \text{and} \quad \|f^{-1}(\boldsymbol{x}) - f^{-1}(\boldsymbol{y})\| = \lambda^{-1} \cdot \|\boldsymbol{x} - \boldsymbol{y}\|,$$

for any $\lambda > 0$ and $\boldsymbol{x}, \mathbf{y} \in A$. Then, an application of Proposition 5.9 to both φ and φ^{-1} yields part (ii). Finally, part (iv) follows immediately from Proposition 2.54 (along with the associated footnote) and the other three parts. □

As with the properties of Lebesgue measure in Proposition 3.70, property (i) above says that Hausdorff measure is translation invariant and property (iii) says that Hausdorff measure is invariant with respect to rotations.

Also, as discussed in Example 3.74 regarding Lebesgue measure, given a set A we seek a *unique* Hausdorff measure which assigns a meaningful value, interpreted intuitively as the 'size' of A. More specifically, consider the following question: Is there a unique t-dimensional Hausdorff measure for which $\mathcal{H}^t(A)$ is even possibly nonzero and finite? In some sense, this question is answered in the affirmative in the next section, even though the answer is perhaps a bit more subtle than one might think, a priori.

5.2. Hausdorff Dimension

The following lemma and definition establish the fact that every $A \subseteq \mathbb{R}^m$ has a unique Hausdorff dimension.

Lemma 5.12. *Fix* $A \subseteq \mathbb{R}^m$. *Then,* $\mathcal{H}^t_\delta(A)$ *is nonincreasing as* $t \geq 0$ *increases. Furthermore, there exists a unique nonnegative real number* σ *such that*

$$\mathcal{H}^t(A) = \begin{cases} +\infty, & \text{if } 0 \leq t < \sigma, \\ 0, & \text{if } \sigma < t. \end{cases} \tag{5.8}$$

[3]Of course, if φ is a *contracting* similarity—as is the case for any member of a self-similar system (see Sections 4.2 and 4.3)—then the scaling ratio λ belongs to $(0, 1)$.

Proof. Let $0 \le s \le t$. As in the proof of Lemma 5.3, suppose $0 < \delta < 1$. Consider a set A and a δ-cover $\{A_j\}_{j=1}^{\infty}$ of A. Then, $\delta^t \le \delta^s$ and for each $j \in \mathbb{N}$, $(\text{diam}(A_j))^t \le (\text{diam}(A_j))^s$. Hence, $\sum_{j=1}^{\infty}(\text{diam}(A_j))^t \le \sum_{j=1}^{\infty}(\text{diam}(A_j))^s$. So, in light of Definition 5.1, taking infima in the latter inequality over all δ-covers of A yields $\mathcal{H}^t_\delta(A) \le \mathcal{H}^s_\delta(A)$. Thus, in light of Definition 5.4, letting $\delta \to 0^+$ in this last inequality yields $\mathcal{H}^t(A) \le \mathcal{H}^s(A)$; i.e., $\mathcal{H}^t(A)$ is nonincreasing as $t \ge 0$ increases.

Now, assume that $\mathcal{H}^\tau(A) < \infty$ and $0 \le \tau < t$. For a δ-cover $\{A_j\}_{j=1}^{\infty}$ of A with $0 < \delta < 1$, we have that

$$\begin{aligned}\sum_{j=1}^{\infty}(\text{diam}(A_j))^t &= \sum_{j=1}^{\infty}\left((\text{diam}(A_j))^{t-\tau} \cdot (\text{diam}(A_j))^\tau\right) \\ &\le \delta^{t-\tau} \cdot \left(\sum_{j=1}^{\infty}(\text{diam}(A_j))^\tau\right).\end{aligned}$$

Hence, $0 \le \mathcal{H}^t_\delta(A) \le \delta^{t-\tau} \cdot \mathcal{H}^\tau_\delta(A)$. Since $\mathcal{H}^t(A) \le \mathcal{H}^\tau(A) < \infty$, we deduce that $\mathcal{H}^t(A) = \lim_{\delta\to 0^+} \mathcal{H}^t_\delta(A) \le \lim_{\delta\to 0^+} \delta^{t-\tau} \cdot \mathcal{H}^\tau_\delta(A) = 0$.

Next, assume that $\mathcal{H}^\gamma(A) > 0$ and $0 \le t < \gamma$. Just as above, for $0 < \delta < 1$, we have $0 \le \mathcal{H}^\gamma_\delta(A) \le \delta^{\gamma-t} \cdot \mathcal{H}^t_\delta(A) \le \mathcal{H}^t_\delta(A)$. Note that since $\gamma - t > 0$, the limit of $\delta^{\gamma-t} \cdot \mathcal{H}^t_\delta(A)$ as $\delta \to 0^+$ exists in $[0, +\infty]$ since $\lim_{\delta\to 0^+} \delta^{\gamma-t} = 0$ (because $0 < \delta < 1$) and also, $\lim_{\delta\to 0^+} \mathcal{H}^t_\delta(A) = \mathcal{H}^t(A)$. Thus, we deduce from (5.3) that $0 < \mathcal{H}^\gamma(A) \le \lim_{\delta\to 0^+} \delta^{\gamma-t} \cdot \mathcal{H}^t_\delta(A)$.

Now, suppose $\lim_{\delta\to 0^+} \mathcal{H}^t_\delta(A) = \mathcal{H}^t(A) < \infty$. Since $\mathcal{H}^\gamma(A) \le \mathcal{H}^t(A)$, we have $\lim_{\delta\to 0^+} \delta^{\gamma-t} \cdot \mathcal{H}^t_\delta(A) = 0$, but this clearly contradicts the fact that $\mathcal{H}^t(A) > 0$. Hence, $\mathcal{H}^t(A) = \lim_{\delta\to 0^+} \mathcal{H}^t_\delta(A) = +\infty$.

The above arguments combine to show that

$$\sup\{t \ge 0 : \mathcal{H}^t(A) = +\infty\} = \inf\{t \ge 0 : \mathcal{H}^t(A) = 0\}. \tag{5.9}$$

Finally, defining σ as either value in (5.9) yields the unique nonnegative real number which satisfies (5.8). □

Definition 5.13 (*Hausdorff Dimension*). Let $A \subseteq \mathbb{R}^m$, with A nonempty. Then, the unique nonnegative real number σ whose existence was established in Lemma 5.12 is called the *Hausdorff dimension* (or *Hausdorff–Besicovitch dimension*) of the set A and is denoted $\dim_H(A)$. *Equivalently, σ is given by* (5.9).

In practice, it is important to remember that, in Definition 5.13, $\sigma := \dim_H(A)$, the Hausdorff dimension of A, is indifferently given by either (5.8) or (5.9). Also, as before, when A is the empty set, we let $\dim_H(\emptyset) := -\infty$.

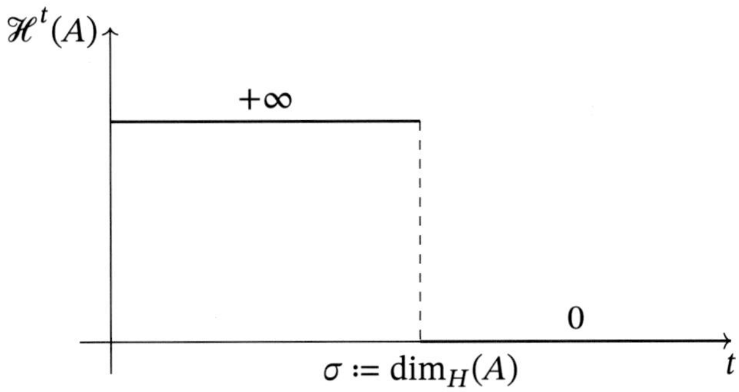

Figure 5.1. Let $A \subseteq \mathbb{R}^m$, with $A \neq \emptyset$, and let $\sigma := \dim_H(A)$ denote the Hausdorff dimension of A. Then, by definition, $\mathcal{H}^t(A) = +\infty$, for $0 \leq t < \sigma$, while $\mathcal{H}^t(A) = 0$, for all $t > \sigma$, where $\mathcal{H}^t(A)$ stands for the t-dimensional Hausdorff measure of A; see Lemma 5.12 and Definition 5.13. Accordingly, in physicists' terminology, $\sigma = \dim_H(A)$ is a *critical exponent.*
In general, we have that $\mathcal{H}^\sigma(A) \in [0, +\infty]$. If $0 < \mathcal{H}^\sigma(A) < \infty$, then A is called a t-set (see Definition 5.14), but, in general, we may also have either that $\mathcal{H}^\sigma(A) = 0$ or $\mathcal{H}^\sigma(A) = +\infty$.

In general, for $t := \dim_H(A)$, then $0 \leq \mathcal{H}^t(A) \leq \infty$ if A is nonempty; that is, one may either have that $\mathcal{H}^t(A) = 0$, or $\mathcal{H}^t(A) = +\infty$, or else $0 < \mathcal{H}^t(A) < \infty$. This observation motivates the next definition.

Definition 5.14 (*t-Set*). If $t := \dim_H(A)$ and $0 < \mathcal{H}^t(A) < \infty$, then A is called a *t-set.*

Observe that it follows from the definition of the Hausdorff dimension (Definition 5.13; see also Lemma 5.12) that if $0 < \mathcal{H}^t(A) < \infty$, for some $t \in \mathbb{R}$, with $t \geq 0$, then we must have that $t = \dim_H(A)$.

Throughout this text, attention is mostly restricted to t-sets that are also Borel sets.

We refer to Figure 5.1 for a graphical illustration of the definition of $\dim_H(A)$ as a *critical exponent* given in Definition 5.13 and Lemma 5.12 above.

Moreover, we refer to Subsection 8.9.2 of Chapter 8 for a discussion of the notions of generalized Hausdorff measure and of exact Hausdorff measure of $A \subseteq \mathbb{R}^m$ with respect to a suitable gauge function (or generalized dimension function, in the sense of Subsection 8.9.1) corresponding to an underlying *scaling law* different from a mere power law.

In addition, we point out that the important notions of Minkowski and box-counting (or box) dimensions will be thoroughly discussed in Chapter 8, especially in Sections 8.1 and 8.3–8.4, along with Subsection 8.6.2, while the notions of (possibly generalized) packing dimension and generalized Minkowski content will be briefly discussed in Section 8.9 (Subsection 8.9.3–8.9.5). Finally, the key inequality comparing the Hausdorff and Minkowski (or, equivalently, the box) dimensions will be provided in Section 8.5.

Exercise 5.15. Let $m \geq 1$ be an arbitrary integer. Find $A \subseteq \mathbb{R}^m$ such that $\mathcal{H}^{\sigma}(A) = 0$ and $B \subseteq \mathbb{R}^m$ such that $\mathcal{H}^{\tilde{\sigma}}(B) = +\infty$, where $\sigma := \dim_H(A)$ and $\tilde{\sigma} := \dim_H(B)$. Also, find a t-set $D \subseteq \mathbb{R}^m$ such that $\dim_H(D) \notin \{0, 1, \dots, m\}$.

Proposition 5.16 (Key Properties of Hausdorff Dimension). *The Hausdorff dimension satisfies the following properties:*

(i) Normalization: $\dim_H(\emptyset) := -\infty$.

(ii) Monotonicity: *If* $A \subseteq B \subseteq \mathbb{R}^m$, *then* $\dim_H(A) \leq \dim_H(B)$.

(iii) Countable Stability:[4] *If* $\{S_n\}_{n=1}^{\infty}$ *is a (finite or) countable collection of subsets of* $\mathbb{R}^m$, *then*

$$\dim_H\left(\bigcup_{n=1}^{\infty} S_n\right) = \sup_{n\in\mathbb{N}}\{\dim_H(S_n)\}. \tag{5.10}$$

(iv) Countable Sets: *If* A *is (finite or) countable, then* $\dim_H(A) = 0$.

Proof. Property (i) holds by convention.

Property (ii) above follows immediately from the normalization and monotonicity properties of Hausdorff measure, viewed as an outer measure; see Remark 5.6 and Definition 3.10.

In order to prove property (iii), note that by monotonicity (ii), we have $\sup_{n\in\mathbb{N}}\{\dim_H(S_n)\} \leq \dim_H\left(\bigcup_{n=1}^{\infty} S_n\right)$. On the other hand, if $t > \dim_H(S_n)$ for each $n \in \mathbb{N}$, then by Lemma 5.12, we have $\mathcal{H}^t(S_n) = 0$. So, by countable subadditivity of the Hausdorff measure, we also have that

$$0 \leq \mathcal{H}^t\left(\bigcup_{n=1}^{\infty} S_n\right) \leq \sum_{n=1}^{\infty} \mathcal{H}^t(S_n) = 0.$$

Hence, it follows that $\mathcal{H}^t\left(\bigcup_{n=1}^{\infty} S_n\right) = 0$. So, in light of either the right-hand side of (5.9) or Definition 5.13 and Lemma 5.12, we deduce that $\dim_H\left(\bigcup_{n=1}^{\infty} S_n\right) \leq t$. But, since $t \geq \sup_{n\in\mathbb{N}}\{\dim_H(S_n)\}$ was arbitrary, a moment's reflection shows that it follows from the above inequality that $\sup_{n\in\mathbb{N}}\{\dim_H(S_n)\} \geq \dim_H\left(\bigcup_{n=1}^{\infty} S_n\right)$, as remained to be shown. And so, we conclude that property (iii) holds, in the countably infinite case.[5]

Note that property (iv) follows immediately from properties (ii) and (iii) along with the fact that $\dim_H((\boldsymbol{x})) = 0$ for any singleton $\{\boldsymbol{x}\} \subseteq \mathbb{R}^m$, since, for any $\varepsilon > 0$, a singleton is contained in a set with diameter less than ε.

[4] In the literature, *countable stability* is also called *σ-stability*.

[5] If $\{S_n\}_{n=1}^{n_0}$ is a finite family, for some $n_0 \in \mathbb{N}$, then (since $\dim_H(\emptyset) = -\infty$) it suffices to let $S_n := \emptyset$, for all $n > n_0$ and to apply property (iii) to the countable family $\{S_n\}_{n=1}^{\infty}$.

This concludes the proof of Proposition 5.16. □

Exercise 5.17. Show that countable stability (property (iii) of Proposition 5.16) implies finite stability, formulated in the obvious manner.

[*Hint*: Use property (i) of Proposition 5.16.]

Proposition 5.18 (Covariance Under Hölder Continuous Maps). *Let $A \subseteq \mathbb{R}^m$ and let $f : A \to \mathbb{R}^n$ be a Hölder continuous function; that is, suppose there exist $c > 0$ and $\alpha > 0$ such that*

$$\|f(\boldsymbol{x}) - f(\boldsymbol{y})\| \leq c \cdot \|\boldsymbol{x} - \boldsymbol{y}\|^{\alpha}, \text{ for every } \boldsymbol{x}, \boldsymbol{y} \in A. \tag{5.11}$$

Then, $\dim_H(f(A)) \leq (1/\alpha) \cdot \dim_H(A)$.

Proof. Let $t > \dim_H(A)$, so that $\mathcal{H}^t(A) = 0$. Then, by Proposition 5.9, we have $\mathcal{H}^{t/\alpha}(f(A)) \leq c^{t/\alpha} \cdot \mathcal{H}^t(A) = 0$. So, $\dim_H(f(A)) \leq t/\alpha$ whenever $t > \dim_H(A)$, in light of Lemma 5.12 and Definition 5.13. Therefore, $\dim_H(f(A)) \leq (1/\alpha) \cdot \dim_H(A)$. □

Corollary 5.19 (Bi-Lipschitz Invariance of Hausdorff Dimension). *Let $A \subseteq \mathbb{R}^m$. If $f : A \to \mathbb{R}^m$ is Lipschitz, then* $\dim_H(f(A)) \leq \dim_H(A)$.

Moreover, let $f : A \to \mathbb{R}^m$ be bi-Lipschitz;[6] *i.e., there are constants $0 < c_1 \leq c_2 < \infty$ such that*

$$c_1 \cdot \|\boldsymbol{x} - \boldsymbol{y}\| \leq \|f(\boldsymbol{x}) - f(\boldsymbol{y})\| \leq c_2 \cdot \|\boldsymbol{x} - \boldsymbol{y}\|, \text{ for every } \boldsymbol{x}, \boldsymbol{y} \in A. \tag{5.12}$$

Then, $\dim_H(f(A)) = \dim_H(A)$. *As a result, the Hausdorff dimension is a* bi-Lipschitz invariant.

Proof. If f is Lipschitz, then taking $\alpha = 1$ in Proposition 5.18 yields $\dim_H(f(A)) \leq \dim_H(A)$.

If f is bi-Lipschitz, then f is invertible and f^{-1} is also bi-Lipschitz. So by taking $\alpha = 1$ and applying Proposition 5.18 to both f and f^{-1}, we deduce that $\dim_H(f(A)) = \dim_H(A)$, as desired. □

Remark 5.20. It follows from Corollary 5.19 that Hausdorff dimension is a *metric invariant*, in the sense that it is invariant under bi-Lipschitz maps. The same will be true of the box (or box-counting) dimension—and hence also of the Minkowski dimension (in Euclidean spaces); see Chapter 8. Contrast this metric invariance (i.e., bi-Lipschitz invariance) property of the Hausdorff, box, and Minkowski dimensions with the fact that the topological dimension is a *topological invariant*, i.e., is invariant under homeomorphisms; see Subsections 8.6.1 and 8.6.2 below.

[6]Note that this hypothesis implies that f is injective and hence is a bijection onto its image $f(A)$. In particular, the inverse map $f^{-1} : f(A) \to A$ exists and is also bi-Lipschitz.

Observe that the metric invariance property of Hausdorff dimension reinforces the fact that Hausdorff measure is a *metric* notion, while the Hausdorff dimension describes certain properties of the underlying space—here, Euclidean space $\mathbb{R}^m$, with $m \in \mathbb{N}$ or, more generally (in the context of Chapters 6 and 7; see, especially, Section 7.5), of arbitrary subsets of a metric space.[7]

Proposition 5.21. *If $A \subseteq \mathbb{R}^m$ and $\dim_H(A) < 1$, then A is totally disconnected.*

Proof. Suppose $\boldsymbol{x}, \boldsymbol{y} \in A$ and $\boldsymbol{x} \neq \boldsymbol{y}$. Define $f : \mathbb{R}^m \to \mathbb{R}$ by $f(\mathbf{z}) = ||\mathbf{z} - \boldsymbol{x}||$, for all $\mathbf{z} \in \mathbb{R}^m$. Note that f is a Lipschitz function with $c = 1$ (i.e., f does not increase distances); so, by Proposition 5.18 (applied with $c := 1$ and $\alpha := 1$ in (5.11)), we have $\dim_H(f(A)) \leq \dim_H(A) < 1$. Thus, by Lemma 5.12 and Remark 5.8, we have $\mathscr{H}^1(f(A)) = 4 \cdot \mathscr{L}^1(f(A)) = 0$. In particular, $f(A)$ can be covered by a countable collection of open intervals with arbitrarily small total length. Now, $\mathbb{R}\backslash f(A)$ is dense since, if it were not, $f(A)$ would contain a nonempty open interval, implying that $\mathscr{H}^1(f(A)) = 4 \cdot \mathscr{L}^1(f(A)) > 0$. So, there is some $r \in \mathbb{R}\backslash f(A)$ such that $0 < r < f(\boldsymbol{y})$. Note that f is continuous; so, the sets

$$f^{-1}((-1,r)) = \{\mathbf{a} \in A : ||\boldsymbol{x} - \mathbf{a}|| < r\} \quad \text{and}$$
$$f^{-1}((r,+\infty)) = \{\mathbf{a} \in A : ||\boldsymbol{x} - \mathbf{a}|| > r\}$$

are open and disjoint. Since $r \notin f(A)$, there is no point $\mathbf{a} \in A$ such that $||\mathbf{x} - \mathbf{a}|| = r$. Therefore, $f^{-1}((-1,r)) \cup f^{-1}((r,+\infty)) = A$, $\boldsymbol{x} \in f^{-1}((-1,r))$, $\boldsymbol{y} \in f^{-1}((r,+\infty))$, and it follows that A is totally disconnected. □

Remark 5.22. In light of Proposition 5.21 (along with its proof) and since $\dim_H(C) = \log_3 2 < 1$, we obtain another proof and a more conceptual explanation of the fact that the classic ternary Cantor set C is totally disconnected.

Exercise 5.23.

(a) Show that if $f : A \subseteq \mathbb{R}^{m'} \to \mathbb{R}^m$ is an isometry—and hence is a bijection from A onto $f(A)$ (since an isometry is always injective)—then, $\dim_H(f(A)) = \dim_H(A)$ but also, with $t := \dim_H(A)$, we have that $\mathscr{H}^t(f(A)) = \mathscr{H}^t(A)$.

(b) Show, by means of an example, that if $t := \dim_H(A)$ and $A \subseteq \mathbb{R}^m$, then we need *not* have that $\mathscr{H}^t(f(A)) = \mathscr{H}^t(A)$, when f is bi-Lipschitz.

5.3. Hausdorff Dimension Calculations

In this section, we compute the Hausdorff dimension and estimate the Hausdorff measure, for several examples.

[7] Note that Hausdorff measure is invariant under bijective (i.e., surjective) isometries, but not, in general, under bi-Lipschitz maps; see Exercise 5.23.

Lemma 5.24. *Let $A \subseteq \mathbb{R}^m$. If there is a constant $c > 0$ such that for every $\delta > 0$ there exists a δ-cover $\{A_j\}_{j=1}^{\infty}$ satisfying $\sum_{j=1}^{\infty}(\operatorname{diam}(A_j))^t \leq c$, then $\dim_H(A) \leq t$.*

Proof. By assumption, $\mathscr{H}_\delta^t(A) \leq c$, for all $\delta > 0$; so, $\mathscr{H}_\delta^t(A) < \infty$. By Lemma 5.12 and Definition 5.13, it then follows that $\dim_H(A) \leq t$. □

Lemma 5.25. *Let $A \subseteq \mathbb{R}^m$. If there is a constant $c > 0$ such that for every $\delta > 0$ and for all δ-covers $\{A_j\}_{j=1}^{\infty}$, we have that $\sum_{j=1}^{\infty}(\operatorname{diam}(A_j))^t \geq c$, then $\dim_H(A) \geq t$.*

Proof. By assumption, $\mathscr{H}_\delta^t(A) \geq c$, for all $\delta > 0$; so, $\mathscr{H}^t(A) > 0$. It then follows from Lemma 5.12 that $\dim_H(A) \geq t$. □

Remark 5.26. By Lemma 5.2 and the well-known fact that open subsets of $\mathbb{R}$ comprise a union of finitely or countably many disjoint open intervals (see Theorem 2.43 in Section 2.2 above or else [**522**], [**571**]),[8] then if $m = 1$ in Lemma 5.25 (i.e, if $A \subseteq \mathbb{R}$), it suffices to consider the case where each A_j in a δ-cover of A is an open interval.

Proposition 5.27 (Unit Interval). *We have that* $\dim_H([0,1]) = 1$.

Proof. Let $\delta > 0$. Choose $n \in \mathbb{N}$ such that $1/n \leq \delta$. Let

$$U_j := \left(\frac{j-1}{3n}, \frac{j+2}{3n}\right), \text{ for } j = 0, \dots, 3n-1. \tag{5.13}$$

Then, $[0,1] \subseteq \bigcup_{j=1}^{3n-1} U_j = \left(\frac{-1}{3n}, 1 + \frac{1}{3n}\right)$ and $\operatorname{diam}(U_j) = \mathscr{L}^1(U_j) = 1/3^n$, for each j as in (5.13). Hence, $\{U_j\}_{j=1}^{3n}$ is a δ-cover of $[0,1]$ comprising $3n$ open intervals. Note that $\sum_{j=1}^{3n-1} \operatorname{diam}(U_j) = 3n\left(\frac{1}{3n}\right) = 1$. By Lemma 5.24, it follows that $\dim_H([0,1]) \leq 1$.

Now, fix $0 \leq t < 1$; let $\varepsilon > 0$ and suppose $\{V_j\}_{j=1}^{\infty}$ is an ε-cover of $[0,1]$. Then, since $t - 1 < 0$,

$$\sum_{j=1}^{\infty}(\operatorname{diam}(V_j))^t = \sum_{j=1}^{\infty}((\operatorname{diam}(V_j))(\operatorname{diam}(V_j))^{t-1}) \geq \varepsilon^{t-1} \cdot \left(\sum_{j=1}^{\infty} \operatorname{diam}(V_j)\right).$$

By Lemma 5.2 and properties of open subsets of the real line,[9] we may take each V_j to be an open interval. So, since $[0,1] \subseteq \bigcup_{j=1}^{\infty} V_j$, we have $\sum_{j=1}^{\infty} \operatorname{diam}(V_j) \geq 1$. Let $a_j := \inf V_j$ and $b_j := \sup V_j$; then $V_j \subseteq (a_j, b_j)$ and $\operatorname{diam}(V_j) = b_j - a_j$, for each $j \in \mathbb{N}$. (If $a_j = 0$, take $V_j = [0, b_j)$, and if $b_j = 1$, take $V_j = (a_j, 1]$.) By the compactness of $[0,1]$, there exists a finite number k of indices $j_1, \dots, j_k$

[8] Interestingly, this classic result was first obtained by Georg Cantor.

[9] See Remark 5.26 above.

such that (after reordering, if necessary) $a_{j_1} = 0, b_{j_k} = 1$, and $a_{j_{l+1}} < b_{j_l}$, for each $l = 1, \dots, k-1$. Hence,

$$\sum_{j=1}^{\infty} \operatorname{diam}(V_j) = \sum_{j=1}^{\infty}(b_j - a_j) \geq \sum_{l=1}^{k-1}(b_{j_l} - a_{j_l})$$

$$> \left(\sum_{l=1}^{k-1}(a_{j_{l+1}} - a_{j_l})\right) + b_{j_k} - a_{j_k} = b_{j_k} - a_{j_1} = 1.$$

Thus, $\sum_{j=1}^{\infty}(\operatorname{diam}(V_j))^t \geq \varepsilon^{t-1}$. Since $t < 1$, we have that $\varepsilon^{t-1} > 1$ when $0 < \varepsilon < 1$. Hence, Lemma 5.25 implies that $\dim_H([0,1]) \geq t$, for all $t < 1$; and so, $\dim_H([0,1]) \geq 1$. Therefore, $\dim_H([0,1]) = 1$, as desired. □

Proposition 5.28 (Classic Cantor Set). *Consider the Cantor set C. Then, $\dim_H(C) =: D_C = \log_3 2$ and $1/2 \leq \mathscr{H}^{D_C}(C) \leq 1$. In particular, C is a D_C-set.*

Proof. Let $C_0 := [0,1]$ and for each $j \in \mathbb{N}_0$, let $C_j := \mathbf{\Phi}_C^j(C_0)$ denote the j-th prefractal approximation of the Cantor set C with respect to C_0; see Example 4.23. Then,

$$C_0 = [0,1], \quad C_1 = [0,1/3] \cup [2/3,1],$$
$$C_2 = [0,1/9] \cup [2/9,1/3] \cup [2/3,7/9] \cup [8/9,1], \text{ etc.}$$

So, for each $j \in \mathbb{N}_0$, C_j comprises of 2^j disjoint compact intervals, each of length 3^{-j}. These intervals can be taken as a 3^{-j}-cover of C which, with $t := \log_3 2$, yields $\mathscr{H}^t_{3^{-j}}(C) \leq 2^j \cdot 3^{-jt} = 1$, for each $j \in \mathbb{N}_0$. Hence, $\mathscr{H}^t(C) \leq 1$ and therefore, $\dim_H(C) \geq t = \log_3 2 = D_C$, as claimed.

Next, let $\{U_j\}_{j=1}^{\infty}$ be a cover of C where, without loss of generality, each U_j is a closed subinterval of $[0,1]$. For each $j \in \mathbb{N}_0$, let $k \in \mathbb{N}_0$ be the unique integer defined by $3^{-(k+1)} \leq \operatorname{diam}(U_j) < 3^{-k}$. Then, U_j intersects at most one of the compact connected components of C_k since each of these components is separated from the others by at least 3^{-k}. If $t := \log_3 2$ and $m \geq k$, then, by the above inequalities, U_j intersects at most

$$2^{m-k} = 2^m \cdot 3^{-tk} \leq 2^m \cdot 3^t \cdot (\operatorname{diam}(U_j))^t$$

of the connected components of C_m. For $m \geq k$ large enough so that $3^{-(m+1)} \leq \operatorname{diam}(U_j)$ for all $j \in \mathbb{N}$, we have $2^m \leq \sum_{j=1}^{\infty}\left(2^m \cdot 3^t \cdot (\operatorname{diam}(U_j))^t\right)$, since the intervals U_j intersect all 2^m of the closed intervals of length 3^{-m} that make up C_m. Hence, $3^{-t} = 1/2 \leq \sum_{j=1}^{\infty}(\operatorname{diam}(U_j))^t$; so, $\mathscr{H}^t(C) \geq 1/2$, and we deduce that $\dim_H(C) \leq t = \log_3 2 = D_C$. Therefore, we conclude that $\dim_H(C) = D_C = \log_3 2$, as desired. □

Example 5.29 (*Smith–Volterra–Cantor Sets*). Given an integer $n \geq 3$, consider the Smith–Volterra–Cantor sets which are denoted $SVC(n)$ and described in Example 4.92. For $n = 3$, we have $SVC(3) = C$, where C is the classic Cantor set considered in Proposition 5.28. So, by Proposition 5.28, $\dim_H(SVC(3)) = \dim_H(C) = D_C = \log_3 2$ and $1/2 \leq \mathscr{H}^{D_C}(SVC(3)) \leq 1$. On the other hand, for $n \geq 4$, $SVC(n)$ has positive length (see part (v) of Example 4.92); for this reason, when $n \geq 4$, $SVC(n)$ is often referred to as a "fat fractal" in the literature. More specifically, in this case,

$$0 < \mathscr{H}^1(SVC(n)) = \mathscr{L}^1(SVC(n)) = \frac{n-3}{n-2} < 1. \tag{5.14}$$

Therefore, we deduce from Lemma 5.12 that $\dim_H(SVC(n)) = 1$, for any $n \geq 4$. In addition, in light of (5.14) and Proposition 5.28, $SVC(n)$ is a t-set for every $n \geq 3$, where $t := D_C = \log_3 2$ if $n = 3$ and $t := 1$ if $n \geq 4$.

The reader may wish to compare the next proposition with Example 4.71 in Chapter 4 above.

Proposition 5.30. *Let A be the attractor of a self-similar system $\mathbf{\Phi} = \{\varphi_j\}_{j=1}^N$, with $N \geq 2$ and positive scaling ratios $\mathbf{r} = (r_1, \dots, r_N)$; furthermore, denote the Hausdorff dimension of A by $\dim_H(A) = D$. Suppose that the sets $\varphi_j(A)$ are pairwise disjoint (for $j = 1, \dots, N$) and assume that $0 < \mathscr{H}^D(A) < \infty$ (i.e., A is a D-set); see Remark* 5.31 *below.*

Then, D is the unique positive real solution[10] *of the* Moran equation

$$\sum_{j=1}^{N} r_j^t = 1, \quad t \in \mathbb{R}. \tag{5.15}$$

In other words, $D = \dim_H(A) = D_S$, where D_S denotes the similarity dimension of the self-similar set A.

Proof. As the attractor of $\mathbf{\Phi}$, A satisfies $A = \mathbf{\Phi}(A) = \bigcup_{j=1}^N \varphi_j(A)$. So by the additivity of Hausdorff measure, $\sum_{j=1}^N \mathscr{H}^D(\varphi_j(A)) = \mathscr{H}^D(A)$. Then, by the scaling property (iv) of Proposition 5.11,

$$\sum_{j=1}^{N} \mathscr{H}^D(\varphi_j(A)) = \sum_{j=1}^{N} \left(r_j^D \cdot \mathscr{H}^D(A)\right) = \left(\sum_{j=1}^{N} r_j^D\right) \cdot \mathscr{H}^D(A) = \mathscr{H}^D(A).$$

Since $0 < \mathscr{H}^D(A) < \infty$, we have $\sum_{j=1}^N r_j^D = 1$. Theorem 4.64 then shows that D is unique, lies in $(0, m]$ (since $N \geq 2$), and coincides with D_S, the similarity dimension of A. □

[10] More specifically, t is the unique real solution of (5.15) and so, necessarily, $t > 0$. In fact, $0 < t \leq m$. (Note that if $N = 1$ were allowed, then we would have $t = 0$.)

Remark 5.31. Note that the assumption $0 < \mathcal{H}^D(A) < \infty$ in Proposition 5.30 is a strong one to make, and its use has not yet been justified in this text. Indeed, even considering the relatively simple examples of the unit interval $[0,1]$ and the Cantor set C, finding a positive lower bound for the corresponding Hausdorff measure is rather difficult. This difficulty is addressed in a rather general setting in the Mass Distribution Principle (Theorem 5.35 in Section 5.4 below). As it turns out, and as is discussed in Moran's Theorem (Theorem 5.40 below), the assumption that $0 < \mathcal{H}^D(A) < \infty$ can be dropped (it holds for any self-similar set A satisfying the OSC) and that the sets $\varphi_j(A)$ (for $j = 1,\dots,N$) are pairwise disjoint can be replaced with the much weaker Open Set Condition (OSC; see Definition 4.65).

Proposition 5.32. *Let $A \subseteq \mathbb{R}^m$ and $t \geq 0$. Define $\mathcal{H}^t_\infty(A)$ by*

$$\mathcal{H}^t_\infty(A) = \inf\left\{ \sum_{j=1}^{\infty} (\operatorname{diam}(U_j))^t \ : \ \{U_j\}_{j=1}^\infty \text{ is a countable cover of } A \right\}.$$

Then, $\dim_H(A) = \inf\{t \geq 0 : \mathcal{H}^t_\infty(A) = 0\}$.

Proof. Suppose $t > \dim_H(A)$. Then, for all $\delta > 0$, we have that $\mathcal{H}^t_\infty(A) \leq \mathcal{H}^t_\delta(A) = 0$. Now, if $\mathcal{H}^t_\infty(A) = 0$, then for each $\delta > 0$ there is a countable cover $\{U_j\}_{j=1}^\infty$ of A such that $\sum_{j=1}^\infty (\operatorname{diam}(U_j))^t \leq \delta^t$. In this case, $\{U_j\}_{j=1}^\infty$ is a δ-cover of A. So, $\mathcal{H}^t_\delta(A) \to 0$, as $\delta \to 0^+$. Thus, $\mathcal{H}^t(A) = 0$ and, hence, we deduce from Lemma 5.12 that $t \geq \dim_H(A)$, as desired. □

5.4. Mass Distribution Principle and Moran's Theorem

Our main goal in this section is to establish Moran's Theorem (Theorem 5.40), which will enable us to calculate the fractal dimensions of a large class of self-similar sets. We begin our discussion with several preliminary results culminating with the very useful Mass Distribution Principle (Theorem 5.35).

Under appropriate hypotheses, upper bounds for the Hausdorff dimension and measure are relatively easy to obtain, as we see next.

Proposition 5.33. *Let A be a set. Given a sequence of positive numbers ε_k such that $\lim_{k\to\infty} \varepsilon_k = 0$, if for each $k \in \mathbb{N}$ the set A can be covered by $N_k \in \mathbb{N}$ sets of diameter at most ε_k, then $\dim_H(A) \leq \liminf_{k\to\infty} \frac{\log N_k}{-\log \varepsilon_k}$.*

Also, if, for some $t \geq 0$, $N_k = O(\varepsilon_k^{-t})$, as $k \to \infty$, then $\mathcal{H}^t(A) < \infty$.

Proof. The stated upper bound for $\dim_H(A)$ follows at once from the definitions of $\mathcal{H}^t_\delta, \mathcal{H}^t$, and $\dim_H$ (see Definitions 5.1, 5.4, and 5.13, respectively). Now, if, for some $t \geq 0$, $N_k = O(\varepsilon_k^{-t})$, as $k \to \infty$, then $\mathcal{H}^t_{\varepsilon_k}(A) \leq N_k \cdot \varepsilon_k^t$ is bounded (uniformly in $k \in \mathbb{N}$). So, as $k \to \infty$, $\mathcal{H}^t_{\varepsilon_k}(A)$ tends to a finite limit, given by $\mathcal{H}^t(A)$. Hence, we obtain the desired upper bound for $\mathcal{H}^t(A)$, guaranteeing that $\mathcal{H}^t(A) < \infty$. □

Definition 5.34 (Mass Distribution). Let $A \subseteq \mathbb{R}^m$. A measure μ is a *mass distribution on* A if $\mathrm{supp}(\mu) \subseteq A$ and $0 < \mu(A) < \infty$.[11]

The following key (but simple) result provides a very useful tool for obtaining a lower bound for the Hausdorff dimension, $\dim_H(A)$, and the (t-dimensional) Hausdorff measure, $\mathcal{H}^t(A)$, of a suitable set $A \subseteq \mathbb{R}^m$.

Theorem 5.35 (Mass Distribution Principle). *Let $A \subseteq \mathbb{R}^m$ and let μ be a mass distribution on A. Assume that, for some $t > 0$, there are positive constants c and ε such that for any $U \subseteq \mathbb{R}^m$ with* $\mathrm{diam}(U) \leq \varepsilon$, *we have*

$$\mu(U) \leq c \cdot (\mathrm{diam}(U))^t. \tag{5.16}$$

Then,

$$(1/c) \cdot \mu(A) \leq \mathcal{H}^t(A) \quad \text{and} \quad t \leq \dim_H(A). \tag{5.17}$$

Proof. Let $\{U_j\}_{j=1}^{\infty}$ be a cover of A. Then, $A \subseteq \bigcup_{j=1}^{\infty} U_j$ and, by inequality (5.16) combined with part (iv) of Proposition 3.29, we have

$$0 < \mu(A) \leq \mu\Big(\bigcup_{j=1}^{\infty} U_j\Big) \leq \sum_{j=1}^{\infty} \mu(U_j) \leq c \cdot \Big(\sum_{j=1}^{\infty} (\mathrm{diam}(U_j))^t\Big).$$

For $\delta > 0$ small enough, we have $\mu(A)/c \leq \mathcal{H}^t_\delta(A)$. It then follows that $\mu(A)/c \leq \mathcal{H}^t_\delta(A)$. Since $\mu(A) > 0$, we deduce from Lemma 5.12 (see also the accompanying Figure 5.1 in Section 5.2) that $t \leq \dim_H(A)$, as desired. □

Note that the Mass Distribution Principle (Theorem 5.35 just above) does not completely remove the difficulty inherent to finding a lower bound for Hausdorff dimension, but it does at least provide an alternative to finding such a lower bound. Another alternative, a local version of the Mass Distribution Principle, is provided in Proposition 5.38 below. To prove the second part of this proposition, the following covering lemma is used.

Lemma 5.36. *Let $\mathcal{B}$ be a family of balls contained in a bounded subset of $\mathbb{R}^m$. Then, there is a finite or countably infinite subcollection $\{B_j\} \subseteq \mathcal{B}$ such that $\bigcup_{B \in \mathcal{B}} B \subseteq \bigcup_j \widetilde{B}_j$, where, for each j, $\widetilde{B}_j$ is the closed ball concentric with B_j and of four times the radius of B_j.*

Proof. For the sake of clarity and brevity, only the finite case is considered in this proof.

[11] See Definition 3.42 for the definition of the support $\mathrm{supp}(\mu)$ of a measure μ.

Let B_1 be a ball of maximal radius out of the finite collection $\mathcal{B}$. For $k \in \mathbb{N}$, suppose $B_1, \dots, B_k$ have been chosen. Define B_{k+1} to be one of the largest balls in $\mathcal{B}$ that does not intersect $B_1, \dots, B_k$. Terminate the process when no such balls remain. If $B \in \mathcal{B}$, then either $B = B_j$, for some j, or else $B \cap B_j \neq \emptyset$, for some j such that $\operatorname{diam}(B) \leq \operatorname{diam}(B_j)$. In either case, $B \subseteq \widetilde{B}_j$ and so $\bigcup_{B \in \mathcal{B}} B \subseteq \bigcup_j \widetilde{B}_j$, as desired. □

Exercise 5.37. Establish the countable case of Lemma 5.36, i.e., the case when $\{B_j\}_{j=1}^{\infty}$ is countably infinite (rather than finite).

Proposition 5.38 (Local Version of the Mass Distribution Principle). *Let μ be a mass distribution on $\mathbb{R}^m$, let $t \geq 0$, let $F \subseteq \mathbb{R}^m$ be a Borel set, and let $0 < c < \infty$ be a constant. Then:*

(i) *If $\limsup_{r \to 0^+} \mu(B_{\boldsymbol{x}}(r))/r^t < c$, for all $\boldsymbol{x} \in F$, then we have that $\mathcal{H}^t(F) \geq \mu(F)/c$.*

(ii) *If $\limsup_{r \to 0^+} \mu(B_{\boldsymbol{x}}(r))/r^t > c$, for all $\boldsymbol{x} \in F$, then we have that $\mathcal{H}^t(F) \leq 8^t \cdot \mu(\mathbb{R}^m)/c$.*

Proof. In order to prove (i), for each $\delta > 0$, let

$$F_\delta := \{\boldsymbol{x} \in F : \mu(B_{\boldsymbol{x}}(r)) < c \cdot r^t, \text{ for all } 0 < r \leq \delta\}. \tag{5.18}$$

Let $\{A_j\}_{j=1}^{\infty}$ be a δ-cover of F; then, it also covers F_δ. If $\boldsymbol{x} \in A_j$, for some $\boldsymbol{x} \in F_\delta$, then $A_j \subseteq B_{\boldsymbol{x}}(\operatorname{diam}(A_j))$. Hence, $\mu(A_j) \leq \mu(B_{\boldsymbol{x}}(\operatorname{diam}(A_j))) \leq c \cdot (\operatorname{diam}(A_j))^t$, and so, $\mu(F_\delta) \leq \sum_{A_j \cap F_\delta \neq \emptyset} \mu(A_j) \leq c \cdot (\sum_{j=1}^{\infty} (\operatorname{diam}(A_j))^t)$. This inequality holds for all $\delta > 0$; so, $\mu(F_\delta) \leq c \cdot \mathcal{H}^t_\delta(F)$, and thus, by Lemma 5.12, we have that $c \cdot \mu(F) \leq \mathcal{H}^t(F)$.

In order to prove (ii), suppose F is bounded. Fix $\delta > 0$ and let $\mathcal{B}$ be given by $\mathcal{B} := \{B_{\boldsymbol{x}}(r) : \boldsymbol{x} \in F,\ 0 < r \leq \delta, \text{ and } \mu(B_{\boldsymbol{x}}(r)) > c \cdot r^t\}$. Then, $F \subseteq \bigcup_{B \in \mathcal{B}} B$ and an application of Lemma 5.36 above yields a sequence $\{B_j\}_{j=1}^{\infty} \subseteq \mathcal{B}$ of pairwise disjoint balls such that $\bigcup_{B \in \mathcal{B}} B \subseteq \bigcup_j \widetilde{B}_j$, where, for each j, $\widetilde{B}_j$ is the closed ball concentric with B_j and of four times the radius. Hence, $\{\widetilde{B}_j\}_{j=1}^{\infty}$ is an 8δ-cover of F. Therefore, since the balls B_j are disjoint, we have successively

$$\begin{aligned} \mathcal{H}^t_{8\delta}(F) &\leq \sum_{j=1}^{\infty} (\operatorname{diam}(\widetilde{B}_j))^t \leq 4^t \cdot \left(\sum_{j=1}^{\infty} (\operatorname{diam}(B_j))^t \right) \\ &\leq (8^t/c) \cdot \left(\sum_{j=1}^{\infty} \mu(B_j) \right) \leq (8^t/c) \cdot \mu(\mathbb{R}^m). \end{aligned} \tag{5.19}$$

Letting $\delta \to 0^+$ in (5.19) yields $\mathcal{H}^t(F) \leq (8^t \cdot \mu(\mathbb{R}^m))/c < \infty$, as desired.

In order to finish the proof of (ii), suppose F is unbounded and $\mathcal{H}^t(F) > (8^t \cdot \mu(\mathbb{R}^m))/c$. Then, the $\mathcal{H}^t$-measure of some bounded subset of F will also exceed $(8^t \cdot \mu(\mathbb{R}^m))/c$, leading to a contradiction.[12] This completes the proof of the proposition. □

One key application of the Mass Distribution Principle (Theorem 5.35) comes in the context of self-similar sets and is expressed fully in Moran's Theorem (Theorem 5.40), which is stated and proved below. The following lemma will be helpful in the proof of this important theorem.

Lemma 5.39. *Let $0 < c_1 < c_2$ and $r > 0$. Assume that $\{V_j\}$ is a collection of disjoint open subsets of $\mathbb{R}^m$ such that each V_j contains a ball of radius $c_1 \cdot r$ and is contained in a ball of radius $c_2 \cdot r$. Then, a ball B of radius r intersects at most $(1 + 2c_2)^n \cdot c_1^{-n}$ of the closures $\overline{V_j}$.*

Proof. If $\overline{V_j}$ intersects B, then $\overline{V_j}$ is contained in the ball concentric with B of radius $(1 + 2c_2) \cdot r$. Suppose q of the sets $\overline{V_j}$ intersect B. Then, by considering the sum of the volumes of the q balls interior to each V_j with radii $c_1 \cdot r$, we have $q(c_1 \cdot r)^n \leq (1 + 2c_2)^n \cdot r^n$. Thus, $q \leq (1 + 2c_2)^n \cdot c_1^{-n}$, as desired. □

The following theorem, known as Moran's Theorem, is a significantly stronger version of Proposition 5.30 in Section 5.3 above. It is essentially due to the probabilist and statistician Patrick Alfred Pierre Moran in the mid-1940s, [**452**]. Furthermore, its proof has been helpful since then in related contexts, such as the study of analysis on fractals or of other nonsmooth spaces.

Theorem 5.40 (Moran's Theorem for Self-Similar Sets: Equality of Hausdorff, Box, Minkowski, and Similarity Dimensions; P. A. Moran, [**452**], 1946). *Let $N \geq 2$ be an integer and let $\mathbf{\Phi} = \{\varphi_j\}_{j=1}^N$ be a self-similar system on $\mathbb{R}^m$ that satisfies the Open Set Condition (OSC, stated in Definition 4.65) and with scaling vector $\mathbf{r} = (r_1, \ldots, r_N)$ comprising positive (but not necessarily distinct) scaling ratios. If*

$$F = \mathbf{\Phi}(F) := \bigcup_{j=1}^{N} \varphi_j(F) \tag{5.20}$$

(that is, if F is the unique attractor of $\mathbf{\Phi}$), then the box-counting (or box) dimension, $\dim_B(F)$, and the Minkowski dimension, $\dim_M(F)$, of F exist and coincide with both the Hausdorff dimension, $\dim_H(F)$, and the similarity dimension, D_S, of F:

$$\dim_H(F) = \dim_B(F) = \dim_M(F) = D_S, \tag{5.21}$$

[12]For instance, let $\{A_n := F \cap B_0(n)\}_{n=1}^{\infty}$ such that $A_n \nearrow F$ and use the monotonicity of the Hausdorff measure.

where D_S is the similarity dimension of F given by the unique real (and hence also positive) solution of the corresponding Moran equation

$$\sum_{j=1}^{N} r_j^t = 1,\ t \in \mathbb{R}. \tag{5.22}$$

Additionally, we have that $0 < \mathscr{H}^{D_S}(F) < \infty$.[13]

Proof. Only the equation $\dim_H(F) = D_S$ is addressed here. The case involving the box-counting dimension $\dim_B(F)$ is addressed in Chapter 8; see Lemma 8.58 and Proposition 8.39, along with Remark 8.59.

Moreover, as we will see in Theorem 8.29 (identity (8.81)) in Section 8.4, we always have that $\dim_B(A) = \dim_M(A)$, for any bounded subset A of $\mathbb{R}^m$ for which one of these fractal dimensions exists. In particular, since $F \subseteq \mathbb{R}^m$ is compact and $\dim_B(F)$ exists, as follows from the aforementioned results in Section 8.4 below, we have that $\dim_B(F) = \dim_M(F)$.

Let $k \in \mathbb{N}$. By iterating the map $\mathbf{\Phi}$ on F, (5.20) yields (see Notation 4.47 and Theorem 4.48) $F = \bigcup_{J\in\mathcal{J}_k} F_J$. Also, for each $J \in \mathcal{J}_k$, the mapping $\varphi_J := \varphi_{j_1}\circ\cdots\circ\varphi_{j_k}$ is a contracting similarity with scaling ratio $r_J = r_{j_1}\cdots r_{j_k}$.[14] Hence, by induction on Moran's equation (5.22), we deduce that (since $t := D_S$)[15]

$$\begin{aligned} (\operatorname{diam}(F))^{D_S} &= \left(\sum\nolimits_{j=1}^{N} r_j^t\right)^k \cdot (\operatorname{diam}(F))^{D_S} \\ &= \sum_{J\in\mathcal{J}_k} \left(r_J^{D_S} \cdot (\operatorname{diam}(F))^{D_S}\right) = \sum_{J\in\mathcal{J}_k} (\operatorname{diam}(F_J))^{D_S}. \end{aligned} \tag{5.23}$$

For each $\delta > 0$, choose k so that $(\operatorname{diam}(F_J))^k \leq (r_{\mathbf{\Phi}})^k = (\max_j r_j)^k \leq \delta$. Then, $\mathscr{H}_\delta^{D_S}(F) \leq (\operatorname{diam}(F))^{D_S}$, for all $\delta > 0$, which implies that $\mathscr{H}^{D_S}(F) \leq (\operatorname{diam}(F))^{D_S} < \infty$, and therefore (by Lemma 5.12), $D_S \geq \dim_H(F)$.

In order to obtain a lower bound for $\dim_H(F)$, we will construct a mass distribution on $\mathcal{J}_\infty$ which can be converted in a natural way to a mass distribution on F via the Carathéodory–Hahn Theorem (Theorem 3.57), allowing for the use of the Mass Distribution Principle (Theorem 5.35).[16]

Let $J \in \mathcal{J}$ (that is, J is a finite word), and let $\mathcal{J}_J \subseteq \mathcal{J}_\infty$ be the so-called *cylinder* of infinite words whose initial terms are given by $J = j_1 \dots j_{|J|}$. That is, if $K \in \mathcal{J}_J$, then $K|_{|J|} = J$. Define a mass distribution μ on $\mathcal{J}_\infty$ such that

[13] Just as in Propositions 5.27 and 5.28, it is hard to find a more explicit lower bound for $\mathscr{H}^{D_S}(F)$.

[14] Here, $J = (j_1, \dots, j_k) \in (\{1, \dots, N\})^k$ where $k = |J|$, the length of the finite word J over the alphabet $\{1, \dots, N\}$.

[15] Clearly, $\operatorname{diam}(F_J) = r_J \cdot \operatorname{diam}(F)$, where $F_J := \varphi_J(F)$.

[16] As we have seen in Section 5.3 and earlier in the present section, obtaining a lower bound for $\mathscr{H}^t(F)$ is not a simple matter, in general.

$\mu(\mathcal{J}_J) = r_J^{D_S}$; see Definition 5.34 above. Since $r_J^{D_S} = \sum_{j=1}^N (r_J \cdot r_j)^{D_S}$,[17] we have $\mu(\mathcal{J}_J) = \sum_{j=1}^N \mu(\mathcal{J}_{Jj})$. Then, by the Carathéodory–Hahn Theorem (Theorem 3.57), we have that μ can be extended to a mass distribution (i.e., a measure) on $\mathcal{J}_\infty$ such that $\mu(\mathcal{J}_\infty) = 1$.

Now, define a measure μ^* supported on F by

$$\mu^*(A) = \mu(\{J : J \in \mathcal{J}_\infty, \boldsymbol{x}_J \in A\}),$$

where $A \subseteq F$. Then, $\mu^*(F) = 1$. Let V be a feasible open set, in the sense of Definition 4.65 for the OSC. Since $\bigcup_{j=1}^N \varphi_j(\overline{V}) = \mathbf{\Phi}(\overline{V}) \subseteq \overline{V}$, the decreasing sequence of compact sets $\mathbf{\Phi}^k(\overline{V})$ converges to F, as $k \to \infty$, in the Hausdorff metric on $\mathbb{R}^m$; see Proposition 4.19. Moreover, $F \subseteq \overline{V}$ and for each finite word $J \in \mathcal{J}$, we have that $F_J \subseteq \overline{V_J}$. Let B be any ball of radius $r < 1$. The value of $\mu^*(B)$ is to be estimated by considering the sets V_J with diameters comparable to B with closures intersecting $F \cap B$.

Next, let $J \in \mathcal{J}_\infty$ and truncate J at the first term j_k for which

$$\big(\min_j r_j\big) \cdot r \leq r_{J|_k} \leq r, \tag{5.24}$$

where the minimum is taken over all $j \in \{1, \dots, N\}$. Let $\mathcal{Q}$ denote the set of all (finite) words obtained in this manner. Then, for each $J \in \mathcal{J}_\infty$, there is exactly one index $k \in \mathbb{N}$ such that $J|_k \in \mathcal{Q}$. By the Open Set Condition, the sets V_j are pairwise disjoint for all $j = 1, \dots, N$. Thus, the sets V_{Kj} are likewise pairwise disjoint, for each $K \in \mathcal{Q}$ and all $j = 1, \dots, N$. Therefore, the collection $\{V_K : K \in \mathcal{Q}\}$ is pairwise disjoint and, moreover, $F \subseteq \bigcup_{K \in \mathcal{Q}} F_K \subseteq \bigcup_{K \in \mathcal{Q}} \overline{V_K}$.

Choose $0 < c_1 < c_2$ such that V contains a ball of radius $c_1 \cdot r$ and is contained in a ball of radius $c_2 \cdot r$. Then, for each $K \in \mathcal{Q}$, the set V_K contains a ball of radius $r_K \cdot c_1$ and thus, also, a ball of radius $(\min_j r_j) \cdot c_1 \cdot r$. Similarly, for each $K \in \mathcal{Q}$, the set V_K is contained in a ball of radius $r_K \cdot c_2$—and thus, also in a ball of radius $c_2 \cdot r$. Let $\mathcal{Q}_1 \subseteq \mathcal{Q}$ denote the words $K \in \mathcal{Q}$ such that B intersects $\overline{V_K}$. We deduce from Lemma 5.39 that there are at most $q = (1 + 2c_2)^n \cdot (c_1 \cdot \min_j r_j)^{-n}$ words in $\mathcal{Q}_1$. Then,

$$\mu^*(B) = \mu^*(F \cap B) \leq \mu\Big(\bigcup_{K \in \mathcal{Q}_1} \mathcal{J}_K\Big), \tag{5.25}$$

since if $J \in \mathcal{Q}$ and $\boldsymbol{x}_J \in F \cap B \subseteq \bigcup_{K \in \mathcal{Q}_1} \overline{V_K}$, there is an integer $k \in \mathbb{N}$ such that $J|_k \in \mathcal{Q}_1$. Therefore, by (5.24) and (5.25), we have that

$$\mu^*(B) \leq \sum_{K \in \mathcal{Q}_1} \mu(\mathcal{J}_K) = \sum_{K \in \mathcal{Q}_1} r_K^{D_S} \leq \sum_{K \in \mathcal{Q}_1} r^{D_S} \leq q \cdot r^{D_S}.$$

[17] Recall from Moran's equation (5.22) that $\sum_{j=1}^N r_j^{D_S} = 1$.

Since any set U is contained in a ball of radius diam(U), we have $\mu^*(U) \leq q \cdot (\mathrm{diam}(U))^{D_S}$. So, the Mass Distribution Principle (Theorem 5.35) applies and yields that $0 < q^{-1} \leq \mathscr{H}^{D_S}(F)$ and $D_S \leq \dim_H(F)$. Hence, $\dim_H(F) = D_S$, as desired.

Since $F \subseteq \mathbb{R}^m$ is bounded, the fact that $\dim_B(F)$ and $\dim_B(M)$ exist and coincide with $\dim_H(F) = D_S$, as claimed in (5.21), is explained in Remark 8.59 of Chapter 8, which makes use of Theorem 8.29 as well as of Lemma 8.58 and Proposition 8.39 in Sections 8.4, 8.5, and 8.7 below.

This concludes the proof of Theorem 5.40. □

We refer back to Chapter 4 (Example 4.69) for various examples of self-similar sets (including the Cantor set, the Sierpinski gasket, the Sierpinski carpet, and the Koch curve) which all satisfy the Open Set Condition—and, hence, to which Moran's Theorem (Theorem 5.40) can be applied in order to calculate the corresponding fractal dimensions, as will be seen in Example 5.43 below, by using the next corollary.

Corollary 5.41. *Assume, in addition to the hypotheses of Moran's Theorem (Theorem 5.40), that the scaling ratios of the self-similar system $\mathbf{\Phi}$ are all equal; namely, $r_1 = \cdots = r_N =: r$, say, with $r \in (0,1)$. Then, the fractal dimensions of the associated self-similar set $F = F_{\mathbf{\Phi}}$ (the attractor of the self-similar system $\mathbf{\Phi}$) coincide with the similarity dimension D_S and are given by*

$$\dim_H(F) = \dim_B(F) = \dim_M(F) = D_S = \frac{\log N}{\log(1/r)} = \log_{r^{-1}} N. \tag{5.26}$$

Proof. In light of Moran's equation (5.22), we have that $Nr^{D_S} = 1$; i.e., $r^{-D_S} = N$ or (upon taking the logarithm of both sides) $D_S = \log N / \log(1/r)$, as claimed. The desired result (5.26) then follows by combining this equality with the identity (5.21) of Theorem 5.40. □

Remark 5.42 (*Similarity Dimension, Self-Similar Set, and OSC*). It follows at once from Moran's Theorem (Theorem 5.40) that *the similarity dimension D_S of a self-similar set F in $\mathbb{R}^m$ is independent of the choice of the self-similar system $\mathbf{\Phi}$ satisfying the OSC and having F as its attractor* (assuming that there is at least one such IFS $\mathbf{\Phi}$).

Indeed, according to identity (5.21) in the conclusion of Moran's Theorem (5.40), D_S must be equal to the common value of $\dim_H(F)$, $\dim_M(F)$, and $\dim_B(F)$, which obviously only depends on F. In this specific sense, the similarity dimension D_S of a self-similar set F satisfying the hypotheses of Moran's Theorem is "intrinsically" well defined: $D_S = D_S(\mathbf{\Phi}) = D_S(\mathbf{\Phi}^*)$, for *any* self-similar system $\mathbf{\Phi}^*$ satisfying the OSC.

Example 5.43 (*Fractal Dimensions of Classic Self-Similar Sets*). In each of the classic examples of self-similar sets $F \subseteq \mathbb{R}^m$ satisfying the Open Set Condition

(OSC), we can apply Theorem 5.40 (Moran's Theorem) and, really, Corollary 5.41, in order to deduce that the fractal dimensions of F are all equal and coincide with the similarity dimension D_S of the corresponding self-similar system $\mathbf{\Phi}$, as in Corollary 5.41 above; i.e., (5.26) holds. We refer the reader to Example 4.69 for a detailed justification of the fact that each of the self-similar sets discussed in (a)–(e) below satisfies the OSC.

By necessity of concision, we will limit ourselves to a few examples of self-similar sets discussed in Chapters 1, 4, and 9 (see, especially, Sections 4.3, 9.5, and 9.6) by considering $F \subseteq \mathbb{R}^m$, with $m = 1, 2$, or 3, and by simply giving the value of D_S, which also coincides with $\dim_H(F)$, and with $\dim_B(F) = \dim_M(F)$ (which both exist, by Theorem 5.40 or Corollary 5.41). The reader should keep in mind, however, that in each case (i.e., in (a)–(e)), identity (5.26) holds.

(a) (*Cantor Set*). Let $F := C$ be the classic (or ternary) Cantor set; see Example 4.23 in Section 4.3. Then, $N = 2$, $r = 1/3$ and thus, $D_S = \log_3 2 \approx 0.631 \in (0, 1)$.

(b) (*Sierpinski Gasket*). Let $F := \mathbf{SG}$ be the Sierpinski gasket; see Example 4.33 in Section 4.3. Then, $N = 3, r = 1/2$ and hence, $D_S = \log_2 3 \approx 1.585 \in (1, 2)$.

(c) (*Koch Curve*). Consider the Koch curve $F := K_C$; see Example 4.31 in Section 4.3. Then, $N = 4, r = 1/3$ and so, $D_S = \log_3 4 \approx 1.262 \in (1, 2)$. As we will see in Section 9.5, exactly the same conclusion holds (i.e., (5.26) holds for this same value of D_S) for the Koch snowflake curve $\mathbf{\Gamma}_{KS}$.

(d) (*Sierpinski Carpet*). Let $F := \mathbf{SC}$ denote the Sierpinski carpet; see Example 4.34 in Section 4.3. Then, $N = 8, r = 1/3$ and hence, $D_S = \log_3 8 \approx 1.893 \in (1, 2)$.

(e) (*Menger Sponge*). Finally, consider the Menger sponge; see Example 4.34 in Section 4.3. Then, $N = 20, r = 1/3$ and so, $D_S = \log_3 20 \approx 2.727 \in (2, 3)$.

Exercise 5.44. (*Higher-Dimensional Sierpinski Carpets*). Generalize the construction of the Sierpinski carpet **SC** and of the Menger sponge **MS** to $\mathbf{SC}^{(m)}$, the corresponding self-similar set in $\mathbb{R}^m$, where $m \geq 2$ is an arbitrary integer, so that $\mathbf{SC}^{(2)} = \mathbf{SC}$ and $\mathbf{SC}^{(3)} = \mathbf{MS}$, as in parts (d) and (e) of Example 5.43 just above. Then, calculate the common value of the similarity, Hausdorff, box-counting, and Minkowski dimensions of $\mathbf{SC}^{(m)}$.

Remark 5.45. In order to motivate the following result which addresses the insight discussed in this remark, consider a self-similar system $\mathbf{\Phi} = \{\varphi_j\}_{j=1}^N$ with attractor F. For each $k \in \mathbb{N}$, the self-similar system given by the N^k contractions of the form φ_J, where $J \in \mathcal{J}_k$, has F as its attractor. In this case, the

Moran equation is given by

$$\sum_{J\in\mathcal{J}_k} r_J^t = \left(\sum_{j=1}^N r_j^t\right)^k = 1,\ t\in\mathbb{R}. \tag{5.27}$$

By Moran's Theorem (Theorem 5.40), $\dim_H(A) = \dim_H(B) = D_S$ is the unique positive real number such that $\sum_{j=1}^N r_j^t = 1$. Therefore, D_S is also the unique positive real solution of (5.27). This insight motivates the following corollary.

Corollary 5.46. *Let $\mathbf{\Phi} = \{\varphi_j\}_{j=1}^N$, with $N \geq 2$, be a self-similar system on $\mathbb{R}^m$ that satisfies the Open Set Condition with scaling vector $\mathbf{r} = (r_1, \ldots, r_N)$, comprising positive scaling ratios. Let F denote the attractor of $\mathbf{\Phi}$ and let D_S denote the similarity dimension of F. Then, in addition to being given by* (5.21) *in Theorem* 5.40*, D_S is also given by the following equality:*

$$D_S = \inf\left\{t\in\mathbb{R} : \sum_{k=0}^{\infty}\left(\sum_{j=1}^N r_j^t\right)^k < \infty\right\}. \tag{5.28}$$

Proof. In light of the equality in (5.27), the result follows from Moran's Theorem (Theorem 5.40) and an application of the Root Test for the series with general term given by $y_k := \left(\sum_{j=1}^N r_j^t\right)^k$, for each $k\in\mathbb{N}$. □

It is also relatively easy to find an upper bound for the Hausdorff dimension of a self-similar set or, more generally, of an iterated function system (IFS), as we next explain.

Proposition 5.47 (Upper Bound for the Hausdorff Dimension of the Attractor of an IFS). *Let $N \geq 2$ and let $\mathbf{\Psi} = \{\psi_j\}_{j=1}^N$ be an iterated function system on $\mathbb{R}^m$ such that for any $\boldsymbol{x}, \boldsymbol{y} \in \mathbb{R}^m$,*

$$\|\psi_j(\boldsymbol{x}) - \psi_j(\boldsymbol{y})\|_m \leq r_j \cdot \|\boldsymbol{x} - \boldsymbol{y}\|_m,$$

where $0 < r_j < 1$ for each $j = 1, \ldots, N$. Then, $\dim_H(F) \leq D$, where D is the unique (and necessarily positive) real number such that $\sum_{j=1}^N r_j^t = 1$.

Exercise 5.48. Give a proof of Proposition 5.47 just above.

In general, an appropriate lower bound for the Hausdorff dimension (and thus also for the lower box-counting dimension or Minkowski dimension) of the attractor of an iterated function system is more difficult to find. However, in the strongly separated case where the attractor is necessarily totally disconnected, a lower bound can be found yet again through an application of the Mass Distribution Principle (Theorem 5.35); see Definition 4.73 (of the Strong Separation Condition) and Remark 4.74.

Proposition 5.49 (Lower Bound for the Hausdorff Dimension of the Attractor of a Suitable IFS). *Assume that $N \geq 2$ and that $\mathbf{\Psi} = \{\psi_j\}_{j=1}^N$ is a strongly separated iterated function system on $\mathbb{R}^m$ (in the sense of Definition 4.73 in Section 4.7 above) such that $c_j \cdot \|\boldsymbol{x} - \boldsymbol{y}\|_m \leq \|\psi_j(\boldsymbol{x}) - \psi_j(\boldsymbol{y})\|_m$, where $0 < c_j < 1$ for each $j = 1, \ldots, N$ and for any $\boldsymbol{x}, \boldsymbol{y} \in \mathbb{R}^m$. Then, $D \leq \dim_H(F)$, where D is the unique (and necessarily positive) real number such that $\sum_{j=1}^N c_j^t = 1$.*

Proof. Note that $\dim_H(F) \leq \underline{\dim}_B(F)$ follows from Proposition 8.39 in Section 8.5 below. In order to show that $D \leq \dim_H(F)$, apply the Mass Distribution Principle (Theorem 5.35).

For each finite word $J \in \mathcal{J}$, let $\mu(F_J) := c_J^D$, where $F_J := \psi_J(F) = \psi_{j_1} \circ \cdots \circ \psi_{j_{|J|}}(F)$ and $c_J := c_{j_1} \cdots c_{j_{|J|}}$. Observe that

$$\sum_{j=1}^{N} \mu(F_{Jj}) = \sum_{j=1}^{N} (c_J \cdot c_j)^D = c_J^D \cdot \left(\sum_{j=1}^{N} c_j^D \right) = \mu(F_J) = \mu\left(\bigcup_{j=1}^{N} F_{Jj} \right).$$

Hence, μ extends to a mass distribution on F such that $\mu(F) = 1$.

Now, let $\boldsymbol{x} \in F$. By Theorem 4.48, there is a unique address given by an infinite word $J \in \mathcal{J}_\infty$ such that $\boldsymbol{x} \in F_{J|_k}$, for all $k \in \mathbb{N}$. For each $0 < r < \delta$, let k denote the smallest integer such that $\delta \cdot c_{J|_k} \leq r < \delta \cdot c_{J|_{k-1}}$. If $K \in \mathcal{J}_\infty$ is such that $J|_k \neq K|_k$, then $F_{J|_k}$ and $F_{K|_k}$ are disjoint and therefore, $r < \delta \cdot c_{J|_{k-1}} \leq \inf\{\|\boldsymbol{x} - \boldsymbol{y}\|_m : \boldsymbol{x} \in F_{J|_k}, \boldsymbol{y} \in F_{K|_k}\}$. So, $F \cap B_{\boldsymbol{x}}(r) \subseteq F_{J|_k}$, and hence, $\mu(F \cap B_{\boldsymbol{x}}(r)) \leq \mu(F_{J|_k}) = c_{J|_k}^D \leq \delta^{-D} \cdot r^D$. If U is a bounded set that intersects F, then $U \subseteq B_{\boldsymbol{x}}(r)$, for some $\boldsymbol{x} \in F$ and where $r := \operatorname{diam}(U)$. Thus, $\mu(U) \leq \delta^{-D} \cdot (\operatorname{diam}(U))^D$, and by the Mass Distribution Principle (Theorem 5.35), we have $\mathcal{H}^D(F) > 0$ and $\dim_H(F) \geq D$. □

A generalization of Moran's Theorem (Theorem 5.40) holds for the attractors of iterated function systems on $\mathbb{R}$ whose contractions are twice differentiable and such that the first derivatives are bounded between 0 and 1, strictly. Indeed, compositions of such contractions are nearly similarity contractions. Also, in order to motivate the choice of the function g made in (5.29) just below, see Remark 5.45 and Corollary 5.46 above.

Theorem 5.50 (Nonlinear IFSs in $\mathbb{R}$ and Moran-Type Equation). *Let $N \geq 2$ and let $\mathbf{\Psi} = \{\psi_j\}_{j=1}^N$ be an iterated function system on $\mathbb{R}$ with attractor F. Suppose $\mathbf{\Psi}$ satisfies the Open Set Condition, such that each contraction ψ_j is twice differentiable on a feasible open set (i.e., here, an open interval) $V \supseteq F$, continuous on $\overline{V}$, and for some constants $0 < a < b < \infty$, we have $a \leq |\psi_j'(x)| \leq b$, for each $j = 1, \ldots, N$ and any $x \in V$.*

Then, for each $t > 0$, *the limit*

$$\lim_{k\to\infty}\Big(\sum_{J\in\mathcal{J}_k}|\psi_J'(x)|^t\Big)^{1/k} =: g(t) \tag{5.29}$$

exists and is independent of $x \in V$.[18] *Moreover,* $g = g(t)$ *is strictly decreasing on* $(0,+\infty)$, *and* $\dim_H(F) = \dim_B(F) = t_0$, *where* t_0 *is the unique positive real solution of the Moran-type equation* $g(t) = 1$, *with* $t \in \mathbb{R}$.

Sketch of the proof. Let $k \in \mathbb{N}$. Note that the iterated function system given by the N^k contractions of the form ψ_J, where $J \in \mathcal{J}_k$, has F as its attractor. Also, since each ψ_j is twice differentiable on U, the compositions ψ_J are also twice differentiable for all $J \in \mathcal{J}_k$. Let

$$a_J := \inf\{|\psi_J'(x)| : x \in V\} \quad \text{and} \quad b_J := \sup\{|\psi_J'(x)| : x \in V\}.$$

Then, for any $u, v \in V$, $a_J \cdot |u - v| \le |\psi_J(u) - \psi_J(v)| \le b_J \cdot |u - v|$. For each $k \in \mathbb{N}$, we apply Propositions 5.47 and 5.49 to the IFSs given by the N^k contractions of the form ψ_J, where $J \in \mathcal{J}_k$, to obtain upper and lower estimates for $\dim_H(F)$ and $\dim_B(F)$ denoted s_k and t_k, respectively. For large enough k, a_J/b_J is approximately 1, and the limit in (5.29) exists for each $t > 0$ and is independent of $x \in V$. In addition, $g(t) = 1$ iff we have $t = \lim_{k\to\infty} s_k = \lim_{k\to\infty} t_k = \dim_H(F) = \dim_B(F)$, as desired. □

Remark 5.51. Under the hypotheses of Theorem 5.50, more is true, in fact; namely, $0 < \mathcal{H}^t(F) < \infty$. We further note that this same theorem also applies to self-similar subsets of the real line.

Remark 5.45 and Corollary 5.46 also motivate an analog of Moran's Theorem (Theorem 5.40) which holds for many self-affine sets, the attractors of iterated function systems with contractions given by affine maps. See Definition 4.39 and Remark 4.40. The key insight amounts to replacing the scaling ratios in Corollary 5.46 with the *singular values* associated with the underlying affine maps. Note that a contracting nonsingular $m \times m$ square matrix has m positive eigenvalues (counting multiplicities). More information on this interesting topic can be found in Falconer's book [**172**], as well as in his earlier research article [**173**].

Definition 5.52 (*Singular Values*). Let $T : \mathbb{R}^m \to \mathbb{R}^m$ be a nonsingular $m\times m$ matrix that is also a contraction on $\mathbb{R}^m$. The *singular values* of T, denoted $1 > \alpha_1 \ge \alpha_2 \ge \cdots \ge \alpha_m > 0$, are the positive square roots of the (necessarily positive) eigenvalues of T^*T, where T^* is the adjoint of T. For $0 \le t \le m$, the *singular value function*, denoted $\sigma^t(T)$, is given by $\sigma^t(T) := \alpha_1 \cdot \alpha_2 \cdots \alpha_{n-1} \cdot \alpha_n^{t-(n-1)}$, where n is the unique positive integer for which $n - 1 < t \le n$.

[18] Here, ψ_J' denotes the derivative of the smooth contraction ψ_J.

Remark 5.53. The singular values $1 > \alpha_1 \geq \alpha_2 \geq \cdots \geq \alpha_m > 0$ of T encode the proportions by which T contracts in different directions. In particular, for a unit ball B in $\mathbb{R}^m$, $T(B)$ is an ellipsoid for which the lengths of the principal semi-axes are given by the singular values of T.

Lemma 5.54. *Let $\{T_j\}_{j=1}^N$ be a finite collection of contracting nonsingular $m \times m$ matrices, with $N \geq 2$. Then, there is a unique real number D_{sing} and called the* singular exponent *of $\{T_j\}_{j=1}^N$, such that $0 \leq D_{\text{sing}} \leq m$ and*

$$\lim_{k\to\infty}\left(\sum_{J\in\mathcal{J}_k}\sigma^{D_{\text{sing}}}(T_J)\right)^{1/k} = 1, \tag{5.30}$$

where, for each $k \in \mathbb{N}_0$, $\mathcal{J}_k$ denotes the set of finite words of length k on the alphabet $S := \{1, \ldots, N\}$. Additionally, D_{sing} is given by

$$D_{\text{sing}} = \inf\left\{t \in \mathbb{R} : \sum_{k=1}^{\infty}\sum_{J\in\mathcal{J}_k}\sigma^t(T_J) < \infty\right\}. \tag{5.31}$$

Proof. Note that for a contracting nonsingular square matrix T, we immediately have that $\sigma^t(T)$ is continuous and strictly decreasing in $t \geq 0$. Fix $t \geq 0$ and let X and Y be contracting nonsingular $m \times m$ matrices. Then,

$$\sigma^t(XY) \leq \sigma^t(X) \cdot \sigma^t(Y); \tag{5.32}$$

that is, σ^t is submultiplicative. Now, let $S_k^t := \sum_{J\in\mathcal{J}_k}\sigma^t(T_J)$, for each $k \in \mathbb{N}_0$. By (5.32), we have for fixed $t \geq 0$ that[19]

$$\begin{aligned} S_{k+n}^t &= \sum_{J\in\mathcal{J}_{k+n}}\sigma^t(T_J) \leq \sum_{J\in\mathcal{J}_{k+n}}\sigma^t(T_{j_1}\circ\cdots\circ T_{j_k})\sigma^t(T_{j_{k+1}}\circ\cdots\circ T_{j_{k+n}}) \\ &= \sum_{J\in\mathcal{J}_{k+n}}\left(\sigma^t(T_{j_1}\circ\cdots\circ T_{j_k})\right)\left(\sigma^t(T_{j_1}\circ\cdots\circ T_{j_n})\right) = S_k^t \cdot S_n^t. \end{aligned}$$

Since the maximum singular value among all of the singular values of the maps T_j lies strictly between 0 and 1, and since S_k^t is submultiplicative, we deduce from the Root Test for series that $\lim_{k\to\infty}(S_k^t)^{1/k} =: S^t$ exists. As a function of t, S^t is decreasing since σ^t is decreasing. Since $S^m \leq 1$, there is a unique value of t, denoted D_{sing}, such that (5.30) holds. Another application of the same Root Test, with terms of the series given by $y_k := S_k^t = \left(\sum_{j=1}^N \sigma^t(T_j)\right)^k$, for all $k \in \mathbb{N}_0$, shows that (5.31) holds. □

The number D_{sing} defined in Lemma 5.54 provides an upper bound for the Hausdorff dimension of self-affine sets, as we see next.

[19] Here, for each $J \in \mathcal{J}_{k+n}$, we write $J = (j_1, \ldots, j_k, j_{k+1}, \ldots, j_{k+n}) \in \mathcal{J}_k \times \mathcal{J}_n$.

Theorem 5.55 (Hausdorff Dimension and Singular Exponent; K. J. Falconer [**173**], 1988). *Let $N \geq 2$ and let $\mathbf{\Psi} = \{\psi_j\}_{j=1}^N = \{T_j(\cdot) + \mathbf{t}_j\}_{j=1}^N$ be a self-affine iterated function system on $\mathbb{R}^m$ with attractor F, where T_j is a nonsingular contracting $m \times m$ matrix and $\mathbf{t}_j \in \mathbb{R}^m$, for each $j = 1, \dots, N$. Then, we have that* $\dim_H(F) \leq D_{\text{sing}}$*; see also Remark* 5.56 *below.*

Proof. Let B be a closed ball in $\mathbb{R}^m$ large enough so that $\psi_j(B) \subseteq B$, for each $j = 1, \dots, N$. For a given $\delta > 0$, choose $k \in \mathbb{N}$ large enough so that $\operatorname{diam}(\psi_J(B)) < \delta$, for all $J \in \mathcal{J}_k$. By Theorem 4.48, $F \subseteq \bigcup_{J \in \mathcal{J}_k} \psi_J(B)$. Note that for each $J \in \mathcal{J}_k$, $\psi_J(B)$ is an ellipsoid with principal semi-axes of lengths $\alpha_1 \cdot b, \dots, \alpha_m \cdot b$, where $b = \operatorname{diam}(B)$ and $\alpha_1, \dots, \alpha_m$ are the singular values of T_J. Therefore, $\psi_J(B)$ is contained in a rectangular parallelepiped R with side lengths $2\alpha_1 \cdot b, \dots, 2\alpha_m \cdot b$.

Now, if $0 \leq t \leq m$ and $n \in \mathbb{N}$ is such that $n - 1 < t \leq n$, then R is covered by at most $4^m \cdot b^m \cdot \alpha_1 \cdots \alpha_{n-1} \cdot \alpha_n^{1-n}$ cubes with side length $\alpha_n \cdot b$. Therefore, $\psi_J(B)$ can be covered by a collection of m-dimensional cubes A_l, where $\operatorname{diam}(A_l) < \alpha_n \cdot \sqrt{m}$, such that

$$\sum_l (\operatorname{diam}(A_l))^t \leq 4^m \cdot b^m \cdot \alpha_1 \cdots \alpha_{n-1} \cdot \alpha_n^{1-n} \cdot (\alpha_n \cdot \sqrt{m})^t \leq (4b\sqrt{m})^m \cdot \sigma^t(T_J).$$

By taking such a $\delta\sqrt{m}$-cover $\{A_l\}$ of $\psi_J(B)$, for each $J \in \mathcal{J}_k$, it follows that $\mathcal{H}^t_{\delta\sqrt{m}}(F) \leq (4b\sqrt{m})^m \cdot \left(\sum_{J \in \mathcal{J}_k} \sigma^t(T_J)\right)$. Note that $\delta \to 0^+$ implies $k \to \infty$. Hence, by Lemma 5.54 (in particular (5.31)), and upon letting $\delta \to 0^+$ in the latter inequality, we deduce that $\mathcal{H}^t(F) = 0$ if $t > D_{\text{sing}}$. Thus, by Lemma 5.12, we conclude that $\dim_H(F) \leq D_{\text{sing}}$, as desired. □

Remark 5.56. In Theorem 5.55, equality holds under the following conditions: If $0 < \operatorname{Lip}(\psi_j) < 1/2$, for each $j = 1, \dots, N$, then (as is shown in [**173**]); see also [**172**]) $\dim_H(F) = D_{\text{sing}}$, *for almost all* $(\mathbf{t}_1, \dots, \mathbf{t}_N)$, in the sense of mN-dimensional Lebesgue measure. That is, if two self-affine iterated function systems comprise contractions given by the same collection of contracting nonsingular matrices $T_1, \dots, T_j$ but possibly different translation vectors, then the attractors have the same Hausdorff dimension given by D_{sing}, unless the translation vectors happen to fall within a set of mN-dimensional Lebesgue measure zero.

Theorem 5.55 above can be applied, for instance, to the (self-affine) Bedford–McMullen carpets (Example 4.44) and their natural generalizations.

5.5. Further Techniques for Calculating Hausdorff Dimension

In this section, we discuss various useful, additional methods for calculating or estimating the Hausdorff dimension of certain classes of fractal sets. We surely do not intend to provide an exhaustive list of such techniques but simply wish to give an introduction to this important topic, having its roots in analysis, measure theory, potential theory, and geometric measure theory, besides fractal geometry.

We begin by discussing the relationship between the length and the 1-dimensional Hausdorff measure of a suitable curve. The corresponding result (Proposition 5.61 below) will be useful in Chapter 9, in particular.

Theorem 5.57. *Let $t \geq 0$ and let E be a Borel subset of $\mathbb{R}^m$ such that $\mathcal{H}^t(E) = +\infty$. Then, there is a compact set $A \subseteq E$ such that $\mathcal{H}^t(A) < \infty$.*

We omit the proof of Theorem 5.57 as it is a bit technical and the corresponding result will not play a key role here, except in some aspects of the latter part of this section.

Definition 5.58 (*Jordan Arc*). Suppose that $a, b \in \mathbb{R}$ (with $a < b$), $m \geq 2$, and that $f : [a, b] \to \mathbb{R}^m$ is a continuous injective function. Then, it is called a *Jordan arc* or, equivalently, a *simple curve* (in $\mathbb{R}^m$).[20] We use the same terminology for the range of f, i.e., for $J = f([a, b])$.

For the purpose of this section, we will not distinguish between a Jordan arc, viewed as a parametrized curve, and its *image* J, defined as the image of $[a, b]$ under f; that is, $J = f([a, b])$. We are justified in doing so here because, according to Definition 5.58, f is injective.

Further discussion, as well as many examples, of Jordan arcs will be provided in Chapter 9. These examples will include the Devil's staircase (or Cantor graph), the Koch curve, the Koch snowflake (which is actually a Jordan curve, in the sense of footnote 20), along with the Sierpinski gasket and carpet; see Sections 9.4–9.7. On the other hand, as we will see in Section 8.6 (Remark 8.48, part (c)), a nontrivial result from topology implies that plane-filling curves (such as the Hilbert and Peano curves) and their higher-dimensional counterparts (called space-filing curves) are *not* Jordan arcs, because they must have multiple points. Moreover, the notion of a continuous (parametrized) curve will be discussed in Sections 9.1 and 9.2.

Remark 5.59. Jordan arcs are connected, compact, and not self-intersecting, and they have two endpoints; see Chapter 9, especially Section 9.1. On the other hand, Jordan curves, as defined in footnote 20, are closed curves and hence have a single endpoint. In the present section, they are viewed as special cases of Jordan arcs. In particular, Definition 5.60 and Proposition 5.61 below both apply to Jordan arcs as well as to Jordan curves.

---

[20]Furthermore, a *Jordan curve* is a closed Jordan arc, i.e., a Jordan arc such that $f(a) = f(b)$. In this case, one only requires that the restriction of f to the *semi-open* interval $[a, b)$ be injective. It can be viewed as a continuous map from S^1 to $\mathbb{R}^2$, where S^1 is the unit circle of $\mathbb{R}^2$ (or of $\mathbb{C}$).

Definition 5.60 (*Length of a Jordan Arc*). Let J be a Jordan arc, viewed as the image of $[a,b]$ under f. Then, the *length of J*, denoted $L(J)$, is given by $L(J) := \sup\left\{\sum_{j=1}^{n} \|f(x_j) - f(x_{j-1})\|\right\}$, where the supremum is taken over all partitions of $[a,b]$, i.e., over finite ordered collections $\{x_j\}_{j=0}^{n}$ of points in $[a,b]$ such that $x_0 := a < x_1 < \cdots < x_{n-1} < x_n := b$.

A Jordan arc J is said to be *rectifiable* if $L(J)$ is finite, and *unrectifiable*, otherwise.[21]

Note that the first part of the next proposition implies that rectifiable Jordan arcs with positive length are 1-sets, in the sense of Definition 5.14.

Proposition 5.61 (Equality of Length and of 1-Dimensional Hausdorff Measure). *If J is a rectifiable Jordan arc in $\mathbb{R}^m$ such that $L(J) > 0$, then*

$$\mathcal{H}^1(J) = L(J). \tag{5.33}$$

Furthermore, if J is unrectifiable (i.e., if $L(J) = +\infty$), then we also have that $\mathcal{H}^1(J) = L(J) = +\infty$.

Proof. Let $f([a,b]) = J$ and let $\boldsymbol{x}, \boldsymbol{y} \in J$. Let $x = f^{-1}(\boldsymbol{x})$ and $y = f^{-1}(\boldsymbol{y})$; clearly, without loss of generality, we may assume that $x < y$. Let $J_{\boldsymbol{x},\boldsymbol{y}} = f([x,y]) \subseteq J$. Then, $J_{\boldsymbol{x},\boldsymbol{y}}$ is a rectifiable Jordan arc. Also, let $\overline{\boldsymbol{xy}}$ denote the line segment connecting $\boldsymbol{x}$ and $\boldsymbol{y}$ in $\mathbb{R}^m$. Then, we clearly have that

$$\|\boldsymbol{x} - \boldsymbol{y}\| = \mathcal{H}^1(\overline{\boldsymbol{xy}}) \leq \mathcal{H}^1(J_{\boldsymbol{x},\boldsymbol{y}}),$$

as the reader can easily verify. So, for any finite ordered collection of points in $[a,b]$ such that $x_0 := a < x_1 < \cdots < x_{n-1} < x_n =: b$, we have that

$$\sum_{j=1}^{n} \|f(x_j) - f(x_{j-1})\| \leq \sum_{j=1}^{n} \mathcal{H}^1(J_{f(x_j),f(x_{j-1})}) \leq \mathcal{H}^1(J).$$

Taking the supremum as in Definition 5.60, over all partitions of $[a,b]$, yields $L(J) \leq \mathcal{H}^1(J)$.

Assuming for now that $L(J) < \infty$, let $g : [0, L(J)] \to J$ denote the function which maps u to the point $\mathbf{u} \in J$ such that $L(J_{f(a),\mathbf{u}}) = u$. Then, $\|g(u) - g(v)\| \leq |u - v|$, for all $u, v \in [0, L(J)]$. Thus, g is Lipschitz and by Proposition 5.9,[22] $\mathcal{H}^1(J) \leq \mathcal{H}^1([0, L(J)]) = L(J)$. Hence, $\mathcal{H}^1(J) = L(J)$, as claimed.

The proof in the unrectifiable case (i.e., when $L(J) = +\infty$) is entirely similar to the above proof, except for the fact that we replace the compact interval $[0, L(J)]$ by $[0, M]$, where $M \in (0, +\infty)$ is arbitrary. One then deduces that $\mathcal{H}^1(J) \geq L(J) = +\infty$—and hence, $\mathcal{H}^1(J) = +\infty$, as required. We leave the details as an exercise for the interested reader. □

[21] One also sometimes requires that $L(J)$ is positive (i.e., nonzero), but we will not do so here.

[22] This proposition is applied with $\alpha := 1$ and $c := 1$.

Proposition 5.62. *Let E be a Borel t-set. Then, there is a positive real number b and a compact set $A \subseteq E$ with $\mathcal{H}^t(A) > 0$ such that*

$$\mathcal{H}^t(A \cap B_{\boldsymbol{x}}(r)) \leq b \cdot r^t, \text{ for all } \boldsymbol{x} \in \mathbb{R}^m \text{ and all } r > 0. \tag{5.34}$$

Proof. Apply Proposition 5.38 to $\mu_E(\cdot) = \mathcal{H}^t(E \cap (\cdot))$, and let E_0 be given by $E_0 := \{\boldsymbol{x} \in \mathbb{R}^m : \limsup_{r\to 0^+} \mathcal{H}^t(E \cap B_{\boldsymbol{x}}(r))/r^t > 2^{1+t}\}$. Then, we have that $\mathcal{H}^t(E_0) \leq 2^t \cdot 2^{-(1+t)} \cdot \mu_E(E) = \mathcal{H}^t(E_0)/2$. Thus, it follows that $0 < \mathcal{H}^t(E)/2 \leq \mathcal{H}^t(E \backslash E_0)$; so, $\limsup_{r\to 0^+} \mathcal{H}^t(E \cap B_{\boldsymbol{x}}(r))/r^t \leq 2^{1+t}$, for all $\boldsymbol{x} \in E\backslash E_0$. By Egoroff's Theorem (see part (v) of Theorem 3.102), there is a compact set $A \subseteq E\backslash E_0$ with $\mathcal{H}^t(A) > 0$ and $r_0 > 0$ such that $\mathcal{H}^t(E \cap B_{\boldsymbol{x}}(r))/r^t \leq 2^{2+t}$, for all $\boldsymbol{x} \in A$ and $0 < r \leq r_0$. However, $\mathcal{H}^t(E \cap B_{\boldsymbol{x}}(r))/r^t \leq \mathcal{H}^t(E)/r_0^t$, for all $r \geq r_0$. Hence, (5.34) holds for all $r > 0$ and b defined accordingly. □

The following corollary is a sort of converse of the Mass Distribution Principle (Theorem 5.35) known in the literature as Frostman's Lemma.

Corollary 5.63 (Frostman's Lemma; O. A. Frostman, [**205**], 1935). *Let E be a Borel set such that $0 < \mathcal{H}^t(E) \leq \infty$. Then, there is a positive real number b and a compact set $A \subseteq E$ with $0 < \mathcal{H}^t(A) < \infty$ such that*

$$\mathcal{H}^t(A \cap B_{\boldsymbol{x}}(r)) \leq b \cdot r^t, \text{ for all } \boldsymbol{x} \in \mathbb{R}^m \text{ and all } r > 0. \tag{5.35}$$

Proof. This follows at once from Theorem 5.57 and Proposition 5.62. □

Remark 5.64. A key consequence of a slight generalization of Frostman's Lemma (Corollary 5.63 just above) is that $\dim_H(A)$, the Hausdorff dimension of a given Borel set $A \subseteq \mathbb{R}^m$, is equal to the supremum of all the values of $t \geq 0$ such that $\mathcal{C}^t(A) > 0$, where $\mathcal{C}^t(A)$ is the t-dimensional (Newtonian) capacity of A, a useful notion from potential theory. For more information about this subject, see, e.g., [**172**] and [**430**].

5.6. Density and Hausdorff Measure of Products

In this section, we introduce and briefly study the notions of Lebesgue and Hausdorff densities, as well as use some of the corresponding results in order to obtain inequalities or equalities concerning fractal dimensions of the Cartesian product of two subsets of Euclidean space $\mathbb{R}^m$. In the latter case, we limit ourselves to the most elementary results.

Definition 5.65 (*Lebesgue's Density*). Let $A \subseteq \mathbb{R}^2$ and $\boldsymbol{x} \in \mathbb{R}^2$. The *Lebesgue density* of A at $\boldsymbol{x}$ is given by

$$\operatorname{dens}(A, \boldsymbol{x}) := \lim_{r\to 0^+} \frac{\mathcal{L}^2(A \cap B_{\boldsymbol{x}}(r))}{\pi r^2}, \tag{5.36}$$

where $\mathcal{L}^2(B_{\boldsymbol{x}}(r)) = \pi r^2$ *is the area of a disk of radius* r.

The following theorem is a classical result of Lebesgue; see, e.g., [**91**], [**193**], [**521**], or [**523**].

Theorem 5.66 (Lebesgue's Density Theorem). *Let $A \subseteq \mathbb{R}^2$ and $\boldsymbol{x} \in \mathbb{R}^2$. Then,* $\operatorname{dens}(A, \boldsymbol{x}) = 1$, *for almost all $\boldsymbol{x} \in A$, and* $\operatorname{dens}(A, \boldsymbol{x}) = 0$, *for almost all $x \notin A$.*

In a similar fashion, and in order to motivate the definition of density with respect to t-dimensional Hausdorff measure, suppose J is a smooth Jordan curve or arc in the plane $\mathbb{R}^2$. By Proposition 5.61, $\dim_H(J) = 1$;[23] so, intuitively, one might expect that $\lim_{r\to 0+} \mathcal{H}^1(J \cap B_{\boldsymbol{x}}(r))/2r = 1$, where $\mathcal{H}^1$ is 1-dimensional Lebesgue measure (i.e., length) and $2r$ is the diameter of the disk $B_{\boldsymbol{x}}(r)$. This is indeed the case. That is, locally, smooth curves have the density of a diameter of a disk.

Definition 5.67 (*Upper and Lower Hausdorff Densities*). Let $A \subseteq \mathbb{R}^m$ be a t-set and let $\boldsymbol{x} \in \mathbb{R}^m$. The *upper* and *lower Hausdorff densities* of A at $\boldsymbol{x}$ are given by

$$\overline{\operatorname{dens}}^t(A, \boldsymbol{x}) := \limsup_{r\to 0+} \frac{\mathcal{H}^t(A \cap B_{\boldsymbol{x}}(r))}{(2r)^t} \quad \text{and} \tag{5.37}$$

$$\underline{\operatorname{dens}}^t(A, \boldsymbol{x}) := \liminf_{r\to 0+} \frac{\mathcal{H}^t(A \cap B_{\boldsymbol{x}}(r))}{(2r)^t}, \tag{5.38}$$

respectively. If $\overline{\operatorname{dens}}^t(A, \boldsymbol{x}) = \underline{\operatorname{dens}}^t(A, \boldsymbol{x})$, then the *density* of A at $\boldsymbol{x}$ is said to *exist* and $\operatorname{dens}^t(A, \boldsymbol{x})$ denotes the common value.

Definition 5.68. Let $A \subseteq \mathbb{R}^m$ be a t-set,[24] and let $\boldsymbol{x} \in \mathbb{R}^m$. If $\overline{\operatorname{dens}}^t(A, \boldsymbol{x}) = \underline{\operatorname{dens}}^t(A, \boldsymbol{x}) = 1$, then $\boldsymbol{x}$ is called a *regular* point of A. Otherwise, $\boldsymbol{x}$ is called an *irregular* point of A. Additionally, the set A is *regular* if $\mathcal{H}^t$-almost all points in A (i.e., all of its points except perhaps for a set of $\mathcal{H}^t$-measure 0) are regular, and A is *irregular* if $\mathcal{H}^t$-almost all points in A are irregular.

We caution the reader that even for a Borel set $A \subseteq \mathbb{R}^m$, 'irregular' does not imply 'not regular'.

Proposition 5.69. *Let $A \subseteq \mathbb{R}^m$ be a t-set. Then, the following two properties hold:*

(i) $\overline{\operatorname{dens}}^t(A, \boldsymbol{x}) = \underline{\operatorname{dens}}^t(A, \boldsymbol{x}) = 0$, *for $\mathcal{H}^t$-almost all $\boldsymbol{x} \notin A$.*

(ii) $2^{-t} \leq \overline{\operatorname{dens}}^t(A, \boldsymbol{x}) \leq 1$, *for $\mathcal{H}^t$-almost all $\boldsymbol{x} \in A$.*

[23] Indeed, by Theorem 9.3 in Section 9.1 below, it follows that J is rectifiable. Hence, by Proposition 5.61, $0 < \mathcal{H}^1(J) = L(J) < \infty$; thus, $\dim_H(J) = 1$.

[24] Recall from Definition 5.14 and the discussion following it that this means that $0 < \mathcal{H}^t(A) < \infty$, from which it follows that $\dim_H(A) = t$.

Proof. For part (i), we prove only the case where A is closed. When A is not closed, the proof is more subtle and involved and is therefore omitted here. Suppose A is closed and $\boldsymbol{x} \notin A$. Then, for all positive r sufficiently small, we have that $A \cap B_{\boldsymbol{x}}(r) = \emptyset$ and so, $\overline{\text{dens}}^t(A, \boldsymbol{x}) = 0$.

In order to prove part (ii), apply Proposition 5.38 to the measure $\mu(\cdot) = \mathcal{H}^t(A \cap (\cdot))$. Let A_0 be given by $A_0 := \{\boldsymbol{x} \in A : \overline{\text{dens}}^t(A, \boldsymbol{x}) < 2^{-t} \cdot c\}$. Then, by part (i) of Proposition 5.38, $\mathcal{H}^t(A_0)/c \leq \mathcal{H}^t(A) \leq \mathcal{H}^t(A_0)$. However, for $0 < c < 1$, this inequality only holds when $\mathcal{H}^t(A_0) = 0$. So, for $\mathcal{H}^t$-almost all $\boldsymbol{x} \in A$, we have $2^{-t} \leq \overline{\text{dens}}^t(A, \boldsymbol{x})$. The upper bound of part (ii) follows from an application of part (ii) of Proposition 5.38 and a similar argument.[25] □

Remark 5.70. Proposition 5.69 leads to many interesting results regarding the structure of t-sets. For a t-set A, if $E \subseteq A$ and $\mathcal{H}^t(E) > 0$, then E is regular (resp., irregular) when A is regular (resp., irregular). Also, if t is not an integer, then A is irregular.

Exercise 5.71. Prove the statements made in Remark 5.70 just above.

Definition 5.72. A 1-set is said to be *arc-like* (resp., *curve-like*) if it is a subset of a countable union of rectifiable Jordan arcs (resp., curves). On the other hand, a 1-set A is said to be *arc-free* (resp., *curve-free*) if $\mathcal{H}^1(A \cap J) = 0$, for all rectifiable Jordan arcs (resp., curves) J.

Remark 5.73. 1-sets cannot only be decomposed into regular and irregular parts which, in turn, can be characterized in terms of *curve-like* and *curve-free*. We refer the interested reader to Section 5.2 of **[172]** for a much more detailed analysis of these topics.

Here are some of the main properties of 1-sets:

(i) A rectifiable Jordan arc or curve is a 1-set; this is a consequence of Proposition 5.61.

(ii) If A is a 1-set, then the set of regular points of A is a regular set, and the set of irregular points of A is an irregular set.

(iii) An arc-like (or a curve-like) 1-set is a regular 1-set.

(iv) A 1-set is irregular iff it is arc-free (or curve-free).

(v) A 1-set is regular iff it is the union of an arc-like (or of a curve-like) set and a set of $\mathcal{H}^1$-measure 0.

We close this section by applying the above results concerning the Hausdorff density in order to obtain an inequality for the Hausdorff measure of a Cartesian product.

[25] Recall that Proposition 5.38 is the local version of the Mass Distribution Principle.

Definition 5.74 (Cartesian Product). For a pair of sets $A \subseteq \mathbb{R}^m$ and $B \subseteq \mathbb{R}^n$, the *Cartesian product* (or simply *product*) $A \times B$ is given by the following expression: $A \times B = \{(\boldsymbol{x}, \boldsymbol{y}) \in \mathbb{R}^{m+n} : \boldsymbol{x} \in A, \boldsymbol{y} \in B\}$.

Example 5.75. Consider the Cartesian product $C \times C \subseteq \mathbb{R}^2$, where C is the classic ternary Cantor set and which corresponds to the choice $A = B := C$. Recall from Exercise 4.27 in Section 4.3 that $C \times C$ is a self-similar set in $\mathbb{R}^2$ and is called the *Cantor dust.*

Theorem 5.76. *Let $A \subseteq \mathbb{R}^m$ and $B \subseteq \mathbb{R}^n$ such that both $\mathcal{H}^s(A) < \infty$ and $\mathcal{H}^t(B) < \infty$, with $s, t \geq 0$. Then, $\mathcal{H}^{s+t}(A \times B) \geq 2^{-(s+t)} \cdot \mathcal{H}^s(A) \cdot \mathcal{H}^t(B)$.*

Proof. If either $\mathcal{H}^s(A) = 0$ or $\mathcal{H}^t(B) = 0$, then the inequality obviously holds. So, assume that A is an s-set and B is a t-set; i.e., $0 < \mathcal{H}^s(A) < \infty$ and $0 < \mathcal{H}^t(B) < \infty$. Let μ be the product measure of $\mathcal{H}^s$ and $\mathcal{H}^t$ on $\mathbb{R}^{m+n}$,[26] supported on $A \times B$ and given for all $E \in \mathcal{B}(\mathbb{R}^m)$ and $F \in \mathcal{B}(\mathbb{R}^n)$ by

$$\mu(E \times F) := \mathcal{H}^s(A \cap E) \cdot \mathcal{H}^t(B \cap F). \tag{5.39}$$

In this case, note that $\mu(\mathbb{R}^{m+n}) = \mathcal{H}^s(A) \cdot \mathcal{H}^t(B)$. By part (ii) of Proposition 5.69,

$$\limsup_{r \to 0^+} \frac{\mathcal{H}^s(A \cap B_{\boldsymbol{x}}(r))}{(2r)^s} \leq 1, \tag{5.40}$$

for $\mathcal{H}^s$-almost all $\boldsymbol{x} \in A$, and

$$\limsup_{r \to 0^+} \frac{\mathcal{H}^t(B \cap B_{\boldsymbol{y}}(r))}{(2r)^t} \leq 1, \tag{5.41}$$

for $\mathcal{H}^t$-almost all $\boldsymbol{y} \in B$.

Note that $B_{(\boldsymbol{x},\boldsymbol{y})}(r) \subseteq B_{\boldsymbol{x}}(r) \times B_{\boldsymbol{y}}(r)$; thus,

$$\begin{aligned} \frac{\mu(B_{(\boldsymbol{x},\boldsymbol{y})}(r))}{(2r)^{s+t}} &\leq \frac{\mu(B_{\boldsymbol{x}}(r) \times B_{\boldsymbol{y}}(r))}{(2r)^{s+t}} \\ &= \frac{\mathcal{H}^s(A \cap B_{\boldsymbol{x}}(r))}{(2r)^s} \cdot \frac{\mathcal{H}^t(B \cap B_{\boldsymbol{y}}(r))}{(2r)^t}. \end{aligned} \tag{5.42}$$

In light of (5.40)–(5.42), we deduce that $\limsup_{r \to 0^+} \frac{\mu(B_{\boldsymbol{x},\boldsymbol{y}}(r))}{(2r)^{s+t}} \leq 1$, for μ-almost all $(\boldsymbol{x}, \boldsymbol{y}) \in A \times B$. By part (i) of Proposition 5.38 and in light of (5.39), we then obtain that $\mathcal{H}^{s+t}(A \times B) \geq 2^{-(s+t)} \cdot \mu(A \times B) = 2^{-(s+t)} \cdot \mathcal{H}^s(A) \cdot \mathcal{H}^t(B)$, as desired. □

Corollary 5.77. *Let $A \subseteq \mathbb{R}^m$ and $B \subseteq \mathbb{R}^n$. Then,*

$$\dim_H(A \times B) \geq \dim_H(A) + \dim_H(B). \tag{5.43}$$

[26]See Exercises 3.125 and 3.128

Proof. Suppose $s < \dim_H(A)$ and $t < \dim_H(B)$, with $s, t \geq 0$. Then, by Lemma 5.12, $\mathcal{H}^s(A) = \mathcal{H}^t(B) = +\infty$. Theorem 5.57 implies that there exist Borel sets $A_0 \subseteq A$ and $B_0 \subseteq B$ such that $0 < \mathcal{H}^s(A_0) < \infty$ and $0 < \mathcal{H}^t(B_0) < \infty$. By Theorem 5.76, it follows that

$$0 < 2^{-(s+t)} \cdot \mathcal{H}^s(A_0) \cdot \mathcal{H}^t(B_0) \leq \mathcal{H}^{s+t}(A_0 \times B_0) \leq \mathcal{H}^{s+t}(A \times B).$$

Hence, $s + t \leq \dim_H(A \times B)$, for any $s < \dim_H(A)$ and any $t < \dim_H(B)$. Therefore, (5.43) holds, as claimed. □

Exercises

Exercise 5.78. Verify that Hausdorff measure is countably additive, as stated in Remark 5.6 above.

Exercise 5.79. For any given $m, n \in \mathbb{N}$, find $A \subseteq \mathbb{R}^m$ and $B \subseteq \mathbb{R}^n$ such that the inequality (5.43) in Corollary 5.77 is strict.

Exercise 5.80 (*Dimension of the Cantor Dust* $C \times C$). Show that we have $\dim_H(C \times C) = \dim_H(C) + \dim_H(C) = D_S = \log 4/\log 3$, where C is the ternary Cantor set.[27]

[*Hint*: Recall from Exercise 4.27 that $F := C \times C$ is self-similar; then, check that F satisfies the OSC and apply Moran's Theorem, Theorem 5.40.]

Exercise 5.81 (*Topologist's Sine Curve*). The *topologist's sine curve* is defined as the closure $\overline{G}$ of the graph G of the function $f(x) := 1/\sin(2\pi x)$, for all $x \in (0, 1]$; i.e., it is equal to G union its limit points, consisting of the closed vertical line segment from $P(0, -1)$ to $Q(0, 1)$.

(i) Show that $\mathcal{H}^1(\overline{G}) = +\infty$. What is $\mathcal{H}^1(G)$ equal to?

[*Hint*: Estimate $\mathcal{H}^1_\delta(G)$ and $\mathcal{H}^1_\delta(\overline{G})$, for any small $\delta > 0$.]

(ii) Calculate $\dim_H(G)$ and $\dim_H(\overline{G})$, as well as $\dim_M(G)$ and $\dim_M(\overline{G})$. Are they all equal? Also compare the 1-dimensional Minkowski contents $\mathcal{M}^1(G)$ and $\mathcal{M}^1(\overline{G})$, provided they exist; see Section 8.1.

[27] See Figure 4.3 in Section 4.3 for a picture of the Cantor dust.

Chapter 6

δ-Approximate Hausdorff Measures, via Carathéodory's Theory

In this chapter and the next one (Chapter 7), we discuss the applications of Carathéodory's theory (studied in detail in Appendix B, Sections B.1–B.4) to the rigorous construction of Hausdorff measure and outer measure on a given metric space, as well as (in the present chapter) of their δ-approximate counterparts.

We first recall the notion of a metric outer measure and prove the corresponding main result (Theorem 6.14) according to which, for such an outer measure, the Borel σ-algebra is contained in the σ-algebra of Carathéodory measurable sets. This key result enables one to also view δ-approximate Hausdorff measures (and later, in Chapter 7, the Hausdorff measure itself) as a Borel measure, as well as to define and "calculate" (at least in principle) or, more realistically, estimate, the Hausdorff dimension and measure of most of the fractals encountered in practice.

In Section 6.1, we provide some additional general information about metric spaces and basic associated concepts, such as the diameter of a bounded subset or the basic notion of equivalent metrics. Then, in Section 6.2, we recall the notion of a metric outer measure and prove that Borel measurable sets are Carathéodory measurable, relative to any given metric outer measure.

In Section 6.3, we use several of the key results of Carathéodory's theory discussed and proved in Chapter 3 and Appendix B in order to construct and study some of the main properties of the δ-approximate Hausdorff measures. This will serve as the basis for the construction and the study in Chapter 7 of many of the key properties of Hausdorff measure, after having let δ tend to zero.

Further information on the subject of Hausdorff measures and δ-approximate Hausdorff measures can be found in, e.g., [**36, 171, 172, 193, 234, 281, 430, 518**] and the many relevant references therein. If one takes into account the information provided in the relevant part of Chapter 3 and the rigorous exposition of Carathéodory's theory provided in Appendix B, our mathematical discussion (in Chapters 6 and 7) of Hausdorff measure and its δ-approximate counterparts, however, is significantly more detailed and complete than in many of those references.

6.1. A Few Generalities About Metric Spaces

We recall or introduce here a few definitions and properties of metric spaces. More information about metric spaces can be found in Chapter 2.

Let (X, d) be a metric space. Recall from Definition 2.12 that a *closed* (resp., an *open*) ball of X (with *center* at $\boldsymbol{x}_0 \in X$ and *radius* r)[1] is denoted[2] by $\overline{B_{\boldsymbol{x}_0}}(r)$ (resp., $B_{\boldsymbol{x}_0}(r)$).[3] If we do not need to specify the radius (or the center), we simply refer to B as a closed (or as an open) ball. As can be easily checked, an open ball is an open subset of X and a closed ball is a closed subset of X.

Remark 6.1. We caution the reader that in spite of the suggestive notation adopted here, the closed ball $\overline{B_{\boldsymbol{x}_0}}(r)$ is not, in general, equal to $\overline{B_{\boldsymbol{x}_0}(r)}$, the closure of the corresponding open ball $B_{\boldsymbol{x}_0}(r)$. In a general metric space X, all we know is that $\overline{B_{\boldsymbol{x}_0}}(r) \supseteq \overline{B_{\boldsymbol{x}_0}(r)}$. If $X := \mathbb{R}^m$ (the Euclidean space), however, then we clearly have the equality $\overline{B_{\boldsymbol{x}_0}}(r) = \overline{B_{\boldsymbol{x}_0}(r)}$.

Given $A \subseteq X$, recall that the *diameter* of A, denoted by $\operatorname{diam}(A)$, is given by (2.6) in Definition 2.8.[4] Note that $\operatorname{diam}(A) \in [0, +\infty]$ but that if A is bounded (i.e., if $A \subseteq B$, for some ball B of X) and nonempty, then $\operatorname{diam}(A) < \infty$. Clearly, the converse is also true. Namely, $\operatorname{diam}(A) < \infty$ iff A is bounded and nonempty; see part (iii) of Exercise 6.2 just below.

[1] "The" center need not be unique; neither does the radius. Indeed, if X is an ultrametric space (e.g., if $X = \mathbb{Q}_p$, the space of p-adic numbers), then any $x_0 \in B$ can be taken as a center of the ball B.

[2] In this chapter (as well as in Chapter 7), for added clarity, we will boldface some of the relevant mathematical symbols.

[3] If no ambiguity may arise about the choice of the underlying metric, we shall refer, for example, to a ball (open or closed) of X or simply to a ball.

[4] It is easy to check that we may take B to be either an open ball or a closed ball, indifferently.

Exercise 6.2.

(i) Show that every (open or closed) ball B is bounded and that if its radius is r, then its diameter satisfies $\operatorname{diam}(B) \leq 2r$.

(ii) Show that if $(X,d) = (\mathbb{R}^m, d_m)$ is Euclidean space, then, $\operatorname{diam}(B) = 2r$, with B as in part (i). Does this equality hold in any metric space? Explain your answer or give a counterexample.

(iii) Show that if X is any metric space and $A \subseteq X$, then $\operatorname{diam}(A) < \infty$ iff A is bounded (and nonempty).

(iv) Show that[5] $A \neq \emptyset$ and $A \subseteq B \Rightarrow \operatorname{diam}(A) \leq \operatorname{diam}(B)$.

Exercise 6.3.

(i) Show that for any $A \subseteq X$, $\operatorname{diam}(A) = \operatorname{diam}(\overline{A})$, where $\overline{A}$ is the closure of A in X.

(ii) Deduce from part (i) that if $X := \mathbb{R}^m$, $B := B_{\boldsymbol{x}_0}(r)$ and $A := \overline{B_{\boldsymbol{x}_0}}(r)$, with $\boldsymbol{x}_0 \in \mathbb{R}^m$ and $r > 0$ arbitrary, then $A = \overline{B}$ (the closure of B in $\mathbb{R}^m$) and $\operatorname{diam}(A) = \operatorname{diam}(B) = 2r$, a result already (partially) stated in part (ii) of Exercise 6.2.

Definition 6.4. A *Borel measure* on X is a measure on the measurable space $(X, \mathcal{B}(X))$, where $\mathcal{B}(X)$ is the Borel σ-algebra, that is, the σ-algebra generated by the open subsets of X (as in Definition 3.20).

Exercise 6.5. Show that $\mathcal{B}(X)$ is the σ-algebra generated by the open balls (and also by the closed balls) of X, as well as by the closed subsets of X.

Definition 6.6 (*Equivalent Metrics*). Two metrics d and $\widetilde{d}$ on a set X are said to be *equivalent* if there exists a positive and finite constant c such that

$$c^{-1} \cdot \widetilde{d}(\boldsymbol{x}, \boldsymbol{y}) \leq d(\boldsymbol{x}, \boldsymbol{y}) \leq c \cdot \widetilde{d}(\boldsymbol{x}, \boldsymbol{y}), \text{ for all } \boldsymbol{x}, \boldsymbol{y} \in X. \tag{6.1}$$

This defines an equivalence relation on the metrics of X.

The following exercise shows, in particular, that given a metric space (X,d), the associated Borel σ-algebra $\mathcal{B}(X)$ does not strictly depend on the metric d itself, but rather it depends on the equivalence class of d for the equivalence relation introduced in Definition 6.6 just above.

Exercise 6.7. Prove that two equivalent metrics d and $\widetilde{d}$ on a set X are equivalent iff they induce the same topology, i.e., iff the metric spaces (X,d) and $(X,\widetilde{d})$ have the same open sets. Deduce that, as a result, the two metric spaces also have the same Borel σ-algebra.

[5] Observe that if the conclusion of part (iv) were true even if $A = \emptyset$, then (since $\operatorname{diam}(\emptyset) := +\infty$) this would imply that the diameter of any subset B of X would have to be infinite.

It is also useful to introduce the following notation: Given $A, B \subseteq X$ (arbitrary subsets of X), *the distance from A to B* is given by

$$\text{dist}(A, B) := \inf\{d(\boldsymbol{x}, \boldsymbol{y}) : \boldsymbol{x} \in A, \boldsymbol{y} \in B\} = \inf\{\text{dist}(\boldsymbol{x}, B) : \boldsymbol{x} \in A\}, \tag{6.2}$$

where for a given $\boldsymbol{x} \in X$, *the distance from $\boldsymbol{x}$ to B* is given by

$$\text{dist}(\boldsymbol{x}, B) := \inf\{d(\boldsymbol{x}, \boldsymbol{y}) : \boldsymbol{y} \in B\}. \tag{6.3}$$

Obviously, provided A and B are bounded and nonempty, $\text{dist}(A, B) \in [0, +\infty)$ and $\text{dist}(A, B) = \text{dist}(B, A)$. The reader may wish to investigate whether the following "Triangle Inequality" is valid, namely, whether $\text{dist}(A, C) \leq \text{dist}(A, B) + \text{dist}(B, C)$, for any $A, B, C \subseteq X$. Also, provided B is bounded and nonempty, we clearly have that $\text{dist}(\boldsymbol{x}, B) \in [0, +\infty)$, for all $\boldsymbol{x} \in X$. The next exercise examines the case when $\text{dist}(\boldsymbol{x}, B) = 0$.

Exercise 6.8. Let $A \subseteq X$ be nonempty. Show that $\text{dist}(\boldsymbol{x}, A) = 0 \Leftrightarrow \boldsymbol{x} \in \overline{A}$, where $\overline{A}$ denotes the closure of A.

Definition 6.9 (Open *ε-Neighborhood*). Given $A \subseteq X$ and $\varepsilon > 0$,

$$A_\varepsilon := \{\boldsymbol{x} \in X : \text{dist}(\boldsymbol{x}, A) < \varepsilon\} \tag{6.4}$$

denotes the *(open) ε-neighborhood* of A.[6]

Exercise 6.10. Let $A \subseteq X$ be nonempty.

(i) Show that the function $\varphi : X \to [0, +\infty)$ given by $\boldsymbol{x} \mapsto \text{dist}(\boldsymbol{x}, A) =: \varphi(\boldsymbol{x})$ is a Lipschitz function (in the sense of Remark 2.45) and hence is continuous. More specifically, show that

$$|\text{dist}(\boldsymbol{x}, A) - \text{dist}(\boldsymbol{y}, A)| \leq d(\boldsymbol{x}, \boldsymbol{y}), \text{ for all } \boldsymbol{x}, \boldsymbol{y} \in X. \tag{6.5}$$

(ii) Show that the open (resp., closed) ε-neighborhoods of A, as given in Definition 6.9 and the corresponding footnote, are indeed open (resp., closed) subsets of X. (Use part(i).)

6.2. Metric Outer Measures and Their Associated Borel Measures

We are now ready to discuss the notion of a metric outer measure μ^* on a given metric space (X, d) and to state as well as establish the main corresponding result; namely, for such an outer measure, $\mathcal{B} \subseteq \mathcal{M}$, where $\mathcal{B} = \mathcal{B}(X)$ is the Borel σ-algebra of X and $\mathcal{M} = \mathcal{M}_{\mu^*}$ is the σ-algebra of μ^*-measurable sets; see Theorem 6.14 below. This key result will then be applied (in Section 6.3) to δ-approximate Hausdorff outer measures and, in particular, will be used to

[6] The *closed ε-neighborhood of A*, denoted $[A]_\varepsilon$, would be defined as in (6.4), except for the strict inequality replaced by a large inequality; see Definition 4.6 in Section 4.1

justify the definition of δ-approximate Hausdorff measures (viewed as Borel measures).

First, it is helpful to recall the definition of a metric outer measure, which was already given in Definition 3.58.

Definition 6.11 (*Metric Outer Measure*). Let (X, d) be a metric space and let μ^* be an outer measure on X, in the sense of Definition 3.10. Then, μ^* is said to be a *metric outer measure* on X if for any $A, B \subseteq X$ such that $\operatorname{dist}(A, B) > 0$, we have $\mu^*(A \cup B) = \mu^*(A) + \mu^*(B)$.

Remark 6.12. Note that, obviously, if A and B are not disjoint, then, in light of (6.2), we have $\operatorname{dist}(A, B) = 0$. Therefore, the condition according to which $\operatorname{dist}(A, B) > 0$ in Definition 6.11 implies that A and B are disjoint.[7] It is natural to say that the subsets A and B are *metrically separated*, when this condition is fulfilled. Observe that it follows that the identity at the very end of Definition 6.11 can be viewed as *a weak form of finite additivity* of the outer measure μ^*.

Remark 6.13. A simple, but important, case when the metric separation condition $\operatorname{dist}(A, B) > 0$ is satisfied is when A and B are *disjoint compact* subsets of X; see Exercise 3.59. Indeed, this follows from the fact that if A and B are nonempty,[8] then the continuous function $\psi(\boldsymbol{x}, \boldsymbol{y}) := d(\boldsymbol{x}, \boldsymbol{y})$ achieves its minimum on the compact subset $A \times B$ of $X \times X$. Hence, the minimum value of ψ must be positive and equal to $\operatorname{dist}(A, B)$.

We can now state and prove the main result of this section (which was also stated, but without proof, in Theorem 3.61). This key result will be established just after we have stated and proved two of its corollaries.

Theorem 6.14 (Metric Outer Measures Induced by Borel Measures). *Let (X, d) be a metric space and let μ^* be a metric outer measure on X, in the sense of Definition* 6.11 *above. Then, the Borel σ-algebra $\mathcal{B}(X)$ of X is contained in $\mathcal{M} = \mathcal{M}_{\mu^*}$, the σ-algebra of μ^*-measurable subsets of X; i.e., $\mathcal{B}(X) \subseteq \mathcal{M}$. Rephrased: Every Borel subset of X is μ^*-measurable.*

Corollary 6.15. *Under the hypothesis of Theorem* 6.14, *every open subset of X, as well as every closed subset of X, is μ^*-measurable. In particular, every compact subset of X is μ^*-measurable (i.e., is Carathéodory measurable with respect to μ^*).*

Proof. This corollary is obvious since, by definition, the σ-algebra $\mathcal{B}(X)$ is generated by the open sets. Hence, it contains all of the closed sets (since $\mathcal{B}(X)$ is stable under complements) and, in particular, all of the compact subsets of X (since such subsets must be both closed and bounded in X). $\square$

[7]The converse is clearly not true, however. For example, consider $A := (0, \frac{1}{2})$ and $B := (\frac{1}{2}, 1)$, viewed as subsets of $\mathbb{R}$. Then, $A \cap B = \emptyset$ and $\operatorname{dist}(A, B) = 0$.

[8]If either A or B is empty, we use the standard convention according to which $\operatorname{dist}(A, B) := +\infty$, since $\inf(\emptyset) := +\infty$.

Corollary 6.16 (Associated Borel Measure). *To every metric outer measure μ^* on a metric space X is naturally associated a Borel measure $\tilde{\mu}$, simply obtained by restricting μ^* to $\mathcal{B}(X)$, the Borel σ-algebra of X. This measure is called* the Borel measure associated with (*the metric outer measure*) μ^*.

Proof. This follows at once from Theorem 6.14, according to which $\mathcal{B}(X) \subseteq \mathcal{M} = \mathcal{M}_{\mu^*}$, and from the Carathéodory Measurability and Restriction Theorem (Theorem B.3 in Appendix B) according to which $\overline{\mu}$, the restriction of the outer measure μ^* to $\mathcal{M}$, is a measure, called the Carathéodory measure associated with μ^*. It then suffices to further restrict μ^* to the Borel σ-algebra $\mathcal{B}(X)$ to obtain the desired measure, $\widetilde{\mu}$. Note that $\widetilde{\mu}$ is a Borel measure since it is a measure on the measurable space $(X, \mathcal{B}(X))$. □

We next give the proof of Theorem 6.14.

Proof of Theorem 6.14. Since, clearly, $\mathcal{B}(X)$ is generated by the closed subsets of X, it suffices to show that each closed set belongs to $\mathcal{M}$, the σ-algebra of μ^*-measurable sets. Indeed, it will then follow that $\mathcal{B}(X) \subseteq \mathcal{M}$, as desired, since $\mathcal{B}(X)$ is the smallest σ-algebra containing all of the closed subsets of X.

Hence, let $E \subseteq X$ be closed and let us show that E is μ^*-measurable. In light of Proposition B.1 in Appendix B, it suffices to prove that for any $A \subseteq X$ with $\mu^*(A) < +\infty$,

$$\mu^*(A) \geq \mu^*(A \cap E) + \mu^*(A \cap E^c). \tag{6.6}$$

The proof of the inequality (6.6) is a little tricky. One can proceed as follows. Consider the sequence $\{D_j\}_{j=1}^{\infty}$ of subsets of X given by

$$D_j := \left\{ \boldsymbol{x} \in A \setminus E \,:\, \operatorname{dist}(\boldsymbol{x}, E) \geq \frac{1}{j} \right\}, \text{ for all } j \geq 1. \tag{6.7}$$

Then, clearly, $D_j \nearrow A \setminus E = A \cap E^c$, as $j \to \infty$; i.e., $D_j \subseteq D_{j+1}$, for all $j \geq 1$, and

$$D := \bigcup_{j=1}^{\infty} D_j = A \setminus E = A \cap E^c. \tag{6.8}$$

In light of (6.7), the inclusions are obvious (since $\frac{1}{j+1} \leq \frac{1}{j}$), and (6.8) results from the fact that E is closed. Indeed, we have successively

$$\begin{aligned} \boldsymbol{x} \in D &\Leftrightarrow \boldsymbol{x} \in A \setminus E \quad \text{and} \quad \operatorname{dist}(\boldsymbol{x}, E) \geq j^{-1}, \text{ for some } j \geq 1 \\ &\Leftrightarrow \boldsymbol{x} \in A \setminus E \quad \text{and} \quad \operatorname{dist}(\boldsymbol{x}, E) > 0 \\ &\Leftrightarrow \boldsymbol{x} \in A \setminus E \quad \text{and} \quad \boldsymbol{x} \notin \overline{E} = E \Leftrightarrow \boldsymbol{x} \in A \setminus E. \end{aligned}$$

Hence, $D = A \setminus E$, as desired. Note that in the third equivalence just above, we have used the fact that E is closed so that $E = \overline{E}$, the closure of E or, by definition, the smallest closed subset containing E, together with the easily checked and well-known relation, valid for any subset of X, $\boldsymbol{x} \in \overline{E} \Leftrightarrow \operatorname{dist}(\boldsymbol{x}, E) = 0$; see Exercise 6.8 above.

Next, we note that since $A \setminus E \supseteq D_j$ and by the monotonicity of the outer measure μ^*, the following inclusions hold, for every $j \geq 1$:

$$A = (A \cap E) \cup (A \cap E^c) = (A \cap E) \cup (A \setminus E) \supseteq (A \cap E) \cup D_j,$$

and they imply that (by the monotonicity of the outer measure μ^*; see part (ii) of Definition 3.10)

$$\mu^*(A) \geq \mu^*(A \cap E) + \mu^*(D_j), \tag{6.9}$$

for every integer $j \geq 1$. We now claim that[9]

$$\mu^*(D_j) \nearrow \mu^*(A \setminus E), \text{ as } j \to \infty. \tag{6.10}$$

The sought for inequality (6.6) will then follow (since $A \cap E^c = A \setminus E$) by letting $j \to \infty$ in the inequality (6.9) and using (6.10). Therefore, we will deduce that E is μ^*-measurable, as desired.

All that remains to show is that the above limit holds, as claimed in (6.10). It helps to first note that by the monotonicity of μ^* and since $D_j \subseteq D_{j+1} \subseteq A \setminus E$, for each $j \geq 1$, the sequence $\{\mu^*(D_j)\}_{j=1}^{\infty}$ is convergent and so, as $j \to \infty$,

$$\mu^*(D_j) \nearrow \alpha, \text{ with } \alpha \leq \mu^*(A \setminus E) < \infty. \tag{6.11}$$

Indeed, $\mu^*(A \setminus E) \leq \mu^*(A) < \infty$, by hypothesis.

Next, construct a sequence $\{F_j\}_{j=1}^{\infty}$ of pairwise disjoint subsets of X, as follows: $D_0 := \emptyset$ and $F_j := D_j \setminus D_{j-1}$, for each $j \geq 1$. In light of (6.8),

$$\bigcup_{j=1}^{\infty} F_j = \bigcup_{j=1}^{\infty} D_j =: D = A \setminus E. \tag{6.12}$$

The key observation is now that, for any integer $j \geq 1$,

$$\text{if } \boldsymbol{x} \in F_{j+1} \text{ and } d(\boldsymbol{x}, \boldsymbol{y}) \leq \frac{1}{j(j+1)}, \text{ then } \operatorname{dist}(\boldsymbol{x}, E) < \frac{2}{j}. \tag{6.13}$$

Indeed, in light of (6.5) in Exercise 6.10, we have the following easily verified inequality: $|\operatorname{dist}(\boldsymbol{x}, E) - \operatorname{dist}(\boldsymbol{y}, E)| \leq d(\boldsymbol{x}, \boldsymbol{y})$; so that, for every integer $j \geq 1$,[10]

$$\operatorname{dist}(\boldsymbol{y}, E) \leq \operatorname{dist}(\boldsymbol{x}, E) + d(\boldsymbol{x}, \boldsymbol{y}) < \frac{1}{j} + \frac{1}{j(j+1)} < \frac{2}{j},$$

which implies (6.13), as claimed.

---

[9]Observe that this does *not* follow in a straightforward way from the fact that $D_j \nearrow A$, as $j \to \infty$. Indeed, μ^* is an outer measure but not a measure. Hence, we cannot simply apply part (v) of Proposition 3.29 in order to deduce (6.10).

[10]Note that $\boldsymbol{x} \in F_{j+1} = D_{j+1} \setminus D_j$ implies that $\boldsymbol{x} \in A \setminus E$ and $\boldsymbol{x} \notin D_j$; hence, in light of (6.7), we have $\operatorname{dist}(\boldsymbol{x}, E) < \frac{1}{j}$.

The reason why (6.13) is significant is that it implies that[11]

$$\operatorname{dist}(F_{j+1}, D_j) \geq \frac{1}{j(j+1)}, \text{ for all } j \geq 1.$$

In particular, we have that[12]

$$\operatorname{dist}(F_{j+1}, D_j) > 0, \text{ for all } j \geq 0, \tag{6.14}$$

and so F_{j+1} and D_j are metrically separated, for every $j \geq 0$. Consequently, since μ^* is a metric outer measure (see Definition 6.11), we deduce from (6.14) that

$$\mu^*(F_{j+1} \cup D_j) = \mu^*(F_{j+1}) + \mu^*(D_j), \tag{6.15}$$

for all $j \geq 0$. (For $j = 0$, this is also valid since $\mu^*(D_0) = \mu^*(\emptyset) = 0$.) Now, since $F_{j+1} = D_{j+1} \setminus D_j$ and $D_j \supseteq D_{j-1}$, we have that $D_{j+1} = F_j \cup D_j \supseteq F_j \cup D_{j-1}$. Thus, by the monotonicity of the outer measure and by (6.15),

$$\mu^*(D_{j+1}) \geq \mu^*(F_j \cup D_{j-1}) = \mu^*(F_j) + \mu^*(D_{j-1}). \tag{6.16}$$

Applying (6.16) repeatedly, we obtain that[13]

$$\mu^*(D_{2n+1}) \geq \sum_{j=1}^{n} \mu^*(F_{2j}) + \mu^*(D_1) \geq \sum_{k=1}^{n} \mu^*(F_{2k}) \quad \text{and}$$

$$\mu^*(D_{2n}) \geq \sum_{k=2}^{n} \mu^*(F_{2k-1}) + \mu^*(D_1) = \sum_{k=1}^{n} \mu^*(F_{2k-1}).$$

Hence, for every integer $n \geq 0$, we have (since $\mu^*(\emptyset) = 0$)

$$\mu^*(D_{2n+1}) \geq \sum_{k=1}^{n} \mu^*(F_{2k}) \quad \text{and} \quad \mu^*(D_{2n}) \geq \sum_{k=1}^{n} \mu^*(F_{2k-1}). \tag{6.17}$$

Since $D_j \subseteq A \setminus E \subseteq A$, we have $\mu^*(D_j) \leq \mu^*(A) < \infty$, for all $j \geq 0$. Thus, the two series (of nonnegative terms) $\sum_{k=1}^{\infty} \mu^*(F_{2k})$ and $\sum_{k=1}^{\infty} \mu^*(F_{2k-1})$ are convergent (and hence also commutatively convergent). Therefore, the full series $\sum_{j=1}^{\infty} \mu^*(F_j)$ is also convergent; in particular, its tail $\sum_{j=n+1}^{\infty} \mu^*(F_j)$ tends to 0, as $n \to \infty$.

We next observe that in light of (6.12) and of the (finite and) countable subadditivity of the outer measure μ^*, as well as because $\bigcup_{j=1}^{n} F_j = D_n$ (since

[11]Indeed, otherwise, $\operatorname{dist}(F_{j+1}, D_j) < 1/j(j+1)$ and in light of the definition of $\operatorname{dist}(F_{j+1}, D_j)$ as an infimum (see (6.2)), there exists $(\boldsymbol{x}, \boldsymbol{y}) \in F_{j+1} \times D_j$ such that $d(\boldsymbol{x}, \boldsymbol{y}) < \frac{1}{j(j+1)}$, which contradicts (6.13).

[12]For $j = 0$, since $D_0 := \emptyset$, we have $\operatorname{dist}(F_1, D_0) = +\infty > 0$.

[13]Note that $D_1 = F_1$ since $D_0 = \emptyset$.

$D_j \subseteq D_{j+1}$, for all $j \geq 1$), we have successively

$$\mu^*(A \setminus E) \leq \mu^*\Big(\bigcup_{j=1}^{n} F_j\Big) + \mu^*\Big(\bigcup_{j=n+1}^{\infty} F_j\Big)$$

$$= \mu^*(D_n) + \mu^*\Big(\bigcup_{j=n+1}^{\infty} F_j\Big) \leq \mu^*(D_n) + \sum_{j=n+1}^{\infty} \mu^*(F_j).$$

Thus, for every $n \geq 1$,

$$\mu^*(A \setminus E) \leq \mu^*(D_n) + \sum_{j=n+1}^{\infty} \mu^*(F_j). \tag{6.18}$$

As we already noted, because it is the tail of a convergent series, the last sum in (6.18) tends to 0, as $n \to \infty$. Therefore, recalling from (6.11) that $\mu^*(D_n) \to \alpha$, as $n \to \infty$, with $\alpha \leq \mu^*(A \setminus E) < \infty$, we deduce by letting $n \to \infty$ in (6.18) that

$$\mu^*(A \setminus E) \leq \alpha \leq \mu^*(A \setminus E), \quad \text{and then} \quad \mu^*(A \setminus E) = \alpha := \lim_{n \to \infty} \mu^*(D_n),$$

which is exactly what remained to be shown; see (6.10) and the discussion surrounding it. This finally concludes the proof of the theorem. □

We will apply Theorem 6.14 in the next section and in the next chapter, when discussing several of the main properties of δ-approximate Hausdorff measures, as well as of Hausdorff measures, respectively.

6.3. Construction and Properties of δ-Approximate Hausdorff Measures

In this section, we will complete and extend to the general setting of metric spaces the discussion of δ-approximate Hausdorff measures provided in Section 5.1 in the case of Euclidean space $\mathbb{R}^m$. In Chapter 7, this will then enable us, in particular, to give detailed proofs of the main statements concerning Hausdorff measures and to establish many of the key properties of those measures (when the proofs do not parallel those sometimes provided in Section 5.1 in the case of $\mathbb{R}^m$).

6.3.1. Construction of δ-Approximate Hausdorff Measures for Various Families. Let $X := (X, d)$ be an arbitrary metric space, and let $t \in \mathbb{R}$, with $t \geq 0$, be given. Then, for a fixed $\delta > 0$, we let (with $\inf(\emptyset) := +\infty$)

$$\mathscr{H}^t_\delta(E) := \inf\Big\{ \sum_{j=1}^{\infty} (\operatorname{diam}(A_j))^t \ : \ \{A_j\}_{j=1}^{\infty} \text{ is a } \delta\text{-cover of } E \Big\}, \tag{6.19}$$

for any $E \subseteq X$, where, by definition, a *δ-cover* of E, also called an *$\mathcal{S}_\delta$-cover* of E below, is a sequence $\{A_j\}_{j=1}^{\infty}$ such that $A_j \subseteq X$ and $\operatorname{diam}(A_j) \leq \delta$, for all $j \geq 1$. (Here, $\operatorname{diam}(A_j)$ denotes the diameter of A_j, defined as in formula (2.6) of Definition 2.8)[14] In other words, with $\mathcal{S} := 2^X$, $\{A_j\}_{j=1}^{\infty}$ is an $\mathcal{S}$-cover consisting of elements of $\mathcal{S}$ of diameter not exceeding δ, the given "cut-off". (Recall that the notion of an $\mathcal{S}$-cover was already introduced in Definition 3.46, for any $\mathcal{S} \subseteq 2^X$.) Equivalently, $\{A_j\}_{j=1}^{\infty}$ is an $\mathcal{S}_\delta$-cover of E, also in the sense of Definition 3.46, where $\mathcal{S}_\delta \subseteq 2^X$ (with $\emptyset \in \mathcal{S}_\delta$) is given by

$$\mathcal{S}_\delta := \mathcal{S} \cap \{A : A \subseteq X, \operatorname{diam}(A) \leq \delta \text{ or } A = \emptyset\}. \tag{6.20}$$

More generally, given $\mathcal{S} \subseteq 2^X$ with $\emptyset \in \mathcal{S}$, and $\mathcal{S}_\delta$ defined by (6.20) just above, it will be convenient to denote by $\mathcal{H}^t_{\delta,\mathcal{S}} = \mathcal{H}^t_{\mathcal{S}_\delta}$ the following obvious counterpart of (6.19) for the family $\mathcal{S}$ of subsets of X (or, equivalently, for $\mathcal{S}_\delta \subseteq 2^X$):[15]

$$\begin{aligned} \mathcal{H}^t_{\delta,\mathcal{S}}(E) &= \mathcal{H}^t_{\mathcal{S}_\delta}(E) \\ &:= \inf\left\{ \sum_{j=1}^{\infty} (\operatorname{diam}(A_j))^t : \{A_j\}_{j=1}^{\infty} \text{ is an } \mathcal{S}_\delta\text{-cover of } E \right\}, \end{aligned} \tag{6.21}$$

for any $E \subseteq X$, where an $\mathcal{S}_\delta$-cover is defined as in Definition 3.46, except for $\mathcal{S}$ replaced by $\mathcal{S}_\delta$. Equivalently, it is an $\mathcal{S}$-cover of E consisting of elements of $\mathcal{S}$ with diameter not exceeding δ; i.e., in light of (6.20), it is a countable collection $\{A_j\}_{j=1}^{\infty}$ of elements of $\mathcal{S}$ such that $E \subseteq \bigcup_{j=1}^{\infty} A_j$ and $\operatorname{diam}(A_j) \leq \delta$, for all $j \geq 1$.

If, much as in Theorem 3.48 (about Carathéodory's construction of outer measures), we consider the following set function $\mu : \mathcal{S}_\delta \to [0, +\infty]$, given by[16]

$$\mu(A) := (\operatorname{diam}(A))^t, \text{ for all } A \in \mathcal{S}_\delta, \tag{6.22}$$

then, clearly, we can rewrite (6.21) as follows:

$$\mathcal{H}^t_{\delta,\mathcal{S}}(E) = \mathcal{H}^t_{\mathcal{S}_\delta}(E) := \inf\left\{ \sum_{j=1}^{\infty} \mu(A_j) : \{A_j\}_{j=1}^{\infty} \text{ is an } \mathcal{S}_\delta\text{-cover of } E \right\}, \tag{6.23}$$

for any $E \subseteq X$, and similarly for $\mathcal{H}^t_\delta(E)$, as given by (6.19). Observe that, by definition, $\mathcal{H}^t_\delta = \mathcal{H}^t_{\mathcal{S}_\delta}$ is the special case when $\mathcal{S} := 2^X$.

Remark 6.17.

(a) In (6.19), (6.21), and (6.23), we could also use finite families $\{A_j\}_{j=1}^{n_0}$, for some arbitrary $n_0 \in \mathbb{N}$. However, in light of (6.23) and Remark 3.47, it suffices to consider countable families $\{A_j\}_{j=1}^{\infty}$ since we can simply let $A_j := \emptyset$, for all $j > n_0$ (because $\mu(\emptyset) := +\infty$).

[14] Recall that $\operatorname{diam}(\emptyset) := +\infty$, since $\operatorname{diam}(\emptyset)$ is given as a supremum; see formula (2.6). It follows that for any $t > 0$, $(\operatorname{diam}(\emptyset))^t = +\infty$; by convention, this is still true even for $t = 0$.

[15] It is especially useful when $\mathcal{S}$ is either the collection of closed balls of X or the collection of open balls of X, or else when $\mathcal{S}$ is a semialgebra (i.e., an elementary family) of X such that the σ-algebra generated by $\mathcal{S}$ contains $\mathcal{B}(X)$, the Borel σ-algebra of X.

[16] Since $\emptyset \in \mathcal{S}$ and in light of (6.20), we have that $\emptyset \in \mathcal{S}_\delta$.

(b) Observe that with the above notation (in (6.19), for $\mathcal{H}^t_\delta$, and in both (6.21) and (6.23), for $\mathcal{H}^t_{\delta,\mathcal{S}} = \mathcal{H}^t_{\mathcal{S}_\delta}$), and with $\mathcal{S} := 2^X$, we have

$$\mathcal{H}^t_\delta = \mathcal{H}^t_{\delta,\mathcal{S}} = \mathcal{H}^t_{\mathcal{S}_\delta}. \tag{6.24}$$

(c) It can be shown that in the definition of $\mathcal{H}^t_\delta$ given in (6.19), we can restrict ourselves to δ-covers by closed sets, as well as by open sets and, when $X := \mathbb{R}^m$ (the m-dimensional Euclidean space), by convex sets, as well as by closed convex sets. However, when $X = \mathbb{R}^m$, one cannot, strictly speaking, restrict one's attention to δ-covers by closed balls or by open balls. Indeed, the latter two families give rise to the same δ-approximate Hausdorff outer measure, called the *δ-approximate spherical Hausdorff outer measure*, and which is bounded from the left and the right by a positive constant times $\mathcal{H}^t_\delta$ but does not coincide with $\mathcal{H}^t_\delta$. All of these facts are established in Theorem 6.24 below; see also Corollary 6.26.

An entirely analogous comment can be made, after having passed to the limit as $\delta \to 0^+$, for the Hausdorff outer measure and measure, as well as for their spherical counterparts (corresponding to covers by closed balls or by open balls); see Section 7.2. From the point of view of the computation of the Hausdorff dimension, however, any of the families $\mathcal{S}$ mentioned in the previous paragraph yields the same result.

6.3.2. δ-Approximate Hausdorff Outer Measures and Measures.

We will next use many of the key results of Carathéodory's theory, along with related results—all established in Appendix B, Chapter 3, and Section 6.2— in order to define and study various incarnations of the δ-approximate Hausdorff outer measures and δ-approximate measures (in the present subsection) and then, after a passage to the limit as $\delta \to 0^+$, of the corresponding outer measure and Hausdorff measure (in Sections 7.1 and 7.2 of Chapter 7) on a general metric space and, in particular, on Euclidean space $\mathbb{R}^m$.

Theorem 6.18 (($\delta, \mathcal{S}$)-Approximate Hausdorff Outer Measure, Carathéodory–Hausdorff Measure, and Borel Measure). *Fix $\delta > 0$. Let $\mathcal{S} \subseteq 2^X$ be any collection of subsets of the metric space X such that $\emptyset \in \mathcal{S}$. Then:*

(i) (Outer Measure, $\mu^*_{\delta,\mathcal{S}} = \mathcal{H}^t_{\delta,\mathcal{S}}$). *The set function $\mu^*_{\delta,\mathcal{S}} := \mathcal{H}^t_{\delta,\mathcal{S}}$ (as defined via Carathéodory's construction specialized in* (6.21)) *is an outer measure on X, called the* ($\delta, \mathcal{S}$)-approximate Hausdorff outer measure (*or the* δ-approximate Hausdorff measure associated with $\mathcal{S}$); *see footnote* 19. *Furthermore, it is a metric outer measure, in the sense of Definition* 6.11.

(ii) (Carathéodory Measure, $\overline{\mu}_{\delta,\mathcal{S}}$). *Also, by Carathéodory's Measurability and Restriction Theorem,*[17] *and still with* $\mu^*_{\delta,\mathcal{S}} = \mathcal{H}^t_{\delta,\mathcal{S}}$, *as in part* (i),

$$\mathcal{M}_{\delta,\mathcal{S}} := \mathcal{M}_{\mu^*_{\delta,\mathcal{S}}} = \{\mu^*_{\delta,\mathcal{S}}\text{-measurable sets}\} \tag{6.25}$$

is a σ*-algebra on* X *and* $\overline{\mu}_{\delta,\mathcal{S}}$, *the Carathéodory measure associated with the outer measure* $\mu^*_{\delta,\mathcal{S}}$ *by restricting* $\mu^*_{\delta,\mathcal{S}}$ *to* $\mathcal{M}_{\delta,\mathcal{S}}$ *(see Theorem* 3.26*), is called the* $(\delta,\mathcal{S})$-approximate Carathéodory–Hausdorff measure *(or the* δ-approximate Carathéodory–Hausdorff measure associated with $\mathcal{S}$*). In addition,* $\overline{\mu}_{\delta,\mathcal{S}}$ *is a complete measure.*

(iii) (Borel Measure, $\widetilde{\mu}_{\delta,\mathcal{S}}$). *Moreover, since* $\mu^*_{\delta,\mathcal{S}}$ *is a metric outer measure (by part* (i)*), and with the* σ*-algebra* $\mathcal{M}_{\delta,\mathcal{S}}$ *defined as in* (6.25) *of part* (ii)*, it follows that* $\mathcal{B}(X) \subseteq \mathcal{M}_{\delta,\mathcal{S}}$,[18] *where* $\mathcal{B}(X)$ *is the Borel* σ*-algebra of* X. *Hence, the restriction of* $\mu^*_{\delta,\mathcal{S}}$ *(or, equivalently, of the Carathéodory measure* $\overline{\mu}_{\delta,\mathcal{S}}$ *from part* (ii)*) to* $\mathcal{B}(X)$*) yields a Borel measure on* X, *denoted by* $\widetilde{\mu}_{\delta,\mathcal{S}}$ *(but, in practice, also by* $\mathcal{H}^t_{\delta,\mathcal{S}}$*) and called the* $(\delta,\mathcal{S})$-approximate Hausdorff–Borel measure *(or the* δ-approximate Hausdorff–Borel measure associated with $\mathcal{S}$*).*[19]

(iv) (Weak Uniqueness of the Extension). *The Carathéodory measure* $\overline{\mu}_{\delta,\mathcal{S}}$, *as a measure extension, is "weakly unique", in the following sense: If* ν *is any other measure on the measurable space* $(X,\mathcal{M}_{\delta,\mathcal{S}})$ *extending the Borel measure* $\widetilde{\mu}_{\delta,\mathcal{S}}$ *from* $\mathcal{B}(X)$ *to the* σ*-algebra* $\mathcal{M}_{\delta,\mathcal{S}}$, *then*

$$\nu(E) \le \overline{\mu}_{\delta,\mathcal{S}}(E), \text{ for all } E \in \mathcal{M}_{\delta,\mathcal{S}}, \tag{6.26}$$

with equality (i.e., $\nu(E) = \overline{\mu}_{\delta,\mathcal{S}}(E)$*) if* $E \in \mathcal{M}_{\delta,\mathcal{S}}$ *is such that* $\overline{\mu}_{\delta,\mathcal{S}}(E) < \infty$.[20]

Corollary 6.19. *With the above notation, the Carathéodory measure* $\overline{\mu}_{\delta,\mathcal{S}}$ *is a complete measure (in the sense of Definition* B.5 *of Appendix* B*).*

Proof. This follows from the fact that the restriction $\overline{\nu}$ of an outer measure ν^* to the σ-algebra $\mathcal{M}_{\nu^*}$ of ν^*-measurable sets is always complete; see the last statement of Theorem B.3 in Appendix B, the Carathéodory Measurability and Extension Theorem. Here, of course, $\overline{\nu}$, which is the Carathéodory measure associated with the outer measure $\nu^* := \mu^*_{\delta,\mathcal{S}}$, coincides with $\overline{\mu}_{\delta,\mathcal{S}}$ and we also have that $\mathcal{M}_{\nu^*} := \mathcal{M}_{\delta,\mathcal{S}}$. □

Corollary 6.20. *All of the results of Theorem* 6.18 *and of Corollary* 6.19 *hold without change when applied to* $\mathcal{H}^t_\delta = \mathcal{H}^t_{\delta,2^X}$ *(corresponding to the important special case of the theorem when* $\mathcal{S} := 2^X$*), where the outer measure* $\mathcal{H}^t_\delta = \mu^*_\delta$ *is*

---

[17] See Theorems 3.19 and 3.26 in Chapter 3 or else Theorem B.3 in Appendix B.

[18] See Theorem 6.14 and Corollary 6.16 in Section 6.2 above.

[19] Clearly, $\mu^*_{\delta,\mathcal{S}}, \overline{\mu}_{\delta,\mathcal{S}}$, and $\widetilde{\mu}_{\delta,\mathcal{S}}$ also depend on $t \ge 0$, but for notational simplicity, we do not indicate it explicitly here. The same comment applies to μ^*_δ, $\overline{\mu}_\delta$, and $\widetilde{\mu}_\delta$ to be considered in Corollary 6.20 and corresponding to the special case when $\mathcal{S} := 2^X$.

[20] Note that the measure $\overline{\mu}_{\delta,\mathcal{S}}$ need not be σ-finite. The analog of this statement will be discussed in Subsection 6.3.3.

given by (6.19). *The associated Carathéodory measure* (*on the measurable space* $(X, \mathcal{M}_\delta := \mathcal{M}_{\mu^*_\delta})$) *and the Borel measure* (*on the measurable space* $(X, \mathcal{B}(X))$) *are then denoted, respectively, by* $\overline{\mu}_\delta$ *and* $\widetilde{\mu}_\delta$.

Definition 6.21. For any $\delta > 0$ and $t \geq 0$, the outer measure $\mu^*_\delta = \mathcal{H}^t_\delta$ in Corollary 6.20 is called the (*t-dimensional*) *δ-approximate Hausdorff outer measure*. Similarly, the associated Carathéodory measure $\overline{\mu}_\delta = \mathcal{H}^t_\delta$ (resp., the Borel measure $\widetilde{\mu}_\delta = \mathcal{H}^t_\delta$) is called the (*t-dimensional*) *δ-approximate Carathéodory–Hausdorff measure* (resp., the *Hausdorff–Borel measure*).

Observe that, for notational simplicity, we denote indifferently by $\mathcal{H}^t_\delta$ either $\mu^*_\delta, \overline{\mu}_\delta$, or $\tilde{\mu}_\delta$. The meaning should be clear from the context.[21]

Next, we give the proof of Theorem 6.18, while postponing to Lemma 6.22 the proof of a key ingredient, namely, the fact that $\mu^*_{\delta,\mathcal{S}}$ is a metric outer measure.

Proof of Theorem 6.18.

(i) The fact that $\mu^*_{\delta,\mathcal{S}}$ is an outer measure follows from Theorem 3.48 since $\mu^*_{\delta,\mathcal{S}} = \mathcal{H}^t_{\delta,\mathcal{S}}$ is defined via the Carathéodory construction in (6.21). Note that in the terminology and the notation of Theorem 3.48, $\mu^*_{\delta,\mathcal{S}}$ is the outer measure associated with the set function $\mu_{\delta,\mathcal{S}} : \mathcal{S} \to [0, +\infty]$ defined by $\mu_{\delta,\mathcal{S}}(A) = (\text{diam}(A))^t$, for all $A \in \mathcal{S}_\delta$, where we recall that $\mathcal{S}_\delta$ stands for the collection of elements $A \in \mathcal{S}$ with $\text{diam}(A) \leq \delta$.

Furthermore, the fact that $\mu^*_{\delta,\mathcal{S}}$ is a metric outer measure (in the sense of Definition 6.11) requires some work and will be established in the proof of Lemma 6.22, following this proof.

(ii) Since, by part (i), $\mu^*_{\delta,\mathcal{S}}$ is an outer measure on X, part (ii) is an immediate consequence of Carathéodory's Measurability and Restriction Theorem, Theorem B.3 in Appendix B (or else Theorems 3.19 and 3.26 in Chapter 3), according to which $\mathcal{M}_{\delta,\mathcal{S}}$ is a σ-algebra and $\overline{\mu}_{\delta,\mathcal{S}}$ is a complete measure on the measurable space $(X, \mathcal{M}_{\delta,\mathcal{S}})$.

(iii) Since, by part (i) (or, really, by Lemma 6.22 below), $\mu^*_{\delta,\mathcal{S}}$ is a *metric* outer measure, part (iii) follows at once from Theorem 6.14 and Corollary 6.16, according to which $\mathcal{B}(X) \subseteq \mathcal{M}_{\delta,\mathcal{S}} := \mathcal{M}_{\mu^*_{\delta,\mathcal{S}}}$ and hence, $\widetilde{\mu}_{\delta,\mathcal{S}}$, obtained by restricting $\overline{\mu}_{\delta,\mathcal{S}}$ to $\mathcal{B}(X)$, is a Borel measure on X.

(iv) Finally, the weak uniqueness statement concerning the measure $\overline{\mu}_{\delta,\mathcal{S}}$, viewed as an extension of $\widetilde{\mu}_{\delta,\mathcal{S}}$ from $\mathcal{B}(X)$ to the σ-algebra $\mathcal{M}_{\delta,\mathcal{S}}$, follows from

[21] In the statement of parts (i)–(iii) of Theorem 6.18, we have adopted a similar terminology and convention for the outer measure $\mu^*_{\delta,\mathcal{S}}$ as well as for the measures $\overline{\mu}_{\delta,\mathcal{S}}$ and $\widetilde{\mu}_{\delta,\mathcal{S}}$, for any family $\mathcal{S} \subseteq 2^X$ (rather than just for $\mathcal{S} := 2^X$).

a somewhat unusual application of the Algebra Version of Carathéodory's Extension Theorem (Theorem B.12 in Appendix B), part (ii). Indeed, with the notation of that theorem, $\mathcal{A} := \mathcal{B}(X)$ (here, the algebra $\mathcal{A}$ is, in fact, a σ-algebra)[22] and the premeasure $\mu_{\delta,\mathcal{S}}$ to be extended from $\mathcal{A}$ to $\mathcal{M}_{\delta,\mathcal{S}}$ (the σ-algebra of $\mu^*_{\delta,\mathcal{S}}$-measurable sets) is, in fact, a countably additive measure (on the measurable space $(X, \mathcal{B}(X))$) and coincides with the Borel measure $\widetilde{\mu}_{\delta,\mathcal{S}}$; see Remark B.10.

In closing, let us point out once again that the reason why we do not have at our disposal here the full strength of part (iv) (under the additional assumption that $\mu_{\delta,\mathcal{S}} = \widetilde{\mu}_{\delta,\mathcal{S}}$ is σ-finite) is that this latter hypothesis is not, in general, satisfied by $\widetilde{\mu}_{\delta,\mathcal{S}}$, in the present context. □

In the course of the proof of step (i) of Theorem 6.18 given just above, we have used the following key lemma.

Lemma 6.22. *The outer measure $\mu^*_{\delta,\mathcal{S}}$ is a metric outer measure on X.*

Proof. According to the definition of a metric outer measure (see Definition 6.11), we must show that if A and B are subsets of X that are metrically separated (i.e., $\operatorname{dist}(A, B) > 0$), then

$$\mu^*_{\delta,\mathcal{S}}(A \cup B) = \mu^*_{\delta,\mathcal{S}}(A) + \mu^*_{\delta,\mathcal{S}}(B). \tag{6.27}$$

First, let us recall that $\operatorname{dist}(A, B) > 0$ implies that A and B are disjoint (since, otherwise, if $A \cap B \neq \emptyset$, then we would have that $\operatorname{dist}(A, B) = 0$).

Next, using once again the notation $\varphi(\{D_j\}) := \sum_{j=1}^{\infty}(\operatorname{diam}(D_j))^t$, we consider any $(\delta, \mathcal{S})$-cover $\{D_j\}$ of $A \cup B$. Since A and B are disjoint, we can clearly write, symbolically, $\{D_j\} = \{D_{j,1}\} \sqcup \{D_{j,2}\}$, where $\{D_{j,1}\}$ corresponds to the D_j's which intersect A (and, hence, do not intersect B), and analogously, $\{D_{j,2}\}$ corresponds to the D_j's which intersect B (and, therefore, do not intersect A). By construction, $\{D_{j,1}\}$ is a $(\delta, \mathcal{S})$-cover of A. A moment's reflection then shows that

$$\varphi(\{D_j\}) = \varphi(\{D_{j,1}\}) + \varphi(\{D_{j,2}\}). \tag{6.28}$$

Hence, by using the definition of $\mu^*_{\delta,\mathcal{S}}(A \cup B)$ as an infimum over all $(\delta, \mathcal{S})$-covers (i.e., over all $\mathcal{S}_\delta$-covers) of $A \cup B$ (and similarly for $\mu^*_{\delta,\mathcal{S}}(A)$ and $\mu^*_{\delta,\mathcal{S}}(B)$),[23] we first deduce from (6.28) that $\varphi(\{D_j\}) \geq \mu^*_{\delta,\mathcal{S}}(A) + \mu^*_{\delta,\mathcal{S}}(B)$, and then, since $\{D_j\}$ was an arbitrary $(\delta, \mathcal{S})$-cover (i.e., $\mathcal{S}_\delta$-cover) of $A \cup B$, we deduce by taking the infimum over all subcovers in the latter inequality that

$$\mu^*_{\delta,\mathcal{S}}(A \cup B) \geq \mu^*_{\delta,\mathcal{S}}(A) + \mu^*_{\delta,\mathcal{S}}(B). \tag{6.29}$$

[22] Consequently, also in the notation of Theorem B.12, $\mathcal{A}$ coincides with the σ-algebra it generates, so that $\mathcal{A} = \mathfrak{M} = \mathcal{B}(X)$.

[23] See the identity (6.21) defining $\mu^*_{\delta,\mathcal{S}}(E)$ for any $E \subseteq X$.

Since $\mu^*_{\delta,\mathcal{S}}$ is an outer measure, it is finitely subadditive (because $\mu^*_{\delta,\mathcal{S}}(\emptyset) = 0$) and therefore, the reverse inequality follows at once and does not make any use of the fact that A and B are disjoint (or, let alone, that $\mathrm{dist}(A, B) > 0$):

$$\mu^*_{\delta,\mathcal{S}}(A \cup B) \leq \mu^*_{\delta,\mathcal{S}}(A) + \mu^*_{\delta,\mathcal{S}}(B). \tag{6.30}$$

Finally, combining (6.29) and (6.30) we obtain the desired equality (6.27). This completes the proof of the lemma. □

6.3.3. Comparison of δ-Approximate Hausdorff Measures for Various Families. The following comments will be useful in the sequel when comparing below the δ-approximate Hausdorff outer measures and measures associated with different families.

Remark 6.23.

(a) In Theorem 6.24 and Corollary 6.26, we will, in particular, state the equality of various $(\delta, \mathcal{S})$-approximate Hausdorff outer measures $\mu^*_{\delta,\mathcal{S}}$ and, consequently, of the associated Carathéodory–Hausdorff measures $\overline{\mu}_{\delta,\mathcal{S}}$ and Borel measures $\widetilde{\mu}_{\delta,\mathcal{S}}$. Actually, in the proof of Theorem 6.24, we will only establish the equality of the outer measures $\mu^*_{\delta,\mathcal{S}} = \mathcal{H}^t_\delta$ (for various families $\mathcal{S}$), which will obviously imply the equalities of the associated measures $\overline{\mu}_{\delta,\mathcal{S}}$ and $\widetilde{\mu}_{\delta,\mathcal{S}}$ obtained by restriction of $\mu^*_{\delta,\mathcal{S}}$ to the σ-algebras $\mathcal{M}_{\delta,\mathcal{S}}$ and $\mathcal{B}(X)$, respectively; see Theorem 6.18 above.[24]

(b) Much as in part (a), when, in Theorem 6.24 or Corollary 6.26 (as well as in Proposition 6.30), we state an inequality between $\mathcal{H}^t_{\delta,\mathcal{S}_1}$ and $\mathcal{H}^t_{\delta,\mathcal{S}_2}$ for $\mathcal{S}_1, \mathcal{S}_2 \subseteq 2^X$, we will prove first the inequality between the outer measures $\mu^*_{\delta,\mathcal{S}_2}$ and $\mu^*_{\delta,\mathcal{S}_1}$,[25] and then, since $\mathcal{B}(X) \subseteq \mathcal{M}_{\delta,\mathcal{S}_1} \cap \mathcal{M}_{\delta,\mathcal{S}_2}$ (because both $\mu^*_{\delta,\mathcal{S}_1}$ and $\mu^*_{\delta,\mathcal{S}_2}$ are metric outer measures), this will imply the exact same inequality between the associated Borel measures $\widetilde{\mu}_{\delta,\mathcal{S}_1}$ and $\widetilde{\mu}_{\delta,\mathcal{S}_2}$, as well as between the Carathéodory–Hausdorff measures $\overline{\mu}_{\delta,\mathcal{S}_1}$ and $\overline{\mu}_{\delta,\mathcal{S}_2}$ on their common σ-subalgebra $\mathcal{M}_{\delta,\mathcal{S}_1} \cap \mathcal{M}_{\delta,\mathcal{S}_2}$.

The reader who is interested in these measure-theoretic subtleties may wish to note the comments in parts (a) and (b) of Remark 6.23 just above, while the other readers may wish to ignore them, at least upon a first reading.

In the sequel, starting with Theorem 6.24 and continuing on with Corollary 6.26 and Proposition 6.30, for example, we will use the following notation. Let $\mathcal{S}_{cl}$ (resp., $\mathcal{S}_{op}$) denote the collection (or family) of all closed (resp., open) subsets of X. Furthermore, let $\mathcal{S}^\#$ (resp., $\mathcal{S}^\times$) be the collection of all closed

[24]The equality of the outer measures clearly implies that of the corresponding σ-algebras of $\mu^*_{\delta,\mathcal{S}_j}$-measurable sets, $\mathcal{M}_{\delta,\mathcal{S}_j}$, for $j = 1, 2$. Also note that, by definition, $\mathcal{B}(X)$ only depends on the topology of X and hence is independent of the choice of both δ and $\mathcal{S}_j$.

[25]This is indeed what is established in the proof of Theorem 6.24.

(resp., open) balls of X. Moreover, when $X := \mathbb{R}^m$ (the m-dimensional Euclidean space), we let $\mathcal{S}_{conv}$ (resp., $\mathcal{S}_{\overline{conv}}$) denote the collection of all convex (resp., closed and convex) subsets of $\mathbb{R}^m$.[26]

For notational simplicity, we also let (for any $\delta > 0$)

$$\mathcal{S}^{\#}_{\delta} := (\mathcal{S}^{\#})_{\delta},\ \mathcal{S}^{\times}_{\delta} := (\mathcal{S}^{\times})_{\delta},\quad \mathcal{S}_{cl,\delta} := (\mathcal{S}_{cl})_{\delta},\ \mathcal{S}_{op,\delta} := (\mathcal{S}_{op})_{\delta}, \text{ etc.} \tag{6.31}$$

In the sequel (starting with Theorem 6.24), given a family $\mathcal{S} \subseteq 2^X$, the notation $\mathcal{H}^t_{\delta} = \mathcal{H}^t_{\delta,2^X}$ (resp., $\mathcal{H}^t_{\delta,\mathcal{S}}$) stands for the δ-approximate (resp., $(\delta,\mathcal{S})$-approximate) Hausdorff outer measure or else measure (by restriction to $\mathcal{B}(X)$, the Borel σ-algebra of X), depending on the context or the goal and interpretation being pursued. Accordingly, given two families $\mathcal{S}_1, \mathcal{S}_2 \subseteq 2^X$, the equality $\mathcal{H}^t_{\delta,\mathcal{S}_1} = \mathcal{H}^t_{\delta,\mathcal{S}_2}$ means that $\mathcal{H}^t_{\delta,\mathcal{S}_1}(E) = \mathcal{H}^t_{\delta,\mathcal{S}_2}(E)$, for all $E \subseteq X$ (resp., for all $E \in \mathcal{B}(X)$), provided that *both* $\mathcal{H}^t_{\delta,\mathcal{S}_1}$ and $\mathcal{H}^t_{\delta,\mathcal{S}_2}$ are viewed as outer measures (resp., Borel measures) on X.

Similarly, the inequality $\mathcal{H}^t_{\delta,\mathcal{S}_1} \leq \mathcal{H}^t_{\delta,\mathcal{S}_2}$ means that $\mathcal{H}^t_{\delta,\mathcal{S}_1}(E) \leq \mathcal{H}^t_{\delta,\mathcal{S}_2}(E)$, for all $E \subseteq X$ (resp., for all $E \in \mathcal{B}(X)$) provided that, as above, *both* $\mathcal{H}^t_{\delta,\mathcal{S}_1}$ and $\mathcal{H}^t_{\delta,\mathcal{S}_2}$ are viewed as outer measures (resp., as Borel measures) on X. Recall that, by construction, $\mathcal{H}^t_{\delta} = \mathcal{H}^t_{\delta,\mathcal{S}}$, with $\mathcal{S} := 2^X$.

The following result (Theorem 6.24) is quite useful for the concrete computation and estimation of $\mathcal{H}^t_{\delta}$, as well as of its limit as $\delta \to 0^+$, $\mathcal{H}^t$, the t-dimensional Hausdorff outer measure or measure on X, to be defined and studied in Chapter 7 below. It is also useful for the computation of the (t-dimensional) Hausdorff dimension, as is defined in Section 5.2 and studied throughout Chapter 5 (when $X = \mathbb{R}^m$) and as is briefly discussed in Section 7.5. Most of it is well known, although possibly not made explicit in most textbooks (such as the precise formalization of Proposition 6.30 and of Theorem 6.24 themselves), but the notion of a "reasonable" metric space and the associated results seem to be new.

In the sequel, by a *"reasonable"* metric space, we mean that there exists $\theta > 0$ such that every (closed or open) ball of X with radius r has diameter θr. Accordingly, if $A := B_{\boldsymbol{x}_0}(r)$ and $B := \overline{B}_{\boldsymbol{x}_0}(r)$, for some $\boldsymbol{x}_0 \in X$ and $r > 0$, then

$$\operatorname{diam}(A) = \operatorname{diam}(B) = \theta r. \tag{6.32}$$

Note that we will show in part (a) of Remark 6.25 that (as is well known) when $X = \mathbb{R}^m$, the Euclidean space, we have $\theta := 2$.

Theorem 6.24 (Comparison of Various δ-Approximate (or $(\delta,\mathcal{S})$-Approximate) Hausdorff Measures and Outer Measures). *Let X be a metric space and fix $\delta > 0$ as well as $t \in \mathbb{R}$, with $t \geq 0$. Then, with the notation and conventions introduced*

[26] Recall that a set $A \subseteq \mathbb{R}^m$ is said to be *convex* if given any two points $\boldsymbol{x}, \boldsymbol{y} \in A$, it contains the *line segment* $\overline{\boldsymbol{xy}}$ connecting $\boldsymbol{x}$ and $\boldsymbol{y}$; namely, $\overline{\boldsymbol{xy}} := \{t\boldsymbol{x} + (1-t)\boldsymbol{y} : t \in [0,1]\} \subseteq A$.

just before the statement of this theorem, we have the following comparison results:

(i) (Restriction to Families of "Nice" δ-Covers: Approximation by Closed or Open Sets). *For an arbitrary metric space X, we have the following equalities and inequalities:*[27]

$$\mathcal{H}^t_\delta = \mathcal{H}^t_{\delta,\mathcal{S}_{cl}} \le \mathcal{H}^t_{\delta,\mathcal{S}_{op}} \quad \text{and} \quad \mathcal{H}^t_{\delta+,\mathcal{S}_{op}} \le \mathcal{H}^t_\delta = \mathcal{H}^t_{\delta,\mathcal{S}_{cl}}. \tag{6.34}$$

(ii) (Restriction to Further Families of "Nice" δ-Covers: Approximation by Convex Sets). *Furthermore, if $X = \mathbb{R}^m$ (the m-dimensional Euclidean space), we also have that $\mathcal{H}^t_\delta = \mathcal{H}^t_{\delta,\mathcal{S}_{conv}} = \mathcal{H}^t_{\delta,\mathcal{S}_{\overline{conv}}}$.*

(iii) (The Spherical Case: Approximation by Closed or Open Balls). *If $X = \mathbb{R}^m$ (the Euclidean space) or, more generally, if the metric space X is "reasonable" (in the sense explained just before the statement of this theorem), then $S^t_\delta := \mathcal{H}^t_{\delta,\mathcal{S}^\#} \le \mathcal{H}^t_{\delta,\mathcal{S}^\times}$, and (with the same meaning for δ^+ as in part* (i) *of this theorem), $\mathcal{H}^t_{\delta+,\mathcal{S}^\times} \le \mathcal{H}^t_{\delta,\mathcal{S}^\#} =: S^t_\delta$, where $S^t_\delta := \mathcal{H}^t_{\delta,\mathcal{S}^\#}$ is called the δ-*approximate spherical Hausdorff outer measure (*or* measure, *depending on the context) and with the same meaning for δ^+ as in part* (i) *of this theorem.*[28]

(iv) (Comparison Between the δ-Approximate Standard and Spherical Hausdorff Outer Measures or Measures). *With the notation $S^t_\delta := \mathcal{H}^t_{\delta,\mathcal{S}^\#}$ adopted in part* (iii)*, and still assuming that $X = \mathbb{R}^m$* (the Euclidean space)*, we have the following inequalities:*[29]

$$\mathcal{H}^t_\delta \le S^t_\delta \quad \text{and} \quad S^t_\delta \le \gamma_{m,t} \cdot \mathcal{H}^t_\delta; \quad \text{i.e.,} \quad \mathcal{H}^t_\delta \le S^t_\delta \le \gamma_{m,t} \cdot \mathcal{H}^t_\delta, \tag{6.36}$$

where the positive constant $\gamma_{m,t}$ satisfies $\gamma_{m,t} > 1$, for every integer $m \ge 2$, and is given, for every integer $m \ge 1$, by

$$\gamma_{m,t} := (2m/(m+1))^{t/2}. \tag{6.37}$$

---

[27] By convention, the second inequality in (6.34) (namely, $\mathcal{H}^t_{\delta+,\mathcal{S}_{op}} \le \mathcal{H}^t_\delta$) means that

$$\mathcal{H}^t_{\delta^*,\mathcal{S}_{op}} \le \mathcal{H}^t_\delta, \text{ for all } \delta^* > \delta. \tag{6.33}$$

Note that $\mathcal{H}^t_{\delta+,\mathcal{S}_{op}} \le \mathcal{H}^t_{\delta,\mathcal{S}_{op}}$; i.e., $\mathcal{H}^t_{\delta^*,\mathcal{S}_{op}} \le \mathcal{H}^t_{\delta,\mathcal{S}_{op}}$, for every $\delta^* > \delta$. This explains why we did *not* combine the two sides of (6.34) into a single string of inequalities and conclude that $\mathcal{H}^t_\delta = \mathcal{H}^t_{\delta,\mathcal{S}_{cl}} = \mathcal{H}^t_{\delta,\mathcal{S}_{op}}$. In fact, of these equalities, only the first one is claimed here.

[28] More specifically, in the last inequality of part (iii), δ^+ stands for any $\delta^* > \delta$. Hence, this inequality means that

$$\mathcal{H}^t_{\delta^*,\mathcal{S}^\times} \le \mathcal{H}^t_{\delta,\mathcal{S}^\#} =: S^t_\delta, \text{ for all } \delta^* > \delta. \tag{6.35}$$

Note that $\mathcal{H}^t_{\delta^*,\mathcal{S}^\times} \le \mathcal{H}^t_{\delta+,\mathcal{S}^\#}$ (i.e., $\mathcal{H}^t_{\delta^*,\mathcal{S}^\times} \le \mathcal{H}^t_{\delta+,\mathcal{S}^\#}$, for every $\delta^* > \delta$), which explains why we could not combine the two identities in part (iii) of Theorem 6.24 into a single string of inequalities (namely, $\mathcal{H}^t_{\delta,\mathcal{S}^\#} \le \mathcal{H}^t_{\delta,\mathcal{S}^\times} \le \mathcal{H}^t_{\delta,\mathcal{S}^\#}$) and thus, we could *not* conclude that $S^t_\delta = \mathcal{H}^t_{\delta,\mathcal{S}^\#} = \mathcal{H}^t_{\delta,\mathcal{S}^\times}$.

[29] More precisely, on the left or right side of (6.36), the first inequality is sharp while the second inequality on the left-hand side of (6.36) is essentially sharp; see Remark 6.35 and Exercise 6.36.

Furthermore, when $m = 1$ *(i.e., if* $X = \mathbb{R}$*, the real line), then* $\gamma_{m,t} = 1$ *and hence, we only have equalities in* (6.36): $\mathcal{H}^t_\delta = S^t_\delta$*, when* $m = 1$.

Remark 6.25.

(a) It is easy to check that any Euclidean space $X := \mathbb{R}^m$ is reasonable, but it is a little more subtle to show that, for example, $X := \mathbb{Q}_p$, the field of p-adic numbers,[30] is reasonable, in the sense of part (iii) of Theorem 6.24 and of the text surrounding formula (6.32). In the Euclidean case, the verification is relatively simple; indeed, given an open ball $B = B_{\boldsymbol{x}_0}(r)$ of $\mathbb{R}^m$ and the associated closed ball $A := \overline{B_{\boldsymbol{x}_0}(r)}$, for some $\boldsymbol{x}_0 \in \mathbb{R}^m$ and $r > 0$, it is easy to check that

$$\text{diam}(A) = \text{diam}(B) = 2r, \tag{6.39}$$

from which it follows that $\mathbb{R}^m$ is reasonable with optimal constant $\theta := 2$; see (6.32).[31]

In order to see why (6.39) holds, it suffices to show that $\text{diam}(A) \geq 2r$. Indeed, $A = B \sqcup \partial A$ (the disjoint union of B and ∂A, the boundary of A) and by the "continuity" of $\mathbb{R}^m$, there are points of B that are arbitrarily close to ∂A, the Euclidean sphere of center $\boldsymbol{x}_0 \in \mathbb{R}^m$ and radius $r > 0$. Hence, since also $\text{diam}(B) \leq \text{diam}(A)$ (because $B \subseteq A$ and B is nonempty), we have that $\text{diam}(A) = \text{diam}(B)$. Furthermore, clearly, by the Triangle Inequality, we also have that $\text{diam}(A) \leq 2r$.

Now, the fact that $\text{diam}(A) \geq 2r$ easily follows from the continuity of $\mathbb{R}^m$ and the fact that A is convex and symmetric with respect to the origin. Indeed, given $\varepsilon > 0$ sufficiently small (say, $0 < \varepsilon < r/2$), there exists a point $\boldsymbol{a} \in A$ such that $d(\boldsymbol{x}_0, \boldsymbol{a}) \geq r - \varepsilon$ (by the continuity of $\mathbb{R}^m$). Then, $\boldsymbol{a}^*$, the symmetric of $\boldsymbol{a}$ with respect to $\boldsymbol{x}_0$ (i.e., $\boldsymbol{a}^* := 2\boldsymbol{x}_0 - \boldsymbol{a}$), satisfies the identity $d(\boldsymbol{x}_0, \boldsymbol{a}^*) = d(\boldsymbol{x}_0, \boldsymbol{a}) \geq r - \varepsilon$. Since A is convex, it contains the line segment $\overline{\boldsymbol{a}\boldsymbol{a}^*}$ connecting $\boldsymbol{a}$ and $\boldsymbol{a}^*$; namely, $\overline{\boldsymbol{a}\boldsymbol{a}^*} = [\boldsymbol{a}, \boldsymbol{a}^*] \subseteq A$. Hence,

$$\text{diam}(A) \geq d(\boldsymbol{a}^*, \boldsymbol{a}) = d(\boldsymbol{a}^*, \boldsymbol{x}_0) + d(\boldsymbol{x}_0, \boldsymbol{a}) \geq 2(r - \varepsilon).$$

Since ε is arbitrary in $(0, r/2)$, we deduce that $\text{diam}(A) \geq 2r$, as desired.

The justification of the statements concerning $\mathbb{Q}_p$ is provided in part (b) just below.

---

[30] In particular, $\mathbb{Q}_p$ is ultrametric. Recall that a metric space (X, d) is said to be *ultrametric* if d satisfies the following strong form of the Triangle Inequality:

$$d(\boldsymbol{x}, \boldsymbol{z}) \leq \max\{d(\boldsymbol{x}, \boldsymbol{z}), d(\boldsymbol{y}, \boldsymbol{z})\}, \text{ for all } \boldsymbol{x}, \boldsymbol{y}, \boldsymbol{z} \in X. \tag{6.38}$$

It follows that the center of an open or closed ball can be chosen anywhere within the ball and, hence, that the diameter of any open or closed ball of radius r is no greater than r (and not $2r$), as can also be deduced directly from the Ultrametric Triangle Inequality (6.38).

[31] Observe that in a general metric space, we only know that $\text{diam}(A) \leq 2r$ and $\text{diam}(B) \leq 2r$, which would be insufficient to conclude.

(b) We claim that $\mathbb{Q}_p$ is unreasonable (i.e., not reasonable), as was stated in part (a) of this remark, but that in spite of this, we have the following inequalities, valid for all $\delta > 0$:

$$\mathscr{H}^t_{\delta,\mathscr{S}^\times} \leq \mathscr{H}^t_{\delta,\mathscr{S}^\#} \quad \text{and} \quad \mathscr{H}^t_{\delta,\mathscr{S}^\#} \leq p^t \cdot \mathscr{H}^t_{\delta,\mathscr{S}^\times}. \tag{6.40}$$

In this sense, $\mathbb{Q}_p$ (the field of p-adic numbers) is an interesting example of unreasonable metric space. Recall that p is a prime number, and hence, $p > 1$.

To see why this is true, let $B := B_{\boldsymbol{x}_0}(r)$, for some $\boldsymbol{x}_0 \in \mathbb{Q}_p$ and $r > 0$, and let $A := \overline{B_{\boldsymbol{x}_0}}(r)$, the closed ball associated with $B_{\boldsymbol{x}_0}(r)$. We then deduce from the ultrametricity of $\mathbb{Q}_p$ that $\operatorname{diam}(A) \leq r$ and hence (since $B \subseteq A$) that[32] $\operatorname{diam}(B) \leq \operatorname{diam}(A) \leq r$. Next, let $n \in \mathbb{Z}$ be the unique integer such that $p^n < r \leq p^{n+1}$. (Recall that distances in $\mathbb{Q}_p$ take their values in the discrete multiplicative group $p^{\mathbb{Z}}$, the valuation group of the nonarchimedean field $\mathbb{Q}_p$.) We claim that

$$\operatorname{diam}(B) = p^n, \text{ whereas } \operatorname{diam}(A) = p^{n+1}, \tag{6.41}$$

which clearly shows that $\mathbb{Q}_p$ cannot be reasonable, since, otherwise, A and B would have the same radius, in particular.

To see why (6.41) holds, first observe that there cannot be any $\boldsymbol{x} \in \mathbb{Q}_p$ with $d(\boldsymbol{x}, \boldsymbol{x}_0) \in (p^n, p^{n+1})$. A moment's reflection then shows that

$$B = \overline{B_{\boldsymbol{x}_0}}(p^n) \quad \text{and} \quad A = \overline{B_{\boldsymbol{x}_0}}(p^{n+1}). \tag{6.42}$$

Hence, since $B \supseteq S_{\boldsymbol{x}_0}(p^n)$ and $A \supseteq \overline{B_{\boldsymbol{x}_0}}(p^{n+1})$ and observing that (again in light of the ultrametricity of $\mathbb{Q}_p$) every sphere of $\mathbb{Q}_p$ of the form $S_{\boldsymbol{x}_0}(\rho)(= \boldsymbol{x}_0 + S_{\boldsymbol{0}}(\rho))$, with $\rho \in p^{\mathbb{Z}}$, is nonempty and has diameter ρ,[33] we deduce that

$$\operatorname{diam}(B) \geq \operatorname{diam}(S_{\boldsymbol{x}_0}(p^n)) = p^n \quad \text{and} \quad \operatorname{diam}(A) \geq \operatorname{diam}(S_{\boldsymbol{x}_0}(p^{n+1})) = p^{n+1}. \tag{6.43}$$

Next, we apply to each of the (closed) balls $B = B_{\boldsymbol{x}_0}(p^n)$ and $A = B_{\boldsymbol{x}_0}(p^{n+1})$ the counterpart of the aforementioned inequality (stated in the text preceding (6.41)) in order to deduce that

$$\operatorname{diam}(B) \leq p^n \quad \text{and} \quad \operatorname{diam}(A) \leq p^{n+1}. \tag{6.44}$$

Combining (6.43) and (6.44), we obtain (6.41), as desired.

Finally, in order to prove the two inequalities in (6.40), we observe that, in light of (6.41), $\operatorname{diam}(A) = p \cdot \operatorname{diam}(B) \leq p \cdot \operatorname{diam}(B)$ and hence that an application of Remark 6.31 below (extending part (iii) of Proposition 6.30) along with part (ii) of the same proposition shows (much as in the proof of part (ii)

---

[32] Note that this is in sharp contrast with the Euclidean case, for which we have $\operatorname{diam}(A) = \operatorname{diam}(B) = 2r$, as was verified in part (a) of this remark.

[33] By contrast, if $\rho \notin p^{\mathbb{Z}}$, $S_{\boldsymbol{x}_0}(\rho) = \emptyset$ and thus, $\operatorname{diam}(S_{\boldsymbol{x}_0}(\rho)) = +\infty$.

of Theorem 6.24) that for every $\delta > 0$,

$$\mathcal{H}^t_{\delta,\mathcal{S}^\#} \leq p^t \cdot \mathcal{H}^t_{\delta,\mathcal{S}^\times} \quad \text{and} \quad \mathcal{H}^t_{\delta,\mathcal{S}^\times} \leq \mathcal{H}^t_{\delta,\mathcal{S}^\#},$$

which is precisely the content of (6.40).

(c) We shall see in Subsection 7.4.6 that it is geometrically natural to suitably normalize $\mathcal{H}^t_{\delta,\mathcal{S}}$ for any family $\mathcal{S} \subseteq 2^X$ (and, in particular, $\mathcal{H}^t_\delta = \mathcal{H}^t_{\delta,2^X}$), by multiplying it by an appropriate positive constant (independent of δ and $\mathcal{S}$). However, since such a modification preserves both inequalities and equalities, the exact counterpart of Theorem 6.24 (and of Corollary 6.26 just below) would still hold for the resulting $(\delta, \mathcal{S})$-approximate (or δ-approximate) Hausdorff outer measures (and measures). After a passage to the limit as $\delta \to 0^+$, the same comment remains valid for the corresponding (t-dimensional) Hausdorff outer measures (and measures) to be defined in Section 7.1.

We postpone the proof of Theorem 6.24 for a while, until after the statement (and proof) of Proposition 6.30, which is key to various parts of the proof. The proof of the theorem is relatively elementary but is a bit tricky and requires a good deal of "mental gymnastic", along with several geometric arguments needed in a few places, especially in order to obtain the results of part (ii) and part (iv) of Theorem 6.24.

We next summarize the main results of Theorem 6.24 in the important special case of Euclidean space $X = \mathbb{R}^m$, which is of most interest to us in this book. Several parts of the original statement simplify considerably in this case.

Corollary 6.26 (Comparison Results: The Case of Euclidean Space $\mathbb{R}^m$). *Fix $\delta > 0$. Then:*

(i) *If $X := \mathbb{R}^m$, the m-dimensional Euclidean space, and with the notation introduced before as well as in the statement of Theorem* 6.24, *then we have the following string of equalities:*

$$\mathcal{H}^t_\delta = \mathcal{H}^t_{\delta,\mathcal{S}_{cl}} = \mathcal{H}^t_{\delta,\mathcal{S}_{conv}} = \mathcal{H}^t_{\delta,\mathcal{S}_{\overline{conv}}}. \tag{6.45}$$

In particular, this fully justifies the statement made in part (c) *of Remark* 6.17 *in this case; see also part* (b) *of Remark* 6.25.

(ii) *Moreover, the δ-approximate standard and spherical Hausdorff outer measures (or measures), $\mathcal{H}^t_\delta$ and $\mathcal{S}^t_\delta$, respectively, are compared as follows:*

$$\mathcal{H}^t_\delta \leq \mathcal{S}^t_\delta \quad \text{and} \quad \mathcal{S}^t_\delta \leq \gamma_{m,t} \cdot \mathcal{H}^t_\delta, \tag{6.46}$$

*where the δ-*approximate spherical outer measure (*or* measure) *is given by*

$$\mathcal{S}^t_\delta := \mathcal{H}^t_{\delta,\mathcal{S}^\#} \quad \text{and where} \quad \gamma_{m,t} := (2m/(m+1))^{t/2}. \tag{6.47}$$

Accordingly, $\gamma_{1,t} = 1$,[34] *and $\gamma_{m,t} > 1$ when $m \geq 2$. Furthermore, the constant $\gamma_{m,t}$ is (essentially) optimal; see Remark* 6.35 *and Exercise* 6.36.

[34] In fact, it follows from (6.46) that $\mathcal{S}^t_\delta = \mathcal{H}^t_\delta$ when $m = 1$.

(iii) *Moreover, we have the following additional equalities and inequalities:*[35]

$$\mathcal{H}^t_\delta = \mathcal{H}^t_{\delta,\mathcal{S}_{cl}} \le \mathcal{H}^t_{\delta,\mathcal{S}_{op}}, \tag{6.49}$$

$$\mathcal{H}^t_{\delta^+,\mathcal{S}_{op}} \le \mathcal{H}^t_{\delta,\mathcal{S}_{cl}} = \mathcal{H}^t_\delta, \tag{6.50}$$

$$\mathcal{S}^t_\delta := \mathcal{H}^t_{\delta,\mathcal{S}^\#} \le \mathcal{H}^t_{\delta,\mathcal{S}^\times}, \quad \text{and} \tag{6.51}$$

$$\mathcal{H}^t_{\delta^+,\mathcal{S}^\times} \le \mathcal{H}^t_{\delta,\mathcal{S}^\#} =: \mathcal{S}^t_\delta. \tag{6.52}$$

Proof. This follows at once from Theorem 6.24 since we are now working in $\mathbb{R}^m$. Note that we use the fact that $\mathbb{R}^m$ is a "reasonable" metric space; see part (a) of Remark 6.25 along with part (iii) of Theorem 6.24 above. □

Remark 6.27. We stress that the statement of Corollary 6.26 will simplify even more upon the passage to the limit as $\delta \to 0^+$, that is, when we are dealing (in Section 7.2) with the ordinary and spherical outer measures (or measures), along with related quantities, instead of with their δ-approximate counterparts.

Parts of the next exercise follow easily from the definitions.

Exercise 6.28.

(i) With the above notation, and if $X := \mathbb{R}^m$, show that for any $\mathcal{S} \in \{\mathcal{S}_{op}, \mathcal{S}_{cl}, \mathcal{S}^\#, \mathcal{S}^\times, \mathcal{S}_{conv}, \mathcal{S}_{\overline{conv}}\}$, the σ-algebra generated by $\mathcal{S}$ coincides with $\mathcal{B}(X)$, the Borel σ-algebra of X.

[*Hint*: Since closed (resp., open) balls are closed (resp., open) and convex subsets of $\mathbb{R}^m$, it suffices to consider the case when $\mathcal{S} \in \{\mathcal{S}^\#, \mathcal{S}^\times\}$ and even, after an easy reduction step, when $\mathcal{S} = \mathcal{S}^\#$.]

(ii) Moreover, find a sufficient condition on the metric space X for the converse of the above result in part (i) to be true when $\mathcal{S} \in \{\mathcal{S}_{op}, \mathcal{S}_{cl}, \mathcal{S}^\#, \mathcal{S}^\times\}$.

Exercise 6.29.

(i) Still when $X := \mathbb{R}^m$, show that $\mathcal{H}^t_\delta = \mathcal{H}^t_{\delta,\mathcal{S}^+}$, where $\mathcal{S}^+$ is the collection of all closed intervals of $\mathbb{R}^m$, given by

$$\mathcal{S}^+ = \left\{ \prod_{k=1}^m [a_k, b_k] \,:\, a_k \le b_k, -\infty \le a_k, b_k \le +\infty \right\}.$$

Furthermore, show that the σ-algebra generated by $\mathcal{S}^+$ coincides with $\mathcal{B}(\mathbb{R}^m)$.

[35] Here, as in Theorem 6.24 (and Proposition 6.30), δ^+ stands for any $\delta^* > \delta$. Accordingly, the inequality in the left-hand side of (6.50) (resp., of (6.52)) means that for every $\delta^* > \delta$,

$$\mathcal{H}^t_{\delta^*,\mathcal{S}_{op}} \le \mathcal{H}^t_{\delta,\mathcal{S}_{cl}} \; (\text{resp., } \mathcal{H}^t_{\delta^*,\mathcal{S}^\times} \le \mathcal{H}^t_{\delta,\mathcal{S}^\#}). \tag{6.48}$$

(ii) Establish the exact analog of part (i) where $\mathcal{S}^+$ is replaced by $\mathcal{S}_{dya}$,[36] the collection of all closed dyadic cubes of $\mathbb{R}^m$.[37]

(iii) Compare $\mathcal{H}^t_\delta$, $\mathcal{H}^t_{\delta,\mathcal{S}^+}$, and $\mathcal{H}^t_{\delta,\mathcal{S}_{dya}}$. Are they all equal? If not, are there inequalities holding between them?

In order to prove Theorem 6.24, it will be convenient to first establish the following proposition and to introduce the following notation:[38]

$$\varphi(\{A_j\}) := \sum_{j=1}^{\infty} \mu(A_j) = \sum_{j=1}^{\infty} (\operatorname{diam}(A_j))^t \tag{6.53}$$

if $A_j \subseteq X$, for all $j \geq 1$. Note that the family $\{A_j\}$ could be finite or countably infinite.

Proposition 6.30 (Comparison Results).

(i) *The map φ (defined as in* (6.53)*) is monotonic, in the following sense:*

$$\varphi(\{A_j\}) \leq \varphi(\{B_j\}), \tag{6.54}$$

provided $\{A_j\} \preccurlyeq \{B_j\}$, i.e., provided that $\operatorname{diam}(A_j) \leq \operatorname{diam}(B_j)$, *for all $j \geq 1$.*[39] *Then, the following statements hold, for every $\delta > 0$:*

(ii) *If $\mathcal{S}_2 \subseteq \mathcal{S}_1 \subseteq 2^X$, then $\mathcal{H}_{\delta,\mathcal{S}_1} \leq \mathcal{H}_{\delta,\mathcal{S}_2}$.*[40]

(iii) (Generalization of (ii)). *If, for any $E \subseteq X$ and given any $\mathcal{S}_2$-cover $\{B_j\}$ of E (by sets B_j of finite diameter), there exists an $\mathcal{S}_1$-cover $\{A_j\}$ of E such that $\{A_j\} \preccurlyeq \{B_j\}$ (in the sense of part* (i)*), then $\mathcal{H}^t_{\delta,\mathcal{S}_1} \leq \mathcal{H}^t_{\delta,\mathcal{S}_2}$.*

(iv) (Further Generalization). *The conclusion of part* (iii) *is slightly modified and is replaced by $\mathcal{H}^t_{\delta^+,\mathcal{S}_1} \leq \mathcal{H}^t_{\delta,\mathcal{S}_2}$ (in the sense of footnote 27) if, in the counterpart of part* (iii) *of this proposition, we assume instead that for every $\varepsilon > 0$ and every $E \subseteq X$, given any $\mathcal{S}_2$-cover $\{B_j\}$ of E (by sets of finite diameter), there exists an $\mathcal{S}_1$-cover $\{A_j\}$ of E such that*

$$\operatorname{diam}(A_j) \leq \operatorname{diam}(B_j) + \varepsilon, \text{ for all } j \geq 1. \tag{6.55}$$

Proof. Fix $\delta > 0$.

(i) This follows at once from the definitions and the fact that for every fixed $t \geq 0$, the function $x \mapsto x^t$ is nondecreasing on $(0, +\infty)$.

[36]There is one difference between $\mathcal{S}^+$ and $\mathcal{S}_{dya}$, however; indeed, $\mathcal{S}_{dya}$ is not a semialgebra, which is unimportant here.

[37]A *dyadic cube* of $\mathbb{R}^m$ is an m-dimensional cube with center having coordinates in the lattice $2^{-k} \cdot \mathbb{Z}^m$ and side lengths 2^{-k}, for some $k \in \mathbb{Z}$; see Subsection 7.4.1.

[38]In the sequel, for notational simplicity, we will most often write $\{A_j\}$ instead of $\{A_j\}_{j=1}^{\infty}$ or $\{A_j\}_{j\geq 1}$.

[39]This is certainly true if $A_j \subseteq B_j$, for all $j \geq 1$. However, typically, we will use the relation $\{A_j\} \preccurlyeq \{B_j\}$ when $B_j \subseteq A_j$, for all $j \geq 1$—and, hence, when $\operatorname{diam}(A_j) = \operatorname{diam}(B_j)$. In this case, if $\{A_j\}$ is a cover of E, so is $\{B_j\}$ because then, $\bigcup_j B_j \subseteq \bigcup_j A_j \subseteq E$. The same comment applies to part (iv) of this proposition.

[40]That is, $\mathcal{H}^t_{\delta,\mathcal{S}_1}(E) \leq \mathcal{H}^t_{\delta,\mathcal{S}_2}(E)$, for all $E \subseteq X$ and all $\delta > 0$.

(ii) This is a special case of (iii) since under the hypothesis of (ii), any $\mathcal{S}_2$-cover is an $\mathcal{S}_1$-cover of $E \subseteq X$.

(iii) In light of the definition of $\mathcal{H}^t_{\delta,\mathcal{S}_2}(E)$ as an infimum (in (6.21)), given $\varepsilon > 0$, there exists an $\mathcal{S}_2$-cover $\{B_j\}$ of E such that $\operatorname{diam}(B_j) \leq \delta$, for all $j \geq 1$ and (with $\varphi(\{B_j\})$ defined as in (6.53)) $\varphi(\{B_j\}) \leq \mathcal{H}^t_{\delta,\mathcal{S}_2}(E) + \varepsilon$. Since, by the hypothesis of part (iii), there exists an $\mathcal{S}_1$-cover $\{A_j\}$ of E such that $\{A_j\} \preccurlyeq \{B_j\}$, it follows from part (i) that $\varphi(\{A_j\}) \leq \varphi(\{B_j\})$. Now, by definition of $\mathcal{H}^t_{\delta,\mathcal{S}_1}(E)$ and φ (in (6.21) and (6.53), respectively), we also have (since, clearly, $\operatorname{diam}(A_j) \leq \operatorname{diam}(B_j) \leq \delta$, for all $j \geq 1$) $\mathcal{H}^t_{\delta,\mathcal{S}_1}(E) \leq \varphi(\{A_j\})$. Next, we combine the above three inequalities in order to obtain that $\mathcal{H}^t_{\delta,\mathcal{S}_1}(E) \leq \mathcal{H}^t_{\delta,\mathcal{S}_2}(E) + \varepsilon$. Since $\varepsilon > 0$ was arbitrary, we conclude that $\mathcal{H}^t_{\delta,\mathcal{S}_1}(E) \leq \mathcal{H}^t_{\delta,\mathcal{S}_2}(E)$, as desired.

(iv) The proof of part (iv) parallels that of part (iii), except for the fact that we can no longer appeal to part (i). Instead (and with the notation of the proof of part (iii)), we can directly estimate $\varphi(\{A_j\})$ in terms of $\varphi(\{B_j\})$, as follows:

$$\varphi(\{A_j\}) \leq \sum_{j=1}^{\infty} (\operatorname{diam}(B_j) + \varepsilon)^t. \tag{6.56}$$

Note that the right-hand side of (6.56) is a nondecreasing function of $\varepsilon > 0$. Therefore, as $\varepsilon \searrow 0$, it admits a limit, equal to $\varphi(\{B_j\}) := \sum_{j\geq 1}(\operatorname{diam}(B_j))^t$. Hence, by letting $\varepsilon \searrow 0$ in the right-hand side of (6.56), we obtain the desired inequality, $\varphi(\{A_j\}) \leq \varphi(\{B_j\})$. The rest of the proof of part (iv) is then the same as that of part (iii). This completes the proof of the proposition. □

The following comment will be useful in the main step of the proof (of part (iv)) of Theorem 6.24.

Remark 6.31. If, in part (iii) of Proposition 6.30, we assume instead of $\{A_j\} \preccurlyeq \{B_j\}$ (i.e., $\operatorname{diam}(A_j) \leq \operatorname{diam}(B_j)$, for all $j \geq 1$) that there exists a constant $c > 0$ such that $\operatorname{diam}(A_j) \leq c \cdot \operatorname{diam}(B_j)$, for all $j \geq 1$, then the conclusion of part (iii) should be replaced by the following inequality: $\mathcal{H}^t_{\delta,\mathcal{S}_1} \leq c^t \cdot \mathcal{H}^t_{\delta,\mathcal{S}_2}$. The corresponding proof is identical to that of part (iii) of Proposition 6.30.

We are now finally ready to give a detailed proof of Theorem 6.24.

Proof of Theorem 6.24. We will use repeatedly the results and the notation of Proposition 6.30. Fix $\delta > 0$.

(i) Let $\mathcal{S} := 2^X$. Since, obviously, $\mathcal{S}_{cl} \subseteq \mathcal{S}$ and $\mathcal{S}_{op} \subseteq \mathcal{S}$, we have (by part (ii) of Proposition 6.30) $\mathcal{H}^t_\delta = \mathcal{H}^t_{\delta,\mathcal{S}} \leq \mathcal{H}^t_{\delta,\mathcal{S}_{cl}}$, and similarly for $\mathcal{H}^t_{\delta,\mathcal{S}_{op}}$ (instead of for $\mathcal{H}^t_{\delta,\mathcal{S}_{cl}}$).

The reverse inequality follows from part (iii) of Proposition 6.30. Indeed, given any $\mathcal{S}$-cover $\{B_j\}$ of $E \subseteq X$ (i.e., any cover of X since $\mathcal{S} := 2^X$, here), we simply let for every $j \geq 1$, $A_j := \overline{B_j}$ (the closure of B_j in X) to deduce (since $\operatorname{diam}(B_j) = \operatorname{diam}(A_j)$, by part (i) of Exercise 6.3) that $\{A_j\} \preccurlyeq \{B_j\}$. Hence, the hypothesis of part (iii) of Proposition 6.30 is satisfied and so we have that $\mathcal{H}^t_{\delta,\mathcal{S}_{cl}} \leq \mathcal{H}^t_{\delta,\mathcal{S}} = \mathcal{H}^t_\delta$. Combining the above two inequalities, we see that $\mathcal{H}^t_\delta = \mathcal{H}^t_{\delta,\mathcal{S}_{cl}}$, as desired.

We next show that $\mathcal{H}^t_{\delta+,\mathcal{S}_{op}} \leq \mathcal{H}^t_{\delta,\mathcal{S}}$. To do so, we apply part (iv) of Proposition 6.30 and thus, we need to verify that its hypothesis is satisfied (which will imply the desired inequality). Accordingly, we let for each $j \geq 1$, $A_j := \{\boldsymbol{x} \in X : \operatorname{dist}(\boldsymbol{x}, B_j) < \varepsilon/2\}$, the $(\varepsilon/2)$-neighborhood of A_j. By part (ii) of Exercise 6.10, A_j is open. Also, it is easy to check that $\operatorname{diam}(A_j) \leq \operatorname{diam}(B_j) + 2 \cdot (\varepsilon/2) = \operatorname{diam}(B_j) + \varepsilon$. Since, clearly, $B_j \subseteq A_j$, for all $j \geq 1$, the sequence $\{A_j\}$ is an $\mathcal{S}_{op}$-cover of E, and hence, the hypothesis of part (iv) of Proposition 6.30 is satisfied. Therefore, $\mathcal{H}^t_{\delta+,\mathcal{S}_{op}} \leq \mathcal{H}^t_\delta = \mathcal{H}^t_{\delta,\mathcal{S}}$, as desired.

It now suffices to combine the above results to deduce that $\mathcal{H}^t_{\delta+,\mathcal{S}_{op}} \leq \mathcal{H}^t_\delta = \mathcal{H}^t_{\delta,\mathcal{S}_{cl}}$. Furthermore, we observe that since $\mathcal{S}_{op} \subseteq \mathcal{S} := 2^X$, it follows from part (i) of Proposition 6.30 that $\mathcal{H}^t_\delta = \mathcal{H}^t_{\delta,\mathcal{S}} \leq \mathcal{H}^t_{\delta,\mathcal{S}_{op}}$. Note that even though $\mathcal{H}^t_{\delta+,\mathcal{S}_{op}} \leq \mathcal{H}^t_{\delta,\mathcal{S}_{op}}$, this does *not* imply that $\mathcal{H}^t_\delta = \mathcal{H}^t_{\delta,\mathcal{S}_{op}} = \mathcal{H}^t_{\delta,\mathcal{S}_{c\ell}}$.

Finally, combining the last two obtained identities, we deduce part (i) of Theorem 6.24; namely, $\mathcal{H}^t_\delta \leq \mathcal{H}^t_{\delta,\mathcal{S}_{op}}$ and $\mathcal{H}^t_{\delta+,\mathcal{S}_{op}} \leq \mathcal{H}^t_\delta = \mathcal{H}^t_{\delta,\mathcal{S}_{cl}}$.

(ii) In order to establish part (ii), we must show that when $X := \mathbb{R}^m$,

$$\mathcal{H}^t_\delta = \mathcal{H}^t_{\delta,\mathcal{S}_{conv}} = \mathcal{H}^t_{\delta,\mathcal{S}_{\overline{conv}}}. \tag{6.57}$$

In essence, the proof of (6.57) relies on the fact that for any $A \subseteq X$, we have that $\operatorname{diam}(A) = \operatorname{diam}(\overline{A})$ (see part (i) of Exercise 6.3) and (when $X := \mathbb{R}^m$, as is the case here) we also have that $\operatorname{diam}(A) = \operatorname{diam}(\operatorname{conv}(A))$; see part (i) of Lemma 6.33 below, especially (6.71) and the text following it.

We now proceed to prove a first string of inequalities connected to (6.57). Indeed, since, obviously, $\mathcal{S}_{\overline{conv}} \subseteq \mathcal{S}_{conv} \subseteq \mathcal{S} := 2^X$, we deduce from a repeated application of part (ii) of Proposition 6.30 that

$$\mathcal{H}^t_\delta = \mathcal{H}^t_{\delta,\mathcal{S}} \leq \mathcal{H}^t_{\delta,\mathcal{S}_{conv}} \leq \mathcal{H}^t_{\delta,\mathcal{S}_{\overline{conv}}}. \tag{6.58}$$

Next, we show that $\mathcal{H}^t_{\delta,\mathcal{S}_{\overline{conv}}} \leq \mathcal{H}^t_\delta$, which, combined with (6.58), will clearly imply the desired string of equalities stated in (6.57). To see why the latter inequality is true , we apply part (iii) of Proposition 6.30 as follows. Given $E \subseteq X$, let $\{B_j\}$ be any $\mathcal{S}$-cover of E (still with $\mathcal{S} = 2^X$) by sets of finite diameter. For each $j \geq 1$, let $A_j := \overline{\operatorname{conv}}(B_j)$, the closed convex hull of B_j (defined just before Lemma 6.33 below). By part (i) of Lemma 6.33,[41] we have that for every

[41] Equivalently, one can apply part (ii) of Lemma 6.33. Indeed, since B_j has finite diameter, so does A_j. Hence, A_j is bounded and thus since it is also closed, it follows that A_j is compact.

$j \geq 1$, $\operatorname{diam}(B_j) = \operatorname{diam}(A_j)$ and, clearly, $B_j \subseteq A_j$. Therefore, it follows from part (ii) of Proposition 6.30 that $\mathcal{H}^t_{\delta,\mathcal{S}_{\overline{conv}}} \leq \mathcal{H}^t_{\delta,\mathcal{S}} = \mathcal{H}^t_\delta$, as desired.

(iii) Next, we establish part (iii) of the theorem. We must therefore show that if the metric space X is reasonable, in the sense of the definition given just before the statement of Theorem 6.24 (which is the case, for example, if $X := \mathbb{R}^m$, according to part (a) of Remark 6.25), then (as in the two identities in part (iii) of Theorem 6.24)

$$\text{(6.59)} \qquad \mathcal{S}^t_\delta := \mathcal{H}^t_{\delta,\mathcal{S}^\#} \leq \mathcal{H}^t_{\delta,\mathcal{S}^\times} \quad \text{and} \quad \mathcal{H}^t_{\delta+,\mathcal{S}^\times} \leq \mathcal{H}^t_{\delta,\mathcal{S}^\#} =: \mathcal{S}^t_\delta.$$

First, note that since X is reasonable, given any open ball B, there exists a closed ball with the same diameter as B and containing A. Therefore, given any $\mathcal{S}^\times$-cover $\{B_j\}$ of $E \subseteq X$, there exists an $\mathcal{S}^\#$-cover $\{A_j\}$ of E such that[42]

$$\text{(6.60)} \qquad \operatorname{diam}(A_j) = \operatorname{diam}(B_j), \text{ for all } j \geq 1.$$

In light of part (iii) of Proposition 6.30 (the hypotheses of which are clearly satisfied), we deduce that

$$\text{(6.61)} \qquad \mathcal{H}^t_{\delta,\mathcal{S}^\#} \leq \mathcal{H}^t_{\delta,\mathcal{S}^\times}.$$

Conversely, given any $\varepsilon > 0$ and any $\mathcal{S}^\#$-cover $\{B_j\}$ of $E \subseteq X$, we claim that there exists an $\mathcal{S}^\times$-cover $\{A_j\}$ of E such that

$$\text{(6.62)} \qquad \operatorname{diam}(A_j) \leq \operatorname{diam}(B_j) + \varepsilon, \text{ for all } j \geq 1.$$

Indeed, if $B_j = \overline{B_{\boldsymbol{x}_j}(r_j)}$ (for some $\boldsymbol{x}_j \in X$ and $r_j > 0$), it suffices to let $A_j := B_{\boldsymbol{x}_j}(r_j + \theta^{-1} \cdot \varepsilon)$, for each $j \geq 1$. Then, clearly, for every $j \geq 1$, $B_j \subseteq A_j$ and[43]

$$\text{(6.63)} \qquad \operatorname{diam}(A_j) = \theta(r_j + \theta^{-1} \cdot \varepsilon) = \theta r_j + \varepsilon = \operatorname{diam}(B_j) + \varepsilon,$$

which implies (6.62). Note that in (6.63), we have used the fact that since X is reasonable, then $\operatorname{diam}(A_j) = \theta(r_j + \theta^{-1} \cdot \varepsilon)$ and $\operatorname{diam}(B_j) = \theta r_j$, since A_j is an open ball of radius $r_j + \theta^{-1} \cdot \varepsilon$ and B_j is a closed ball of radius r_j.

Now, it follows from part (iv) of Proposition 6.30 that

$$\text{(6.64)} \qquad \mathcal{H}^t_{\delta+,\mathcal{S}^\times} \leq \mathcal{H}^t_{\delta,\mathcal{S}^\#};$$

i.e., $\mathcal{H}^t_{\delta^*,\mathcal{S}^\times} \leq \mathcal{H}^t_{\delta,\mathcal{S}^\#}$, for every $\delta^* > \delta$.

The desired result, (6.59), now follows by combining (6.61) and (6.64). This concludes the proof of part (iii) of the theorem.

---

[42]Indeed, if $B_j = B_{\boldsymbol{x}_0}(r_j)$ (for some $\boldsymbol{x}_0 \in X$ and $r_j > 0$), it suffices to let $A_j := \overline{B_j}$. Then, $A_j \supseteq B_j$ and (by the reasonableness of X) $\operatorname{diam}(A_j) = \theta r_j = \operatorname{diam}(B_j)$. Hence, $\bigcup_j A_j \supseteq \bigcup_j B_j \supseteq E$, so that $\{A_j\}$ is an $\mathcal{S}^\#$-cover of E satisfying (6.60), as desired.

[43]Recall that according to Remark 6.25, when $X := \mathbb{R}^m$, we have $\theta := 2$, whereas when $X := \mathbb{Q}_p$, the field of p-adic numbers, no such $\theta \geq 1$ exists.

(iv) Finally, we establish part (iv) of Theorem 6.24. Accordingly, we must show that (with $S^t_\delta := \mathcal{H}^t_{\delta,\mathcal{S}^\#}$) the inequalities in (6.36) hold; namely, with $\gamma_{m,t} > 0$ given by (6.37), we have that

$$\mathcal{H}^t_\delta \le S^t_\delta \quad \text{and} \tag{6.65}$$

$$S^t_\delta \le \gamma_{m,t} \cdot \mathcal{H}^t_\delta. \tag{6.66}$$

From now on, we assume that $X := \mathbb{R}^m$ (the m-dimensional Euclidean space), as in parts (iii) and (iv) of Theorem 6.24, and we fix an integer $m \ge 1$ and a real number $t \ge 0$. We want to compare S^t_δ (the δ-approximate *spherical* Hausdorff outer measure or measure) and $\mathcal{H}^t_\delta$ (the δ-approximate *standard* or classic Hausdorff outer measure or measure).

Since $\mathcal{H}^t_\delta := \mathcal{H}^t_{\delta,\mathcal{S}}$, with $\mathcal{S} := 2^X$, one inequality (namely, (6.65)) is easy to obtain. Indeed, since $\mathcal{S}^\# \subseteq \mathcal{S}$, it follows from part (ii) of Proposition 6.30 that $\mathcal{H}^t_\delta := \mathcal{H}^t_{\delta,\mathcal{S}} \le \mathcal{H}^t_{\delta,\mathcal{S}^\#} =: S^t_\delta$, which is precisely (6.65) above.

Next, we aim at proving a suitable form of a reverse inequality, namely, at bounding S^t_δ in terms of a suitable positive constant times $\mathcal{H}^t_\delta$. This reverse inequality will turn out to coincide with (6.66), with the positive constant $\gamma_{m,t}$ given by (6.37).

For this purpose, we will appeal to Lemma 6.34 (in the form of the Jung Sharp Inequality (6.73), stated in Subsection 6.3.4 below), according to which, given any nonempty compact subset K of $\mathbb{R}^m$, there exists an "optimal" closed ball A containing K and with diameter satisfying the following inequality (see inequality (6.73) in Lemma 6.34 below):

$$\operatorname{diam}(A) \le (2m/(m+1))^{1/2} \cdot \operatorname{diam}(K). \tag{6.67}$$

In order to be able to apply this result, we first need to recall from part (i) of Theorem 6.24 that $\mathcal{H}^t_\delta = \mathcal{H}^t_{\delta,\mathcal{S}_{cl}}$; see the first equality in (6.34).

Now, let $E \subseteq X$ be arbitrary and let $\{B_j\}$ be any $\mathcal{S}_{cl}$-cover of E (composed of sets of finite diameter). Then, for each integer $j \ge 1$, in light of Lemma 6.34 (Jung's Theorem) applied to the compact set $K := B_j$,[44] there exists a closed ball A_j (i.e., $A_j \in \mathcal{S}^\#$) such that $B_j \subseteq A_j$ and (see the inequality (6.67) above)

$$\operatorname{diam}(A_j) \le (2m/(m+1))^{1/2} \cdot \operatorname{diam}(B_j). \tag{6.68}$$

Clearly, it then follows that $\{A_j\}$ is an $\mathcal{S}^\#$-cover of E satisfying the inequality

[44] Since B_j has finite diameter, it is nonempty and bounded. Indeed, $B_j \subseteq B_{\boldsymbol{x}_j}(d_j)$, where $d_j := \operatorname{diam}(B_j)$ and $\boldsymbol{x}_j$ is the center of the ball; see also part (iii) of Exercise 6.2. Hence, since B_j is also closed, we deduce that B_j is compact, as required in the statement of Lemma 6.34. This is the reason why we are working here with the family $\mathcal{S}_{cl}$ instead of with $\mathcal{S} := 2^X$.

(6.68), for each $j \geq 1$.[45] Therefore, in light of the variant of part (iii) of Proposition 6.30, which is stated in Remark 6.31 (and with $c := (2m/(m+1))^{1/2}$, in light of (6.68)), we deduce that

$$\mathscr{H}^t_{\delta,\mathcal{S}^\#} \leq c^t \cdot \mathscr{H}^t_{\delta,\mathcal{S}_{cl}} = (2m/(m+1))^{t/2} \cdot \mathscr{H}^t_{\delta,\mathcal{S}_{cl}}. \tag{6.69}$$

Since $S^t_\delta := \mathscr{H}^t_{\delta,\mathcal{S}^\#}$, by definition, and since $\mathscr{H}^t_\delta = \mathscr{H}^t_{\delta,\mathcal{S}_{cl}}$ (by part (i) of Theorem 6.24; see the first equality in (6.34)), we conclude that (6.69) is precisely equivalent to the remaining desired inequality (see (6.66) above); namely, $S^t_\delta \leq \gamma_{m,t} \cdot \mathscr{H}^t_\delta$, where $\gamma_{m,t} := (2m/(m+1))^{t/2}$, as in (6.37). This completes the proof of the main statement of part (iv) of the theorem, namely, of the inequalities (6.65) and (6.66) above.

The remaining statements are very easy to check. First, in light of (6.37), $\gamma_{m,t} > 1$ for every $m \geq 1$, as desired. Second, if $m = 1$ (i.e., if $X := \mathbb{R}$), then, obviously, $\gamma_{1,t} = 1$ (independently of $t \geq 0$) and hence, we can combine (6.65) and (6.66) to obtain that $\mathscr{H}^t_\delta \leq S^t_\delta \leq \mathscr{H}^t_\delta$; i.e., $\mathscr{H}^t_\delta = S^t_\delta$, as stated at the very end of part (iv).

This finally completes the proof of part (iv) of the theorem and, consequently, of the entire theorem (Theorem 6.24). □

Exercise 6.32 (*The Case of the Real Line*). Assume that $m = 1$; i.e., $X := \mathbb{R}$, the real line.

(i) Show directly that $S^t_\delta = \mathscr{H}^t_\delta$; see part (iv) of Theorem 6.24.

(ii) Obtain a direct proof of Jung's Theorem (Lemma 6.34 below) in this case. Namely, show that given any nonempty compact set $K \subseteq \mathbb{R}$, there exists a closed and bounded (i.e., compact) interval A such that $\operatorname{diam}(A) = \operatorname{diam}(K)$. Is this closed interval (ball) A unique? What is a possible choice of A when $K = C$, the classic Cantor set?

[*Hint*: (a) *For the existence*: Let $A := \overline{\operatorname{conv}}(K)$. (b) *For the presumed uniqueness*: If A' is a closed interval containing K, then $A' \supseteq \overline{\operatorname{conv}}(K)$.]

In the course of the very last step of the proof of part (iv) of Theorem 6.24, given just above, namely, the proof of (6.66), we have used the following two results (Lemma 6.33 and Lemma 6.34), which are of a geometric nature and, more specifically, belong to the realm of convex geometry and of discrete geometry.

In Lemma 6.33, given a convex subset B of $\mathbb{R}^m$, we denote by $\operatorname{conv}(B)$ (resp., $\overline{\operatorname{conv}}(B)$) its *convex hull* (resp., *closed convex hull*), i.e., the smallest convex (resp., closed convex) subset of $\mathbb{R}^m$ containing B.

---

[45] Indeed, by construction, $A_j \in \mathcal{S}^\#$ and $B_j \subseteq A_j$, for all $j \geq 1$. Hence, $E \subseteq \bigcup_j B_j \subseteq \bigcup_j A_j$.

Lemma 6.33 (Convex Hull).

(i) *If* $A \subseteq \mathbb{R}^m$, *then (with the above notation)*[46]

$$
\text{(6.70)}\qquad \begin{aligned} \operatorname{conv}(A) &= \left\{ \sum_{k=1}^{n} \alpha_k \cdot \boldsymbol{a}_k \,:\, n \in \mathbb{N},\ \boldsymbol{a}_k \in A,\ \alpha_k \geq 0 \text{ and } \sum_{k=1}^{n} \alpha_k = 1 \right\} \\ &= \{\text{ convex combinations of elements of } A\}. \end{aligned}
$$

It follows from (6.70) *that (with* $B := \operatorname{conv}(A)$*),*

$$
\text{(6.71)}\qquad \operatorname{diam}(B) = \operatorname{diam}(A) \quad \text{and} \quad \overline{B} = \overline{\operatorname{conv}(A)} = \overline{\operatorname{conv}}(A).
$$

Hence, we also have that $\operatorname{diam}(B) = \operatorname{diam}(\overline{B}) = \operatorname{diam}(A)$.

(ii) *Furthermore, if* $A \subseteq \mathbb{R}^m$ *is compact (i.e., is closed and bounded),*[47] *then so is* $\operatorname{conv}(A)$. *In this case, by the second part of* (6.71), *the convex hull and the closed convex hull of A coincide:* $\operatorname{conv}(A) = \overline{\operatorname{conv}}(A)$*; i.e., the convex hull of A is closed. Also, their diameters are the same and are both equal to* $\operatorname{diam}(A)$.

Proof. The detailed proof of the lemma is left as an exercise for the reader, who may want to consult a standard textbook on convex geometry; see, e.g., [**517**]. We only give here a few key indications.

First, the identity (6.70) is easy to establish. Once this is done, the fact that (6.71) holds is an easy consequence of (6.70). Recall, in particular, that

$$
\boldsymbol{x} \in \overline{A} \Leftrightarrow \boldsymbol{x} = \lim_{n \to \infty} \boldsymbol{x}_n, \text{ with } \boldsymbol{x}_n \in A, \text{ for all } n \geq 1.
$$

The last identity in part (i) follows from (6.71) and the fact that the diameter of a set and of its closure coincide; see part (i) of Exercise 6.3.

For part (ii), recall that since A, $\operatorname{conv}(A)$, and $\overline{\operatorname{conv}}(A)$ have the same diameter, according to part (i), it follows from part (iii) of Exercise 6.2 that $\operatorname{conv}(A)$ has finite diameter and hence is bounded.[48] But since $\operatorname{conv}(A)$ is also closed (because A is closed), in light of the second part (6.71), we deduce that A is compact. The remaining statements in part (ii) follow from part (i). □

6.3.4. Jung's Sharp Inequality and Related Results. We will admit without proof the following result, which is also known as "Jung's Theorem" [**278**, **279**] in the literature; see, e.g., [**130**].

[46] It turns out that a well-known result due to Carathéodory (once again, in this chapter) states that in (6.70), the integer n can be chosen to be no greater than $m+1$: $n \leq m+1$. This is proved by descending induction on n. It is not needed in the sequel, however.

[47] Equivalently, when A is nonempty, this is the case if A is closed and has finite diameter; see part (iii) of Exercise 6.2.

[48] The case when A is empty is trivial and needs not be discussed here.

Lemma 6.34 (Jung's Sharp Inequality for Smallest Enclosing Balls; H. W. E. Jung, [**278,279**]). *Let $K \subseteq \mathbb{R}^m$ be compact and nonempty, with $m \in \mathbb{N}$ arbitrary. Then, there exists a closed ball A such that $K \subseteq A$ and with radius r satisfying*

$$r \leq d \cdot \sqrt{m/2(m+1)}, \tag{6.72}$$

where $d := \operatorname{diam}(K)$. Equivalently, A has diameter $\operatorname{diam}(A) = 2r$ satisfying the following inequality:[49]

$$\operatorname{diam}(A) \leq d \cdot \sqrt{2m/(m+1)}. \tag{6.73}$$

Remark 6.35. The ball A is called a *minimum enclosing ball* for K. The constant $\sqrt{m/2(m+1)}$ in (6.72) (or, equivalently, its analog $\sqrt{2m/(m+1)}$ in (6.73)) is optimal. For example, when K is a closed equilateral triangle in $\mathbb{R}^2$, then the minimum enclosing disk (or enclosing circle) has radius $\sqrt{1/3}\cdot d = \sqrt{2/2(2+1)}\cdot d$. Moreover, note that for any integer $m \geq 2$, the radius r of A is strictly greater than $d/2$; equivalently, the diameter $\operatorname{diam}(A) = 2r$ of A is strictly greater than $d = \operatorname{diam}(K)$. (The only case where we have an equality is when $m = 1$.) Observe that, in general, a minimum enclosing ball is not unique.

This shows that even when $X = \mathbb{R}^m$, there is no hope to improve the inequality comparing $\mathcal{H}^t_\delta$ (the δ-approximate Hausdorff outer measure) and the δ-approximate Hausdorff spherical outer measure, $S^t_\delta := \mathcal{H}^t_{\delta,\mathcal{S}^\#} = \mathcal{H}^t_{\delta,\mathcal{S}^\times}$, in part (iv) of Theorem 6.24 (or in the corresponding proof), as well as in Corollary 6.26. An entirely analogous comment can be made (after passing to the limit as $\delta \to 0^+$) about the Hausdorff measure $\mathcal{H}^t$ and the spherical Hausdorff measure $\mathcal{H}^t_{\mathcal{S}^\#} = \mathcal{H}^t_{\mathcal{S}^\times}$ to be discussed in Sections 7.1 and 7.2. It is unimportant, however, since most of the time, all one cares about, in practice, is to know whether $\mathcal{H}^t$ (or its spherical counterpart, $\mathcal{H}^t_{\mathcal{S}^\#} = \mathcal{H}^t_{\mathcal{S}^\times}$) is finite, equal to zero, or else lies strictly between 0 and $+\infty$. The obtained inequalities show that in either case (that of the Hausdorff measure or of its spherical counterpart), the answer to this question is the same. As an important consequence, the Hausdorff dimensions associated with either the standard Hausdorff measure or its spherical counterpart coincide.

Let $X := \mathbb{R}^m$, the m-dimensional Euclidean space. In the next, open-ended, exercise, the reader is asked to investigate, among other things, the optimality of the positive constant appearing in the inequalities stated in part (iv) of Theorem 6.24 and following from Jung's Theorem (Lemma 6.34 above):

$$S^t_\delta \leq \gamma_{m,t} \cdot \mathcal{H}^t_\delta, \tag{6.74}$$

while the first inequality in the aforementioned result stated that

$$\mathcal{H}^t_\delta \leq S^t_\delta. \tag{6.75}$$

[49]Note that since $K \subseteq A$, we also have $d \leq \operatorname{diam}(A)$, for any $m \geq 1$. In particular, when $m = 1$ (i.e., when $X = \mathbb{R}$), (6.73) (resp., (6.72)) becomes an equality: $\operatorname{diam}(A) = d$ (resp., $r = d/2$).

Here, in (6.74), $\gamma_{m,t} := (2m/(m+1))^{t/2}$, as given by (6.37). Hence, $\gamma_{m,t} > 1$ for all $m \geq 2$, while $\gamma_{1,t} = 1$.

Accordingly, given an integer $m \geq 1$, along with real numbers $t \geq 0$ and $\delta > 0$, we let

$$\Lambda_{m,t,\delta} := \inf\{c > 0 : S^t_\delta \leq c \cdot \mathcal{H}^t_\delta\}. \tag{6.76}$$

Since we will also, and especially, be interested in the exact counterpart of (6.74) and (6.75) (as well as, more generally, of Theorem 6.24), for the spherical and standard Hausdorff measure or outer measure S^t and $\mathcal{H}^t$ (obtained, respectively, as the limit of S^t_δ and $\mathcal{H}^t_\delta$, as $\delta \to 0^+$; see Sections 7.1 and 7.2 below), we also let for a given integer $m \geq 1$ and real number $t \geq 0$,

$$\Lambda_{m,t} := \inf\{c > 0 : S^t \leq c \cdot \mathcal{H}^t\}. \tag{6.77}$$

The main question is to determine whether or not $\Lambda_{m,t,\delta} = \gamma_{m,t}$ and, similarly, $\Lambda_{m,t} = \gamma_{m,t}$. Clearly, in light of (6.74) and (6.75) (obtained in part (iv) of Theorem 6.24), we have that

$$1 \leq \Lambda_{m,t,\delta} \leq \gamma_{m,t} \quad \text{and} \quad 1 \leq \Lambda_{m,t} \leq \gamma_{m,t}, \tag{6.78}$$

with equalities when $m = 1$ since $\gamma_{1,t} = 1$ (i.e., $\Lambda_{1,t,\delta} = \Lambda_{1,t} = 1$), because in this case, $\mathcal{H}^t_\delta = S^t_\delta$, an immediate consequence of (6.74) and (6.75) when $m = 1$ and hence also $\mathcal{H}^t = S^t$ (after having let $\delta \to 0^+$). Therefore, we may assume from now on that $m \geq 2$.

Exercise 6.36 (*Optimality of* $\gamma_{m,t}$*?*). Let $X := \mathbb{R}^m$ and let $m \geq 2$, $t \geq 0$ and (for part (i)) $\delta > 0$. Furthermore, let $\Lambda_{m,t,\delta}$ and $\Lambda_{m,t}$ be given by (6.76) and (6.77), respectively.

(i) (*Optimality for the δ-Approximate Spherical and Standard Hausdorff Measures?*) Investigate whether or not $\Lambda_{m,t,\delta} = \gamma_{m,t}$.

(ii) (*Optimality for the Spherical and Standard Hausdorff Measures?*) Similarly, determine whether or not $\Lambda_{m,t} = \gamma_{m,t}$.

(iii) Show that in (i) (resp., (ii)), it does not matter whether $\mathcal{H}^t_\delta$ and S^t_δ (resp., S^t and $\mathcal{H}^t$) are both viewed as outer measures or both as Borel measures on $\mathbb{R}^m$.

[*Hint*: In part (i) and part (ii), respectively, the main issue is whether given any $\varepsilon > 0$ sufficiently small, one can find a Borel subset E of $\mathbb{R}^m$ such that $S^t_\delta(E) > (\gamma_{m,t} - \varepsilon) \cdot \mathcal{H}^t_\delta(E)$, and similarly for $S^t(E)$ and $\mathcal{H}^t(E)$, although for a possibly different Borel set E.]

Exercise 6.37.

(a) Show that in a general metric space X, the best inequalities we can expect in the statement of a counterpart of Jung's Theorem (Lemma 6.34) are given by $d/2 \leq r \leq d$. Indeed, if X is uniform,[50] then we always have that $r = d$. On the other hand, if X is the set $\mathbb{R}^m$ equipped with the "taxicab" or "Manhattan" metric, given by $d_1(\mathbf{x}, \mathbf{y}) := \sum_{j=1}^{m} |x_j - y_j|$, for any $\mathbf{x} = (x_1, \dots, x_m)$, $\mathbf{y} = (y_1, \dots, y_m)$ in $\mathbb{R}^m$, then a moment's reflection shows that we always have that $r = d/2$.

(b) Find a "non-Euclidean geometry" analog of Jung's Theorem in the case when $X = S^{m-1}$, the unit sphere of $\mathbb{R}^m$ (i.e., when X is "spherical"), and in the case when X is hyperbolic (i.e., when X is the hyperbolic m-dimensional space).[51]

(c) Show that in Euclidean space $X = \mathbb{R}^{m+1}$ and for every integer $m \geq 0$, Jung's Sharp Inequality (6.72) in Lemma 6.34 is achieved (i.e., for this choice of K, it becomes an equality) for the standard m-simplex $K = K_m$ given by

$$K_m := \left\{ \mathbf{x} = (x_0, \dots, x_m) \in \mathbb{R}^{m+1} : \sum_{j=0}^{m} x_j = 1, x_j \geq 0, \text{ for } j = 0, \dots, m \right\},$$

or, more generally (but equivalently), for any m-simplex in $\mathbb{R}^{m+1}$. Are these the only (nonempty) compact subsets K of $\mathbb{R}^{m+1}$ for which Jung's Sharp Inequality is achieved?

Exercise 6.38. Looking ahead, one could also use the so-called *Isodiametric Inequality*[52] rather than the Sharp Jung Inequality, in order to improve the estimate comparing S^t_δ and $\mathcal{H}^t_\delta$ (and, eventually, S^t and $\mathcal{H}^t$, the t-dimensional spherical and standard Hausdorff measures, respectively). This open-ended exercise consists in asking the reader to examine whether the just-mentioned estimates can indeed be improved in this case and to state the corresponding estimates, as well as to consider the exact counterpart (in this new context) of Exercise 6.36 above.

Often, one does not care about obtaining an optimal (or close to optimal) estimate but one simply needs to obtain a (nontrivial) bound. In such situations, it is advantageous to use the following tautological (and, thus, very simple) result as a substitute for Jung's Sharp Inequality (or Theorem), stated in Lemma 6.34. We refer to Remark 6.42 below for a use of this simple result.

[50] That is, if the distance between any two distinct points is equal to 1.

[51] For a complete answer to this question, see **[123]**.

[52] See, e.g., (7.160) at the beginning of the proof of Theorem 7.66, along with the text preceding the statement of Theorem 7.66, for a statement and discussion of this inequality. The *Isodiametric Inequality* states that for any compact set $K \subseteq \mathbb{R}^m$ (with diameter $d := \operatorname{diam}(K)$),

$$\mathcal{L}^m(K) \leq (d/2)^m \cdot \omega_m, \tag{6.79}$$

where ω_m is the volume of the unit ball in $\mathbb{R}^m$. This highly nontrivial inequality is clearly better than the simple inequality stated in (6.80) of Lemma 6.39. Without loss of generality, one may assume that K is compact because $\operatorname{diam}(A) = \operatorname{diam}(\overline{A})$ and if K is unbounded, then $d = +\infty$ so that (6.79) is then trivially satisfied.

Lemma 6.39 (Weak Form of Jung's Theorem). *Let K be a nonempty compact set in Euclidean space $\mathbb{R}^m$, with m an arbitrary positive integer, and let $d := \operatorname{diam}(K)$ denote its diameter. Then, there is a closed ball B of radius $r := d$ (and, hence, of diameter equal to $2d$) such that $K \subseteq B$. Thus, with ω_m denoting the volume of the unit ball in $\mathbb{R}^m$, we have that*

$$\mathcal{L}^m(K) \leq (2d)^m \cdot \omega_m. \tag{6.80}$$

Proof. Since K is compact and nonempty, its diameter d is finite. Let $\boldsymbol{x}_0 \in K$ be arbitrary. Then, clearly, the closed ball $B = B_{\boldsymbol{x}_0}(d)$ satisfies the stated condition $B \supseteq K$. Indeed, if $\boldsymbol{x} \in K$, then, by definition of the diameter and since $\boldsymbol{x}_0, \boldsymbol{x} \in K$, $d(\boldsymbol{x}_0, \boldsymbol{x}) \leq d := \operatorname{diam}(K)$. Hence, $\boldsymbol{x} \in B$, as required. This completes the proof of the lemma. □

As a consequence of Lemma 6.39, one obtains the following estimate, which (in the case when $X = \mathbb{R}^m$) is clearly worse than the one obtained in (6.74) (or, equivalently, in (6.36) of part (iv) of Theorem 6.24) or, certainly, than the one we would obtain by using the Isodiametric Inequality (see Exercise 6.38 above), but presents the advantage of having an elementary and completely self-contained proof and of also being valid in any metric space (rather than just in Euclidean space $\mathbb{R}^m$); see Remark 6.42 below.

Lemma 6.40. *For any $t \geq 0$, the (t-dimensional) δ-approximate spherical and Hausdorff measures, $\mathcal{H}^t_\delta$ and S^t_δ, on X satisfy the following estimate:*

$$\mathcal{H}^t_\delta \leq S^t_\delta \leq 2^t \cdot \mathcal{H}^t_\delta. \tag{6.81}$$

Proof. The second inequality in (6.81) is proved exactly as in the corresponding estimate in (6.36) of part (iv) of Theorem 6.24, except that one uses inequality (6.80) of Lemma 6.39 rather than Jung's inequality (6.73) of Lemma 6.34. As is pointed out in Remark 6.42 below, the first part of Lemma 6.39 remains valid without change in any metric space (X, d), rather than just in Euclidean space $\mathbb{R}^m$.

Finally, note that the first inequality in (6.81) follows at once from the definitions of the δ-approximate Hausdorff measure as an infimum, in light of the fact that $S^t_\delta := \mathcal{H}_{\delta, \mathcal{S}^\#}$ and $\mathcal{S}^\# \subseteq 2^X$. □

Remark 6.41 (*Optimality when $m = 1$*). If $m = 1$, the result stated in Lemma 6.39 is optimal. Indeed, in this case, we should take $B := \operatorname{conv}(K)$ (or if, more generally, K is a nonempty bounded subset of $\mathbb{R}^m$, $B := \overline{\operatorname{conv}}(K)$, the smallest closed interval containing K); then, by Lemma 6.33, part (i), $\operatorname{diam}(B) = \operatorname{diam}(K) =: d$ instead of $2d$. For example, if $K = C$, the ternary Cantor set, then $B = [0, 1]$, which can be viewed as the ball of center $1/2 \in K$ and radius $1/2$ (in $\mathbb{R}$) or, equivalently, diameter $d = 1$, which is the diameter of C.

Remark 6.42 (*Extension to Arbitrary Metric Spaces*). Obviously, except for the estimate (6.80), which is specific to Euclidean space $\mathbb{R}^m$, Lemma 6.39 remains valid for any (nonempty) compact subset of X (or, more generally, for any nonempty bounded subset of X), where X is an arbitrary metric space. This will enable us, for example, to obtain self-contained proofs of new estimates valid in any metric space, such as those given in parts of Chapter 7, including, for example, the ones comparing the spherical and standard Hausdorff measures; see Corollary 7.17.

6.3.5. Translation Invariance and Scale Covariance. We close this section by stating a simple proposition and an exercise, along with related comments.

Proposition 6.43. *Let* $X := \mathbb{R}^m$. *Then,* $\mathcal{H}^t_\delta := \mathbb{R}^m$ *satisfies the following two key properties:*

(i) (Translation Invariance). $\mathcal{H}^t_\delta(E + \boldsymbol{x}) = \mathcal{H}^t_\delta(E)$, *for all* $E \subseteq \mathbb{R}^m$ *and* $\boldsymbol{x} \in \mathbb{R}^m$, *and where* $E + \boldsymbol{x} := \{\boldsymbol{y} + \boldsymbol{x} : \boldsymbol{y} \in E\}$.

(ii) (Scale Covariance). $\mathcal{H}^t_{\lambda\delta}(\lambda E) = \lambda^t \cdot \mathcal{H}^t_\delta(E)$, *for all* $E \subseteq \mathbb{R}^m$ *and* $\lambda > 0$, *and where* $\lambda E := \{\lambda \boldsymbol{y} : \boldsymbol{y} \in E\}$.

Proof. This follows easily from the definition of $\mathcal{H}^t_\delta(E)$ given in (6.19) but, in light of the importance of this result (especially for $\mathcal{H}^t$, after a passage to the limit as $\delta \to 0^+$), we provide the details.

Indeed, clearly, in light of the definition of the diameter of a set, we have that $\operatorname{diam}(A + \boldsymbol{x}) = \operatorname{diam}(A)$ and $\operatorname{diam}(\lambda A) = \lambda \cdot \operatorname{diam}(A)$, since the metric $d = d_m$ on $\mathbb{R}^m$ is translation invariant and "scale covariant"; i.e., $d(\boldsymbol{a} + \boldsymbol{x}, \boldsymbol{b} + \boldsymbol{x}) = d(\boldsymbol{a}, \boldsymbol{b})$ and $d(\lambda \boldsymbol{a}, \lambda \boldsymbol{b}) = \lambda \cdot d(\boldsymbol{a}, \boldsymbol{b})$, for all $\boldsymbol{a}, \boldsymbol{b} \in \mathbb{R}^m$.

Since $\{B_j\}$ is a δ-cover of $E + \boldsymbol{x}$ iff $\{A_j := B_j - \boldsymbol{x}\}$ is a δ-cover of E and $\varphi(\{A_j + \boldsymbol{x}\}) = \varphi(\{A_j\})$, we have successively

$$\begin{aligned}\mathcal{H}^t_\delta(E + \boldsymbol{x}) &= \inf\left\{ \sum_{j\geq 1} (\operatorname{diam}(A_j + \boldsymbol{x}))^t \ : \ \{A_j + \boldsymbol{x}\} \text{ is a } \delta\text{-cover of } E + \boldsymbol{x} \right\} \\ &= \inf\left\{ \sum_{j\geq 1} (\operatorname{diam}(A_j))^t \ : \ \{A_j\} \text{ is a } \delta\text{-cover of } E \right\} = \mathcal{H}^t_\delta(E).\end{aligned}$$

Furthermore, since, in light of the scale covariance, $\{B_j\}$ is a $(\lambda\delta)$-cover of λE iff $\{A_j := \lambda^{-1} \cdot B_j\}$ is a δ-cover of E and $\varphi(\{\lambda A_j\}) = \lambda^t \cdot \varphi(\{A_j\})$, we have successively

$$\begin{aligned}\mathcal{H}^t_{\lambda\delta}(\lambda E) &= \inf\left\{\sum_{j\geq 1}(\operatorname{diam}(\lambda A_j))^t \ : \ \{\lambda A_j\} \text{ is a } (\lambda\delta)\text{-cover of } \lambda E\right\}\\ &= \inf\left\{\lambda^t \cdot \left(\sum_{j\geq 1}(\operatorname{diam}(A_j))^t\right) \ : \ \{A_j\} \text{ is a } \delta\text{-cover of } E\right\}\\ &= \lambda^t \cdot \inf\left\{\sum_{j\geq 1}(\operatorname{diam}(A_j))^t \ : \ \{A_j\} \text{ is a } \delta\text{-cover of } E\right\} = \lambda^t \cdot \mathcal{H}^t_\delta(E).\end{aligned}$$

This concludes the proof of the proposition. □

Remark 6.44. Naturally, Proposition 6.43 has an obvious counterpart for the associated Borel measure $\widetilde{\mu}_\delta$ on $(\mathbb{R}^m, \mathcal{B}(\mathbb{R}^m))$ and for the measure $\overline{\mu}_\delta$ on $(\mathbb{R}^m, \mathcal{M}_\delta)$, where $\mathcal{M}_\delta$ denotes the σ-algebra of μ^*_δ-measurable sets (in the notation of Theorem 6.18 and Corollary 6.20).

Exercises

Our final exercise, which could clearly be extended to any family $\mathcal{S} \subseteq 2^X$ (with $\emptyset \in \mathcal{S}$), is important because it answers a natural question concerning "intrinsic" and "extrinsic" (δ-approximate) Hausdorff outer measures and measures. Much as in Remark 6.44 just above, we continue using the notation of Theorem 6.18, as well as of Corollary 6.20 and its extension.

Exercise 6.45. Let $X = (X, d)$ be a metric space and given $D \subseteq X$, consider the associated metric space $D = (D, d_D)$, with d_D denoting here the restriction to D of the metric d on X. The point of this exercise is to check that we can start from X and then restrict to D, or consider D intrinsically. Indeed, the associated Hausdorff outer measures and measures coincide with their counterparts.

(i) Let $D \subseteq X$. Show that the restriction $\mu^*_{\delta,D}$ to D of the outer measure $\mu^*_\delta = \mathcal{H}^t_\delta$ on X coincides with the *intrinsic* δ-approximate Hausdorff outer measure on D.

(ii) Similarly, let $D \in \mathcal{M}_\delta$ (i.e., D is μ^*_δ-measurable); then, the restriction $\overline{\mu}_{\delta,D}$ to D (i.e., to the σ-algebra $\mathcal{M}_{\delta,D} = \mathcal{M}_\delta \cap D := \{A \cap D : A \in \mathcal{M}_\delta\}$) of the Carathéodory measure on X (i.e., on the measurable space $(X, \mathcal{M}_\delta)$) coincides with the *intrinsic* Carathéodory measure on D (i.e., on the measurable space $(D, \mathcal{M}_{\delta,D})$, obtained by restricting the δ-approximate Hausdorff outer measure on D).

(iii) State and establish the counterpart for the following Borel measures on X, $\widetilde{\mu}_\delta$, and its restriction $\widetilde{\mu}_{\delta,D}$ on D (i.e., on the Borel σ-algebra $\mathcal{B}(D) = \mathcal{B}(X) \cap D = \{A \cap D : A \in \mathcal{B}(X)\}$), as well as for the *intrinsic* Borel measure on $(D, \mathcal{B}(D))$ associated with the metric outer measure $\mu^*_{\delta,D}$ on D.

Chapter 7

Construction and Properties of Hausdorff Measure

Now that, in Chapter 6, we have constructed and studied in detail the δ-approximate Hausdorff outer measures $\mathcal{H}^t_{\delta,\mathcal{S}}$ (and $\mathcal{H}^t_{\delta}$), along with the associated δ-approximate Hausdorff measures, it is a relatively simple matter to define and determine the main properties of the corresponding Hausdorff measures $\mathcal{H}^t_{\mathcal{S}}$ and, in particular, $\mathcal{H}^t$, for any $t \geq 0$. In short, it suffices to pass to the limit as $\delta \to 0^+$ in order to derive $\mathcal{H}^t$ from the family $\{\mathcal{H}^t_{\delta}\}_{\delta>0}$.

After having constructed Hausdorff measure in Section 7.1, along the lines described just above, and compared in Section 7.2 the values of the Hausdorff measures corresponding to different underlying families, we establish in Section 7.3 several of the key geometric properties of Hausdorff measure, including its bi-Lipschitz invariance and—in the important special case of Euclidean space $\mathbb{R}^m$—its invariance under translations and rotations, its covariance under scaling, as well as (both in Sections 7.3 and 7.4) the equality of (suitably normalized) m-dimensional Hausdorff and Lebesgue measures, $\mathcal{H}^m$ and $\mathcal{L}^m$. Finally, in Section 7.5, we briefly review the definition and basic properties of Hausdorff dimensions in general metric spaces (rather than in Euclidean spaces, as in Section 5.2).

We note that the notions of Hausdorff measure and dimension were introduced by Felix Hausdorff in 1918 in [**234**].

7.1. Construction of Hausdorff Measure, via Its δ-Approximate Counterparts

As in Chapter 6, let $\delta > 0$, $t \geq 0$ $(t \in \mathbb{R})$ be arbitrary, and given any metric space (X, d), denote by $\mathcal{H}^t_{\delta,\mathcal{S}}$ or $\mu^*_{\delta,\mathcal{S}}$, indifferently, the corresponding (t-dimensional) δ-approximate Hausdorff outer measure studied in Subsection 6.3.3 and associated with a family $\mathcal{S} \subseteq 2^X$ such that $\emptyset \in \mathcal{S}$.

We also adopt the same notation as in Subsection 6.3.2 for the (t-dimensional) δ-approximate Hausdorff–Carathéodory measure $\mathcal{H}^t_{\delta,\mathcal{S}} = \overline{\mu}_{\delta,\mathcal{S}}$, defined by restricting $\mu^*_{\delta,\mathcal{S}}$ to the σ-algebra $\mathcal{M}^*_{\delta,\mathcal{S}} := \mathcal{M}$ of Carathéodory measurable sets, as well as for the (t-dimensional) δ-approximate Hausdorff–Borel measure $\mathcal{H}^t_\delta = \widetilde{\mu}_{\delta,\mathcal{S}}$, obtained by further restricting $\mu^*_{\delta,\mathcal{S}}$ (or, equivalently, by restricting $\overline{\mu}_{\delta,\mathcal{S}}$) to the Borel σ-algebra $\mathcal{B}(X) \subseteq \mathcal{M}^*_{\delta,\mathcal{S}}$; see Theorem 6.18 and Definition 6.21.

In the special case when $\mathcal{S} := 2^X$, we write, indifferently, $\mathcal{H}^t_\delta$ or μ^*_δ for the standard (t-dimensional) δ-approximate Hausdorff outer measure, and similarly for the associated Hausdorff–Carathéodory and Hausdorff–Borel measures $\overline{\mu}_\delta$ and $\widetilde{\mu}_\delta$, respectively; see Corollary 6.20. We then have the following simple lemma.

Lemma 7.1. *Let $t \geq 0$, with $t \in \mathbb{R}$. Then, the outer measure $\mathcal{H}^t_{\delta,\mathcal{S}} := \mu^*_{\delta,\mathcal{S}}$ can be viewed as a nonincreasing function of $\delta > 0$; i.e., for any $E \subseteq X$, the $[0, +\infty]$-valued function $\delta \mapsto \mu^*_{\delta,\mathcal{S}}(E)$ is nonincreasing on $(0, +\infty)$.*

Proof. Let $E \subseteq X$. In light of the definition of $\mu^*_{\delta,\mathcal{S}}(E) = \mathcal{H}^t_{\delta,\mathcal{S}}(E)$ provided in (6.21), given $0 < \delta_1 \leq \delta_2$, the family of (countable) δ_1-covers of E is clearly contained in the family of (countable) δ_2-covers of E. Hence, since $\mu^*_{\delta_2,\mathcal{S}}(E)$ (resp., $\mu^*_{\delta_1,\mathcal{S}}(E)$) is given by an infimum taken over the former (resp., over the latter) family, we have that $\mu^*_{\delta_2,\mathcal{S}}(E) \leq \mu^*_{\delta_1,\mathcal{S}}(E)$, as desired. □

Since a nonincreasing $[0, +\infty]$-valued function on $(0, +\infty)$ always has a limit (in $[0, +\infty]$) as $\delta > 0$ tends to zero, given by the supremum of this function on $(0, +\infty)$,[1] the following definition is entirely justified, in light of Lemma 7.1.

Definition 7.2. Given $t \in \mathbb{R}$ with $t \geq 0$, we let[2]

$$\mu^*_{\mathcal{S}} = \mathcal{H}^t_{\mathcal{S}} := \lim_{\delta \to 0^+} \mathcal{H}^t_{\delta,\mathcal{S}} = \sup_{\delta > 0} \mathcal{H}^t_{\delta,\mathcal{S}} \tag{7.1}$$
$$\left(= \lim_{\delta \to 0^+} \mu^*_{\delta,\mathcal{S}} = \sup_{\delta > 0} \mu^*_{\delta,\mathcal{S}}\right);$$

[1]See Proposition A.5 in Appendix A for a counterpart of this fact for sequences rather than for functions of a continuous variable.

[2]Both in (7.1) and in (7.2), the supremum and the limit exist in $[0, +\infty]$, the extended nonnegative half-line.

i.e., for every $E \subseteq X$, we let

$$(7.2)\qquad \begin{aligned} \mu_{\mathcal{S}}^*(E) = \mathcal{H}^t_{\mathcal{S}}(E) &:= \lim_{\delta \to 0^+} \mathcal{H}^t_{\delta,\mathcal{S}}(E) = \sup_{\delta>0} \mathcal{H}^t_{\delta,\mathcal{S}}(E) \\ &(= \lim_{\delta \to 0^+} \mu^*_{\delta,\mathcal{S}}(E) = \sup_{\delta>0} \mu^*_{\delta,\mathcal{S}}(E)). \end{aligned}$$

Note that, as was pointed out in Section 5.1,[3] and since $\delta \mapsto \mathcal{H}^t_{\delta,S}$ is a nonincreasing function of δ on $(0, +\infty)$, in the sense of Lemma 7.1, each of the suprema over $\delta > 0$ in identities (7.1) and (7.2) of Definition 7.2 (as well as in parts (i) and (ii) of Proposition 7.7 below) can be replaced indifferently by the corresponding supremum over $\delta \in [0, 1]$.

In the sequel, unless explicitly specified otherwise, $t \in \mathbb{R}$ is such that $t \geq 0$ is arbitrary, but fixed. In particular, $\mu_{\mathcal{S}}^* = \mathcal{H}^t_{\mathcal{S}}$ depends implicitly on t, and similarly for $\overline{\mu}_{\mathcal{S}}$ and $\widetilde{\mu}_{\mathcal{S}}$ (as well as for μ^*, $\overline{\mu}$, and $\widetilde{\mu}$, corresponding to $\mathcal{S} := 2^X$), to be introduced below. Of course, an entirely analogous comment could have been made about the δ-approximate counterparts of these notions; namely, $\mu^*_{\delta,\mathcal{S}} = \mathcal{H}^t_{\delta,\mathcal{S}}$, $\overline{\mu}_{\delta,\mathcal{S}}$, and $\widetilde{\mu}_{\mathcal{S}}$, as well as for μ^*_{δ}, $\overline{\mu}_{\delta}$, and $\widetilde{\mu}_{\delta}$ (corresponding to the important special case when $\mathcal{S} := 2^X$).

We can now state and prove the following key result.

Theorem 7.3 (Outer Measure and Metric Outer Measure). *Let $t \geq 0$, with $t \in \mathbb{R}$, and let (X, d) be an arbitrary metric space. Then, the following two properties hold:*

(i) *$\mu_{\mathcal{S}}^* = \mathcal{H}^t_{\mathcal{S}}$ is an outer measure on X.*

(ii) *In addition, $\mu_{\mathcal{S}}^* = \mathcal{H}^t_{\mathcal{S}}$ is a metric outer measure, in the sense of Definition 6.11.*

Proof.

(i) Let $\delta > 0$. Consider $\{E_j\}_{j=1}^{\infty}$, a countable collection of subsets of X, and let $E := \bigcup_{j=1}^{\infty} E_j$. Then, we have that

$$(7.3)\qquad \mu^*_{\delta,\mathcal{S}}(E) \leq \sum_{j=1}^{\infty} \mu^*_{\delta,\mathcal{S}}(E_j) \leq \sum_{j=1}^{\infty} \mu^*_{\mathcal{S}}(E_j),$$

where the first inequality in (7.3) follows from the fact that $\mu^*_{\delta,\mathcal{S}}$ is an outer measure[4] and hence is countably subadditive, while the second inequality is a consequence of (7.2) in Definition 7.2 which implies that

$$\mu^*_{\delta,\mathcal{S}}(E_j) \leq \mu^*_{\mathcal{S}}(E_j), \text{ for all } j \geq 1.$$

[3]See the proof of Lemma 5.3, along with Definition 5.4.

[4]See Theorem 6.18, part (i).

It now follows from (7.3) that

$$\mu^*_{\delta,\mathcal{S}}(E) \le \sum_{j=1}^{\infty} \mu^*_{\mathcal{S}}(E_j), \quad \text{for all } \delta > 0. \tag{7.4}$$

Since the right-hand side of (7.4) is independent of $\delta > 0$, we deduce once again from (7.2) that

$$\mu^*_{\mathcal{S}}(E) = \sup_{\delta>0} \mu^*_{\delta,\mathcal{S}}(E) \le \sum_{j=1}^{\infty} \mu^*_{\mathcal{S}}(E_j), \tag{7.5}$$

which yields the countable subadditivity of $\mu^*_{\mathcal{S}}$. We leave it as a simple exercise for the interested reader to deduce from (7.2) in Definition 7.2 that the set function $\mu^*_{\mathcal{S}}$ is monotonic, in the sense of part (ii) of Definition 3.10. In addition, since clearly, in light of (7.5), the set function $\mu^*_{\mathcal{S}}$ is $[0, +\infty]$-valued and (because $\mu^*_{\delta,\mathcal{S}}$, being an outer measure, satisfies $\mu^*_{\delta,\mathcal{S}}(\emptyset) = 0$), we have that

$$\mu^*_{\mathcal{S}}(\emptyset) = \lim_{\delta\to 0^+} \mu^*_{\delta,\mathcal{S}}(\emptyset) = \lim_{\delta\to 0^+} 0 = 0.$$

In light of Definition 3.10 for an outer measure, we conclude that $\mu^*_{\mathcal{S}}$ is an outer measure. This completes the proof of part (i) of the theorem.

(ii) Let us next show that $\mu^*_{\mathcal{S}}$ is a *metric* outer measure. For this purpose, let $A, B \subseteq X$ such that $\operatorname{dist}(A, B) > 0$, where, as in (6.2), $\operatorname{dist}(A, B)$ denotes the distance between A and B; see Definition 6.11. Then, since for each $\delta > 0$, $\mu^*_{\delta,\mathcal{S}}$ is a metric outer measure (see Lemma 6.22 above), we have that

$$\mu^*_{\delta,\mathcal{S}}(A \cup B) = \mu^*_{\delta,\mathcal{S}}(A) + \mu^*_{\delta,\mathcal{S}}(B), \quad \text{for all } \delta > 0. \tag{7.6}$$

Therefore, it suffices to let $\delta \to 0^+$ in (7.6) in order to deduce from (7.2) that

$$\mu^*_{\mathcal{S}}(A \cup B) = \mu^*_{\mathcal{S}}(A) + \mu^*_{\mathcal{S}}(B),$$

which shows that $\mu^*_{\mathcal{S}}$ is a metric outer measure and completes the proof of part (ii) as well as of the entire theorem. □

In light of part (i) of Theorem 7.3, the following definition is fully justified.

Definition 7.4 (*t-Dimensional Hausdorff Outer Measure*). Given $t \ge 0$, $\mu^*_{\mathcal{S}} := \mathcal{H}^t_{\mathcal{S}}$, as introduced in Definition 7.2, is called the (*t-dimensional*) *Hausdorff outer measure* (on X) *associated with the family* $\mathcal{S}$.

In particular, when $\mathcal{S} := 2^X$, the power set of X, then $\mu^* = \mathcal{H}^t$ is called the (*t-dimensional*) *Hausdorff outer measure* (on X), or sometimes, the *standard outer measure* on X.

According to the Carathéodory Measurability and Restriction Theorem (Theorem B.3 in Appendix B), we can now restrict $\mu^*_{\mathcal{S}} = \mathcal{H}^t_{\mathcal{S}}$ to the σ-algebra

$\mathcal{M}_{\mathcal{S}} := \mathcal{M}_{\mu^*_{\mathcal{S}}}$ of $\mu^*_{\mathcal{S}}$-measurable sets in order to obtain the Carathéodory measure $\overline{\mu}_{\mathcal{S}} = \mathcal{H}^t_{\mathcal{S}}$ associated with the outer measure $\mu^*_{\mathcal{S}}$.[5]

Similarly, since (by part (ii) of Theorem 7.3) $\mu^*_{\mathcal{S}}$ is a metric outer measure on X, we know from Theorem 6.14 (and Corollary 6.16) that the Borel σ-algebra $\mathcal{B}(X)$ is contained in $\mathcal{M}_{\mathcal{S}}$—and hence that we can restrict $\overline{\mu}_{\mathcal{S}}$ (or, equivalently, $\mu^*_{\mathcal{S}}$) to $\mathcal{B}(X)$. The resulting measure is, by definition, a Borel measure and is denoted by $\widetilde{\mu}_{\mathcal{S}}$.

Lemma 7.5 (Borel vs. Carathéodory Measurable σ-Algebras). *We have the following two inclusions of σ-algebras (of X):*

$$\mathcal{B}(X) \subseteq \mathcal{N}_{\mathcal{S}} := \bigcap_{\delta>0} \mathcal{M}_{\delta,\mathcal{S}} \subseteq \mathcal{M}_{\mathcal{S}}, \tag{7.7}$$

where, as before, $\mathcal{M}_{\mathcal{S}}$ is the σ-algebra of $\mu^*_{\mathcal{S}}$-measurable sets and as in (6.25), *for each $\delta > 0$, $\mathcal{M}_{\delta,\mathcal{S}}$ is the σ-algebra of $\mu^*_{\delta,\mathcal{S}}$-measurable sets.*

Proof. According to parts (i) and (iii) of Theorem 6.18, respectively, we know that for each $\delta > 0$, $\mu^*_{\delta,\mathcal{S}}$ is a metric outer measure; hence, we have that $\mathcal{B}(X) \subseteq \mathcal{M}_{\delta,\mathcal{S}}$; see Theorem 6.14. This implies the first inclusion in (7.7).

Next, in order to see why the second inclusion of (7.7) holds, we proceed as follows. Let $E \in \mathcal{N}_{\mathcal{S}}$; i.e., $E \in \mathcal{M}_{\delta,\mathcal{S}}$, for all $\delta > 0$. We must show that $E \in \mathcal{M}_{\mathcal{S}}$. Since, by assumption, E is $\mu^*_{\delta,\mathcal{S}}$-measurable, for every $\delta > 0$, we have for all $A \subseteq X$ and all $\delta > 0$,[6] $\mu^*_{\delta,\mathcal{S}}(A) = \mu^*_{\delta,\mathcal{S}}(A \cap E) + \mu^*_{\delta,\mathcal{S}}(A \cap E^c)$. We can thus let $\delta \to 0^+$ in the latter identity and then deduce that $\mu^*_{\mathcal{S}}(A) = \mu^*_{\mathcal{S}}(A \cap E) + \mu^*_{\mathcal{S}}(A \cap E^c)$, which, again by Definition 3.11, means that E is $\mu^*_{\mathcal{S}}$ -measurable (since $A \subseteq X$ is arbitrary); i.e., $E \in \mathcal{M}_{\mathcal{S}}$, as desired. □

The following definition completes Definition 7.4 of $\mu^*_{\mathcal{S}}$, viewed as the Hausdorff outer measure associated with $\mathcal{S}$.[7]

---

[5]*Caution*: Strictly speaking, this is *not* the Carathéodory measure, in the sense of Definition 3.50, because the outer measure $\mu^*_{\mathcal{S}}$ is not obtained directly via Theorem 3.48, but only via a two-step process. However, by an abuse of language, we will still refer to $\overline{\mu}_{\mathcal{S}}$ as *the Carathéodory measure associated with* $\mu^*_{\mathcal{S}}$.

[6]Recall the original definition of ν^*-measurability, where ν^* is an outer measure; see, especially, identity (3.6) in Definition 3.11.

[7]A sample list of references on (or related to) the Hausdorff dimension and/or measure includes, e.g., [7], [11], [19], [23–26], [28–30], [34], [35], [36], [40–42], [43–45], [47], [50], [54], [59], [61], [72], [74], [84], [85], [86], [92], [98], [101,102], [103–105], [106–117], [118], [119], [120], [121,543], [401], [145], [149–151], [152–155], [157], [160], [167], [169], [170], [171–173], [175], [181–183], [193], [194], [198–200], [205], [207], [208], [316], [219], [226, 227], [231], [232], [233], [234], [235], [245, 246], [250] [253], [254], [258], [261], [265–270], [281], [282], [286], [289], [293], [294], [295], [308], [310], [312], [317], [318–326], [329,330], [332], [333–338], [352–354], [356], [366], [368], [370], [372], [374], [375], [380], [382–385], [386–388], [390, 391], [399,400], [402], [415–420], [423–426], [427], [430], [431], [434], [439,440], [622], [452], [454], [463], [468], [471–473], [483], [485–487], [491,492], [497], [503], [506], [507], [516], [518], [534], [542], [546], [547], [555], [557], [558,559], [572–574], [575], [576–580], [581], [590,591], [593,594], [585], [611], [619], [620], [621], and [629], along with the relevant references therein.

Definition 7.6 (*Hausdorff–Carathéodory and Hausdorff–Borel or Hausdorff Measures*). We have the following different kinds of Hausdorff measures, depending on the σ-algebras on which they act:

(i) (*t-Dimensional Hausdorff–Carathéodory Measure*). The measure $\overline{\mu}_{\mathcal{S}} = \mathcal{H}^t_{\mathcal{S}}$ on the σ-algebra $\mathcal{M}_{\mathcal{S}}$, obtained by restricting the outer measure $\mu^*_{\mathcal{S}} = \mathcal{H}^t_{\mathcal{S}}$ to $\mathcal{M}_{\mathcal{S}}$, is called the *($t$-dimensional) Hausdorff–Carathéodory measure associated with the family* $\mathcal{S}$.[8]

(ii) (*t-Dimensional Hausdorff–Borel Measure*). Moreover, the Borel measure $\widetilde{\mu}_{\mathcal{S}}$ obtained by further restriction $\mu^*_{\mathcal{S}}$ to the smaller σ-algebra $\mathcal{B}(X)$ (or, equivalently, by restricting $\overline{\mu}_{\mathcal{S}}$ to $\mathcal{B}(X)$) is called the *(t-dimensional) Hausdorff–Borel measure* (or, frequently, just the *Hausdorff measure*) *associated with the family* $\mathcal{S}$.

(iii) (*Standard t-Dimensional Hausdorff Measure*). In particular, in the important special case when $\mathcal{S} := 2^X$, $\overline{\mu} := \overline{\mu}_{2^X}$ (resp., $\widetilde{\mu} = \widetilde{\mu}_{2^X}$) is called the (standard) *Hausdorff–Carathéodory measure* (resp., the *Hausdorff–Borel measure*, or simply the *Hausdorff measure*) on X. It is also sometimes referred to as the *standard Hausdorff measure* on X.

Depending on the context, we denote by $\mathcal{H}^t$ either $\overline{\mu}$ or $\widetilde{\mu}$, just as we denote by $\mathcal{H}^t_{\mathcal{S}}$ either $\overline{\mu}_{\mathcal{S}}$ or $\widetilde{\mu}_{\mathcal{S}}$.

Proposition 7.7 (Hausdorff Measures as Limits of δ-Approximate Measures). *The Hausdorff measure can be expressed as the limit as $\delta \to 0^+$ of the corresponding δ-approximate Hausdorff measures, as follows:*

(i) $\overline{\mu}_{\mathcal{S}}$, *restricted to the σ-algebra $\mathcal{N}_{\mathcal{S}} := \bigcap_{\delta>0} \mathcal{M}_{\delta,\mathcal{S}}$,*[9] *is the limit as $\delta \to 0^+$ of the $(\delta, \mathcal{S})$-approximate Hausdorff–Carathéodory measures $\overline{\mu}_{\delta,\mathcal{S}}$; i.e., for every $E \in \mathcal{N}_{\mathcal{S}}$, we have that*

$$\overline{\mu}_{\mathcal{S}}(E) = \lim_{\delta\to 0^+} \overline{\mu}_{\delta,\mathcal{S}}(E) = \sup_{\delta>0} \overline{\mu}_{\delta,\mathcal{S}}(E). \tag{7.8}$$

(ii) *The Hausdorff–Borel measure (or Hausdorff measure) $\widetilde{\mu}_{\mathcal{S}}$ is the limit as $\delta \to 0^+$ of the corresponding $(\delta, \mathcal{S})$-approximate measures $\widetilde{\mu}_{\delta,\mathcal{S}}$; i.e., for every $E \in \mathcal{B}(X)$, we have that*

$$\widetilde{\mu}_{\mathcal{S}}(E) = \lim_{\delta\to 0^+} \widetilde{\mu}_{\delta,\mathcal{S}}(E) = \sup_{\delta>0} \widetilde{\mu}_{\delta\mathcal{S}}(E). \tag{7.9}$$

Proof. In light of Definition 7.2 for the Hausdorff outer measure $\mu^*_{\mathcal{S}}$ associated with $\mathcal{S}$ (see, especially, (7.1) and (7.2)), this follows at once from Lemma 7.5 and Definition 7.6, where $\overline{\mu}_{\mathcal{S}}$ and $\widetilde{\mu}_{\mathcal{S}}$ are the restriction of $\mu^*_{\mathcal{S}}$ to the σ-algebra $\mathcal{M}_{\mathcal{S}} \supseteq \mathcal{N}_{\mathcal{S}}$ and to the Borel σ-algebra $\mathcal{B}(X)$, respectively. □

[8] Recall from Lemma 7.5 that the inclusions stated in (7.7) hold.

[9] Recall from Lemma 7.5 that $\mathcal{N}_{\mathcal{S}}$ is a σ-subalgebra of $\mathcal{M}_{\mathcal{S}}$ and contains $\mathcal{B}(X)$ as a σ-subalgebra.

The following uniqueness result is the counterpart of part (iv) of Theorem 6.18 but does not follow from it, although its proof is entirely parallel to the proof of part (iv).[10]

Proposition 7.8 (Weak Uniqueness of the Extension). *The Hausdorff–Carathéodory measure $\overline{\mu}_{\mathcal{S}}$, viewed as a measure extension of the Borel measure $\widetilde{\mu}_{\mathcal{S}}$, is "weakly unique", in the following sense: If ν is any other measure on the measurable space $(X, \mathcal{M}_{\mathcal{S}})$ extending the Borel measure $\widetilde{\mu}_{\mathcal{S}}$ from $\mathcal{B}(X)$ to all of $\mathcal{M}_{\mathcal{S}}$, then $\nu(E) \leq \overline{\mu}_{\mathcal{S}}(E)$,* for all $E \in \mathcal{M}_{\mathcal{S}}$, *with equality (i.e., $\nu(E) = \overline{\mu}_{\mathcal{S}}(E)$) whenever $E \in \mathcal{M}_{\mathcal{S}}$ is such that $\overline{\mu}_{\mathcal{S}}(E) < \infty$.*[11]

Proof. As was alluded to just above, the proof parallels that of part (iv) of Theorem 6.18. In short, we apply (the uniqueness part of) the Algebra Version of the Carathéodory Extension Theorem, Theorem B.12 (part (ii)) in Appendix B, with $\mathcal{A} := \mathcal{B}(X)$, which is a σ-algebra rather than a mere algebra, and with the set function $\mu := \widetilde{\mu}_{\mathcal{S}}$, a measure on the Borel σ-algebra $\mathcal{B}(X)$ rather than a mere premeasure on an algebra. We omit the details and refer instead to the proof of part (ii) of Theorem B.12 which the reader will easily be able to adapt to the present situation. □

Proposition 7.9. *The Hausdorff–Carathéodory measure $\overline{\mu}_{\mathcal{S}} = \mathcal{H}^t_{\mathcal{S}}$ is a complete measure, in the sense of Definition* B.5 *in Appendix* B.

Proof. This follows at once from the last statement of Theorem B.3 in Appendix B (the Measurability and Restriction Theorem), according to which the Carathéodory measure (here, $\overline{\mu}_{\mathcal{S}}$) associated to an outer measure (here, $\mu^*_{\mathcal{S}}$) is always complete. □

Remark 7.10. Note that we are not claiming here (and let alone proving) that $\overline{\mu}_{\mathcal{S}}$ is always the completion of the Borel measure $\widetilde{\mu}_{\mathcal{S}}$, in the sense of Theorem B.6. As an interesting and creative exercise, the reader may wish to investigate the following question, even in the important special case when $\mathcal{S} := 2^X$: *Is the measure $\overline{\mu}_{\mathcal{S}}$ the completion of the Borel measure $\widetilde{\mu}_{\mathcal{S}}$?*

The above question may be closely connected with the fact that Hausdorff measure is, in general, not σ-finite, as well as with the weak (as opposed to strong) uniqueness statement of Proposition 7.8; see Exercise 7.52 below. Compare with Theorem B.6 in Appendix B, where the uniqueness statement is, however, phrased in terms of complete (rather than σ-finite) measures.

Corollary 7.11. *All of the results of Theorem* 7.3, *Lemma* 7.5, *as well as of Propositions* 7.7, 7.8, *and* 7.9, *are valid, in particular, in the important special case*

[10]In light of Proposition 7.7, if we tried to deduce Proposition 7.8 from part (iv) of Theorem 6.18, we would obtain a weaker uniqueness result about the restriction of $\overline{\mu}_{\mathcal{S}}$ to $\mathcal{M}_{\mathcal{S}}$ instead of about $\overline{\mu}_{\mathcal{S}}$ itself.

[11]We point out that the measure $\overline{\mu}_{\mathcal{S}}$ is *not* σ-finite, in general, as will be further discussed later in this subsection. This fact explains why one cannot obtain a stronger uniqueness statement here.

when $\mathcal{S} := 2^X$ *and, hence, for the Hausdorff outer measure* $\mathcal{H}^t = \mu^*$ *along with the Hausdorff–Carathéodory and the Hausdorff–Borel (or simply Hausdorff) measures* $\mathcal{H}^t = \overline{\mu}$ *and* $\mathcal{H}^t = \widetilde{\mu}$, *respectively.*

This statement holds true for any metric space (X, d) *and, in particular, for the key special case of Euclidean space* $(\mathbb{R}^m, d_m)$, *equipped with its natural Euclidean metric* d_m.

Remark 7.12 (*Hausdorff Dimension in a General Metric Space*)**.** Now that we have defined Hausdorff measure in a precise manner, we can define Hausdorff dimension in exactly the same way as we did in Section 5.2 in the case of Euclidean space $\mathbb{R}^m$ (see, especially, Definition 5.13 and the text surrounding it). Recall, however, that in the present chapter (and unless explicitly stated otherwise), we work with an arbitrary metric space X instead of with $\mathbb{R}^m$. We refer the interested reader to Section 7.5 at the end of this chapter, for more precision about the definition of the Hausdorff dimension in an arbitrary metric space (X, d); see, especially, Definition 7.75.

7.2. Comparison of Hausdorff Measures for Various Families

The following comments will be quite useful in the sequel when comparing below the Hausdorff measures and outer measures associated with different families.

Remark 7.13. The exact analogs (of parts (a) and (b)) of Remark 6.23 and of the unnumbered comment preceding Theorem 6.24 hold for $\mathcal{H}^t_{\mathcal{S}} \in \{\mu^*_{\mathcal{S}}, \overline{\mu}_{\mathcal{S}}, \widetilde{\mu}_{\mathcal{S}}\}$, instead of for $\mathcal{H}^t_{\delta,\mathcal{S}} \in \{\mu^*_{\delta,\mathcal{S}}, \overline{\mu}_{\delta,\mathcal{S}}, \widetilde{\mu}_{\delta,\mathcal{S}}\}$, respectively, and for $\mathcal{M}_{\mathcal{S}}$ in place of $\mathcal{M}_{\delta,\mathcal{S}}$, as well as with Theorem 7.14 and Corollary 7.15 below substituted for Theorem 6.24 and Corollary 6.26 from Subsection 6.3.3.

In particular, given $\mathcal{S}_1, \mathcal{S}_2 \subseteq 2^X$ and with $\mathcal{H}^t_{\mathcal{S}_j} := \mu^*_{\mathcal{S}_j}$, for $j = 1, 2$, the equality $\mathcal{H}^t_{\mathcal{S}_1} = \mathcal{H}^t_{\mathcal{S}_2}$ (resp., the inequality $\mathcal{H}^t_{\mathcal{S}_1} \leq \mathcal{H}^t_{\mathcal{S}_2}$) means that $\mathcal{H}^t_{\mathcal{S}_1}(E) = \mathcal{H}^t_{\mathcal{S}_2}(E)$ (resp., $\mathcal{H}^t_{\mathcal{S}_1}(E) \leq \mathcal{H}^t_{\mathcal{S}_2}(E)$), for all $E \subseteq X$ (or, equivalently, for all $E \in 2^X$); i.e., $\mu^*_{\mathcal{S}_1}(E) = \mu^*_{\mathcal{S}_2}(E)$ (resp., $\mu^*_{\mathcal{S}_1}(E) \leq \mu^*_{\mathcal{S}_2}(E)$), for all $E \subseteq X$, where $\mu_{\mathcal{S}_1}$ and $\mu_{\mathcal{S}_2}$ are the Hausdorff outer measures respectively associated with $\mathcal{S}_1$ and $\mathcal{S}_2$.

This is true, analogously, for the Hausdorff–Carathéodory measures $\overline{\mu}_{\mathcal{S}_1}$ and $\overline{\mu}_{\mathcal{S}_2}$ (except for $E \in 2^X$ replaced by any $E \in \mathcal{M}_{\mathcal{S}_1} \cap \mathcal{M}_{\mathcal{S}_2} =: \mathcal{M}_{\mathcal{S}_1,\mathcal{S}_2}$),[12] as well as for the Hausdorff–Borel (or simply Hausdorff) measures $\widetilde{\mu}_{\mathcal{S}_1}$ and $\widetilde{\mu}_{\mathcal{S}_2}$ (except for $E \in 2^X$ now replaced by any $E \in \mathcal{B}(X)$).[13]

---

[12]It is immediate to check that $\mathcal{M}_{\mathcal{S}_1,\mathcal{S}_2}$ is a σ-algebra of X containing $\mathcal{B}(X)$.

[13]Clearly, by restriction to the appropriate σ-algebras, the statements concerning $\overline{\mu}_{\mathcal{S}_1}, \overline{\mu}_{\mathcal{S}_2}$ and for $\widetilde{\mu}_{\mathcal{S}_1}, \widetilde{\mu}_{\mathcal{S}_2}$ follow at once from the above ones about $\mu^*_{\mathcal{S}_1}$ and $\mu^*_{\mathcal{S}_1}$.

The following theorem (Theorem 7.14) is the counterpart for Hausdorff measures (and outer measures) of Theorem 6.24. Observe that it takes a somewhat simpler form than in Theorem 6.24, due to the fact that in the limit as $\delta \to 0^+$, several inequalities in the latter theorem become significantly simpler and, in several cases, even become equalities. For the same reasons, the exact same comment can be made for Corollary 7.15 of Theorem 7.14, in comparison to Corollary 6.26 of Theorem 6.24.

Theorem 7.14 (Comparison of Various Hausdorff Measures (and Outer Measures)). *Let X be a metric space and fix $t \in \mathbb{R}$ such that $t \geq 0$. Then, with the notation and the conventions introduced in the present subsection,*[14] *we have the following comparison results:*

(i) (Families of Closed Sets and of Open Sets). *For any metric space X, we have the following equalities:* $\mathcal{H}^t = \mathcal{H}^t_{\mathcal{S}_{cl}} = \mathcal{H}^t_{\mathcal{S}_{op}}$.

(ii) (Families of Convex Sets and of Closed Convex sets). *If $X := \mathbb{R}^m$ (the m-dimensional Euclidean space), then we have the following additional equalities:* $\mathcal{H}^t = \mathcal{H}^t_{\mathcal{S}_{conv}} = \mathcal{H}^t_{\mathcal{S}_{\overline{conv}}}$.

(iii) (Spherical Hausdorff Measures: Families of Closed Balls and of Open Balls). *If $X := \mathbb{R}^m$ (the m-dimensional Euclidean space) or, more generally, if the metric space X is "reasonable",*[15] *then we have the following additional equalities:* $\mathcal{H}^t_{\mathcal{S}^\times} = \mathcal{H}^t_{\mathcal{S}^\#} =: S^t$, *where S^t is the (t-dimensional)* spherical Hausdorff measure (*or* outer measure).

(iv) (Comparison Between the Standard Spherical Measures (or Outer Measures)). *If $X := \mathbb{R}^m$ (the m-dimensional Euclidean space) and with the notation S^t introduced for the (t-dimensional) spherical measure (or outer measure), we also have the following inequalities (with S^t defined as in part* (iii) *just above):*

$$\mathcal{H}^t \leq S^t \leq \gamma_{m,t} \cdot \mathcal{H}^t, \tag{7.10}$$

where, as in part (iv) *of Theorem 6.24, the positive constant $\gamma_{m,t}$ satisfies $\gamma_{m,t} > 1$ for every integer $m \geq 2$ and is given for every integer $m \geq 1$ by*

$$\gamma_{m,t} := (2m/(m+1))^{t/2}. \tag{7.11}$$

Furthermore, when $m = 1$ (i.e., in the case when $X := \mathbb{R}$, the real line), $\gamma_{1,t} = 1$ and hence, we only have equalities in (7.10): $\mathcal{H}^t = S^t$, if $m = 1$.

Proof. In essence, this follows easily (after a passage to the limit as $\delta \to 0^+$) from Theorem 6.24, the counterpart of the present theorem for δ-approximate Hausdorff measures (and outer measures). First, recall from Remark 7.13 that it suffices to establish the theorem for the Hausdorff outer measures.

[14] See, in particular, Remark 7.13 above.

[15] See the text preceding the statement of Theorem 6.24 for the notion of a reasonable metric space.

(i) By part (i) of Theorem 6.24, we have, for all $\delta > 0$ and $E \subseteq X$,

$$\mathscr{H}^t_\delta(E) = \mathscr{H}^t_{\delta,\mathcal{S}_{cl}}(E) \leq \mathscr{H}^t_{\delta,\mathcal{S}_{op}}(E) \quad \text{and} \tag{7.12}$$

$$\mathscr{H}^t_{\delta^+,\mathcal{S}_{op}} \leq \mathscr{H}^t_\delta(E) = \mathscr{H}^t_{\delta,\mathcal{S}_{cl}}(E); \tag{7.13}$$

see (6.34).[16] In particular, (7.13) yields

$$\mathscr{H}^t_{2\delta,\mathcal{S}_{op}}(E) \leq \mathscr{H}^t_\delta(E) = \mathscr{H}^t_{\delta,\mathcal{S}_{cl}}(E), \tag{7.14}$$

which is significantly less sharp than (7.13) but is sufficient for our purpose.

Next, we simply let $\delta \to 0^+$ in (7.12) and (7.14) and deduce, in light of Definition 7.2 (see, especially, identity (7.2)), that

$$\mathscr{H}^t(E) = \mathscr{H}^t_{\mathcal{S}_{cl}}(E) \leq \mathscr{H}^t_{\mathcal{S}_{op}}(E) \quad \text{and} \quad \mathscr{H}^t_{\mathcal{S}_{op}}(E) \leq \mathscr{H}^t(E) = \mathscr{H}^t_{\mathcal{S}_{cl}}(E).$$

Clearly, these two inequalities, once combined, yield the desired equalities: $\mathscr{H}^t(E) = \mathscr{H}^t_{\mathcal{S}_{cl}}(E) = \mathscr{H}^t_{\mathcal{S}_{op}}(E)$, as claimed.

(ii) Part (ii) follows immediately from part (ii) of Theorem 6.24, according to which $\mathscr{H}^t_\delta(E) = \mathscr{H}^t_{\delta,\mathcal{S}_{conv}}(E) = \mathscr{H}^t_{\delta,\mathcal{S}_{\overline{conv}}}(E)$, for every $\delta > 0$ and every $E \subseteq X$. Holding $E \subseteq X$ fixed but letting $\delta \to 0^+$ in the latter identity, we deduce (once again from Definition 7.2, identity (7.2)) that $\mathscr{H}^t(E) = \mathscr{H}^t_{\mathcal{S}_{conv}}(E) = \mathscr{H}^t_{\mathcal{S}_{\overline{conv}}}(E)$, as desired.

(iii) Part (iii) follows from part (iii) of Theorem 6.24 in much the same way as part (i) followed from part (i) of Theorem 6.24. Indeed, assuming that X is reasonable, by part (iii) of Theorem 6.24, we have that

$$S^t_\delta(E) := \mathscr{H}^t_{\delta,\mathcal{S}^\#}(E) \leq \mathscr{H}^t_{\delta,\mathcal{S}^\times}(E) \quad \text{and} \tag{7.15}$$

$$\mathscr{H}^t_{\delta^+,\mathcal{S}^\times}(E) \leq \mathscr{H}^t_{\delta,\mathcal{S}^\#}(E) =: S^t_\delta(E), \tag{7.16}$$

for all $\delta > 0$ and $E \subseteq X$. Then, we replace δ^+ by 2δ (say) in (7.16) and after having let $\delta \to 0^+$ in (7.15) and the resulting substitute for (7.16) (while holding $E \subseteq X$ fixed), we deduce that

$$S^t(E) := \mathscr{H}^t_{\mathcal{S}^\#}(E) \leq \mathscr{H}^t_{\mathcal{S}^\times}(E) \leq \mathscr{H}^t_{\mathcal{S}^\#}(E) = S^t(E),$$

which yields the equalities $S^t(E) = \mathscr{H}^t_{\mathcal{S}^\#}(E) = \mathscr{H}^t_{\mathcal{S}^\times}(E)$, as desired.

(iv) Finally, this part follows routinely from part (iv) of Theorem 6.24, according to which (when $X = \mathbb{R}^m$ and with $\gamma_{m,t}$ given by (7.11)) we have that $\mathscr{H}^t_\delta(E) \leq S^t_\delta(E) \leq \gamma_{m,t} \cdot \mathscr{H}^t_\delta(E)$, for all $\delta > 0$ and all $E \subseteq X$; see (6.36). Holding E fixed while letting $\delta \to 0^+$ in the latter inequality, we deduce from identity (7.2) in Definition 7.2 that $\mathscr{H}^t(E) \leq S^t(E) \leq \gamma_{m,t} \cdot \mathscr{H}^t(E)$, which is equivalent to (7.10), the main conclusion of part (iv). Note that when $m = 1$, it follows at once from (7.11) that $\gamma_{1,t} = 1$ (independently of $\delta > 0$) and hence, the first inequality of (7.10) becomes the equality $\mathscr{H}^t(E) = S^t(E)$, as claimed in the last

[16] Recall from inequality (6.33) in footnote 27 of Chapter 6 that, in (7.13), the symbol δ^+ means that the corresponding inequality holds for any $\delta^* > \delta$ (and thus, for example, for $\delta^* = 2\delta$).

statement of part (iv). This completes the proof of part (iv) and thus also of the entire theorem. □

The next result, which is the counterpart of Corollary 6.26 for Hausdorff measures (or outer measures), follows at once from Theorem 7.14 (and its proof) specialized to the case when $X := \mathbb{R}^m$. Recall from part (a) of Remark 6.25 that Euclidean spaces are reasonable metric spaces, in the sense specified just before Theorem 6.24.

Corollary 7.15 (The Case of Euclidean Space $\mathbb{R}^m$). *Assume that $X = \mathbb{R}^m$, the m-dimensional Euclidean space (with $m \geq 1$). Then:*

(i) *We have the following equalities:*

$$\mathcal{H}^t = \mathcal{H}^t_{\mathcal{S}_{cl}} = \mathcal{H}^t_{\mathcal{S}_{op}} = \mathcal{H}^t_{\mathcal{S}_{conv}} = \mathcal{H}^t_{\mathcal{S}_{\overline{conv}}}. \tag{7.17}$$

(ii) *In addition, we have the equalities*

$$S^t := \mathcal{H}^t_{\mathcal{S}^\#} = \mathcal{H}^t_{\mathcal{S}^\times}, \tag{7.18}$$

while the spherical and standard Hausdorff measures (or outer measures), S^t and $\mathcal{H}^t$, respectively, are compared via the following string of inequalities:

$$\mathcal{H}^t \leq S^t \leq \gamma_{m,t} \cdot \mathcal{H}^t, \tag{7.19}$$

with $\gamma_{m,t} \geq 1$ given by $\gamma_{m,t} = (2m/(m+1))^{t/2}$. In particular, $\gamma_{m,t} > 1$ for any integer $m \geq 2$, $\gamma_{1,t} = 1$ when $m = 1$—and hence, in the latter case, (7.19) *becomes an equality*: $S^t = \mathcal{H}^t$, if $m = 1$.

Example 7.16 (*Field of p-Adic Numbers*). As we have seen in part (b) of Remark 6.25, $X := \mathbb{Q}_p$ (the field of p-adic numbers, where p is an arbitrary prime) is not a reasonable metric space. However, in light of (6.40), we have the following inequalities, for every $\delta > 0$:

$$\mathcal{H}^t_{\delta,\mathcal{S}^\times} \leq \mathcal{H}^t_{\delta,\mathcal{S}^\#} =: S^t_\delta \quad \text{and} \quad S^t_\delta := \mathcal{H}^t_{\delta,\mathcal{S}^\#} \leq p^t \cdot \mathcal{H}^t_{\delta,\mathcal{S}^\times}. \tag{7.20}$$

Applying these inequalities to a fixed (but arbitrary) $E \subseteq X$, we obtain, after having passed to the limit as $\delta \to 0^+$,

$$\mathcal{H}^t_{\mathcal{S}^\times} \leq \mathcal{H}^t_{\mathcal{S}^\#} =: S^t \quad \text{and} \quad S^t := \mathcal{H}^t_{\mathcal{S}^\#} \leq p^t \cdot \mathcal{H}^t_{\mathcal{S}^\times}. \tag{7.21}$$

Since $p > 1$, this is in sharp contrast with the case of Euclidean space $X = \mathbb{R}^m$ discussed in part (ii) of Corollary 7.15 above; see (7.18) and (7.19).

The following result, Corollary 7.17, which is the exact counterpart for Hausdorff measures of Lemma 6.40 and Remark 6.42, provides worse estimates than the ones given in part (iv) of Theorem 7.14 in the special case of Euclidean space $\mathbb{R}^m$. However, it presents the advantage of having an elementary and completely self-contained proof, as well as of being valid for any metric space, rather than just for $\mathbb{R}^m$.

Let us also mention that, in practice, one typically only needs to know that S^t and $\mathcal{H}^t$ are comparable, without the specific knowledge of an optimal (or nearly optimal) estimate. In particular, it follows from Corollary 7.17 just below that for any $A \in \mathcal{B}(X)$, or, more generally, for any $A \subseteq X$, if $\mathcal{H}^t$ and S^t are interpreted as being outer measures rather than Borel measures,

$$\mathcal{H}^t(A) > 0 \Leftrightarrow S^t(A) > 0, \quad \mathcal{H}^t(A) < \infty \Leftrightarrow S^t(A) < \infty, \text{ and} \tag{7.22}$$

$$0 < \mathcal{H}^t(A) < \infty \Leftrightarrow 0 < S^t(A) < \infty. \tag{7.23}$$

Corollary 7.17 (Comparison Between Spherical and Standard Hausdorff Measures). *Let $t \geq 0$, with $t \in \mathbb{R}$, and let (X, d) be any metric space. Then, the spherical and standard Hausdorff measures $S^t := \mathcal{H}^t_{S^\#}$ and $\mathcal{H}^t$ can be compared via the following simple estimates: $\mathcal{H}^t \leq S^t \leq 2^t \cdot \mathcal{H}^t$.*

Proof. This follows at once from Lemma 6.40, itself a consequence of the Weak Form of Jung's Inequality (Lemma 6.39) for the δ-approximate spherical and Hausdorff measures, which, in light of Remark 6.42, is valid in any metric space. Indeed, it suffices to let $\delta \to 0^+$ in (6.81), in order to deduce that $\mathcal{H}^t \leq S^t \leq 2^t \cdot \mathcal{H}^t$, as desired. □

In Chapter 5, we have seen (without proof)[17] that $\mathcal{H}^1$, $\mathcal{H}^2$, and $\mathcal{H}^3$ are natural generalizations of length, area, and volume, in $\mathbb{R}$, $\mathbb{R}^2$, and $\mathbb{R}^3$, respectively. In Section 7.4, we will actually fully justify and extend this result to $\mathcal{H}^m$, the m-dimensional Hausdorff measure on $\mathbb{R}^m$, which will be shown to coincide (up to a multiplicative positive constant) with m-dimensional volume $\mathcal{L}^m$, the Lebesgue measure on $\mathbb{R}^m$; see Theorem 7.53 and Corollary 7.58, which are valid for any integer $m \geq 1$.

In the next exercise, the reader is asked, in particular, to show that $\mathcal{H}^0$, the 0-dimensional Hausdorff measure on a metric space X, coincides with the counting measure on X.

Exercise 7.18. For $t \in \mathbb{R}$, $t \geq 0$, let $\mathcal{H}^t$ be the t-dimensional Hausdorff measure (or outer measure) on a metric space (X, d).

(i) Show that $\mathcal{H}^0 = \rho$ is the counting measure on X. Namely, for all $E \subseteq X$,

$$\mathcal{H}^0(E) = \rho(E) := \begin{cases} \#E, & \text{if } E \text{ is finite,} \\ +\infty, & \text{if } E \text{ is infinite,} \end{cases} \tag{7.24}$$

where $\#E$ denotes the number of points in the finite set E (that is, the cardinality of E).

[17] For the case of the length of a suitable curve, however, see Proposition 5.61, which is proved in Section 5.5.

(ii) Show that if $t > 0$, then $\mathcal{H}^t(\{x\}) = 0$, for all $x \in X$; i.e., $\mathcal{H}^t$ is a *nonatomic* (or *diffuse*) measure.

(iii) Deduce from parts (i) and (ii) that for every fixed $t \geq 0$, any point-set or singleton $\{x\}$ in X (i.e., with $x \in X$) has t-dimensional Hausdorff dimension zero (i.e., $\dim_H(\{x\}) = 0$, for any $x \in X$) and hence that for every (at most) countable set $E \subseteq X$, we have $\dim_H(E) = 0$.[18]

[*Hint*: Use a covering argument and then the σ-stability of Hausdorff dimension in order to deduce that $\dim_H(E) = 0$.]

7.3. Key Geometric Properties of Hausdorff Measure

Let (X, d) be a metric space and let $t \in \mathbb{R}$ such that $t \geq 0$. In the sequel, we let $\mathcal{H}^t = \widetilde{\mu}$ be the Borel measure associated with the Hausdorff outer measure μ^* on X; that is, $\mathcal{H}^t = \widetilde{\mu}$ is the (t-dimensional) Hausdorff–Borel (or Hausdorff measure, in short), defined on the measurable space $(X, \mathcal{B}(X))$.

We leave it as an exercise for the interested reader to extend, whenever possible—and, if needed, suitably adjust—some of the results discussed below to the Hausdorff–Carathéodory outer measure μ^* (defined on all 2^X), as well as to the Hausdorff–Carathéodory measure, defined on the measurable space $(X, \mathcal{M} := \mathcal{M}_{\mu^*})$.

In the sequel, for notational simplicity, if X, Y are metric spaces, we shall denote by d the metric on X and on Y indifferently (instead of, say, by d_X and d_Y, respectively), and similarly for other notions depending on the metric, such as $\mathcal{H}^t$, the t-dimensional Hausdorff measure on X or on Y, and $\operatorname{diam}(A)$, the diameter of A, for any subset A of X or of Y (instead of $\mathcal{H}^t_X$ or $\mathcal{H}^t_Y$ and of $\operatorname{diam}_X(A)$ or of $\operatorname{diam}_Y(A)$, say). We adopt an analogous simplifying notation for $\dim_H(A)$, the Hausdorff dimension of A.

7.3.1. Translation and Isometry Invariance, Scaling, and Similarity Covariance.

We are now ready to state the first new basic properties of Hausdorff measure; they deal with the behavior of $\mathcal{H}^t$ under the action of Lipschitz and bi-Lipschitz maps.

Proposition 7.19 (Behavior of $\mathcal{H}^t$ Under the Action of Lipschitz Maps).[19] *Let X, Y be metric spaces and let $f : X \to Y$ be a Lipschitz map, with Lipschitz*

[18]The notion of Hausdorff dimension was introduced in Chapter 5 in the case of Euclidean space $\mathbb{R}^m$ and will be briefly revisited in Section 7.5 in the general case of metric spaces. In short, $\dim_H$ is defined exactly as in Section 5.2, where X was assumed to be equal to $\mathbb{R}^m$.

[19]In light of the proof of Proposition 7.19 and of the beginning of Remark 7.28 below, Proposition 7.19 along with each of its corollaries, Corollaries 7.20, 7.21, and 7.25, remains valid for any $A \subseteq X$, provided $\mathcal{H}^t$ is then interpreted as μ^*, the (t-dimensional) Hausdorff outer measure on X. The same exact comment applies, for example, to Proposition 7.31 and Corollary 7.33 below.

constant $\mathrm{Lip}(f) \le M$ *(where* $0 \le M < \infty$*).*[20] *Then,*[21]

$$\mathcal{H}^t(f(A)) \le M^t \cdot \mathcal{H}^t(A), \text{ for all } A \in \mathcal{B}(X). \tag{7.25}$$

Proof. The proof is divided into the following two steps:

Step 1: For any $\delta > 0$, we first obtain a suitable counterpart of (7.25) for $\mathcal{H}^t_\delta$, interpreted here as being the (t-dimensional) δ-approximate Hausdorff measure. Namely, we show that for any fixed $\delta > 0$ and for every $\varepsilon > 0$,

$$\mathcal{H}^t_{M\delta}(f(A)) \le M^t \cdot \mathcal{H}^t_\delta(A), \text{ for all } A \subseteq X. \tag{7.26}$$

Indeed, fix $\delta > 0$ and let $\varepsilon > 0$ be arbitrary. Then, by the definition of $\mathcal{H}^t_\delta(A)$ as an infimum (see (6.21)), there exists a countable δ-cover $\{A_j\}$ of A such that

$$\mathcal{H}^t_\delta(A) + \varepsilon \ge \varphi(\{A_j\}) := \sum_{j=1}^{\infty} (\mathrm{diam}(A_j))^t. \tag{7.27}$$

Clearly, $\{B_j := f(A_j)\}$ is a countable δM-cover of $B := f(A)$; see (7.28) just below, which itself follows from inequality (7.43) in Lemma 7.27. Also, since, as can be easily verified (see again Lemma 7.27 below),

$$\mathrm{diam}(B_j) \le M \cdot \mathrm{diam}(A_j) \ (\le M\delta), \text{ for all } j \ge 1, \tag{7.28}$$

it follows that

$$\varphi(\{B_j\}) \le M^t \cdot \varphi(\{A_j\}). \tag{7.29}$$

By definition of $\mathcal{H}^t_{M\delta}(B)$, we have that

$$\mathcal{H}^t_{M\delta}(B) \le \varphi(\{B_j\}). \tag{7.30}$$

Hence, by combining (7.27), (7.29), and (7.30), and upon multiplication of (7.27) by M^t, we deduce from (7.29) that

$$M^t \cdot (\mathcal{H}^t_\delta(A) + \varepsilon) \ge \mathcal{H}^t_{M\delta}(B) = \mathcal{H}^t_{M\delta}(f(A)). \tag{7.31}$$

Since $\varepsilon > 0$ was arbitrary, we can replace it by $M^{-t} \cdot \varepsilon$ (or simply let $\varepsilon \to 0^+$) in (7.31) in order to obtain (7.26).

Step 2: It now suffices to let $\delta \to 0^+$ in (7.26) in order to deduce that (7.25) holds, as desired. This concludes the proof of the proposition; see, however, Remark 7.28 below. □

---

[20] Recall from Definition 2.44 and formula (2.13) that the inequality $\mathrm{Lip}(f) \le M$ means that $d(f(\boldsymbol{x}), f(\boldsymbol{y})) \le M \cdot d(\boldsymbol{x}, \boldsymbol{y})$, for all $\boldsymbol{x}, \boldsymbol{y} \in X$.

[21] Strictly speaking, at least on the left-hand side of (7.25), we should interpret $\mathcal{H}^t$ as being Hausdorff outer measure. Indeed, $f(A)$ need not be Borel if A is Borel, even when f is a Lipschitz map and let alone a continuous map; see Remark 7.28 below. This comment does not apply to Corollaries 7.21 and 7.22, however, in light of the statements of those results.

In the last statement of each of the following corollaries, we use the notion of Hausdorff dimension, $\dim_H$, introduced in Euclidean spaces in Section 5.2 and to be briefly revisited in Section 7.5 below, in the case of arbitrary metric spaces.[22]

Corollary 7.20. *Assume that the hypotheses of Proposition* 7.19 *are satisfied. Then, for any* $A \in \mathcal{B}(X)$ *and* $t \geq 0$, *we have*

$$\mathcal{H}^t(f(A)) > 0 \Rightarrow \mathcal{H}^t(A) > 0, \tag{7.32}$$

$$\mathcal{H}^t(A) < \infty \Rightarrow \mathcal{H}^t(f(A)) < \infty, \text{ and also} \tag{7.33}$$

$$\mathcal{H}^t(A) = 0 \Rightarrow \mathcal{H}^t(f(A)) = 0. \tag{7.34}$$

In particular, we have the following inequality between the Hausdorff dimensions of $f(A)$ *and of* A:

$$\dim_H(f(A)) \leq \dim_H(A). \tag{7.35}$$

Proof. The first three statements (in (7.32), (7.33), and (7.34)) follow at once from (7.25) in Proposition 7.19. The fourth (and last) statement, inequality (7.35), follows from the definition of Hausdorff dimension, given in Section 5.2; see Lemma 5.12, formula (5.9), and Definition 5.13. Indeed, if $t > \dim_H(A)$, then $\mathcal{H}^t(A) = 0$. Hence, in light of (7.34), it follows that $\mathcal{H}^t(f(A)) = 0$, which in turn implies—again according to the aforementioned definition of the Hausdorff dimension—that $\dim_H(f(A)) \leq t$. Since $t > \dim_H(A)$ was arbitrary, we conclude that $\dim_H(f(A)) \leq \dim_H(A)$, as stated in (7.35). This concludes the proof of the corollary. □

Corollary 7.21 (Behavior of $\mathcal{H}^t$ Under Bi-Lipschitz Maps—Bi-Lipschitz Invariance of the Hausdorff Dimension $\dim_H$). *Assume that* $f : X \to Y$ *is a bi-Lipschitz map with* $\operatorname{Lip}(f) \leq M$ *and* $\operatorname{Lip}(f^{-1}) \leq m^{-1}$,[23] *for some finite and positive constants m and M. Then, for any* $A \in \mathcal{B}(X)$,

$$m^t \cdot \mathcal{H}^t(A) \leq \mathcal{H}^t(f(A)) \leq M^t \cdot \mathcal{H}^t(A), \text{ for all } t \geq 0. \tag{7.37}$$

Consequently, $\mathcal{H}^t(f(A))$ *and* $\mathcal{H}^t(A)$ *are simultaneously* > 0, *finite (i.e.,* $< \infty$), *in* $(0, +\infty)$, *or else equal to* 0.

It follows that the Hausdorff dimension is invariant under bi-Lipschitz maps:

$$\dim_H(f(A)) = \dim_H(A). \tag{7.38}$$

One thus says that the Hausdorff dimension is a bi-Lipschitz invariant.

[22] It suffices to know for now that the definition of $\dim_H$ is exactly the same in a general metric space as in Euclidean space $\mathbb{R}^m$, which is the setting of Section 5.2.

[23] More specifically, recall that this statement means that

$$m \cdot d(\boldsymbol{x}, \boldsymbol{y}) \leq d(f(\boldsymbol{x}), f(\boldsymbol{y})) \leq M \cdot d(\boldsymbol{x}, \boldsymbol{y}), \text{ for all } \boldsymbol{x}, \boldsymbol{y} \in X. \tag{7.36}$$

This clearly implies that f is injective and hence that it is a bijection from X onto $f(X)$, with inverse map denoted by f^{-1}.

Proof. This follows from Proposition 7.19 applied to $f : X \to Y$ (with $\operatorname{Lip}(f) \leq M$) and to $A \in \mathcal{B}(X)$, as well as to its inverse map $f^{-1} : f(X) \subseteq Y \to X$ (with $\operatorname{Lip}(f^{-1}) \leq m^{-1}$) and $B := f(A) \in \mathcal{B}(Y)$, where $A \in \mathcal{B}(X)$; see Remark 7.28 below. The first application of Proposition 7.19 yields the second inequality in (7.37), in light of (7.25), while the aforementioned second application yields the first inequality in (7.37), as follows:

$$\mathscr{H}^t(f^{-1}(B)) \leq \operatorname{Lip}(f^{-1}) \cdot \mathscr{H}^t(B) \leq m^{-t} \cdot \mathscr{H}^t(B);$$

i.e., since $f^{-1}(B) = A$ and $B = f(A)$, we have $m^t \cdot \mathscr{H}^t(A) \leq \mathscr{H}^t(f(A))$, as desired. This concludes the proof of the corollary. □

Corollary 7.22 (Independence of the Hausdorff Dimension on the Choice of an Equivalent Metric). *Let (X, d) be a metric space. Then, given any fixed $A \in \mathcal{B}(X)$, whether $\mathscr{H}^t(A) > 0$, $\mathscr{H}^t(A) < \infty$, $0 < \mathscr{H}^t(A) < \infty$, or $\mathscr{H}^t(A) = 0$, the value of $\mathscr{H}^t(A)$, the t-dimensional Hausdorff measure of A, is independent of the choice of an equivalent metric d' (to d) on X, in the sense of Definition* 6.6.

In particular, the value of the Hausdorff dimension $\dim_H(A)$ is independent of this choice; i.e., it depends only on the equivalence class of the metric d, in the sense of Definition 6.6.

Proof. This follows immediately from Corollary 7.21 because, as can be easily verified, the fact that a metric d' is equivalent to d means precisely that the identity map from (X, d') to (X, d) is bi-Lipschitz. □

The next corollary is more specific than Corollary 7.22 (and also implies it at once). Furthermore, it is essentially a consequence of Corollary 7.21 (and of (7.36) in footnote 23) about the action of bi-Lipschitz maps. However, we will instead provide a direct proof of this result.

Corollary 7.23 (Equivalence of Hausdorff Measures Under Equivalent Metrics). *Assume that two metrics d and d' on the same topological (metrizable) space X are equivalent; i.e., there exists $c \geq 1$ such that $c^{-1} \cdot d(\mathbf{x}, \mathbf{y}) \leq d'(\mathbf{x}, \mathbf{y}) \leq c \cdot d(\mathbf{x}, \mathbf{y})$, for all $\mathbf{x}, \mathbf{y} \in X$. Then, the corresponding Hausdorff measures $\mathscr{H}^t$ and $(\mathscr{H}^t)'$ are "equivalent", in the following sense: for every $t \geq 0$, we have $c^{-t} \cdot (\mathscr{H}^t)' \leq \mathscr{H}^t \leq c^t \cdot (\mathscr{H}^t)'$.*

In particular, the associated outer measures satisfy the exactly analogous inequalities and so do the associated Hausdorff–Carathéodory measures (with $\mathcal{M} = \mathcal{M}'$, the corresponding σ-algebras of measurable sets), as well as the associated Borel (or Hausdorff–Borel) measures.

Proof. If $\operatorname{diam}(A)$ and $\operatorname{diam}'(A)$ denote the diameter of $A \subseteq X$ with respect to the metrics d and d', respectively, then

$$c^{-1} \cdot \operatorname{diam}'(A) \leq \operatorname{diam}(A) \leq c \cdot \operatorname{diam}'(A). \tag{7.39}$$

Hence, we have that $\{A_j\}$ is a δ-cover of $E \subseteq X$ (relative to d') iff $\{A_j\}$ is a $c\delta$-cover of E (relative to d). One can then easily deduce from (7.39) that the corresponding δ-approximate Hausdorff outer measures ($\mathcal{H}^t_\delta$ and $(\mathcal{H}^t_\delta)'$, respectively) satisfy the following inequalities, for all $E \subseteq X$:

$$c^{-t} \cdot (\mathcal{H}^t_\delta)'(E) \le \mathcal{H}^t_{c\delta}(E) \le c^t \cdot (\mathcal{H}^t_\delta)'(E). \tag{7.40}$$

Upon letting $\delta \to 0^+$ in (7.40), we obtain—since the constant c is independent of $\delta > 0$ and since $t \ge 0$—that the associated Hausdorff outer measures satisfy the following inequalities:[24] $c^{-t} \cdot (\mathcal{H}^t)'(E) \le \mathcal{H}^t(E) \le c^t \cdot (\mathcal{H}^t)'(E)$, for all $E \subseteq X$. This proves the stated inequalities, as desired.

The remaining statements in the corollary are now easy to verify. We leave the corresponding verification as a simple exercise for the reader. □

Example 7.24. We can apply Corollaries 7.22 and 7.23 to any of the usual equivalent metrics on $\mathbb{R}^m$, namely, $d_{m,p}$ (with $1 \le p \le \infty$), given by $d_{m,p}(\mathbf{x}, \mathbf{y}) := \left(\sum_{j=1}^m |x_j - y_j|^p\right)^{1/p}$ (for $p < \infty$) and $d_{m,\infty}(\mathbf{x}, \mathbf{y}) := \max_{1\le j\le m} |x_j - y_j|$, respectively, where $\mathbf{x} = (x_1, \dots, x_m) \in \mathbb{R}^m$ and $\mathbf{y} = (y_1, \dots, y_m) \in \mathbb{R}^m$.[25] In particular, the Hausdorff dimension of a set $E \subseteq \mathbb{R}^m$ is independent of the choice of the metric $d_{m,p}$, with $1 \le p \le \infty$. Moreover, whether $\mathcal{H}^t(E)$ is > 0, $< \infty$, or in $(0, +\infty)$ is also independent of the choice of p. Clearly, the $p = 2$ case corresponds to the usual Euclidean space $\mathbb{R}^m$.

Corollary 7.25 (Invariance of $\mathcal{H}^t$ Under Isometries). *Let $J : X \to Y$ be an isometry from X to Y.*[26] *Then, for every $A \in \mathcal{B}(X)$, we have that*

$$\mathcal{H}^t(J(A)) = \mathcal{H}^t(A), \text{ for all } t \ge 0. \tag{7.41}$$

It follows that

$$\dim_H(J(A)) = \dim_H(A). \tag{7.42}$$

Proof. This follows at once from Corollary 7.21 with $f := J$, $f^{-1} = J^{-1}$, $\mathrm{Lip}(f) = \mathrm{Lip}(f^{-1}) = 1$, and $A \in \mathcal{B}(X)$. □

Exercise 7.26. Give a *direct* proof of Corollary 7.25, based on the original definition of $\mathcal{H}^t_\delta$ for each $\delta > 0$, as well as on that of $\mathcal{H}^t$.

[*Hint*: When J is an isometry, there is a very simple bijective relationship between the δ-covers $\{A_j\}$ of A and the δ-covers $\{B_j\}$ of $B := J(A)$, for any $A \subseteq X$ and, in particular, for any $A \in \mathcal{B}(X)$.]

[24] For notational simplicity, we omit here the stars normally used to denote outer measures.

[25] In essence, as is well known, the equivalence of any two of these metrics—and here, also, of the associated norms—follows from the compactness of the closed unit ball of $\mathbb{R}^m$.

[26] Recall that this means that $d(J(\boldsymbol{x}), J(\boldsymbol{y})) = d(\boldsymbol{x}, \boldsymbol{y})$, for all $\boldsymbol{x}, \boldsymbol{y} \in X$. The latter identity implies that J is injective and hence is a bijection from X onto $J(X) \subseteq Y$.

In the course of the proof of Proposition 7.19 (more specifically, in order to obtain (7.28)), we have used the following simple lemma.

Lemma 7.27. *Under the hypotheses of Proposition* 7.19, *we have that*

$$\operatorname{diam}(f(D)) \le M \cdot \operatorname{diam}(D), \quad \text{for all } D \subseteq X. \tag{7.43}$$

Proof. This follows at once from the definition of the diameter of a set. Indeed,

$$\begin{aligned}\operatorname{diam}(f(D)) &= \sup\{d(f(\boldsymbol{x}), f(\boldsymbol{y})) : \boldsymbol{x}, \boldsymbol{y} \in D\} \le \sup\{M \cdot d(\boldsymbol{x}, \boldsymbol{y}) : \boldsymbol{x}, \boldsymbol{y} \in D\}\\ &= M \cdot \sup\{d(\boldsymbol{x}, \boldsymbol{y}) : \boldsymbol{x}, \boldsymbol{y} \in D\} = M \cdot \operatorname{diam}(D),\end{aligned}$$

where in the first inequality, we have used the fact that $\operatorname{Lip}(f) \le M$; see footnote 20 of Proposition 7.19. □

The nonexpert reader may wish to omit the following technical remark and move on directly to Exercise 7.30 and Proposition 7.31.[27]

Remark 7.28 (*The Continuous Image of a Borel Set Need Not Be Borel*). Strictly speaking, in the proof of Proposition 7.19 provided above, we have shown both a stronger and a weaker result. Namely, we have proved that (with $\mu^* = \mathcal{H}^t$ denoting the t-dimensional Hausdorff outer measure) the following inequality holds: $\mu^*(f(A)) \le M^t \cdot \mu^*(A)$, for all $A \subseteq X$. In order to obtain the desired result (7.25) stated in Proposition 7.19 and since μ^* restricts to $\mathcal{H}^t$ on $\mathcal{B}(X)$ (or on $\mathcal{B}(Y)$), respectively, we need to show that under the hypotheses of the proposition, $f(A) \in \mathcal{B}(Y)$ if $A \in \mathcal{B}(X)$. But this is *not* true, in general, as we now explain.

As was stated in footnote 19, we should interpret $\mathcal{H}^t$ on the left-hand side of (7.25) as standing for the (t-dimensional) Hausdorff outer measure on Y. Indeed, in general, it is *not* true that if X and Y are metric spaces and $f : X \to Y$ is a continuous map, then $f(A) \in \mathcal{B}(Y)$ if $A \in \mathcal{B}(X)$.[28] In fact, it is not even true if X and Y are Euclidean spaces (possibly of different dimensions) and f is Lipschitz continuous. Actually, there are well-known counterexamples for which $X := \mathbb{R}^m, Y := \mathbb{R}$, and f is the projection of $\mathbb{R}^m$ onto $\mathbb{R}$, sending $\mathbf{x} = (x_1, \ldots, x_m) \in \mathbb{R}^m$ to the first coordinate (say), $x_1 =: f(\mathbf{x})$; here, the integer $m \ge 2$ is arbitrary.

It follows that, in general, in Proposition 7.19, we should assume that $A \in \mathcal{B}(X)$ is such that $f(A) \in \mathcal{B}(Y)$ in (7.25).

However, if X and Y are Euclidean spaces (for example)—or, more generally, if X and Y are Polish spaces (as defined in footnote 29)—and $f : X \to Y$ is

[27] Observe that we do not need to use the first part of Remark 7.28 (or Exercise 7.29) in order to prove either Corollary 7.21 or Corollary 7.25. Indeed, since for those corollaries the inverse map f^{-1} exists, is Lipschitz and hence continuous, as well as, a fortiori, Borel measurable, $f(A) = (f^{-1})^{-1}(A)$ belongs to $\mathcal{B}(Y)$ whenever A belongs to $\mathcal{B}(X)$.

[28] This incorrect belief was held by Lebesgue himself, as is well known among historians of mathematics—and, especially, of modern analysis. The error in Lebesgue's argument was found by Suslin.

continuous *and injective*, then $A \in \mathcal{B}(X)$ implies that $f(A) \in \mathcal{B}(Y)$:[29] see Exercise 7.29 just below. For this reason, in Proposition 7.19, $\mathcal{H}^t$ can be interpreted as being the (t-dimensional) Hausdorff–Borel measure on both sides of (7.25). (Indeed, a Lipschitz map is necessarily continuous and injective.) Moreover, because a bi-Lipschitz map from X to Y is a homeomorphism from X onto its image (and hence maps a Borel subset of X onto a Borel subset of Y), it follows from Corollaries 7.21–7.23 and 7.25 that the same statement as just above (and adapted in the obvious way) is true, but now for X and Y arbitrary metric spaces.

The following exercise supplements Remark 7.28 just above.

Exercise 7.29 (*A Continuous Injective Map Sends a Borel Set Onto a Borel Set*). Show that if $f : X \to Y$ is an injective continuous map, where X and Y are arbitrary Euclidean spaces, then, for any $A \in \mathcal{B}(X)$, we have that $f(A) \in \mathcal{B}(Y)$.

[*Hint*: Consider $\mathcal{T} \subseteq \mathcal{B}(X)$ given by $\mathcal{T} := \{A \in \mathcal{B}(X) : f(A) \in \mathcal{B}(Y)\}$. Then, show that $\mathcal{T}$ is a σ-algebra of X containing all open subsets of X.[30] (Note that, because f is injective, we have that $Y \setminus f(A) = f(X \setminus A)$, for any $A \subseteq X$ since, by definition, $\mathcal{B}(X)$ is generated by the open sets of X.) Deduce, in particular, that $\mathcal{T} = \mathcal{B}(X)$, as a σ-algebra.]

In the following easy exercise, the reader is asked, in essence, to extend Proposition 7.19 and Corollary 7.20 from Lipschitz maps to Hölder continuous maps.

Exercise 7.30. Extend Proposition 5.18 about the action of Hölder continuous maps to the case of general metric spaces. That is, show that if $f : (X_1, d_1) \to (X_2, d_2)$ is Hölder continuous of order α, for some $\alpha \in (0, 1]$,[31] then for any $A \subseteq X_1$, we have $\dim_H(f(A)) \leq (1/\alpha) \cdot \dim_H(A)$.

Furthermore, still for any $A \subseteq X_1$, estimate $\mathcal{H}^t(f(A))$ in terms of $\mathcal{H}^t(A)$, where $t \geq 0$ is arbitrary.

The following simple, but very important, result is central to the proof of the fact that when $X = \mathbb{R}^m$, the m-dimensional Euclidean space, $\mathcal{H}^m$ is proportional to $\mathcal{L}^m$ (Lebesgue measure on $\mathbb{R}^m$); see in Section 7.4 below the proof of Proposition 7.49, especially in the case when $n < 0$.

---

[29] More generally, such a result (due to Suslin and Lusin) is valid if X and Y are Polish spaces, i.e., are complete and separable metric spaces.

[30] In order to see why $\mathcal{T}$ contains every open set, one may use, for example, Lemma 7.35 in Subsection 7.4.1 below, according to which any open set of Euclidean space is a countable union of compact sets, along with the fact that the continuous image of a compact set is compact—and hence is also a Borel set.

[31] Recall from Chapter 3 that this statement means that there exists a constant $c > 0$ such that $d_2(f(\boldsymbol{x}), f(\boldsymbol{y})) \leq c \cdot (d_1(\boldsymbol{x}, \boldsymbol{y}))^\alpha$, for all $\boldsymbol{x}, \boldsymbol{y} \in X_1$.

Proposition 7.31 (Key Geometric Properties of Hausdorff Measure). *Let $X := \mathbb{R}^m$ be the m-dimensional Euclidean space. Then, $\mathcal{H}^t$ satisfies the following key properties:*[32]

(i) (Translation Invariance). *For every $A \in \mathcal{B}(X)$ and $\boldsymbol{x} \in \mathbb{R}^m$,*

$$\mathcal{H}^t(A + \boldsymbol{x}) = \mathcal{H}^t(A), \text{for all } t \geq 0. \tag{7.44}$$

Hence,

$$\dim_H(A + \boldsymbol{x}) = \dim_H(A). \tag{7.45}$$

(ii) (Scale Covariance or "Scaling Property"). *For every $A \in \mathcal{B}(X)$ and $\lambda > 0$,*

$$\mathcal{H}^t(\lambda A) = \lambda^t \cdot \mathcal{H}^t(A), \text{ for all } t \geq 0. \tag{7.46}$$

Hence,

$$\dim_H(\lambda A) = \dim_H(A). \tag{7.47}$$

Proof. The identities (7.44) and (7.46) follow at once from the respective identities obtained in part (i) and part (ii) of Proposition 6.43 about the δ-approximate Hausdorff measure $\mathcal{H}^t_\delta$. Accordingly, for every $\boldsymbol{x} \in \mathbb{R}^m$ and $\lambda > 0$, we have that $\mathcal{H}^t_\delta(A+\boldsymbol{x}) = \mathcal{H}^t_\delta(A)$ and $\mathcal{H}^t_\delta(\lambda A) = \lambda^t \cdot \mathcal{H}^t_\delta(A)$. It then suffices to let $\delta \to 0^+$ in each part of the latter identity to deduce the corresponding identity for $\mathcal{H}^t$, namely, (7.44) and (7.46), respectively. Finally, (7.45) and (7.47) are easy consequences of the latter identities. □

Remark 7.32. Since a translation of $\mathbb{R}^m$ is clearly a bijective isometry (from $\mathbb{R}^m$ to itself), the translation invariance of $\mathcal{H}^t$ (i.e., part (i) of Proposition 7.31) also follows from Corollary 7.25.

The following result is very important in fractal geometry. In fact, in many informal introductions to fractal geometry, it is often used in order to "calculate" heuristically the similarity dimension of a self-similar set in $\mathbb{R}^m$.

Corollary 7.33 (Behavior Under the Action of Similarities: Similarity Covariance). *If S is a similarity transformation of the Euclidean space $X := \mathbb{R}^m$, with scaling ratio $r > 0$, then for every $A \in \mathcal{B}(\mathbb{R}^m)$, we have that*

$$\mathcal{H}^t(S(A)) = r^t \cdot \mathcal{H}^t(A), \text{ for all } t \geq 0. \tag{7.48}$$

Thus,

$$\dim_H(S(A)) = \dim_H(A). \tag{7.49}$$

[32] Here, $A+\boldsymbol{x}$, the translate of A by $\boldsymbol{x} \in \mathbb{R}^m$ (resp., λA, the scaled version of A by the scaling factor $\lambda > 0$), is defined in part (i) (resp., in part (ii)) of Proposition 6.43.

Proof. Since a similarity map of scaling ratio r is the composition of a translation, rotation (which are thus bijective isometries of $\mathbb{R}^m$), and a scaling transformation of scaling ratio r, as well as (possibly) a reflection, the result follows directly from Corollary 7.25, combined with Proposition 7.31. □

Example 7.34 (*Computation of the Similarity and Hausdorff Dimensions of a Self-Similar Set*). One of the main goals of this example is to illustrate the use of the above geometric properties of Hausdorff measure by discussing and justifying the semi-heuristic argument typically provided in order to compute the similarity dimension D_S, along with the Hausdorff dimension $\dim_H(K)$ of a self-similar set K.[33]

Let $N \geq 2$ be any integer, and let $\mathcal{S} := \{S_j\}_{j=1}^N$ be a finite set of N contractive similitudes (or a self-similar system) of $\mathbb{R}^m$ with each S_j having similarity ratio r_j $(0 < r_j < 1)$, for $j = 1, \ldots, N$; thus, $\mathcal{S}$ is a self-similar system, in the sense of Definition 4.26. Let K be the associated self-similar set; i.e., K is the attractor of $\mathcal{S}$, according to the terminology used in much of the literature on this subject. Hence, K is the unique nonempty compact subset of $\mathbb{R}^m$ such that

$$K = \bigcup_{j=1}^{N} S_j(K); \tag{7.50}$$

see Sections 4.2 and 4.3. Then, hoping for the best—that is, assuming that the images of K under $\mathcal{S}$, the subsets $S_j(K)$ of $\mathbb{R}^m$, for each $j = 1, \ldots, N$, are sufficiently separated so that the finite union appearing on the right-hand side of (7.50) is a "disjoint union", *up to a set of $\mathcal{H}^t$-measure zero*[34]—we can apply $\mathcal{H}^t$ to both sides of the fixed point equation (7.50) and obtain successively[35]

$$\begin{aligned} \mathcal{H}^t(K) = \mathcal{H}^t\left(\bigcup_{j=1}^{N} S_j(K)\right) &= \sum_{j=1}^{N} \mathcal{H}^t(S_j(K)) \\ = \sum_{j=1}^{N} \left(r_j^t \cdot \mathcal{H}^t(K)\right) &= \left(\sum_{j=1}^{N} r_j^t\right) \cdot \mathcal{H}^t(K). \end{aligned} \tag{7.51}$$

Here, in the second equality of (7.51), we have used the hoped-for *"separation condition", according to which, in particular, the images $S_j(K)$ are disjoint, up*

[33]For convenience, we use here somewhat different notation (and, occasionally, terminology) than in the relevant parts of this book; compare, e.g., Section 2.3, Chapters 4, 5, and 8.

[34]This is one of the reasons for introducing some sort of "separation condition" on $\mathcal{S}$. One other technical, but crucial, reason is to guarantee that $0 < \mathcal{H}^t(K) < \infty$, as is also required in order to justify the present informal computation of the similarity dimension D_S.

[35]A sufficient condition guaranteeing such a conclusion is that the *Open Set Condition* (OSC) holds; namely, there exists a nonempty bounded open set V of $\mathbb{R}^m$ such that $\bigcup_{j=1}^N S_j(V) \subseteq V$ and the sets $S_j(V)$ are pairwise disjoint, for $j = 1, \ldots, N$; see, e.g., [**172**, Section 9.2], along with Chapters 4 and 5 above (Definition 4.65 and Theorem 5.40).

to a set of Hausdorff measure zero; namely, $\mathcal{H}^t(S_j(K) \cap S_k(K)) = 0$, for $j, k = 1, \ldots, N$, with $j \neq k$. Furthermore, in the third equality of (7.51), we have used Corollary 7.33 applied to each similarity transformation S_j with similarity ratio r_j, for $j = 1, \ldots, N$, which yields (since K and $S_j(K)$, being compact and, hence, closed, certainly belong to $\mathcal{B}(\mathbb{R}^m)$)

$$\mathcal{H}^t(S_j(K)) = r_j^t \cdot \mathcal{H}^t(K), \text{ for } j = 1, \ldots, N. \tag{7.52}$$

Next, in light of (7.51), we have that $\mathcal{H}^t(K) \cdot \left(1 - \sum_{j=1}^N r_j^t\right) = 0$, which—*provided that the "separation condition" on $\mathcal{S}$ is sufficient in order to guarantee that* $0 < \mathcal{H}^t(K) < \infty$—implies that $t \in \mathbb{R}$, $t \geq 0$, must satisfy the following *Moran equation*:

$$\sum_{j=1}^N r_j^t = 1. \tag{7.53}$$

Since the scaling ratios r_j satisfy $0 < r_j < 1$, for $j = 1, \ldots, N$ (that is, since the similitudes S_j are contraction maps, for $j = 1, \ldots, N$), it is easy to check (as is done in Chapter 5; see Proposition 5.30) that the Moran equation (7.53) has a unique *real* solution t[36] and that t necessarily satisfies the inequalities $0 < t \leq m$. Moreover, by definition, this unique number t is equal to D_S, the *similarity dimension* of the self-similar set K. In addition, since $0 < \mathcal{H}^t(K) < \infty$, t must also coincide with $\dim_H(K)$, the Hausdorff dimension of K. Thus, we conclude that

$$\dim_H(K) = D_S = t, \tag{7.54}$$

the unique (real) solution of the Moran equation (7.53).

Note that in the important special case (often encountered in the applications; see, e.g., [**419**]) when all the similitudes have the same scaling ratio r (i.e., $r_j = r$, for all $j = 1, \ldots, N$), with necessarily $0 < r < 1$, then (7.54) takes the following well-known and more explicit form:

$$\dim_H(K) = D_S = t = \frac{\log N}{\log(1/r)}; \tag{7.55}$$

i.e., the unique (real) solution $t = D_S$ of the Moran equation (7.53) coincides with the Hausdorff dimension of K, $\dim_H(K)$, and is given by

$$t = \frac{\log N}{\log(1/r)} = \log_{1/r} N. \tag{7.56}$$

We should point out that—according to Moran's Theorem (Theorem 5.40) and if the self-similar system $\mathcal{S}$ satisfies the Open Set Condition (OSC)—the above

[36] When we study the *complex* solutions of (7.53), the resulting equation has infinitely many solutions, with a beautiful quasiperiodic structure (in the nonlattice case). We refer the reader to [**371**], [**375**, Chapters 2 and 3], along with the sequel to this book, [**119**], as well as [**376**], for a lot more information about the so-called *complex dimensions* of self-similar fractal strings and the complex solutions of the Moran equation.

conclusions (see (7.53)–(7.56)) are fully justified[37] and D_S also coincides with $\dim_M(K)$, the Minkowski (or box) dimension of K, discussed in Chapter 8 below and which will be key to the sequel of this book, [**119**], as well as to large parts of Chapter 9.

This concludes our semi-heuristic discussion of the computation of the similarity dimension and of the Hausdorff dimension of a (suitable) self-similar set, illustrating the key scaling property along with the isometry (and hence also translation) invariance of the Hausdorff measure, as used in an essential manner in Corollary 7.33 (and relying on Proposition 7.19 and its corollaries, as well as on Proposition 7.31).

A lot more information can be found in the mathematical literature (see, e.g., [**172, 175, 193, 356, 375, 430**], along with the many related references therein) and in appropriate places in this book (see, e.g., Chapters 4, 5, and 8) regarding, in particular, the type of separation condition (on $\mathcal{S}$),[38] enabling us to rigorously justify this semi-heuristic discussion. In the present book, as was mentioned above, the precise justification and hypothesis (the "Open Set Condition") of Example 7.34 are provided in Theorem 5.40 (Moran's Theorem) and its proof; see Section 5.4.

7.4. Lebesgue Measure vs. m-Dimensional Hausdorff Measure on $\mathbb{R}^m$

Our main goal in this section is to establish the equality between $\mathcal{L}^m$ (the m-dimensional Lebesgue measure on $\mathbb{R}^m$) and $\mathcal{H}^m_{nor}$, a suitably normalized version of the m-dimensional Hausdorff measure on $\mathbb{R}^m$. See, especially, Theorem 7.53 and Proposition 7.49 along with Theorem 7.50.

We will provide, in particular, a seemingly novel proof (due to the first author) of the corresponding well-known main result. It is based, in particular, on transfinite induction and the Borel hierarchy enabling one to exhaust the Borel σ-algebra of $\mathbb{R}^m$, as well as on the use of tessellations of Euclidean space $\mathbb{R}^m$ by dyadic cubes.

7.4.1. Tessellations by Dyadic Cubes and Their Applications. The following technical lemma (Lemma 7.35) is very useful in many areas of mathematics, including harmonic analysis, spectral theory, geometric measure theory, and fractal geometry. It will have as a consequence, in particular, a key aspect of the proof of the (Borel) regularity of the Hausdorff measure $\mathcal{H}^m$ and the proof of the fact that $0 < \mathcal{H}^m(Q_0) < \infty$ and $\mathcal{H}^m(\partial Q_0) = 0$, where Q_0 is

[37] The corresponding proof, however, is significantly more involved than the above semi-heuristic computation; compare with the proof of Theorem 5.40.

[38] The most prominent of these separation conditions—which, in the context of self-similar systems, is equivalent to the "Open Set Condition"—is the "Strong Open Set Condition" discussed, in particular, in Chapter 4; see Definition 4.65 and the comment following it.

the unit cube of $\mathbb{R}^m$, and similarly for any dyadic cube Q in $\mathbb{R}^m$; see Theorem 7.41. It will follow that $\mathscr{H}^m$ is a nonzero Radon measure on $\mathbb{R}^m$ (i.e., is finite on every compact subset of $\mathbb{R}^m$) and that every open set of $\mathbb{R}^m$ is σ-compact; see Corollary 7.39. Eventually, this will provide us with a rather direct, but perhaps not completely elementary, proof of the fact that for every integer $m \geq 1$, $\mathscr{H}^m$ (the m-dimensional Hausdorff measure on $\mathbb{R}^m$)—and, more generally, any (nondegenerate) translation-invariant measure on $\mathbb{R}^m$—is equal to $\mathcal{L}^m$ (the m-dimensional Lebesgue measure on $\mathbb{R}^m$), up to a positive multiplicative constant.

Before stating the lemma, we need to introduce some convenient notation concerning the tessellations of the Euclidean space $\mathbb{R}^m$ by dyadic cubes. Let

$$\mathcal{K} = \bigcup_{n\in\mathbb{Z}} \mathcal{K}_n \tag{7.57}$$

be the countable collection of dyadic cubes of $\mathbb{R}^m$ tessellating $\mathbb{R}^m$ and defined as follows (see Figure 7.1):

For each $n \in \mathbb{Z}$, $\mathcal{K}_n$ is the countable collection of (closed and hence also compact) dyadic cubes (of $\mathbb{R}^m$) of level n, that is, of side lengths 2^{-n} and with vertices all belonging to the lattice (of mesh 2^{-n} and of rank m) $\Lambda_n := 2^{-n} \cdot \mathbb{Z}^m$. Accordingly, a cube Q belongs to $\mathcal{K}_n$ iff

$$Q = \prod_{j=1}^{m} [a_j, b_j], \tag{7.58}$$

with

$$a_j := \frac{k_j}{2^n} \quad \text{and} \quad b_j := \frac{k_j+1}{2^n}, \text{ for some } k_j \in \mathbb{Z}, \tag{7.59}$$

so that $b_j - a_j = 2^{-n} \in 2^{-n} \cdot \mathbb{Z}$, for all $j = 1, \dots, m$, as required.

By construction, all of the cubes in $\mathcal{K}_n$, with $n \in \mathbb{Z}$ arbitrary, have disjoint interiors and hence can only meet along their common boundaries. Also, note that by bisecting the sides (and still for $n \in \mathbb{Z}$ arbitrary), each cube in the $(n-1)$-th generation $\mathcal{K}_{n-1}$ can be subdivided into 2^m cubes of the n-th generation $\mathcal{K}_n$. Moreover, these 2^m cubes of $\mathcal{K}_n$ can only meet along their boundaries, more specifically, along a common face; see Figures 7.1 and 7.2.

Lemma 7.35 (Tessellation of Open Sets by Dyadic Cubes). *Every (nonempty) open set U in the m-dimensional Euclidean space $\mathbb{R}^m$ is a countable union of dyadic compact cubes $\{Q_\alpha\}_{\alpha\in R(U)}$ having pairwise disjoint interiors:*[39]

$$U = \bigcup_{\alpha\in R(U)} Q_\alpha, \tag{7.60}$$

[39] Such cubes are said to be *nonoverlapping*; i.e., they can only intersect along their common boundaries (or faces, here).

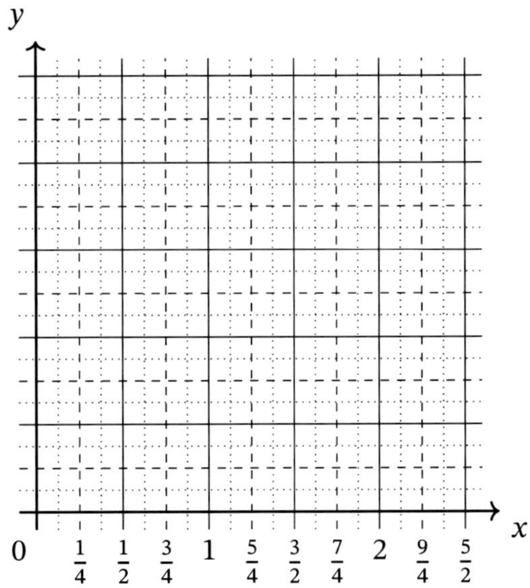

Figure 7.1. Tessellation of the Euclidean plane $\mathbb{R}^2$ by dyadic cubes.

where $R(U) \subseteq \mathbb{Z}^m$ is a countable set and $\mathring{Q}_\alpha \cap \mathring{Q}_\beta = \emptyset$ for $\alpha \neq \beta$, with $\alpha, \beta \in R(U)$. Here, for example, $\mathring{Q}_\alpha$ denotes the interior of the closed (and hence also compact) dyadic cube Q_α.[40]

Proof. Let us use the notation concerning the standard tessellation of $\mathbb{R}^m$ by dyadic cubes introduced in the text surrounding (7.57)–(7.59) above.

For each integer $n \in \mathbb{Z}$, let[41]

$$T(n, U) := \{Q \in \mathcal{K}_n : Q \subseteq U\}, \tag{7.61}$$

and then,

$$R(n, U) := T(n, U) \setminus T(n-1, U) = \{Q \in \mathcal{K}_n \setminus \mathcal{K}_{n-1} : Q \subseteq U\}. \tag{7.62}$$

It is not hard to show (see Exercise 7.36 and the key observation below) that

$$U = \bigcup_{Q \in R(U)} Q, \tag{7.63}$$

where

$$R(U) := \bigcup_{n=-\infty}^{\infty} R(n, U), \tag{7.64}$$

[40]For notational convenience, in the statement of Lemma 7.35 (see, especially, identity (7.60)) and its proof, we identify the countable set $\{Q : Q \in R(U)\}$ with the countable family of dyadic cubes $\{Q_\alpha\}_{\alpha \in R(U)}$.

[41]Note that $T(n-1, U) \subseteq T(n, U)$, for each $n \in \mathbb{Z}$.

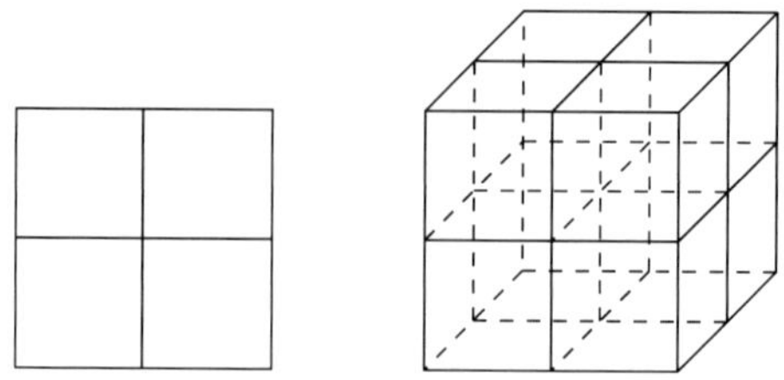

(a) (b)

Figure 7.2. The dyadic cubes in $\mathcal{K}_n$ are obtained from the dyadic cubes in $\mathcal{K}_{n-1}$ by bisecting the sides of the cubes in $\mathcal{K}_{n-1}$. Accordingly, via this process, a cube in $\mathcal{K}_{n-1}$ yields 2^m cubes in $\mathcal{K}_n$. Note that the number 2^m is independent of $n \in \mathbb{Z}$ and depends only on m, the dimension of the underlying Euclidean space $\mathbb{R}^m$. Part (a) (resp., part (b)) of the figure depicts the $m = 2$ (resp., $m = 3$) case; see Exercise 7.36.

with each $R(n, U)$ defined as in (7.62).[42] Clearly, by construction, $R(U) \subseteq \mathcal{K} = \bigcup_{n\in\mathbb{Z}} \mathcal{K}_n$—and hence, $R(U)$ is countable (as a countable union of countable sets), since for each $n \in \mathbb{Z}$, $\mathcal{K}_n$ is countable. This yields (7.60), along with the other claims in the lemma, as desired.

Note that the key reason for why (7.63) holds is given by the following simple observation, with U^c denoting the complement of U in $\mathbb{R}^m$:

$$x \in U \Leftrightarrow \operatorname{dist}(x, U^c) > 0. \tag{7.65}$$

Indeed, U^c is closed (since U is open) and hence, $x \in U^c \Leftrightarrow \operatorname{dist}(x, U^c) = 0$, according to Exercise 6.8. Now, in order to show that any given $x \in U$ belongs to some $Q \in R(U)$ or, equivalently (in light of (7.64)), to $R(n, U)$, for some $n \in \mathbb{Z}$, it suffices to choose n large enough so that

$$\sqrt{m} \cdot 2^{-n} < \operatorname{dist}(x, U^c), \tag{7.66}$$

the distance from x to U^c. Note that since $x \in U$, it follows from (7.65) that $\operatorname{dist}(x, U^c) > 0$.

This completes the proof of the lemma. □

Exercise 7.36. Rephrase in your own words both the proof of Lemma 7.35 and the construction of the standard dyadic tessellations of $\mathbb{R}^m$ outlined just before the statement of Lemma 7.35. In particular, convince yourself, possibly by the help of a picture, that (7.63) holds, because $x \in U$ belongs to some cube $Q \in R(n, U)$, provided inequality (7.66) holds.

[*Hint*: For the latter assertion, it is useful to observe that if $Q \in \mathcal{K}_n$, then every point q of Q is within a distance from $x \in Q$ at most equal to $\sqrt{m} \cdot 2^{-n}$; indeed, $\operatorname{diam}(Q) = \sqrt{m} \cdot 2^{-n}$.]

[42]Obviously, $U \supseteq \bigcup_{Q\in R(U)} Q$, in light of (7.61)–(7.64).

Remark 7.37 (*Whitney Cubes*). In the literature, the cubes Q in the countable set $R(U)$ given by (7.64) (and (7.62)) are often called *Whitney cubes* — or, more specifically, *the Whitney cubes associated with the open set* U (or, rather, *the Whitney cubes associated with the closed set* U^c, the complement of U in $\mathbb{R}^m$); see, e.g., [**564**], [**429**], or [**318**].

Exercise 7.38. Show that, in the terminology of Remark 7.37 just above, each of the Whitney cubes associated with the open set U has side lengths (or, equivalently, diameter)[43] comparable to the distance of its center (or, more generally, of any of its points) to the boundary ∂U of U. Provide, in particular, a quantitative version of this statement.

Since each dyadic cube is compact—because it is both a closed and bounded subset of $\mathbb{R}^m$—the following result is an immediate consequence of Lemma 7.35.

Corollary 7.39 (σ-Compactness of $\mathbb{R}^m$). *Every nonempty open set U in $\mathbb{R}^m$ is σ-compact; i.e., here, U is a countable union of compact sets in $\mathbb{R}^m$.*

Since the empty set, $\emptyset$, is compact, we may restrict our attention to countable (rather than to at most countable) unions here. Note that a finite union of compact sets is compact—and hence is closed and bounded. However, in the present case of $\mathbb{R}^m$—and except for $\mathbb{R}^m$ itself (which is unbounded)—no nonempty subset of $\mathbb{R}^m$ can be both closed and open (i.e., clopen). Hence, in Corollary 7.39, as well as in Lemma 7.35, the stated union is necessarily countably infinite.

Naturally, in the special case when $U := \mathbb{R}^m$ itself, a direct proof of Corollary 7.39 can be given by simply observing that $\mathbb{R}^m = \bigcup_{n=0}^{\infty}[-n, n]^m$, for example.

The next result could also be considered as a corollary of Lemma 7.35 (and its proof), but its own proof has enough subtlety (especially in the case of Hausdorff measure $\mathscr{H}^m$) that it may deserve to be viewed as an independent proposition.

Proposition 7.40 (Disjoint Union, Up to an $\mathscr{H}^m$-Null Set). *With the notation of Lemma* 7.35, *the following properties hold*:

The union in (7.60) *of Lemma* 7.35 *is disjoint, up to a set of Hausdorff (resp.,* Lebesgue) *measure zero, i.e., up to a set of $\mathscr{H}^m$-measure (resp., $\mathcal{L}^m$-measure) zero, also called an $\mathscr{H}^m$-null (resp., $\mathcal{L}^m$-null) set.*[44] *Thus,*

$$\mathscr{H}^m(U) = \sum_{\alpha \in R(U)} \mathscr{H}^m(Q_\alpha) \quad \left(\text{resp., } \mathcal{L}^m(U) = \sum_{\alpha \in R(U)} \mathcal{L}^m(Q_\alpha)\right). \tag{7.67}$$

[43]Note that if a cube Q of $\mathbb{R}^m$ has side lengths 2^{-n} (for example, if $Q \in \mathcal{K}_n$), for some $n \in \mathbb{Z}$, then its diameter is given by $\operatorname{diam}(Q) = \sqrt{m} \cdot 2^{-n}$.

[44]Equivalently, $U = \bigcup_{\alpha \in R(U)} \mathring{Q}_\alpha$, up to an $\mathcal{H}^m$-null (resp., an $\mathcal{L}^m$-null) set. Note that the latter (countable) union of *open* cubes is disjoint, by construction.

Proof. The proof is divided into the following three steps:

Step 1. *For any m-dimensional cube Q in $\mathbb{R}^m$, we have that $\mathscr{H}^m(\partial Q) = 0$; i.e., its boundary ∂Q is an $\mathscr{H}^m$-null set (or is a set of $\mathscr{H}^m$-measure zero).*

Indeed, this key fact is established in Theorem 7.41 (part (iii) and case (b)) below.[45] The exact analog (and much simpler result) for Lebesgue measure $\mathcal{L}^m$ is also proved in that same theorem; see the end of cases (a) and (b) in ibid; see also Remark 7.42 for an alternative proof in the case of $\mathcal{L}^m$. This concludes the proof of Step 1.

Step 2. The first identity of (7.67) holds.

The proof of this step is straightforward. Indeed, we may clearly assume U to be nonempty. Then, since, in light of identity (7.60) in Lemma 7.35, the open set U of $\mathbb{R}^m$ is a *countable* union of cubes with pairwise disjoint interiors, it follows from Step 1 that the union is disjoint, up to a set A of $\mathscr{H}^m$-measure zero. In order to show that, we can choose (still with the notation of Lemma 7.35) the Borel set $A := \bigcup_{\alpha\in R(U)} \partial Q_\alpha$, which is a countable union of $\mathscr{H}^m$-null sets and hence an $\mathscr{H}^m$-null set itself, as we now explain: $0 \leq \mathscr{H}^m(A) \leq \sum_{\alpha\in R(U)} \mathscr{H}^m(\partial Q_\alpha) = 0$, by the σ-subadditivity of $\mathscr{H}^m$; thus, $\mathscr{H}^m(A) = 0$, as desired. This concludes the proof of Step 2, as well as of the case of the m-dimensional Hausdorff measure $\mathscr{H}^m$ in Proposition 7.40.

Step 3. The second identity of (7.67) holds.

Indeed, the proof of Step 3 mimics that of Step 2, except for $\mathscr{H}^m$ replaced with $\mathcal{L}^m$. We have already recalled in Step 1 that $\mathcal{L}^m(\partial Q) = 0$, for every dyadic cube Q. For a direct proof in the case of $\mathcal{L}^m$, see part (iii), case (b), of Theorem 7.41 or else see Remark 7.42. This is the key ingredient needed in order to mimic the proof of Step 2, where $\mathcal{L}^m$ is now substituted for $\mathscr{H}^m$. In fact, as we have noted earlier (in footnote 45), these arguments are valid more generally for any nondegenerate translation-invariant Borel measure on $\mathbb{R}^m$; see Theorem 7.50 or, simply, Remark 7.48.

This completes the proof of Step 3—and hence also of the entire proposition. □

In the course of the proof of Proposition 7.40 (Step 1), we have used the following key (and deceptively obvious) result, which will be used repeatedly, in various forms as well as explicitly or implicitly, throughout the remainder of this chapter.

[45] For an alternative proof, valid for any (nondegenerate) translation-invariant Borel measure on $\mathbb{R}^m$, see Theorem 7.50 or just Remark 7.48.

Theorem 7.41 (m-Dimensional Hausdorff Measure of the Unit Cube Q_0 of $\mathbb{R}^m$ and of Its Boundary ∂Q_0). *Let $Q_0 = [0,1]^m$ be the (closed) m-dimensional unit cube of the Euclidean space $\mathbb{R}^m$ and let ∂Q_0 denote its boundary. Then:*

(a) *The following inequalities, and equalities, hold:*[46]

(i) $$\mathcal{H}^m(Q_0) < \infty,$$

(ii) $$\mathcal{H}^m(Q_0) > 0, \text{ as well as}$$

(iii) $$\mathcal{H}^m(\partial Q_0) = 0, \text{ and hence also}$$

(iv) $$0 < \mathcal{H}^m(\mathring{Q}_0) = \mathcal{H}^m(Q_0) < \infty,$$

where ∂Q_0 and $\mathring{Q}_0$ denote the boundary and the interior of Q_0, respectively.

Furthermore, the same statements as in (i) *and* (iii)–(iv) *are true if $\mathcal{H}^m$ is replaced by the m-dimensional Lebesgue measure on $\mathbb{R}^m$, $\mathcal{L}^m = \mathrm{vol}^m$. (Clearly, the analog of* (ii) *also holds since $\mathcal{L}^m(Q_0) = 1$.)*

(b) (Discrete Scale Covariance). *Moreover, for any dyadic cube Q of side lengths 2^{-n}, with $n \in \mathbb{Z}$ arbitrary, the exact counterpart of* (i)–(iv) *holds for Q, instead of just for Q_0. In fact, we have the following "discrete scale-covariance" relation:*

(7.68) $$\mathcal{H}^m(Q) = 2^{-nm} \cdot \mathcal{H}^m(Q_0)$$

and similarly

(7.69) $$\mathcal{L}^m(Q) = 2^{-nm} \cdot \mathcal{L}^m(Q_0) = 2^{-nm}.$$

Proof. In essence, this follows from the self-similarity of Q_0 (or, more generally, of any dyadic cube Q),[47] along with the translation invariance of $\mathcal{H}^m$ and $\mathcal{L}^m$ (see Proposition 7.31 and part (i) of Proposition 3.70, respectively) which, by induction on $|n|$, with $n \in \mathbb{Z}$ (or on $n \in \mathbb{N}_0$), implies the discrete scale covariance of $\mathcal{H}^m$ and $\mathcal{L}^m$, as expressed in (7.68) and (7.69), respectively; see the proof of Proposition 7.49 below, especially for the case when $n < 0$.

We will focus on the estimates concerning $\mathcal{H}^m$ which are, by far, the most interesting and delicate ones. Indeed, as was already alluded to earlier, either for Q_0 (part (a)) or for any dyadic cube Q (part (b)), part (ii) (resp., part (iii)) is valid for any translation-invariant Borel measure ν on $\mathbb{R}^m$ (resp., such that $\nu(Q_0) < \infty$),[48] including $\nu := \mathcal{L}^m$, while part (i) is trivial for $\mathcal{L}^m$ in the case of the unit cube Q_0 (since $\mathcal{L}^m(Q_0) = 1$) and follows (still for $\mathcal{L}^m$ or, more

[46] We state the properties (i) and (ii) separately because they have rather different proofs. Furthermore, as we will see, (iii) (resp., (i)) extends to any (resp., nontrivial) translation-invariant measure on $\mathbb{R}^m$, while (ii) does not, at least not in an obvious way.

[47] Indeed, observe that $Q_0 = \bigcup_{j=1}^{2^m} S_j(Q_0)$, where $\{S_j\}_{j=1}^{2^m}$ is a suitable family of 2^m similarity transformations of $\mathbb{R}^m$ with the same scaling ratio 1/2; see Example 4.96, in Section 4.10.

[48] We shall even see that if a translation-invariant Borel measure on $\mathbb{R}^m$ is such that $\nu(Q_0) = +\infty$, then $\nu(A) = +\infty$, for any Borel subset A of $\mathbb{R}^m$.

generally, for any translation-invariant Borel measure ν such that $\nu(Q_0) < \infty$) by translation invariance—and hence also by discrete scale covariance with scaling ratio 1/2, in light of part (iii) for Q_0.

We next proceed to establish part (a), and then part (b), for $\mathscr{H}^m$.

Part (a): We prove successively inequalities (i) and (ii), as well as the equality (iii), from which (iv) will easily follow.

(i) *We claim that* $\mathscr{H}^m(Q_0) < \infty$.

Indeed, let us repeatedly bisect the sides of the unit cube Q_0, n times, with $n \in \mathbb{N}$ arbitrary. We then obtain $2^{nm} = (2^m)^n = (2^n)^m$ dyadic cubes $Q_{j,n}$ (for $j = 1, \ldots, 2^{nm}$), of side lengths 2^{-n}. By construction, the finite collection $\{Q_{j,n}\}_{j=1}^{2^{nm}}$ is a δ_n-cover of Q_0, with $\delta_n := 2^{-n} \cdot \sqrt{m}$.[49] Therefore, by definition of $\mathscr{H}^m_\delta(Q_0)$ (see (6.19)),[50] we obtain that

$$\mathscr{H}^m_{\delta_n}(Q_0) \leq \sum_{j=1}^{2^{nm}} (\operatorname{diam}(Q_{j,n}))^m = 2^{nm} \cdot 2^{-nm} \cdot m^{m/2} = m^{m/2}; \tag{7.70}$$

so,

$$\mathscr{H}^m_{\delta_n}(Q_0) \leq m^{m/2}, \text{ for every integer } n \geq 1. \tag{7.71}$$

Since $\delta_n := 2^{-n} \cdot \sqrt{m} \to 0^+$, as $n \to \infty$, we deduce from the definition of $\mathscr{H}^m(Q_0)$ as the limit of $\mathscr{H}^m_\delta(Q_0)$ as $\delta \to 0^+$ that (7.71) yields the following inequalities, upon taking the limit as $n \to \infty$: $\mathscr{H}^m(Q_0) \leq m^{m/2} < \infty$. This implies the desired estimate, namely, $\mathscr{H}^m(Q_0) < \infty$, and concludes the proof of (i) in part (a) of the theorem.[51]

(ii) *We next claim that* $\mathscr{H}^m(Q_0) > 0$, *which proves inequality* (ii) *in part* (a) *of the theorem.*

In fact, we will establish the stronger result according to which

$$\mathscr{H}^m(Q_0) \geq c_m \cdot \mathcal{L}^m(Q_0) = c_m > 0, \tag{7.72}$$

where we have used the fact that $\mathcal{L}^m(Q_0) = 1$ and where c_m is a finite and positive constant depending only on the dimension m of the ambient space

[49] Since the dyadic cubes in the collection $\{Q_{j,n}\}_{j=1}^{2^{nm}}$ are all translates of one another, they clearly all have the same diameter given, via elementary Euclidean geometry, by the following expression: $\operatorname{diam}(Q_{j,n}) = 2^{-n} \cdot \sqrt{m} =: \delta_n$, for $j = 1, \ldots, 2^{nm}$.

[50] A priori, one should use the δ_n-approximate *outer measure* of Q_0, but since $Q_0 \in \mathcal{B}(\mathbb{R}^m)$, the latter value coincides with $\mathscr{H}^m_{\delta_n}(Q_0)$.

[51] Clearly, the exact same argument applied to any dyadic cube Q of side lengths 2^{-k}, with $k \in \mathbb{Z}$, leads to the inequality $\mathscr{H}^m(Q) \leq m^{m/2} \cdot 2^{-km} = (2^{-k} \cdot \sqrt{m})^m$, which implies that $\mathscr{H}^m(Q) < \infty$, as claimed in part (b) of the theorem.

$\mathbb{R}^m$.[52] The precise value of the constant c_m is not important for our purposes. In Remark 7.45, we will indicate how to obtain (7.72) and a non-optimal value of c_m via completely elementary means. However, in the proof of (ii) given below, we will also explain how to improve the value of the constant c_m by using Jung's Inequality, also called Jung's Theorem (Lemma 6.34).

Finally, in part (iii), we will give an entirely self-contained, and partly new, proof of (7.72)—and hence also of the inequality in part (ii) of the theorem.[53]

Let us now proceed to prove (7.72)—and hence also the inequality stated in part (ii).

Fix $\delta > 0$. Then, given any $\varepsilon > 0$, and in light of the definition of $\mathscr{H}^m_{\delta,\mathcal{S}^\#}(Q_0)$ as an infimum (see (6.21)), there exists a δ-cover of Q_0 by a countable collection of closed balls $\{B_j\}_{j=1}^\infty$ of $\mathbb{R}^m$ (with $\operatorname{diam}(B_j) \leq \delta$, for all $j \geq 1$, by definition) such that

$$\sum_{j=1}^{\infty}(\operatorname{diam}(B_j))^m \leq \mathcal{S}^m_\delta(Q_0) + \varepsilon = \mathscr{H}^m_{\delta,\mathcal{S}^\#}(Q_0) + \varepsilon, \tag{7.73}$$

where, as in Theorem 6.24 and Corollary 6.26, $\mathcal{S}^\#$ denotes the family of all closed balls of $\mathbb{R}^m$.

Recall from part (iv) of Theorem 6.24 (see (6.36) and (6.37))—or from its present special case when $X := \mathbb{R}^m$, the m-dimensional Euclidean space (see part (ii) of Corollary 6.26, identity (6.46))—that the δ-approximate standard and spherical Hausdorff measures $\mathscr{H}^m_\delta$ and $\mathcal{S}^m_\delta$ are comparable. In particular, by letting $t := m$ in the second inequality of (6.46) and specializing accordingly the value $\gamma_{m,t}$ given by (6.37), we obtain that for any $E \subseteq X$,

$$\mathcal{S}^m_\delta(E) \leq \gamma_m \cdot \mathscr{H}^m_\delta(E), \tag{7.74}$$

where

$$\gamma_m (= \gamma_{m,m}) = (2m/(m+1))^{m/2}. \tag{7.75}$$

Hence, applying this inequality to $E := Q_0$, and combining the resulting inequality with (7.73), we obtain that

$$\sum_{j=1}^{\infty}(\operatorname{diam}(B_j))^m \leq \gamma_m \cdot \mathscr{H}^m_\delta(Q_0) + \varepsilon \leq \gamma_m \cdot \mathscr{H}^m(Q_0) + \varepsilon, \tag{7.76}$$

where, in the last inequality of (7.76), we have used the fact that, by construction, $\mathscr{H}^m(Q_0) \geq \mathscr{H}^m_\delta(Q_0)$.

[52]The exact same proof will also show that, actually, Q_0 can be replaced by any set $E \subseteq \mathbb{R}^m$ in a suitable analog of the first inequality in (7.72). Namely, $\mathscr{H}^m(E) \geq c_m \cdot \mathcal{L}^m(E)$, provided $\mathscr{H}^m$ and $\mathcal{L}^m$ are interpreted as the m-dimensional Hausdorff and Lebesgue outer measures on $\mathbb{R}^m$, respectively; see Remark 7.45. Note that if $E \in \mathcal{B}(\mathbb{R}^m)$, then $\mathscr{H}^m$ and $\mathcal{L}^m$ are the usual m-dimensional Hausdorff and Lebesgue Borel measures on $\mathbb{R}^m$, respectively.

[53]The possibly novel aspect of the proof of part (ii) (and of (7.72)) are discussed in some detail in Remarks 7.45 and 7.46 below.

Next, we observe that for a closed ball B of radius $r > 0$ (and hence also of diameter $2r$), it is immediate to check the so-called Isodiametric Inequality; namely,

$$\mathcal{L}^m(B) \leq \omega_m \cdot ((\text{diam}(B))/2)^m = \omega_m \cdot r^m, \tag{7.77}$$

where ω_m is the volume of the closed unit ball $B_0 := \overline{B_0(1)}$ in $\mathbb{R}^m$:

$$\omega_m := \mathcal{L}^m(B_0) = \text{vol}^m(B_0). \tag{7.78}$$

Indeed, by the translation invariance of $\mathcal{L}^m$ (see Proposition 3.70, part (i)), we may assume that B is centered at the origin. Then, if $r = 1$ (i.e., if $B := B_0$, the closed unit ball of $\mathbb{R}^m$), (7.77) is a tautology since $\mathcal{L}^m(B_0) = \omega_m$, by (7.78). Moreover, if $B = \overline{B_0(r)}$, then the scale covariance of $\mathcal{L}^m$ (see Proposition 3.70, part (ii)), clearly yields the following equalities, as stated earlier: $\mathcal{L}^m(B) = r^m \cdot \mathcal{L}^m(B_0) = \omega_m \cdot r^m$.

Now, applying the inequality (7.77) to each closed ball B_j (of diameter $\text{diam}(B_j)$), for $j \geq 1$, we obtain that

$$\mathcal{L}^m(B_j) \cdot \omega_m^{-1} \cdot 2^m \leq (\text{diam}(B_j))^m, \text{ for all } j \geq 1. \tag{7.79}$$

Upon summation of the inequalities (7.79) over all integers $j \geq 1$, we obtain, after combining the resulting inequalities with (7.76), that

$$\omega_m^{-1} \cdot 2^m \cdot \left(\sum_{j=1}^{\infty} \mathcal{L}^m(B_j) \right) \leq \sum_{j=1}^{\infty} (\text{diam}(B_j))^m \leq \gamma_m \cdot \mathcal{H}^m(Q_0) + \varepsilon. \tag{7.80}$$

We then note that since, by construction, $Q_0 \subseteq \bigcup_{j=1}^{\infty} B_j$, we deduce from the monotonicity and subadditivity of the measure (or of the outer measure) $\mathcal{L}^m$ that $\mathcal{L}^m(Q_0) \leq \sum_{j=1}^{\infty} \mathcal{L}^m(B_j)$. Upon multiplication of the latter inequality by the positive constant $\omega_m^{-1} \cdot 2^m$, we deduce from the resulting inequality, combined with (7.80), that

$$2^m \cdot \omega_m^{-1} \cdot \mathcal{L}^m(Q_0) \leq \gamma_m \cdot \mathcal{H}^m(Q_0) + \varepsilon.$$

Since $\varepsilon > 0$ is arbitrary, we deduce, as desired, that

$$0 < c_m \cdot \mathcal{L}^m(Q_0) \leq \mathcal{H}^m(Q_0), \tag{7.81}$$

where

$$c_m := 2^m \cdot \omega_m^{-1} \cdot \gamma_m^{-1} = 2^m \cdot (\omega_m \cdot \gamma_m)^{-1}, \tag{7.82}$$

and with ω_m and γ_m given by (7.78) and (7.75), respectively.

This completes the proof of claim (ii) of part (a), as well as of (7.72).

We refer the interested reader to Remark 7.45 for a completely elementary and self-contained proof of the fact that $\mathcal{H}^m(Q_0) > 0$ and, hence, of claim (ii)—no longer relying on Jung's Inequality but instead, on a very simple substitute for it—which implies that $\mathcal{H}^m(Q_0) \geq \omega_m^{-1} \cdot \mathcal{L}^m(Q_0) = \omega_m^{-1} > 0$, a worse, but sufficient for our purpose, estimate than the one provided in (7.81) and (7.82).

(iii) We next establish claim (iii) of part (a) of the theorem; namely, we prove that (iii) holds: $\mathcal{H}^m(\partial Q_0) = 0$.

Now that we have shown that (i) and (ii) hold, we know that for every integer $k \geq 1$,

$$0 < \mathcal{H}^k(Q_0^k) < \infty, \tag{7.83}$$

where $Q_0^k := [0,1]^k$ is the k-dimensional cube in $\mathbb{R}^k$. We will use this key fact in order to provide a quick proof of claim (iii). It has the merit of being conceptually very simple and of relying on a few elementary properties of the notion of Hausdorff dimension.[54]

Beforehand, let us observe that ∂Q_0, the boundary of the unit cube $Q_0 = Q_0^m$ in $\mathbb{R}^m$, consists of 2^m faces, each of which can be identified (via an appropriate choice of the coordinate axes) with $Q_0^{m-1} = [0,1]^{m-1}$, the $(m-1)$-dimensional (closed) unit cube in $\mathbb{R}^{m-1}$. Since a finite, and even a countable, union of Borel measurable $\mathcal{H}^m$-null sets is $\mathcal{H}^m$-null (i.e., is of $\mathcal{H}^m$-measure zero), it suffices to show that $\mathcal{H}^m(\Phi) = 0$, where $\Phi := \Phi_0^{m-1}$ is the bottom (horizontal and closed) face of $Q_0 = Q_0^m$.

We chose the coordinate axes $x_1, \ldots, x_m$ so that $\Phi = \Phi_0^{m-1}$ lies in the hyperplane of $\mathbb{R}^m$ of equation $x_m = 0$ and can therefore be identified with $[0,1]^{m-1} \times \{0\}$ and, hence, via an isometry (and thus also a bi-Lipschitz map) with $Q_0^{m-1} = [0,1]^{m-1}$, the $(m-1)$-dimensional cube of $\mathbb{R}^{m-1}$.

Since Hausdorff dimension is invariant under bi-Lipschitz maps (see Corollary 7.21, identity (7.38)), we deduce from the above discussion and from (7.83), applied with $k := m-1$, that $\dim_H(\Phi_0^{m-1}) = m-1$. Thus, by definition, this fact implies that $\mathcal{H}^t(\Phi_0^{m-1}) = 0$, for all $t > m-1$. Hence, $\mathcal{H}^m(\Phi_0^{m-1}) = \mathcal{H}^m(\Phi) = 0$, from which it follows that $\mathcal{H}^m(\partial Q_0) = 0$, as was already explained above.

This concludes the proof of claim (iii) in part (a) of the theorem.

(iv) Since, in light of (iii), we know that $\mathcal{H}^m(\partial Q_0) = 0$ and since Q_0 is the disjoint union of its interior and of its boundary, $Q_0 = \mathring{Q}_0 \sqcup \partial Q_0$, we have, as claimed, that $\mathcal{H}^m(Q_0) = \mathcal{H}^m(\mathring{Q}_0) + \mathcal{H}^m(\partial Q_0) = \mathcal{H}^m(\mathring{Q}_0)$.

This completes the proof of claim (iv) in part (a) of the theorem. It also concludes the proof of the entire part (a) of the theorem because the corresponding statements for $\mathcal{L}^m$ (instead of for $\mathcal{H}^m$) are obtained in the same (and, in fact, in a simpler) manner.

We next proceed to establish part (b), which follows from part (a) by a simple substitution argument.

Part (b): This part follows at once from the method of proof of part (a) by simply replacing the unit cube Q_0 by the given dyadic cube Q, of side lengths 2^{-n}, for some arbitrary $n \in \mathbb{Z}$. In order to prove the counterpart of claim (i),

[54] Alternatively, one could provide a proof of claim (iii) based on a discrete scaling argument and relying solely on the translation invariance of $\mathcal{H}^m$, along with the fact that $\mathcal{H}^m(Q_0) < \infty$, established in (i), but it would be significantly longer and more involved. As a result, we omit it.

as was already observed in footnote 51, we then obtain that[55] $\mathcal{H}^m(Q) \leq m^m \cdot 2^{-mn} = (m2^{-n})^m < \infty$, as desired, while for establishing the analog of claim (ii), we obtain that, with $c_m > 0$ given by (7.82),

$$\mathcal{H}^m(Q) \geq c_m \cdot \mathcal{L}^m(Q) = c_m \cdot 2^{-mn} \cdot \mathcal{L}^m(Q_0) = c_m \cdot 2^{-mn} > 0.$$

Finally, for proving the counterpart of claim (iii), the exact same argument as in the proof of claim (iii) in part (a), but still with Q_0 replaced by Q, shows that any $(m-1)$-dimensional face Φ of Q has Hausdorff dimension $m-1$; i.e., $\dim_H(\Phi) = m-1$. It then follows from the definition of $\dim_H$ that $\mathcal{H}^m(\Phi) = 0$.[56] Hence, since ∂Q is covered by the finite union of its 2^m $(m-1)$-dimensional faces, we deduce that $\mathcal{H}^m(\partial Q) = 0$, as desired.

Consequently, since the counterpart (iv) is now obvious, we have established all of the statements of the theorem (i.e., Theorem 7.41) concerning Hausdorff measure $\mathcal{H}^m$. As a last comment, we refer to Remark 7.42 and Exercise 7.43 for further discussion of the case of Lebesgue measure $\mathcal{L}^m$ (instead of Hausdorff measure $\mathcal{H}^m$) on $\mathbb{R}^m$. We observe for now that only case (iii) of part (a) (i.e., the proof of the fact that $\mathcal{L}^m(\partial Q_0) = 0$, for the unit cube Q_0 of $\mathbb{R}^m$) requires some discussion since, in light of the scale covariance of $\mathcal{L}^m$ (see part (ii) of Proposition 7.31), we have that, for any dyadic cube Q of side lengths 2^{-n}, for some arbitrary $n \in \mathbb{Z}$, $0 < \mathcal{L}^m(Q) = (2^{-n})^m \cdot \mathcal{L}^m(Q_0) = 2^{-nm} \cdot 1 = 2^{-nm} < \infty$.

This finally completes the proof of Theorem 7.41 □

Remark 7.42. Here is another direct proof of the fact that if Q_0 is the unit cube in $\mathbb{R}^m$, then $\mathcal{L}^m(\partial Q_0) = 0$, as claimed in part (iii) of case (a) of Theorem 7.41.[57]

In light of Exercise 3.126 and with the notation of that exercise about products of Lebesgue measures, we have $\mathcal{L}^m_{\mathbb{R}^m} = \mathcal{L}^{m-1}_{\mathbb{R}^{m-1}} \times \mathcal{L}^1_{\mathbb{R}}$, where for $k \in \{1, \dots, m\}$, $\mathcal{L}^k_{\mathbb{R}^k}$ stands for the k-dimensional Lebesgue measure on $\mathbb{R}^k$. In short, $\mathcal{L}^m = \mathcal{L}^{m-1} \times \mathcal{L}^1$, the product measure of $\mathcal{L}^{m-1}$ and of $\mathcal{L}^1$, in the sense of Exercise 3.125.

Thus, since as was noted earlier, the bottom (horizontal) face Φ_0^{m-1} of the unit cube Q_0 of $\mathbb{R}^m$ can be identified with $[0,1]^{m-1} \times \{0\} \subseteq \mathbb{R}^{m-1} \times \mathbb{R} = \mathbb{R}^m$, by labeling the coordinates $x_1, \dots, x_m$ of $\mathbb{R}^m$ so that Φ_0^{m-1} lies in the hyperplane of $\mathbb{R}^m$ of equation $x_m = 0$, we deduce from the definition of a product measure

[55] Naturally, one could also invoke instead the scale covariance of $\mathcal{H}^m$ (see part (ii) of Proposition 7.31), according to which $\mathcal{H}^m(Q) = (2^{-n})^m \cdot \mathcal{H}^m(Q_0) = 2^{-nm} \cdot \mathcal{H}^m(Q_0)$, and then invoke claims (i) and (ii) of part (a) in order to deduce that $0 < \mathcal{H}^m(Q) < \infty$, as desired. However, we prefer to emphasize here the simplicity of this substitution argument.

[56] The same argument would show that any of the k-dimensional faces Φ^k of the cube Q (with $0 \leq k \leq m-1$) has Hausdorff dimension $k < m$ and hence that $\mathcal{H}^m(\Phi^k) = 0$.

[57] The fact that, for any dyadic cube Q, we have that $\mathcal{L}^m(\partial Q) = 0$, follows by simply substituting Q for Q_0 (exactly as in the case of $\mathcal{H}^m$ in part (b)) and by using the scale covariance of $\mathcal{L}^{m-1}$ (part (ii) of Proposition 7.31).

(see identity (3.43) of Exercise 3.125)[58] that

$$\mathcal{L}^m(\Phi_0^{m-1}) = \mathcal{L}^{m-1}([0,1]^{m-1}) \cdot \mathcal{L}^1(\{0\}) = 1 \cdot 0 = 0.$$

Here, we have used the fact that $\mathcal{L}^{m-1}([0,1]^{m-1}) = 1$ (since $[0,1]^{m-1}$ is the unit cube of $\mathbb{R}^{m-1}$), along with the easily verifiable fact according to which $\mathcal{L}^1(\{0\}) = 0$, as was noted and proved in Lemma 3.28.[59]

Finally, as in the proof of claim (iii) in part (a) of Theorem 7.41, we deduce from the translation invariance of $\mathcal{L}^m$ that $\mathcal{L}^m(\Phi) = \mathcal{L}^m(\Phi_0^{m-1}) = 0$, for any $(m-1)$-dimensional face of the unit cube Q_0 of $\mathbb{R}^m$. Also, since ∂Q_0 is contained in (and is, in fact, equal to) the union of the 2^m closed $(m-1)$-dimensional faces of Q_0, we deduce from the finite subadditivity of $\mathcal{L}^m$ that $\mathcal{L}^m(\partial Q_0) = 0$, as desired.

Exercise 7.43. Provide all of the details of the proof outlined in Remark 7.42 just above, and according to which $\mathcal{L}^m(\partial Q_0) = 0$, where $Q_0 = [0,1]^m$.

Remark 7.44. Here is a slightly different way of proving the fact that $\mathcal{L}^m(\Phi_0^{m-1}) = 0$—and hence that $\mathcal{L}^m(\partial Q_0) = 0$, as well as $\mathcal{L}^m(\partial Q) = 0$, for any dyadic cube Q of $\mathbb{R}^m$. View $\Phi := \Phi_0^{m-1}$ as a subset of $[0,1]^{m-1} \times [-\varepsilon, \varepsilon]$, for any $\varepsilon > 0$. Therefore, by the monotonicity of $\mathcal{L}^m$ and since $\mathcal{L}^m = \mathcal{L}^{m-1} \times \mathcal{L}^1$, we have successively

$$\begin{aligned} 0 \le \mathcal{L}^m(\Phi) &\le \mathcal{L}^m([0,1]^{m-1} \times [-\varepsilon, \varepsilon]) = \mathcal{L}^{m-1}([0,1]^{m-1}) \cdot \mathcal{L}^1([-\varepsilon, \varepsilon]) \\ &= (2\varepsilon) \cdot \mathcal{L}^{m-1}([0,1]^{m-1}) = 2\varepsilon, \end{aligned} \tag{7.84}$$

where, in the next to last equality, we used the scaling property of $\mathcal{L}^1$ and the fact that $\mathcal{L}^k([0,1]^k) = 1$, for every integer $k \geq 1$.

It now suffices to let $\varepsilon \to 0^+$ in (7.84) in order to conclude that $0 \le \mathcal{L}^m(\Phi) \le 0$; i.e., $\mathcal{L}^m(\Phi) = 0$, as desired.

Remark 7.45 (*Self-Contained Proof of Part* (ii) *of Theorem* 7.41). We give here a fully self-contained proof of part (ii) of Theorem 7.41. Recall from the proof of part (iv) of Theorem 6.24 that, in the present case of the Euclidean space $\mathbb{R}^m$, the aforementioned results obtained in part (iv) of Theorem 6.24 or in part (ii) of Corollary 6.26—more specifically, the inequality (7.74)—depend on the sharp Jung Inequality (Lemma 6.34).

However, for our present purpose—namely, in order to show that $\mathcal{H}^m(Q_0) > 0$ and hence also that $\mathcal{H}^m(Q) > 0$, for all dyadic cubes Q of $\mathbb{R}^m$—it would suffice to use a weak form of Jung's Theorem (Lemma 6.34), which is very simple to prove and according to which any compact set $K \subseteq \mathbb{R}^m$ is contained in a

[58]Note that $[0,1]^{m-1} \times \{0\}$ is a (degenerate) rectangle of $\mathbb{R}^{m-1} \times \mathbb{R}$.

[59]Indeed, for every integer $k \geq 1$, and by the translation invariance of $\mathcal{L}^k$ (see part (i) of Proposition 3.29), we have $\mathcal{L}^k(\{\text{point}\}) = \mathcal{L}^1(\{0\})$ and since $\infty > \mathcal{L}^k([0,1]^k) \geq \mathcal{L}^k(\mathbb{Q} \cap [0,1]^k) = +\infty$ (by the monotonicity and subadditivity of $\mathcal{L}^k$), unless $\mathcal{L}^k(\{\text{point}\}) = 0$, we deduce that $\mathcal{L}^k(\{x\}) = \mathcal{L}^k(\{0\}) = 0$, for all $x \in \mathbb{R}^k$. One shows similarly that for any integer $k \geq 1$, $\mathcal{H}^k(\{x\}) = \mathcal{H}^k(\{0\}) = 0$, for all $x \in \mathbb{R}^k$.

closed ball of radius $r := \operatorname{diam}(K)$ and hence also of diameter $d := 2\operatorname{diam}(K)$; see Lemma 6.39 and its proof. It then follows from this weak (and trivial) form of Jung's Theorem that the following weaker form of the inequality (7.74) holds:[60]

$$S_\delta^m \leq 2^m \cdot \mathcal{H}_\delta^m. \tag{7.85}$$

It compares the (δ-approximate) spherical and standard Hausdorff measures (or outer measures), S_δ^m and $\mathcal{H}_\delta^m$, respectively.[61] One then deduces from (7.85), exactly as in the proof of (ii) of part (a) of Theorem 7.41, that we have $\mathcal{H}^m(Q_0) \geq \omega_m^{-1} \cdot \mathcal{L}^m(Q_0) = \omega_m^{-1} \cdot 1 = \omega_m^{-1}$ or, more generally, according to Remark 7.46 just below, that for any $E \subseteq \mathbb{R}^m$, $\mathcal{H}^m(E) \geq \omega_m^{-1} \cdot \mathcal{L}^m(E)$.

Remark 7.46 (*Extension of the Result of Claim* (ii) *of Theorem* 7.41). Close inspection of the derivation of the estimate (7.72) in the proof of claim (ii) in part (a) of Theorem 7.41 shows that Q_0 itself did not play any role—except for the fact that $\mathcal{L}^m(Q_0) = 1 > 0$, enabling us to deduce from (7.72) or, equivalently, from (7.81), that $\mathcal{H}^m(Q_0) > 0$. Actually, the exact same proof as in claim (ii) of part (a) of the theorem shows that we can replace Q_0 by any subset E of $\mathbb{R}^m$:

$$\mathcal{H}^m(E) \geq c_m \cdot \mathcal{L}^m(E), \text{ for all } E \subseteq \mathbb{R}^m, \tag{7.86}$$

with c_m still given by (7.82). In general, in (7.86), one should interpret $\mathcal{H}^m$ and $\mathcal{L}^m$ as the (m-dimensional) Hausdorff and Lebesgue outer measures. However, if $E \in \mathcal{B}(\mathbb{R}^m)$, then $\mathcal{H}^m$ and $\mathcal{L}^m$ in (7.86) stand for the Hausdorff measure and the Lebesgue measure, respectively.

In fact, in Theorem 7.53 and the comments following it, along with Theorem 7.66, we will see that

$$\mathcal{H}^m(E) = (2^m \cdot \omega_m^{-1}) \cdot \mathcal{L}^m(E), \text{ for any } E \subseteq \mathbb{R}^m, \tag{7.87}$$

where $\omega_m := \mathcal{L}^m(B_0)$, the m-dimensional volume of the unit ball in $\mathbb{R}^m$ (as in (7.78)). More specifically, in Theorem 7.53, we show that $\mathcal{H}^m(E) = \beta_m \cdot \mathcal{L}^m(E)$, for some constant $\beta_m > 0$, whereas the precise determination of β_m, as given by (7.87), would essentially require a suitable form of the Isoperimetric Inequality; see the next comment, Remark 7.47.

Remark 7.47 (*Possible Novelty of the Proof of Claim* (ii) *of Theorem* 7.41). We explain here the main new aspect of our proof of the fact that $\mathcal{H}^m(Q_0) > 0$, as stated in claim (ii) of Theorem 7.41—or, more generally, of the fact that $\mathcal{H}^m(Q_0) \geq c_m \cdot \mathcal{L}^m(Q_0)$, with $c_m > 0$, and similarly with Q_0 replaced by any $E \subseteq \mathbb{R}^m$ and, in particular, by any dyadic cube Q as in (ii) of part (b) of Theorem 7.41. The novelty in question is that, contrary to what is stated (or implied) by many textbooks and expository papers related to this subject (see, e.g., [**182**],

---

[60] See Lemma 6.40, along with Lemma 6.39.

[61] To see why inequality (7.85) is worse than inequality (7.74), $S_\delta^m \leq \gamma_m \cdot \mathcal{H}_\delta^m$, with γ_m given as in (7.75), it suffices to note that for any $m \in \mathbb{N}$, $2m/(m+1) \leq 4$, from which it follows that $\gamma_m \leq 2^m$.

[**430**], [**193**], [**36**]), one does *not* have to use the general *Isodiametric Inequality*, according to which

$$\mathcal{L}^m(A) \leq \omega_m \cdot ((\mathrm{diam}(A))/2)^m, \text{ for all } A \in \mathcal{B}(\mathbb{R}^m), \tag{7.88}$$

where ω_m denotes the volume of the unit ball in $\mathbb{R}^m$. The latter inequality (7.88) is itself intimately connected with the celebrated *Isoperimetric Inequality* and also a consequence of the *Brunn–Minkowski Inequality* from convex geometry (see, e.g., [**540**]) which states that

$$\mathcal{L}^m(A_1 + A_2)^{1/m} \geq \mathcal{L}^m(A_1)^{1/m} + \mathcal{L}^m(A_2)^{1/m}, \text{ for all } A_1, A_2 \in \mathcal{B}(\mathbb{R}^m), \tag{7.89}$$

where $A_1 + A_2 := \{a_1 + a_2 : a_1 \in A_1 \text{ and } a_2 \in A_2\}$ is the *Minkowski sum* of the sets A_1 and A_2.

The way we avoid having to use the general Isodiametric Inequality is twofold:

(i) We first replace the standard δ-approximate Hausdorff measure by the spherical Hausdorff measure defined in terms of δ-covers by closed balls instead of by arbitrary subsets of $\mathbb{R}^m$, thanks to Theorem 6.24 (part (iv)) and Corollary 6.26.

(ii) We then observe that the Isodiametric Inequality (7.88) is trivially satisfied for any closed ball A.

Finally, we point out—as was alluded to at the end of Remark 7.45—that in order to obtain the precise relationship given in the equality (7.87) above, one must use an appropriate form of the celebrated Isoperimetric Inequality, typically, the Isodiametric Inequality (7.88). For most practical purposes, however, the precise value of the proportionality constant between $\mathcal{H}^m$ and $\mathcal{L}^m$ is unimportant.

Remark 7.48 (*Scaling Argument Valid for All Translation-Invariant Borel Measures*). Here is a slightly different way to prove that ∂Q_0—or, more generally, the boundary ∂Q of any dyadic cube Q—has measure zero. It does not use the notion of Hausdorff measure and is based instead on a closely related, but different, scaling argument. For this reason, it presents the advantage of being valid for any translation-invariant Borel measure ν on $\mathbb{R}^d$ such that $\nu(Q_0) < \infty$.[62] As we have seen in the proof of claim (iii) in part (a) of Theorem 7.41, given any $n \in \mathbb{N}_0$,[63] one can bisect n consecutive times the sides of the unit cube Q_0 so as to obtain 2^{nm} closed dyadic cubes of side lengths 2^{-n} and covering Q_0. A moment's reflection then shows that, of those cubes, only $2^{(n-1)(m-1)}$

[62]We shall see later, by using transfinite induction (and hence also the Axiom of Choice), that a translation-invariant Borel measure on $\mathbb{R}^m$ which is not identically equal to $+\infty$ (i.e., which is *nontrivial*) must be such that $\nu(Q_0) < \infty$ and must therefore be such that $\nu(\{x\}) = 0$, for all $x \in \mathbb{R}^m$; see Proposition 7.49 and Theorem 7.50, along with their method of proof.

[63]An entirely similar argument can be used if $n \in \mathbb{Z}$, with $n < 0$, except that one must then go "backward" instead of "forward"; see the proof of Proposition 7.49 below.

are needed to cover the bottom horizontal and closed $(m-1)$-dimensional face $\Phi_0^{m-1} = [0,1]^{m-1} \times \{0\}$ of Q_0; namely, these are the above dyadic cubes which touch Φ_0^{m-1}. Note that, as in the aforementioned proof, we choose the coordinates $\{x_j\}_{j=1}^m$ of $\mathbb{R}^m$ so that Φ_0^{m-1} lies in the hyperplane of equation $x_n = 0$.

Now, one small but important technical point is that in the present proof, we replace all of the covering dyadic cubes by the associated half-closed, half-open dyadic cubes (with the same centers and side lengths).[64] Then, in the notation of footnote 64 and still with $n \in \mathbb{N}_0$, Φ_0^{m-1} is still covered by the same $2^{(n-1)(m-1)}$ half-closed, half-open dyadic cubes $Q^\#$ (in the bottom face), while $Q_0^\#$ (the unit cube without its top face) is precisely covered by the same 2^{nm} dyadic (half-closed, half-open) dyadic cubes $Q^\#$. Since ν is translation invariant, all of these dyadic cubes $Q^\#$ have the same "ν-volume", $\nu(Q^\#) =: v_n$, because they have the same side lengths 2^{-n} and hence are translates of one another. Furthermore, since, by construction, all of these dyadic cubes are now pairwise disjoint, we deduce that[65,66]

$$\nu(Q_0^\#) = 2^{nm} \cdot v_n \quad \text{and} \tag{7.90}$$
$$0 \le \nu(\Phi_0^{m-1}) \le 2^{(n-1)(m-1)} \cdot v_n = 2^{1-m} \cdot 2^{-n} \cdot (2^{nm} \cdot v_n)$$

or, equivalently, in light of (7.90),

$$0 \le \nu(\Phi_0^{m-1}) \le 2^{1-m} \cdot 2^{-n} \cdot \nu(Q_0^\#). \tag{7.91}$$

Since $\nu(Q_0^\#) \le \nu(Q_0) < \infty$ (because, obviously, $Q_0^\# \subseteq Q_0$), we deduce by letting $n \to \infty$ in (7.91) that $\nu(\Phi_0^{m-1}) = 0$, as desired.

Just as in the proof of case (iii) in part (a) of Theorem 7.41, we then deduce that for each face Φ of Q_0, $\nu(\Phi) = 0$ and since ∂Q_0 can be covered by finitely many (really, 2^m) faces, that $\nu(\partial Q_0) = 0$, as claimed.

Finally, we observe, as in part (b) of the proof of Theorem 7.41, that there is nothing special about the unit cube Q_0. We could have just as well started with any closed dyadic cube Q (instead of with Q_0) and used the same reasoning in order to deduce that $\nu(\partial Q) = 0$.

---

[64] More specifically, if the original closed dyadic cube is of the form $Q := \prod_{j=1}^m [a_j, b_j]$, then we replace it by $Q^\# := \prod_{j=1}^m [a_j, b_j)$. In particular, the closed unit cube $Q_0 = [0,1]^m$ is replaced by $Q_0^\# = [0,1)^m$, which is the unit cube Q_0 from which the top (horizontal) face Φ_1^{m-1} has been removed.

[65] Indeed, in order to obtain (7.90), it is crucial to observe here that, by design since the cubes are half-closed, half-open, the disjoint union of the corresponding dyadic cubes is precisely equal to $Q_0^\#$ and does not just cover $Q_0^\#$.

[66] The first author (Michel L. Lapidus) would like to thank his colleague David Weisbart for having suggested to try to use half-closed, half-open (rather than closed) dyadic cubes in this setting.

The next result (Proposition 7.49), combined with Remark 7.48 just above, will enable us (in Theorem 7.50) to extend Theorem 7.41 to all (nondegenerate) translation-invariant Borel measures on $\mathbb{R}^m$.[67] In its proof, we will anticipate on the use of a possibly novel technique in the present context, based on transfinite induction—and hence also on the use of the Axiom of Choice or, equivalently, of the Well-Ordering Principle[68]—to be introduced in Step 4 of the second proof of Theorem 7.53. Therefore, we will not give all of the details but instead refer the interested reader to the just-mentioned theorem and its proof, along with Exercises 7.54 and 7.55 for further information, and leave the easy completion of the proof to the reader.

Proposition 7.49 (Nondegenerate Translation-Invariant Borel Measures on $\mathbb{R}^m$). *Let ν be a translation-invariant measure on $\mathbb{R}^m$. Then, either $\nu = c_0 \cdot \rho$, where ρ is the counting measure on $\mathbb{R}^m$ and $c_0 := \nu(\{0\})$ (in which case $\nu(Q_0) = +\infty$, unless $c_0 = 0$ and hence also $\nu = 0$) or else*

$$0 < \nu(Q_0) < \infty. \tag{7.92}$$

Also, if $\nu(Q_0) < \infty$, then ν is a nonatomic *(or* diffuse*) measure; i.e.,*

$$\nu(\{\boldsymbol{x}\}) = 0, \text{ for all } \boldsymbol{x} \in \mathbb{R}^m. \tag{7.93}$$

Rephrased: If ν is nonzero, then $\nu(Q_0) > 0$ and if ν is nontrivial (i.e., if $\nu(E) < \infty$, for some nonempty $E \in \mathcal{B}(\mathbb{R}^m)$), then $\nu(Q_0) < \infty$, in which case ν is nonatomic *or* diffuse *(i.e., (7.93) holds).*

Finally, all of the above statements extend without change with Q_0 (the unit cube in $\mathbb{R}^m$) replaced by any closed dyadic cube Q in $\mathbb{R}^m$.

Proof.

(i) *Assume that $\nu(Q_0) = 0$. Then, we claim that $\nu = 0$, the zero measure.*

Indeed, clearly then, because they are translates of one another and since ν is translation invariant, all dyadic cubes $Q_{(n)}$ of side lengths 2^{-n}, for some $n \in \mathbb{Z}$, have the same ν-volume; i.e., $v_n := \nu(Q_{(n)})$, with $v_n \in [0, +\infty]$, depending only on $n \in \mathbb{Z}$.

Now, for a fixed $n \geq 0$, since Q_0 is equal to the disjoint union of finitely many (more specifically 2^{nm}) half-closed, half-open dyadic cubes of side lengths 2^{-n} (see Remark 7.48 above), we clearly have $v_n = 0$. Indeed,

$$0 = v_0 = \nu(Q_0) = 2^{nm} \cdot \nu(Q_{(n)}) = 2^{nm} \cdot v_n = 0;$$

so $v_n = 0$. It follows that $\nu(Q_{(n)}) = 0$, for all $n \geq 0$.

[67] Here, a Borel measure ν is said to be *nondegenerate* if ν is neither the zero measure nor the trivial measure identically equal to $+\infty$ (except at the empty set, where it must be equal to 0).

[68] We note that, in the present case, the significantly weaker Axiom of Countable Choice is sufficient; see, e.g., [**273**].

Also, if $n < 0$, then any dyadic cube $Q_{(n)}$ of side lengths $2^{-n} = 2^{|n|}$ can be covered by $2^{|n|m}$ dyadic cubes of side lengths 1 and, thus, of ν-volume $v_0 = \nu(Q_0) = 0$. Consequently, by the (finite) subadditivity of ν, and since we have $0 \leq \nu(Q_{(n)}) \leq 2^{|n|m} \cdot \nu(Q_0) = 0$, we also have that $\nu(Q_{(n)}) = 0$, for any $n < 0$. Hence, we conclude that $v_n = \nu(Q_{(n)}) = 0$, for all $n \in \mathbb{Z}$; i.e., $\nu(Q) = 0$, for any dyadic cube Q.

Next, in light of Proposition 7.40, since any open set U is an at most countable union of dyadic cubes, we have that $\nu(U) = 0$; indeed, a Borel measurable subset of a countable union of ν-null sets is ν-null, by the countable subadditivity of ν. We therefore deduce that

$$\nu(U) = 0, \text{ for all open sets } U \text{ in } \mathbb{R}^m. \tag{7.94}$$

Anticipating on the proof of Theorem 7.53 (and the classic terminology used therein), we deduce by repeated application of (7.94) that any G_δ set, and then any $G_{\delta\sigma}$ set, then any $G_{\delta\sigma\delta}$ set, etc., is ν-null, i.e., has ν-measure zero.[69]

By transfinite induction and by using the Borel hierarchy of the Borel σ-algebra $\mathcal{B}(\mathbb{R}^m)$ (see Exercises 7.54 and 7.55 below), one deduces, much as in the proof of Theorem 7.53 below and in the aforementioned exercises, that $\nu(B) = 0$, for all $B \in \mathcal{B}(\mathbb{R}^m)$; i.e., $\nu = 0$, the zero measure.

(ii) *Next, assume that* $\nu(Q_0) < \infty$. *Then, we claim that* (7.93) *holds; i.e.,* $\nu(\{x\}) = 0$, *for all* $x \in \mathbb{R}^m$.

Indeed, by the translation invariance of ν, we have that

$$\nu(\{x\}) = \nu(\{0\}), \text{ for all } x \in \mathbb{R}^m. \tag{7.95}$$

Furthermore, observe that $\mathbb{Q}^m$ is a Borel set, since it is the countable union of its points $\{x\}$, with $x \in \mathbb{Q}^m$. (Note that each singleton $\{x\}$ is closed and hence is a Borel set.)

Now, if $\nu(\{0\}) > 0$,[70] then[71]

$$\nu(Q_0) \geq \nu(\mathbb{Q}^m \cap Q_0) = \nu(\{0\}) \cdot (+\infty) = +\infty. \tag{7.96}$$

Thus, $\nu(Q_0) = +\infty$, which contradicts our assumption. We conclude that $\nu(\{0\}) = 0$ and hence also, in light of (7.95), that $\nu(\{x\}) = 0$, for all $x \in \mathbb{R}^m$, as desired.[72]

---

[69] As we will see in the proof of Theorem 7.53, a G_δ set is a countable intersection of open sets, a $G_{\delta\sigma}$ set is a countable union of G_δ sets, a $G_{\delta\sigma\delta}$ set is a countable intersection of $G_{\delta\sigma}$ sets, etc.

[70] We allow here the case when $\nu(\{0\}) = +\infty$, although it will follow from the present argument that this situation cannot occur.

[71] In light of identity (7.95), the first equality in (7.96) follows since $\mathbb{Q}^m \cap Q_0 = (\mathbb{Q} \cap [0,1])^m$ is countably infinite and ν is countably additive.

[72] We note that the exact same proof as in (ii) shows, simply by substituting $E \in \mathcal{B}(\mathbb{R}^m)$ for Q_0, where $\nu(E) < \infty$ (assuming such a Borel set E exists), that we also reach the same conclusion, even in this more general case.

(iii) *Assume that* $\nu(Q_0) = +\infty$ *and that* ν *is nontrivial. We then claim that* $\nu = c_0 \cdot \rho$, *where* $c_0 := \nu(\{0\})$, *with* $0 \leq c_0 < \infty$, *and* ρ *is the counting measure on* $\mathbb{R}^m$. *Note that* $c_0 > 0$ *unless* $\nu = 0$.

Indeed, first of all, we note that $0 \leq c_0 < \infty$ because assuming that $c_0 = +\infty$ would clearly imply (since $\nu(\{\boldsymbol{x}\}) = c_0$, for all $\boldsymbol{x} \in \mathbb{R}^m$) that for any $B \in \mathcal{B}(\mathbb{R}^m)$, $B \neq \emptyset$, we have that $\nu(B) = +\infty$; i.e., ν is trivial, which contradicts our hypothesis.

Next, we note that if $B \in \mathcal{B}(\mathbb{R}^m)$ is finite or countable, then B is the finite or countable disjoint union of its points $\{\boldsymbol{x}\}$, with $\boldsymbol{x} \in B$. Therefore, by the (finite or) countable additivity of the measure ν, we have that

$$\nu(B) = \sum_{x \in B} \nu(\{\boldsymbol{x}\}). \tag{7.97}$$

Hence, since $\nu(\{\boldsymbol{x}\}) = \nu(\{0\}) =: c_0 < \infty$, by identity (7.95), which is valid independently of whether $\nu(Q_0) < \infty$ or $\nu(Q_0) = +\infty$, we deduce that (with $\#B$ denoting the number of points of the finite set B)

$$\nu(B) = \begin{cases} +\infty, & \text{if } B \text{ is countably infinite,} \\ c_0 \cdot (\#B), & \text{if } B \text{ is finite,} \end{cases} \tag{7.98}$$

unless $c_0 := \nu(\{0\}) = 0$, in which case $\nu(B) = 0$, by (7.97).

Moreover, if B is an uncountably infinite set, then $B \supseteq B_1$, for some countably infinite Borel set; for example, we can let $B_1 := \mathbb{Q}^m \cap B$. (Clearly, $B \in \mathcal{B}(\mathbb{R}^m)$, since B and $\mathbb{Q}^m$ belong to the σ-algebra $\mathcal{B}(\mathbb{R}^m)$.) Therefore, $\nu(B) \geq \nu(B_1) = +\infty$, and thus, $\nu(B) = +\infty$. In light of (7.98), applied to the countable set B_1, this implies once again that

$$\nu(B) = \begin{cases} +\infty, & \text{if } B \text{ is infinite,} \\ c_0 \cdot (\#B), & \text{if } B \text{ is finite,} \end{cases} \tag{7.99}$$

which (since it is now valid for any $B \in \mathcal{B}(\mathbb{R}^m)$) is exactly equivalent to stating that $\nu = c_0 \cdot \rho$, as claimed above.

This completes the proof of the proposition. □

The next theorem will enable us, in particular, to show that any nondegenerate translation-invariant Borel measure on $\mathbb{R}^m$ coincides, up to a positive and finite multiplicative constant depending only on m, with the m-dimensional Lebesgue measure $\mathcal{L}^m$ on $\mathbb{R}^m$—and hence also (by Theorem 7.53 below) with the m-dimensional Hausdorff measure $\mathcal{H}^m$ on $\mathbb{R}^m$.

Theorem 7.50 (Extension of Theorem 7.41 to Nondegenerate Translation Borel Measures—Discrete Scale Covariance). *Let* ν *be a translation-invariant Borel measure on* $\mathbb{R}^m$ *which is* nondegenerate.[73] *Then, the exact analog of part* (b) *of*

[73] Recall that a Borel measure ν is *nondegenerate* if ν is neither the zero measure nor the trivial $+\infty$ measure for which, by definition, $\nu(\emptyset) = 0$ and $\nu(B) = +\infty$, for every nonempty Borel set B.

Theorem 7.41 *holds for* ν, *instead of just for the special case of* $\mathcal{H}^m$. *In particular, for any dyadic cube Q, we have that*

$$0 < \nu(\mathring{Q}) = \nu(Q) < \infty, \ \nu(\partial Q) = 0, \tag{7.100}$$

and, in fact, if Q has side lengths 2^{-n}, *for some* $n \in \mathbb{Z}$, *then*

$$\nu(Q) = 2^{-nm} \cdot \nu(Q_0), \tag{7.101}$$

with $0 < \nu(\mathring{Q}_0) = \nu(Q_0) < \infty$ *and* $\nu(\partial Q_0) = 0$. *As before, we refer to the property stated in* (7.101) *as* discrete scale covariance *(of* ν*). Furthermore,* ν *is* nonatomic *(or* diffuse*); i.e.,* $\nu(\{\boldsymbol{x}\}) = 0$, *for all* $\boldsymbol{x} \in \mathbb{R}^m$.

Proof. This follows at once by combining Remark 7.48 and Proposition 7.49. □

Exercise 7.51. In light of the forthcoming discussion (in Subsection 7.4.3) of transfinite induction and of the Borel hierarchy, provide the missing details in the proof of part (i) of the proof of Proposition 7.49 above, using transfinite induction and according to which, if ν is a nonzero translation-invariant measure on $\mathbb{R}^m$, then $\nu(Q_0) > 0$.

[*Hint*: Compare the proofs of Theorems 7.53 and 7.57; also, refer to the information provided in Exercises 7.54 and 7.55 in Subsection 7.4.3 below.]

Exercise 7.52.

(a) Given $t \in \mathbb{R}$, $t \geq 0$, consider the t-dimensional Hausdorff–Borel measure $\mathcal{H}^t$ on $\mathbb{R}^m$. Is $\mathcal{H}^t$ a Radon measure? Also, is it σ-finite?

[*Hint*: Distinguish (at least) two different cases: $t = m$ and $t \neq m$.]

(b) Deduce from part (a) that the following statement is *wrong*, in general: "$\mathcal{H}^t$ is a σ-finite, translation-invariant Borel measure on $\mathbb{R}^m$. Therefore, it must be proportional to Lebesgue measure on $\mathbb{R}^m$."

Explain your answer.[74] Does this contradict, in particular, Theorem 7.50 above or Corollary 7.58 below?

7.4.2. Equality of $\mathcal{L}^m$ and $\mathcal{H}^m_{nor}$, via Transfinite Induction. The next result is important, in practice, because for any integer $m \geq 1$, it explicitly identifies, up to a multiplicative normalization constant, the m-dimensional Hausdorff measure $\mathcal{H}^m$ on $\mathbb{R}^m$ with the classic m-dimensional volume on $\mathbb{R}^m$, interpreted as $\mathcal{L}^m$, the standard (m-dimensional) Lebesgue measure on $\mathbb{R}^m$. Accordingly, Hausdorff measure $\mathcal{H}^m_{nor}$, suitably normalized so that $\mathcal{H}^m_{nor}(Q_0) = \mathcal{L}^m(Q_0) = 1$, coincides with $\mathcal{L}^m$. This suggests to think heuristically of $\mathcal{H}^t$, the t-dimensional Hausdorff measure, for any $t \geq 0$ (or of its suitably normalized version, $\mathcal{H}^t_{nor}$, to be briefly discussed further on), as a natural t-dimensional generalization of length ($t = 1$), area ($t = 2$), volume ($t = 3$), ..., m-dimensional

[74]For $t = 0$, see Exercise 7.18 according to which $\mathcal{H}^0 = \rho$, the counting measure on $\mathbb{R}^m$.

volume ($t = m$, for any integer $m \geq 1$), and even, of counting measure (which corresponds to $t = 0$).

Theorem 7.53 (Equality of $\mathcal{H}^m$ and $\mathcal{L}^m$, Up to a Multiplicative Constant). *The m-dimensional Hausdorff measure $\mathcal{H}^m$ on Euclidean space $\mathbb{R}^m$ is proportional to the m-dimensional Lebesgue measure on $\mathbb{R}^m$. More specifically,*

$$\mathcal{H}^m = \beta_m \cdot \mathcal{L}^m, \tag{7.102}$$

where $\beta_m := \mathcal{H}^m(Q_0)$ is a positive and finite constant which depends only on the positive integer m and can be explicitly computed,[75] *and where $Q_0 := [0,1]^m$ is the m-dimensional unit cube in $\mathbb{R}^m$.*[76]

Proof. We will give two different proofs of this theorem. Namely, we will provide a classic one (the "first proof"), relying on the uniqueness, up to a positive multiplicative constant, of a Haar measure (i.e., a translation-invariant Radon measure) on an abelian (say) locally compact Hausdorff group. Moreover, we will give a direct, as well as possibly new, proof (the "second proof"), which makes an essential use of the above discussion concerning dyadic tessellations of $\mathbb{R}^m$, along with their applications (see, especially, Lemma 7.35 and Proposition 7.40), and relies on a suitable use of the hierarchy of the class of Borel sets via transfinite induction (Subsection 7.4.3).

First Proof. The proof of this theorem usually relies on a classic theorem according to which a nonzero Radon measure (i.e., a regular Borel measure which is finite on compact sets)[77] that is left translation invariant on $\mathbb{R}^m$ (or, more generally, which is left translation invariant on a locally compact and metrizable group G)[78] is unique, up to a finite and positive constant.[79] Here, the (additive) group $G := \mathbb{R}^m$ is abelian and hence, there is no need to distinguish between left and right translation-invariant Radon measures (i.e., Haar measures) in the present context. The translation invariance of Hausdorff measure $\mathcal{H}^m$ is precisely the content of part (i) of Proposition 7.31 above.

[75] See formula (7.150) in Theorem 7.66 below for the explicit value of β_m, for any integer $m \geq 0$.

[76] It is clear that if the proportionality constant exists, then it is uniquely determined. Indeed, since $\mathcal{L}^m(Q_0) = 1$, we must have (in light of (7.102) read from right to left) that it is given by $\beta_m = \beta_m \cdot \mathcal{L}^m(Q_0) = \mathcal{H}^m(Q_0)$.

[77] In Subsection 7.4.5, we will give the definition of Borel regularity (Definition 7.61), along with a proof of a key part of the regularity of $\mathcal{H}^m$ (and similarly for $\mathcal{L}^m$); see Theorem 7.62 and Exercise 7.64.

[78] It turns out that every open subset of a locally compact and metrizable topological space G is σ-compact, which is the case of $G := \mathbb{R}^m$ here (see Corollary 7.39 above). Then, a Borel measure on G is a Radon measure (and hence is regular) iff it is finite on all of the compact subsets of G. This result (see, e.g., **[193]**) does not require any underlying (topological) group structure on G.

[79] In Theorem 7.57 (and Corollary 7.58), we will give a direct proof of this theorem in the special case of the abelian group $\mathbb{R}^m$ of interest to us here, in the spirit of the second proof of Theorem 7.53 to be given below. Note that in the general case, the group need not be abelian; then, one works with either left or right Haar measures. Such a nonzero (left or right) translation-invariant Radon measure is called a (left or right) Haar measure on G.

An entirely similar argument would establish the Borel regularity (which also comes for free, according to footnote 77 on page 279) and the translation invariance, a well-known key fact in Lebesgue's integration theory on $\mathbb{R}^m$.

It then remains to verify that $\mathcal{H}^m(K) < \infty$, for every compact set K in $\mathbb{R}^m$. Since such a set K is contained in a large enough cube Q, and in light of the translation and rotation invariance as well as of the scale covariance of $\mathcal{H}^m$ (see Proposition 7.31 and Corollary 7.25 or, alternatively, Corollary 7.33 about the covariance of $\mathcal{H}^m$ under similarity transformations), this follows from the monotonicity of the measure $\mathcal{H}^m$ and the fact (established in part (i) of case (a) of Theorem 7.41) that $\mathcal{H}^m(Q_0) < \infty$, where, as before, Q_0 is the unit cube in $\mathbb{R}^m$.[80]

This concludes our first proof of the theorem (Theorem 7.53), which is the standard one found in the literature.

Second Proof. This is a direct proof of the theorem, not relying on any classic, but somewhat complicated, result concerning the uniqueness of Haar measures on locally compact (abelian) groups, except for the well-known (but perhaps somewhat forgotten nowadays) description of Borel sets, itself relying on transfinite induction and, in particular, on the Axiom of Choice or, rather, its equivalent form, the Well-Ordering Principle, which are frequently used in advanced analysis and functional analysis; see, e.g., [**193**, **521**, **522**, **524**] along with Subsection 7.4.3. It is apparently a new proof (due to the first author) of this theorem. At least, the authors could not find it in the literature. It can easily be adapted to prove related results, including the uniqueness of a Haar (i.e., a translation-invariant Radon) measure on $\mathbb{R}^m$ (or on another suitable topological group).

Step 1 (Outline of the Proof). *One first recalls that*

$$0 < \mathcal{H}^m(Q_0) < \infty, \text{ where } Q_0 := [0,1]^m. \tag{7.103}$$

This statement follows from parts (i) and (ii) of Theorem 7.41 above. It enables us to normalize $\mathcal{H}^m$ as follows, so that (like $\mathcal{L}^m$) it is a probability measure when restricted to Q_0; i.e., with the notation introduced just below $\mathcal{H}^m_{nor}(Q_0) = \mathcal{L}^m(Q_0) = 1$. More specifically, let the *normalized m-dimensional Hausdorff measure* $\mathcal{H}^m_{nor}$ be given by

$$\mathcal{H}^m_{nor} = \mathcal{H}^m(Q_0)^{-1} \cdot \mathcal{H}^m; \tag{7.104}$$

[80]The fact that $\mathcal{L}^m$ is finite on compact sets of $\mathbb{R}^m$ follows exactly in the same way (by using the translation and rotation invariance, along with the scale covariance of $\mathcal{L}^m$), except for the simplifying fact that, by construction of $\mathcal{L}^m$, we have $\mathcal{L}^m(Q_0) = \mathcal{L}^m([0,1]^m) = 1^m = 1$.

i.e.,

$$\mathcal{H}^m_{nor}(A) = \mathcal{H}^m(Q_0)^{-1} \cdot \mathcal{H}^m(A), \tag{7.105}$$

for any Borel set A in $\mathbb{R}^m$ (i.e., for any $A \in \mathcal{B}(\mathbb{R}^m)$).

In order to preserve the symmetry of our notation, we also let

$$\mathcal{L}^m_{nor} = \mathcal{L}^m; \tag{7.106}$$

indeed, since $\mathcal{L}^m(Q_0) = 1$, there is no need to further normalize $\mathcal{L}^m$.

Our goal is to show that

$$\mathcal{H}^m_{nor} = \mathcal{L}^m_{nor}; \tag{7.107}$$

i.e.,

$$\mathcal{H}^m_{nor}(A) = \mathcal{L}^m_{nor}(A), \text{ for any } A \in \mathcal{B}(\mathbb{R}^m). \tag{7.108}$$

We will first show that (7.108) holds for every (dyadic) cube of $\mathbb{R}^m$ (Step 2), then for any open set U in $\mathbb{R}^m$ (Step 3), and, finally, for every Borel set A in $\mathbb{R}^m$ (Step 4), as desired.

Step 2 (Arbitrary Cubes). *We have that*[81]

$$\mathcal{H}^m_{nor}(Q) = \mathcal{L}^m_{nor}(Q), \text{ for every cube } Q \text{ in } \mathbb{R}^m. \tag{7.109}$$

Since, by definition (see (7.105)), we have $\mathcal{H}^m_{nor}(Q_0) = \mathcal{L}^m_{nor}(Q_0)(= 1)$, the identity (7.109) follows at once from the similarity covariance of $\mathcal{H}^m$ (see Corollary 7.33) and also, as is well known and the reader can easily verify, of $\mathcal{L}^m$. Equivalently, it follows from the translation and rotation invariance as well as from the scale covariance of $\mathcal{H}^m$ (see Proposition 7.31) and their exact counterpart for $\mathcal{L}^m$.[82]

Step 3 (Arbitrary Open Sets). *We have that*

$$\mathcal{H}^m_{nor}(U) = \mathcal{L}^m_{nor}(U), \text{ for every open set } U \text{ in } \mathbb{R}^m. \tag{7.110}$$

Indeed, if U is an open subset of $\mathbb{R}^m$ and in light of Proposition 7.40, we have (with the notation of Lemma 7.35 and Proposition 7.40)

$$\mathcal{H}^m_{nor}(U) = \sum_{\alpha \in R(U)} \mathcal{H}^m_{nor}(Q_\alpha). \tag{7.111}$$

Similarly, in light of (7.106) and Proposition 7.40, we have that

$$\mathcal{L}^m_{nor}(U) = \sum_{\alpha \in R(U)} \mathcal{L}^m_{nor}(Q_\alpha). \tag{7.112}$$

[81] In the proof of Step 1 of Theorem 7.57, we will see that it suffices to establish (7.109) for all dyadic cubes, in which case *only the translation invariance of $\mathcal{H}^m$ and of $\mathcal{L}^m$ is needed.* It follows that in order to establish Step 3 of the present proof, we only need to use the translation invariance of $\mathcal{H}^m$ and $\mathcal{L}^m$.

[82] Actually, since in light of the proof of Steps 3 and 4 below, we only need to know that (7.109) holds for all dyadic cubes and since the faces of dyadic cubes are parallel to the coordinate hyperplanes, we would not need to invoke here the invariance under isometries (and, in particular, under rotations) of $\mathcal{H}^m$ and $\mathcal{L}^m$.

Recall that the index set $R(U)$ appearing in both (7.111) and (7.112) is countable, so that the corresponding sums are ordinary series of positive terms, which are either convergent or divergent, depending on whether $\mathcal{H}^m(U)$ or $\mathcal{L}^m(U)$ is finite or equal to $+\infty$, respectively.

Now, it suffices to apply the result of Step 2, specialized to dyadic cubes (see (7.109)) and according to which, in particular, $\mathcal{H}^m_{nor}(Q_\alpha) = \mathcal{L}^m_{nor}(Q_\alpha)$, for all $\alpha \in R(U)$, to deduce from (7.111) and (7.112) that $\mathcal{H}^m_{nor}(U) = \mathcal{L}^m_{nor}(U)$, as desired. This concludes the proof of Step 3.

Step 4 (Arbitrary Borel Sets—Borel Hierarchy and Transfinite Induction). *For all $A \in \mathcal{B}(\mathbb{R}^m)$, we have that $\mathcal{H}^m_{nor}(A) = \mathcal{L}^m_{nor}(A)$ or, equivalently (in light of* (7.104) *and* (7.105)*),*

$$\mathcal{H}^m(A) = \mathcal{H}^m(Q_0) \cdot \mathcal{L}^m(A), \tag{7.113}$$

as claimed in the statement of the theorem; see identity (7.102) *and the text following it.*

In order to establish (7.113), we use the classification of Borel sets described in Exercise 7.54 below. Accordingly, a Borel subset A of $\mathbb{R}^m$ is either an open set in $\mathbb{R}^m$ or a G_δ (i.e., a countable intersection of open sets of $\mathbb{R}^m$) or a $G_{\delta\sigma}$ (i.e., a countable union of G_δ sets) or else a $G_{\delta\sigma\delta}$ (i.e., a countable intersection of $G_{\delta\sigma}$ sets), and so on, but with one important caveat. Namely, one does not have to continue "only" countably many times but, instead, one must repeat the process uncountably many times, more specifically, up to (but strictly less than) ω, the first uncountable ordinal.

We stress that in order to properly define this ascending family of subsets of $\mathcal{B}(\mathbb{R}^m)$—namely, $G_\delta, G_{\delta\sigma}, G_{\delta\sigma\delta}, \ldots$, and well beyond—one must use *transfinite induction*, based on the *well-ordered set*

$$\Omega_1 := \{\kappa : \kappa \text{ ordinal}, \kappa < \omega\}. \tag{7.114}$$

For the nonexpert reader, let us mention that in carrying out such an induction, one must distinguish between "standard" ordinals (analogous to the integers in ordinary induction, based on $\mathbb{N}$, since each integer n has a successor, namely, $n + 1$) and the so-called *limit ordinals* (which do not have any counterpart among the integers); see, e.g., [**82, 83, 288, 561**].[83] But except for this technical difference, such a transfinite induction is carried out in much the same way as a standard (countable) induction.

[83] If τ is a limit ordinal in the above *well-ordered* set Ω_1 in (7.114), then, essentially by definition and by using the natural partial order of Ω_1, $\lambda := \sup\{\tau : \tau \text{ ordinal}, \tau < \lambda\}$, which is the key property making transfinite induction both relatively easy to carry out and different from an ordinary (countable) induction. Note that, by contrast, for any $n \in \mathbb{N}$, we have that $n + 1 \neq \sup\{k \in \mathbb{N} : k < n + 1\} = n$.

We point out that if we denote by $\mathcal{U}$, $\mathcal{G}_\delta$, $\mathcal{G}_{\delta\sigma}$, $\mathcal{G}_{\delta\sigma\delta}$, ... the class of open sets, G_δ sets, $G_{\delta\sigma}$ sets, $G_{\delta\sigma\delta}$ sets, ..., respectively, we clearly have (since $\emptyset, X \in \mathcal{U}$), the following family of inclusions:

$$\mathcal{U} \subseteq \mathcal{G}_\delta \subseteq \mathcal{G}_{\delta\sigma} \subseteq \mathcal{G}_{\delta\sigma\delta} \subseteq \cdots . \tag{7.115}$$

In addition, $\emptyset, X$ belong to each of these classes. Moreover, $\mathcal{U}$ is closed under (finite or) countable unions as well as under finite intersections and so are $\mathcal{G}_{\delta\sigma}$, $\mathcal{G}_{\delta\sigma\delta\sigma}$, etc., whereas $\mathcal{F}$ (the collection of all closed sets) is closed under (finite or) countable intersections as well as under finite unions and so are $\mathcal{G}_\delta$, $\mathcal{G}_{\delta\sigma\delta}$, $\mathcal{G}_{\delta\sigma\delta\sigma\delta}$, etc.

The above statement is obvious for $\mathcal{U}$, by definition. Furthermore, in the course of the proof of Case 2 and Case 3 below, respectively, we will show that $\mathcal{G}_\delta$ and $\mathcal{G}_{\delta\sigma\delta}$ satisfy the just-stated properties. The other classes are dealt with in exactly the same manner. In the sequel, it will be helpful to keep the above comments in mind.

Next, in agreement with the aforementioned classification of the Borel sets, we proceed to show that (7.113) holds for all $A \in \mathcal{U}$ (this is just Step 3 above), then for all $A \in \mathcal{G}_\delta$, next for all $A \in \mathcal{G}_{\delta\sigma}$, ..., and so on, uncountably many times, as explained below.

Upon an application of transfinite induction,[84] based on the ordinals strictly less than ω (the first uncountable ordinal; see (7.114)) and which we leave to the expert reader to carry out, one concludes that the identity (7.113) holds for all $A \in \mathcal{B}(\mathbb{R}^m)$, as desired.

As was already pointed out, the fact that (7.113) holds for all $A \in \mathcal{U}$ (i.e., for all open sets in $\mathbb{R}^m$) was precisely the content of Step 3 above; see (7.110).

Case 1 (*A is a G_δ Set in $\mathbb{R}^m$*): Let $A \in \mathcal{G}_\delta$; i.e., let A be a G_δ set of $\mathbb{R}^m$. Accordingly,

$$A = \bigcap_{n=1}^{\infty} A_n, \tag{7.116}$$

where $A_n \in \mathcal{U}$ (i.e., A_n is open in $\mathbb{R}^m$) for all integers $n \geq 1$.[85] First, note that if there exists $n_0 \in \mathbb{N}$ such that $\mathcal{H}^m_{nor}(A_n) < \infty$—and hence also $\mathcal{L}^m(A_n) < \infty$, since $\mathcal{H}^m_{nor}(A_n) = \mathcal{L}^m_{nor}(A_n)$, by Step 3—for all $n \geq n_0$, then it is easy to deduce the desired conclusion from Step 3.

[84] See, e.g., [**82, 83, 288, 309, 521, 523, 561**] for the notion of transfinite induction and for the Borel hierarchy of Borel sets; see also Exercises 7.54 and 7.55 in Subsection 7.4.3 below.

[85] Clearly, since X is open, we can restrict our attention to countable intersections in (7.116).

Indeed, in this case, by part (vi) of Proposition 3.29, applied successively to the positive measures $\mathcal{H}^m_{nor}$ and $\mathcal{L}^m_{nor}$, we have that[86]

$$\mathcal{H}^m_{nor}(A) = \lim_{n\to\infty} \mathcal{H}^m_{nor}(A_n) \quad \text{and} \tag{7.117}$$
$$\mathcal{L}^m_{nor}(A) = \lim_{n\to\infty} \mathcal{L}^m_{nor}(A_n).$$

However, since each A_n is open (i.e., $A_n \in \mathcal{U}$), we know that

$$\mathcal{H}^m_{nor}(A_n) = \mathcal{L}^m_{nor}(A_n), \text{ for all } n \geq 1. \tag{7.118}$$

Therefore, it clearly follows by combining (7.117) and (7.118) that $\mathcal{H}^m_{nor}(A) = \mathcal{L}^m_{nor}(A)$, as we wanted to show.

Next, we consider the case when $\mathcal{H}^m(A_n)$ (and hence also $\mathcal{L}^m(A_n)$) is equal to $+\infty$, for infinitely many integers $n \geq 1$. Then, we modify the above argument as follows:

By Lemma 7.35, applied to the open set $\mathbb{R}^m$, we have that

$$\mathbb{R}^m = \bigcup_{\alpha\in R(\mathbb{R}^m)} Q_\alpha, \tag{7.119}$$

a countable union of dyadic cubes. But since according to Step 1 of the proof of Proposition 7.40, the boundary of each dyadic cube Q_α is both $\mathcal{H}^m$-null and $\mathcal{L}^m$-null, we also have that, up to both an $\mathcal{H}^m$-null and $\mathcal{L}^m$-null set,

$$\mathbb{R}^m = \bigcup_{\alpha\in R(\mathbb{R}^m)} \mathring{Q}_\alpha, \tag{7.120}$$

the countable union of the family of open dyadic cubes $\{\mathring{Q}_\alpha\}_{\alpha\in R(\mathbb{R}^m)}$, in contrast to the exact equality (7.119), which holds set-theoretically and not just measure-theoretically. See also footnote 44 to Proposition 7.40 on page 263.

In the remainder of this argument, for notational simplicity, we identify the countable set $R(\mathbb{R}^m)$ with $\mathbb{N}$ and write accordingly $\{\mathring{Q}_j\}_{j=1}^\infty$ instead of $\{\mathring{Q}_\alpha\}_{\alpha\in R(\mathbb{R}^m)}$.

In light of (7.120), and with the above convention, we have that

$$A = A \cap \mathbb{R}^m = \bigcup_{j=1}^\infty (A \cap \mathring{Q}_j),$$

up to a set of both $\mathcal{H}^m$-measure zero and $\mathcal{L}^m$-measure zero. Hence, by part (v) of Proposition 3.29, applied successively to the positive measures $\mathcal{H}^m_{nor}$ and

[86] Note that, without loss of generality, we may assume that $\{A_n\}_{n=1}^\infty$ is nonincreasing. Indeed, it suffices to replace each $A_n \in \mathcal{U}$ by $\bigcap_{k=1}^n A_k \in \mathcal{U}$. Then, (7.116) still holds and $A_n \subseteq A_{n+1}$, for all $n \geq 1$. For simplicity, here, we do not change the notation for the new A_n's.

$\mathcal{L}^m_{nor}$, we obtain that[87]

$$\mathcal{H}^m_{nor}(A) = \mathcal{H}^m_{nor}\left(\bigcup_{j=1}^{\infty}(A \cap \mathring{Q}_j)\right) = \lim_{j\to\infty} \mathcal{H}^m_{nor}(A \cap \mathring{Q}_j) \tag{7.121}$$

and, similarly (since $\mathcal{L}^m_{nor} = \mathcal{L}^m$), that

$$\mathcal{L}^m_{nor}(A) = \mathcal{L}^m_{nor}\left(\bigcup_{j=1}^{\infty}(A \cap \mathring{Q}_j)\right) = \lim_{j\to\infty} \mathcal{L}^m_{nor}(A \cap \mathring{Q}_j). \tag{7.122}$$

Now, by (7.116), we have that for each fixed integer $j \geq 1$,

$$A \cap \mathring{Q}_j = \bigcap_{n=1}^{\infty}(A_n \cap \mathring{Q}_j).$$

Note that according to our current construction, $\mathcal{H}^m_{nor}(A_n \cap \mathring{Q}_j) < \infty$ and $\mathcal{L}^m_{nor}(A_n \cap \mathring{Q}_j) < \infty$, for every integer $n \geq 1$.[88] Hence, we can once again apply part (vi) of Proposition 3.29 to each of the positive measures $\mathcal{H}^m_{nor}$ and $\mathcal{L}^m_{nor}$ in order to obtain that, still for a fixed $j \geq 1$,[89]

$$\mathcal{H}^m_{nor}(A \cap \mathring{Q}_j) = \lim_{n\to\infty} \mathcal{H}^m_{nor}(A_n \cap \mathring{Q}_j) \tag{7.123}$$

and similarly

$$\mathcal{L}^m_{nor}(A \cap \mathring{Q}_j) = \lim_{n\to\infty} \mathcal{L}^m_{nor}(A_n \cap \mathring{Q}_j). \tag{7.124}$$

Since A_n and $\mathring{Q}_j$ are open subsets of $\mathbb{R}^m$, $A_n \cap \mathring{Q}_j$ is also open in $\mathbb{R}^m$, for every $n \geq 1$, and hence, it follows from Step 3 above (see (7.110)) that, for every $n \geq 1$ and still for a fixed $j \geq 1$,

$$\mathcal{H}^m_{nor}(A_n \cap \mathring{Q}_j) = \mathcal{L}^m_{nor}(A_n \cap \mathring{Q}_j). \tag{7.125}$$

Combining (7.123)–(7.125) and remembering that $j \geq 1$ was fixed but arbitrary, we deduce at once that $\mathcal{H}^m_{nor}(A \cap \mathring{Q}_j) = \mathcal{L}^m_{nor}(A \cap \mathring{Q}_j)$, for all $j \geq 1$.

Finally, we combine (7.121), (7.122), and the latter identity to deduce that $\mathcal{H}^m_{nor}(A) = \lim_{j\to\infty} \mathcal{H}^m_{nor}(A \cap \mathring{Q}_j) = \lim_{j\to\infty} \mathcal{L}^m_{nor}(A \cap \mathring{Q}_j) = \mathcal{L}^m_{nor}(A)$, as desired.

---

[87]Strictly speaking, in the remainder of the proof of this step, we should replace each $\mathring{Q}_j$ by $\mathring{\widetilde{Q}}_j$, where $\widetilde{Q}_j := \bigcup_{k=1}^{j} Q_k$—and hence also $\mathring{\widetilde{Q}}_j = \bigcup_{k=1}^{j} \mathring{Q}_k$—for all $j \geq 1$. Then, $\{\mathring{\widetilde{Q}}_j\}_{j=1}^{\infty}$ is nondecreasing (so that part (v) of Proposition 3.29 can be applied), $\mathcal{H}^m_{nor}(A \cap \mathring{Q}_j)$ and $\mathcal{L}^m_{nor}(A \cap \mathring{Q}_j)$ are both finite, for every $j \geq 1$, and $A = \bigcup_{j=1}^{\infty}(A \cap \mathring{\widetilde{Q}}_j)$. For notational simplicity, however, we continue to write $\mathring{Q}_j$ instead of $\mathring{\widetilde{Q}}_j$ throughout this argument. Observe that (just like $\mathring{Q}_j \in \mathcal{U}$) $\mathring{\widetilde{Q}}_j \in \mathcal{U}$, for all $j \geq 1$.

[88]Indeed, $A_n \cap \mathring{Q}_j \subseteq Q_j$ and, as we have already shown towards the beginning of the first proof of this theorem, $\mathcal{H}^m$ and $\mathcal{L}^m$ are finite on compact sets; so that $\mathcal{H}^m(A_n \cap \mathring{Q}_j) \leq \mathcal{H}^m(Q_j) < \infty$, and entirely similarly for $\mathcal{L}^m$.

[89]Recall from footnote 86 that, without loss of generality, $\{A_n\}_{n=1}^{\infty}$ may be assumed to be nonincreasing (so that part (vi) of Proposition 3.29 can be applied).

In summary, we have shown that

$$\mathcal{H}^m_{nor}(A) = \mathcal{L}^m_{nor}(A), \quad \text{for all } A \in \mathcal{G}_\delta. \tag{7.126}$$

This concludes the proof of the case when $A \in \mathcal{G}_\delta$ (Case 1), i.e., whenever A is a G_δ set in $\mathbb{R}^m$.

Case 2 (*A is a $G_{\delta\sigma}$ Set in $\mathbb{R}^m$*): The case when $A \in \mathcal{G}_{\delta\sigma}$ (i.e., A is a $G_{\delta\sigma}$ set in $\mathbb{R}^m$) is easier to handle than Case 1, where A was assumed to be a G_δ set. Indeed, let A be a $G_{\delta\sigma}$ set in $\mathbb{R}^m$. This means that

$$A = \bigcup_{n=1}^{\infty} B_n, \tag{7.127}$$

where $B_n \in \mathcal{G}_\delta$ (i.e., B_n is a G_δ set in $\mathbb{R}^m$), for every integer $n \geq 1$. We can therefore once again apply part (v) of Proposition 3.29 to the positive measures $\mathcal{H}^m_{nor}$ and $\mathcal{L}^m_{nor}$ in order to obtain that[90]

$$\mathcal{H}^m_{nor}(A) = \lim_{n\to\infty} \mathcal{H}^m_{nor}(B_n) \tag{7.128}$$

and similarly

$$\mathcal{L}^m_{nor}(A) = \lim_{n\to\infty} \mathcal{L}^m_{nor}(B_n).$$

But, by construction, B_n is a G_δ set for each integer $n \geq 1$. Hence, by Case 1 (see (7.126)), we know that $\mathcal{H}^m_{nor}(B_n) = \mathcal{L}^m_{nor}(B_n)$, for all $n \geq 1$.

In light of (7.127), we can now combine (7.128) and the latter identity to deduce that

$$\mathcal{H}^m(A) = \mathcal{L}^m(A), \quad \text{for all } A \in \mathcal{G}_{\delta\sigma}. \tag{7.129}$$

This concludes the proof of the case when $A \in \mathcal{G}_{\delta\sigma}$ (Case 2), i.e., when A is a $G_{\delta\sigma}$ set.

Case 3 (*A is a $G_{\delta\sigma\delta}$ Set in $\mathbb{R}^m$*): The case when $A \in \mathcal{G}_{\delta\sigma\delta}$ (i.e., when A is a $G_{\delta\sigma\delta}$ set or, equivalently, a countable intersection of $G_{\delta\sigma}$ sets in $\mathbb{R}^m$) is handled in much the same way as Case 1 where $A \in \mathcal{G}_\delta$. Indeed, we now have, by hypothesis,

$$A = \bigcap_{n=1}^{\infty} D_n, \tag{7.130}$$

where D_n belongs to $\mathcal{G}_{\delta\sigma}$, for every $n \in \mathbb{N}$.[91] We are not going to repeat entirely the argument used in Case 1, but we will go over most of it in order to help

[90] Much as in Case 1, we may assume without loss of generality that $\{B_n\}_{n=1}^{\infty}$ is nondecreasing, so that we can apply part (v) of Proposition 3.29, and (7.127) still holds, with $B_n \in \mathcal{G}_\sigma$, for all $n \geq 1$. Indeed, it suffices to replace each B_n by $\bigcup_{k=1}^{n} B_k$, for all $n \geq 1$, and to use the fact that $\mathcal{G}_\sigma$ is clearly closed under finite unions.

[91] Here, much as in footnote 86, we should replace each $D_n \in \mathcal{G}_{\delta\sigma}$ by $\bigcap_{k=1}^{n} D_k \in \mathcal{G}_{\delta\sigma}$, for all $n \geq 1$, so that $\{D_n\}_{n=1}^{\infty}$ is nonincreasing and part (vi) of Proposition 3.29 can be applied.

convince the reader. We simply outline the main points and highlight several new aspects of this situation.

Since $\mathbb{R}^m = \bigcup_{j=1}^{\infty} \mathring{Q}_j$, up to an $\mathcal{H}^m_{nor}$-null set and an $\mathcal{L}^m_{nor}$-null set (see (7.120) and the text following it), we have that $A = A \cap \mathbb{R}^m = \bigcup_{j=1}^{\infty}(A \cap \mathring{Q}_j)$.[92] Since, because each $\mathring{Q}_j$ is bounded, $\mathcal{H}^m_{nor}(A \cap \mathring{Q}_j) < \infty$ (resp., $\mathcal{L}^m_{nor}(A \cap \mathring{Q}_j) < \infty$), for all $j \geq 1$, we can apply part (v) of Proposition 3.29 in order to obtain that for each fixed integer $j \geq 1$,

$$\mathcal{H}^m_{nor}(A) = \lim_{j \to \infty} \mathcal{H}^m_{nor}(A \cap \mathring{Q}_j) \tag{7.131}$$

(resp., $\mathcal{L}^m_{nor}(A) = \lim_{j \to \infty} \mathcal{L}^m_{nor}(A \cap \mathring{Q}_j)$). But, in light of (7.130), we have that (by Proposition 3.29, part (v), and since, then, $A \cap \mathring{Q}_j = \bigcap_{n \geq 1}(D_n \cap \mathring{Q}_j)$)

$$\mathcal{H}^m_{nor}(A \cap \mathring{Q}_j) = \lim_{n \to \infty} \mathcal{H}^m_{nor}(D_n \cap \mathring{Q}_j) \tag{7.132}$$

(resp., $\mathcal{L}^m_{nor}(A \cap \mathring{Q}_j) = \lim_{n \to \infty} \mathcal{L}^m_{nor}(D_n \cap \mathring{Q}_j)$), for each fixed integer $j \geq 1$.

Now, we know from Case 2 above (see (7.129)) that

$$\mathcal{H}^m_{nor}(D_n \cap \mathring{Q}_j) = \mathcal{L}^m_{nor}(D_n \cap \mathring{Q}_j), \text{ for all } n \geq 1 \text{ and } j \geq 1. \tag{7.133}$$

Indeed, $D_n \in \mathcal{G}_{\delta\sigma}$ and $\mathring{Q}_j \in \mathcal{U}$ but, by (7.115), $\mathcal{U} \subseteq \mathcal{G}_\delta \subseteq \mathcal{G}_{\delta\sigma}$. Hence, $\mathring{Q}_j \in \mathcal{G}_{\delta\sigma}$ and so, $D_n \cap \mathring{Q}_j \in \mathcal{G}_{\delta\sigma}$, because $\mathcal{G}_{\delta\sigma}$ is closed under finite intersections.

In order to see why this last statement is true, consider $E, F \in \mathcal{G}_{\delta\sigma}$, with $E = \bigcup_{n \geq 1} E_n$ and $F = \bigcup_{n \geq 1} F_n$, where $E_n, F_n \in \mathcal{G}_\delta$ for all $n \in \mathbb{N}$. Now, clearly, we have that $E \cap F = \bigcup_{n,\ell \geq 1}(E_n \cap F_\ell)$. Note that $E_n \cap F_\ell \in \mathcal{G}_\delta$ since, by definition, $\mathcal{G}_\delta$ is closed under finite, and even countable, intersections. Since, by construction, $\mathcal{G}_{\delta\sigma}$ is closed under countable unions, it follows from the above identity that $E \cap F \in \mathcal{G}_{\delta\sigma}$.

Next, we can conclude as follows. In light of (7.133) and (7.132), we have that $\mathcal{H}^m_{nor}(A \cap \mathring{Q}_j) = \mathcal{L}^m_{nor}(A \cap \mathring{Q}_j)$, for all $j \geq 1$. Finally, in light of the latter equality, we deduce at once from (7.131) that $\mathcal{H}^m_{nor}(A) = \mathcal{L}^m_{nor}(A)$. Since $A \in \mathcal{G}_{\delta\sigma\delta}$ was arbitrary, the proof of Case 3 is now complete.

One then proceeds in exactly the same manner as for Case 1 (where $A \in \mathcal{G}_\delta$) or for Case 3 (where $A \in \mathcal{G}_{\delta\sigma\delta}$), and one finishes the proof by using transfinite

[92]Here, as in footnote 87, we should replace each $\mathring{Q}_j$ by $\mathring{\tilde{Q}}_j$, where $\mathring{\tilde{Q}}_j$ is defined exactly as in footnote 87, so that $\{A \cap \mathring{\tilde{Q}}_j\}_{j=1}^{\infty}$ is nondecreasing and part (v) of Proposition 3.29 can be applied.

induction, as was indicated earlier in Step 3 of this "second proof" of the theorem.[93] This finally concludes the proof of Step 4, as well as of the second proof of Theorem 7.53. □

7.4.3. Transfinite Induction and Borel Hierarchy.

In the next exercise, and in the follow-up exercise (Exercise 7.55), we provide further information, from different points of view, concerning transfinite induction and the Borel hierarchy. The reader interested in learning more deeply these beautiful subjects from classic set theory, analysis, and topology, may wish to consult the following references: [**82**], [**83**, especially pp. 299–300], [**288**], [**309**, Section 27, Chapter VIII], [**521, 523**], and [**561**, especially pp. 115–117].

Exercise 7.54 (*Borel Hierarchy Revisited, Transfinite Induction, and Completion of Step* 4)**.** The goal of this exercise is to describe a convenient way of completing the details of the proof of Step 4 in the second proof of Theorem 7.53 above (henceforth, abbreviated as "Step 4").[94]

As before, let $\Omega_1 := \{\lambda : \lambda < \omega_1\}$ denote the well-ordered set of countable ordinals, where, by definition, ω_1 is the first uncountable ordinal. Let $\mathcal{G}^{(0)} := \mathcal{U}$, the collection of all open sets in X, where $X := \mathbb{R}^m$. Then, given any $\lambda \in \Omega_1$, let

$$\mathcal{G}^{(\lambda)} := \bigcup_{\beta<\lambda} (\mathcal{G}^{(\beta)})_{\delta\sigma}, \tag{7.134}$$

where, much as in Step 4 and for all $\beta < \lambda$,

$$(\mathcal{G}^{(\beta)})_{\delta\sigma} := ((\mathcal{G}^{(\beta)})_{\delta})_{\sigma} \tag{7.135}$$

denotes the collection of all countable unions of countable intersections of elements of $\mathcal{G}^{(\beta)}$.

(i) (*Exhausting the Borel σ-Algebra*). Use transfinite induction on $\lambda \in \Omega_1$ in order to show, when $X := \mathbb{R}^m$ (or, more generally, when (X, d) is an arbitrary metric space), that $\mathcal{G}^{(\lambda)}$ is well defined for any $\lambda \in \Omega_1$ and that the family $\{\mathcal{G}^{(\lambda)}\}_{\lambda\in\Omega_1}$ is an ascending chain of subsets of $\mathcal{B}(X)$, the Borel σ-algebra of X:[95]

$$\mathcal{G}^{(\lambda)} \subseteq \mathcal{G}^{(\lambda')} \subseteq \mathcal{B}(X), \text{ whenever } \lambda < \lambda' < \omega_1. \tag{7.136}$$

---

[93] See Exercises 7.54 and 7.55 for a lot more information and guidance about the corresponding transfinite induction steps and about the associated exhaustion of $\mathcal{B}(\mathbb{R}^m)$ by a suitable nonincreasing family $\{\mathcal{G}^{(\lambda)}\}_{\lambda<\omega_1}$, where ω_1 is the first uncountable ordinal; in other words, the family $\{\mathcal{G}^{(\lambda)}\}_{\lambda<\omega_1}$ is indexed by the well-ordered set Ω_1 of all countable ordinals.

[94] Throughout this exercise, "Step 4" refers to Step 4 of the second proof of Theorem 7.53, which begins near (7.113).

[95] When $\lambda \in \Omega_1$ is a *successor ordinal* (i.e., when $\lambda = \beta + 1$, for some ordinal β necessarily in Ω_1), then we can replace the definition of $\mathcal{G}^{(\lambda)}$ given in (7.134) by $\mathcal{G}^{(\lambda)} := (\mathcal{G}^{(\beta)})_{\delta\sigma} := ((\mathcal{G}^{(\beta)})_{\delta})_{\sigma}$, with $\lambda := \beta + 1$. However, when $\lambda \in \Omega_1$ is a *limit ordinal* (i.e., when $\lambda = \sup\{\tau : \tau < \lambda\}$), then we must truly use (7.134) in order to define $\mathcal{G}^{(\lambda)}$. Note that if τ is an ordinal such that $\tau < \lambda$, then $\tau \in \Omega_1$.

Moreover, show, also by using transfinite induction on $\lambda \in \Omega_1$, that, still when $X := \mathbb{R}^m$ (or, more generally, when (X, d) is a suitable metric space), the family $\{\mathcal{G}^{(\lambda)}\}_{\lambda\in\Omega_1}$ exhausts the Borel σ-algebra $\mathcal{B}(X)$, in the following sense:[96]

$$\mathcal{B}(X) = \bigcup_{\lambda\in\Omega_1} \mathcal{G}^{(\lambda)}. \tag{7.137}$$

[*Hint*: Clearly, in light of (7.136), it suffices to show that any Borel set $A \subseteq X$ belongs to some $\mathcal{G}^{(\lambda)}$, with $\lambda < \omega_1$. In essence, this statement follows, in particular, from the fact that in a metric space, every open set is an F_σ set (i.e., a countable union of an increasing sequence of open sets in X), combined with the elementary fact that any countable union of countable sets is still countable and that, by definition, ω_1 is the *first uncountable ordinal*.[97] Also, observe that for any limit ordinal $\lambda \in \Omega_1$, $\mathcal{G}^{(\lambda)}$ is closed under complementation.]

(ii) (*Completion of the Proof of Step* 4). Let $X := \mathbb{R}^m$, for simplicity. Given $\lambda \in \Omega_1$, let $P(\lambda)$ be the property according to which, in the notation of Theorem 7.53 and of its second proof,

$$\mathcal{H}^m_{nor}(A) = \mathcal{L}^m_{nor}(A), \text{ for all } A \in \mathcal{G}^{(\lambda)}, \tag{7.138}$$

where $\mathcal{G}^{(\lambda)}$ is defined by (7.134).

Use Step 3 (identity (7.110)) of the second proof of Theorem 7.53 in order to begin the transfinite induction and hence to verify that $P(0)$ holds. (This statement follows at once from Step 3 and the definition of $\mathcal{G}^{(0)}$ in part (i), according to which $\mathcal{G}^{(0)} := \mathcal{U}$.) Then, use transfinite induction on $\lambda \in \Omega_1$ in order to show that the property $P(\lambda)$ (i.e., the identity (7.138)) holds for every $\lambda \in \Omega_1$. Deduce from the "Exhaustion Identity" (7.137) in part (i) that, still with $X := \mathbb{R}^m$, $\mathcal{H}^m_{nor}(A) = \mathcal{L}^m_{nor}(A)$, for all $A \in \mathcal{B}(X)$, which completes the proof of Step 4 as well as of the entire Theorem 7.53.

[*Hint*: Suitably adapt or combine the arguments involved in establishing Case 1 (near (7.116)), the case of G_δ sets, then Case 2 (near (7.127)), the case of $G_{\delta\sigma}$ sets, and finally, Case 3 (near (7.130)), the case of $G_{\delta\sigma\delta}$ sets. Also, review the text of the footnotes 95–97 associated with part (i).]

---

[96] A property $P(\lambda)$, with $\lambda \in \Omega_1$, say, can be established by transfinite induction as follows. Firstly, show that $P(0)$ holds, i.e., that it is true. Secondly, given any $\lambda \in \Omega_1$, assume that $P(\beta)$ holds for all $\beta < \lambda$ and then show that $P(\lambda)$ holds. Observe that $\beta < \lambda$ with $\lambda \in \Omega_1$ implies that $\beta \in \Omega_1$.

Alternatively, one can distinguish between the two types of ordinals: If $\lambda = \beta + 1 \in \Omega_1$ (necessarily with $\beta \in \Omega_1$) is a *successor ordinal*, then it suffices to check that $P(\beta + 1)$ holds whenever $P(\beta)$ holds, with $\beta \in \Omega_1$; this is akin, of course, to a standard induction. On the other hand, if λ is a *limit ordinal* (i.e., $\lambda = \sup\{\beta \in \Omega_1 : \beta < \lambda\}$), then we really have to check that $P(\lambda)$ holds whenever $P(\beta)$ does, with $\beta < \lambda$. (See also footnote 95 on page 288.)

[97] It is useful to recall that an ordinal λ can be viewed as a well-ordered set $(\lambda, \leq)$ such that for all $x \in \lambda$ (i.e., x is a proper subset of λ), we have that $x = \{y \in \lambda : y < x\}$. Alternatively, a set λ is well-ordered iff it is strictly well-ordered by set membership and every element of λ is itself (or can be identified with) a subset of λ. (These two alternative and equivalent definitions of an ordinal are due to John von Neumann.) As a consequence, Ω_1 (or, more generally, any set of ordinals) is totally ordered and well-ordered, and every subset of λ (with $\lambda \in \Omega_1$, say) is itself an ordinal. Also, every family of ordinals in the set Ω_1 has a supremum in Ω_1.

In the next exercise, we use some of the same notation (e.g., Ω_1, etc.) and definitions (concerning ordinals, etc.) as in Exercise 7.54 just above and as in Step 4 of the second proof of Theorem 7.53.

Exercise 7.55 (*The Borel Hierarchies, Continued*). Let (X, d) be a metric space. Furthermore, let $\Sigma^{(0)} := \mathcal{U}$ and $\Pi^{(0)} := \mathcal{F}$ denote, respectively, the collection of all open sets and that of all closed sets in X. Then, define simultaneously and by transfinite induction (on $\lambda \in \Omega_1$) two families of subsets of 2^X (and also, as we shall see in part (ii), of $\mathcal{B}(X)$):[98]

For every $\lambda \in \Omega_1$ (i.e., for every finite or countable ordinal), let[99]

$$\Sigma^{(\lambda)} := \left(\bigcap_{\beta<\lambda} \Pi^{(\beta)} \right)_\sigma \quad \text{and} \quad \Pi^{(\lambda)} := \left(\bigcup_{\beta<\lambda} \Sigma^{(\beta)} \right)_\delta . \tag{7.139}$$

Also, still for any $\lambda \in \Omega_1$, we let $\Delta^{(\lambda)} := \Sigma^{(\lambda)} \cap \Pi^{(\lambda)}$.

For any $\lambda \in \Omega_1$, the classes $\Sigma^{(\lambda)}$, $\Pi^{(\lambda)}$, and $\Delta^{(\lambda)}$ are called, respectively, the *additive, multiplicative*, and *ambiguous classes* of X.

(i) (*Exhaustion of the Borel Class*). Verify that the Borel σ-algebra $\mathcal{B}(X)$ can be written in any of the following ways:

$$\mathcal{B}(X) = \bigcup_{\lambda\in\Omega_1} \Sigma^{(\lambda)} = \bigcup_{\lambda\in\Omega_1} \Pi^{(\lambda)} = \bigcup_{\lambda\in\Omega_1} \Delta^{(\lambda)}, \tag{7.140}$$

where, as before, $\Omega_1 := \{\tau : \tau \text{ ordinal}, \tau < \omega_1\}$ is the set of (finite or) countable ordinals.[100]

(ii) Show that the additive, multiplicative, and ambiguous classes $\Sigma^{(\lambda)}$, $\Pi^{(\lambda)}$, and $\Delta^{(\lambda)}$ satisfy the inclusion relations described in Figure 7.3 and which the interested reader may wish to further specify when $\lambda \in \Omega_1$ is a limit ordinal.

Furthermore, still for $\lambda \in \Omega_1$, show that the additive classes $\Sigma^{(\lambda)}$ are closed under countable unions, while the multiplicative classes $\Pi^{(\lambda)}$ are closed under countable intersections.

Moreover, prove that, for every $\lambda \in \Omega_1$, the complements of the elements of $\Pi^{(\lambda)}$ are the elements of $\Sigma^{(\lambda)}$. Is the counterpart of this statement true if the roles of $\Sigma^{(\lambda)}$ and $\Pi^{(\lambda)}$ are interchanged? If not, what is a correct analogous statement, if any?

[98] In (7.139), much as in (7.134) and (7.135) above, $\mathcal{R}_\sigma$ (resp., $\mathcal{R}_\delta$) denotes the collection of all at most countable intersections (resp., all at most countable unions) of elements of $\mathcal{R}$, where $\mathcal{R} \subseteq 2^X$. Note that if X (resp., $\emptyset$) belongs to $\mathcal{R}$, then we may simply take countably infinite intersections (resp., countably infinite unions) in the definition of $\mathcal{R}_\sigma$ (resp., $\mathcal{R}_\delta$).

[99] If, for some $\beta \in \Omega_1$, $\lambda := \beta + 1$ is a successor ordinal, then we can equivalently let

$$\Sigma^{(\lambda)} := \left(\Pi^{(\beta)}\right)_\sigma \quad \text{and} \quad \Pi^{(\lambda)} := \left(\Sigma^{(\beta)}\right)_\delta .$$

[100] Here, Ω_1 starts at 0 instead of at 1, as is usually done. The notation adopted here in this exercise is more in keeping with the one used in Exercise 7.54 and in Step 4 of the second proof of Theorem 7.53.

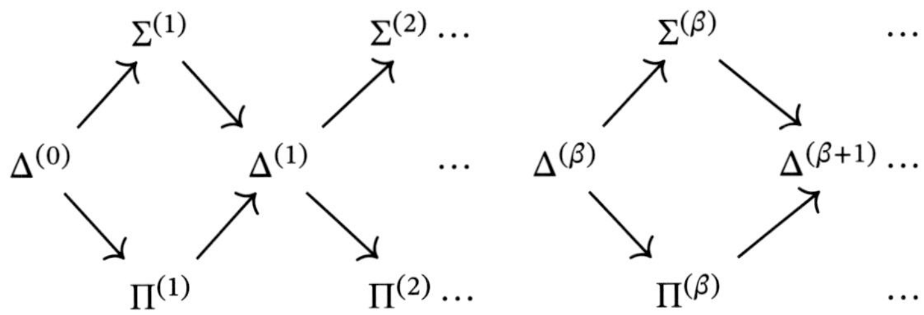

Figure 7.3. The Hierarchy Classes of Borel sets. Here, an arrow $A \to B$ means that $A \subseteq B$.

Finally, show that for every $\lambda \in \Omega_1$, the ambiguous class $\Delta^{(\lambda)}$ is an algebra over X, in the sense of measure theory; see Definition 3.2.

(iii) Show that $\Delta^{(0)}$ is the collection of all clopen (i.e., both closed and open) sets in X. Moreover, check that $\Sigma^{(1)} = \mathcal{F}_\sigma$ is the collection of all F_σ sets (i.e., the countable unions of closed sets in X), while $\Pi^{(1)} = \mathcal{G}_\delta$ is the collection of all G_δ sets (i.e., the countable intersections of open sets) in X.

(iv) Similarly, in the notation of Step 4 and of Exercise 7.54, show that $\Sigma^{(2)} = \mathcal{G}_{\delta\sigma} =: \mathcal{G}^{(1)}$ (resp., $\Pi^{(2)} = \mathcal{F}_{\sigma\delta}$, the collection of all countable intersections of elements of $\Sigma^{(1)} = \mathcal{F}_\sigma$), $\Sigma^{(3)} = \mathcal{F}_{\sigma\delta\sigma}$, $\Pi^{(3)} = \mathcal{G}_{\delta\sigma\delta}$, $\Sigma^{(4)} = \mathcal{F}_{\sigma\delta\sigma\delta}$, $\Pi^{(4)} = \mathcal{G}_{\delta\sigma\delta\sigma} =: \mathcal{G}^{(2)}$, and more generally, $\Pi^{(2n)} = \mathcal{G}^{(n)}$, for all integers $n \geq 0$, etc.

(v) Compare $\Pi^{(2\lambda)}$ and $\mathcal{G}^{(\lambda)}$, for any $\lambda \in \Omega_1$. Are they equal for all successor ordinals as well as for all limit ordinals and, hence, for all $\lambda \in \Omega_1$?

(vi) (*Existence of Non-Borel Sets*). Let $X := \mathbb{R}^m$. Show that for every $\lambda \in \Omega_1$, each class $\Sigma^{(\lambda)}$, $\Pi^{(\lambda)}$, or $\Delta^{(\lambda)}$ (in (7.139) and in the text following it, as well as the class $\mathcal{G}^{(\lambda)}$ (in (7.134) of Exercise 7.54) has at most the cardinality of the continuum, $c = 2^{\aleph_0}$. Deduce from (7.140) in part (i) that the Borel σ-algebra $\mathcal{B}(X)$ has cardinality at most $\aleph_1 \cdot 2^{\aleph_0} = \aleph_1 \cdot c = c$, because $\aleph_1 \leq c$, where $\aleph_1$ is the *first uncountable cardinal* (or *ordinal*).[101]

Try to show (or else admit) that $\mathcal{M} = \mathcal{M}(\mathbb{R}^m)$, the σ-algebra of all Lebesgue measurable sets on $\mathbb{R}^m$, has cardinality $2^c > c$. Deduce from this discussion that *there exist many Lebesgue measurable sets which are* not *Borel measurable.*

Finally, independently of the computation of the cardinality of $\mathcal{M}$, show that there exist many subsets of $\mathbb{R}^m$ which are not Borel measurable; indeed, the cardinality of $2^{\mathbb{R}^m}$ is 2^c, which, by the Cantor Diagonal Argument, is strictly greater than c.[102]

[101] In fact, it is not difficult to check that the cardinality of the Borel class $\mathcal{B}(X)$ is equal to c.

[102] Observe that we are implicitly using here the Axiom of Choice (or, equivalently, the Well-Ordering Principle) since the Exhaustion Identity (7.140) is proved by using transfinite induction. Actually, it can be shown that the Axiom of Choice is needed in order to prove the existence of subsets of $\mathbb{R}^m$ that are not Borel measurable—and hence, a fortiori, not Lebesgue measurable.

Exercise 7.56. As before, let $\mathscr{H}^m_{nor} = \mathscr{H}^m(Q_0)^{-1} \cdot \mathscr{H}^m$ denote the normalized m-dimensional Hausdorff measure, and let $\mathcal{L}^m_{nor} = \mathcal{L}^m$ denote the m-dimensional Lebesgue measure on $\mathbb{R}^m$; see (7.104) and (7.106).

(i) Investigate whether or not one can *directly* show that

$$\mathscr{H}^m_{nor}(F) = \mathcal{L}^m_{nor}(F), \text{ for all closed subsets } F \text{ of } \mathbb{R}^m.$$

(ii) If possible, use part (i) to give a simpler proof of Theorem 7.53, assuming Step 3 of the second proof of that theorem (see (7.110)), and arguing that $\{B \in \mathcal{B}(\mathbb{R}^m) : \mathscr{H}^m_{nor}(B) = \mathcal{L}^m_{nor}(B)\}$ is then a σ-algebra containing $\mathcal{U}$ (the class of all open sets in $\mathbb{R}^m$) and hence coincides with $\mathcal{B}(\mathbb{R}^m)$.

7.4.4. Uniqueness of Translation-Invariant Borel Measures. Next, we turn our attention to giving a direct proof of a variant, and a stronger form, of the uniqueness of a Haar measure (i.e., a nonzero translation-invariant Radon measure) on the abelian group $\mathbb{R}^m$, up to a multiplicative constant.

Theorem 7.57 (Uniqueness of Haar Measure on $\mathbb{R}^m$). *Let ν be a translation-invariant Borel measure on $\mathbb{R}^m$ such that $\nu(Q_0) = 1$, where Q_0 is the unit cube in $\mathbb{R}^m$. Then, ν coincides with the m-dimensional Lebesgue measure on $\mathbb{R}^m$: $\nu = \mathcal{L}^m$.*

Before providing the proof of Theorem 7.57, we state its most important corollary, from the perspective of this book.

Since $\mathscr{H}^m$ is translation invariant and $0 < \mathscr{H}^m(Q_0) < \infty$, the following result is an immediate consequence of Theorem 7.57, applied to the translation-invariant and *normalized* measure on $\mathbb{R}^m$ given by $\nu := \mathscr{H}^m_{nor}$.

Corollary 7.58 (Equality of Normalized m-Dimensional Hausdorff and Lebesgue Measures on $\mathbb{R}^m$). *With the notation used in the second proof of Theorem 7.53 (i.e., $\mathscr{H}^m_{nor} := \mathscr{H}^m(Q_0)^{-1} \cdot \mathscr{H}^m$ and $\mathcal{L}^m_{nor} := \mathcal{L}^m$), we have that $\mathscr{H}^m_{nor} = \mathcal{L}^m_{nor}$, from which Theorem 7.53 follows, with $\beta_m := \mathscr{H}^m(Q_0)$; namely, $\mathscr{H}^m = \beta_m \cdot \mathcal{L}^m$.*

The proof of Theorem 7.57 given below follows along the same lines as the second proof of Theorem 7.53 given above. Because of that, we only stress the new aspects of the proof, which also show that in the second proof of Theorem 7.53, we only needed to know that $\mathscr{H}^m$ and $\mathcal{L}^m$ were translation invariant. This proof seems to be new in the present context of $\mathbb{R}^m$.

Proof of Theorem 7.57. The main new observation is that in order to establish this theorem, it suffices to use the translation invariance of $\mathcal{L}^m$, as well as of ν, of course.

Step 1 (Discrete Scale Covariance). *For all dyadic cubes Q of $\mathbb{R}^m$, we have that*

$$\nu(Q) = \nu(\mathring{Q}) = \mathcal{L}^m(Q) = \mathcal{L}^m(\mathring{Q}) \quad \text{and} \tag{7.141}$$

$$\nu(\partial Q) = \mathcal{L}^m(\partial Q) = 0. \tag{7.142}$$

Indeed, since both ν and $\mathcal{L}^m$ are nondegenerate translation-invariant measures, this follows at once from Theorem 7.50, according to which, in particular, if Q is a cube of side lengths 2^{-n}, for some $n \in \mathbb{Z}$, then the discrete scale-covariance property of ν implies that $\nu(Q) = 2^{-nm} \cdot \nu(Q_0)$, and similarly for $\mathcal{L}^m$; compare (7.68) and (7.69) in Theorem 7.41. Recall that $\nu(Q_0) = 1$, by hypothesis, and that $\mathcal{L}^m(Q_0) = 1$, by construction.

The rest of this proof parallels the remainder of the proof of Theorem 7.53, except for the fact that we do not have to normalize ν since $\nu(Q_0) = \mathcal{L}^m(Q_0) = 1$.

Step 2 (Equality on Open Sets). *For every open set U in $\mathbb{R}^m$, we have that $\nu(U) = \mathcal{L}^m(U)$.*

This follows exactly as in Step 3 (see (7.110)) of the second proof of Theorem 7.53. Indeed, by Step 1 just above (see the first and the second equalities in (7.141)) and since $\nu(\partial Q) = 0$, for all dyadic cubes (see (7.142)), we deduce from Proposition 7.40 and its proof that we have the following disjoint union (recall that $R(U)$ is a countable set): $U = \bigsqcup_{\alpha \in R(U)} \mathring{Q}_\alpha$, up to a set of ν-measure (and $\mathcal{L}^m$-measure) zero. Hence, by the countable additivity of both ν and $\mathcal{L}^m$, we have that

$$\nu(U) = \sum_{\alpha \in R(U)} \nu(\mathring{Q}_\alpha) = \sum_{\alpha \in R(U)} \mathcal{L}^m(\mathring{Q}_\alpha) = \mathcal{L}(U),$$

as required. This completes the proof of Step 2.

Step 3 (Borel Hierarchy and Transfinite Induction). *For all $A \in \mathcal{B}(\mathbb{R}^m)$, we have that $\nu(A) = \mathcal{L}^m(A)$.*

This follows exactly as in the proof of Step 4 of the second proof of Theorem 7.53; see identity (7.113) from Step 4, along with the classification of Borel subsets of $\mathbb{R}^m$ via transfinite induction, as discussed in Exercise 7.54 above. It suffices to replace $\mathcal{H}^m_{nor}$ by ν and to recall that $\mathcal{L}^m_{nor} = \mathcal{L}^m$ in the aforementioned proof.

This concludes the proof of Theorem 7.57. □

In the next corollary (of Theorems 7.50 and 7.57), we refine the assumptions of Theorem 7.57 by no longer assuming that $\nu(Q_0) = 1$.

Corollary 7.59. *Let ν be a translation-invariant Borel measure which is nondegenerate.*[103] *Then, there exists a positive and finite constant $\alpha_{m,\nu}$, depending only*

[103] Recall that this means that ν is a nonzero measure and is not identically equal to $+\infty$.

on the integer m and on ν *such that (with* $\alpha_{m,\nu} := \nu(Q_0)$*)*

$$\nu = \alpha_{m,\nu} \cdot \mathcal{L}^m. \tag{7.143}$$

Proof. According to Theorem 7.50 (identity (7.100)), we know that $0 < \nu(Q_0) < \infty$. Therefore, we can normalize ν by letting $\nu_{nor} := \nu(Q_0)^{-1} \cdot \nu$. Then, ν_{nor} is a translation-invariant Borel measure such that $\nu_{nor}(Q_0) = 1$. Hence, by Theorem 7.57, applied to ν_{nor}, we conclude that $\nu_{nor} = \mathcal{L}^m$; i.e., (7.143) holds, with $\alpha_{m,\nu} = \nu(Q_0) \in (0, +\infty)$, as claimed. □

Corollary 7.60. *Under the assumptions of Corollary* 7.59, *the following statements are equivalent, for any* $E \in \mathcal{B}(\mathbb{R}^m)$:

(i) $\nu(E) > 0 \Leftrightarrow \mathcal{L}^m(E) > 0$;

(ii) $\nu(E) < \infty \Leftrightarrow \mathcal{L}^m(E) < \infty$; *and*

(iii) $0 < \nu(E) < \infty \Leftrightarrow 0 < \mathcal{L}^m(E) < \infty$.

Furthermore, in (iii), *if either one of the two equivalent conditions is satisfied, then* $\dim_H(E) = m$.

Note that this result applies, in particular, to $\nu := \mathcal{H}^m$ *on* $\mathbb{R}^m$.

Proof. This is an immediate consequence of Corollary 7.59 and the observation according to which if one of the two equivalent conditions of (iii) is satisfied, then $0 < \mathcal{L}^m(E) < \infty$ and hence, by Theorem 7.53, we also have that $0 < \mathcal{H}^m(E) < \infty$. Consequently, in light of the definition of the Hausdorff dimension, we must have that $\dim_H(E) = m$, as desired. □

7.4.5. Borel Regularity of the m-Dimensional Hausdorff and Lebesgue Measures $\mathcal{H}^m$ and $\mathcal{L}^m$. In this subsection, we define the notion of Borel regularity and establish at least part of the proof of the regularity of $\mathcal{H}^m$—or, equivalently, of $\mathcal{L}^m$, in light of Theorem 7.53.

Definition 7.61 (*Borel Regularity*)**.** Let ν be a Borel measure on a metric space X (or, more generally, on a topological space X). Then, ν is said to be *regular* (or *Borel regular*) if, for every open set U in X,

$$\nu(U) = \sup\{\nu(K) : K \subseteq U,\ K \text{ compact}\} \tag{7.144}$$

and, for every $A \in \mathcal{B}(\mathbb{R}^m)$,

$$\nu(A) = \inf\{\nu(U) : A \subseteq U,\ U \text{ open}\}. \tag{7.145}$$

More specifically, the property stated in (7.144) (resp., (7.145)) is referred to as *inner regularity* (resp., *outer regularity*).

We can now state the following result.

Theorem 7.62. *Let $X = \mathbb{R}^m$ be the m-dimensional Euclidean space. Then, the Hausdorff measure $\mathscr{H}^m$ (or, equivalently, the m-dimensional Lebesgu measure $\mathcal{L}^m$) is Borel regular.*

Furthermore, it is a Radon measure on $\mathbb{R}^m$; i.e., it is a regular Borel measure on $\mathbb{R}^m$ which is finite on every compact subset of $\mathbb{R}^m$.

Proof. The claim that $\mathscr{H}^m$ (or, equivalently, $\mathcal{L}^m$, by Theorem 7.53) is finite on every compact set $K \subseteq \mathbb{R}^m$ follows from the fact that K is contained in some large dyadic cube Q (of side lengths $2^{|n|}$, for some $n \in \mathbb{Z}$, say) and then (with $\nu := \mathscr{H}^m$, for example), $\nu(K) \leq \nu(Q) = 2^{|n|m} \cdot \nu(Q_0)$, where $\nu(Q_0) < \infty$. Note that in the last equality, we have used the (discrete) scale covariance of ν; see (7.68) in Theorem 7.41.

Next, we prove the inner regularity of $\mathscr{H}^m$; i.e., we show that (7.144) holds. This follows easily from Proposition 7.40. Indeed, in light of (7.67) of Proposition 7.40, we have that for any open set U in $\mathbb{R}^m$,

$$\mathscr{H}^m(U) = \sum_{\alpha \in R(U)} \mathscr{H}^m(Q_\alpha), \tag{7.146}$$

where $R(U)$ is a suitable countable index set.

Now, for notational simplicity, let us rewrite (7.146) in the following form: $\mathscr{H}^m(U) = \sum_{j=1}^{\infty} \mathscr{H}^m(Q^{(j)})$, where each $Q^{(j)}$ is a (compact) dyadic cube. Thus, clearly (by convention, we let $k := 0$ if $U = \emptyset$),

$$\mathscr{H}^m(U) = \sup\left\{\sum_{j=1}^{k} \mathscr{H}^m(Q^{(j)}) \,:\, k \in \mathbb{N}_0\right\}. \tag{7.147}$$

But since $K_k := \bigcup_{j=1}^{k} Q^{(j)}$ is a disjoint union, up to a set of $\mathscr{H}^m$-measure zero,[104] we can substitute $\mathscr{H}^m(K_k)$ for the finite sum appearing on the right-hand side of (7.147). Hence, we can approximate $\mathscr{H}^m(U)$ arbitrarily closely by $\mathscr{H}^m(K_k)$, for some large enough $k_0 \in \mathbb{N}_0$. Since K_{k_0} is compact, and since, clearly, $\mathscr{H}^m(K) \leq \mathscr{H}^m(U)$ whenever $K \subseteq U$, this implies that (7.144) in Definition 7.61 holds for every open subset U of $\mathbb{R}^m$.

We leave the proof of the outer regularity of $\mathscr{H}^m$ (see (7.145) in Definition 7.61) as an exercise for the interested reader; see Exercise 7.65 below.

In closing this proof, we note that the fact that $\mathscr{H}^m$ is a Radon measure now follows immediately from the fact that, in light of Theorem 7.53, $\mathscr{H}^m$ is also (like $\mathcal{L}^m$) a Borel regular measure which is finite on every compact set in $\mathbb{R}^m$. □

[104]Indeed, in light of part (b) of Theorem 7.41, we have that $\mathscr{H}^m(Q^{(j)}) = \mathscr{H}^m(\mathring{Q}^{(j)})$, for all $j \geq 1$.

The following open-ended exercise asks the reader to explore the issue of uniqueness of (normalized) Haar measure in a broader setting than the one considered thus far.

Exercise 7.63.

(i) Suitably extend and adapt the statements of Proposition 7.49 and Theorems 7.53 and 7.57 to a metrizable, locally compact topological group $X := G$ which is equipped with a translation-invariant metric d,[105] under appropriate additional hypotheses on G (e.g., G could be assumed to be separable and/or σ-compact), which the reader will determine.[106]

[*Advice*: First consider the easier and more familiar case when the group G is abelian.]

(ii) In particular, deduce from part (i) that under the hypotheses found in part (i) and if G is compact, G admits a unique (normalized) *left Haar measure* ν_H,[107] such that $\nu_H(G) = 1$; and similarly if "left" is replaced by "right".[108]

Exercise 7.64. Show that with $\nu := \mathcal{H}^m$ (or, equivalently, with $\nu := \mathcal{L}^m$), the inner regularity property (7.144) in Definition 7.61 is still valid for any Borel set $A \subseteq \mathbb{R}^m$.

[*Hint*: Recall from Theorem 7.62 that $\nu := \mathcal{H}^m$ is a Radon measure.]

Exercise 7.65. Provide a proof of the outer regularity of $\mathcal{H}^m$ (or, equivalently, of $\mathcal{L}^m$, by Theorem 7.53) which was omitted in the proof of Theorem 7.62 just above; i.e., show that (7.145) holds in Definition 7.61 (with $\nu := \mathcal{H}^m$).

[*Hint*: It suffices to show that $\nu(A) \geq \inf\{\nu(U) : A \subseteq U,\ U \text{ open}\}$, for any $A \in \mathcal{B}(\mathbb{R}^m)$. To see why this is true, use a method similar to the one leading to the proof of Lemma 7.35 and Proposition 7.40, but now also by considering the cubes which intersect ∂A, the boundary of A.]

7.4.6. Isodiametric Inequality and Exact Value of the Proportionality Constant Between $\mathcal{H}^m$ and $\mathcal{L}^m$. We close this discussion of the main properties of the m-dimensional Hausdorff measure on Euclidean space $\mathbb{R}^m$ by providing the specific value of the positive proportionality constant β_m in Theorem 7.53. We do not give a complete proof—which is somewhat complicated and relies on the Isodiametric Inequality (stated in (7.148) in footnote 109 of this chapter),[109] itself a consequence of the Brunn–Minkowski Inequality (7.89)

---

[105] In the abelian case, this means that $d(\boldsymbol{x}+\boldsymbol{z}, \boldsymbol{y}+\boldsymbol{z}) = d(\boldsymbol{x}, \boldsymbol{y})$, for all $\boldsymbol{x}, \boldsymbol{y}, \boldsymbol{z} \in G$. In the nonabelian case, one must distinguish between left invariant and right invariant metrics, in the obvious sense.

[106] By "separable", we mean that G contains a dense and at most countable subset.

[107] A *left Haar measure* ν on G is a Radon measure on G such that $\nu(g \cdot A) = \nu(A)$, for all $g \in G$ and $A \in \mathcal{B}(G)$, and similarly for a right Haar measure.

[108] Of course, it is well known that Theorem 7.57 (the existence and uniqueness of a left (resp., right) Haar measure, up to a positive multiplicative constant) holds for any locally compact topological group.

[109] Recall that this inequality states that for every Borel subset E of $\mathbb{R}^m$,

$$\mathcal{L}^m(E) \leq 2^{-m} \cdot \omega_m \cdot (\operatorname{diam}(E))^m. \tag{7.148}$$

from convex geometry. In essence, what is needed in order to determine the precise value of β_m is a suitable (geometric measure-theoretic) version of the Isoperimetric Inequality.[110] See Remark 7.46 above, along with [**182**, Section 3.2.3], [**183, 557**], or [**167**, especially Section 3.3].

We postpone for a little while the proof of Theorem 7.66, which will be self-contained except for the fact that we will assume without further justification the Isodiametric Inequality.

Theorem 7.66. *In Theorem 7.53, the positive and finite proportionality constant β_m is given, for every integer $m \geq 1$, by*[111]

$$\beta_m = 2^m \cdot \omega_m^{-1}, \tag{7.150}$$

where ω_m is the volume of the closed unit ball of $\mathbb{R}^m$.

Accordingly, the m-dimensional Hausdorff and Lebesgue measures on $\mathbb{R}^m$ are connected via the following precise identity:[112]

$$\mathcal{H}^m = 2^m \cdot \omega_m^{-1} \cdot \mathcal{L}^m. \tag{7.151}$$

We can now state the following useful corollary of Theorem 7.66, as well as of other results previously established in this chapter.

Corollary 7.67 (m-Dimensional Volume of Cubes and Balls in $\mathbb{R}^m$). *The following properties hold:*

(i)
$$\mathcal{H}^m(Q_0) = 2^m \cdot \omega_m^{-1},$$

where Q_0 is the closed (or, equivalently, is the open) unit cube in $\mathbb{R}^m$.

(ii)
$$\mathcal{H}^m(Q) = 2^{m(1-n)} \cdot \omega_m^{-1},$$

---

[110]The general geometric-theoretic version of the Isoperimetric Inequality can be stated as follows: Let Ω be a nonempty bounded open set in $\mathbb{R}^m$, with boundary $\partial\Omega$. Assume that $\partial\Omega$ is Minkowski measurable—and hence also necessarily of Minkowski dimension $m-1$—and with positive and finite Minkowski content (see Section 8.1) $\mathcal{M} = \mathcal{A}_{m-1}(\partial\Omega)$, a generalization of the $(m-1)$-dimensional volume of $\partial\Omega$, denoted $|\partial\Omega|_{m-1}$; note that when $\partial\Omega$ is smooth, say, C^1—or piecewise smooth—then $\mathcal{M} = |\partial\Omega|_{m-1}$. The *general Isoperimetric Inequality* states that

$$\mathcal{M} = \mathcal{A}_{m-1}(\partial\Omega) \geq m\omega_m^{1/m} \cdot \mathcal{L}^m(\Omega)^{(m-1)/m}, \tag{7.149}$$

with equality iff Ω is an open ball in $\mathbb{R}^m$.

[111]The specific, and well-known, value of ω_m is recalled in part (a) of Remark 7.68 below.

[112]This means, of course, that $\mathcal{H}^m(E) = 2^m \cdot \omega_m^{-1} \cdot \mathcal{L}^m(E)$, for any $E \subseteq \mathbb{R}^m$, provided $\mathcal{H}^m$ and $\mathcal{L}^m$ are both interpreted as the m-dimensional Hausdorff and Lebesgue outer measure, and for any $E \in \mathcal{B}(\mathbb{R}^m)$ (or, more generally, for any $E \in \mathcal{M}$, the σ-algebra of Lebesgue measurable sets in $\mathbb{R}^m$) provided $\mathcal{H}^m$ and $\mathcal{L}^m$ are both interpreted as the m-dimensional Hausdorff and Lebesgue measure on $(\mathbb{R}^m, \mathcal{B}(\mathbb{R}^m))$ (or on $(\mathbb{R}^m, \mathcal{M})$, respectively). Note that it follows from Theorem 7.53 and the relevant definitions that $\mathcal{M}$ can be viewed, indifferently, as the σ-algebra of Lebesgue or of Hausdorff measurable sets in $\mathbb{R}^m$.

if Q is any dyadic cube in $\mathbb{R}^m$ *of side lengths* 2^{-n}, *with* $n \in \mathbb{Z}$.[113]

(iii) $$\mathcal{H}^m(B_\mathbf{0}) = 2^m,$$

where $B_\mathbf{0} = \overline{B_{\mathbf{0}(1)}}$ *is the closed (or, equivalently, is the open) unit ball.*

(iv) $$\mathcal{H}^m(B_{\boldsymbol{x}_0,r}) = (2r)^m = d^m,$$

where $B_{\boldsymbol{x}_0,r} = \overline{B_{\boldsymbol{x}_0}(r)}$ *is any closed (or, equivalently, is any open) ball in* $\mathbb{R}^m$, *with center* $\boldsymbol{x}_0 \in \mathbb{R}^m$ *and radius* $r > 0$*—and, thus, with diameter* $d := \mathrm{diam}(B_r) = 2r$.

With the notation of (i)–(iv) *just above, we have that, for any* $n \in \mathbb{Z}$, $r > 0$, *and* $\boldsymbol{x}_0 \in \mathbb{R}^m$,

(v) $$\mathcal{H}^m(\partial Q_0) = \mathcal{H}^m(\partial Q) = \mathcal{H}^m(\partial B_\mathbf{0}) = \mathcal{H}^m(\partial(B_{\boldsymbol{x}_0,r})) = 0.$$

Proof.

(i) follows at once from (7.151) since $\mathcal{L}^m(Q_0) = 1$.

(ii) follows from (i) and the discrete scale-covariance property of $\mathcal{H}^m$ and $\mathcal{L}^m$ (see Theorem 7.41, formulas (7.68) and (7.69)), according to which $\mathcal{H}^m(Q) = 2^{-nm} \cdot \mathcal{H}^m(Q_0)$, and hence, as desired, we have that $\mathcal{H}^m(Q) = 2^{-nm} \cdot 2^m \cdot \omega_m^{-1} = 2^{m(1-n)} \cdot \omega_m^{-1}$.

(iii) follows from the definition of ω_m and the identity (7.151). Indeed, as required, we have that $\mathcal{H}^m(B_\mathbf{0}) = 2^m \cdot \omega_m^{-1} \cdot \mathcal{L}^m(B_\mathbf{0}) = 2^m \cdot \omega_m^{-1} \cdot \omega_m = 2^m$.

(iv) follows from (iii), along with the translation-invariance and the (general) scale-covariance property of $\mathcal{H}^m$ (see parts (i) and (ii) of Proposition 5.11, respectively), according to which (since $B_{\boldsymbol{x}_0,r} = \boldsymbol{x}_0 + rB_\mathbf{0}$), we have that $\mathcal{H}^m(B_{\boldsymbol{x}_0,r}) = \mathcal{H}^m(B_{\mathbf{0},r}) = r^m \cdot \mathcal{H}^m(B_\mathbf{0}) = (2r)^m$, as claimed.

(v) We know from part (b) of Theorem 7.41 that $\mathcal{H}^m(\partial Q) = \mathcal{L}^m(\partial Q) = 0$. Since $B_{\boldsymbol{x}_0,r} = \boldsymbol{x}_0 + r \cdot B_{\mathbf{0},1}$ and hence also $\partial(B_{\boldsymbol{x}_0,r}) = \boldsymbol{x}_0 + r \cdot \partial(B_{\mathbf{0},1})$, we deduce from the translation invariance and the (general) scale covariance of $\mathcal{H}^m$ (parts (i) and (ii) of Proposition 5.11) that (with $B_{\mathbf{0},1} := \overline{B_\mathbf{0}(1)}$) we have $\mathcal{H}^m(\partial(B_{\boldsymbol{x}_0,r})) = \mathcal{H}^m(\partial(B_{\mathbf{0},r})) = r^m \cdot \mathcal{H}^m(\partial(B_{\mathbf{0},1}))$, where $\partial(B_{\mathbf{0},1}) = S_\mathbf{0}^{m-1}$, the $(m-1)$-dimensional sphere in $\mathbb{R}^m$.

We leave it to the reader to verify that[114] $0 < \mathcal{H}^{m-1}(S_\mathbf{0}^{m-1}) < \infty$. Hence, as expected, we deduce that the $(m-1)$-dimensional sphere $S_\mathbf{0}^{m-1}$ has Hausdorff

[113] More generally, by using the translation and (general) scale-covariance properties of $\mathcal{H}^m$ (see parts (i) and (ii) of Proposition 7.31), one could also, of course, obtain the value of $\mathcal{H}^m(Q_{\boldsymbol{x}_0,\eta})$, where $Q_{\boldsymbol{x}_0,\eta}$ is an arbitrary cube of center $\boldsymbol{x}_0 \in \mathbb{R}^m$ and side lengths $\eta > 0$. Accordingly, we would find that (independently of $\boldsymbol{x}_0 \in \mathbb{R}^m$) $\mathcal{H}^m(Q_{\boldsymbol{x}_0,\eta}) = (2\eta)^m \cdot \omega_m^{-1}$.

[114] As is well known, a simple computation—based on the Change of Variables Formula—shows that the $(m-1)$-dimensional volume of $S_\mathbf{0}^{m-1}$ is given by

(7.152) $$\mathrm{vol}^{m-1}(S_\mathbf{0}^{m-1}) = m \cdot \mathrm{vol}^m(B_\mathbf{0}) = m\omega_m = 2\pi^{m/2}/\Gamma(m/2),$$

where we have used (7.154) in part (a) of Remark 7.68, along with the functional equation for Γ according to which $\Gamma(1 + (m/2)) = (m/2)\Gamma(m/2)$.

dimension equal to $m-1$:

$$\dim_H(S_{\mathbf{0}}^{m-1}) = \dim_H(\partial(B_{\mathbf{0},1})) = m-1. \tag{7.153}$$

It follows from the definition of $\dim_H$ that (since $m > m-1 = \dim_H(S_{\mathbf{0}}^{m-1})$) $\mathcal{H}^m(S_{\mathbf{0}}^{m-1}) = \mathcal{H}^m(\partial(B_{\mathbf{0},1})) = 0$. Therefore, by combining (7.153) and the latter identity, we obtain, as desired, that $\mathcal{H}^m(\partial(B_{\mathbf{x}_0,r})) = 0$.[115]

Finally, we note that in (i)–(iv), the claim that the given cubes or balls could be chosen to be closed or open, indifferently, follows from the fact that the boundaries of the cubes (or balls) have m-dimensional Hausdorff measure equal to zero. This completes the proof of the corollary. □

Remark 7.68.

(a) (*Value of ω_m, the Volume of the Unit Ball in $\mathbb{R}^m$*). As is well known, the Change of Variables Formula in $\mathbb{R}^m$ combined with the use of spherical coordinates yields the following specific value of $\omega_m := \mathrm{vol}^m(\overline{B_{\mathbf{0}}(1)})$, the volume of the closed unit ball in $\mathbb{R}^m$:

$$\omega_m = \pi^{m/2}/\Gamma(1+(m/2)) = \pi^{m/2}/(m/2)!\,, \tag{7.154}$$

where Γ denotes the classic gamma function; i.e.,

$$\omega_m = \pi^k/k!\,, \text{ if } m = 2k \quad \text{is even} \quad (k \geq 1), \text{ and} \tag{7.155}$$

$$\omega_m = \pi^k \cdot k!\,/(2k+1)!\,, \text{ if } m = 2k+1 \quad \text{is odd} \quad (k \geq 0). \tag{7.156}$$

(b) (*Normalization of Hausdorff Measure $\mathcal{H}^t$ on $\mathbb{R}^m$*). In light of Theorem 7.66, footnote 112, and of part (a) of the present remark (see, especially, (7.154)), it is natural to normalize $\mathcal{H}^t$, the t-dimensional Hausdorff measure on $\mathbb{R}^m$, in the following way, for any $t \in \mathbb{R}$, $t \geq 0$:[116]

$$\mathcal{H}^t_{nor} := 2^{-t} \cdot \omega_t \cdot \mathcal{H}^t, \tag{7.157}$$

where, by mimicking (7.154) with the integer $m \geq 1$ replaced by the real number $t \geq 0$,[117]

$$\omega_t := \pi^{t/2}/\Gamma(1+(t/2)) = 2\pi^{t/2}/t\Gamma(t/2) = 2\pi^{t/2}/(t/2)!\,. \tag{7.158}$$

Of course, by design, when $m \geq 1$ is an integer, we recover the fact, used earlier in this chapter, that the normalized Hausdorff measure $\mathcal{H}^m_{nor}$ (i.e., the probability measure associated with $\mathcal{H}^m$) coincides with Lebesgue measure—and hence also with the m-dimensional volume—on $\mathbb{R}^m$:

$$\mathcal{H}^m_{nor} = \mathcal{L}^m = \mathrm{vol}^m(\cdot). \tag{7.159}$$

---

[115]An alternative, and perhaps simpler, argument would consist in suitably adapting and refining our earlier proof of the fact that $\mathcal{H}^m(\partial Q_0) = 0$; see the proof of part (iii) of Theorem 7.41.

[116]See, e.g., **[172, 182, 430, 557]** and also **[352, 353]**, where a similar normalization was used for the Minkowski content.

[117]Note that according to Exercise 7.18, we have that $\mathcal{H}^0_{nor} := \mathcal{H}^0 = \rho$, the counting measure on $\mathbb{R}^m$, because $\omega_0 = 1$, in light of the first equality in (7.158). Note that in the second equality of (7.158), we use this natural convention.

Proof of Theorem 7.66. As was mentioned earlier, we will assume without proof the Isodiametric Inequality,[118]

$$\mathcal{L}^m(A) \leq \omega_m \cdot (\operatorname{diam}(A)/2)^m = \mathcal{L}^m(\overline{B_{\mathbf{0}}(r(A))}), \text{ for all } A \in \mathcal{B}(\mathbb{R}^m) \tag{7.160}$$

(with $r(A) := \operatorname{diam}(A)/2$), which asserts geometrically that among all of the Borel subsets of $\mathbb{R}^m$ with the same diameter $d := \operatorname{diam}(A)$, a ball B of diameter d (i.e., of radius $r := d/2$) has the largest volume (given by the right-hand side of the inequality (7.160), with $A := B$).

Step 1 (Isodiametric Equality for Balls). *The Isodiametric Inequality becomes an equality for any (closed or open) ball $B := B_{\boldsymbol{x}_0}(r)$ of arbitrary center $\boldsymbol{x}_0 \in \mathbb{R}^m$ and radius $r > 0$; more specifically,*

$$\mathcal{L}^m(B) = \operatorname{vol}^m(B) = \omega_m \cdot r^m = 2^{-m} \cdot \omega_m \cdot (\operatorname{diam}(B))^m \tag{7.161}$$

or, rather (with $B_{\mathbf{0}} := \overline{B_{\mathbf{0}}(1)}$),

$$\mathcal{L}^m(B) = r^m \cdot \operatorname{vol}^m(B_{\mathbf{0}}) = 2^{-m} \cdot \omega_m \cdot (\operatorname{diam}(B))^m. \tag{7.162}$$

Indeed, this well-known fact follows at once from part (a) of Remark 7.68 and, hence, ultimately, from the Change of Variables Formula.

Note, however, that for the purpose of the present proof of Step 1 of Theorem 7.66, all we need to know is that (7.162) holds, with $\omega_m := \operatorname{vol}^m(B_0) = \mathcal{L}^m(B_0)$; in particular, we do not care about the specific value of ω_m given by (7.154)–(7.156) in part (a) of Remark 7.68.

Indeed, (7.162) follows immediately from the translation invariance and scale covariance of $\mathcal{L}^m$ (see parts (i) and (ii) of Proposition 3.70) according to which (since $B_{\boldsymbol{x}_0,r} = \boldsymbol{x}_0 + rB_{\mathbf{0}}$)

$$\begin{aligned}\mathcal{L}^m(B) = \mathcal{L}^m(B_{\boldsymbol{x}_0,r}) &= \mathcal{L}^m(\overline{B_{\boldsymbol{x}_0}(r)}) = \mathcal{L}^m(rB_{\mathbf{0}}) \\ &= r^m \cdot \mathcal{L}^m(B_{\mathbf{0}}) = 2^{-m} \cdot \omega_m \cdot (\operatorname{diam}(B))^m.\end{aligned}$$

This concludes the easy proof of Step 1.

The rest of the proof is divided into two steps (Steps 2 and 3), each corresponding to one of the inequalities associated with the main claim of Theorem 7.66, namely, the equality (7.151), which can be equivalently written as follows: $\mathcal{L}^m = 2^{-m} \cdot \omega_m \cdot \mathcal{H}^m$.

Step 2. *We claim that*

$$\mathcal{L}^m(A) \leq 2^{-m} \cdot \omega_m \cdot \mathcal{H}^m(A), \textit{ for all } A \in \mathcal{B}(\mathbb{R}^m). \tag{7.163}$$

[118]See [**167**, Theorem 2.4, p. 89] for a proof of this inequality.

Indeed, let us fix $A \in \mathcal{B}(\mathbb{R}^m)$ and show that for any $\delta > 0$,

$$\mathcal{L}^m(A) \leq 2^{-m} \cdot \omega_m \cdot \mathcal{H}^m_\delta(A), \tag{7.164}$$

from which the inequality (7.163) follows at once by letting $\delta \to 0^+$.

Now, let us fix $\delta > 0$ and consider an arbitrary countable δ-cover $\{A_j\}_{j=1}^\infty$ of A, so that $A \subseteq \bigcup_{j=1}^\infty A_j$ and $\mathrm{diam}(A_j) \leq \delta$, for all $j \geq 1$. Then, by the monotonicity and countable subadditivity of $\mathcal{L}^m$ (used to justify the first equality of (7.165)) combined with the Isodiametric Inequality (7.160) applied to each A_j, with $j \geq 1$ (in order to deduce the second inequality of (7.165)), we obtain that

$$\begin{aligned}\mathcal{L}^m(A) \leq \sum_{j=1}^\infty \mathcal{L}^m(A_j) &\leq \sum_{j=1}^\infty \left(2^{-m} \cdot \omega_m \cdot (\mathrm{diam}(A_j))^m\right) \\ &= 2^{-m} \cdot \omega_m \cdot \left(\sum_{j=1}^\infty (\mathrm{diam}(A_j))^m\right).\end{aligned} \tag{7.165}$$

By taking the infimum of the right-hand side of (7.165) over all δ-covers $\{A_j\}_{j=1}^\infty$ of A, we obtain the desired inequality (7.164), from which (7.163) follows (by then letting $\delta \to 0^+$), as was already noted. This completes the proof of Step 2.

Step 3. *We claim that the following reverse inequality of* (7.163) *holds*:

$$\mathcal{L}^m(A) \geq 2^{-m} \cdot \omega_m \cdot \mathcal{H}^m(A), \text{ for all } A \in \mathcal{B}(\mathbb{R}^m). \tag{7.166}$$

Indeed, we may assume that $\mathcal{L}^m(A) < \infty$, because the inequality (7.166) is obvious when $\mathcal{L}^m(A) = +\infty$.[119] Let us fix $\varepsilon > 0$. Now, by definition of $\mathcal{L}^m(A)$ as an infimum (see Section 3.4, Definition 3.65), given any $\delta > 0$, there exists a δ-cover $\{A_j\}_{j=1}^\infty$ of A by rectangles (i.e., parallelepipeds of $\mathbb{R}^m$) A_j of $\mathbb{R}^m$ of diameter not exceeding $\delta > 0$ and such that

$$\sum_{j=1}^\infty \mathrm{vol}^m(A_j) = \sum_{j=1}^\infty \mathcal{L}^m(A_j) \leq \mathcal{L}^m(A) + \varepsilon. \tag{7.167}$$

In light of Exercise 7.69 below, each rectangle A_j can be written as the disjoint union of a countable family $\{B_{j,n}\}_{n=1}^\infty$ of (disjoint, open) balls, up to an $\mathcal{L}^m$-null—and hence also up to an $\mathcal{H}^m$-null— set (since by Theorem 7.53, $\mathcal{H}^m = \beta_m \cdot \mathcal{L}^m$, with $0 < \beta_m < \infty$).[120] Therefore, since a countable union of $\mathcal{L}^m$-null sets is $\mathcal{L}^m$-null, and similarly with $\mathcal{H}^m$ instead of $\mathcal{L}^m$, we have that,

[119] Although it is not needed here, note that it then follows from Theorem 7.53 that $\mathcal{H}^m(A) = \beta_m \cdot \mathcal{L}^m(A) = +\infty$, since $\beta_m > 0$.

[120] All we would need to know here is that $\mathcal{L}^m \geq d_m \cdot \mathcal{H}^m$, for some finite constant $d_m > 0$, an inequality for which the reader is asked to find a direct, self-contained proof in Exercise 7.70 below, via several helpful hints.

up to an $\mathcal{L}^m$-null (and hence also $\mathcal{H}^m$-null) subset of $\mathbb{R}^m$,

$$\bigsqcup_{j,n\geq 1} B_{j,n} = \bigcup_{j\geq 1} A_j \supseteq A. \tag{7.168}$$

Note that in light of (7.168) and the text surrounding it, $\{B_{j,n}\}_{j,n\geq 1}$ is a δ-cover of A, up to an $\mathcal{L}^m$-null (and hence also up to an $\mathcal{H}^m$-null) Borel set.[121] Thus, a moment's reflection shows that—since this implies that $\sum_{j=1}^{\infty} \mathcal{L}^m(A_j) = \sum_{j,n=1}^{\infty} \mathcal{L}^m(B_{j,n})$—(7.167) combined with Step 1 yields successively

$$\begin{aligned} \mathcal{L}^m(A) + \varepsilon \geq \sum_{j,n=1}^{\infty} \mathcal{L}^m(B_{j,n}) &= \sum_{j,n=1}^{\infty} \big(2^{-m} \cdot \omega_m \cdot (\mathrm{diam}(B_{j,n}))^m\big) \\ &= 2^{-m} \cdot \omega_m \cdot \left(\sum_{j,n=1}^{\infty} (\mathrm{diam}(B_{j,n}))^m \right), \end{aligned} \tag{7.169}$$

where in the first equality of (7.169), we have used the translation invariance and scaling covariance of $\mathcal{L}^m$ applied to each ball $B_{j,n}$, with $j, n \in \mathbb{N}^*$; accordingly, $\mathcal{L}^m(B_{j,n}) = 2^{-m} \cdot \omega_m \cdot (\mathrm{diam}(B_{j,n}))^m$, for all $j, n \geq 1$.

Since $\{B_{j,n}\}_{j,n=1}^{\infty}$ is a δ-cover of A, up to an $\mathcal{H}^m$-null Borel set (and hence also up to an $\mathcal{H}_\delta^m$-null set, because $0 \leq \mathcal{H}_\delta^m \leq \mathcal{H}^m$), a further moment's reflection shows that (by definition of $\mathcal{H}_\delta^m(A)$ as an infimum taken over all δ-covers of A) $\mathcal{H}_\delta^m(A) \leq \sum_{j,n=1}^{\infty} (\mathrm{diam}(B_{j,n}))^m$.

By combining the latter inequality (multiplied through by $2^{-m} \cdot \omega_m$) and (7.169), we obtain the following inequality: $\mathcal{L}^m(A) + \varepsilon \geq 2^{-m} \cdot \omega_m \cdot \mathcal{H}_\delta^m(A)$. Since $\delta > 0$ is arbitrary, we finally deduce from this last inequality by letting $\delta \to 0^+$ that $\mathcal{L}^m(A) + \varepsilon \geq 2^{-m} \cdot \omega_m \cdot \mathcal{H}^m(A)$.

Since $\varepsilon > 0$ is arbitrary, we obtain by letting $\varepsilon \to 0^+$ in the latter inequality that $\mathcal{L}^m(A) \geq 2^{-m} \cdot \omega_m \cdot \mathcal{H}^m(A)$, which is precisely the desired inequality (7.166) and concludes the proof of Step 3 as well as of Theorem 7.66, in light of the reverse inequality (7.164) obtained in Step 2 above. □

The result of the following "exercise" was used in a crucial way in the proof of Step 3 of Theorem 7.66; see, e.g., (7.168) and the discussion surrounding it.

Exercise 7.69. Show that each rectangle (i.e., parallelepiped) R in $\mathbb{R}^m$ can be written as a disjoint union of a countable family of (open, disjoint) balls in $\mathbb{R}^m$, up to a (Borel) set of m-dimensional Lebesgue measure zero (i.e., up to an $\mathcal{L}^m$-null set).

The result of the next exercise can be used as a simple and self-contained substitute for the use of the more sophisticated Theorem 7.53 in the proof of Step 3 of Theorem 7.66; see footnote 120 on page 301.

[121]See footnote 120, along with the discussion surrounding the associated footnote indicator.

Exercise 7.70. The goal of this exercise is to give a direct proof of the fact that there exists a finite and positive constant d_m, depending only on $m \in \mathbb{N}$, such that

$$\mathcal{L}^m(A) \geq d_m \cdot \mathcal{H}^m(A), \quad \text{for all } A \in \mathcal{B}(\mathbb{R}^m). \tag{7.170}$$

Observe that this inequality implies that an $\mathcal{L}^m$-null (Borel) set is also $\mathcal{H}^m$-null.

(i) Assume that $\mathcal{L}^m(A) < \infty$. Fix $\varepsilon > 0$. By covering A by a sequence of congruent cubes (e.g., dyadic cubes of the same generation) $\{A_j\}_{j=1}^\infty$, show that $\mathcal{L}^m(A) + \varepsilon \geq m^{-m/2} \cdot \mathcal{H}^m_\delta(A)$, where δ is the common side length of the cubes A_j.

[*Hint*: Note that $\mathrm{vol}^m(A_j) = \mathcal{L}^m(A_j) = \delta^m = \big(\mathrm{diam}(A_j)/\sqrt{m}\big)^m$.]

(ii) First, let $\delta \to 0^+$ in order to obtain the counterpart of the inequality obtained in part (i) just above, with $\mathcal{H}^m(A)$ substituted for $\mathcal{H}^m_\delta(A)$. Then, since $\varepsilon > 0$ is arbitrary, deduce that the following inequality holds: $\mathcal{L}^m(A) \geq m^{-m/2} \cdot \mathcal{H}^m(A)$, for all $A \in \mathcal{B}(\mathbb{R}^m)$ with $\mathcal{L}^m(A) < \infty$.

(iii) Use the σ-finiteness of $\mathcal{L}^m$ (which is very easy to verify directly) in order to prove that the inequality obtained in part (ii) above continues to hold even if $\mathcal{L}^m(A) = +\infty$. Conclude that (7.170) holds for all $A \in \mathcal{B}(\mathbb{R}^m)$, where $d_m := m^{-m/2}$.

Exercise 7.71. Show directly, by using the definitions of $\mathcal{H}^1$ and $\mathcal{L}^1$ on $\mathbb{R}$, that $\mathcal{H}^1 = \mathcal{L}^1$, as outer measures (and hence also as Borel and as Carathéodory measures) on $\mathbb{R}^m$.

Exercise 7.72. Use a simple geometric argument in order to verify directly that the Isodiametric Inequality is valid for an arbitrary cube Q in $\mathbb{R}^m$.

[*Hint*: Consider a ball circumscribed to Q and with the same center as the cube.]

Exercise 7.73. Use the Change of Variables Formula for (Lebesgue) integrable functions on $\mathbb{R}^m$ in order to show that $\omega_m := \mathcal{L}^m(\overline{B_{\mathbf{0}}(1)})$ is given by formulas (7.154)–(7.156) and to deduce that $\mathrm{vol}^{m-1}(S^{m-1}_{\mathbf{0}})$ is given by formula (7.152) in footnote 114.

Exercise 7.74. Show in a few lines, and without any computation, that for a ball B of radius $r > 0$ (and hence of diameter $d := 2r$), the Isodiametric Inequality becomes an equality: $\mathcal{L}^m(B) = \mathrm{vol}^m(B) = \omega_m \cdot r^m = \omega_m \cdot (d/2)^m$.

[*Hint*: Use the translation invariance and scale covariance of $\mathcal{L}^m$; see part (i) and part (ii) of Proposition 3.70, respectively.]

7.5. Hausdorff Dimension in General Metric Spaces

We close this chapter by briefly extending the notion of Hausdorff dimension (introduced in Section 5.2 for any Euclidean space $\mathbb{R}^m$) to the case of an arbitrary metric space. The corresponding definition and justification are the same as in Section 5.2; see, especially, Definition 5.13, Figure 5.1 in Section 5.2, along with the identities (5.8), (5.9) and the text surrounding them.

Definition 7.75 (*Hausdorff Dimension in an Arbitrary Metric Space*)**.** Let (X, d) be a metric space, with t-dimensional Hausdorff measure denoted $\mathcal{H}^t$, for $t \geq 0$. Furthermore, let $A \subseteq X$ and $A \neq \emptyset$. Then, the *Hausdorff dimension* of A, denoted $\dim_H(A)$, is the unique extended nonnegative real number σ (i.e., $\sigma = \dim_H(A) \in [0, +\infty]$) such that[122]

$$\mathcal{H}^t(A) = \begin{cases} +\infty, & \text{if } 0 \leq t < \sigma, \\ 0, & \text{if } \sigma < t. \end{cases} \tag{7.171}$$

Equivalently, $\dim_H(A)$ is the unique extended nonnegative real number (i.e., $\dim_H(A) \in [D, +\infty]$) defined by the following identities:[123]

$$\begin{aligned} \dim_H(A) &= \sup\{t \geq 0 : \mathcal{H}^t(A) = +\infty\} \\ &= \inf\{t \geq 0 : \mathcal{H}^t(A) = 0\} = \inf\{t \geq 0 : \mathcal{H}^t(A) < +\infty\}. \end{aligned} \tag{7.172}$$

Remark 7.76. Naturally, since (as the reader can easily verify) the restriction of $\mathcal{H}^t$ to $A \subseteq X$ coincides with the (intrinsic) t-dimensional Hausdorff measure of A—viewed as a metric space of its own (equipped with the induced metric, obtained by restricting d to $A \times A$)—it would have sufficed to define $\dim_H(X)$, for any metric space (X, d).

Exercise 7.77. Justify the statements of both parts of Definition 7.75. In particular, prove the existence and the uniqueness of $\sigma \in [0, +\infty]$ such that (7.171) holds, in the first part of this definition. Also, prove that the equalities in identity (7.172) hold, in the second part of the definition.

The reader is referred to Section 7.2 above in order to understand the statement of the next exercise.

Exercise 7.78.

(a) Show that, in Definition 7.75 and when $X := \mathbb{R}^m$ (Euclidean space), we can replace $\mathcal{H}^t$ by any of the other versions of (t-dimensional) Hausdorff measures introduced in Section 7.2 (for example, the spherical measure S^t); see Theorem 7.14 and Corollary 7.15.

---

[122] For an illustration, see Figure 5.1 in Section 5.2 (with $\mathbb{R}^m$ replaced by X in its caption).

[123] As usual, we let $\dim_H(\emptyset) := -\infty$.

(b) In the case of a general metric space X, precisely formulate and then establish a result analogous to the one stated in part (a), but now for a narrower class of Hausdorff measures; see, especially, Theorem 7.14, part (i) and—when X is a "reasonable" metric space—part (iii), in Section 7.2 above. In the case when X is *not* reasonable, determine whether or not the Hausdorff dimension remains uncharged when we replace $\mathscr{H}^t$ by S^t, the (t-dimensional) spherical measure (see, e.g., part (iv) of Theorem 7.14); if the answer is "no", provide an explicit counterexample.

Our final exercise provides a simple extension to the present general setting of Proposition 5.16 about the basic properties of Hausdorff dimension.

Exercise 7.79 (*Key Properties of Hausdorff Dimension*). Prove the exact counterpart (for a general metric space X) of Proposition 5.16 in Section 5.2 regarding, in particular, the monotonicity and the countable stability of Hausdorff dimension $\dim_H(.)$.

Chapter 8

Minkowski Content and Minkowski Dimension

For a complex natural shape, dimension is relative. It varies with the observer. The same object can have more than one dimension, depending on how you measure it and what you want to do with it. And dimension need not be a whole number; it can be fractional. Now an ancient concept, dimension, becomes thoroughly modern.

Benoît B. Mandelbrot

In this chapter, we introduce, in Section 8.1, the important notions of Minkowski dimension [**445**] (or Minkowski–Bouligand dimension, [**56**]), Minkowski content, Minkowski nondegeneracy, and Minkowski measurability, which play a key role in many applications of fractal geometry to dynamical systems, harmonic analysis, mathematical physics, probability theory, operator algebras and noncommutative geometry, spectral theory, and number theory (see, e.g., [**106–115, 172, 175, 182, 235, 238, 286, 318, 320, 324, 326, 328–330, 356, 370, 374, 375, 419, 429, 430, 491, 593**]), as we will see in Chapter 9 and, especially, in the sequels to this book [**119**], [**116**], [**117**], as well as to condensed matter physics, computational physics, chemistry, mathematical biology, geology, computer science, engineering, economics and finance, along with other scientific and technical fields.

In Section 8.2, we illustrate these notions by means of two different examples, the a-string and the Cantor string, which are two important special cases of fractal strings (or 1-dimensional fractal drums), a notion also discussed in the latter part of this same section; see, e.g., [**235, 238, 318, 326, 328, 329, 340,**

353, 356, 370, 374, 375]. We also provide a brief introduction—especially in the 1-dimensional case (i.e., for fractal strings)—to the notion of complex (fractal) dimensions (see, e.g., [**370, 374, 375**], [**356**], [**326**], [**238**], [**329**], [**330**], [**353**]), which significantly and naturally extends the (upper) Minkowski (or box) dimension to the complex plane via the use of suitable geometric or fractal zeta functions and will play a central role in the sequels to this book ([**119**], [**116**] and [**117**]), both in one and higher dimensions, along with its many geometric, dynamical, spectral, and number-theoretic applications. We mention that the notions of Minkowski dimension and complex dimensions will also be useful in Chapter 9 (on fractal curves), while a brief introduction to the higher-dimensional theory of complex dimensions will be provided in Section 9.3.

In Section 8.3, we discuss the notion of box-counting dimension (or box dimension, in short)—also known in the literature as capacity dimension or Kolmogorov dimension, among many other names. In particular, we show that, as is well known, the Minkowski dimension coincides with the box-counting dimension, for any bounded subset of Euclidean space $\mathbb{R}^m$. The latter fact explains in part why the Minkowski dimension is so useful in the physical, engineering, biological, and numerical applications.

In Section 8.4, we provide some of the basic properties of the Minkowski (or box) dimension. Furthermore, in Section 8.5, we compare the Minkowski and Hausdorff dimensions, while in Section 8.6, we compare the topological and Hausdorff (along with Minkowski) dimensions (Subsection 8.6.2), as well as state some of the basic properties of the topological dimension (Subsection 8.6.1).

Moreover, in Section 8.7, we discuss an application to the attractors of certain dynamical systems and, in Section 8.8, we give an explicit example for which the lower and upper box-counting—and hence also Minkowski—dimensions are different, i.e., for which the Minkowski dimension does not exist. It turns out, however, that the upper Minkowski dimension is the more natural notion, from the mathematical point of view (especially when dealing with problems from harmonic analysis, spectral theory, and number theory) and from the perspective of the theory of complex fractal dimensions (see, e.g., [**318, 370, 374, 375**], [**356**], [**329, 330**]), as will be seen in the sequels to this book, [**119**], [**116**], and [**117**].

Finally, in Section 8.9, we discuss scaling laws (or gauge functions) that are different from the usual power functions (Subsection 8.9.1). This leads us to naturally define the useful notions of generalized Hausdorff measures (Subsection 8.9.2) and generalized Minkowski contents (Subsection 8.9.5), along with the associated notions of generalized dimensions. We also briefly discuss the notions of packing dimension and modified box-counting dimension (Subsections 8.9.3 and 8.9.4) and compare them with the Hausdorff and Minkowski (or box) dimensions. Note that, like the Hausdorff dimension—but unlike the Minkowski (or box) dimension, which is associated with a content—the packing dimension is associated with a measure, the so-called packing measure.

8.1. Minkowski Content and Dimension

The Hausdorff dimension of a set A is defined in terms of δ-covers of A which, for a given $\delta > 0$, are countable collections of sets with diameter *at most* δ whose union contains A (see Chapters 5–7). If we restrict the δ-covers to sets with diameter *equal to* δ, much of the difficulty inherent to calculating the corresponding dimension is removed, but this benefit comes at a price. With this restriction, the corresponding function that provides a value for the 'size' of the set is no longer a measure. In particular, such a function fails to be even finitely additive. This framework is made precise in this chapter with a discussion of *Minkowski content* and *Minkowski dimension*, beginning with this section (as well as of box-counting dimension, beginning with Section 8.3).

Notation 8.1. For $A \subseteq \mathbb{R}^m$ and a point $\boldsymbol{x} \in \mathbb{R}^m$, let $\operatorname{dist}(\boldsymbol{x}, A)$ denote the distance between $\boldsymbol{x}$ and A given by

$$\operatorname{dist}(\boldsymbol{x}, A) := \inf\{\|\boldsymbol{x} - \mathbf{a}\|_m \ : \ \mathbf{a} \in A\}, \tag{8.1}$$

where $\|\cdot\|_m$ denotes the usual m-dimensional Euclidean norm. The notation $d_m = d_{m,2}$ is used to denote m-dimensional Euclidean metric. Let $\varepsilon > 0$. The *open ε-neighborhood* of A, denoted A_ε, is the set of points of $\mathbb{R}^m$ which are within ε of A. Specifically,

$$A_\varepsilon := \{\boldsymbol{x} \in \mathbb{R}^m \ : \ \operatorname{dist}(\boldsymbol{x}, A) < \varepsilon\}. \tag{8.2}$$

Consider subsets of $\mathbb{R}^3$ and let $\varepsilon > 0$. If A is a line segment with length ℓ, then A_ε, the open ε-neighborhood of A, is a 'sausage' (a tube-like object, also called the ε-tube of A) such that $\mathscr{L}^3(A_\varepsilon) \sim \pi\varepsilon^2\ell$, as $\varepsilon \to 0^+$. Note that, in this case, $\mathscr{L}^3(A_\varepsilon) \sim c \cdot \varepsilon^{3-t}$, as $\varepsilon \to 0^+$, where $t := 1$ plays the role of a dimension, which happens to coincide with the Hausdorff dimension of A in this case, and $c := \pi\ell$ assigns a value as the measure of A with respect to the dimension played by $t = 1$. Now, for a fractal subset of $\mathbb{R}^3$ such as an embedded copy of the Sierpinski gasket $\mathbf{SG} = A$, it turns out that we still obtain $\mathscr{L}^3(A_\varepsilon) \sim c \cdot \varepsilon^{3-t}$, as $\varepsilon \to 0^+$, where $t := \dim_H(\mathbf{SG}) = \log_2 3$, but in general c is not assigned by a measure. In the development of Minkowski content and dimension presented in this chapter, these intuitive notions are carefully refined and the corresponding ramifications are investigated.

Definition 8.2 (*Upper and Lower t-Dimensional Contents*)**.** Let $t \geq 0$. The *upper* and *lower t-dimensional Minkowski contents* of a (nonempty) bounded

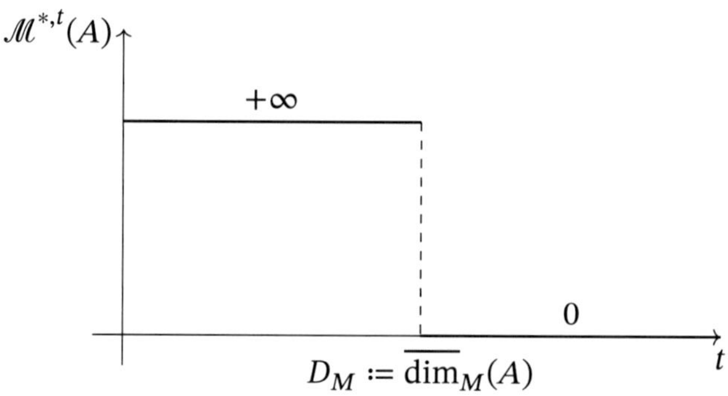

Figure 8.1. The upper Minkowski dimension of the nonempty bounded set $A \subseteq \mathbb{R}^m$, $D_M := \overline{\dim}_M(A)$, is the unique real number $D_M \geq 0$ such that—with $\mathcal{M}^{*,t}(A)$ denoting the t-dimensional upper Minkowski content of A, as in (8.4) of Definition 8.3—$\mathcal{M}^{*,t}(A) = +\infty$, for $0 \leq t < D_M$ and $\mathcal{M}^{*,t}(A) = 0$, for $t > D_M$. For $t = D_M := \overline{\dim}_M(A)$, anything permissible can happen; i.e., $\mathcal{M}^* = \mathcal{M}^{*,D_M}$ can take any value in the extended half-line $[0, +\infty]$. In physicists' terminology, the number $D_M := \overline{\dim}_M(A)$ is called a *critical exponent.* An entirely similar figure could be provided with $\overline{\dim}_M(A)$ and $\mathcal{M}^{*,t}(A)$ respectively replaced by $\underline{\dim}_M(A)$ (the lower Minkowski dimension of A) and $\mathcal{M}_*^t(A)$ (the t-dimensional lower Minkowski content of A).

set $A \subseteq \mathbb{R}^m$ are, respectively, given by

$$\mathcal{M}^{*,t}(A) := \limsup_{\varepsilon \to 0^+} \frac{\mathcal{L}^m(A_\varepsilon)}{\varepsilon^{m-t}} \quad \text{and} \quad \mathcal{M}_*^t(A) := \liminf_{\varepsilon \to 0^+} \frac{\mathcal{L}^m(A_\varepsilon)}{\varepsilon^{m-t}}, \tag{8.3}$$

where the upper and lower limits in (8.3) hold in the extended real half-line $[0, +\infty]$.

Definition 8.3 (*Upper and Lower Minkowski Dimensions—Existence of the Minkowski Dimension*). The *upper* and *lower Minkowski dimensions* of a (nonempty) bounded set A, denoted $\overline{\dim}_M(A)$ and $\underline{\dim}_M(A)$, are respectively given by (see Figure 8.1)

$$\overline{\dim}_M(A) := \inf\{t \geq 0 : \mathcal{M}^{*,t}(A) = 0\} = \sup\{t \geq 0 : \mathcal{M}^{*,t}(A) = +\infty\}, \tag{8.4}$$

$$\underline{\dim}_M(A) := \inf\{t \geq 0 : \mathcal{M}_*^t(A) = 0\} = \sup\{t \geq 0 : \mathcal{M}_*^t(A) = +\infty\}. \tag{8.5}$$

When $\overline{\dim}_M(A) = \underline{\dim}_M(A)$, the common value is called the *Minkowski dimension* of A, denoted as simply $\dim_M(A)$. One then says that *the Minkowski dimension of A, $\dim_M(A)$, exists.*[1]

[1]If $A := \emptyset$, the empty set, then we may choose, by convention, to let $\overline{\dim}_M(\emptyset) = \underline{\dim}_M(\emptyset) := -\infty$; i.e., $\dim_M(\emptyset)$ exists and is equal to $-\infty$.

Note that the equalities on the right of (8.4) and (8.5) are indeed justified. For instance, suppose $t < \overline{\dim}_M(A)$ and let $u \in (t, \overline{\dim}_M(A))$. By definition, $\mathcal{M}^{*,u}(A) \in (0, +\infty]$ and thus, $\mathcal{M}^{*,t}(A) = +\infty$. Similarly, given $t > \overline{\dim}_M(A)$, there is some $u \in (\overline{\dim}_M(A), t)$ such that $\mathcal{M}^{*,u}(A) = 0$ and therefore, $\mathcal{M}^{*,t}(A) = 0$.

From now on, most of the time, we will assume implicitly that the bounded set A is nonempty. (Recall that we let $\dim_M(\emptyset) := -\infty$.)

The following useful result, stated in the next exercise, is easily proved as a consequence of Definition 8.3 and the justification following it.

Exercise 8.4. Show that we also have that[2]

$$\overline{\dim}_M(A) = \inf\{t \geq 0 : \mathcal{M}^{*,t}(A) < \infty\} = \sup\{t \geq 0 : \mathcal{M}^{*,t}(A) > 0\}, \tag{8.6}$$

$$\underline{\dim}_M(A) = \inf\{t \geq 0 : \mathcal{M}^{t}_{*}(A) < \infty\} = \sup\{t \geq 0 : \mathcal{M}^{t}_{*}(A) > 0\}. \tag{8.7}$$

[*Hint*: Observe that $0 < \mathcal{M}^{*,t}(A)$ (with possibly $\mathcal{M}^{*,t}(A) = +\infty$) implies that $t \leq \overline{\dim}_M(A)$ and, similarly, $\mathcal{M}^{*,t}(A) < \infty$ (with possibly $\mathcal{M}^{*,t}(A) = 0$) implies that $t \geq \overline{\dim}_M(A)$. Then, reason analogously for $\mathcal{M}^{t}_{*}(A)$ and $\underline{\dim}_M(A)$ instead of $\mathcal{M}^{*,t}(A)$ and $\overline{\dim}_M(A)$, respectively.]

Remark 8.5.

(a) Clearly, since the lower limit does not exceed the upper limit (of the same function), and since the lower limit of a nonnegative quantity is nonnegative, we have that, for any $t \geq 0$,

$$0 \leq \mathcal{M}^{t}_{*}(A) \leq \mathcal{M}^{*,t}(A) \leq \infty. \tag{8.8}$$

(b) We therefore always have that

$$0 \leq \underline{\dim}_M(A) \leq \overline{\dim}_M(A) \leq m. \tag{8.9}$$

Indeed, the first inequality in (8.9) is obvious in light of the first equality in (8.5), for example. Furthermore, in light of (8.4) in Definition 8.3, the second inequality in (8.9) clearly follows from part (a) just above—or, more precisely, from the middle inequality of (8.8).

Finally, the third (and last) inequality of (8.9) can be proved as follows. Since A is bounded and, obviously, $A_\varepsilon \subseteq A_1$ for $0 < \varepsilon \leq 1$, we have that (since A_1 is clearly also bounded) A_1—and thus also A—is contained in a ball of center the origin and radius $R > 0$ in $\mathbb{R}^m$. Hence, with $V_A(\varepsilon) := \mathcal{L}^2(A_\varepsilon)$, there exists a constant $c > 0$ such that

$$V_A(\varepsilon) = \frac{V_A(\varepsilon)}{\varepsilon^{m-m}} \leq c \cdot R^m < \infty, \text{ for all } \varepsilon \in (0, 1];$$

[2]Here, when we write, for example, $\mathcal{M}^{*,t}(A) > 0$, we allow for $\mathcal{M}^{*,t}(A)$ to be equal to $+\infty$.

and so, upon taking the upper limit as $\varepsilon \to 0^+$, we deduce from the first equality in (8.3) of Definition 8.2 that

$$\mathcal{M}^{*,m}(A) = \limsup_{\varepsilon \to 0^+} \frac{V_A(\varepsilon)}{\varepsilon^{m-m}} \,(\leq c \cdot R^m) < \infty.$$

Therefore, in light of identity (8.6) in Exercise 8.4 above, we have that $\overline{\dim}_M(A) \leq m$, as desired.

(c) If, in addition to being bounded, $A \subseteq \mathbb{R}^m$ is Borel (or, more generally, Lebesgue) measurable and $\mathcal{L}^m(A) > 0$ (i.e., if A is *not* of Lebesgue measure zero), then $\dim_M(A)$ exists and is equal to m: $\dim_M(A) = m$. Moreover, A is Minkowski measurable, with Minkowski content given by $\mathcal{M}^m(A) = \mathcal{L}^m(A)$; see Definition 8.6 below.

Indeed, clearly, $V_A(\varepsilon) := \mathcal{L}^m(A_\varepsilon) \geq \mathcal{L}^m(A) > 0$, for all $\varepsilon > 0$. Hence, we have that $\mathcal{M}_*^m(A) = \liminf_{\varepsilon \to 0^+} V_A(\varepsilon)/\varepsilon^{m-m} \,(\geq \mathcal{L}^m(A)) > 0$; therefore, $m \leq \underline{\dim}_M(A)$, in light of the second equality in identity (8.7) of Exercise 8.4 above. But then, it follows from the last two inequalities of (8.9) in part (b) of the present remark that $m \leq \underline{\dim}_M(A) \leq \overline{\dim}_M(A) \leq m$, and so, $\underline{\dim}_M(A) = \overline{\dim}_M(A) = m$; i.e., $\dim_M(A)$ exists and equals m, as stated above.

We next prove the assertion regarding the Minkowski measurability and content of A. Clearly, in light of Definitions 8.2 and 8.3, we have that

$$\mathcal{L}^m(A) \leq \mathcal{M}_*^m(A) \leq \mathcal{M}^{*,m}(A) \leq \mathcal{L}^m(A).$$

Hence, since $0 < \mathcal{L}^m(A) < \infty$ (by hypothesis and because A is bounded), we have that $0 < \mathcal{M}_*^m(A) = \mathcal{M}^{*,m}(A) = \mathcal{L}^m(A) < \infty$; i.e., in the notation of Definition 8.6 below, $0 < \mathcal{M}_* = \mathcal{M}^* = \mathcal{L}^m(A) < \infty$; and so, by Definition 8.3 and Exercise 8.4, we recover the fact that $\dim_M(A)$ exists and equals m. Also, in light of Definition 8.6 below, we conclude from the latter identity that A is Minkowski measurable with Minkowski content $\mathcal{M} = \mathcal{M}^m(A) = \mathcal{L}^m(A)$, as claimed.

(d) This last comment (in part (c)), along with part (b), will be useful in understanding the formulation of Theorem 8.10 below. Observe that we have that $\infty > V_A(\varepsilon) := \mathcal{L}^m(A_\varepsilon) > 0$, for all $\varepsilon > 0$. Indeed, since A is nonempty and the distance function in (8.1) is continuous, it follows from the definition of A_ε in (8.2) that, for *any* bounded subset A of $\mathbb{R}^m$, A_ε is an *open* and *nonempty* (as well as *bounded*) subset of $\mathbb{R}^m$. Thus, $V_A(\varepsilon) = \mathcal{L}^m(A_\varepsilon) > 0$, as claimed just above.[3]

Definition 8.6. Let A be a bounded subset of $\mathbb{R}^m$ and let $D_M = \overline{\dim}_M(A)$. Then, the *upper* and *lower Minkowski contents* of A are, respectively, given by

$$\mathcal{M}^* = \mathcal{M}^*(A) := \mathcal{M}^{*,D_M}(A) \quad \text{and} \quad \mathcal{M}_* = \mathcal{M}_*(A) := \mathcal{M}_*^{D_M}(A). \tag{8.10}$$

[3]Note that, in light of part (c) of this remark, it follows that $\dim_M(A_\varepsilon) = m$ and A_ε is Minkowski measurable, with Minkowski content $\mathcal{M}^m(A_\varepsilon) = \mathcal{L}^m(A_\varepsilon)$; see Definition 8.6.

Clearly, we always have that

$$0 \leq \mathcal{M}_*(A) \leq \mathcal{M}^*(A) \leq \infty. \tag{8.11}$$

If $0 < \mathcal{M}_*(A)(\leq)\mathcal{M}^*(A) < \infty$, then A is said to be *Minkowski nondegenerate.* If $\mathcal{M}_*(A) = \mathcal{M}^*(A)$, then the common value, denoted $\mathcal{M} = \mathcal{M}(A)$, is called the *Minkowski content* of A and is necessarily given by the following limit (which exists in $[0, +\infty]$):

$$\mathcal{M}(A) = \lim_{\varepsilon \to 0^+} \frac{\mathcal{L}^m(A_\varepsilon)}{\varepsilon^{m-D_M}}. \tag{8.12}$$

If, in addition, $0 < \mathcal{M}(A) < \infty$, then A is said to be *Minkowski measurable*, with *Minkowski content* $\mathcal{M} = \mathcal{M}(A)$.[4]

Note that if $0 < \mathcal{M}_*^t(A) \leq \mathcal{M}^{*,t}(A) < \infty$, for some $t \geq 0$, then $\dim_M(A)$ exists and is equal to t. Hence, A is Minkowski nondegenerate and $t = \dim(A)$. As a result, if A is Minkowski nondegenerate—and, in particular, if A is Minkowski measurable—then the Minkowski dimension, $\dim_M(A)$, of A exists; see Exercise 8.7.

A moment's reflection shows that the above assertion follows from the definitions of $\overline{\dim}_M(A)$ and $\underline{\dim}_M(A)$ in (8.4) and (8.5) of Definition 8.3.

Exercise 8.7. Prove that if the bounded set $A \subseteq \mathbb{R}^m$ is Minkowski nondegenerate—and, in particular, if A is Minkowski measurable—then the Minkowski dimension, $\dim_M(A)$, exists.

The next exercise enables us to answer a very natural question.

Exercise 8.8. Show by means of concrete examples that all the possibilities allowed by the string of inequalities given by (8.11) can actually occur.

Example 8.9. Let $\varepsilon > 0$, $E := [0,1] \times \{\mathbf{0}\} \times \{\mathbf{0}\}$ (a line segment with unit length embedded in $\mathbb{R}^3$), $F := [0,1] \times [0,1] \times \{\mathbf{0}\}$ (a unit square embedded in $\mathbb{R}^3$), and let $G := [0,1] \times [0,1] \times [0,1]$ (the unit cube in $\mathbb{R}^3$). Let $\varepsilon > 0$ and recall that the volume of a 3-dimensional sphere of radius ε is $(4/3)\pi\varepsilon^3$. Then,

$$\begin{aligned}
\mathcal{L}^3(E_\varepsilon) &= (4/3)\pi\varepsilon^3 + \pi\varepsilon^2, \\
\mathcal{L}^3(F_\varepsilon) &= (4/3)\pi\varepsilon^3 + 2\pi\varepsilon^2 + 2\varepsilon, \text{ and} \\
\mathcal{L}^3(G_\varepsilon) &= (4/3)\pi\varepsilon^3 + 2\pi\varepsilon^2 + 6\varepsilon + 1.
\end{aligned} \tag{8.13}$$

[4] A sample list of references about (or related to) the Minkowski dimension and/or content—or else about the box-dimension—includes [7], [19], [34], [46], [56], [59], [63], [72], [92], [95], [103–105], [106–117], [118], [119], [122], [126–129], [136], [137], [147], [168], [160], [170], [171], [172], [174], [175], [182], [191], [194], [196], [199], [200], [202], [209], [210–212], [227], [235], [238–242], [254], [286], [290], [299], [302], [303–305], [313–315], [318–342], [346–367], [369–379], [390, 391], [397–399], [418–421], [427], [428], [429], [430], [434], [445], [446], [449–451], [463], [469, 470], [480], [481], [483], [491], [492], [498], [499], [502], [503], [505], [508], [516], [535], [542], [562], [590–593], [598], [600, 601], [606], [619], [620], [621], [628], [629], and [631, 632], along with the relevant references therein.

In the case of the line segment E, we have

$$\mathcal{M}^{*,t}(E) = \mathcal{M}_*^t(E) = \begin{cases} +\infty, & \text{if } 0 \leq t < 1, \\ \pi, & \text{if } t = 1, \\ 0, & \text{if } 1 < t \leq m. \end{cases} \tag{8.14}$$

Hence, $\dim_M(E)$ exists and equals 1, and E is Minkowski measurable with Minkowski content $\mathcal{M} = \mathcal{M}^1(E) = \pi$. In a similar fashion, it follows that $\dim_M(F) = 2$, with $\mathcal{M} = \mathcal{M}^2(F) = 2$, and $\dim_M(G) = 3$, with $\mathcal{M} = \mathcal{M}^3(G) = 1 = \mathcal{L}^3(G)$. In particular, each of F and G is also Minkowski measurable, with the indicated respective Minkowski content.

We point out that, much more generally, it is shown in [**356**, Section 4.7] that the (upper) Minkowski dimension as well as (when it exists) the Minkowski content (of a given bounded set $A \subseteq \mathbb{R}^m$)[5]—along with the associated complex dimensions (see [**375**] and [**356**]), to be briefly discussed in Sections 8.2 and 9.3 below and, in much greater detail, in [**119**], the sequel to this book—do *not* depend on the dimension of the Euclidean space $\mathbb{R}^{m'}$ (with $m' \geq m$) in which the given bounded set $A \subseteq \mathbb{R}^m$ is embedded.

The following result is very useful in calculating the Minkowski dimension of many fractal curves, including, in particular, the Weierstrass nowhere differentiable curve (and its natural generalizations), to be studied in Section 9.7 and, in much greater detail, in the relevant part of [**116**].

Theorem 8.10 (Explicit Expressions for the Upper and Lower Minkowski Dimensions). *The upper and lower Minkowski dimensions of a (nonempty) bounded set $A \subseteq \mathbb{R}^m$ are equivalently given, respectively, by*

$$\overline{\dim}_M(A) = \limsup_{\varepsilon \to 0^+} \left(m - \frac{\log \mathcal{L}^m(A_\varepsilon)}{\log \varepsilon} \right) = m - \liminf_{\varepsilon \to 0^+} \frac{\log \mathcal{L}^m(A_\varepsilon)}{\log \varepsilon}, \tag{8.15}$$

$$\underline{\dim}_M(A) = \liminf_{\varepsilon \to 0^+} \left(m - \frac{\log \mathcal{L}^m(A_\varepsilon)}{\log \varepsilon} \right) = m - \limsup_{\varepsilon \to 0^+} \frac{\log \mathcal{L}^m(A_\varepsilon)}{\log \varepsilon}. \tag{8.16}$$

Consequently, if $\dim_M(A)$ *exists—which is the case, for example, if A is Minkowski nondegenerate (or, in particular, if A is Minkowski measurable)—then the following limit exists and equals* $\dim_M(A)$:

$$\dim_M(A) = \lim_{\varepsilon \to 0^+} \left(m - \frac{\log \mathcal{L}^m(A_\varepsilon)}{\log \varepsilon} \right) = m - \lim_{\varepsilon \to 0^+} \frac{\log \mathcal{L}^m(A_\varepsilon)}{\log \varepsilon}. \tag{8.17}$$

[5]The statement concerning the Minkowski content was first proved (in lesser generality than in [**356**, Section 4.7]) by M. Kneser in [**299**]; see also M. Resman, [**505**].

Proof. We first note that by using the basic properties of the upper and lower limits (see Appendix A), we have that

$$\limsup_{\varepsilon\to 0^+}\left(m - \frac{\log \mathscr{L}^m(A_\varepsilon)}{\log\varepsilon}\right) = m + \limsup_{\varepsilon\to 0^+}\left(-\frac{\log \mathscr{L}^m(A_\varepsilon)}{\log\varepsilon}\right)$$
$$= m - \liminf_{\varepsilon\to 0^+}\left(\frac{\log \mathscr{L}^m(A_\varepsilon)}{\log\varepsilon}\right),$$

and similarly with the roles of the upper and lower limits interchanged. Consequently, it suffices to show that the first equality in both (8.15) and (8.16) is true.

We will first focus on proving the first equality of (8.15) and will then leave the verification of the first equality of (8.16) for the interested reader because its proof is entirely similar.

For pedagogical reasons, we begin the proof of the first equality in (8.15) by a preliminary step which is not logically necessary but which may be helpful to the reader who is unfamiliar with the manipulation of upper and lower limits.

Step 0 (The Minkowski Measurable Case). *If A is Minkowski measurable, then the limit in* (8.15) *exists in* $(0, +\infty)$ *and is equal to* $\dim_M(A)$; *i.e.,* (8.15) *holds.*

Indeed, by hypothesis, A is Minkowski measurable—and hence, by Exercise 8.7, $\underline{\dim}_M(A) = \overline{\dim}_M(A) = \dim_M(A) =: D_A$. Then, the Minkowski content $\mathcal{M}$ of A exists in $(0, +\infty)$ and is given by the following limit: $\mathcal{M} = \lim_{\varepsilon\to 0^+} V_A(\varepsilon)/\varepsilon^{m-D_A}$. *(Here, and thereafter, we let $V_A(\varepsilon) := \mathscr{L}^2(A_\varepsilon)$, for notational simplicity.)*

Since the logarithm is a continuous function on $(0, +\infty)$, it follows from the above expression for $\mathcal{M}$ that the following limit exists and equals $\log \mathcal{M} \in \mathbb{R}$:

$$\log \mathcal{M} = \lim_{\varepsilon\to 0^+} \log\left(\frac{V_A(\varepsilon)}{\varepsilon^{m-D_A}}\right) = \lim_{\varepsilon\to 0^+} (\log V_A(\varepsilon) - (m - D_A)\log\varepsilon);$$

i.e., with $o(1)$ standing for a function of $\varepsilon > 0$ tending to 0, as $\varepsilon \to 0^+$,

$$\log V_A(\varepsilon) - (m - D_A)\log\varepsilon = (\log \mathcal{M})(1 + o(1))$$

and, upon division by $\log\varepsilon < 0$ (assuming, as we can do without loss of generality, that $0 < \varepsilon < 1$),

$$\text{(8.18)} \qquad \frac{\log V_A(\varepsilon)}{\log\varepsilon} - (m - D_A) = \left(\frac{\log \mathcal{M}}{\log\varepsilon}\right)(1 + o(1)) \to 0, \quad \text{as } \varepsilon \to 0^+,$$

since $\log \mathcal{M}/\log\varepsilon \to 0$, as $\varepsilon \to 0^+$ (because $1/\log\varepsilon \to 0$, as $\varepsilon \to 0^+$, and $\log \mathcal{M} \in \mathbb{R}$).

Therefore, (8.18) shows that the following limit exists and equals D_A (which proves (8.15) as needed): $D_A = \lim_{\varepsilon\to 0^+}(m - (\log V_A(\varepsilon)/\log\varepsilon))$.[6]

Step 1. *For any bounded set* $A \subseteq \mathbb{R}^m$, *we have that* $\overline{\beta} \le \overline{\dim}_M(A)$, *where* $\overline{\beta}$ *denotes the upper limit on the right-hand side of* (8.15); *i.e.,*

$$\overline{\beta} := \limsup_{\varepsilon\to 0^+}\left(m - \frac{\log V_A(\varepsilon)}{\log\varepsilon}\right). \tag{8.19}$$

Indeed,[7] let $d \ge 0$ be an arbitrary real number such that $d < \overline{\beta}$.[8] Then (see Appendix A), by the definition of the upper limit, there exists a sequence of positive numbers $\{\varepsilon_n\}_{n=1}^{\infty}$ such that $\varepsilon_n \to d^+$, as $n\to\infty$ (and $0<\varepsilon_n<1$, for all $n\ge 1$), as well as $\log V_A(\varepsilon_n)/\log\varepsilon_n \to m-d$; i.e.,

$$\frac{\log V_A(\varepsilon_n)}{(m-d)\log\varepsilon_n} = \frac{\log V_A(\varepsilon_n)}{\log(\varepsilon_n^{m-d})} \to 1, \quad \text{as } n\to\infty.$$

Hence, given any $\delta \in (0,1)$, there exists $N\in\mathbb{N}$ such that $n\ge N$ implies that $\log V_A(\varepsilon_n)/\log(\varepsilon_n^{m-d}) \ge \delta > 0$; i.e., $\log V_A(\varepsilon_n) \le \delta\log(\varepsilon_n^{m-d}) = \log(\varepsilon_n^{(m-d)\delta})$. Upon exponentiating both sides of the latter identity, we deduce that for all $n \ge N$, we have that $V_A(\varepsilon_n) \le \varepsilon_n^{(m-d)\delta} = \varepsilon_n^{m-(d\delta+m(1-\delta))}$; thus, $\mathcal{M}^{*,d\delta+m(1-\delta)} = \limsup_{\varepsilon\to 0^+}(V_A(\varepsilon)/\varepsilon^{m-(d\delta+m(1-\delta))}) \ge 1 > 0$, by the definition of the upper limit.[9] Now, since $\mathcal{M}^{*,d\delta+m(1-\delta)} > 0$, it follows from the definition of $\overline{\dim}_M(A)$ (see (8.4) in Definition 8.3) that $d\delta + m(1-\delta) \le \overline{\dim}_M(A)$. Since $\delta\in(0,1)$ was chosen to be arbitrary, we can let $\delta\to 1^-$ in the latter inequality to deduce that $d \le \overline{\dim}_M(A)$. Finally, since d itself was chosen to be an arbitrary real number such that $0 \le d < \overline{\beta}$, we can also let $d \to \overline{\beta}^-$ to obtain that $\overline{\beta} \le \overline{\dim}_M(A)$, as claimed. This concludes the proof of Step 1.

Step 2. *For any* (*nonempty*) *set* $A \subseteq \mathbb{R}^m$ (*with* $\overline{\beta}$ *given by* (8.19), *as in Step* 1), we have that $\overline{\dim}_M(A) \le \overline{\beta}$.

Indeed, let us reason by contradiction ("ad absurdum") and assume, therefore, that $\overline{\dim}_M(A) > \overline{\beta}$. Let $d > 0$ be such that[10]

$$\overline{\dim}_M(A) > d > \overline{\beta}. \tag{8.20}$$

[6] As was alluded to just before the statement of Step 0, the true beginning of the proof of Theorem 8.10 starts with Step 1.

[7] We leave it to the reader to verify that $\overline{\beta} \ge 0$ (i.e., $\liminf_{\varepsilon\to 0^+}(\log V_A(\varepsilon)/\log\varepsilon) \le m$), a fact which is not necessary if we simply let $d\in\mathbb{R}$ (instead of $d\ge 0$, with $d < \overline{\dim}_M(A)$) and then use the result of Step 2 below, according to which $\overline{\beta} \ge \overline{\dim}_M(A) \ge 0$.

[8] Note that since $d < \overline{\beta}$, d must be a (finite) real number even if $\overline{\beta} = +\infty$—a possibility which will be excluded by Step 2, anyway; also, $d < m$. (Observe that the proof of Step 2 is independent of Step 1.)

[9] Note that, once again, we have reversed the sense of the inequality because, for all $n \ge 1$, $\varepsilon_n^{m-d} < 1$, and hence, $\log(\varepsilon_n^{m-d})$ is negative.

[10] Since $\overline{\dim}_M(A) \ge 0$, we can always choose d close enough to $\overline{\dim}_M(A)$, unless $\overline{\dim}_M(A) = 0$, a case which can easily be dealt with separately, as the reader is invited to verify.

Then, since $d < \overline{\dim}_M(A)$, it follows from the definition of $\overline{\dim}_M(A)$ in (8.4) that $\mathcal{M}^{*,d}(A) := \limsup_{\varepsilon\to 0^+} \left(V_A(\varepsilon)/\varepsilon^{m-d}\right) = +\infty$. Hence, given any sequence $\{\varepsilon_n\}_{n=1}^\infty$ of positive numbers such that $\varepsilon_n \to 0^+$,[11] and given any real number $B > 1$, there exists $N \in \mathbb{N}$ such that $n \geq N$ implies that $V_A(\varepsilon_n)/\varepsilon_n^{m-d} \geq B$; i.e., $\log V_A(\varepsilon_n) - (m-d)\log\varepsilon_n \geq \log B$, or (upon dividing both sides of the latter inequality by $-\log\varepsilon_n > 0$, since $0 < \varepsilon_n < 1$)

$$(8.21)\qquad (m-d) - \frac{\log V_A(\varepsilon_n)}{\log\varepsilon_n} \geq \frac{-\log B}{\log\varepsilon_n}, \text{ for all } n \geq N.$$

Letting $n \to \infty$ in (8.21) (while assuming that $n \geq N$), we deduce that (since $-\log B/\log\varepsilon_n \to 0^+$)

$$\lim_{n\to\infty}\left((m-d) - \frac{V_A(\varepsilon_n)}{\log\varepsilon_n}\right) = \lim_{n\to\infty}\left(m - \frac{V_A(\varepsilon_n)}{\log\varepsilon_n}\right) - d \geq 0;$$

i.e., $\lim_{n\to\infty}(m - (V_A(\varepsilon_n)/\log\varepsilon_n)) \geq d$. Since $\{\varepsilon_n\}_{n=1}^\infty$ is an arbitrary sequence tending to 0^+, it follows that $\overline{\beta} := \limsup_{\varepsilon\to 0^+}(m - (\log V_A(\varepsilon)/\log\varepsilon)) \geq d$. But since, in light of (8.20), $d > \overline{\beta}$, we deduce from the latter inequality that $\overline{\beta} > \overline{\beta}$, which is an obvious contradiction. Hence, $\overline{\dim}_M(A) \leq \overline{\beta}$, which completes the proof of Step 2.

Step 3. *For any (nonempty) bounded subset A of $\mathbb{R}^m$, we have that $\overline{\dim}_M(A) = \overline{\beta}$; i.e., the identity* (8.15) *in Theorem* 8.10 *holds, as desired.*

Indeed, this follows at once by combining Step 1 and Step 2.

Steps 4 and 5 (Counterparts of Steps 2 and 3). *For any (nonempty) bounded subset A of $\mathbb{R}^m$, we have that $\underline{\beta} \leq \underline{\dim}_M(A)$* and *$\underline{\dim}_M(A) \leq \underline{\beta}$; i.e., the exact analog of Step* 3 *holds*:

$$(8.22)\qquad \underline{\dim}_M(A) := \underline{\beta}, \quad \text{where } \underline{\beta} := \liminf_{\varepsilon\to 0^+}\left(m - \frac{\log V_A(\varepsilon)}{\log\varepsilon}\right).$$

Indeed, the proof is entirely analogous to that of Steps 2 and 3, except for the upper limits (in various expressions) replaced by the corresponding lower limits.

Step 6 (The Minkowski Nondegenerate Case). *If A is Minkowski nondegenerate, then* $\dim_M(A)$ *exists and equals the following limit, which also exists*:

$$(8.23)\qquad \dim_M(A) = \lim_{\varepsilon\to 0^+}\left(m - \frac{\log V_A(\varepsilon)}{\log\varepsilon}\right) =: \beta.$$

[11] Clearly, without loss of generality, we may also assume that $0 < \varepsilon_n < 1$, for all $n \geq 1$.

Indeed, in light of Step 3 (see (8.15)) and of Steps 4 and 5 (see (8.22)), we have that $\overline{\dim}_M(A) = \overline{\beta}$ and $\underline{\dim}_M(A) = \underline{\beta}$. But, since A is Minkowski nondegenerate we know that $\dim_M(A)$ exists (i.e., $\underline{\dim}_M(A) = \overline{\dim}_M(A) =: \dim_M(A)$) and so, we deduce from the last two equalities the following identity: $\dim_M(A) = \underline{\beta} = \overline{\beta}$. Hence, the lower and upper limits of the expression $m-(\log V_A(\varepsilon)/\log \varepsilon)$ as $\varepsilon \to 0^+$ coincide, which precisely means that the corresponding limit exists (it is called β in (8.23) above) and coincides with $\dim_M(A)$; i.e., (8.23) holds, as desired.

This completes the proof of Step 6, as well as of the entire theorem (i.e., Theorem 8.10). □

Exercise 8.11. Prove Steps 4 and 5, which are the exact analogs of Steps 2 and 3 (taken together) in the proof of Theorem 8.10 just above.

8.2. Examples: The a-String and the Cantor String (or Set)

In this section, we illustrate the above concepts—namely, the Minkowski dimension, the (upper, lower) Minkowski content, along with the notion of Minkowski nondegeneracy and of Minkowski measurability—by means of two main examples, the a-string (Example 8.12) and the Cantor string—along with its boundary, the Cantor set (Example 8.17); see, especially, Example 8.12 (as well as Remark 8.13, Theorem 8.15), and Example 8.17, respectively.

In the process, we will briefly introduce the important and useful notions of a fractal string and of its associated complex dimensions. The theory of fractal strings, their complex dimensions—along with their higher-dimensional counterparts, relative fractal drums and, in particular, arbitrary bounded subsets of $\mathbb{R}^m$, and their complex dimensions—will be the central object of the sequel to the present book [**119**]—based, especially, on [**370, 374, 375**] and [**356**] (see also [**329**], [**330**]), as will be further explained below.

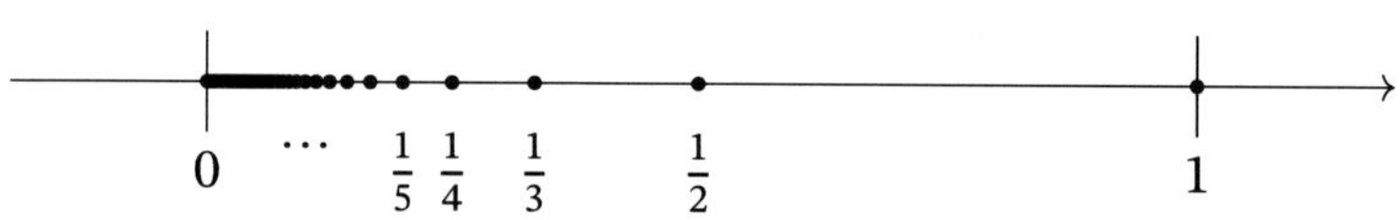

Figure 8.2. The a-string $\mathcal{L}_a$, as described in Example 8.12, with geometric realization $\Omega_a := \bigcup_{j=1}^{\infty}((j+1)^{-a}, j^{-a})$ and with boundary $\partial\Omega_a = \{j^{-a} : j \in \mathbb{N}\}$. Here, $a = 1$.

Example 8.12 (*From a-Strings to General Fractal Strings and Their Complex Dimensions*). Fix $a > 0$ and let Ω_a denote the bounded open subset of $\mathbb{R}$ composed of the disjoint open intervals $(1/(j+1)^a, 1/j^a)$, where $j = 1, 2, \ldots$.[12]

$$\Omega_a := \bigsqcup_{j=1}^{\infty} \left(\frac{1}{(j+1)^a}, \frac{1}{j^a} \right); \tag{8.24}$$

see Figure 8.2. Clearly, $\Omega_a \subseteq (0, 1]$ and 0 is a limit point of Ω_a. Let $\partial\Omega_a$ be the boundary of Ω_a; namely,

$$\partial\Omega_a := \left\{ \frac{1}{j^a} \ : \ j \geq 1 \right\} \cup \{0\}. \tag{8.25}$$

Note that $0 \in \partial\Omega_a$ because the boundary $\partial\Omega_a$ is closed and 0 is a limit point of $\{j^{-a} \ : \ j \geq 1\}$, the set of endpoints of the intervals $((j+1)^{-a}, j^{-a})$, the connected components of Ω_a. Clearly, $\partial\Omega_a$ is a compact subset of $\mathbb{R}$ since it is closed and bounded.

The *a-string* Ω_a was introduced by M. L. Lapidus in [**318**, Example 5.1] as a first explicit example of a fractal string and has been used by him to test various conjectures, such as the "modified Weyl–Berry Conjecture" (or MWB Conjecture, in short), also introduced in [**318**]. Its boundary $\partial\Omega_a$, even though it is a countable set, has a nontrivial Minkowski dimension, D_a:

$$D_a := \dim_M(\partial\Omega_a) = \frac{1}{a+1}. \tag{8.26}$$

Furthermore, it is Minkowski measurable, with Minkowski content, $\mathcal{M}_a$, given by

$$\mathcal{M}_a := \mathcal{M}^{D_a} = \left(\frac{2^{1-D_a}}{1-D_a} \right) \cdot a^{D_a}. \tag{8.27}$$

In particular, the Minkowski dimension of $\partial\Omega_a$ exists and equals D_a, as given by (8.26).

These results were obtained by the first author in [**318**, Example 5.1 and Appendix C] and were later generalized by the first author and C. Pommerance in [**353**] (announced in [**352**]) in order to characterize the Minkowski measurability of fractal strings. The general notion of a fractal string was also introduced in [**352, 353**] in order to solve the 1-dimensional case of the aforementioned MWB Conjecture (from [**318**]) and to discover in the process unsuspected connections with the Riemann zeta function.[13]

---

[12]We caution the reader that throughout this discussion in Example 8.12 as well as in Remark 8.13 and Theorem 8.15, Ω_a (e.g., in (8.24)) and $\partial\Omega_a$ (e.g., in (8.25)) do *not* stand for the a-neighborhood of Ω_a and $\partial\Omega_a$, respectively. In fact, in order to avoid any confusion, the ε-neighborhood of $\partial\Omega_a$ will be denoted by $(\partial\Omega_a)_\varepsilon$ in the proof of Theorem 8.15.

[13]See also [**354**] where the notion of a *fractal spray*, an exact higher-dimensional analog of a fractal string, was introduced.

Shortly thereafter, in [**340**] (announced in [**339**]), M. L. Lapidus and H. Maier obtained a new spectral reformulation of the celebrated Riemann Hypothesis [**515**],[14] by considering and completely solving a suitable inverse spectral problem for fractal strings, a natural converse of the 1-dimensional MWB Conjecture (now a theorem, in light of [**352, 353**]). This provided, in particular, a natural geometric interpretation of the *critical strip* for the Riemann zeta function, namely, $\{s \in \mathbb{C} : 0 \le \mathrm{Re}(s) \le 1\}$, which is the receptacle for the (principal) complex dimensions of a fractal string—then only heuristically understood or conjectured to exist, in [**352, 353**] and in [**319, 320**]—namely, for the complex dimensions with maximal real part, necessarily equal to the (upper) Minkowski dimension $D_{\mathcal{L}}$ of (the boundary of) the fractal string.[15]

Accordingly, the question à la Mark Kac [**280**]—but now interpreted quite differently — *Can one hear the shape of a fractal string*? — or, more precisely, *Can one hear the Minkowski measurability of a fractal string*? (of Minkowski dimension $D = D_{\mathcal{L}} \in (0,1)$)—is shown to be intimately connected with the Riemann Hypothesis. In fact, this *inverse spectral problem* is shown in [**340**] to have an affirmative answer (for all fractal strings of a given dimension $D \in (0,1)$) iff $\zeta(s) \neq 0$, for $\mathrm{Re}(s) = D$ — and, therefore, it has an affirmative answer for all fractal strings of dimension $D \neq 1/2$ (with $D \in (0,1)$ arbitrary) iff the Riemann Hypothesis is true.

These results—along with other results and conjectures from [**318–320**], [**353, 354**], [**339, 340**], and [**235**], in particular—helped motivate the development of the rigorous theory of complex dimensions for fractal strings (essentially, the 1-dimensional theory of complex fractal dimensions) and discover, in the process, a number of additional connections between fractal geometry and number theory. See the books by M. L. Lapidus and M. van Frankenhuijsen [**370, 374**] and [**375**]—along with many accompanying related works,[16] including the papers [**327, 328**], the survey article [**329**], and the books [**238, 326**] and [**330**].

Eventually—building on the 1-dimensional theory but going well beyond in its scope—the higher-dimensional theory of complex fractal dimensions was developed by M. L. Lapidus, G. Radunović, and D. Žubrinić in the book [**356**] and in a series of accompanying papers, [**357–365**] and [**355**]. It can be applied to any bounded (or, equivalently, compact) subset of $\mathbb{R}^m$ and, more generally,

---

[14] Recall that the Riemann Hypothesis states that $\zeta(s) = 0$, with $s \in \mathbb{C}$, and $0 \le \mathrm{Re}(s) \le 1$ implies that $\mathrm{Re}(s) = 1/2$. See also, e.g., [**124, 125**], [**222, 223**] and [**156**], [**262**], [**264**], [**331**], [**477**], [**588**], along with [**370**], [**375**], [**326**], [**238**], [**330**].

[15] See Exercise 8.14, which states that $0 \le D_{\mathcal{L}} \le 1$, for any (infinite) fractal string $\mathcal{L}$.

[16] See, e.g., [**64, 72, 84, 87, 88, 106–120, 126–129, 131, 135–137, 160, 162–166, 172, 174, 175, 227, 235, 239–242, 290, 297, 303–305, 318–324, 327–329, 331–340, 347–367, 369, 371–373, 376–378, 398, 427, 449–451, 463, 469–471, 480, 481, 498, 499, 508, 516, 586, 587, 600, 601, 631, 632**].

to any relative fractal drum (RFD) of $\mathbb{R}^m$, with $m \geq 1$ being an arbitrary positive integer.[17] It is based on the use of new *fractal zeta functions* (namely, the *distance* and *tube zeta functions*, among others) very significantly extending the geometric zeta functions associated with fractal strings.[18]

Most of these topics will be discussed in a lot more detail in the sequel to the present book, [**119**], which is largely dedicated to providing an accessible exposition of the theory of complex fractal dimensions, both for fractal strings (i.e., in one dimension, which presents many interesting special features) and in higher dimensions, i.e., in the much more general context of arbitrary relative fractal drums (RFDs) in m-dimensional space $\mathbb{R}^m$—including arbitrary bounded subsets of $\mathbb{R}^m$—(and other suitable metric measure spaces, [**606**], [**379**]), where $m \geq 1$ is any integer, along with its geometric, spectral, dynamical, and number-theoretic applications.

Very briefly, a *fractal string* is a (nonempty) bounded open set $\Omega \subseteq \mathbb{R}$ or, more generally, an open set $\Omega \subseteq \mathbb{R}$ with finite length: $\mathcal{L}^1(\Omega) < \infty$. Hence, it can be written as a countable disjoint union of its connected components:[19]

$$\Omega := \bigsqcup_{j=1}^{\infty} I_j, \tag{8.28}$$

where the sets I_j are (necessarily) bounded open intervals of length ℓ_j, for $j = 1, 2, \ldots$. Equivalently, a fractal string is given by a sequence $\mathcal{L} := \{\ell_j\}_{j=1}^{\infty}$ of positive numbers, called the *lengths* of $\mathcal{L}$, tending to zero. Since $\mathcal{L}^1(\Omega) = \sum_{j=1}^{\infty} \ell_j < \infty$, we may assume that $\mathcal{L}$ is written in nonincreasing order (and taking into account the multiplicities):

$$0 < \ell_1 \leq \ell_2 \leq \cdots \leq \ell_j \leq \cdots, \quad \text{with } \ell_j \searrow 0, \text{ as } j \to \infty. \tag{8.29}$$

Then, Ω is called a *geometric realization* of the fractal string $\mathcal{L}$; furthermore, $\partial\Omega$, the boundary of Ω, is called the *boundary* of the fractal string. Even though $\partial\Omega$ changes with Ω, its (upper) Minkowski dimension—and, more generally, its set of complex dimensions—does not depend on the geometric realization Ω of $\mathcal{L}$; i.e., it only depends $\mathcal{L} = \{\ell_j\}_{j=1}^{\infty}$.[20]

---

[17] In this general context, (ordinary) fractal strings can be viewed as a special case of relative fractal drums in $\mathbb{R}$ (hence, $m = 1$), while their boundaries are arbitrary compact subsets of $\mathbb{R}$.

[18] See Subsections 9.3.8 and 9.3.2–9.3.7 below for a brief introduction to RFDs and to fractal zeta functions, respectively.

[19] The case when Ω is a finite disjoint union of intervals is trivial, so that we leave it aside here.

[20] This is why it is also called the (upper) Minkowski dimension of $\mathcal{L}$. Henceforth, it is important to keep in mind that for any fractal string $\mathcal{L}$, in the definition of the ε-neighborhood (as well as of $D_{\mathcal{L}}$), we replace $(\partial\Omega)_\varepsilon$ by $\widetilde{\partial\Omega}_\varepsilon := (\partial\Omega)_\varepsilon \cap \Omega = \{x \in \Omega : d(x, \partial\Omega) < \varepsilon\}$, the *inner ε-neighborhood* of $\partial\Omega$ (relative to Ω)—or, equivalently, the ε-neighborhood of the RFD $(\partial\Omega, \Omega)$, in the sense of Subsection 9.3.8.

Moreover, the *geometric zeta function*, $\zeta_{\mathcal{L}} = \zeta_{\mathcal{L}}(s)$, of $\mathcal{L} = \{\ell_j\}_{j=1}^{\infty}$ is given, for all $s \in \mathbb{C}$ with $\mathrm{Re}(s)$ sufficiently large, by the Dirichlet series

$$\zeta_{\mathcal{L}}(s) := \sum_{j=1}^{\infty} \ell_j^s. \tag{8.30}$$

Then, the *upper Minkowski dimension*, $D_{\mathcal{L}}$, of $\mathcal{L}$ coincides with the abscissa of convergence of $\zeta_{\mathcal{L}}$ (see [**319, 320, 370, 374, 375**]):[21] $D_{\mathcal{L}} = \sigma_{\mathcal{L}}$, so that (8.30) holds for all $s \in \mathbb{C}$ with $\mathrm{Re}(s) > D_{\mathcal{L}}$, and $D_{\mathcal{L}}$ is the best possible (i.e., smallest) such real number.

If, in addition, we assume that the geometric zeta function $\zeta_{\mathcal{L}}$ admits a (necessarily unique) meromorphic extension to a connected open set U strictly containing the half-plane of (absolute) convergence

$$\Pi_{\mathcal{L}} := \{s \in \mathbb{C} : \mathrm{Re}(s) > D_{\mathcal{L}}\}, \tag{8.32}$$

then the *set of visible complex dimensions*, $\mathcal{D}_{\mathcal{L}}(U)$, of $\mathcal{L}$ (*relative to* U) is the set of poles of $\zeta_{\mathcal{L}}$ belonging to U:

$$\mathcal{D}_{\mathcal{L}}(U) := \{\omega \in U : \zeta_{\mathcal{L}}(\omega) = \infty\}. \tag{8.33}$$

If, furthermore, $\zeta_{\mathcal{L}}$ admits a (necessarily unique) meromorphic extension to all of $\mathbb{C}$, then $\mathcal{D}_{\mathcal{L}} := \mathcal{D}_{\mathcal{L}}(\mathbb{C})$ is called the *set of complex dimensions* of $\mathcal{L}$.

It turns out that $\zeta_{\mathcal{L}}$ (being given by a Dirichlet series with nonnegative terms for $\mathrm{Re}(s) > D_{\mathcal{L}}$) is always holomorphic for $\mathrm{Re}(s) > D_{\mathcal{L}}$ (see, e.g., [**331, 496, 545**]) and that $\Pi_{\mathcal{L}} = \{\mathrm{Re}(s) > D_{\mathcal{L}}\}$ is the largest (open, right) half-plane to which $\zeta_{\mathcal{L}}$ can be holomorphically continued, because $\zeta_{\mathcal{L}}$ always has a singularity at $s = D_{\mathcal{L}}$; see loc. cit. Hence, $\mathcal{D}_{\mathcal{L}} \subseteq \{\mathrm{Re}(s) \leq D_{\mathcal{L}}\} \cap U$—and, in particular, when $\zeta_{\mathcal{L}}$ can be meromorphically extended to all of $\mathbb{C}$ (and $U := \mathbb{C}$), then $\mathcal{D}_{\mathcal{L}} = \mathcal{D}_{\mathcal{L}}(\mathbb{C}) \subseteq \mathbb{C}\backslash\Pi_{\mathcal{L}} = \{\mathrm{Re}(s) \leq D_{\mathcal{L}}\}$.

Let us now return to our example of the a-string, Ω_a (as given by (8.24)) or, equivalently, $\mathcal{L}_a = \{\ell_j\}_{j=1}^{\infty}$, where

$$\ell_j := (j+1)^{-a} - j^{-a}, \text{ for } j = 1, 2, \ldots. \tag{8.34}$$

In this case, it is shown in [**375**, Theorem 6.21, p. 205] that $\zeta_{\mathcal{L}}$ admits a meromorphic continuation to all of $\mathbb{C}$ and that its set of complex dimensions is given by

$$\mathcal{D}_a = \mathcal{D}_{\mathcal{L}_a}(\mathbb{C}) = \{D_a, 0, -D_a, -2D_a, \ldots\} = \{D_a\} \cup \{-kD_a : k \in \mathbb{N}_0\}, \tag{8.35}$$

where each complex dimension is simple (i.e., is a simple pole of $\zeta_{\mathcal{L}}$). Here, $D_a = 1/(a+1)$ is the Minkowski dimension of $\mathcal{L}_a$ (i.e., of $\partial\Omega_a$), as given by

---

[21] By definition (see, e.g., [**545**], [**496**], [**375**], [**331**], or [**356**, Subsection 2.1.3]),

$$\sigma_{\mathcal{L}} := \inf\left\{\alpha \in \mathbb{R} : \sum_{j=1}^{\infty} \ell_j^{\alpha} < \infty\right\} = \inf\left\{\mathrm{Re}(s) : \sum_{j=1}^{\infty} \ell_j^{s} \text{ converges absolutely}\right\}. \tag{8.31}$$

It follows that $0 \leq D_{\mathcal{L}} \leq 1$. (Note that if $\mathcal{L}$ is finite, then $\sigma_{\mathcal{L}} = -\infty$, so that $D_{\mathcal{L}} = \max(\sigma_{\mathcal{L}}, 0)$.)

(8.26), and is the complex dimension on the real axis which has the maximal real part.[22]

The example of the a-string will be further discussed (from the point of view of the complex dimensions) in the sequel to this book, [**119**], along with many other examples of fractal stings (including self-similar strings) and of higher-dimensional relative fractal drums (in $\mathbb{R}^m$, with $m \geq 2$), as well as of bounded subsets of $\mathbb{R}^m$—studied both from the point of the Minkowski dimension, the (lower, upper) Minkowski content, and of the complex dimensions. For the complete theory of complex dimensions in dimension 1 and $m \geq 1$, respectively, we refer the interested reader to [**375**] and [**356**].

Remark 8.13. Observe that because $\partial\Omega_a$ is countable, the Hausdorff dimension of $\partial\Omega_a$ must be equal to zero: $H_a := \dim_H(\partial\Omega_a) = 0$, in contrast to the Minkowski dimension, which, since $D_a = 1/(a+1)$, takes any possible value in the open interval $(0, 1)$, as the parameter a varies in $(0, +\infty)$.

Note, moreover, that as was pointed out in [**318**]—in connection to the first author's partial resolution of the Weyl–Berry Conjecture and the sharpness of the remainder estimates obtained in loc. cit.—via examples in $\mathbb{R}^m$ based on the a-string, for any integer $m \geq 1$, that the fact that the Hausdorff dimension of a countable set is equal to zero is directly related to the good mathematical properties of the Hausdorff measure—namely, its countable stability (see part (iii) of Proposition 5.16), itself a consequence of the fact that the notion of Hausdorff dimension is associated to a (countably additive) measure.

This is in sharp contrast with the fact that the (upper) Minkowski content is only finitely subadditive and hence, the (upper) Minkowski dimension is only finitely stable, which allows it to take nonzero values on countable sets and, more importantly (but closely related to that), enables all of the new connections between fractal geometry and many aspects of number theory briefly mentioned or alluded to below and discovered, in particular, in [**238–242, 318–320, 326–330, 339, 340, 352–354, 356, 367, 369, 370, 374, 375**].

Exercise 8.14.

(a) Show that for any fractal sting $\mathcal{L}$, we have that

$$0 \leq D_{\mathcal{L}} \leq 1, \tag{8.36}$$

where $D_{\mathcal{L}}$ is the upper Minkowski dimension of $\mathcal{L}$ (and also coincides with $\sigma_{\mathcal{L}}$, the abscissa of convergence of $\mathcal{L}$, if $\mathcal{L} := \{\ell_j\}_{j=1}^{\infty}$ is infinite).

[*Hint*: Recall that if Ω is a geometric realization of $\mathcal{L}$, then $\mathcal{L}^1(\Omega) < \infty$.]

[22] More precisely, (8.35) yields the set of possible complex dimensions, because for certain values of a, it could be that the residue of $\zeta_{\mathcal{L}}$ at one of the possible complex dimensions $-kD_a$ (with $k \in \mathbb{N}_0$) vanishes. However, for any $a > 0$, D_a is a true complex dimension; i.e., it is a pole of $\zeta_{\mathcal{L}}$.

(b) Deduce from part (a) that the principal complex dimensions of any fractal sting $\mathcal{L}$ (i.e., the complex dimensions with real part $D_{\mathcal{L}}$) must belong to the critical strip (for the Riemann zeta function), $\{0 \leq \operatorname{Re}(s) \leq 1\}$.

We next establish the results concerning the a-string towards the beginning of Example 8.12; see, in particular, (8.26) and (8.27).

Theorem 8.15 (Minkowski Dimension and Content of the a-String; [**318**]). *Given any real number $a > 0$, let Ω_a be the a-string defined as in* (8.24), *Example* 8.12, *and let $\partial\Omega_a$ be its boundary, as given in* (8.25).

Then, $D_a := \dim_M(\partial\Omega_a)$ exists and is given by $D_a = 1/(a+1)$.

Moreover, $\partial\Omega_a$ (i.e., the a-string $\mathcal{L}_a$ as introduced in the latter part of Example 8.12) *is (inner) Minkowski measurable and its (inner) Minkowski content $\mathcal{M}_a$ is given by*

$$\mathcal{M}_a = \mathcal{M}^{D_a}(\partial\Omega_a) = \left(\frac{2^{1-D_a}}{1-D_a}\right) \cdot a^{D_a}. \tag{8.37}$$

Proof. We will follow the proof given in [**318**, Appendix C]. For any $\varepsilon > 0$, let $(\widetilde{\partial\Omega_a})_\varepsilon := (\partial\Omega_a)_\varepsilon \cap \Omega_a = \{x \in \Omega_a : d(x, \partial\Omega_a) < \varepsilon\}$ denote the (inner) ε-neighborhood of $\partial\Omega_a$. We must show that for $d = 1/(a+1)$, the limit of

$$\tilde{L}_d(\varepsilon) := \varepsilon^{-(1-d)} \cdot V_a(\varepsilon) \tag{8.38}$$

exists in $(0, +\infty)$, with $V_a(\varepsilon) := \mathcal{L}^1((\widetilde{\partial\Omega_a})_\varepsilon)$. Of course, once this is established, we deduce that $D_a = 1/(a+1)$ is, in fact, the (inner) Minkowski dimension of $\mathcal{L}_a$ (or, equivalently, of $\partial\Omega_a$), and that the latter exists. For pedagogical reasons, however, we will also show how to "discover" (without prior knowledge) that $d := 1/(a+1)$ is the value of d yielding the (inner) Minkowski dimension of $\mathcal{L}_a$ (or of $\partial\Omega_a$).

Since $\Omega_a = \bigsqcup_{j=1}^{\infty} I_j$, with $I_j = ((j+1)^{-a}, j^{-a})$, for all $j \geq 1$, a moment's reflection shows that—with $\ell_j := (j+1)^{-a} - j^{-a}$ and $(\widetilde{\partial I}_j)_\varepsilon := (\partial I_j)_\varepsilon \cap \Omega_a$, for all $j \geq 1$, as well as for any $\varepsilon \in (0, 1/4)$—

$$\begin{aligned} V(\varepsilon) &= \sum_{j=1}^{\infty} (\widetilde{\partial I}_j)_\varepsilon = \sum_{j:\ell_j \geq 2\varepsilon} 2\varepsilon + \sum_{j:\ell_j < 2\varepsilon} \ell_j \\ &= 2\varepsilon N_{\mathcal{L}_a}\left(\frac{1}{2\varepsilon}\right) + \sum_{j:\ell_j < 2\varepsilon} \ell_j =: A(\varepsilon) + B(\varepsilon), \end{aligned} \tag{8.39}$$

where (with $\mathcal{L}_a := \{\ell_j\}_{j=1}^{\infty}$), for each $x > 0$,

$$N_{\mathcal{L}_a}(x) := \sum_{\ell_j^{-1} \leq (2\varepsilon)^{-1}} 1 = \#(\{j \geq 1 : \ell_j^{-1} \leq x\}) \tag{8.40}$$

denotes the *geometric counting function* of the fractal string $\mathcal{L}_a$, counting the number of reciprocal lengths not exceeding x.

Observe that, in the present case,

$$\ell_j = j^{-a} - (j+1)^{-a} \sim a \cdot j^{-(a+1)} = a \cdot j^{-1/D_a}, \text{ as } j \to \infty, \tag{8.41}$$

where $D_a := 1/(a+1)$. Indeed,

$$\begin{aligned}\ell_j &= j^{-a} \cdot \left(1 - \left(\frac{j+1}{j}\right)^{-a}\right) = j^{-a} \cdot \left(1 - \left(1 + \frac{1}{j}\right)^{-a}\right) \\ &= j^{-a} \cdot \left(1 - \left(1 - \frac{a}{j}\right) + o(1)\right) = a \cdot j^{-(a+1)} \cdot (1 + o(1)),\end{aligned}$$

where $o(1)$ stands for a sequence tending to zero as $j \to \infty$. Thus, (8.41) holds.

From (8.41), it is easy to infer that $N_{\mathcal{L}_a}(x)$, as defined by (8.40), satisfies the following asymptotics:

$$N_{\mathcal{L}_a}(x) \sim (ax)^{D_a} = a^{D_a} \cdot x^{D_a}, \text{ as } x \to +\infty, \tag{8.42}$$

and hence also the first term on the right-hand side of the last equality of (8.39) satisfies

$$\begin{aligned}A(\varepsilon) &:= 2\varepsilon N_{\mathcal{L}_a}\left(\frac{1}{2\varepsilon}\right) \sim 2^{1-D_a} \cdot a^{D_a} \cdot \varepsilon^{1-D_a} \\ &= 2^{a/(a+1)} \cdot a^{1/(a+1)} \cdot \varepsilon^{a/(a+1)}, \text{ as } \varepsilon \to 0^+,\end{aligned} \tag{8.43}$$

while the second term on the right-hand side of the last equality of (8.39) satisfies

$$\begin{aligned}B(\varepsilon) &:= \sum_{j:\ell_j<2\varepsilon} \ell_j = \sum_{j>N_{\mathcal{L}_a}(1/2\varepsilon)} \ell_j \sim \left(\frac{2\varepsilon}{a}\right)^{aD_a} \\ &= \left(\frac{2\varepsilon}{a}\right)^{a/(a+1)}, \text{ as } \varepsilon \to 0^+.\end{aligned} \tag{8.44}$$

Indeed, in light of (8.41), and then of (8.42), we have successively

$$\begin{aligned}B(\varepsilon) &\sim \int_{N_{\mathcal{L}_a}(1/2\varepsilon)}^{+\infty} at^{-(a+1)}dt = \frac{a}{1-(a+1)}[t^{-a}]_{t=N_{\mathcal{L}_a}(1/2\varepsilon)}^{+\infty} \\ &= (N_{\mathcal{L}_a}(1/2\varepsilon))^{-a} \sim \left(\frac{2\varepsilon}{a}\right)^{aD_a}, \text{ as } \varepsilon \to 0^+.\end{aligned} \tag{8.45}$$

Thus, (8.44) holds.

Now, putting (8.43) and (8.44) together, we deduce from (8.39), upon multiplication by $\varepsilon^{-(1-d)}$, that for any $d \geq 0$,

$$\varepsilon^{-(1-d)} \cdot V_a(\varepsilon) \sim 2^{a/(a+1))} \cdot a^{1/(a+1)} \cdot \varepsilon^{a/(a+1)-(1-d)},$$

and since $1 - D_a = a/(a+1) = aD_a$,[23]

$$\varepsilon^{-(1-d)} \cdot V_a(\varepsilon) \sim \left(\frac{2^{1-D_a}}{1-D_a}\right) \cdot a^{D_a} \cdot \varepsilon^{d-D_a}, \quad \text{as } \varepsilon \to 0^+. \tag{8.46}$$

[23]This identity also implies that $1 + (1/a) = (a+1)/a = 1/(1-D_a)$.

Indeed, we have successively

$$(8.47)\qquad \begin{aligned} \varepsilon^{-(1-d)} \cdot V_a(\varepsilon) &\sim 2^{1-D_a} \cdot a^{D_a} \cdot \varepsilon^{d-D_a} + 2^{1-D_a} \cdot a^{-(1-D_a)} \cdot \varepsilon^{d-D_a} \\ &= 2^{1-D_a} \cdot a^{D_a} \cdot \left(1 + \frac{1}{a}\right) \cdot \varepsilon^{d-D_a} \\ &= \left(\frac{2^{1-D_a}}{1-D_a}\right) \cdot a^{D_a} \cdot \varepsilon^{d-D_a}, \end{aligned}$$

where we have used (8.45), in the last equality.

Observe that in (8.47), as is pointed out in [**318**], the contributions of $A(\varepsilon)$ and $B(\varepsilon)$—that is, of the first sum and of the second sum in (8.39)—are precisely equal iff $a = 1$, i.e., iff we are in the *midfractal case* when $D_a = 1/2$, in the terminology of [**318**].[24] Indeed,

$$a^{D_a} = a^{-(1-D_a)} = a^{D_a - 1} \Leftrightarrow a^{-1} = 1 \Leftrightarrow a = 1.$$

It now suffices to let $d := D_a = 1/(a+1)$ in (8.46) in order to deduce at once that

$$(8.48)\qquad \varepsilon^{-(1-D_a)} \cdot V_a(\varepsilon) \sim \left(\frac{2^{1-D_a}}{1-D_a}\right) \cdot a^{D_a}, \quad \text{as } \varepsilon \to 0^+;$$

i.e., since the right-hand side of (8.48) is independent of ε, the following limit exists and is given as follows:

$$\mathcal{M}_a = \mathcal{M}_a^{D_a} = \lim_{\varepsilon \to 0^+} \varepsilon^{-(1-D_a)} \cdot V_a(\varepsilon) = \left(\frac{2^{1-D_a}}{1-D_a}\right) \cdot a^{D_a}.$$

This means that $\mathcal{L}_a$ (i.e., $\partial\Omega_a$) is (inner) Minkowski measurable and that its (inner) Minkowski content $\mathcal{M}_a$ is equal to $(2^{1-D_a}/(1-D_a)) \cdot a^{D_a}$, as desired. Hence, (8.37) holds. Finally, since $\mathcal{L}_a$ is Minkowski measurable, if follows that the Minkowski dimension of $\mathcal{L}_a$ exists and coincides with $D_a = 1/(a+1)$. This completes the proof of the theorem. □

Remark 8.16. Recall that as the parameter a varies in $(0, +\infty)$, the Minkowski dimension takes on all of the values in the open interval $(0, 1)$.

Furthermore, observe, as is pointed out in [**318**], that if a is replaced by its reciprocal $1/a$, then $D_a = 1/(a+1)$ is transformed into $1 - D_a = a/(a+1)$, with equality iff we are in the midfractal case when $D_a = 1/2$; i.e., here iff $a = 1$; see also the comment following (8.47) above.

Even though it may be fortuitous, this fact is significant in light of the comments between fractal strings and their spectra and the Riemann zeta function as well as with the Riemann Hypothesis; see, in particular, [**238, 318–320, 328, 329, 339, 340, 353, 354, 369, 370, 374, 375**] (along with [**119**], for an

[24] Recall from Exercise 8.14 that, for any fractal string $\mathcal{L}$, we have that $0 \le D_{\mathcal{L}} \le 1$, where $D_{\mathcal{L}}$ is the (upper) Minkowski dimension of $\mathcal{L}$.

exposition). Recall that the functional equation for the Riemann zeta function connects $\zeta(s)$ and $\zeta(1-s)$ and that the Riemann Hypothesis states that $\zeta(s)=0,\ 0\le \mathrm{Re}(s)\le 1 \Rightarrow \mathrm{Re}(s)=1/2$; i.e., the critical zeros of ζ are located on the critical line $\mathrm{Re}(s)=1/2$, the vertical line of abscissa $1/2$.

Example 8.17 (*The Cantor String and Set*; [**354**,**370**,**375**]). Consider the *Cantor String*, the canonical geometric realization of which is given as follows. Let $\Omega_{Can} = \bigsqcup\{\text{deleted intervals}\}$ denote the disjoint union of all of the deleted intervals in the construction of the ternary Cantor set, namely,

$$\begin{aligned}
&\left(\frac{1}{3},\frac{2}{3}\right),\ \left(\frac{1}{9},\frac{2}{9}\right),\ \left(\frac{7}{9},\frac{8}{9}\right),\\
&\left(\frac{1}{27},\frac{2}{27}\right),\ \left(\frac{7}{27},\frac{8}{27}\right),\ \left(\frac{19}{27},\frac{20}{27}\right),\ \left(\frac{25}{27},\frac{26}{27}\right),\dots \qquad (8.49)\\
&\dots 2^{n-1} \quad \text{intervals of the same length } \frac{1}{3^n},\dots,\ \text{for } n=1,2,3,\dots.
\end{aligned}$$

Accordingly, the corresponding fractal string $\mathcal{L}_{Can}$ consists of the following lengths:

$$\begin{aligned}
&\frac{1}{3}\ (\text{with multiplicity } 1),\ \frac{1}{9}\ (\text{with multiplicity } 2),\\
&\frac{1}{27}\ (\text{with multiplicity } 4), \text{ and, generally,} \qquad (8.50)\\
&\frac{1}{3^n}\ (\text{with multiplicity } 2^{n-1}),\ \text{for } n=1,2,3,\dots.
\end{aligned}$$

Note that the boundary of Ω_{Can} coincides with the ternary Cantor set: $\partial\Omega_{Can} = C$.

It is shown in [**354**] that the Cantor string and the Cantor set are (inner) Minkowski nondegenerate, with Minkowski dimension (which must therefore exist) equal to $\log_3 2$, as expected:

$$D_{Can} = \dim_M(\partial\Omega_{Can}) = \dim_M(\mathcal{L}_{Can}) = \log_3 2 = \frac{\log 2}{\log 3}, \qquad (8.51)$$

which, as we know, is also equal to the Hausdorff dimension of $\partial\Omega_{Can} = C$ (the Cantor set). Furthermore, the (inner) lower and upper Minkowski contents of the Cantor string and of the Cantor set are precisely computed in [**354**] and shown to be given by (with $D_{Can} = \log_3 2$, as in (8.51))

$$\mathcal{M}_* = \mathcal{M}_{*,Can} = 2^{2-D_{Can}} = 2.5830\dots \quad \text{and} \qquad (8.52)$$

$$\mathcal{M}^* = \mathcal{M}^*_{Can} = 2^{1-D_{Can}} \cdot D_{Can}^{-D_{Can}} \cdot (1-D_{Can})^{1-D_{Can}} = 2.4950\cdots. \qquad (8.53)$$

It follows that the Cantor string and the Cantor set are not Minkowski measurable. We refer the interested reader to the sequel to this book, [**119**], for an exposition of these results and of their proofs (which have been greatly extended in different directions—as well as explained, conceptually—in [**370, 374, 375**], [**356**]).

Here, we limit ourselves to providing the easy, but enlightening, computation of the geometric zeta function ζ_{Can} and of the complex dimensions of the Cantor string.

In light of (8.50) and of the definition (given in (8.30)) of the geometric zeta function of a fractal string $\mathcal{L} = \{\ell_j\}_{j=1}^{\infty}$, we have successively

$$\begin{aligned}\zeta_{Can}(s) &= \sum_{j=1}^{\infty} \ell_j^s = \sum_{n=1}^{\infty} 2^{n-1} \cdot (3^{-n})^s \\ &= \frac{1}{2}\sum_{n=1}^{\infty}(2 \cdot 3^{-s})^n = \frac{1}{2}(2 \cdot 3^{-s}) \sum_{n=0}^{\infty}(2 \cdot 3^{-s})^n \qquad (8.54)\\ &= 3^{-s}\frac{1}{1-2\cdot 3^{-s}} = \frac{3^{-s}}{1-2\cdot 3^{-s}} = \frac{1}{3^s-2},\end{aligned}$$

where these equalities hold for $|2 \cdot 3^{-s}| = 2 \cdot 3^{-\operatorname{Re}(s)} < 1$ (i.e., for $\operatorname{Re}(s) > \log 2/\log 3$), in order for the geometric series $\sum_{n=0}^{\infty}(2 \cdot 3^{-s})^n$ to converge.

In summary, in light of (8.54) and the text following it, we have shown that the Dirichlet series $\sum_{j=1}^{\infty} \ell_j^s$—initially defining $\zeta_{\mathcal{L}}(s)$—converges (absolutely) for $\operatorname{Re}(s) > D_{Can} = \log_3 2$ and that D_{Can} is the largest real number α such that this series converges for $\operatorname{Re}(s) > \alpha$.

Therefore, we have verified directly for this example that the following identity holds: $\log_3 2 = D_{Can} = \sigma_{Can}$, the abscissa of convergence of the Dirichlet series $\sum_{j=1}^{\infty} \ell_j^s$ (as defined in (8.31) of footnote 21 on page 322; see the discussion following (8.30)). We have also shown that ζ_{Can} admits a (necessarily unique) meromorphic continuation to all of $\mathbb{C}$, given by

$$\zeta_{Can}(s) = \frac{3^{-s}}{1-2\cdot 3^{-s}} = \frac{1}{3^s-2}, \quad \text{for all } s \in \mathbb{C}. \qquad (8.55)$$

Indeed, in light of (8.54) and the text following it, the identity (8.55) holds for all $s \in \mathbb{C}$ with $\operatorname{Re}(s) > D_{Can} = \log_3 2$, but since the right-hand side of (8.55) is clearly a meromorphic function in all of $\mathbb{C}$, it follows from the Principle of Analytic (i.e., here, Meromorphic) Continuation that ζ_{Can} can be (uniquely) meromorphically extended to the entire complex plane and that, also, the identity (8.55) still holds for all $s \in \mathbb{C}$, as claimed.

We next calculate the complex dimensions of the Cantor string as follows. Since complex dimensions are the poles of ζ_{Can}, we deduce at once from (8.55) that (by using the properties of the principal branch of the complex logarithm)

$$
\begin{aligned}
\omega \in \mathcal{D}_{Can} = \mathcal{D}_{Can}(\mathbb{C}) &\Leftrightarrow 3^s = 2 \\
&\Leftrightarrow s = \frac{\log 2}{\log 3} + \frac{2i\,n\pi}{\log 3}, \quad \text{with } n \in \mathbb{Z} \text{ arbitrary; i.e.,}
\end{aligned}
$$

$$
\begin{aligned}
\mathcal{D}_{Can} = \mathcal{D}_{Can}(\mathbb{C}) &= \left\{\log_3 2 + i\,n\frac{2\pi}{\log 3} \;:\; n \in \mathbb{Z}\right\} \\
&= \{D_{Can} + i\,n\mathbf{p}_{Can} \;:\; n \in \mathbb{Z}\},
\end{aligned} \tag{8.56}
$$

where $\mathcal{D}_{Can} = \log_3 2$ is the Minkowski dimension of the Cantor string $\mathcal{L}_{Can}$ (as well as of its boundary the Cantor set $\partial\Omega_{Can}$) and $\mathbf{p}_{Can} = 2\pi/\log 3$ is called the *oscillatory period* of the lattice, self-similar fractal string $\mathcal{L}_{Can}$ (or of C)—for reasons that are explained in detail in [**370**, **375**] and which will also be provided in [**119**]. It is not difficult to show that, in the present case, $\mathcal{D}_{Can}$ is also the set of complex dimensions of the Cantor set $C = \partial\Omega_{Can}$, the boundary of the Cantor string; see Figure 8.3.

As an example of application of the notion of complex dimensions, we provide (without proof) the *fractal tube formula* for the Cantor string, which expresses the volume (i.e., length, here, since $m = 1$) of the sufficiently small (inner) ε-neighborhoods of the Cantor string. More specifically, if

$$
V(\varepsilon) = V_{Can}(\varepsilon) = \mathcal{L}^1((\widetilde{\partial\Omega}_{Can})_\varepsilon), \quad \text{where } (\widetilde{\partial\Omega}_{Can})_\varepsilon := (\partial\Omega_{Can})_\varepsilon \cap \Omega_{Can}, \tag{8.57}
$$

then (see [**370**, Subsection 1.1.2] or [**375**, Subsection 1.1.2, pp. 12–16]), for all $\varepsilon \in (0, 1/2]$,

$$
V(\varepsilon) = V_{Can}(\varepsilon) = (2\varepsilon)^{1-D} \cdot \left(\left(\frac{1}{2}\right)^{\{-\log_3(2\varepsilon)\}} + \left(\frac{3}{2}\right)^{\{-\log_3(2\varepsilon)\}}\right) - 2\varepsilon, \tag{8.58}
$$

where $\{y\} := y - [y] \in [0, 1)$ denotes the fractional part of $y \in \mathbb{R}$ and $[y] \in \mathbb{Z}$ is the integer part of y. Here and henceforth, we let, for notational simplicity,

$$
D := D_{Can} = \log_3 2 = \frac{\log 2}{\log 3} \quad \text{and} \quad \mathbf{p} := \mathbf{p}_{Can} = \frac{2\pi}{\log 3} \tag{8.59}
$$

be, respectively, the Minkowski dimension and the oscillatory period of the Cantor string Ω_{Can} or $\mathcal{L}_{Can}$ (as well as of the Cantor set $C = \partial\Omega_{Can}$, here).

Observe that the function between parentheses in (8.58) is bounded, nonconstant, and multiplicatively periodic (with multiplicative period 3, or, equivalently, 1/3). More specifically, it takes the same value at ε and $\varepsilon/3$; see Figure 8.4.

We next use the following Fourier series expansion for the bounded, nonconstant, 1-periodic function $x \mapsto b^{-\{x\}}$, where $b > 0$, $b \neq 1$ is arbitrary (see

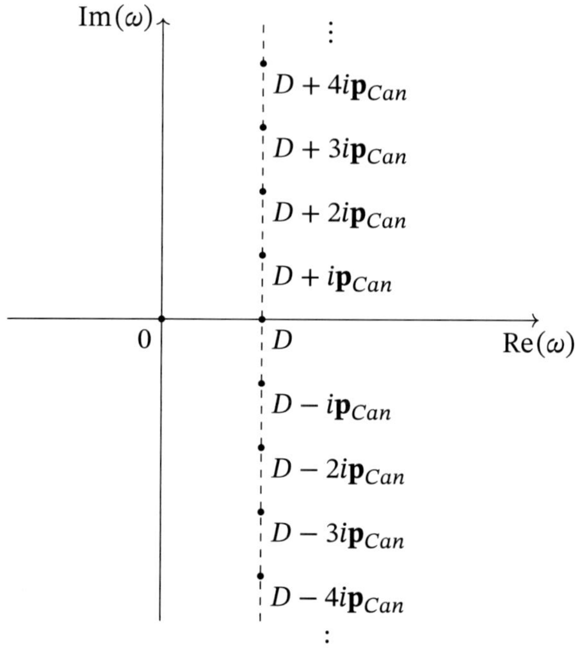

Figure 8.3. The set of complex dimensions of the Cantor string $\mathcal{L}_{Can}$ (and of the Cantor set C), as given by (8.56). Here $D = \log_3 2$ and $\mathbf{p}_{Can} = 2\pi/\log 3$, are, respectively, the Minkowski dimension and the oscillatory period of $\mathcal{L}_{Can}$ (as well as of $C = \partial\mathcal{L}_{Can}$).

[**375**, equation (1.13), p. 14]):

$$b^{-\{x\}} = \left(\frac{b-1}{b}\right) \cdot \sum_{n\in\mathbb{Z}} \frac{e^{2\pi i n x}}{\log b + 2\pi i n}. \tag{8.60}$$

Applying (8.60) to both $b := 1/2$ and $b := 3/2$ and substituting the result into the right-hand side of (8.58), we deduce from (8.58) that

$$V(\varepsilon) = V_{Can}(\varepsilon) = \frac{1}{2\log 3} \cdot \sum_{n\in\mathbb{Z}} \frac{(2\varepsilon)^{1-D-in\mathbf{p}}}{(D+in\mathbf{p})(1-D-in\mathbf{p})} - 2\varepsilon. \tag{8.61}$$

In light of (8.56) and (8.55), which give the complex dimensions and the geometric zeta function of the Cantor string (as well as of the Cantor set), we can rewrite (8.61) just above in the following form:[25]

$$V(\varepsilon) = V_{Can}(\varepsilon) = \sum_{\omega\in\mathcal{D}_{Can}} \operatorname{res}(\zeta_{Can}, \omega) \cdot \frac{2^{1-\omega} \cdot \varepsilon^{1-\omega}}{\omega(1-\omega)} - 2\varepsilon, \tag{8.62}$$

[25] We need to observe, in particular, that, for any $\omega \in \mathcal{D}_{Can}$, the residue at $s := \omega$ of $\zeta_{Can}(s)$ (as expressed in (8.55)) is given by $\operatorname{res}(\zeta_{Can}, \omega) = \lim_{s\to\omega}(s-\omega)\zeta_{Can}(s) = 1/2\log 3$.

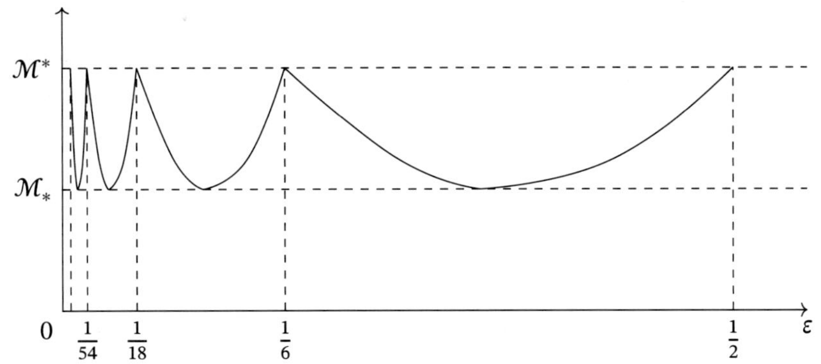

(a)

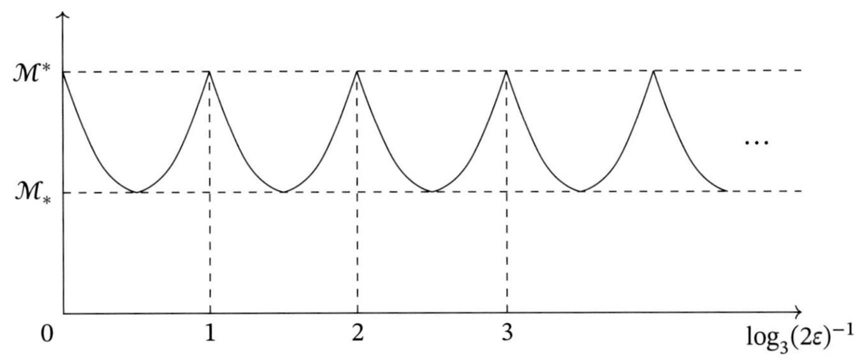

(b)

Figure 8.4. (a) The function $\varepsilon \mapsto \varepsilon^{-(1-D)} \cdot (V_{Can}(\varepsilon) + 2\varepsilon)$ viewed multiplicatively. Here, $\mathcal{M}_*$ and $\mathcal{M}^*$, the (inner) lower and upper Minkowski content of the Cantor string (or Cantor set) are, respectively, given by (8.52) and (8.53). (b) The same function as in (a), now viewed additively, i.e., as a function of $\log_3(2\varepsilon)^{-1}$.

which is the general form of the fractal tube formula for fractal strings, in the present case of simple complex dimensions; i.e., all of the poles of the geometric zeta function ζ_{Can} are simple; see [**375**, Chapter 8].

Observe that the fractal tube formula in (8.62) *expresses the tube function* $V_{Can} = V_{Can}(\varepsilon)$ *as a* "fractal power series" *with exponents the fractal complex codimensions* $1 - \omega$ *of the fractal string* Ω_{Can} *(or of the Cantor set), where* $\omega \in \mathcal{D}_{Can}$. Note that $\Omega_{Can} \subseteq \mathbb{R}$ and $C = \partial\Omega_{Can} \subseteq \mathbb{R}$, so that the "embedding dimension" is $m := 1$ and hence, $1 - \omega$ is indeed a *complex codimension.*

Note that because nonreal complex dimensions come in complex conjugate pairs, $V_{Can}(\varepsilon)$—as given by the right-hand side of (8.61) or, equivalently, of (8.62)—is real-valued. This is called the *Reality Principle* in [**375**].

We close this discussion by pointing out that the complex dimensions of the Cantor string Ω_{Can} (or of the Cantor set, $\partial\Omega_{Can} = C$) are in vertical arithmetic progression (with ratio $i\mathbf{p}_{Can} = i2\pi/\log 3$) and therefore, in light of (8.61), give rise to logarithmic periodic—or, equivalently, multiplicatively periodic—oscillations, as follows (see also Figure 8.4):

$$V_{Can}(\varepsilon) + 2\varepsilon = \varepsilon^{1-D} \cdot G_{Can}(\log_3(\varepsilon^{-1})), \tag{8.63}$$

where G_{Can} is a continuous, bounded, and nonconstant 1-periodic function given for all $x \in \mathbb{R}$ by

$$G_{Can}(x) := \frac{1}{2\log 3} \cdot \sum_{n\in\mathbb{Z}} \frac{2^{1-D-in\mathbf{p}_{Can}}}{(D + in\mathbf{p}_{Can})(1 - D - in\mathbf{p}_{Can})} e^{i2\pi nx}. \tag{8.64}$$

Also, G_{Can} is bounded away from zero and infinity. More specifically,

$$0 < \mathcal{M}_* = \min_{x\in[0,1]} G_{Can}(x) < \max_{x\in[0,1]} G_{Can}(\boldsymbol{x}) = \mathcal{M}^* < \infty, \text{ for all } x \in \mathbb{R}, \tag{8.65}$$

where $\mathcal{M}_*$ and $\mathcal{M}^*$ are, respectively, the lower and upper (inner) Minkowski content of the Cantor string (or Cantor set), as given by (8.52) and (8.53).

The details of the computation leading to the above fractal tube formula (8.64) can be found in loc. cit., as well as in the sequel to this book [**119**].

The following exercise is intended for the reader who is familiar with Fourier series.

Exercise 8.18. Justify the Fourier series expansion stated in (8.60) and show that the series is conditionally convergent for all $x \in \mathbb{R}$; i.e., it is convergent but not absolutely convergent.

Exercise 8.19. Define the logarithmic *average Minkowski content* of the Cantor string (or set) by the following Cesàro average of $V(\varepsilon)/\varepsilon^{1-D}$:

$$\widetilde{\mathcal{M}}_{Can} = \mathcal{M}_{av,Can} := \lim_{t\to\infty} \frac{1}{\log t} \int_{1/t}^{1} \frac{V(\varepsilon)}{\varepsilon^{1-D}} \frac{d\varepsilon}{\varepsilon}. \tag{8.66}$$

Show that $\mathcal{M}_{av,Can}$ is, indeed, well defined and calculate its value. More specifically show that $\mathcal{M}_{av,Can} = (1/1) \int_0^1 G_{Can}(\tau) d\tau$.

[*Hint*: Use (8.61) or (8.63) and (8.64).]

Many additional examples of computations of Minkowski content—or, if it does not exist, of suitably averaged content, as in Exercise 8.19 just above—or else of lower and upper Minkowski contents, as well as of (upper) Minkowski dimension and of complex dimensions, in one and in higher dimensions, can be found in [**375**], [**356**], along with [**119**].

8.3. Box-Counting Dimension and Content

A well-known equivalent formulation of Minkowski dimension is *box-counting dimension* or *box dimension*, in short. Note that, at least intuitively, for a bounded nonempty set $A \subseteq \mathbb{R}^m$, we have

$$\mathscr{L}^m(A_\varepsilon) \approx c \cdot \varepsilon^m \cdot N_B(A, \varepsilon^{-1}), \tag{8.67}$$

which is to say that the volume of the open ε-neighborhood of a set A is approximately equal to the product of $N_B(A, \varepsilon^{-1})$, the maximum number of disjoint closed balls of radius ε with centers in A, and the "size" of each ball, given by $c \cdot \varepsilon^m$, where c is an appropriately defined positive constant.

Definition 8.20 (*Box-Counting Function*). Let A be a bounded subset of $\mathbb{R}^m$. The *box-counting function* of A is the map $N_B(A, \cdot)\colon (0, +\infty) \to \mathbb{N}_0$, where $N_B(A, x)$ denotes the maximum number of disjoint closed balls $B_{\boldsymbol{a}}(x^{-1})$ with centers $\boldsymbol{a} \in A$ of radius $x^{-1} > 0$.

Remark 8.21. Although it may seem unnatural to define the box-counting function in terms of the reciprocal of the radius $\varepsilon := x^{-1}$, this perspective is used later in [**119**] in order to connect to the theory of complex dimensions of fractal strings (as is done in [**367**] and [**535**]); thus, the convention is adopted here. Note that the geometric aspects of the box-counting function are still understood in this context, since $N_B(A, \varepsilon^{-1})$ denotes the maximum number of disjoint closed balls $B_{\boldsymbol{a}}(\varepsilon)$ with centers $\boldsymbol{a} \in A$ and radii $\varepsilon := x^{-1}$. Also, ε tends to 0 iff x tends to $+\infty$.

Definition 8.22 (*Lower and Upper Box-Counting Dimensions*). Given any (nonempty) bounded set $A \subseteq \mathbb{R}^m$, the *lower* and *upper box-counting dimensions* of A are given, respectively, by

(8.68)

$$\underline{\dim}_B(A) = \liminf_{x \to +\infty} \frac{\log N_B(A, x)}{\log x} \quad \text{and} \quad \overline{\dim}_B(A) = \limsup_{x \to +\infty} \frac{\log N_B(A, x)}{\log x}.$$

When $\underline{\dim}_B(A) = \overline{\dim}_B(A)$, the following limit exists and is called the *box-counting dimension* of A, denoted $\dim_B(A)$:

$$\dim_B(A) = \lim_{x \to +\infty} \frac{\log N_B(A, x)}{\log x}. \tag{8.69}$$

In this case, the *box-counting dimension* (or *box dimension*, in short) of A is said to *exist*. (As usual, we let $\dim_B(\emptyset) = -\infty$.)

Note that in the literature, as well as in many places in this book and its sequels ([**119**],[**116**], [**117**]), the box-counting dimension is called more briefly the *box dimension*, and similarly for the lower and upper box-counting dimensions.

Clearly, in light of Definition 8.22, we always have that

$$0 \le \underline{\dim}_B(A) \le \overline{\dim}_B(A) \le m. \tag{8.70}$$

As noted throughout the literature, the types of “boxes” counted by a box-counting function can vary. Given $\varepsilon > 0$, $N_B(A, \varepsilon^{-1})$ may be taken to be the minimum number of sets of diameter ε required to cover A. Another commonly used notion of box-counting function is provided by taking $N_B(A, \varepsilon^{-1})$ to be the number of ε-mesh cubes that intersect A. The *ε-mesh* (or *ε-grid*) is the family of m-dimensional cubes of the form

$$[n_1\varepsilon, n_1(1+\varepsilon)] \times \cdots \times [n_k\varepsilon, n_k(1+\varepsilon)], \text{ where } n_1, \dots, n_k \text{ are in } \mathbb{Z}.$$

Proposition 8.23. *For a set $A \subseteq \mathbb{R}^m$ and $\varepsilon > 0$, the value of the (upper and lower) box-counting dimension(s) in Definition* 8.22 *remain unchanged when the box-counting function $N_B(A, \varepsilon^{-1})$ is given by any of the following:*

(i) *the maximum number of disjoint closed balls of radius ε with centers in A, denoted $N_1(A, \varepsilon^{-1})$;*

(ii) *the minimum number of sets of diameter ε required to cover A, denoted $N_2(A, \varepsilon^{-1})$;*

(iii) *the total number of ε-mesh cubes that intersect A, denoted $N_3(A, \varepsilon^{-1})$.*

Proof. In order to prove the equivalence (i) $\Leftrightarrow$ (ii), let $\{B_j\}_{j=1}^{N_1(A,\varepsilon^{-1})}$ denote a disjoint collection of closed balls with radii ε and centers in A. If $\boldsymbol{a} \in A$, then $\operatorname{dist}(\boldsymbol{a}, B_j) < \varepsilon$, for some j; otherwise, another ball of radius ε centered at $\boldsymbol{a}$ could be added to form a larger collection of disjoint balls. So, a 4ε-cover of A is obtained by taking the $N_1(A, \varepsilon^{-1})$ balls concentric with the B_j's and radii 2ε (thus, diameter 4ε). Hence, $N_2(A, (4\varepsilon)^{-1}) \le N_1(A, \varepsilon^{-1})$. Now, suppose that $\{B_j\}_{j=1}^{N_1(A,\varepsilon^{-1})}$ denotes a disjoint collection of closed balls with radii $\varepsilon \in (0,1)$ and centers in A. Let $\{A_n\}_{n=1}^{k}$ be an ε-cover of A. Since the center of each ball B_j is contained in some A_n, each B_j contains at least one A_n. Also, since the balls B_j are disjoint, there are at least as many A_n's as B_j's. Thus, $N_1(A, (4\varepsilon)^{-1}) \le N_2(A, \varepsilon^{-1})$. So, combining the above two inequalities and dividing by $-\log \varepsilon > 0$ yields

$$\frac{\log N_2(A, (4\varepsilon)^{-1})}{-\log 4\varepsilon + \log 4} \le \frac{\log N_1(A, \varepsilon^{-1})}{-\log \varepsilon} \le \frac{\log N_2(A, \varepsilon^{-1})}{-\log \varepsilon}.$$

Letting $\varepsilon \to 0^+$—or, equivalently, with $x := \varepsilon^{-1}$, letting $x \to +\infty$—yields

$$\limsup_{x\to+\infty} \frac{\log N_2(A,x)}{\log x} \le \limsup_{x\to+\infty} \frac{\log N_1(A,x)}{\log x} \le \limsup_{x\to+\infty} \frac{\log N_2(A,x)}{\log x}.$$

Hence, the value of the upper box-counting dimension in Definition 8.22 remains the same when $N_B(A,x)$ is replaced by either $N_1(A,x)$ or $N_2(A,x)$. By a similar proof, the same can be said for the lower box-counting dimension.

Now, to prove (ii) ⇔ (iii), note that a set of diameter at most ε is contained in 3^m ε-mesh cubes. Indeed, an ε-mesh cube containing a point of A taken along with its $3^m - 1$ neighboring ε-mesh cubes provides an ε-cover of A. So, $N_3(A, \varepsilon^{-1}) \leq 3^m \cdot N_2(A, \varepsilon^{-1})$. On the other hand, the $N_3(A, \varepsilon^{-1})$ ε-mesh cubes which intersect A provide a collection of $N_3(A, \varepsilon^{-1})$ sets of diameter $\varepsilon\sqrt{m}$ that covers A. Thus, $N_2(A, (\varepsilon\sqrt{m})^{-1}) \leq N_3(A, \varepsilon^{-1})$. So, combining the last two inequalities and then dividing by $-\log \varepsilon > 0$ yields

$$\frac{\log N_2(A, (\varepsilon\sqrt{m})^{-1})}{-\log(\varepsilon\sqrt{m}) + \log\sqrt{m}} \leq \frac{\log N_3(A, \varepsilon^{-1})}{-\log \varepsilon} \leq \frac{\log 3^m + \log N_2(A, \varepsilon^{-1})}{-\log \varepsilon}.$$

Following an argument similar to the one provided in the proof of the equivalence (i) ⇔ (ii) above, we show that the values of the upper and lower box-counting dimensions in Definition 8.22 remain the same when $N_B(A, x)$ is replaced by either $N_2(A, x)$ or $N_3(A, x)$. □

The type of calculation provided in Example 8.24 below to find the box-counting dimension of the Cantor set C is fairly standard in fractal geometry. It features a pair of arguments, one to find an upper bound and another to find a lower bound, where each argument has a geometric step followed by an analytic step.

Example 8.24 (*Cantor Set*). Consider the Cantor set C and suppose that $3^{-n} < \varepsilon \leq 3^{-(n-1)}$, for some $n \in \mathbb{N}$. Then, there are 2^n disjoint closed intervals of length 3^{-n} which provide an ε-cover of C; see Example 1.1. Hence, $N_B(C, \varepsilon^{-1}) \leq 2^n$, where N_B is given by the minimum number of sets in an ε-cover of C, and we have

$$\overline{\dim}_B(A) = \limsup_{x \to +\infty} \frac{\log N_B(A, x)}{\log x} \leq \limsup_{n \to \infty} \frac{\log 2^n}{-\log 3^{-(n-1)}} = \log_3 2. \tag{8.71}$$

Now, an interval of length ε, where $3^{-(n+1)} < \varepsilon \leq 3^{-n}$, intersects at most one of the 2^n closed intervals of length 3^{-n} in the n-th prefractal approximation in the construction of C; see Example 1.1. So, at least 2^n such intervals of length ε are required to cover C. Hence, $N_B(C, \varepsilon^{-1}) \geq 2^n$, and therefore,

$$\underline{\dim}_B(A) = \liminf_{x \to +\infty} \frac{\log N_B(A, x)}{\log x} \leq \liminf_{n \to \infty} \frac{\log 2^n}{-\log 3^{-(n+1)}} = \log_3 2. \tag{8.72}$$

Exercise 8.25.

(a) (*Generalized Cantor Sets*). Generalize Example 8.24 to the class of Smith–Volterra–Cantor (SVC) sets described in Example 4.92 of Section 4.9.

(b) (*Other Fractal Sets*). Obtain the counterpart of the results of Example 8.24 for the Koch curve K_C, the Sierpinski gasket **SG**, and the Sierpinski carpet **SC** described, respectively, in Examples 4.31, 4.33, and 4.34 of Section 4.3.

The concept underlying the following proposition generalizes the technique exhibited in Example 8.24 and shows that one need only consider certain discrete sequences of scales which tend to zero in order to determine the box-counting dimension of a set.

Proposition 8.26. *Let* $0 < \lambda < 1$ *and let* A *be a bounded subset of* $\mathbb{R}^m$. *Then* (*with* $k \in \mathbb{N}$),

$$\underline{\dim}_B(A) = \liminf_{k\to\infty} \frac{\log N_B(A,\lambda^k)}{-\log \lambda^k}, \quad \text{and} \tag{8.73}$$

$$\overline{\dim}_B(A) = \limsup_{k\to\infty} \frac{\log N_B(A,\lambda^k)}{-\log \lambda^k}. \tag{8.74}$$

Proof. Only the upper limit is considered in this proof since the case of the lower limit follows by a similar argument. If $\lambda^{k+1} \leq \varepsilon < \lambda^k$, then

$$\frac{\log N_B(A,\varepsilon^{-1})}{-\log \varepsilon} \leq \frac{\log N_B(A,\lambda^{-(k+1)})}{-\log \lambda^k} = \frac{\log N_B(A,\lambda^{-(k+1)})}{-\log \lambda^{k+1} + \log \lambda}. \tag{8.75}$$

Therefore, with $x := \varepsilon^{-1}$,

$$\limsup_{x\to+\infty} \frac{\log N_B(A,x)}{\log x} \leq \limsup_{k\to\infty} \frac{\log N_B(A,\lambda^{-k})}{-\log \lambda^k}. \tag{8.76}$$

The reverse inequality is immediate. □

The terminology "box-counting content" does not seem to appear in the literature. It is naturally motivated by the characterization of the Minkowski measurability (and of Minkowski nondegeneracy) obtained for fractal strings (i.e., when $m = 1$) by the first author and C. Pomerance in [**353**] (announced in [**352**]); see also [**319, 320, 329, 354**], along with [**370, 374, 375**]. This characterization will be discussed in detail in the sequel to this book, [**119**].

A very brief mention of the Minkowski Measurability Criterion obtained in [**353**] states that an ordinary fractal string $\mathcal{L}$ with geometric counting function $N_{\mathcal{L}}$ is Minkowski measurable (resp., Minkowski nondegenerate), with Minkowski dimension $D \in (0,1)$ iff $N_{\mathcal{L}}(x) \sim C \cdot x^D$, for some constant $C \in (0,+\infty)$ (resp., $N_{\mathcal{L}}(x) \asymp x^D$), as $x \to +\infty$.[26]

Furthermore, the notion of box-counting content (see Definition 8.27 below) provides a natural setting for a very useful result of S. P. Lalley in [**313**] (see also [**314, 315**]) and it allows for content to be studied in the context of box-counting functions. Note that for a bounded set $A \subseteq \mathbb{R}^m$, we have, at least loosely,

$$\mathscr{L}^m(A_\varepsilon) \approx \varepsilon^m \cdot N_B(A,\varepsilon^{-1}). \tag{8.77}$$

[26]See Remark 2.27 (resp., part (iii) of Notation 2.26) for the meaning of the symbol $\sim$ (resp., $\asymp$) in this context.

That is, the volume of the open ε-neighborhood of a set A is approximately equal to the product of the maximum number of disjoint closed balls of radius ε with centers in A and the "size" of each ball given by ε^m.

Definition 8.27 (*Upper and Lower Box-Counting Contents*). Let $A \subseteq \mathbb{R}^m$ be a bounded set such that $D_B = \dim_B(A)$ exists. Then, the *upper* and *lower box-counting contents* of A are respectively denoted by

$$\mathscr{B}^*(A) := \limsup_{x \to +\infty} \frac{N_B(A,x)}{x^{D_B}} \quad \text{and} \quad \mathscr{B}_*(A) := \liminf_{x \to +\infty} \frac{N_B(A,x)}{x^{D_B}}. \tag{8.78}$$

If $\mathscr{B}^*(A) = \mathscr{B}_*(A)$, the common value, denoted $\mathscr{B}(A)$, is called the *box-counting content* of A and is given by

$$\mathscr{B}(A) = \lim_{x \to +\infty} \frac{N_B(A,x)}{x^{D_B}}. \tag{8.79}$$

Additionally, if $0 < \mathscr{B}^*(A) = \mathscr{B}_*(A) < \infty$, then A is said to be *box-counting measurable.*

Example 8.28. For the closed unit interval $[0,1]$, it is readily shown that $\dim_B([0,1]) = \dim_M([0,1]) = 1$, $[0,1]$ is Minkowski measurable, and, in fact, $\mathscr{M}([0,1]) = 1$. It is also true that $[0,1]$ is box-counting measurable. In this case, $\mathscr{B}([0,1]) = 1/2$.

According to the aforementioned Minkowski Measurability Criterion obtained in [**353**], the Cantor set C is neither Minkowski measurable nor box-counting measurable. In fact, as was first shown in [**352**] and [**353**] (see also [**369, 370, 374, 375**] for significant extensions), we have that $0 < \mathscr{M}_*(C) < \mathscr{M}^*(C) < \infty$ and, similarly, $0 < \mathscr{B}_*(C) < \mathscr{B}^*(C) < \infty$.

8.4. Properties of Minkowski and Box-Counting Dimensions

The following theorem is a classic result in dimension theory. It says that upper and lower Minkowski and box-counting dimensions are equivalent.

Theorem 8.29 (Equality of Box and Minkowski Dimensions). *Let A be a nonempty bounded subset of $\mathbb{R}^m$. Then,*

$$\underline{\dim}_B(A) = \underline{\dim}_M(A) \quad \text{and} \quad \overline{\dim}_B(A) = \overline{\dim}_M(A). \tag{8.80}$$

In particular, if either of the fractal dimensions $\dim_B(A)$ *or* $\dim_M(A)$ *exists, then so does the other one and we have the following equality:*

$$\dim_B(A) = \dim_M(A); \tag{8.81}$$

i.e., the Minkowski and box dimensions coincide.

Proof. Recall that, by convention, $\dim_B(\emptyset) = \dim_M(\emptyset) = -\infty$. Hence, without loss of generality, we may assume that A is nonempty.

In this proof, $N_B(A, \varepsilon^{-1})$ is taken to be given by $N_1(A, \varepsilon^{-1})$, the maximum number of disjoint closed ball of radius ε with centers in A and $\underline{\dim}_M$ is as in Theorem 8.10. Also, only the lower dimensions are considered here, as the proof in the case of the upper dimensions follows by a similar argument.

Suppose $0 < \varepsilon < 1$. Then, the union of $N_B(A, \varepsilon^{-1})$ disjoint closed balls with centers in A is contained in $[A]_\varepsilon$. Hence,

$$N_B(A, \varepsilon^{-1}) \cdot \varepsilon^m \cdot \omega_m \le \mathscr{L}^m(A_\varepsilon),$$

where ω_m is the volume of the unit ball in $\mathbb{R}^m$. Thus,

$$\frac{\log N_B(A, \varepsilon^{-1}) + \log(2^m \cdot \omega_m) + m \log \varepsilon}{-\log \varepsilon} \le \frac{\log \mathscr{L}^m(A_\varepsilon)}{-\log \varepsilon}.$$

Taking $\varepsilon \to 0^+$—or, equivalently, letting $x \to +\infty$, with $x := \varepsilon^{-1}$—yields

$$\underline{\dim}_B(A) - m \le \liminf_{\varepsilon \to 0^+} \frac{\log \mathscr{L}^m(A_\varepsilon)}{-\log \varepsilon}. \tag{8.82}$$

Again, assume that $0 < \varepsilon < 1$. Then, A_ε is contained in the union of the $N_B(A, \varepsilon^{-1})$ closed balls concentric with those that determine the value of $N_B(A, \varepsilon^{-1})$ but with radii 2ε. So, $\mathscr{L}^m(A_\varepsilon) \le N_B(A, \varepsilon^{-1})(2\varepsilon)^m \cdot \omega_m$, and thus, $\frac{\log \mathscr{L}^m(A_\varepsilon)}{-\log \varepsilon} \le \frac{\log N_B(A,\varepsilon^{-1}) + \log(2^m \cdot \omega_m) + m \log \varepsilon}{-\log \varepsilon}$. Hence, taking $\varepsilon \to 0^+$—or, equivalently, letting $x \to +\infty$—yields

$$\liminf_{\varepsilon \to 0^+} \frac{\log \mathscr{L}^m(A_\varepsilon)}{-\log \varepsilon} \le \underline{\dim}_B(A) - m. \tag{8.83}$$

We conclude by combining the inequalities (8.82) and (8.83) that the first equality in (8.80) holds. As was alluded to above, the proof of the second equality in (8.80) is entirely similar.

Finally, we note that if $\dim_B(A)$ exists (i.e., if $\underline{\dim}_B(A) = \overline{\dim}_B(A)$), then, in light of the two equalities in (8.80), we have that $\underline{\dim}_M(A) = \overline{\dim}_M(A)$; thus, $\dim_M(A)$ exists and equals $\dim_B(A)$. The reasoning is entirely analogous if we interchange the roles of the box and the Minkowski dimensions. This completes the proof of the theorem. □

The Minkowski dimension—and, equivalently, the box-counting dimension, in light of Theorem 8.29—exhibit some of the same properties as the Hausdorff dimension such as bi-Lipschitz invariance (see Proposition 8.36 below). However, there are some significant differences as well.

Proposition 8.30. *Let $\overline{A}$ denote the closure of $A \subseteq \mathbb{R}^m$. Then,*

$$\underline{\dim}_M(\overline{A}) = \underline{\dim}_M(A) \quad \text{and} \quad \overline{\dim}_M(\overline{A}) = \overline{\dim}_M(A). \tag{8.84}$$

In particular, if $\dim_M(A)$ *exists, then so does* $\dim_M(\overline{A})$ *and then,* $\dim_M(A) = \dim_M(\overline{A})$.

Proof. For every $\varepsilon > 0$, we have that A_ε, the open ε-neighborhood of A, is equal to the open ε-neighborhood of $\overline{A}$. The results follow immediately from Theorem 8.10 (or simply from Definitions 8.2 and 8.3), together with the elementary (and easily verified) fact, already used in several places in this book and according to which $\operatorname{dist}(\boldsymbol{x}, A) = \operatorname{dist}(\boldsymbol{x}, \overline{A})$, for all $\boldsymbol{x} \in \mathbb{R}^m$. □

Remark 8.31. An immediate consequence of Proposition 8.30 is that in order to find the Minkowski (or box-counting) dimension of a set A, one may instead find the Minkowski (or box-counting) dimension of any dense subset of A, even if such a dense subset is countable. One may also replace A by its closure $\overline{A}$.

Example 8.32. Consider the partition of the unit interval into disjoint sets of rational and irrational numbers: $[0,1] = ([0,1] \cap \mathbb{Q}) \cup ([0,1] \backslash \mathbb{Q})$. Note that for any point $x \in \mathbb{R}$, we have $\dim_H(\{x\}) = 0$. So, since Hausdorff dimension is countably stable and $[0,1] \cap \mathbb{Q}$ is a countable set, we have that $\dim_H([0,1] \cap \mathbb{Q}) = 0$ (see also part (iv) of Proposition 5.16). On the other hand, $\overline{[0,1] \cap \mathbb{Q}} = \overline{[0,1] \backslash \mathbb{Q}} = [0,1]$. Therefore, by Proposition 8.30, we also have that $\dim_M([0,1] \cap \mathbb{Q}) = \dim_M([0,1] \backslash \mathbb{Q}) = \dim_M[0,1] = 1$.

Even sparse countable sets, so sparse that they are *not* dense in any set with positive 1-dimensional Hausdorff measure, may have positive box-counting dimension.

Example 8.33. Let $A := \{0\} \cup \{1/j\}_{j=1}^\infty = \{0, 1, 1/2, \ldots\}$. Since A is countable, $\dim_H(A) = 0$. However, $\dim_B(A) = 1/2$.

Suppose $0 < \varepsilon < 1/2$ and let n be the unique positive integer such that $1/(n(n+1)) \leq \varepsilon < 1/((n-1)n)$. The set of points $\{0\} \cup \{1/j\}_{j=n}^\infty$ is contained in the union of $n+1$ intervals of length ε, and $n-1$ intervals of length ε cover the remaining set $\{1/j\}_{j=1}^{n-1}$. Hence, $N_B(A, \varepsilon^{-1}) \leq 2n$, where $N_B(A, \varepsilon^{-1})$ is the minimum number of intervals in an ε-cover of A, and hence,

$$\frac{\log N_B(A, \varepsilon^{-1})}{-\log \varepsilon} \leq \frac{\log 2n}{\log((n-1)n)} = \frac{\log 2 + \log n}{2 \log n + \log(1 - 1/n)}.$$

The rightmost expression tends to $1/2$, as $n \to \infty$. Thus, $\overline{\dim}_B(A) \leq 1/2$.

Now, note that the distance between consecutive points in A is $1/k - 1/(k+1) = 1/(k(k+1))$, where $k \in \mathbb{N}$. As above, suppose $0 < \varepsilon < 1/2$ and let n be the unique positive integer such that $1/(n(n+1)) < \varepsilon \leq 1/((n-1)n)$.

If $\mathrm{diam}(U) \leq \varepsilon \leq 1/((n-1)n)$, then U contains at most one of the points $\{1/j\}_{j=1}^{n}$. Thus, at least n sets of diameter ε are required to cover A. Hence, $N_B(A, \varepsilon^{-1}) \geq n$, and so,

$$\frac{\log N_B(A,\varepsilon^{-1})}{-\log \varepsilon} \geq \frac{\log n}{\log(n(n+1))} = \frac{\log n}{2\log n + \log(1+1/n)}. \tag{8.85}$$

The rightmost expression tends to $1/2$ as $n \to \infty$. Thus, $\underline{\dim}_B(A) \geq 1/2$.

Proposition 8.34 (Basic Properties of Box and Minkowski Dimensions). *The box-counting (or box) dimension—and, equivalently, in light of Theorem* 8.29 *above, the Minkowski dimension—satisfy the following properties:*[27]

(i) Monotonicity: *If $A \subseteq B \subseteq \mathbb{R}^m$, with B bounded, then* $\underline{\dim}_B(A) \leq \underline{\dim}_B(B)$ *and* $\overline{\dim}_B(A) \leq \overline{\dim}_B(B)$.

(ii) Finite Stability (of the Upper Box-Counting Dimension): *For any two bounded sets $A, B \subseteq \mathbb{R}^m$,*

$$\overline{\dim}_B(A \cup B) = \max\{\overline{\dim}_B(A), \overline{\dim}_B(B)\}.$$

(iii) Range of Values: *If $A \subseteq \mathbb{R}^m$ is bounded and nonempty, then*

$$0 \leq \underline{\dim}_B(A) \leq \overline{\dim}_B(A) \leq m.$$

(iv) Open Sets: *If Ω is a nonempty bounded open subset of $\mathbb{R}^m$, then* $\dim_B(\Omega) = m$.

(v) Finite Sets: *If A is finite and nonempty, then* $\dim_B(A) = 0$.

Proof. In this proof, let A and B denote nonempty bounded subsets of $\mathbb{R}^m$ and let $N_B(A, \varepsilon^{-1})$ denote the number of ε-mesh cubes that intersect A (as in part (iii) of Proposition 8.23).

To prove property (i), note that $A \subseteq B$ implies that $N_B(A,\varepsilon^{-1}) \leq N_B(B,\varepsilon^{-1})$, and so $\underline{\dim}_B(A) \leq \underline{\dim}_B(B)$ and $\overline{\dim}_B(A) \leq \overline{\dim}_B(B)$.

To prove property (ii), for any two sets A and B we have both $A \subseteq A \cup B$ and $B \subseteq A \cup B$.[28] By monotonicity, it follows that $\max\{\overline{\dim}_B(A), \overline{\dim}_B(B)\} \leq \overline{\dim}_B(A \cup B)$. On the other hand, for $\varepsilon > 0$, we have that

$$\begin{aligned} N_B(A \cup B, \varepsilon^{-1}) &\leq N_B(A,\varepsilon^{-1}) + N_B(B,\varepsilon^{-1}) \\ &\leq 2\max\{N_B(A,\varepsilon^{-1}), N_B(B,\varepsilon^{-1})\}. \end{aligned} \tag{8.86}$$

Taking the logarithm on both sides of (8.86) and interchanging the logarithms and max (since log is an increasing function), then dividing by $-\log\varepsilon > 0$ (for

[27] In the statement—or in the ensuing proof—of Proposition 8.34, the subscript B stands for "box", whereas the other uses of B are clearly of a different nature.

[28] Since $\dim_B(\emptyset) = -\infty$, it clearly suffices to assume that A and B are both nonempty.

$0 < \varepsilon < 1$) yields

$$\text{(8.87)} \quad \frac{\log N_B(A \cup B, \varepsilon^{-1})}{-\log \varepsilon} \le \frac{\log 2}{-\log \varepsilon} + \max\left\{\frac{\log N_B(A, \varepsilon^{-1})}{-\log \varepsilon}, \frac{\log N_B(B, \varepsilon^{-1})}{-\log \varepsilon}\right\}.$$

Taking upper limits in (8.87) as $\varepsilon \to 0^+$ then yields the desired reverse inequality (since max is a continuous function):

$$\overline{\dim}_B(A \cup B) \le \max\left\{\overline{\dim}_B(A), \overline{\dim}_B(B)\right\}.$$

To prove property (iii), let $\varepsilon \in (0,1)$. Since A is bounded, it is contained in a cube U which, in turn, is contained in a finite number of cubes of side lengths ε. Hence, $N_B(A, \varepsilon^{-1}) \le N_B(U, \varepsilon^{-1}) \le n \cdot \varepsilon^m$, for some $n \in \mathbb{N}$. Therefore, we have that $0 \le \underline{\dim}_B(A) \le \overline{\dim}_B(A) \le m$; this follows by taking the logarithm, dividing by $-\log \varepsilon > 0$, and then taking the upper limit in the previous string of inequalities.

To prove property (iv), let Ω be a (nonempty) bounded open subset of $\mathbb{R}^m$. Then, Ω contains a cube U and so, for $0 < \varepsilon < 1$, we have that

$$\text{(8.88)} \quad c \cdot \varepsilon^m \le N_B(U, \varepsilon^{-1}) \le N_B(\Omega, \varepsilon^{-1}),$$

where $c > 0$ is independent of ε. Therefore, after taking the logarithm, dividing by $-\log \varepsilon$, and then taking the lower limit accordingly, as well as combining with property (iii) (applied to $\Omega := A$ and which yields $\overline{\dim}_B(\Omega) \ge \underline{\dim}_B(\Omega) \ge m$), we deduce that

$$\text{(8.89)} \quad \underline{\dim}_B(\Omega) = \overline{\dim}_B(\Omega) = m, \text{ and so, } \dim_B(\Omega) = m.$$

Finally, in order to prove property (v), if A is finite and contains $n > 0$ points, then $N_B(A, \varepsilon^{-1}) = n$, for all sufficiently small positive ε. It readily follows that $\underline{\dim}_B(A) = \overline{\dim}_B(A) = 0$, as desired.

This concludes the proof of the proposition. □

Remark 8.35. Note that the counterpart of property (ii) (finite stability) does *not* hold for the lower box-counting (or Minkowski) dimension. The delicate and complicated construction of an example that exhibits different upper and lower box-counting dimension is the focus of Section 8.8 below.

Also, neither the upper nor the lower box-counting dimensions is *countably* stable. For example, consider the countable set $A = \{0\} \cup \{1/j\}_{j=1}^{\infty} \subseteq \mathbb{R}$ described in Example 8.33. By property (v) of Proposition 8.34 regarding finite sets, we have $\dim_B\{1/j\} = \dim_B\{0\} = 0$, for every $j \in \mathbb{N}$. And yet, we have that

$$0 = \sup_{j \in \mathbb{N}} \{\dim_B\{0\}, \dim_B\{1/j\}\} < \dim_B\left(\{0\} \cup \{1/j\}_{j=1}^{\infty}\right) = 1/2.$$

Despite the fact that the box-counting dimension and, equivalently, the Minkowski dimension are not countably stable, it is still the case that box-counting dimension is preserved by bi-Lipschitz transformations—and, more generally, that a Lipschitz transformation does not strictly increase box-counting dimension.

Note that, naturally, in light of Theorem 8.29 above, the next proposition has an exact counterpart for the lower and upper Minkowski dimensions (as well as, when it exists, the Minkowski dimension).

Proposition 8.36 (Box Dimension and Hölder Continuity). *Let A be a bounded subset of $\mathbb{R}^m$ and let $f : A \to \mathbb{R}^n$. Then:*

(i) *If f is Hölder continuous of exponent α, that is, if there are some $\alpha, c > 0$ such that*[29]

$$\|f(\boldsymbol{x}) - f(\boldsymbol{y})\|_n \le c \cdot \|\boldsymbol{x} - \boldsymbol{y}\|_m^\alpha, \text{ for all } \boldsymbol{x}, \boldsymbol{y} \in A, \tag{8.90}$$

then

$$\underline{\dim}_B(f(A)) \le (1/\alpha) \cdot \underline{\dim}_B(A) \quad \text{and} \quad \overline{\dim}_B(f(A)) \le (1/\alpha) \cdot \overline{\dim}_B(A). \tag{8.91}$$

(ii) *If f is bi-Lipschitz, that is, if there are $0 < c_1 < c_2$ such that*

$$c_1 \cdot \|\boldsymbol{x} - \boldsymbol{y}\|_m \le \|f(\boldsymbol{x}) - f(\boldsymbol{y})\|_n \le c_2 \cdot \|\boldsymbol{x} - \boldsymbol{y}\|_m, \text{ for all } \boldsymbol{x}, \boldsymbol{y} \in A, \tag{8.92}$$

then $\underline{\dim}_B(f(A)) = \underline{\dim}_B(A)$ and $\overline{\dim}_B(f(A)) = \overline{\dim}_B(A)$.

In particular, if $\dim_B(A)$ exists, then $\dim_B(f(A))$ also exists and

$$\dim_B(f(A)) = \dim_B(A). \tag{8.93}$$

We therefore say that the box-counting—and hence also the Minkowski dimension—is a bi-Lipschitz invariant.

Proof. In this proof, we let $N_B(A, \varepsilon^{-1})$ denote the minimum number of sets in an ε-cover of A.

In order to prove part (i), let $\{A_j\}$ be an ε-cover of A. Then, $\{A \cap A_j\}$ is also an ε-cover of A. Moreover, by (8.90),

$$\operatorname{diam}(f(A \cap A_j)) \le c \cdot (\operatorname{diam}(A \cap A_j))^\alpha \le c \cdot (\operatorname{diam}(A_j))^\alpha \le c \cdot \varepsilon^\alpha.$$

Hence, $\{f(A \cap A_j)\}$ is a $c \cdot \varepsilon^\alpha$-cover of $f(A)$. So,

$$N_B(f(A), (c\varepsilon^\alpha)^{-1}) \le N_B(A, \varepsilon^{-1})$$

and, for all $\varepsilon > 0$ sufficiently small, we have that

$$\frac{\log N_B(f(A), (c\varepsilon^\alpha)^{-1})}{-\log(c\varepsilon^\alpha) + \log c} \le \frac{\log N_B(A, \varepsilon^{-1})}{\alpha(-\log \varepsilon)}.$$

Taking the upper and lower limits as $\varepsilon \to 0^+$ in this last inequality yields the first and the second equalities in (8.91), respectively.

[29]Here, for any $k \in \mathbb{N}$, $\|\cdot\|_k$ denotes the Euclidean norm in $\mathbb{R}^k$.

In order to prove part (ii), note that, due to part (i), it suffices to prove only the lower bound (the left side of (8.92)) when f is bi-Lipschitz. Suppose f is bi-Lipschitz and let $\boldsymbol{a}, \boldsymbol{b} \in f(A)$. Then, f is a bijection with inverse map $f^{-1} : f(A) \to A$ such that

$$c_1 \cdot \|f^{-1}(\boldsymbol{a}) - f^{-1}(\boldsymbol{b})\|_m \le \|f(f^{-1}(\boldsymbol{a})) - f(f^{-1}(\boldsymbol{b}))\|_n = \|\boldsymbol{a} - \boldsymbol{b}\|_n.$$

Hence, the map f^{-1} is Lipschitz. So, part (i) applied to f^{-1} yields

$$\underline{\dim}_B(A) = \underline{\dim}_B(f^{-1}(f(A))) \le \underline{\dim}_B(f(A)) \quad \text{and}$$
$$\overline{\dim}_B(A) = \overline{\dim}_B(f^{-1}(f(A))) \le \overline{\dim}_B(f(A)),$$

which completes the proof of the proposition. □

The next exercise provides a simple, and often useful, alternate way of computing the Minkowski (or, equivalently, the box-counting) dimension; see, e.g., [**370, 374, 375**] and [**356**].

Exercise 8.37. Let A be a (nonempty) bounded set and let $V_A(\varepsilon) = \mathcal{L}^m(A_\varepsilon)$, for any $\varepsilon > 0$. Then:

(i) Show that

$$\overline{\dim}_M(A) = \overline{\dim}_B(A) = \inf\{t \ge 0 : V_A(\varepsilon) = O(\varepsilon^{m-t}), \text{ as } \varepsilon \to 0^+\},$$

where the implied constant is allowed to depend on A and t.

(ii) Find a corresponding expression for $\underline{\dim}_M(A) = \underline{\dim}_B(A)$.

[*Hint*: For part (ii), you may have to introduce a symbol which is often used in analytic number theory, but rarely so in other parts of mathematics.]

Exercise 8.38.

(i) For the lower and upper Minkowski dimensions, show directly the exact counterpart of parts (i)–(v) stated for the box dimension in Proposition 8.34.[30] In particular, prove the finite stability of the upper Minkowski dimension, $\overline{\dim}_M$, the analog of part (ii) in Proposition 8.34.

(ii) Deduce from the finite stability of $\overline{\dim}_M$ (see part (i) just above) that any polygonal curve in $\mathbb{R}^m$ on a compact interval $[a, b]$ has Minkowski dimension equal to 1.

[*Hint*: Use, in particular, the fact that the topological dimension does not exceed the upper Minkowski dimension; see Subsection 8.6.2 below. Also, observe that a polygonal curve is nothing but a finite union of line segments.]

[30] Of course, in light of Theorem 8.29 in Section 8.4, this statement follows at once from Proposition 8.34, but this is clearly not what is asked here.

8.5. Hausdorff versus Minkowski

In this section, we compare, in particular, the Hausdorff and box-counting—or, equivalently, in light of Theorem 8.29 above, Minkowski—dimensions. In the latter part of this section, we also study the fractal dimensions of the Cartesian products of certain subsets of $\mathbb{R}^m$.

Proposition 8.39 (Comparison of Hausdorff and Box Dimensions). *Let A be a bounded subset of $\mathbb{R}^m$.*[31] *Then,*

$$\dim_H(A) \leq \underline{\dim}_B(A) \leq \overline{\dim}_B(A). \tag{8.94}$$

Proof. Clearly, since, $\underline{\dim}_B(A) \leq \overline{\dim}_B(A)$, it suffices to prove the first inequality in (8.94) and to assume that $A \neq \emptyset$. Now, suppose that, for some $t \geq 0$, $1 < \mathcal{H}^t(A) = \lim_{\delta\to 0^+} \mathcal{H}^t_\delta(A)$. Then, $1 < \mathcal{H}^t_\delta(A) \leq N_B(A, \delta^{-1}) \cdot \delta^t$, where $N_B(A, \delta^{-1})$ is the minimum number of sets of diameter equal to δ required to cover A and δ is sufficiently small (hence, $0 < \delta < 1$). So,

$$0 < \log(N_B(A, \delta^{-1}) \cdot \delta^t) = \log N_B(A, \delta^{-1}) + t \log \delta. \tag{8.95}$$

Therefore, solving for t and letting $\delta \to 0^+$ in (8.95) yields the following inequality: $t \leq \liminf_{\delta\to 0^+}(\log N_B(A, \delta^{-1})/ - \log \delta)$. The result then follows by appealing to Lemma 5.12. □

In light of Theorem 8.29 in Section 8.4, we obtain at once the following corollary of Proposition 8.39.

Corollary 8.40 (Comparison of Hausdorff and Minkowski Dimensions). *Let A be a nonempty bounded subset of $\mathbb{R}^m$. Then,*

$$\dim_H(A) \leq \underline{\dim}_M(A) \leq \overline{\dim}_M(A). \tag{8.96}$$

In particular, if $\dim_M(A)$ exists (or, equivalently, if $\dim_B(A)$ exists), then we have that

$$\dim_H(A) \leq \dim_M(A)\ (= \dim_B(A)). \tag{8.97}$$

Example 8.41 (*Smith–Volterra–Cantor Sets*). As shown in Example 5.29, for the Smith–Volterra–Cantor set $SVC(n)$ (described in Example 4.92), where the integer n satisfies $n \geq 4$, we have that

$$0 < \mathcal{H}^1(SVC(n)) = \mathcal{L}^1(SVC(n)) = \frac{n-3}{n-2} < 1,$$

and, therefore, $\dim_H SVC(n) = 1$. So, by Theorem 8.29, Proposition 8.39, and by property (iii) of Proposition 8.34, we also have that

$$1 = \dim_H(SVC(n)) = \dim_B(SVC(n)) = \dim_M(SVC(n)).$$

[31] Clearly, the result is still true if $A = \emptyset$ since all of the terms of (8.94) are then equal to $-\infty$.

Now, in the present case where $m = 1$ (i.e., on the real line), the Minkowski content of $SVC(n)$ turns out to be given by

$$0 < \mathscr{M}^1(SVC(n)) = \mathscr{L}^1(SVC(n)) = \frac{n-3}{n-2}. \tag{8.98}$$

To partially justify (8.98), note that $D = \dim_M SVC(n) = m = 1$. To streamline notation, let $S = SVC(n)$. Then, assuming that the limit exists, we have that

$$\mathscr{M}^1(S) = \lim_{\varepsilon\to 0^+} \mathscr{L}^1(S_\varepsilon)/\varepsilon^{m-D} = \lim_{\varepsilon\to 0^+} \mathscr{L}^1(S_\varepsilon) = (n-3)/(n-2).$$

The countable (and hence also finite) additivity of Lebesgue measure yields

$$\mathscr{L}^1(S_\varepsilon) = \mathscr{L}^1(S) + \mathscr{L}^1(S_\varepsilon \backslash S) = \frac{n-3}{n-2} + \mathscr{L}^1(S_\varepsilon \backslash S).$$

So, it remains to show that $\lim_{\varepsilon\to 0^+} \mathscr{L}^1(S_\varepsilon\backslash S) = 0$, as can be proved by the interested reader; in fact, much more can be shown. Indeed, it turns out that

$$\mathscr{L}^1(S_\varepsilon\backslash S) = O(\varepsilon^{1-\alpha}), \text{ as } \varepsilon \to 0^+,$$

where, for some real number α such that $D < \alpha < 1$, D denotes the *inner Minkowski dimension* of S (as defined in Section 8.2 above) and $0 < D < 1$. Moreover, the corresponding *inner Minkowski content*—as defined in Section 8.2 above by means of $S_\varepsilon \cap S^c$, the inner ε-neighborhood of S (instead of by means of S_ε, the two-sided neighborhood of S) does not exist.

In the case where $n = 3$, we have $SVC(3) = C$ (the classic Cantor set) and $0 < \dim_H(C) = \dim_M(C) = \log_3 2 < 1$; see Proposition 5.28 and Example 8.24. So, by Lemma 5.12, we immediately have $\mathscr{L}^1(C) = 0$. However, a more direct approach shows that, with $m = 1$ and $n = 3$,

$$\mathscr{L}^1(C_\varepsilon) = \frac{n-3}{n-2} + \mathscr{L}^1(C_\varepsilon\backslash C) = \mathscr{L}^1(C_\varepsilon\backslash C), \text{ as } \varepsilon \to 0^+.$$

The set $C_\varepsilon\backslash C$ is very closely related to the Cantor string Ω_C; see Example 8.17. In this case, we obtain that

$$\mathscr{L}^1(C_\varepsilon\backslash C) = O(\varepsilon^{1-\log_3 2}), \text{ as } \varepsilon \to 0^+.$$

Unlike the sets $SVC(n)$ for $n \geq 4$, the Cantor set $SVC(3) = C$ is *not* Minkowski measurable (see Example 8.17) and, in fact, the upper and lower Minkowski contents of C are equal to the *inner* upper and lower Minkowski contents of C, respectively. The notions of inner Minkowski content and dimension play important roles in the development of the theory of ordinary fractal strings; see, especially, [**329**, **356**, **370**, **374**, **375**] and the sequel to this book, [**119**], along with Section 8.2 above.

Recall from Corollary 5.77 that for sets A and B, we have a lower bound for the Hausdorff dimension of the Cartesian product $A \times B$ given by

$$\dim_H(A \times B) \geq \dim_H(A) + \dim_H(B).$$

The following proposition shows that for the upper box-counting dimension, the inequality is reversed.

Proposition 8.42 (Cartesian Product and Upper Box Dimension). *If $A \subseteq \mathbb{R}^m$ and $B \subseteq \mathbb{R}^n$ are bounded with $m, n \in \mathbb{N}$, then*

$$\overline{\dim}_B(A \times B) \leq \overline{\dim}_B(A) + \overline{\dim}_B(B). \tag{8.99}$$

Proof. Let $\varepsilon \in (0,1)$, and let $N_B(A, \varepsilon^{-1})$ and $N_B(B, \varepsilon^{-1})$ denote the number of ε-mesh m-dimensional cubes that intersect A and the number of ε-mesh n-dimensional cubes that intersect B, respectively. Then, $A \times B$ intersects at most $N_B(A, \varepsilon^{-1}) \cdot N_B(B, \varepsilon^{-1})$ ε-mesh $(m+n)$-dimensional cubes. Thus,

$$\frac{\log N_B(A \times B, \varepsilon^{-1})}{-\log \varepsilon} \leq \frac{\log N_B(A, \varepsilon^{-1}) + \log N_B(B, \varepsilon^{-1})}{-\log \varepsilon},$$

where $N_B(A \times B, \varepsilon^{-1})$ denotes the number of ε-mesh $(m+n)$-dimensional cubes that intersect $A \times B$. Letting $\varepsilon \to 0^+$ and taking upper limits yields the desired inequality (8.99). □

The following proposition provides an upper bound for the Hausdorff dimension of a product of subsets of the real line which involves both the Hausdorff dimension of one of the sets and the upper box-counting dimension of the other.

Proposition 8.43. *If $A \subseteq \mathbb{R}$ is arbitrary and $B \subseteq \mathbb{R}$ is bounded, then*

$$\dim_H(A \times B) \leq \dim_H(A) + \overline{\dim}_B(B). \tag{8.100}$$

Proof. Let $s, t \in \mathbb{R}$ be such that $s > \dim_H(A)$ and $t > \overline{\dim}_B(B)$. Also, let $N_B(B, \varepsilon^{-1})$ denote the minimum number of intervals of length ε required to cover B. There exists $\delta_0 > 0$ such that B is covered by at most δ^{-t} intervals of length δ, for all $0 < \delta \leq \delta_0$. For each such δ, let $\{A_j\}$ be a δ-cover of A comprising intervals such that $\sum_{j=1}^{\infty} (\operatorname{diam}(A_j))^s < 1$. Next, for each j, let $\{B_{j,k}\}$ be a cover of B comprising $N_B(B, \delta^{-1})$ intervals of length $\operatorname{diam}(A_j)$. Now, $A_j \times B$ is covered by $N_B(B, (\operatorname{diam}(A_j))^{-1})$ squares $\{A_j \times B_{j,k}\}$ with side lengths $\operatorname{diam}(A_j)$. Hence, $A \times B \subseteq \bigcup_j \bigcup_k (A_j \times B_{j,k})$, and so,

$$\begin{aligned}
\mathcal{H}^{s+t}_{\delta\sqrt{2}}(A \times B) &\leq \sum_j \sum_k (\operatorname{diam}(A_j \times B_{j,k}))^{s+t} \\
&\leq \sum_j \left(N_B(B, (\operatorname{diam}(A_j))^{-1}) \cdot 2^{(s+t)/2} \cdot (\operatorname{diam}(A_j))^{s+t}\right) \\
&\leq 2^{(s+t)/2} \cdot \sum_j \left((\operatorname{diam}(A_j))^{-t} \cdot (\operatorname{diam}(A_j))^{s+t}\right) < 2^{(s+t)/2}.
\end{aligned}$$

Thus, letting $\delta \to 0^+$ yields $\mathcal{H}^{s+t}(A \times B) < \infty$. Therefore, by Lemma 5.12, we obtain that (8.100) holds. □

Exercise 8.44. Show that if $A \subseteq \mathbb{R}^m$ is arbitrary and $B \subseteq \mathbb{R}^n$, with B bounded, then $\dim_H(A \times B) \leq \dim_H(A) + \overline{\dim}_B(B)$.

An immediate consequence of Proposition 8.43 (or, really, when $m \geq 2$, of Exercise 8.44) and of Corollary 5.77, is a sufficient condition under which the Hausdorff dimension of the Cartesian product of a pair of sets is equal to the sum of the Hausdorff dimensions of the underlying sets.[32]

Corollary 8.45 (Hausdorff Dimension of a Cartesian Product, Revisited). *If $A \subseteq \mathbb{R}^m$ is arbitrary and $B \subseteq \mathbb{R}^n$, with B bounded, and if $\overline{\dim}_B(B) = \dim_H(B)$, then we have the following equality:*

$$\dim_H(A \times B) = \dim_H(A) + \dim_H(B). \tag{8.101}$$

8.6. Topological versus Hausdorff and Minkowski Dimensions

In this section, we first consider (in Subsection 8.6.1) the purely topological properties of the topological dimension, $\dim_T$, and we then compare (in Subsection 8.6.2) the topological ($\dim_T$) and the Hausdorff ($\dim_H$) dimensions. In short, we show that $\dim_T \leq \dim_H$. Moreover, under mild hypotheses, it turns out that the topological dimension $\dim_T(X)$ of X is the minimum value of the Hausdorff dimensions of the metric spaces Z which are homeomorphic to X.

8.6.1. Properties of the Topological Dimension. We provide here, without proof (with one exception), several of the basic properties of the topological dimension—also called the (Lebesgue) covering dimension in the literature because it coincides with it within the present context of separable metric spaces.[33] For simplicity, we restrict ourselves to suitable subsets of Euclidean space $\mathbb{R}^m$, with $m \in \mathbb{N}$ arbitrary, instead of considering the more general context of separable metric spaces. Also for simplicity, we certainly do not give all of the key properties and we state their hypotheses in their simplest form; see, however, part (a) of Remark 8.48.

General references on this subject (including for the proof of Theorem 8.46 in a more general context, in particular) are the classic book by W. Hurewicz and H. Wallman, [**260**], along with the more recent book by R. Engelking, [**161**], both entitled *Dimension Theory*. (See also, e.g., [**184**], [**440**], and [**459**].)

We refer the reader to Subsection 1.3.2 in the introductory chapter of the present book for the definition (Definition 1.18) of the topological dimension $\dim_T$ and the intuition underlying this concept. Recall, in particular, that $\dim_T$ takes only integer values, $0, 1, 2, \ldots$, on nonempty subsets of $\mathbb{R}^m$. In fact, as we

---

[32] Recall from Proposition 8.39 at the beginning of this section that we always have $\dim_H(B) \leq \overline{\dim}_B(B)$.

[33] Recall that a topological space X is said to be separable if there exists a finite or countable subset D of X which is dense in X; i.e., $\overline{D} = X$.

shall see in part (d) of Remark 8.48, the range of $\dim_T$ is $\{0, 1, \dots, m\}$ on such subsets of $\mathbb{R}^m$; further recall that $\dim_T(\emptyset) := -\infty$.

Theorem 8.46 (Basic Properties of the Topological Dimension). *The topological dimension* $\dim_T$, *defined on the subsets of Euclidean space* $\mathbb{R}^m$, *satisfies the following properties*:

(i) (Subspace Property). *If* $A \subseteq B \subseteq \mathbb{R}^m$, *then* $\dim_T(A) \leq \dim_T(B)$.

(ii) (Countable Basis Property). *Let* $A \subseteq \mathbb{R}^m$ *be closed (e.g., compact). Then,* $\dim_T(A) \leq n$ *(with* $n \in \mathbb{N}$*) iff* A *admits a countable basis* **B** *of open sets of* A *such that* $\dim_T(U) \leq n-1$, *for every* $U \in \mathbf{B}$.

(iii) (Sum Property). *Let* $A \subseteq \mathbb{R}^m$ *be closed. If* $A = \bigcup_{j \geq 1} F_j$ *is a finite or countable union of closed subsets* F_j *such that* $\dim_T(F_j) \leq n$, *for some* $n \in \mathbb{N}_0$ *and for all* $j \geq 1$, *then* $\dim_T(A) \leq n$.

(iv) (Addition Property). *For any two nonempty closed subsets* A, B *of* $\mathbb{R}^m$, *we have that* $\dim_T(A \cup B) \leq \dim_T(A) + \dim_T(B) + 1$.

(v) (Cartesian Product Property). *If* A, B *are closed subsets of* $\mathbb{R}^m$,[34] *then* $\dim_T(A \times B) \leq \dim_T(A) + \dim_T(B)$.

(vi) (Open Submanifold Property). *Let* U *be a nonempty open subset of* $\mathbb{R}^m$. *Then,* $\dim_T(U) = m$. *In particular,* $\dim_T(\mathbb{R}^m) = m$.

(vii) (Topological Invariance Property). *The topological dimension is invariant under homeomorphisms; i.e., if* f *is a homeomorphism from* A *onto* B,[35] *with* $A \subseteq \mathbb{R}^{m_1}$ *and* $B \subseteq \mathbb{R}^{m_2}$ *(so that* $B := f(A)$*), then* $\dim_T(B) = \dim_T(A)$. *Consequently, the topological dimension* $\dim_T$ *is a* topological invariant.

Exercise 8.47. By using Definition 1.18 in Subsection 1.3.2 prove the topological invariance (property (vii) in Theorem 8.46 just above) of the topological dimension.

Remark 8.48.

(a) (*More General Hypotheses*). We briefly indicate some of the more general hypotheses under which several of the above properties stated in Theorem 8.46 hold. For instance, (ii) and (iii) hold if A is any separable metric space, while (iv) holds for any two separable subsets A, B of a given metric space. Finally, (v) holds if A and B are any two separable metric spaces, while the topological invariance of $\dim_T$ is valid in any topological space.

(b) (*Distinct Euclidean Spaces Are* Not *Homeomorphic*). The fact that the topological dimension is a topological invariant (property (vii) in Theorem 8.46) enables us to resolve easily several basic problems of 19th-century mathematics. For example, it follows at once from it that $\mathbb{R}^{m_1}$ and $\mathbb{R}^{m_2}$ (with $m_1, m_2 \in \mathbb{N}$)

[34] Recall that $\dim_T(\emptyset) := -\infty$. Also, $\dim_T(A) \geq 0$, for any nonempty subset A of $\mathbb{R}^m$.

[35] This means, of course, that f is a bijection from A onto B and that both $f : A \to B$ and its inverse map, $f^{-1} : B \to A$, are continuous. If A and B are compact, then f^{-1} is automatically continuous.

are homeomorphic iff $m_1 = m_2$.[36] More generally, in light of property (vi), if U_j is a nonempty open subset of $\mathbb{R}^{m_j}$, for $j = 1, 2$, then U_1 is homeomorphic to U_2 iff $m_1 = m_2$. Also, if Q_j is an m_j-dimensional closed cube of $\mathbb{R}^{m_j}$ (i.e., $Q_j = [0,1]^{m_j}$, the unit cube in $\mathbb{R}^{m_j}$), for $j = 1, 2$, then Q_1 and Q_2 are homeomorphic iff $m_1 = m_2$.

In particular, $[0,1]$ (the unit interval of $\mathbb{R}$), $[0,1]^2$ (the unit square of $\mathbb{R}^2$), and $[0,1]^3$ (the unit cube of $\mathbb{R}^3$) are pairwise non-homeomorphic.

(c) (*Consequence for Plane and Space-Filling Curves*). It follows from part (b) that a plane-filling curve (such as the Peano curve or the Hilbert curve; see Subsection 1.4.3 above and Remark 9.63 below)—or, more generally, any space-filling curve—must have multiple points.

Indeed, for example, if $\gamma : [0,1] \to [0,1]^2$ is a continuous parametrization of a plane-filling curve such that $\gamma([0,1]) = [0,1]^2$, and if γ did not have any multiple point (i.e., if γ were an injective map), then γ would be a continuous bijection and its inverse map $\gamma^{-1} : [0,1]^2 \to [0,1]$ would also be continuous (as well as a bijection).[37] Hence, γ would thus be a homeomorphism between $[0,1]$ and $[0,1]^2$— which, as we have seen, in light of part (b) of this remark, is impossible.

Of course, the exact same reasoning can be applied to space-filling curves with range equal to $[0,1]^m$, for any integer $m \geq 2$.

(d) (*Range of* $\dim_T$). Since $\dim_T(\mathbb{R}^m) = m$ (see property (vi)), it follows from property (i) (also in Theorem 8.46) that for any nonempty subset A of $\mathbb{R}^m$, we have

$$0 \leq \dim_T(A) \leq m. \tag{8.102}$$

Furthermore, we claim that the range of $\dim_T$ on the class of nonempty subsets of $\mathbb{R}^m$ is precisely $\{0, 1, \ldots, m\}$. To see this, given any fixed $p \in \mathbb{N}_0$, apply properties (vi) and (vii) to $(0,1)^p$—first viewed as a nonempty open subset of $\mathbb{R}^p$, which implies that $\dim_T((0,1)^p) = p$, and then viewed as being homeomorphic to $E_p := (0,1)^p \times \{0\}_{m-p} \subseteq \mathbb{R}^m \simeq \mathbb{R}^p \times \mathbb{R}^{m-p}$, where $\{0\}_{m-p}$ is the zero vector of $\mathbb{R}^{m-p}$.[38] We conclude that $\dim_T(E_p) = p$, as desired.

Exercise 8.49. Show that any closed p-dimensional cube of $\mathbb{R}^m$, with $0 \leq p \leq m$, has topological dimension p. In particular, $\dim_T([0,1]^m) = m$.

---

[36]Indeed, for $j = 1, 2$, $\dim_T(\mathbb{R}^{m_j}) = m_j$, by property (vi) of Theorem 8.46.

[37]To see this, let $F \subseteq [0,1]^2$ be closed and hence compact, since $[0,1]^2$ is compact. We must check that $(\gamma^{-1})^{-1}(F) = \gamma(F)$ is closed (in $([0,1])$). Indeed, it is the case because $\gamma(F)$ is compact, as the continuous image of a compact set, and thus, $\gamma(F)$ is closed, as a compact subset of $[0,1]^2$ (as well as of $\mathbb{R}^2$).

[38]Note that, unless $p = m$, $(0,1)^p \times \{0\}_{m-p}$ is not an open subset of $\mathbb{R}^m$, which does not affect the reasoning. Also, if $p = 0$, we identify $\mathbb{R}^p$ and $(0,1)^p$ to a point, which necessarily has topological dimension zero.

8.6.2. Comparison Between Topological, Hausdorff, and Minkowski Dimensions. Remarkably, even though the topological dimension $\dim_T$ is of a topological nature whereas the Hausdorff measure $\dim_H$ is of a metric nature, and $\dim_T$ and $\dim_H$ are defined extremely differently, they can always be compared. This important fact was first observed in 1931 by G. Nöbeling [**464**] for subsets of Euclidean space (see Theorem 8.50 below) and the corresponding result was then extended in 1937 in a very elegant manner by E. Szpilrajn [**581**] to arbitrary subsets of any metric space X (see Theorem 8.53 and Corollary 8.55 below).

Theorem 8.50 (Comparison of $\dim_T$ and $\dim_H$: Case of Euclidean Spaces; G. Nöbeling, [**464**], 1931). *For any subset $A \subseteq \mathbb{R}^m$, we have the following inequality between the topological and Hausdorff dimensions of A:*

$$\dim_T(A) \leq \dim_H(A). \tag{8.103}$$

Hence, in light of Proposition 8.39 *and Theorem* 8.29, *if $A \subseteq \mathbb{R}^m$ is nonempty, we have that*

$$\begin{aligned} 0 \leq \dim_T(A) \leq \dim_H(A) \leq \underline{\dim}_B(A) = \underline{\dim}_M(A) \\ \leq \overline{\dim}_B(A) = \overline{\dim}_M(A) \leq m. \end{aligned} \tag{8.104}$$

Proof. First, recall that $\dim_T(\emptyset) = \dim_H(\emptyset) = -\infty$ and that, by part (iii) of Proposition 8.34 and by Theorem 8.29, we have that $\underline{\dim}_B = \underline{\dim}_M \leq m$.[39] Therefore, (8.104) follows from (8.103) along with Proposition 8.39.

Next, we proceed exactly as in [**464**] and prove (8.103) by induction, using the fact that the topological dimension is defined recursively; see Definition 1.18 in Subsection 1.3.2. More specifically, given any $l \in \mathbb{N}_0$, we show by induction on $l \in \mathbb{N}_0$ that if A is a (necessarily nonempty) subset of $\mathbb{R}^m$ with $\dim_T(A) \geq l$, then $\dim_H(A) \geq l$. A moment's reflection shows that this claim implies (8.103), as desired.

First, we note that the claim is obviously true for $l = 0$ since we must have that $\dim_H(A) \geq 0$ (because $A \neq \emptyset$).

Next, assume that the claim is true for $l \in \mathbb{N}_0$ and let us prove that it is also true for $l + 1$. Hence, consider $B \subseteq \mathbb{R}^m$, with $\dim_T(B) \geq l + 1$; we want to show that $\dim_H(B) \geq l + 1$.

Since $\dim_T(B) \geq l + 1$, there must exist $\boldsymbol{x} \in B$ and $R_0 > 0$ such that for every $(m-1)$-dimensional sphere S_R (centered at $\boldsymbol{x}$ with radius $R \leq R_0$),[40] the

[39] For the latter inequality, we could also use inequality (8.102) in part (d) of Remark 8.48.

[40] Accordingly, S_R is the boundary of the m-dimensional ball of center $\boldsymbol{x}$ and radius R.

following inequality holds:[41]

$$\dim_T(B \cap S_R) \geq l. \tag{8.105}$$

Now, we use the above information (in (8.105)) in order to show that $\dim_H(B) \geq l+1$, as required. To this aim, given any $n \in \mathbb{N}$, let $\{T_{n_j}\}_{j=1}^{\infty}$ be a cover of B by balls of diameter $d_{n_j} < 1/n$, for every integer $j \geq 1$.

Given any positive real number $\rho < l+1$ (i.e., $l > \rho - 1$) and $r \in (0, R]$, we let (for each $j \geq 1$)

$$g_{n_j}(r) := \begin{cases} (d_{n_j})^{\rho-1}, & \text{if } S_r \cap T_{n_j} \neq \emptyset, \\ 0, & \text{otherwise,} \end{cases} \tag{8.106}$$

and

$$\sigma_n(r) := \sum_{j=1}^{\infty} g_{n_j}(r) = \sum_{j\geq 1,\, S_r \cap T_{n_j} \neq \emptyset} (d_{n_j})^{\rho-1} \in [0, +\infty]. \tag{8.107}$$

Recalling the definition of the (t-dimensional) δ-approximate Hausdorff outer measure $\mathscr{H}^t_\delta$ in Subsection 6.3.1 (see formula (6.19), with $t := \rho - 1$ and $\delta = 1/n$, here), we deduce at once from (8.107) that for each $r \in (0, R]$, we have that[42]

$$\sigma_n(r) \geq \mathscr{H}^{\rho-1}_{1/n}(S_r \cap B), \text{ for all } n \in \mathbb{N}. \tag{8.108}$$

Since, by construction, $\dim_T(S_r \cap B) \geq l > \rho - 1$, we must have, in light of the induction hypothesis (applied to $A := S_r \cap B$), that

$$\dim_H(S_r \cap B) \geq l > \rho - 1. \tag{8.109}$$

Therefore, by definition of the Hausdorff dimension (see Definition 5.13) and since (8.109) implies that $\dim_H(S_r \cap B) > \rho - 1$, we must have that

$$\mathscr{H}^{\rho-1}(S_r \cap B) = +\infty. \tag{8.110}$$

Since by definition (see formula (5.3) in Definition 5.4 of Section 5.1)

$$\mathscr{H}^{\rho-1}(S_r \cap B) = \lim_{\delta \to 0^+} \mathscr{H}^{\rho-1}_\delta(S_r \cap B), \tag{8.111}$$

we also have that $\mathscr{H}^{\rho-1}(S_r \cap B) = \lim_{n\to\infty} \mathscr{H}^{\rho-1}_{1/n}(S_r \cap B)$; we deduce from (8.108) and (8.110)–(8.111) that the following limit exists and is given by

$$\lim_{n\to\infty} \sigma_n(r) = +\infty, \text{ still for all } r \in (0, R]. \tag{8.112}$$

[41]Indeed, otherwise, every point of B would have arbitrarily small neighborhoods V with $\dim_T(V) \leq l-1$. Hence, by Definition 1.18 of the topological dimension, we would have that $\dim_T(B) \leq l$, which contradicts our induction hypothesis.

[42]Indeed, for every fixed $n \in \mathbb{N}$, because $\{T_{n_j}\}_{j=1}^{\infty}$ is an n^{-1}-cover of B, it is also an n^{-1}-cover of $S_r \cap B$. Hence, $\{T_{n_j} : j \in \mathbb{N},\ S_r \cap T_{n_j} \neq \emptyset\}$ is also an n^{-1}-cover of $S_r \cap B$, from which (8.108) follows since $\mathscr{H}^{\rho-1}_{1/n}(S_r \cap B)$ is an infimum of quantities including $\sigma_n(r)$ itself; see formula (5.2) in Definition 5.1.

In light of the definition of σ_n in (8.107), $\{\sigma_n\}_{n=1}^{\infty}$ is a nondecreasing sequence of nonnegative functions on $(0, R]$; it therefore follows from the Monotone Convergence Theorem (see part (iv) of Theorem 3.116 in Section 3.6)—combined with the Fubini–Tonelli Theorem—that

$$\sum_{j=1}^{\infty} \int_0^R g_{n_j}(r)dr = \int_0^R \sigma_n(r)dr \to +\infty, \quad \text{as } n \to \infty. \tag{8.113}$$

Fix $n \in \mathbb{N}$, for now. Next, we observe that, by construction (see, in particular, (8.106) above) for every $j \in \mathbb{N}$, $g_{n_j}(r) = 0$, for all $r \notin I_j$ and $g_{n_j}(r) = (d_{n_j})^{\rho-1}$, for all $r \in I_j$, where I_j is an interval of length at most $d_{n_j} = \operatorname{diam}(I_{n_j})$. Therefore, in light of (8.106) and (8.107), we have that

$$\sum_{j=1}^{\infty} (d_{n_j})^{\rho} = \sum_{j=1}^{\infty} \int_0^{d_{n_j}} (d_{n_j})^{\rho-1} dr \geq \sum_{j=1}^{\infty} \int_0^R \sigma_n(r)dr. \tag{8.114}$$

We now let $n \to \infty$ in (8.114) in order to deduce from (8.113) that

$$\sum_{j=1}^{\infty} (d_{n_j})^{\rho} = \sum_{j=1}^{\infty} \left(\operatorname{diam}(I_{n_j})\right)^{\rho} \to +\infty, \quad \text{as } n \to \infty. \tag{8.115}$$

But since for each fixed $n \in \mathbb{N}$, $\{T_{n_j}\}_{j=1}^{\infty}$ was chosen to be an *arbitrary* n^{-1}-cover of B by balls of diameter less than $1/n$ (i.e., $B \subseteq \bigcup_{j\geq 1} T_{n_j}$ and $\operatorname{diam}(T_{n_j}) = d_{n_j} < 1/n$, for all $j \geq 1$), and since the exact counterpart of (8.110) and (8.111) holds (except for $\rho - 1$ and $S_r \cap B$ respectively replaced by ρ and B), a moment's reflection shows that (8.115) implies that

$$\mathscr{H}^{\rho}(B) = \lim_{n\to\infty} \mathscr{H}^{\rho}_{1/n}(B) = \sup_{n\geq 1} \mathscr{H}^{\rho}_{1/n}(B) = +\infty. \tag{8.116}$$

By definition of the Hausdorff dimension $\dim_H(B)$ (see Definition 5.13), it follows at once from (8.116) that $\dim_H(B) \geq \rho$. And, finally, since we recall that ρ was chosen to be an arbitrary positive number such that $\rho < l + 1$, we conclude that $\dim_H(B) \geq l+1$. This completes the proof of the induction—and hence also of the entire theorem. □

Exercise 8.51. Fully justify (8.116) in light of the surrounding text and of (8.115), in particular.

Remark 8.52. In the above proof of Theorem 8.50 (Nöbeling's Theorem), we have closely followed the outline of Nöbeling's proof given in 1931 [**464**], which truly deserves to be better known. We gave, however, considerably more details, when we felt that it could be helpful to the reader. In particular, the above proof of Theorem 8.50 provides an excellent review of the basic definitions involving the topological dimension and the Hausdorff dimension, along with the δ-approximate Hausdorff measure and the Hausdorff measure themselves.

We next state, without proof, a remarkable complement and generalization of Theorem 8.50, valid for any metric space—more specifically, for any metrizable space (contained in a separable metric space, such as $\mathbb{R}^m$)[43] and, in particular, for any subset of $\mathbb{R}^m$.

Theorem 8.53 (Characterization of the Topological Dimension in Terms of the Hausdorff Dimension; E. Szpilrajn (E. Marczewski) [**581**], 1937). [44] *Let X be a metrizable space contained in a separable metric space W.*[45] *Then, the topological dimension of X, $\dim_T(X)$, is equal to the Hausdorff dimension of at least one metric space Z (contained in W) which is homeomorphic to X. Furthermore, $\dim_T(X)$ is the smallest (i.e., the minimum) such value. Therefore,*

$$\dim_T(X) = \min_Y \dim_H(Y), \tag{8.117}$$

where the minimum is attained[46] *and is taken over all metric spaces Y (contained in W) which are homeomorphic to X.*[47]

Remark 8.54. Note that it is not at all obvious, a priori, that, first, the infimum corresponding to the right-hand side of (8.117) in Theorem 8.53 is attained (i.e., is actually a minimum), and, secondly, that it is an integer (necessarily in $\mathbb{N}_0$ if X is nonempty).

Next, we explain how one can very simply recover Nöbeling's Theorem (Theorem 8.50) from Szpilrajn's Theorem (Theorem 8.53).

Corollary 8.55. *For any $A \subseteq \mathbb{R}^m$, we have that*

$$\dim_T(A) \le \dim_H(A), \tag{8.118}$$

where A is equipped with the Euclidean metric on $\mathbb{R}^m$.

Proof. Apply Theorem 8.53 to $X := A \subseteq \mathbb{R}^m =: W$. Then, by (8.117), with $Y := A \subseteq \mathbb{R}^m$, the metric space (contained in the separable metric space $\mathbb{R}^m$) equipped with the Euclidean metric, we must have that the inequality (8.118) holds, as desired. □

Remark 8.56. A key result established in [**581**] in order to prove Theorem 8.53 is the following one:

Let $n \in \mathbb{N}_0$. A metrizable space X (contained in a separable metric space W) has topological dimension $\le n$ (i.e., $\dim_T(X) \le n$) iff it is homeomorphic

---

[43] A *metrizable space* X is a topological space the topology of which can be induced by a metric on X.

[44] E. Szpilrajn actually stands for Edward Marczewski, who used E. Szpilrajn as a pseudonym in an attempt to escape the Nazis.

[45] For example, we can choose $W = \mathbb{R}^m$, the m-dimensional Euclidean space, where $m \in \mathbb{N}$.

[46] In other words, the underlying infimum is actually a minimum.

[47] Alternatively, we may assume in (8.117) that $X \subseteq Y \subseteq W$ and that Y induces the topology on X.

to a Lebesgue null set Y in the $(n+1)$-dimensional unit cube $[0,1]^{n+1} \subseteq \mathbb{R}^{n+1}$; i.e., $\mathscr{L}^{n+1}(Y) = 0$ or, equivalently,[48] $\mathscr{H}^{n+1}(Y) = 0$.

The following and final comment supplements Remark 8.56 just above.

Remark 8.57 (Sketch of the Proof of Theorem 8.53). In essence, the proof of Theorem 8.53 given in [**581**]—and, in particular, of the key result stated in Remark 8.56—goes as follows:

(i) First, in agreement with the "only if" part of the result stated in Remark 8.56, one shows that if a metrizable space X (contained in a separable metric space W) is such that $\mathscr{H}^{n+1}(X) = 0$, then $\dim_T(X) \leq n$.

(ii) With X as in part (i), and thus such that $\dim_T(X) \leq n$, and given any $t > n$, show that there exists $Z \subseteq [0,1]^{n+1} \subseteq \mathbb{R}^{n+1}$, with Z homeomorphic to X and $\mathscr{H}^t(Z) = 0$. Therefore, $\dim_H(Z) \leq t$.

(iii) Strongly improve the conclusion of (ii) but by now only assuming that $\dim_T(X) \leq n$ (and $X \subseteq W$, where W is a separable metric space) and then by showing that X is homeomorphic to some $Z \subseteq [0,1]^{n+1} \subseteq \mathbb{R}^{n+1}$, such that $\mathscr{H}^t(Z) = 0$, for any $t > n$. Hence, $\dim_H(Z) \leq n$.[49]

(iv) Conclude from (iii) that Z (as obtained in part (iii)) satisfies the desired property; namely, $\dim_H(Z) = n = \dim_T(X)$. Hence, $Z \subseteq [0,1]^{n+1}$ achieves the minimum on the right-hand side of (8.117), as required in the statement of Theorem 8.53.

8.7. Dimensions of Attractors

We briefly study here the fractal dimensions of the attractors of certain dynamical systems (here, self-similar systems or, more generally, suitable IFSs). For a lot more information about this subject, see [**172, 175**], and, especially for a variety of examples corresponding to nonlinear (and, often, chaotic) dynamical systems, see Ya. B. Pesin's book, [**491**].

Lemma 8.58. *Let* $\mathbb{N} \geq 2$ *and let* $\boldsymbol{\Phi} = \{\varphi_j\}_{j=1}^N$ *be a self-similar system on* $\mathbb{R}^m$ *that satisfies the Open Set Condition with scaling vector* $\mathbf{r} = (r_1, \dots, r_N)$ *comprising positive scaling ratios. If* $F = \boldsymbol{\Phi}(F) := \bigcup_{j=1}^N \varphi_j(F)$*, that is, if* F *is the attractor of* $\boldsymbol{\Phi}$*, then* $\overline{\dim}_B(F) \leq D_S$*, where* D_S *is the similarity dimension of* F *given by the unique* (and necessarily positive) *real solution of the corresponding Moran equation* (5.22): $\sum_{j=1}^N r_j^t = 1$, $t \in \mathbb{R}$.

[48] Recall that, for any $n \in \mathbb{N}_0$ and when the underlying space is $\mathbb{R}^{n+1}$, then $\mathscr{H}^{n+1} = (c_{n+1})^{-1} \cdot \mathscr{L}^{n+1}$, for some finite and positive (explicit) constant; see Theorems 7.53 and 7.66 in Section 7.4.

[49] Indeed, we must have that $\dim_H(Z) \leq t$, for all $t > n$.

Proof. Recall the proof of Moran's Theorem (Theorem 5.40), which deals with the Hausdorff dimension of self-similar sets. As in that proof, let V denote a feasible open set; that is, V denotes a nonempty open set such that $\varphi_j(V) \subseteq V$ for each j and the images $\varphi_j(V)$ are pairwise disjoint. Also, given $0 < r < 1$, let Ω be defined by the finite set of finite words which stem from curtailing infinite words $J \in \mathfrak{J}_\infty$ at the first term j_k for which we have $(\min_j r_j) \cdot r \leq r_{J|_k} \leq r$; see (5.24). Then, for each $J \in \mathfrak{J}_\infty$, there is exactly one index $k \in \mathbb{N}$ such that $J|_k = j_1 \ldots j_k \in \Omega$.[50] By induction on Moran's equation (5.22), we have

$$\sum_{\Omega} r_{J|_k}^{D_S} = \sum_{\Omega} (r_{j_1} r_{j_2} \cdots r_{j_k})^{D_S} = 1.$$

Then, by Lemma 5.39, Ω contains at most $(\min_j r_j)^{-D_S} \cdot r^{-D_S}$ words. For each $J|_k \in \Omega$, we have $\operatorname{diam}(\overline{V}_{J|_k}) = r_{J|_k} \cdot \operatorname{diam}(\overline{V}) \leq r \cdot \operatorname{diam}(\overline{V})$; so, the family $\{\overline{V}_{J|_k}\}_{J|_k \in \Omega}$ is a cover of F comprising at most $(\min_j r_j)^{-D_S} \cdot r^{-D_S}$ sets of diameter equal to $r \cdot \operatorname{diam}(\overline{V})$. Taking the box-counting function $N_B(F, \varepsilon^{-1})$ to be the minimum number of sets of diameter ε required to cover F, we have $\overline{\dim}_B(F) \leq D_S$. □

Remark 8.59. Note that Lemma 8.58 and Proposition 8.39 combine to complete the proof of Moran's Theorem (Theorem 5.40). In particular, if F is the attractor of a self-similar system $\mathbf{\Phi} = \{\varphi_j\}_{j=1}^N$ on $\mathbb{R}^m$ that satisfies the Open Set Condition with scaling vector $\mathbf{r} = (r_1, \ldots, r_N)$ comprising positive scaling ratios, then $\dim_B(F) = D_S$, where D_S is the similarity dimension of F given by the unique (and necessarily positive) real solution of the corresponding Moran equation. Indeed, by Lemma 8.58, Proposition 8.39, and Moran's Theorem (Theorem 5.40, applied to the Hausdorff dimension), we have $D_S = \dim_H(F) \leq \underline{\dim}_B(F) \leq \overline{\dim}_B(F) \leq D_S$. Thus, $\dim_B(F)$ exists and $D_S = \dim_H(F) = \dim_B(F)$. It then follows from Theorem 8.29 in Section 8.4 above that $\dim_M(F)$ exists and equals $\dim_B(F)$, as desired.

Example 8.60. Assume that $A \subseteq \mathbb{R}^m$ and $B \subseteq \mathbb{R}^n$, where A is a self-similar set satisfying the Open Set Condition and B is bounded. Then,

$$\dim_H(A \times B) = \dim_H(A) + \dim_H(B) = D_S + \dim_H(B), \tag{8.119}$$

where D_S is the similarity dimension of A. Indeed, by Moran's Theorem (Theorem 5.40) we have $\dim_H(A) = \dim_B(A) = D_S$; so, by Corollary 8.45 the equalities in (8.119) hold.

Let $A, B \subseteq \mathbb{R}$, where A and B are Smith–Volterra–Cantor sets, say, $A = SVC(n_1)$ and $B = SVC(n_2)$, for some $n_1, n_2 \in \mathbb{N}$, where $n_1, n_2 \geq 3$. As shown

[50] We use here the notation pertaining to symbolic dynamics introduced in Section 4.5 above.

in Examples 4.92, 5.29, and 8.41 (and since $\mathcal{H}^1 = \mathcal{L}^1$), we have

$$\dim_H(A) = \dim_B(A) = \dim_H(B) = \dim_B(B) = 1,$$

$$\mathcal{H}^1(A) = \frac{n_1 - 3}{n_1 - 2}, \quad \text{and} \quad \mathcal{H}^1(B) = \frac{n_2 - 3}{n_2 - 2}.$$

So, by Corollary 8.45, we have $\dim_H(A \times B) = 2$. Moreover, $A \times B$ is totally disconnected (see Exercise 8.63 below) and yet, by Theorem 5.76, $A \times B$ has positive area (2-dimensional Hausdorff measure) such that

$$\mathcal{H}^2(A \times B) \geq \frac{n_1 - 3}{n_1 - 2} \cdot \frac{n_2 - 3}{n_2 - 2}.$$

The upper and lower bounds for the Hausdorff dimensions of the attractors of certain iterated function systems also hold for the upper and lower box-counting (equivalently, Minkowski) dimensions.[51]

Proposition 8.61. *Let $\Psi = \{\psi_j\}_{j=1}^N$ be an iterated function system on $\mathbb{R}^m$ such that $N \geq 2$ and $\|\psi_j(\boldsymbol{x}) - \psi_j(\boldsymbol{y})\|_m \leq r_j \cdot \|\boldsymbol{x} - \boldsymbol{y}\|_m$, for any $\boldsymbol{x}, \boldsymbol{y} \in \mathbb{R}^m$, where $0 < r_j < 1$, for each $j = 1, \ldots, N$. Then,*

$$\dim_H(F) \leq \overline{\dim}_B(F) \leq D_1,$$

where D_1 is the unique (and necessarily positive) real number t such that $\sum_{j=1}^N r_j^t = 1$.

If $\Psi = \{\psi_j\}_{j=1}^N$ is a strongly separated iterated function system on $\mathbb{R}^m$ (in the sense of Definition 4.73 in Section 4.7 above) such that

$$c_j \cdot \|\boldsymbol{x} - \boldsymbol{y}\|_m \leq \|\psi_j(\boldsymbol{x}) - \psi_j(\boldsymbol{y})\|_m, \quad \text{for any } \boldsymbol{x}, \boldsymbol{y} \in \mathbb{R}^m,$$

where $0 < c_j < 1$, for each $j = 1, \ldots, N$. Then,

$$D_2 \leq \dim_H(F) \leq \underline{\dim}_B(F),$$

where D_2 is the unique (and necessarily positive) real number t such that $\sum_{j=1}^N c_j^t = 1$.

8.8. Distinct Upper and Lower Box-Counting (or Minkowski) Dimensions

The sole purpose of this section is to prove the following theorem, which asserts the existence of a subset of the real line with distinct upper and lower box-counting (and hence also Minkowski) dimensions.

Theorem 8.62. *Let $0 < \alpha \leq \beta < 1$. Then, there exists a countable compact set $E \subseteq \mathbb{R}$ such that*

$$\dim_H(E) = 0, \quad \underline{\dim}_B(E) = \alpha, \quad \text{and} \quad \overline{\dim}_B(E) = \beta. \tag{8.120}$$

[51] The interested reader may wish to review Proposition 5.49 (and related results) in Section 5.4.

Loosely speaking, the goal of the proof of this theorem is to construct a set E such that, for each $k \in \mathbb{N}$, there exist $\varepsilon_{\alpha,k}, \varepsilon_{\beta,k} > 0$ such that

$$\left| \frac{\log N_B(E, \varepsilon_{\alpha,k}^{-1})}{-\log \varepsilon_{\alpha,k}} \right| < \frac{1}{2^{k+1}} \quad \text{and} \quad \left| \frac{\log N_B(E, \varepsilon_{\beta,k}^{-1})}{-\log \varepsilon_{\beta,k}} \right| < \frac{1}{2^{k+1}},$$

which would produce the desired result.

Proof. For each $n \in \mathbb{N}$, let $a_n := e^{-n}$. Note that $\lim_{n\to\infty} a_n = 0$. The desired set E will be constructed as an increasing sequence, starting at 0, followed by a collection E_1 of points equally separated by length $a_1 = e^{-1}$, followed by a collection E_2 of points equally separated by length $a_2 = e^{-2}$, and so forth. Let

$$\begin{aligned}
E_0 &= \{0\}, \\
E_1 &= \{a_1, 2a_1, \ldots, b_1 a_1\}, \\
E_2 &= \{b_1 a_1 + a_2, b_1 a_1 + 2a_2, \ldots, b_1 a_1 + b_2 a_2\}, \\
&\vdots \\
E_{n+1} &= \{S_n + a_{n+1}, S_n + 2a_{n+1}, \ldots, S_n + b_{n+1} a_{n+1}\}, \\
&\vdots
\end{aligned}$$

where, for all $n \in \mathbb{N}$, $S_n := \sum_{k=1}^{n} a_k b_k$, and $\{b_n\}_{n=1}^{\infty}$ is the sequence of non-negative integers constructed below using a four-step recursive process. Let $S := \lim_{n\to\infty} S_n$ (a limit which will be shown to exist). Then, the desired set E will be given by

$$E := \{S\} \cup \left(\bigcup_{n=0}^{\infty} E_n \right). \tag{8.121}$$

The key difficulty in this proof lies in the construction of the sequence $\{b_n\}_{n=1}^{\infty}$. To start the four-step recursive process, note that $P_n := \sum_{k=1}^{n} [e^{\alpha k}]$ grows like $e^{\alpha n}$. Define n_1 to be the smallest positive integer such that[52]

$$\left| \frac{1}{n_1} \log P_n - \alpha \right| < \frac{1}{2}.$$

Then, for $n = 1, \ldots, n_1$, let $b_n := [e^{\alpha n}]$, where (as before) $[x]$ denotes the unique integer part of $x \in \mathbb{R}$.

Now, choose c such that $e^{\beta} < c < e$. Define n_2 to be the unique positive integer such that $b_{n_1} \cdot c^{n_2 - n_1} \leq e^{\beta n_2} < b_{n_1} \cdot c^{n_2 - n_1 + 1}$. For $n = n_1 + 1, \ldots, n_2$, let $b_n := b_{n_1} \cdot c^{n - n_1}$.

[52]Throughout the remainder of this proof, the word "integer" refers to a positive integer.

Next, choose $n_3 > n_2$ to be the smallest integer such that $n_2 < n_3$ and

$$\left| \frac{1}{n_3} \log\left(\sum_{k=1}^{n_2} b_k + \sum_{k=n_2+1}^{n_3} [e^{\beta k}] \right) - \beta \right| < \frac{1}{2}.$$

For $n = n_2 + 1, \ldots, n_3$, let $b_n := [e^{\beta n}]$.

For the last step in the initial four-step process, choose $n_4 > n_3$ to be the unique integer such that $[e^{\alpha n_4}] \leq [e^{\beta(n_4 - n_3)}] < [e^{\alpha n_4}] + 1$. Then, for $n = n_3 + 1, \ldots, n_4$, let b_n be given by a constant; namely, $b_n := [e^{\beta n_3}]$.

In order to proceed from here, define B_n to be the partial sums of the numbers b_k given by $B_n := \sum_{k=1}^{n} b_k$. Iterate the procedure defined above so that for each $k \in \mathbb{N}_0$:

(i) $b_n := [e^{\alpha n}]$, for $n = n_{4k} + 1, \ldots, n_{4k+1}$, where n_{4k+1} is the smallest integer such that

$$\left| \frac{1}{n_{4k+1}} \log B_{n_{4k+1}} - \alpha \right| < \frac{1}{2^{k+1}}; \tag{8.122}$$

(ii) $b_n := b_{n_{4k+1}} \cdot c^{n - n_{4k+1}}$, for $n = n_{4k+1} + 1, \ldots, n_{4k+2}$, where the integer n_{4k+2} is uniquely defined by the inequalities

$$b_{n_{4k+1}} \cdot c^{n_{4k+2} - n_{4k+1}} < e^{\beta n_{4k+1}} \leq b_{n_{4k+1}} \cdot c^{n_{4k+2} - n_{4k+1} + 1};$$

(iii) $b_n := [e^{\beta n}]$, for $n = n_{4k+2} + 1, \ldots, n_{4k+3}$, where n_{4k+3} is the smallest (positive) integer such that

$$\left| \frac{1}{n_{4k+3}} \log B_{n_{4k+3}} - \beta \right| < \frac{1}{2^{k+1}}; \quad \text{and} \tag{8.123}$$

(iv) $b_n := [e^{\beta n_{4k+3}}]$, for $n = n_{4k+3} + 1, \ldots, n_{4k+4}$, where n_{4k+4} is uniquely determined by the inequalities

$$[e^{\alpha n_{4k+4}}] \leq [e^{\beta(n_{4k+4} - n_{4k+3})}] < [e^{\alpha n_{4k+4}}] + 1.$$

Note that, as defined, the sequence $\{b_n\}_{n=1}^{\infty}$ is an increasing sequence of positive integers and $b_n \to +\infty$, as $n \to \infty$. Also, since $a_n = e^{-n}$ and $b_n \leq e^{\beta n}$, for each $n \in \mathbb{N}$, we have

$$S = \sum_{n=1}^{\infty} a_n b_n \leq \sum_{n=1}^{\infty} e^{(\beta-1)n} = \frac{e^{\beta-1}}{1 - e^{\beta-1}} < \infty.$$

By (8.122) and the definition of n_{4k+1}, we have

$$\liminf_{n\to\infty} \frac{\log B_n}{n} \leq \lim_{k\to\infty} \frac{\log B_{n_{4k+1}}}{n_{4k+1}} = \alpha.$$

Also, the following inequality holds: $B_n \geq \sum_{k=1}^{n} e^{\alpha k} = (e^{\alpha(n+1)} - 1)/(e^{\alpha} - 1)$. So, after taking logarithms and dividing by n, we obtain that

$$\frac{\log B_n}{n} \geq \frac{\log(e^{\alpha(n+1)} - 1)}{n} - \frac{\log(e^{\alpha} - 1)}{n}.$$

Now, since

$$\lim_{n\to\infty} \left(\frac{\log(e^{\alpha(n+1)} - 1)}{n} - \frac{\log(e^{\alpha} - 1)}{n} \right) = \alpha,$$

we deduce that $\liminf_{n\to\infty}(\log B_n / n) \geq \alpha$. Therefore,

$$\liminf_{n\to\infty} \frac{\log B_n}{n} = \alpha. \tag{8.124}$$

By a similar argument, which uses (8.122) and the definition of n_{4k+3}, we also obtain that

$$\limsup_{n\to\infty} \frac{\log B_n}{n} = \beta. \tag{8.125}$$

Next, since $b_{n+1}/b_n \leq c$ and $B_n \geq b_n$, for all $n \in \mathbb{N}$, we have successively

$$\begin{aligned} \frac{S - S_n}{a_n B_n} &\leq \sum_{k=n+1}^{\infty} \frac{e^{-k} \cdot b_k}{e^{-n} \cdot b_n} \leq \sum_{k=n+1}^{\infty} \left(\frac{c}{e}\right)^{k-n} \\ &= \sum_{k=1}^{\infty} \left(\frac{c}{e}\right)^{k} = \frac{c/e}{1 - (c/e)} < \infty. \end{aligned} \tag{8.126}$$

Let E be the set defined by (8.121). Note that (8.126) implies that E is bounded. Let $N_B(E, \varepsilon^{-1})$ denote the minimum number of open intervals of length ε required to cover E. Given $\varepsilon > 0$, choose n to be the unique integer such that $e^{-(n+1)} < \varepsilon \leq e^{-n}$. By construction, the first $B_n = \sum_{k=1}^{n} b_k$ points in E are separated by a distance of at least e^{-n}. Hence, $N_B(E, \varepsilon^{-1}) \geq B_n$.

Moreover, by (8.126), the number of open intervals of length ε required to cover $[S_n, S]$ is at most $(S - S_n)/e^{-(n+1)} = (S - S_n)/e a_n \leq C B_n / e$, where $C := (c/e)/(1 - (c/e))$. So, $N_B(E, \varepsilon^{-1}) \leq (1 + C/e) B_n$ and thus,

$$\frac{\log B_n}{n+1} \leq \frac{\log N_B(E, \varepsilon^{-1})}{-\log \varepsilon} \leq \frac{\log(1 + C/e) + \log B_n}{n}. \tag{8.127}$$

By our choice of $n = n(\varepsilon)$, we have that $\varepsilon \to 0^+$ implies $n \to \infty$. Hence, (8.124), (8.125), and (8.127) combine to yield $\underline{\dim}_B(E) = \alpha$ and $\overline{\dim}_B(E) = \beta$, as desired. Finally, we have that $\dim_H(E) = 0$ since E is countable. This completes the proof of (8.120)—and hence also of Theorem 8.62. □

Exercise 8.63. Show that if A and B are Smith–Volterra–Cantor sets, then their Cartesian product $A \times B$ is totally disconnected.

Exercise 8.64. Compute (via a direct calculation) $\dim_B(C)$, the box dimension of the (classic, ternary) Cantor set C. Show that $\dim_B(C)$ exists and satisfies $\dim_B(C) = D_C := \inf\{t \in \mathbb{R} : \sum_{n=0}^{\infty}(2^{n-1}/3^n) < \infty\} = \log_3 2$. Generalize these results to Smith–Volterra–Cantor sets.

8.9. Generalized Hausdorff, Packing, and Minkowski Dimensions

Our goal in this section is to give a brief overview of several useful generalizations of several of the notions discussed thus far, namely, the notions of Minkowski contents, Hausdorff measures, and a new type of measure called the "packing measure", along with the associated fractal dimension, called the "packing dimension" and which coincides with the so-called "modified box-counting dimension". We do not plan to be exhaustive in any way and choose typically the simplest (rather than the most general) assumptions. We also omit most of the proofs, for which the reader is referred to the relevant literature.

8.9.1. Gauge Functions (or Scaling Laws). Let $\boldsymbol{\Psi}$ denote the class of functions $h\colon (0,\eta) \to (0,+\infty)$, for some $\eta > 0$ sufficiently small, which are right-continuous, monotonically increasing (i.e., nondecreasing), such that $h(0^+) = 0$,[53] and satisfy the following condition, often referred to as the *doubling condition* in the literature: there exists $M > 0$ such that

$$h(2x) \leq M \cdot h(x), \text{ for all } x \in (0,\eta/2). \tag{8.128}$$

The elements h of $\boldsymbol{\Psi}$ are called *gauge functions* (or *dimension functions*) in the mathematical literature, while they are often referred to as *scaling laws* in physics.[54]

Example 8.65 (*Scaling Laws or Gauge Functions*). We give here various examples of gauge functions h, i.e., of elements of $\boldsymbol{\Psi}$. The verification is left to the interested reader.

(a) (*Power Laws*). Fix $t > 0$. A simple, but key, example of a gauge function is then given by $h(x) = h_t(x) := x^t$, for all $x > 0$, which corresponds to the standard power laws.

(b) (*Examples of Logarithmic Scaling Laws*). The following examples of gauge functions are of particular significance in probability theory and, especially, in the theory of Brownian motion, as we shall see in the next two subsections: $h(x) = x^2 \cdot \log(\log(x^{-1}))$, for all $x \in (0,\eta)$; $h(x) = x^2 \cdot \log(x^{-1}) \cdot \log(\log(\log(x^{-1}))) = x^2 \cdot \log(x^{-1}) \cdot \log^{\circ 3}(x^{-1})$, for all $x \in (0,\eta)$, where for any $n \in \mathbb{N}$, $\log^{\circ n}$ denotes the n-th *iterated logarithm*, obtained as the composition (when it makes sense) of log with itself n times (so that $\log^{\circ 1} = \log$); and

[53] Hence, h can be extended by continuity to all of $[0,\eta)$ by letting $h(0) := 0$.

[54] The latter terminology is especially justified when h is exact, in a sense to be specified later.

$h(x) = x^2/\log^{\circ 2}(x^{-1})$, for all $x \in (0, \eta)$. As was already stated, each of the above functions h belong to $\boldsymbol{\Psi}$.

(c) (*More General Scaling Laws*). More general and exotic examples of gauge functions (or scaling laws) can be found in the literature, including, for example, in C. Tricot's thesis [**591**] and in the memoir by C. Q. He and M. L. Lapidus [**235**]. The examples in the latter reference significantly extend those discussed in part (b) just above.

Exercise 8.66. Verify that all of the functions h given in part (b) of Example 8.65 are, indeed, gauge functions, i.e., belong to $\boldsymbol{\Psi}$, for suitable values of $\eta > 0$ and $M > 0$ which you will specify in each case. Generalize these examples of gauge functions.

8.9.2. Generalized Hausdorff Measures and Dimension Functions. Let $h \in \boldsymbol{\Psi}$; i.e., let h be a gauge function, as in Subsection 8.9.1 just above. Given an arbitrary $A \subseteq \mathbb{R}^m$, with $m \in \mathbb{N}$, we let (much as in Chapters 3, 5, and 7)[55]

$$(8.129)\qquad \begin{aligned} \mathscr{H}^h(A) &= \lim_{\delta \to 0^+} \mathscr{H}^h_\delta(A) \\ &:= \lim_{\delta \to 0^+} \inf \left\{ \sum_{j=1}^{\infty} h(\operatorname{diam}(A_j)) \,:\, \{A_j\}_{j=1}^{\infty} \text{ is a } \delta\text{-cover of } A \right\}. \end{aligned}$$

The extended number $\mathscr{H}^h(A) \in [0, +\infty]$ is well defined—as the limit as $\delta \to 0^+$ of the nondecreasing family $\{\mathscr{H}^h_\delta(A)\}_{\delta>0}$, called the *$\delta$-approximate h-Hausdorff* measures of A. For the latter discussion, as well as in the present Subsection 8.9.2, we could restrict ourselves to covering by open (or closed) balls of $\mathbb{R}^m$, for example, as we did in the standard case in Chapters 6 and 7. In any case, the resulting measures will be equivalent—in the sense that, for a given choice of h, any such measure is between two finite and positive constants times the other measure.

Much as in Section 5.1, and by using the corresponding results stated in Chapter 5 (as well as proved in Chapters 6 and 7), one verifies that $\mathscr{H}^h$ is a metric outer measure (see Section 6.2)—and hence that the σ-algebra of $\mathscr{H}^h$-measurable sets contains $\mathcal{B}(\mathbb{R}^m)$, the Borel σ-algebra of $\mathbb{R}^m$. For notational simplicity, we will also denote by $\mathscr{H}^h$ the restriction of the outer measure $\mathscr{H}^h$ to $\mathcal{B}(\mathbb{R}^m)$, which (by Carathéodory's Restriction Theorem, Theorem 3.26 or

[55]By necessity of concision, we use a more compact notation in order to be able to define $\mathscr{H}^h$ in one step instead of in two steps. The same comment applies to the definition of $\mathscr{P}^h$, the h-packing measure; see Subsection 8.9.3

Theorem B.3 in Appendix B) is, in fact, a Borel measure on $\mathbb{R}^m$. Then, $\mathcal{H}^h$ is called the *h-(generalized) Hausdorff measure* or the *(generalized) Hausdorff measure with respect to (or associated with) the gauge function* (or *dimension function*) *h*.

Definition 8.67 (*Exact Hausdorff Dimension Function*). Let $A \in \mathcal{B}(\mathbb{R}^m)$. If, for some $h \in \Psi$, we have that $0 < \mathcal{H}^h(A) < \infty$, then h is said to be an *exact Hausdorff dimension function* (or an *exact gauge function*) *for A*, while $\mathcal{H}^h(A)$ is said to be an *exact Hausdorff measure for A*. Clearly, such a gauge function h is not unique; of course, the same is true for $\mathcal{H}^h$.

Example 8.68. (a) (*Standard Power Laws*). If, for some $t > 0$, we have that $0 < \mathcal{H}^t(A) < \infty$ (i.e., A is a t-set), then the power function $h(x) := x^t$ (as in part (a) of Example 8.65) is an exact dimension function for A and the standard Hausdorff dimension of A (here, $\dim_H(A) = t$)—as given by the common value in (5.9) and by Definition 5.13 in Section 5.2—is an appropriate way to "measure" the roughness of A.

Even if either $\mathcal{H}^t(A) = 0$ or $\mathcal{H}^t(A) = +\infty$, the t-dimensional Hausdorff dimension of A, as defined in Section 5.2 (Definition 5.13), precisely corresponds to the choice of the gauge function $h(x) = h_t(x) = x^t$, and then, $\mathcal{H}^{h_t}(A) = \mathcal{H}^t(A)$, the t-dimensional (outer measure) or Hausdorff measure of A, as defined in Definition 5.4 and Theorem 5.5 of Section 5.1. But, of course in either case (i.e., if we have $\mathcal{H}^t(A) = 0$ or $\mathcal{H}^t(A) = +\infty$), h_t is no longer an exact dimension function for A and $\mathcal{H}^t$ is no longer an exact Hausdorff measure for A.

(b) (*Brownian Paths in* $\mathbb{R}^m$). Almost surely (i.e., with probability 1 with respect to Wiener's probability measure), a sample Brownian path in the Euclidean plane $\mathbb{R}^2$ has Hausdorff dimension 2 but its 2-dimensional Hausdorff or Lebesgue measure is equal to zero (or, equivalently, its area is equal to zero). However, a significantly more precise result was obtained by S. J. Taylor in [**584**]. Namely, almost surely (i.e., with probability 1), "the" *exact* dimension function for a Brownian path (with values in $\mathbb{R}^2$) is given by the gauge function $h(x) := x^2 \cdot \log(x^{-1}) \cdot \log^{\circ 3}(x^{-1})$, defined in part (b) of Example 8.65.

Two years earlier, in 1962, Z. Ciesielski and S. J. Taylor had shown in [**85**] that a typical Brownian path (with values) in $\mathbb{R}^m$, for $m \geq 3$, admits as an *exact Hausdorff dimension function* $h(x) := x^2 \cdot \log(\log(x^{-1}))$, as given in part (b) of Example 8.65. *Rephrased*: With probability 1 (i.e., almost surely), and for such a choice of h, a Brownian path A in $\mathbb{R}^m$ (with $m \geq 3$) satisfies $0 < \mathcal{H}^h(A) < \infty$.[56]

[56] For an accessible and essentially self-contained introduction to Brownian motion in $\mathbb{R}^m$, and to the corresponding Wiener probability measure on the space of continuous paths from $[0,1]$ (or $[0,+\infty)$) to $\mathbb{R}^m$, see, e.g., [**274**, Chapters 2–4] and the relevant references therein. See also, e.g., [**454**] and [**506**].

In Example 8.72 of Subsection 8.9.3, we will see an analog of this statement for the packing measure, for which the exact dimension function now is given by $h(x) := x^2/\log^{\circ 2}(x^{-1})$, also introduced in part (b) of Example 8.65.

Remark 8.69. The notions of generalized Hausdorff measure and dimension function (or gauge function) were already essentially introduced in Felix Hausdorff's pioneering 1918 paper, [**234**]. It has been used and extended by many authors since then, including Abram Besicovitch (and his collaborators), for example in [**44–47**], for the study, in particular, of number-theoretically defined subsets of $\mathbb{R}$ and of fractal curves. As is clear from part (b) of Example 8.68 above and from its counterpart in Subsection 8.9.3 below, this notion—along with its analogue for packing measures—is also very useful in the study of analysis, probability, and stochastic processes. See, e.g., the 2004 survey articles by Y. Xiao [**621**], as well as the books by K. J. Falconer [**171, 172, 175**] and by C. J. Bishop and Y. Peres [**50**] or by P. Mörters and Y. Peres [**454**], along with the relevant references therein.

8.9.3. Generalized Packing Measures and Dimension Functions. We briefly discuss here the "dual" notions, in some sense, of the (generalized) Hausdorff measure and Hausdorff dimension (function), namely, the (generalized) packing measure and dimension (function). These notions were first introduced by D. Sullivan in the late 1970s and early 1980s [**576–578**] for topological, geometric group-theoretic, and dynamical systems reasons—as well as, independently, precisely defined and studied in the early-to-mid 1980s from an analytical, measure-theoretic, and probabilistic point of view by C. Tricot in [**590**] and by the latter author and S. J. Taylor in [**585**]. We choose here to present these notions in the simplest possible way because they are technically somewhat involved and, possibly, a little cumbersome. Just like the Hausdorff dimension (and unlike the Minkowski and box-counting dimension), the packing dimension—along with the associated "modified box-counting dimension" —are usually very difficult to compute, in practice. However, they are rather useful theoretical tools.

Let h be a gauge function; i.e., $h \in \boldsymbol{\Psi}$, where $\boldsymbol{\Psi}$ was defined in Subsection 8.9.1. Given $A \subseteq \mathbb{R}^m$, let

(8.130)

$$\mathcal{P}^h_{pr}(A) := \lim_{\delta \to 0^+} \sup \left\{ \sum_{j=1}^{\infty} h(2\rho_j) \, : \, \overline{B_{\boldsymbol{x}_j}}(\rho_j) \text{ are disjoint}, \boldsymbol{x}_j \in A, 0 \leq \rho_j < \delta \right\}.$$

This defines a set function, $\mathcal{P}^h_{pr}$, on $\mathbb{R}^m$ which is *not* an outer measure; it is, however, a premeasure (in the sense of Section 3.3) and, therefore, can be extended to an outer measure (via Theorem 3.48) which is a metric outer measure (in the sense of Definition 3.58) and, therefore, a Borel measure on $\mathbb{R}^m$ (via Definition 3.50 and Theorem 3.61), denoted $\mathcal{P}^h$ and called the *h-(generalized) packing measure* or the (*generalized*) *packing measure with respect to the gauge function* (or *dimension function*) h.

Accordingly, for any Borel subset A of $\mathbb{R}^m$ (i.e., for any $A \in \mathcal{B}(\mathbb{R}^m)$), we have that (see formula (3.12) in Theorem 3.48)

$$\mathcal{P}^h(A) = \inf\left\{ \sum_{j=1}^{\infty} \mathcal{P}^h_{pr}(A_j) \, : \, \{A_j\}_{j=1}^{\infty} \text{ is a cover of } A \right\}, \tag{8.131}$$

where $\mathcal{P}^h_{pr}$ is defined by (8.130) just above.

The next definition is the precise analogue for packing measure of Definition 8.67 for Hausdorff measure in Subsection 8.9.2 above.

Definition 8.70 (*Exact Packing Dimension Function*). Let $A \in \mathcal{B}(\mathbb{R}^m)$. If, for some $h \in \Psi$, we have that $0 < \mathcal{P}^h(A) < \infty$, then h is said to be an *exact packing dimension function* (or an *exact gauge function*) *for* A, while $\mathcal{P}^h$ is said to be an *exact packing measure for* A.

Remark 8.71 (*Power Laws: Ordinary Packing Dimension*). The precise counterpart of part (a) of Example 8.68 in Subsection 8.9.2 above holds here, with Hausdorff dimension and measure respectively replaced by packing dimension and packing measure. Accordingly, in particular, for any fixed $t \geq 0$ and for the associated *power law* gauge function (or *power law dimension function*) $h(x) := h_t(x) = x^t$, for all $x > 0$, we recover—and, in fact, define, in the present case—the (ordinary) *packing dimension*, denoted $\dim_P$, and the *t-dimensional packing measure* $\mathcal{P}^t := \mathcal{P}^{h_t}$. More specifically, for any $A \subseteq \mathbb{R}^m$, we have that

$$\dim_P(A) := \sup\{t \geq 0 \, : \, \mathcal{P}^t(A) = +\infty\} = \inf\{t \geq 0 \, : \, \mathcal{P}^t(A) = 0\}, \tag{8.132}$$

while by convention, $\dim_P(\emptyset) := -\infty$.

We can now state an analog for packing (instead of Hausdorff) measures of the case $m \geq 3$ of part (b) of Example 8.68 in Subsection 8.9.2 above.

Example 8.72 (*Exact Packing Dimension of Brownian Paths in* $\mathbb{R}^m$, $m \geq 3$). Almost surely (i.e., with probability 1 relative to the Wiener probability measure), a Brownian motion path in $\mathbb{R}^m$ admits as an *exact packing dimension function* the gauge function $h(x) := x^2/\log^{\circ 2}(x^{-1})$, given in part (b) of Example 8.65 above. This result, obtained in 1985, is due to S. J. Taylor and C. Tricot in [**585**]. It is worth recalling from part (b) of Example 8.68 that, by contrast, the exact Hausdorff dimension function of a typical Brownian path in $\mathbb{R}^m$ (with $m \geq 3$) is given instead by $h(x) = x^2 \cdot \log^{\circ 2}(x^{-1})$, a 1962 result due to Z. Ciesielski and S. J. Taylor in [**85**].

8.9.4. Packing and Modified Box-Counting Dimensions. Just as is the case for Hausdorff measures (see Subsection 8.9.2) when we choose for the gauge functions (or dimension functions) the power functions $h(x) := h_t(x) = x^t$, for all $x > 0$, where $t \geq 0$ is arbitrary, we obtain the *(standard)* t-dimensional *packing dimension*, defined exactly as for the Hausdorff dimension (see formula (5.9) and Definition 5.13 in Section 5.2) as follows:

For any set $A \subseteq \mathbb{R}^m$—and, in particular, for any $A \in \mathcal{B}(\mathbb{R}^m)$—the *packing dimension* of A, denoted $\dim_P(A)$, is given by (8.132), where $\mathcal{P}^t := \mathcal{P}^{h_t}$ denotes, respectively, the *packing outer measure* (when $A \subseteq \mathbb{R}^m$ is arbitrary) or the *packing measure* (when $A \subseteq \mathbb{R}^m$ is a Borel set). In other words, $\dim_P(A)$ is the unique real number $\sigma \geq 0$ such that $\mathcal{P}^t(A) = +\infty$, for $0 \leq t < \sigma$ and $\mathcal{P}^t(A) = 0$, for $t > \sigma$; compare with formula (5.8) of Lemma 5.12 in Section 5.2. We stress that for $t := \sigma = \dim_P(A)$, we may have either $\mathcal{P}^\sigma(A) = 0$ or $\mathcal{P}^\sigma(A) = +\infty$ or else $0 < \mathcal{P}^\sigma(A) < \infty$. In the latter case (and assuming that $A \in \mathcal{B}(\mathbb{R}^m)$), the power function $h_\sigma(x) = x^\sigma$ is an exact dimension function for A, while the associated packing measure $\mathcal{P}^\sigma = \mathcal{P}^{h_\sigma}$ is an exact packing measure for A, in the sense of Definition 8.70 in Subsection 8.9.3 above.

Let A be a bounded subset of $\mathbb{R}^m$. Recall from Section 8.3 (Definition 8.22), that $\underline{\dim}_B(A)$ and $\overline{\dim}_B(A)$ denote, respectively, the lower and upper box-counting dimensions of A, as given by (8.68) in Definition 8.22. Further recall (from Definition 8.3 in Section 8.1 and Theorem 8.29 in Section 8.4) that $\underline{\dim}_B(A) = \underline{\dim}_M(A)$ and $\overline{\dim}_B(A) = \overline{\dim}_M(A)$, where $\underline{\dim}_M(A)$ and $\overline{\dim}_M(A)$ denote, respectively, the lower and upper Minkowski dimension of A, as defined in (8.5) and (8.4).

Then, the *lower* and *upper modified box-counting* (or, *box*, in short) *dimensions*—denoted by $\underline{\dim}_{MB}$ and $\overline{\dim}_{MB}$, respectively—are defined, respectively, by

$$\underline{\dim}_{MB}(A) = \inf\left\{ \sup_{j\geq 1} \underline{\dim}_B(A_j) \,:\, A \subseteq \bigcup_{j=1}^{\infty} A_j \right\} \text{ and} \tag{8.133}$$

$$\overline{\dim}_{MB}(A) = \inf\left\{ \sup_{j\geq 1} \overline{\dim}_B(A_j) \,:\, A \subseteq \bigcup_{j=1}^{\infty} A_j \right\}. \tag{8.134}$$

It is not difficult to check that, essentially by construction (see (8.133) and (8.134)), the lower and upper modified box-counting dimensions are countably stable unlike the upper and lower box-counting or Minkowski dimensions, which are only finitely stable: If $A = \bigcup_{j=1}^{\infty} A_j$, then

$$\underline{\dim}_{MB}(A) = \sup_{j\geq 1} \underline{\dim}_{MB}(A_j),$$

and similarly for $\overline{\dim}_{MB}$.

In addition, we say that the modified box dimension of A *exists* if $\underline{\dim}_{MB}(A) = \overline{\dim}_{MB}(A)$ and then denote by $\dim_{MB}(A)$ this common value, as well as calling it the *modified box-counting* (or *box*) *dimension* of A.

Clearly, in light of (8.133) and (8.134), by choosing the cover $\{A_j\}_{j\geq 1}$ of A so that $A_1 := A$ and $A_j := \emptyset$, for all $j \geq 2$, we deduce that[57]

$$\underline{\dim}_{MB}(A) \leq \underline{\dim}_B(A) \quad \text{and} \quad \overline{\dim}_{MB}(A) \leq \overline{\dim}_B(A). \tag{8.135}$$

Also, since $\underline{\dim}_B \leq \overline{\dim}_B$, it follows at once from (8.133) and (8.134) that

$$\underline{\dim}_{MB}(A) \leq \overline{\dim}_{MB}(A). \tag{8.136}$$

Furthermore, it was shown by C. Tricot in [**590**] that $\overline{\dim}_{MB}(A) = \dim_P(A)$, the packing dimensions of A (as given by (8.132) in Subsection 8.9.3 above). Moreover, it can be checked (see the proof of Theorem 8.73 below) that, since $\dim_H \leq \underline{\dim}_B$, we also have that $\dim_H(A) \leq \underline{\dim}_{MB}(A)$. Since, in addition, $\dim_T \leq \dim_H$, we can summarize the above discussion as follows.

Theorem 8.73 (Comparison of Topological, Minkowski, Packing, Hausdorff, Box and Modified Box Dimensions). *For any nonempty bounded subset A of $\mathbb{R}^m$, we have that $\underline{\dim}_{MB}(A) \leq \underline{\dim}_B(A)$ and*

$$\begin{aligned} 0 \leq \dim_T(A) \leq \dim_H(A) \leq \underline{\dim}_{MB}(A) \leq \overline{\dim}_{MB}(A) \\ = \dim_P(A) \leq \overline{\dim}_B(A) = \overline{\dim}_M(A) \leq m. \end{aligned} \tag{8.137}$$

Proof. In light of the discussion just above the statement of the theorem, all we have to do is show that $\dim_H(A) \leq \underline{\dim}_{MB}(A)$. This inequality follows by combining (8.133), the monotonicity and countable stability of $\dim_H$ (see parts (ii) and (iii) of Proposition 5.16 in Section 5.2 above), and the fact that, on bounded subsets of $\mathbb{R}^m$, $\dim_H \leq \underline{\dim}_B$ (by Proposition 8.39 in Section 8.5 above).

Indeed, if $A \subseteq \bigcup_{j\geq 1} A_j$, then (since, $\dim_H(A_j) \leq \underline{\dim}_B(A_j)$, for all $j \geq 1$), $\dim_H(A) \leq \dim_H(\bigcup_{j\geq 1} A_j) = \sup_{j\geq 1} \dim_H(A_j) \leq \sup_{j\geq 1} \underline{\dim}_B(A_j)$. By taking the infimum on both sides over all countable covers $\{A_j\}_{j\geq 1}$ of A, we obtain—in light of (8.133) and since $\dim_H(A)$ does not depend on the covering—$\dim_H(A) \leq \underline{\dim}_{MB}(A)$, as desired. $\square$

Exercise 8.74. Show by means of examples that we can have strict inequalities whenever we have an inequality and not just an equality in the statement of Theorem 8.73 and, in particular, in (8.137).

[57] Note that $\dim_{MB}(\emptyset)$ exists and is equal to $-\infty$, since the same is true for $\dim_B(\emptyset)$.

[*Hint*: You may use other results in Chapter 5 and in previous subsections in the present chapter, including Example 8.75 just below.]

The following classic example is interesting for its own sake and provides a rich source of counterexamples to various potential inequalities or equalities between fractal dimensions.

Example 8.75. Let $\{L_n\}_{n=0}^{\infty}$ be a sequence of numbers in $(0,1]$ such that $L_0 := 1$ and $0 < L_{n+1} < (1/2)L_n$, for all $n \geq 1$, so that $M_n := L_n - 2L_{n+1} > 0$, for every $n \geq 1$. Then, much as in the construction of the classic Cantor set, define a decreasing sequence $\{K_n\}_{n=0}^{\infty}$ of nonempty compact subsets of $[0,1]$ as follows. Let $K_0 := [0,1]$, and for every connected component of K_n (which must be a closed interval contained in $[0,1]$), delete the corresponding (open) middle interval of length $M_n = L_n - 2L_{n+1}$, thereby creating two closed intervals (of $[0,1]$) of length L_{n+1},[58] which become connected components of K_{n+1}. Note that all of the connected components (necessarily closed, and thus compact, intervals, here) of K_{n+1} are obtained in this manner. Next, let $K := \bigcap_{n=0}^{\infty} K_n$, which is a nonempty compact subset of $[0,1]$. In fact, K is a *topological Cantor set*; i.e., it is a (nonempty) compact, totally disconnected, and perfect (that is, without isolated points) topological space. It follows that $K \subseteq [0,1]$ is infinite and, in fact, has the cardinality of the continuum (i.e., of $[0,1]$ or, equivalently, of $\mathbb{R}$).

The very interesting fact about K is that its Hausdorff and packing dimensions are given, respectively, by (note that $\log 2^n = n \log 2$, for all $n \geq 1$)

$$\dim_H(K) = \liminf_{n\to\infty} \frac{n \log 2}{\log(1/L_n)} \quad \text{and} \quad \dim_P(K) = \limsup_{n\to\infty} \frac{n \log 2}{\log(1/L_n)},$$

also equal to $\overline{\dim}_{MB}(K)$ by the discussion preceding Theorem 8.73. It follows that, for many choices of the sequence $\{L_n\}_{n=0}^{\infty}$, we have $\dim_H(K) < \dim_P(K) = \overline{\dim}_{MB}(K)$.

Note that, of course, if we choose $L_n := 3^{-n}$, for all integers $n \geq 0$, then $K = C$, the classic ternary Cantor set, and so $\dim_H(C) = \dim_P(C) = \dim_{MB}(C) = \dim_B(C) = \log 2/\log 3$, the similarity dimension of C (viewed as a self-similar subset of $\mathbb{R}$).

Exercise 8.76.

(a) Justify the construction of the topological Cantor set K and all of the statements concerning K (in particular, the identities for $\dim_H(K)$ and $\dim_P(K)$) in Example 8.75 just above.

(b) Deduce from Example 8.75 that given any real numbers α and β, with $0 \leq \alpha \leq \beta \leq 1$, there exists a compact set $K \subseteq [0,1]$ such that $\dim_H(K) = \alpha$ and $\dim_P(K) = \beta = \overline{\dim}_{MB}(K)$.

[58] The reader is invited to illustrate this construction by means of a suitable figure.

(c) Generalize the result of part (b) to the m-dimensional Euclidean space. Namely, replace 1 by m (in part (b)) above and given $0 \leq \alpha \leq \beta \leq m$, with $m \in \mathbb{N}$, find a compact set $K \subseteq [0,1]^m$ such that $\dim_H(K) = \alpha$ and $\dim_P(K) = \beta = \overline{\dim}_{MB}(K)$.

[*Hint*: Simply, consider the Cartesian product $K = \widetilde{K} \times [0,1]^p$, for a suitable value of $p \in \{0, \ldots, m-1\}$, where $\widetilde{K} \subseteq [0,1]$ is obtained in part (b), via the construction provided in Example 8.75 and for appropriate values of $\widetilde{\alpha}$ and $\widetilde{\beta}$ such that $0 \leq \widetilde{\alpha} \leq \widetilde{\beta} \leq 1$.]

(d) Rephrase the construction of Example 8.75 in the language of fractal strings (see [**370, 375**], along with Section 8.2 above) and calculate, if possible, the upper Minkowski (or box) dimension of K. Address the counterpart of parts (b) and (c) just above, suitably adapted and extended.

Exercise 8.77. (a) Let $A := \mathbb{Q} \cap [0,1]$, the set of rationals between 0 and 1. Show that $\dim_T(A) = \dim_H(A) = \dim_P(A) = \dim_{MB}(A) = 0$, whereas $\dim_M(A) = \dim_B(A) = 1$.

(b) Show by means of examples that there exist bounded subsets A of $\mathbb{R}^m$ such that $\dim_P(A) < \dim_P(\overline{A})$, and similarly for $\dim_H(A)$.

8.9.5. Generalized Minkowski Contents and Dimension Functions. We briefly discuss here the generalized upper and lower Minkowski content, as introduced and at length studied by C. Q. He and M. L. Lapidus in [**235**], especially in the context of fractal strings and fractal drums. For simplicity, we do not give here the precise hypotheses on the underlying gauge function h. Assume, for example, that for some real number $D \geq 0$, the function f given by $f(x) := h(x)/x^D$, for $x > 0$, belongs to $\boldsymbol{\Psi}$, as defined in Subsection 8.9.1 above.[59] For various purposes, these hypotheses can be significantly refined by using the notion of function of slow growth (near 0^+) or of slow decay (near $+\infty$, as is done in Subsection 6.1.1.2 of [**356**], on pages 544–545 and used in several places in loc. cit., but by necessity of concision and in order to avoid being overly technical, we omit the corresponding discussion and refer the interested reader to loc. cit. for the relevant information. See also [**158**] (building on parts of [**235**]), where such gauge functions were classified in more detail from the point of view of regularly varying functions.

For instance, when $m = 2$, $h(x) := \log(x^{-1}) \cdot \log^{\circ 2}(x^{-1})$ or $h(x) := \log(x^{-1}) \cdot \log^{\circ 3}(x^{-1})$ or else $h(x) := x^2 / \log^{\circ 2}(x^{-1})$ are admissible choices of the gauge

[59] Note that, following [**235**], we call the *gauge function* in this subsection, the function $h(x)$ rather than the function $f(x) = h(x)/x^D$, which we call the associated *dimension function*, by consistency with the rest of this section (i.e., of Section 8.9).

function h, corresponding, respectively, to the admissible choices of (generalized Minkowski) dimension functions in part (b) of Example 8.65 in Subsection 8.9.1.[60]

With the gauge function h given as above, we define the *D-dimensional upper* and *lower (generalized) Minkowski contents with respect to the gauge function h* (or *h-Minkowski contents*, in short)—and denote them, respectively, by $\mathcal{M}^{*,h}(A)$ and $\mathcal{M}_*^h(A)$—as follows:[61]

$$\mathcal{M}^{*,h}(A) = \limsup_{\varepsilon \to 0^+} \frac{\mathscr{L}^m(A_\varepsilon)}{\varepsilon^{m-D} \cdot h(\varepsilon)} \tag{8.138}$$

and similarly for $\mathcal{M}_*^h(A)$, with a lower (rather than an upper) limit, as $\varepsilon \to 0^+$. Clearly, $0 \leq \mathcal{M}_*^h(A) \leq \mathcal{M}^{*,h}(A) \leq \infty$. Under mild hypotheses on h, we have that $\dim_M(A) = D$. Furthermore, A is said to be *h-Minkowski nondegenerate* if $0 < \mathcal{M}_*^h(A) \leq \mathcal{M}^{*,h}(A) < \infty$, in which case the gauge function and the generalized lower and upper h-Minkowski contents are said to be *exact* while the *dimension function* $f(x) := h(x)/x^D$ is also said to be *exact*.

Moreover, if $\mathcal{M}_*^h(A) = \mathcal{M}^{*,h}(A)$, we then call this common value (in $[0, +\infty]$) the *(generalized) h-Minkowski content* of A and denote it by $\mathcal{M}^h(A)$. As usual, A is said to be *h-Minkowski measurable* if, in addition, $0 < \mathcal{M}^h(A) < \infty$, i.e., if A is *h-Minkowski nondegenerate* and the following limits exist (necessarily in $(0, +\infty)$):

$$\lim_{\varepsilon \to 0^+} \frac{\mathscr{L}^m(A_\varepsilon)}{\varepsilon^{m-D} \cdot h(\varepsilon)} = \lim_{\varepsilon \to 0^+} \frac{\mathscr{L}^m(A_\varepsilon)}{\varepsilon^m \cdot f(\varepsilon)} =: \mathcal{M}^h(A), \tag{8.139}$$

the *h-Minkowski content* of A.

Observe that when $h \equiv 1$, we recover the usual notions of (upper, lower) D-dimensional Minkowski contents, Minkowski nondegeneracy, and Minkowski measurability introduced in Section 8.1 above.

Finally, we note that with the standard precautions, we can use as an admissible gauge function h the n-th iterated logarithm or its reciprocal. Both types of gauge functions are useful in the applications, as illustrated in [**235**, **356**] and [**355**, **365**] (see also [**63**]) to which the interested reader is referred for a number of additional results involving the notions of (upper, lower) generalized h-Minkowski contents, h-Minkowski nondegeneracy, and h-Minkowski measurability.

[60] Recall that in Subsections 8.9.1–8.9.4, h was referring to the full dimension function, including the part corresponding to the power function x^m (here, x^2). However, for any integer $m \geq 1$, a much greater variety of (generalized, Minkowski) dimension functions f—and hence also of underlying functions h—is allowed here.

[61] Note that in terms of the *dimension function* $f(x) := h(x)/x^D$, we have that $\mathcal{M}_*^h(A) = \liminf_{\varepsilon \to 0^+} (\mathscr{L}^m(A_\varepsilon)/\varepsilon^m f(\varepsilon))$—and similarly for $\mathcal{M}^{*,h}(A)$, with an upper (rather than a lower) limit, as $\varepsilon \to 0^+$.

Part 3

Fractal Curves and Their Complex Dimensions

Chapter 9

Epilogue: A Primer of Fractal Curves and Their Complex Dimensions

The will is infinite
and the execution confined
the desire is boundless
and the act a slave to limit

Ernest Cesàro, 1905, [**78**].

I turn away with fright and horror from this lamentable plague of continuous functions that have no derivatives.

Charles Hermite, in a letter to Thomas Stieltjes, 1893, [**243**].

It is a case where it is truly natural to think of these continuous but nowhere differentiable functions which mathematicians have imagined, and which were wrongly regarded as being mere mathematical curiosities, because experiments actually suggest their existence.

Jean Perrin, discussing the geometry of Brownian paths.[1]

[1] Jean Perrin was awarded the 1926 Nobel Prize in Physics for his experimental work on Brownian motion and the corresponding confirmation of the Atomic Hypothesis; see, e.g., his 1913 book, entitled *The Atoms*, [**488**].

Our goal in this last chapter is to provide a relatively brief introduction to the notion of a fractal curve and the notion of complex (fractal) dimensions. Both subjects will be central topics in the sequels to this book (including, especially, [**119**] and [**116**]). The aim here is to wet the reader's appetite and give a taste of these subjects, by means of several examples, but not at all to be exhaustive or fully rigorous. A much more detailed discussion and complete proofs for most of the key stated properties will be provided in the respective volumes, [**119**] and [**116**] as well as in [**117**].

The plan of this chapter is as follows:

In Section 9.1, we discuss the notions of (continuous) parametrized curves and their polygonal approximations, along with their lengths and (bounded or unbounded) total variations.

In Section 9.2, we discuss several theorems providing different sets of hypotheses under which lengths and limits (in the sense of the uniform convergence or of the Hausdorff metric) can be interchanged.

In Section 9.3, we provide a very brief overview of the notions of complex dimensions and fractal zeta functions of bounded subsets of $\mathbb{R}^m$—or, more generally, of relative fractal drums (RFDs) in $\mathbb{R}^m$—with $m \in \mathbb{N}$ arbitrary. This subject will be key to [**119**], the sequel to this volume, where it will be discussed in much greater detail.

In the remainder of Chapter 9, we discuss in some detail various properties of several important classes of fractal curves, including (in Section 9.4), the Devil's staircase (i.e., the graph of the Cantor function), the Koch curve and the Koch snowflake curve (in Section 9.5), the Sierpinski gasket and carpet, along with their higher-dimensional analogs (in Section 9.6), the family of Weierstrass nowhere differentiable curves (in Section 9.7)—as well as, much more briefly, plane- (or space)-filling curves (in Remark 9.63 at the end of Section 9.4), building on Subsection 1.4.3, such as the Peano and the Hilbert curves.

These classes of examples—along with other fractal curves—will be studied in a lot more depth (as well as by using new research tools associated with the theory of complex dimensions) in [**116**] and [**117**], in addition to [**119**].

9.1. Continuous Curves: Their Lengths and Polygonal Approximations

We begin by providing a few definitions, most of which may be familiar to students of advanced calculus.

Given an integer $m \geq 2$, a *parametrized curve* is a continuous map $\gamma : [a, b] \to \mathbb{R}^m$, where $[a, b]$ is a compact (i.e., closed and bounded) interval of the real line $\mathbb{R}$, with $a, b \in \mathbb{R}$ and $a < b$.[2] All of the parametrized curves discussed here will be assumed to be continuous, sometimes implicitly.

In standard calculus, one usually considers curves which are differentiable —that is, geometrically, which admit a (single) tangent line at every point— or which, at least, are piecewise differentiable, i.e., such that the restriction to each open interval (t_{j-1}, t_j) in a (finite) partition of $[a, b]$,

$$\sigma : a =: t_0 < t_1 < \cdots < t_{n-1} < t_n := b, \tag{9.1}$$

is differentiable, and for which γ has a right and a left derivative at every point t_j of the partition, for $j = 0, \ldots, n$—other than at $t_0 := a$ and $t_n := b$, respectively, where only the corresponding one-sided derivative is required to exist.

Given $k \in \mathbb{N}$, the parametrized curve γ is said to be C^k (or *of class* C^k) if γ is k times continuously differentiable. By convention, the case when $k = 0$ corresponds to a continuous (but not necessarily differentiable) curve: γ is of class C^0 iff γ is continuous.

The *range* (or *image*) of γ is given by $\mathbf{\Gamma} := \gamma([a, b]) \subseteq \mathbb{R}^m$. Then, $\mathbf{\Gamma}$ is called the *curve* associated with the parametrized curve γ.[3] Since γ is continuous and $[a, b]$ is a nonempty interval—and hence is connected and compact in $\mathbb{R}$—$\mathbf{\Gamma}$ is a nonempty compact and connected subset of $\mathbb{R}^m$.

Often, in geometric topology, one takes a less stringent view of what a "curve" is. In that broader context, one considers a *curve* (in $\mathbb{R}^m$) to be simply a (nonempty) compact and connected subset of $\mathbb{R}^m$, also called an *m-dimensional continuum*. If $m = 2$, $\mathbf{\Gamma}$ is said to be a *planar curve*, while if $m = 3$, $\mathbf{\Gamma}$ is called a *space curve*. This more general point of view is especially useful when a parametrization of the underlying "curve" is either not available or not relevant to the problem under investigation. This is the case, for example, for the Sierpinski gasket (a plane curve) or for its 3-dimensional analog, the Sierpinski tetrahedron (a space curve), as well as for their higher-dimensional counterparts (when $m \geq 3$); see Figure 9.22 (or Figure 4.6 in Section 4.3) and Figure 9.25 in Section 9.6 below, for the $m = 2$ and $m = 3$ cases, respectively. And, similarly, it is also a convenient point of view for the Sierpinski carpet (a plane curve) or for its 3-dimensional analog, the Menger sponge (a space curve), as well as for their higher-dimensional counterparts (when $m \geq 3$); see Figure 9.24 (or Figure 4.7 in Section 4.3) and Figure 9.26 in Section 9.6 below, for the $m = 2$ and $m = 3$ cases, respectively.

---

[2]Occasionally, we may wish to consider other (possibly unbounded) intervals of $\mathbb{R}$, such as $[0, +\infty)$ or $\mathbb{R} = (-\infty, +\infty)$ itself. An example of the latter situation is the Weierstrass parametrized curve, which is initially defined on $\mathbb{R}$ but can then be restricted to $[0, 1]$, by 1-periodicity (for integer values of the underlying parameter).

[3]*Caution*: The curve $\mathbf{\Gamma} \subseteq \mathbb{R}^m$ is, of course, not to be mistaken with the graph of the map γ; namely, $Gr(\gamma) := \{(t, \gamma(t)) : a \leq t \leq b\} \subseteq \mathbb{R}^{m+1}$, which is the range $\rho([a, b])$ of the map given by $\rho(t) := (t, \gamma(t))$, for all $t \in [a, b]$.

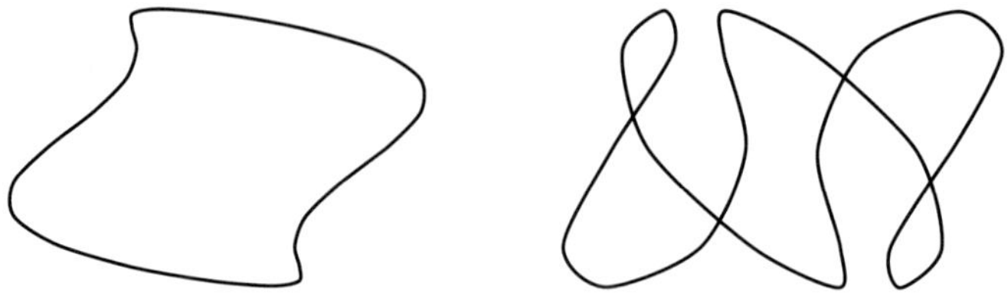

(a) (b)

Figure 9.1. (a) A simple closed curve: a Jordan curve. (b) A closed, but not simple, curve; indeed, this curve has self-intersections.

The above, more general, point of view is also suitable for the Koch curve and the Koch snowflake curve, for which, however, an underlying (continuous) parametrization can be calculated explicitly; see Figure 4.4 in Section 4.3 above and Figures 9.15 and 9.19 in Section 9.5 below for a depiction of the Koch curve and of the (Koch) snowflake curve, respectively.

For now, however, we continue considering the case of a continuous parametrized curve $\gamma : [a,b] \to \mathbb{R}^m$ and of its image (or range), the curve $\mathbf{\Gamma} = \gamma([a,b])$, viewed as a compact, connected, and nonempty subset of $\mathbb{R}^m$.

In the literature, $\mathbf{\Gamma}$ is sometimes identified with its underlying parametrized curve γ. Doing this is convenient in certain situations but may also be misleading (and lead to unwanted errors), especially if the map γ is *not* injective and hence also not a bijection from $[a,b]$ onto the curve $\mathbf{\Gamma}$.

If $\gamma(a) = \gamma(b)$—i.e., if the endpoints $A := \gamma(a)$ and $B := \gamma(b)$ of the curve $\mathbf{\Gamma}$ coincide—then $\mathbf{\Gamma}$ is said to be a *closed curve*; otherwise (i.e., if $A \neq B$), it is an *open curve*.[4]

If $\mathbf{\Gamma}$ is a closed curve, it is said to be *simple* if the restriction of γ to the open interval (a,b) is injective, i.e., if $\mathbf{\Gamma}$ does not have any multiple point other than its endpoint $A = B$. In this case, $\mathbf{\Gamma}$ is called a *Jordan curve*, as in footnote 20 of Section 5.5 above; see Figure 9.1.

If, on the other hand, $\mathbf{\Gamma}$ is an open curve (i.e., if $\gamma(a) \neq \gamma(b)$), then $\mathbf{\Gamma}$ is said to be a *simple curve* (or *simple*, in short) if γ is injective (on all of $[a,b]$).[5] It is then called a *Jordan arc*, as in Definition 5.58 of Section 5.5 above; see Figures 9.2 and 9.3.

Given a partition σ of $[a,b]$, as in (9.1) above, the associated *polygonal approximation* $\mathbf{\Gamma}_{Pol}$ to the curve $\mathbf{\Gamma}$ is the range of the piecewise linear (or affine)

[4]An entirely similar terminology is used in string theory (from theoretical physics); in that context, one then talks about closed strings and open strings, respectively.

[5]It then follows from elementary topology that γ is a homeomorphism from $[a,b]$ onto $\mathbf{\Gamma} = \gamma([a,b])$; i.e., $\gamma : [a,b] \to \mathbb{R}^m$ is a continuous bijection onto its range $\mathbf{\Gamma}$, with a continuous inverse map $\gamma^{-1} : \mathbf{\Gamma} \to [a,b]$.

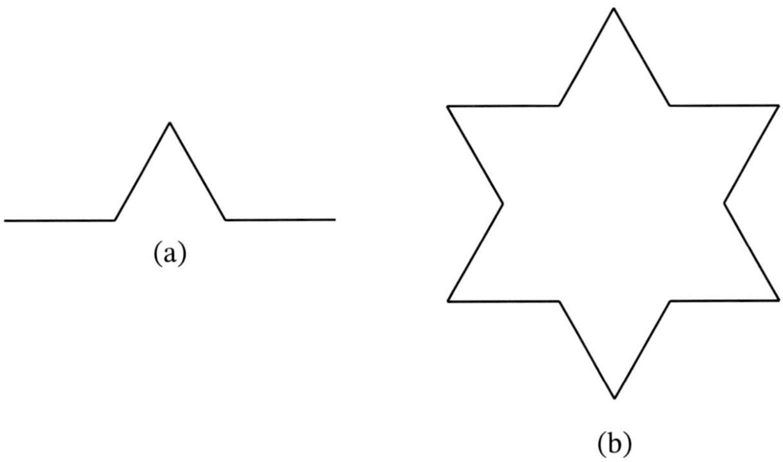

Figure 9.2. (a) An example of a Jordan arc. (b) An example of a Jordan curve. The curve in (a) (resp., in (b)) depicts the first nontrivial (prefractal) polygonal approximation to the Koch curve (resp., to the Koch snowflake curve).

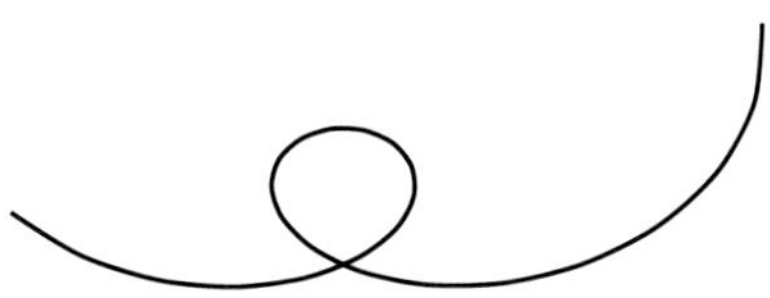

Figure 9.3. An open curve which is *not* simple. It is therefore *not* a Jordan arc; indeed, $\gamma(c) = \gamma(d)$, with $a < c < d < b$.

map $\gamma_{Pol} : [a, b] \to \mathbb{R}^m$ such that

(9.2) $$\gamma_{Pol}(t_j) = \gamma(t_j), \text{ for } j = 0, 1, \ldots, n.$$

Accordingly, $\mathbf{\Gamma}_{Pol}$ is the polygonal curve going through the points $P_j := \gamma(t_j)$ of the curve $\mathbf{\Gamma}$, for $j = 0, 1, \ldots, n$; in particular, A and B, the endpoints of $\mathbf{\Gamma}$, belong to $\mathbf{\Gamma}_{Pol}$. Observe that $\mathbf{\Gamma}_{Pol}$ is the polygonal curve consisting precisely of the line segments $\overline{P_{j-1}P_j}$, for $j = 1, \ldots, n$, in that order; see Figure 9.4.

Then, much as in Definition 5.60 of Section 5.5, *the length of* $\mathbf{\Gamma}$, denoted $L(\mathbf{\Gamma})$, is the supremum (in $[0, +\infty]$) of the lengths $L(\mathbf{\Gamma}_{Pol})$ of all of its polygonal approximations $\mathbf{\Gamma}_{Pol}$. Because γ is continuous—and hence also uniformly continuous—on $[a, b]$, it can be shown (see Theorem 9.20 in Section 9.2 below, along with the text preceding it) that $L(\mathbf{\Gamma})$ is also equal to the limit (which then exists in $[0, +\infty]$) of the lengths $L(\mathbf{\Gamma}_{Pol})$, *as the mesh of* $\mathbf{\Gamma}_{Pol}$ (i.e., *as the mesh of the associated partition* σ, *namely,* $\text{mesh}(\sigma) := \max_{j=1,\ldots,n} |t_j - t_{j-1}|$) *tends to zero*:

(9.3) $$L(\mathbf{\Gamma}) = \sup L(\mathbf{\Gamma}_{Pol}) = \lim L(\mathbf{\Gamma}_{Pol}),$$

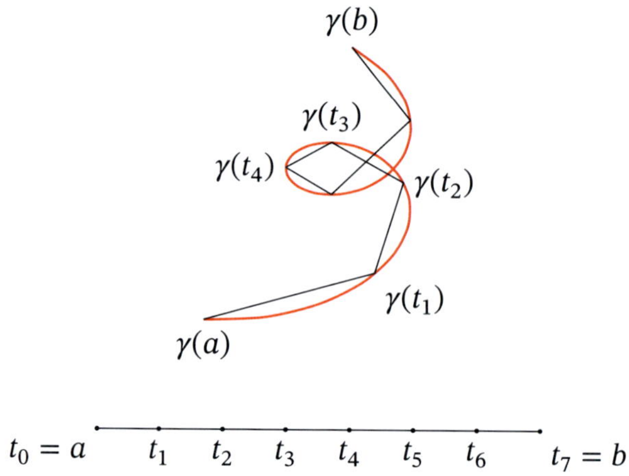

Figure 9.4. Top: the polygonal approximation, $\mathbf{\Gamma}_{Pol}$, to a given continuous curve $\mathbf{\Gamma}$, and associated with the partition $\sigma = \{t_j\}_{j=0}^n$ of $[a, b]$ given by (9.1). Bottom: the partition of $[a, b]$, $\sigma = \{t_j\}_{j=0}^n$, with $t_0 := a$ and $t_n := b$, also as in (9.1). Here, $n := 7$.

where the supremum is taken over all polygonal approximations of $\mathbf{\Gamma}$ (or, equivalently, over all partitions σ of $[a, b]$) and the limit is taken as mesh$(\sigma) \to 0^+$.[6] Here, the *length of* $\mathbf{\Gamma}_{Pol}$ is given by[7]

$$L(\mathbf{\Gamma}_{Pol}) := \sum_{j=1}^{n} |\gamma(t_j) - \gamma(t_{j-1})| = \sum_{j=1}^{n} |\overline{P_{j-1}P_j}| =: V_\sigma(\gamma), \tag{9.4}$$

which is just the sum of the lengths of the line segments $\overline{P_{j-1}P_j}$ of which $\mathbf{\Gamma}_{Pol}$ is comprised, for $j = 1, \ldots, n$.

One can say that $\mathbf{\Gamma}_{Pol}$ *is the polygonal approximation of* $\mathbf{\Gamma}$ *adapted to the partition* σ—which, implicitly, implies that $\mathbf{\Gamma}_{Pol}$ is parametrized by a continuous map $\gamma_{Pol} : [a, b] \to \mathbb{R}^m$, defined on the same interval $[a, b]$ as for γ itself.

Remark 9.1. We note that $L(\mathbf{\Gamma}_{Pol})$, as given by (9.4), is also called the *total variation of* γ *associated with the partition* σ, as given by (9.1), and adapted to the polygonal approximation $\mathbf{\Gamma}_{Pol}$ to $\mathbf{\Gamma}$. On the other hand, $L(\mathbf{\Gamma})$, the length of $\mathbf{\Gamma}$—as given initially by the supremum of $L(\mathbf{\Gamma}_{Pol})$ over all polygonal approximations $\mathbf{\Gamma}_{Pol}$ to $\mathbf{\Gamma}$ (i.e., over all partitions σ of $[a, b]$), in the first equality of (9.3)—is also called the *total variation of* γ (on $[a, b]$).

[6]We point out that the first equality in (9.3) is the actual definition of the length $L(\mathbf{\Gamma})$ of $\mathbf{\Gamma}$, whereas the second equality in (9.3) is really a theorem (see Theorem 9.20 in Section 9.2 below, along with the text preceding it), which is a consequence of this definition, as well as of the uniform continuity of the continuous function γ on the compact internal $[a, b]$. The proof of (9.3)—which is classical and is omitted for now, in order not to interrupt the flow of the exposition—will be included in the proof of Theorem 9.20. In spirit, it is similar to the proof of some of the key properties of the Riemann integral.

[7]The notation $V_\sigma(\gamma)$ will turn out to be useful when defining the total variation $V(\gamma)$ of the continuous map γ; see identities (9.5)–(9.7), further on.

Exercise 9.2 (*The Case of a Polygonal Curve*). Show that if $\gamma = \gamma_{Pol}$—that is, if γ itself is a polygonal curve—then formulas (9.3) and (9.4) are compatible. In other words, one can use either (9.4) or (9.3) in order to define or obtain the length of a polygonal curve. This observation can be applied, in particular, to the polygonal approximations $\mathbf{\Gamma}_{Pol}$ to a given (parametrized) curve $\mathbf{\Gamma}$.

[*Hint*: First, consider the case when the partition σ used in (9.3) or (9.4) is adapted to the polygonal curve γ_{Pol}. Then, in the general case, refine the given partition accordingly.]

By definition, $L(\mathbf{\Gamma}) \in [0, +\infty]$. If, in addition, $\mathbf{\Gamma}$ has finite length (i.e., if $L(\mathbf{\Gamma}) < \infty$), then $\mathbf{\Gamma}$ is said to be a *rectifiable curve*. Otherwise (i.e., if $L(\mathbf{\Gamma}) = +\infty$), then $\mathbf{\Gamma}$ is called an *unrectifiable curve*.

Completely similarly, the continuous map $\gamma : [a, b] \to \mathbb{R}^m$ is said to be of *bounded variation* if $V(\gamma) < \infty$, where

$$V(\gamma) := \sup_{\sigma} V_\sigma(\gamma) \tag{9.5}$$

is called *the total variation of* γ. Here, on the right-hand side of (9.5), the supremum (in $[0, +\infty]$) is taken over all partitions σ of $[a, b]$. Furthermore, $V_\sigma(\gamma)$ is defined by the last two equalities of (9.4) above.

Note that, in light of the definition of $L(\mathbf{\Gamma})$ given by the first equality of (9.3), we simply have

$$V(\gamma) = L(\mathbf{\Gamma}), \tag{9.6}$$

where $\mathbf{\Gamma} := \gamma([a, b])$. Accordingly, the curve $\mathbf{\Gamma}$ is rectifiable iff the parametrized curve γ is of bounded variation. Also, in light of (9.4), we have

$$V_\sigma(\gamma) = L(\mathbf{\Gamma}_{Pol}), \tag{9.7}$$

where $\mathbf{\Gamma}_{Pol}$ is the polygonal approximation to $\mathbf{\Gamma}$ adapted to the partition σ.

Contrary to what is stated in many places (for example, in [**593**, Chapter 11]), *a fractal curve need* not *have an infinite length*. For example, the Devil's staircase (i.e., the graph of the Cantor function) has finite length but it is clearly "fractal", whatever meaning one wants to give to this term; see Subsection 9.4.7 below. On the other hand, the Koch curve has infinite length (i.e., it is not rectifiable). Therefore, a fractal curve can be either rectifiable or unrectifiable: It can have finite length or infinite length. Either situation can arise. Actually, we will see further on in this chapter, as well as in the sequel to this book, [**119**], that for any geometric object, it is the presence of *nonreal* complex dimensions which is the true signature of "fractality"—or, at least, seems to be a much better signature of fractality than its counterpart in earlier definitions, as is discussed in the books [**326, 356, 370, 374, 375**], the survey article [**329**] and will be expanded upon in the sequels to this book, especially [**119**] and [**116**]. Accordingly, as will be shown in Subsection 9.4.7 below, the Cantor graph (or Devil's staircase) is fractal, because it has nonreal complex dimensions with real part $\log_3 2$, the fractal dimension of the Cantor set.

We now come to a well-known result in elementary calculus (at least in the special case discussed in Example 9.6 below when $m := 1$)—although one which is not always proved very rigorously in such courses. We will provide a complete and detailed proof in [**116**] but for now, we precisely state the result and give a sketch of the proof based on a standard physical argument.

Theorem 9.3 (Length of a Smooth Curve). *Let $m \in \mathbb{N}$. If $\mathbf{\Gamma}$ is a smooth curve (i.e., here, if its underlying parametrization $\gamma : [a,b] \to \mathbb{R}^m$ is of class C^1), then its length $L(\mathbf{\Gamma})$—or, equivalently, the total variation of γ—is finite (i.e., $\mathbf{\Gamma}$ is rectifiable) and is given by the following formula:*

$$L(\mathbf{\Gamma}) = \int_a^b |\gamma'(t)|dt, \tag{9.8}$$

where $\gamma'(t) \in \mathbb{R}^m$ denotes the derivative of γ at any $t \in [a,b]$ (or the velocity vector $\vec{v}(t)$ of the associated motion at time t) and $|\gamma'(t)|$ denotes the magnitude (or Euclidean norm) of $\gamma'(t)$—or else the speed (at time t), $v(t)$, of the corresponding motion.

Proof. (*Sketch*) This heuristic proof is divided into two steps:

Step 1 (*Physical Argument*): Since γ is smooth (i.e., continuously differentiable, here) over an infinitesimal interval of length Δt—say, the subinterval $[t, t+\Delta t]$ of $[a,b]$—the velocity vector $\vec{v}(\tau) = \gamma'(\tau) \in \mathbb{R}^m$ (and hence also the speed $v(\tau) = |\gamma'(\tau)|$) is approximately constant with value $\vec{v}(t)$ (resp., $v(t)$) during the time interval $[t, t+\Delta t]$. It follows that the corresponding particle (or body) moves (approximately) by the distance (or arclength) given by $\Delta \boldsymbol{s} = v(t)\Delta t$ during this time interval $[t, t+\Delta t]$, along the curve $\mathbf{\Gamma}$. Alternatively, using the suggestive Leibniz notation, we can write that

$$d\boldsymbol{s} = v(t)dt = |\gamma'(t)|dt. \tag{9.9}$$

Step 2 (*Riemann Integration*): To obtain (9.8), we simply integrate from a to b the "infinitesimal relation" (9.9) in order to obtain the desired result (with $\boldsymbol{s} = \boldsymbol{s}(t)$ denoting the arclength function of $\mathbf{\Gamma}$, for all $t \in [a,b]$), as stated in (9.8): $L(\gamma) = \boldsymbol{s}(b) = \int_a^b d\boldsymbol{s} = \int_a^b v(t)dt = \int_a^b |\gamma'(t)|dt$.

This completes the sketch of the heuristic proof of the theorem. □

Exercise 9.4. Give a rigorous proof of Steps 1 and 2 in the above "heuristic proof" of Theorem 9.3.

[*Hint*: (i) For Step 1, appeal to the Mean Value Theorem. (ii) For Step 2, use the Riemann sum approximation to $\int_a^b |\gamma'(t)|dt$ corresponding to an arbitrary polygonal approximation $\mathbf{\Gamma}_{Pol}$ to $\mathbf{\Gamma}$.]

By applying Theorem 9.3 to the restriction of γ to $[a, t]$, where $t \in (a, b]$ is fixed, we obtain the following result, which yields Theorem 9.3 by letting $\tau := b$. Recall that the *arclength function* $\mathbf{s} = \mathbf{s}(\tau)$ measures the *arclength* (i.e., the length) of the portion of the curve $\mathbf{\Gamma}$ along the arc $\widehat{AP}$, where $A := \gamma(a)$ and $P = P(\tau) := \gamma(\tau)$, for any $\tau \in [a, b]$.

Corollary 9.5 (Arclength Function of a Smooth Curve). *Under the same hypotheses as in Theorem 9.3, and with the notation introduced in Step 1 of the proof of that theorem, the arclength function $\mathbf{s} = \mathbf{s}(\tau)$ of the C^1 smooth curve $\mathbf{\Gamma}$ is given by (with $\widehat{AP}$ denoting the arc of $\mathbf{\Gamma}$ between $A := \gamma(a)$ and $P := \gamma(\tau)$)*

$$L(\widehat{AP}) = \mathbf{s}(\tau) = \int_a^\tau |\gamma'(t)|dt = \int_a^\tau v(t)dt, \text{ for all } \tau \in [a, b]. \tag{9.10}$$

Example 9.6 (*The Special Case of Graphs*). Let $m \in \mathbb{N}$ be arbitrary. If $f : [a, b] \to \mathbb{R}^m$ is a C^1 smooth function, then, clearly, its graph (still denoted by $\mathbf{\Gamma}$),

$$\mathbf{\Gamma} = \mathrm{Gr}(f) := \{(t, f(t)) : t \in [a, b]\} \subseteq \mathbb{R}^{m+1} \simeq \mathbb{R} \times \mathbb{R}^m, \tag{9.11}$$

can be viewed as the curve $\mathbf{\Gamma} = \gamma([a, b])$ in $\mathbb{R}^{m+1}$ associated with the continuously differentiable (i.e., C^1) parametrized curve $\gamma : [a, b] \to \mathbb{R}^m$ given by $\gamma(t) := (t, f(t))$, for all $t \in [a, b]$. It then follows from Theorem 9.3 that the length of $\mathbf{\Gamma}$ is finite (i.e., $\mathbf{\Gamma}$ is rectifiable) and is given by

$$L(\mathbf{\Gamma}) = \int_a^b \sqrt{1 + |f'(t)|^2}\, dt, \tag{9.12}$$

where f' denotes the derivative of the vector-valued (i.e., $\mathbb{R}^m$-valued, here) function f and $|\cdot|$ denotes the Euclidean norm in $\mathbb{R}^m$.

In particular, when $m = 1$—i.e., when f is real-valued and hence, $\mathbf{\Gamma}$ is a planar curve, we recover the following well-known formula from elementary calculus:

$$L(\mathbf{\Gamma}) = \int_a^b \sqrt{1 + (f'(t))^2}\, dt = \int_a^b \sqrt{1 + \left(\frac{df}{dt}(t)\right)^2}\, dt. \tag{9.13}$$

Both (9.12) and (9.13) follow from formula (9.8) in Theorem 9.3, along with the obvious fact that, in light of the above definition of γ,

$$\gamma'(t) = (1, f'(t)) \in \mathbb{R} \times \mathbb{R}^m \simeq \mathbb{R}^{m+1}, \text{ for all } t \in [a, b], \text{ and so} \tag{9.14}$$

$$|\gamma'(t)| = \sqrt{1 + |f'(t)|^2} \left(= \sqrt{1 + (f'(t))^2}\,, \text{ when } m = 1 \right), \text{ for all } t \in [a, b].$$

Naturally, (9.12) and (9.13) have a counterpart expressing the arclength function as an integral, just as in formula (9.10) of Corollary 9.5 above.

Example 9.7 (*Cautionary Comment About Fractal Curves*). It is important to note that neither Theorem 9.3 nor the kinematic relation (9.9)—and hence also neither Corollary 9.5 nor Example 9.6—has an appropriate counterpart for fractal curves (instead of for C^1 smooth curves)—expressed, for instance, in terms of ordinary velocities or speeds.

For example, for the Devil's staircase $\mathbf{\Gamma} = \mathbf{\Gamma}_{DS}$ (the graph of the Cantor function), we are in the $m = 1$ case of Example 9.6 just above. Accordingly, if the analog of Theorem 9.3 (and hence also the analog of Example 9.6) continued to hold in this situation, then the length of $\mathbf{\Gamma}$ would have to be equal to one (i.e., $L(\mathbf{\Gamma}) = 1$), which is clearly not true, as we next explain.

In fact, without having to calculate the explicit (and finite) value of $L(\mathbf{\Gamma})$, we simply can bound $L(\mathbf{\Gamma})$ from below by the length $L(\mathbf{\Gamma}_{Pol})$ of the n-th (natural *piecewise linear*) polygonal approximation $\mathbf{\Gamma}_{n,Pol}$ to $\mathbf{\Gamma}$ (with $n \in \mathbb{N}$ arbitrary), which is strictly greater than 1 and tends to 2, as $n \to \infty$.

Indeed (see Figure 9.5 for the $n = 2$ and the $n = 3$ cases), a moment's reflection shows that it follows from (9.4) that for every integer $n \geq 1$,

$$\begin{aligned} L(\mathbf{\Gamma}_{n,Pol}) &= h_n + \left(\frac{1}{3} + \frac{2}{9} + \frac{4}{27} + \cdots + \frac{2^{n-1}}{3^n} \right) \\ &= h_n + \left(\frac{1}{3} \sum_{j=0}^{n-1} \left(\frac{2}{3} \right)^j \right), \end{aligned} \tag{9.15}$$

where $h_n \to 1^-$, as $n \to \infty$. Note that the piecewise linear curve $\mathbf{\Gamma}_{n,Pol}$—the n-th prefractal approximation to the Devil's staircase $\mathbf{\Gamma}$—is comprised of both horizontal line segments or "horizontal steps" (the sum of the lengths of which is equal to $\frac{1}{3} + \frac{2}{9} + \cdots + \frac{2^{n-1}}{3^n}$) and of oblique line segments or "oblique steps" (the sum of the lengths of which is equal to h_n, by definition of h_n). Furthermore, as $n \to \infty$, the oblique steps become steeper and steeper, until in the limit as $n \to \infty$, they become vertical (and the total length of their projections along the y-axis is equal to 1, as $n \to \infty$; see Figure 9.5 and its caption. This explains both why (9.15) holds and $h_n \to 1^-$, as $n \to \infty$.

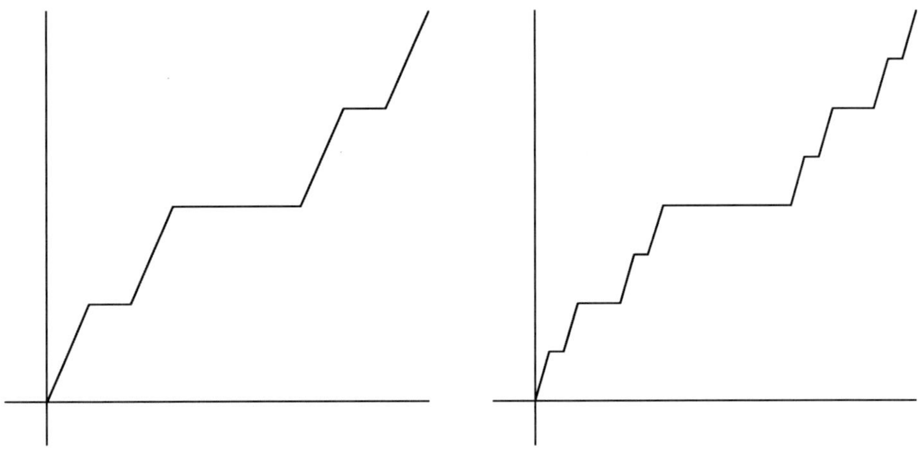

Figure 9.5. The second and third *piecewise linear* prefractal approximations (or natural *piecewise linear* polygonal approximations), $\mathbf{\Gamma}_{2,Pol}$ (left) and $\mathbf{\Gamma}_{3,Pol}$ (right), to the Devil's staircase $\mathbf{\Gamma} = \mathbf{\Gamma}_{DS}$. If we were to draw the n-th prefractal approximation, $\mathbf{\Gamma}_{n,Pol}$, to $\mathbf{\Gamma}$, for $n \in \mathbb{N}$ very large, then the oblique steps of $\mathbf{\Gamma}_{n,Pol}$ would become increasingly vertical, casting a shadow along the y-axis of total length arbitrarily close to 1, while the horizontal steps would cast a shadow along the x-axis of total length arbitrarily close to 1 as well (since the Cantor set C has zero length: $\mathcal{L}^1(C) = 0$).

It now suffices to let $n \to \infty$ in (9.15) in order to deduce that, since $L(\mathbf{\Gamma}) \geq L(\mathbf{\Gamma}_{n,Pol})$, for every $n \in \mathbb{N}$ (by the first equality in (9.3)),

$$L(\mathbf{\Gamma}) \geq \left(\lim_{n\to\infty} h_n\right) + \frac{1}{3}\sum_{j=1}^{\infty}\left(\frac{2}{3}\right)^j = 1 + \frac{1}{3}\left(\frac{1}{1-\frac{2}{3}}\right) = 1 + 1 = 2. \tag{9.16}$$

Thus, $L(\mathbf{\Gamma}) \geq 2 > 1$ and so, $L(\mathbf{\Gamma}) > 1$, as desired.[8] This contradicts our claim according to which $L(\mathbf{\Gamma}) = 1$ if the counterpart of Theorem 9.3—and hence also of Example 9.6—holds in the present context.

We next explain why the latter claim is true. Indeed, the reason why $L(\mathbf{\Gamma})$ would have to be equal to one is that if $f : [0,1] \to \mathbb{R}$ is the Cantor function, then, as we shall soon review (in Section 9.4), its derivative f' exists and is equal to zero on the open set $[0,1]\backslash C$. Indeed, by construction, f is constant on the complement of the Cantor set C, a set of 1-dimensional Lebesgue measure zero (i.e., $\mathcal{L}^1(C) = 0$; see Figure 1.8 in Subsection 1.4.1, along with Example 1.32). Therefore, the derivative of f is almost everywhere equal to zero on $[0,1]$:

$$f'(t) = 0, \text{ for } \mathcal{L}^1\text{-a.e. } t \in [0,1]. \tag{9.17}$$

Since the Lebesgue and Riemann integrals of the continuous and real-valued function $(1 + (f')^2)^{1/2}$ on the compact interval $[0,1]$ must coincide, we thus

[8] Actually it can be shown with a little more effort that $L(\mathbf{\Gamma}) = 2$, a fact which is not needed in the present discussion; for a rigorous proof, see Proposition 9.40 and Remark 9.41 in Subsection 9.4.3 below.

deduce from (9.17) that, if the exact counterpart of (9.12) in Example 9.6 were to hold for this fractal curve, then we would have successively

$$L(\mathbf{\Gamma}) = \int_0^1 \sqrt{1+(f'(t))^2}\,dt = \int_0^1 \sqrt{1+0^2}\,dt = \int_0^1 dt = 1, \tag{9.18}$$

which we have shown above to be impossible since $L(\mathbf{\Gamma}) > 1$.[9]

A significant philosophical and physical implication of this discussion is that motion along a fractal curve—such as the Devil's staircase, the Koch curve, or else a (random) Brownian path, for example—cannot be dealt with in the same way as in ordinary (i.e., smooth) calculus and in standard classical mechanics or kinematics. In particular, the notions of velocity and speed—as well as, a fortiori, of acceleration and the associated Newton's Second Law—are typically useless for motion along a fractal curve. Consequently, all of the intuition and experience developed during one's mathematics and physics training must be completely modified in ways that are yet to be fully discovered.

A suitable (and yet to be developed) version of the so-called fractional calculus, in conjunction with the theory of complex fractal dimensions, should likely be used in this context. In other words, an entirely new *fractal calculus* must be invented. More discussion of this very important topic may be provided in the sequels to the present book, especially [**116**] and [**117**].

Remark 9.8. We point out that, according to a well-known result of M. Shishikura [**547**], deterministic examples of fractal—but not plane-filling—planar curves $\mathbf{\Gamma}$ of maximum possible Hausdorff, Minkowski, and box-counting dimension (i.e., here, $\dim_H(\mathbf{\Gamma}) = \dim_M(\mathbf{\Gamma}) = \dim_B(\mathbf{\Gamma}) = 2$) include the boundary of the Mandelbrot set, along with certain Julia sets (of quadratic polynomials of a complex variable, as in Subsection 1.6.1).

9.2. Interchange of Lengths and Uniform or Hausdorff Limits

In this section, we begin by comparing the uniform convergence of continuous parametrized curves and the Hausdorff convergence (i.e., the convergence with respect to the Hausdorff metric) of the corresponding curves. Furthermore, we discuss conditions under which lengths and uniform limits of continuous parametrized curves (resp., Hausdorff limits of continuous curves) can be interchanged. We also caution the reader that, as is well known, these interchanges of limits are not always possible. Complete proofs of these and related results (when they are not already included here) will be provided with more details in a sequel to this book, [**116**].

---

[9] An entirely similar argument would show that the counterpart for fractal curves of Corollary 9.5 cannot hold either.

We first recall the notion of uniform convergence, in the context of interest to us. Let $[a, b]$ be a *compact* interval of $\mathbb{R}$, with $a < b$. For each integer $n \geq 0$, let $\gamma_n : [a, b] \to \mathbb{R}^m$ be continuous. Then, we say that the sequence $\{\gamma_n\}_{n=0}^{\infty}$ *converges uniformly to* $\gamma : [a, b] \to \mathbb{R}^m$ if

$$\|\gamma_n - \gamma\|_\infty := \sup_{a \leq t \leq b} |\gamma_n(t) - \gamma(t)| \to 0, \text{ as } n \to \infty; \tag{9.19}$$

i.e., for every $\varepsilon > 0$, there exists an integer $n_0 \geq 0$ such that for all $n \geq n_0$, $|\gamma_n(t) - \gamma(t)| \leq \varepsilon$, for all $t \in [a, b]$. Henceforth, we denote by $C([a, b]) = C([a, b], \mathbb{R}^m)$ the space of continuous functions from $[a, b]$ to $\mathbb{R}^m$, equipped with the *metric of uniform convergence* (or "sup metric"), given by

$$\delta_\infty(f, g) := \|f - g\|_\infty = \sup_{a \leq t \leq b} |f(t) - g(t)|, \text{ for all } f, g \in C([a, b]), \tag{9.20}$$

where the associated norm, $\|\cdot\|_\infty$, is given by $\|h\|_\infty := \sup_{a \leq t \leq b} |h(t)|$, for all $h \in C([a, b])$.

A key, but elementary, fact is that the metric space $(C([a, b]), \delta_\infty)$ is complete; i.e., every sequence $\{\gamma_n\}_{n=0}^{\infty}$ of elements of $C([a, b])$ which is *uniformly Cauchy* (i.e., which is a Cauchy sequence in the metric space $(C([a, b]), \delta_\infty)$) is uniformly convergent to some (necessarily unique) $\gamma \in C([a, b])$.

Theorem 9.9 (Uniform Convergence of Continuous Functions). *If a sequence $\{\gamma_n\}_{n=1}^{\infty}$ of elements of $C([a, b])$ converges uniformly to a function $\gamma : [a, b] \to \mathbb{R}^m$, then $\gamma \in C([a, b])$. In short, a uniformly convergent sequence of continuous functions is continuous.*

Corollary 9.10 (Completeness of $C([a, b])$). *The metric space $(C([a, b]), \delta_\infty)$ of continuous parametrized curves on the compact interval $[a, b]$—equipped with the metric of uniform convergence δ_∞ given by* (9.20) *above—is complete.*

Exercise 9.11. Prove Theorem 9.9 and Corollary 9.10.

Exercise 9.12. Show that if $f \in C([a, b])$, then f is uniformly continuous on $[a, b]$; i.e., for every $\varepsilon > 0$, there exists $\delta > 0$ such that for all $t, \tilde{t} \in [a, b]$, we have that $|t - \tilde{t}| \leq \delta \Rightarrow |f(t) - f(\tilde{t})| \leq \varepsilon$.

Let $\{\gamma_n\}_{n=0}^{\infty}$ be a sequence of continuous parametrized curves on $[a, b]$ (i.e., $\gamma_n \in C([a, b])$, for all $n \geq 0$) and, for each $n \geq 0$, let $\mathbf{\Gamma}_n = \gamma_n([a, b]) \subseteq \mathbb{R}^m$ be the continuous curve associated with γ_n. The next result shows that the uniform convergence of $\{\gamma_n\}_{n=0}^{\infty}$ to γ implies the Hausdorff convergence of $\{\mathbf{\Gamma}_n\}_{n=0}^{\infty}$ to $\mathbf{\Gamma}$.

Theorem 9.13 (Uniform Convergence Implies Convergence in the Hausdorff Metric). *With the above notation, assume that $\{\gamma_n\}_{n=0}^{\infty}$ converges uniformly to γ; i.e., according to Theorem* 9.9 *and Corollary* 9.10 *above, assume that $\gamma_n \to \gamma$, in*

$C([a,b])$. Then, the curves $\mathbf{\Gamma}_n := \gamma_n([a,b])$ converge to the curve $\mathbf{\Gamma} := \gamma([a,b])$, in the Hausdorff metric (as defined in Section 4.1):

$$d_H(\mathbf{\Gamma}_n, \mathbf{\Gamma}) \to 0, \text{ as } n \to \infty; \tag{9.21}$$

i.e.,

$$\mathbf{\Gamma}_n \to \mathbf{\Gamma}, \text{ as } n \to \infty, \text{ in } (\mathbb{K}(\mathbb{R}^m), d_H). \tag{9.22}$$

Proof. First, observe that since $\mathbf{\Gamma}_n := \gamma_n([a,b])$ and $\mathbf{\Gamma} := \gamma([a,b])$, $\mathbf{\Gamma}_n$ (for any $n \geq 0$) and $\mathbf{\Gamma}$ are compact (since γ_n and γ are continuous) and nonempty subsets of $\mathbb{R}^m$. Thus, $\mathbf{\Gamma}_n$ (for any $n \geq 0$) and $\mathbf{\Gamma}$ belong to $\mathbb{K}(\mathbb{R}^m)$.

Next, assume that $\|\gamma_n - \gamma\|_\infty \to 0$, as $n \to \infty$. Then, given $\varepsilon > 0$, there exists an integer $N \geq 1$ such that $n \geq N$ implies that $|\gamma_n(t) - \gamma(t)| \leq \varepsilon$, for all $t \in [a,b]$.

We claim that for all $n \geq N$, we have that $d_H(\mathbf{\Gamma}_n, \mathbf{\Gamma}) \leq \varepsilon$, which will prove that $\{\mathbf{\Gamma}_n\}_{n=1}^\infty$ converges to $\mathbf{\Gamma}$ in $(\mathbb{K}(\mathbb{R}^m), d_H)$, as desired. Now, according to the equivalence (4.6) in Lemma 4.7,

$$d_H(\mathbf{\Gamma}_n, \mathbf{\Gamma}) \leq \varepsilon \Leftrightarrow \mathbf{\Gamma}_n \subseteq [\mathbf{\Gamma}]_\varepsilon \text{ and } \mathbf{\Gamma} \subseteq [\mathbf{\Gamma}_n]_\varepsilon. \tag{9.23}$$

where, for example, $[\mathbf{\Gamma}]_\varepsilon$ denotes the closed ε-neighborhood of Γ given in (4.4) of Definition 4.6. Since

$$d_H(\mathbf{\Gamma}_n, \mathbf{\Gamma}) = d_H(\mathbf{\Gamma}, \mathbf{\Gamma}_n) \quad \text{and} \quad \|\gamma_n - \gamma\|_\infty = \|\gamma - \gamma_n\|_\infty, \tag{9.24}$$

it clearly suffices to show that $\mathbf{\Gamma}_n \subseteq [\mathbf{\Gamma}]_\varepsilon$. In order to prove the latter inclusion, we argue as follows:

Fix $n \geq N$ and let $P_n \in \mathbf{\Gamma}_n$ be an arbitrary point in $\mathbf{\Gamma}_n$. Then, $P_n = \gamma_n(\tau)$, for some $\tau \in [a,b]$. Define $P \in \mathbf{\Gamma}$ by the equality $P = \gamma(\tau)$.

Since, by construction, and because $n \geq N$ here, we have that $|P_n - P| = |\gamma_n(\tau) - \gamma(\tau)| \leq \varepsilon$. But clearly, since $P_n \in \mathbf{\Gamma}_n$ and $P \in \mathbf{\Gamma}$,

$$\operatorname{dist}(P_n, \mathbf{\Gamma}) = \inf_{Q \in \mathbf{\Gamma}} |P_n - Q| \leq |P_n - P| \leq \varepsilon.$$

Therefore, $P_n \in [\mathbf{\Gamma}]_\varepsilon$, in light of the definition of $[\mathbf{\Gamma}]_\varepsilon$ given in (4.4) of Definition 4.6. This proves the required inclusion (namely, $\mathbf{\Gamma}_n \subseteq [\mathbf{\Gamma}]_\varepsilon$) and concludes the proof of the theorem. □

Exercise 9.14. With the same notation as in Theorem 9.13, show that if $\{\gamma_n\}_{n=0}^\infty$ is a uniform Cauchy sequence (i.e., a Cauchy sequence in $C([a,b])$), then $\mathbf{\Gamma}_n = \gamma_n([a,b])$ is a Cauchy sequence in $(\mathbb{K}(\mathbb{R}^m), d_H)$. Deduce that, under these hypotheses and as $n \to \infty$, γ_n converges uniformly to some (unique) $\gamma \in C([a,b])$ and that $\mathbf{\Gamma}_n$ converges to $\mathbf{\Gamma} := \gamma([a,b])$, in the Hausdorff metric.

The question naturally arises to know whether or not—under the above hypotheses (in Theorem 9.13) or under a different set of hypotheses (but with the same notation)—the length of $\mathbf{\Gamma}_n$ converges to the length of $\mathbf{\Gamma}$. In other words, does $L(\mathbf{\Gamma}_n)$ converge to $L(\mathbf{\Gamma})$, as $n \to \infty$?

The following example, provided in Exercise 9.15 below (see also [**81**, V1-3-x11]), shows that the answer to the above question is "no", in general, under the hypotheses of Theorem 9.9—and hence also under the assumptions of Theorem 9.13. Accordingly, if $\gamma_n \to \gamma$, uniformly (as in Theorem 9.9)—and let alone (by Theorem 9.13), if $\mathbf{\Gamma}_n \to \mathbf{\Gamma}$, in the Hausdorff metric—then $L(\mathbf{\Gamma}_n)$ need not converge to $L(\mathbf{\Gamma})$, in general.

Example 9.15. For each $n \in \mathbb{N}$, let $f_n : [0,1] \to \mathbb{R}$ be given for all $t \in (0,1]$ by $f_n(t) := n^{-1} \cdot \sin(n\pi t)$, and let $f_0(t) := 1$, also for all $t \in (0,1)$. Furthermore, for any $n \in \mathbb{N}_0$, let $\gamma_n(t) := (t, f_n(t))$, for all $t \in [0,1]$, and let $\mathbf{\Gamma}_n := \gamma_n([0,1])$ be the graph of f_n or, equivalently, the image (or range) of γ_n. Then, show that the hypotheses of Theorem 9.9 are satisfied, with $\gamma(t) := (t,0)$, for all $t \in [0,1]$, and $\mathbf{\Gamma} := \gamma([0,1])$; i.e., $f(t) := 0$, for all $t \in [0,1]$, and $\mathbf{\Gamma}$, the graph of f (or, equivalently, the image of γ) is the following horizontal line segment: $\mathbf{\Gamma} = [0,1] \times \{0\}$.

Also, note that since, clearly, $\gamma_n \to \gamma$, uniformly on $[0,1]$, we have that $\mathbf{\Gamma}_n \to \mathbf{\Gamma}$, in $(\mathbb{K}(\mathbb{R}^2), d_H)$, by Theorem 9.13 or else by direct verification. However, the reader is asked to check via a direct computation that (in light of Theorem 9.3, along with Corollary 9.5 and Example 9.6)

$$L(\mathbf{\Gamma}_n) = \int_0^1 \sqrt{1 + (f_n'(t))^2}\, dt \nrightarrow L(\mathbf{\Gamma}) = 1, \text{ as } n \to \infty. \tag{9.25}$$

In fact, $L(\mathbf{\Gamma}_n) \to l > 1$, as $n \to \infty$, which shows that (9.25) holds and hence that we cannot interchange lengths and uniform limit—and let alone (by Theorem 9.13) Hausdorff limit—in the present case.

Exercise 9.16. Justify all of the statements in Example 9.15 just above.

The next result—which can also be found in [**593**] (but with a less detailed proof)—provides a sufficient condition under which the sequence $\{\mathbf{\Gamma}_{n,Pol}\}_{n=0}^{\infty}$ of polygonal approximations to a given *simple* curve $\mathbf{\Gamma}$, converges to $\mathbf{\Gamma}$ in the Hausdorff metric (i.e., in $(\mathbb{K}(\mathbb{R}^m), d_H)$)—and then, in addition, the lengths and the limit can be interchanged.

Theorem 9.17 (Polygonal Approximations and Hausdorff Metric: Interchange of Lengths and Limit). *Let $\mathbf{\Gamma}$ be a (continuous) simple curve (i.e., a Jordan arc or else a Jordan curve), and let $\{\delta_n\}_{n=0}^{\infty}$ be a sequence of positive numbers tending to 0 as $n \to \infty$. Furthermore, let $\{\mathbf{\Gamma}_{n,Pol}\}_{n=0}^{\infty}$ be a sequence of polygonal approximations to $\mathbf{\Gamma}$ which is* adapted to $\{\delta_n\}_{n=0}^{\infty}$; *i.e.,* $\text{mesh}(\mathbf{\Gamma}_{n,Pol}) \leq \delta_n$, *for all $n \geq 0$.*[10]

[10]This statement means here that, for each $n \in \mathbb{N}_0$, all of the line segments out of which $\mathbf{\Gamma}_{n,Pol}$ is built have a length not exceeding δ_n.

Then,

$$d_H(\mathbf{\Gamma}_{n,Pol}, \mathbf{\Gamma}) \to 0, \text{ as } n \to \infty; \tag{9.26}$$

i.e., with the notation of Section 4.1, *we have that*

$$\mathbf{\Gamma}_{n,Pol} \to \mathbf{\Gamma}, \text{ in } (\mathbb{K}(\mathbb{R}^m), d_H). \tag{9.27}$$

Moreover, we can interchange lengths and Hausdorff limit in this case:

$$L(\mathbf{\Gamma}) = \lim_{n\to\infty} L(\mathbf{\Gamma}_{n,Pol}) = \sup_{n\geq 0} L(\mathbf{\Gamma}_n). \tag{9.28}$$

Here, $L(\mathbf{\Gamma})$ may be finite or infinite. Also, it follows from (9.28) *that the length of $\mathbf{\Gamma}$ is independent of the choice of the adapted sequence of polygonal approximations to $\mathbf{\Gamma}$.*

Proof. First, observe that $\mathbf{\Gamma}$ and each polygonal approximation $\mathbf{\Gamma}_{n,Pol}$ are nonempty compact subsets of $\mathbb{R}^m$, as the continuous images of the nonempty compact set $[a, b]$. Furthermore, as will be explained in Remark 9.18 below and since $\mathbf{\Gamma}$ is a simple curve, both γ and γ^{-1} are uniformly continuous maps. In particular, given $\varepsilon > 0$, there exists $\eta > 0$ such that

$$|t_1 - t_2| \leq \eta \text{ (with } t_1, t_2 \in [a, b]) \Rightarrow |\gamma(t_1) - \gamma(t_2)| \leq \varepsilon, \tag{9.29}$$

where $|\cdot|$ denotes the Euclidean norm (or length) in $\mathbb{R}^m$.[11] In addition, given $\varepsilon' > 0$, there exists $\eta' > 0$ such that

$$|\boldsymbol{x}_1 - \boldsymbol{x}_2| \leq \eta' \text{ (with } \boldsymbol{x}_1, \boldsymbol{x}_2 \in \mathbf{\Gamma}) \Rightarrow |\gamma^{-1}(\boldsymbol{x}_1) - \gamma^{-1}(\boldsymbol{x}_2)| \leq \varepsilon'. \tag{9.30}$$

Next, note that since $\delta_n \to 0^+$, as $n \to \infty$, we can find $N_0 \in \mathbb{N}$ such that

$$\delta_n \leq \eta', \text{ for all } n \geq N_0. \tag{9.31}$$

From now on, let us fix $n \geq N_0$, so that (9.31) holds. Let

$$a =: t_{1,n} < t_{2,n} < \cdots < t_{k_n,n} := b$$

be a (finite) partition of $[a, b]$ such that the points $\gamma(t_{j,n}) = P_{j,n}$, with $j = 0, 1, \ldots, k_n$, are the vertices of the n-th polygonal curve $\mathbf{\Gamma}_{n,Pol}$ approximating $\mathbf{\Gamma}$.

In light of (9.30) (the uniform continuity of γ^{-1}) and (9.31), as well as of the corresponding assumption of the theorem, we have that

$$t_{j,n} - t_{j-1,n} = |t_{j,n} - t_{j-1,n}| \leq \varepsilon', \text{ for } j = 1, \ldots, k_n. \tag{9.32}$$

Since each given $t \in [a, b]$ belongs to a subinterval $[t_{j-1,n}, t_{j,n}]$ of $[a, b]$, for some $j \in \{1, \ldots, k_n\}$, we deduce from (9.32) and from (9.31), applied with $\varepsilon' := \eta$, that $t_{j,n} - t = |t - t_{j,n}| \leq t_{j,n} - t_{j-1,n} \leq \eta$, still for the same index

[11] Here and thereafter, for notational simplicity, we use the notation $|\boldsymbol{x} - \boldsymbol{y}|$ instead of $d_m(\boldsymbol{x}, \boldsymbol{y})$ in order to denote the Euclidean distance between $\boldsymbol{x}$ and $\boldsymbol{y}$, with $\boldsymbol{x}, \boldsymbol{y} \in \mathbb{R}^m$.

$j \in \{1, \ldots, k_n\}$. We now deduce from the latter identity combined with (9.29) (the uniform continuity of γ) that, with $P := \gamma(t)$ and $P_{j,n} := \gamma(t_{j,n})$,

$$|P - P_{j,n}| = |\gamma(t) - \gamma(t_{j,n})| \leq \varepsilon, \text{ for some } j \in \{1, \ldots, k_n\}. \tag{9.33}$$

(Here, $|\cdot|$ still denotes the Euclidean norm in $\mathbb{R}^m$.) Since, obviously, the point $P_{j,n}$ lies on the polygonal curve $\mathbf{\Gamma}_{n,Pol}$, it follows from (9.33) that every point P of the curve $\mathbf{\Gamma}$ lies within a distance not exceeding ε from $\mathbf{\Gamma}_{n,Pol}$:

$$\operatorname{dist}(P, \mathbf{\Gamma}_{n,Pol}) = \inf\{|P - Q| : Q \in \mathbf{\Gamma}_{n,Pol}\} \leq \varepsilon; \tag{9.34}$$

in other words, every point P of $\mathbf{\Gamma}$ belongs to $[\mathbf{\Gamma}_{n,Pol}]_\varepsilon$, the closed ε-neighborhood of $\mathbf{\Gamma}_{n,Pol}$ (see (4.4) in Definition 4.6):

$$\mathbf{\Gamma} \subseteq [\mathbf{\Gamma}_{n,Pol}]_\varepsilon, \text{ for all } n \geq N_0. \tag{9.35}$$

Similarly, but more simply, we next establish the counterpart of (9.35) just above, with the roles of $\mathbf{\Gamma}$ and $\mathbf{\Gamma}_{n,Pol}$ interchanged. To this aim, we note that any point Q of $\mathbf{\Gamma}_{n,Pol}$ belongs to one of the line segments $\overline{P_{j-1,n}P_{j,n}}$ of which the polygonal curve $\mathbf{\Gamma}_{n,Pol}$ is composed; i.e., with the above notation,

$$Q \in \overline{P_{j-1,n}\, P_{j,n}}, \text{ for some } j \in \{1, \ldots, k_n\}. \tag{9.36}$$

Hence, it follows from (9.36) that given any $Q \in \mathbf{\Gamma}_{n,Pol}$, we have that

$$|Q - P_{j,n}| \leq |\overline{P_{j-1,n}\, P_{j,n}}|, \text{ for some } j \in \{1, \ldots, k_n\}. \tag{9.37}$$

Next, by hypothesis of the theorem, all of the line segments of $\mathbf{\Gamma}_{n,Pol}$ have length $\leq \delta_n$; furthermore, $\delta_n \to 0^+$, as $n \to \infty$. Consequently, there exists $N_1 \geq N_0$ such that $n \geq N_1 \Rightarrow \delta_n \leq \varepsilon$. By combining (9.37) and the latter implication, we obtain that, provided that $n \geq N_1$ (which will be assumed from now on),

$$|Q - P_{j,n}| \leq \varepsilon, \text{ for some } j \in \{1, \ldots, k_n\}. \tag{9.38}$$

But since, by construction of the polygonal approximation $\mathbf{\Gamma}_{n,Pol}$, the point $P_{j,n}$ also belongs to the curve $\mathbf{\Gamma}$, we deduce from (9.38) (and from (4.1) in Definition 4.1, where d_m has been replaced by the simpler notation $|\cdot|$) that, for all $n \geq N_1$, $\operatorname{dist}(Q, \mathbf{\Gamma}) = \inf\{|Q-P| : P \in \mathbf{\Gamma}\} \leq \varepsilon$. Since Q is an arbitrary point of $\mathbf{\Gamma}_{n,Pol}$, the latter inequality precisely means that for all $n \geq N_1$, $\mathbf{\Gamma}_{n,Pol} \subseteq [\mathbf{\Gamma}]_\varepsilon$, the closed ε-neighborhood of $\mathbf{\Gamma}$.

It now suffices to combine the latter inclusion and the inclusion in (9.35) to deduce that (since $N_1 \geq N_0$, by construction) for every integer $n \geq N_1$,

$$\mathbf{\Gamma} \subseteq [\mathbf{\Gamma}_{n,Pol}]_\varepsilon \quad \text{and} \quad \mathbf{\Gamma}_{n,Pol} \subseteq [\mathbf{\Gamma}]_\varepsilon. \tag{9.39}$$

Consequently, in light of Lemma 4.7 (see (4.6)), we deduce from (9.39) that for all $n \geq N_1$, the Hausdorff distance between $\mathbf{\Gamma}_{n,Pol}$ and $\mathbf{\Gamma}$ satisfies the following inequality: $d_H(\mathbf{\Gamma}_{n,Pol}, \Gamma) \leq \varepsilon$, still for all $n \geq N_0$. We have therefore established that $d_H(\boldsymbol{\Gamma}_{n,Pol}, \Gamma) \to 0$, as $n \to \infty$; i.e., (9.26) (or, equivalently, (9.27)) holds, as needed to be shown.

The fact that (9.28)—the statement about lengths—also holds follows at once from the identity (9.3) characterizing (and defining) $L(\boldsymbol{\Gamma})$, the length of $\boldsymbol{\Gamma}$; see also Theorem 9.20 below and the text preceding it. The easy proof is left to the interested reader.

This concludes the proof of the theorem (i.e., Theorem 9.17). □

Remark 9.18. We claim that—as was required towards the beginning of the proof of Theorem 9.17—γ is uniformly continuous and that, since the curve $\boldsymbol{\Gamma}$ is simple, the inverse map γ^{-1} is well defined and is uniformly continuous. Indeed, γ is continuous on the compact metric space $[a,b]$—and hence, it is also uniformly continuous on $[a,b]$; clearly, this part does not require that $\boldsymbol{\Gamma}$ be simple. Furthermore, if, in addition, γ is injective (i.e., if $\boldsymbol{\Gamma}$ is simple), then the map $\gamma^{-1} : \boldsymbol{\Gamma} \to [a,b]$ is continuous (and hence also uniformly continuous, since $\boldsymbol{\Gamma}$ is a compact metric space), as we now explain. Indeed, $\boldsymbol{\Gamma} = \gamma([a,b])$ is compact and, for any closed (and thus also compact) subset K of $[a,b]$, $(\gamma^{-1})^{-1}(K) = \gamma(K)$ is compact (as the continuous image of a compact set) and therefore, it is also closed (in $[a,b]$). Hence, γ^{-1} is continuous, as desired.

The following interesting example was studied, in particular, by Henri Lebesgue, the discoverer of the "Lebesgue integral", briefly discussed in Section 3.6.

Example 9.19 (*A Fractal-Like Example*). We start with an equilateral triangle, $A_0A_1A_2$, with side lengths 1. Then, one constructs a sequence of (simple) polygonal curves $\{\boldsymbol{\Gamma}_n\}_{n=0}^{\infty}$ contained within the closure of the above equilateral triangle and such that $\{\boldsymbol{\Gamma}_n\}_{n=0}^{\infty}$ converges in the Hausdorff metric (i.e., in $(\mathbb{K}(\mathbb{R}^2), d_H)$ to a simple curve $\boldsymbol{\Gamma}$, specifically, the horizontal line segment $\boldsymbol{\Gamma} = \overline{A_1A_2}$ (the base of the triangle), of length l and such that $l := \lim_{n\to\infty} L(\Gamma_n)$ exists and is finite but

$$L(\boldsymbol{\Gamma}_n) \nrightarrow L(\boldsymbol{\Gamma}) = 1, \text{ as } n \to \infty. \tag{9.40}$$

In fact, here, by construction, we have that $L(\boldsymbol{\Gamma}_n) = 2$, for all integers $n \geq 1$ and so, l obviously exists and equals $2 \neq 1$.

The construction of $\{\boldsymbol{\Gamma}_n\}_{n=0}^{\infty}$ goes as follows. Let $\boldsymbol{\Gamma}_0 := \overline{A_1A_2}$, as was mentioned earlier. Furthermore, let $\boldsymbol{\Gamma}_1$ denote the polygonal curve (of length 2) consisting of the line segments $\overline{A_2A_0}$ and $\overline{A_0A_1}$, in that order. Moreover, let A_3, A_4, and A_5 be, respectively, the midpoints of the line segments $\overline{A_2A_0}$, $\overline{A_0A_1}$, and $\overline{A_1A_2}$. Then, $\boldsymbol{\Gamma}_2$ is the polygonal curve (also of length $2 = 4(1/2)$) consisting of the $4 = 2^2$ line segments (of length 1/2) $\overline{A_2A_3}$, $\overline{A_3A_5}$, $\overline{A_5A_4}$, and $\overline{A_4A_1}$, in that order; see Figure 9.6; and so on .... Accordingly, at the n-th stage of the

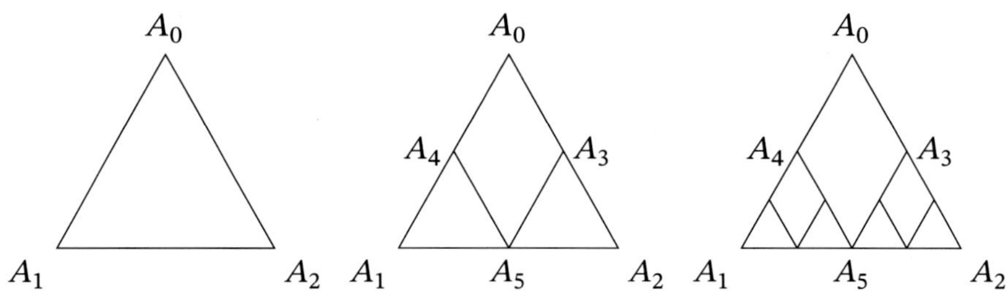

Figure 9.6. The first few steps of the construction of the sequence of (simple) polygonal curves $\{\mathbf{\Gamma}_n\}_{n=0}^{\infty}$ in Example 9.19 converging to $\mathbf{\Gamma} := \overline{A_1A_2}$, the base of the unit equilateral triangle $A_0A_1A_2$. For any $n \geq 1$, each curve $\mathbf{\Gamma}_n$ has length 2, while $\mathbf{\Gamma}$ has length 1. Hence, we have that $1 = L(\mathbf{\Gamma}) \neq \lim_{n\to\infty} L(\mathbf{\Gamma}_n) = 2$, whereas $d_H(\mathbf{\Gamma}_n, \mathbf{\Gamma}) \to 0$, as $n \to \infty$. Note that the polygonal curves are clearly not polygonal approximations to the horizontal line segment $\mathbf{\Gamma} = \overline{A_1A_2}$, in the sense of Section 9.1.

construction, where $n \geq 1$, the polygonal curve $\mathbf{\Gamma}_n$ consists of 2^n line segments of length $1/2^{n-1}$ and hence also has length $2 = 2^n/2^{n-1}$.

It is easy to check that $d_H(\mathbf{\Gamma}_n, \mathbf{\Gamma}) \to 0$, as $n \to \infty$, for example by using Lemma 4.7 and Remark 4.8 in Section 4.1.

Note that, in agreement with the lower semi-continuity (stated in Theorem 9.22 below) of the length function, and by (9.40) we have that

$$1 = L(\mathbf{\Gamma}) \leq \underline{\lim}_{n\to\infty} L(\mathbf{\Gamma}_n) = 2. \tag{9.41}$$

Also, observe that the continuous curves $\mathbf{\Gamma}_n$ are polygonal curves but obviously *not* polygonal approximations to $\mathbf{\Gamma}$, in the sense of Section 9.1 above.

In light of the identities (9.5)–(9.7) in Section 9.1 above, the following result justifies both the identity (9.3)—and, more specifically, the second equality of (9.3) along with the statement of footnote 6 in Section 9.1 above. Its proof, which is elementary but somewhat technical, can be omitted upon a first reading.

Theorem 9.20 (Total Variation, Length, and Limits). *Let the map $\gamma : [a, b] \to \mathbb{R}^m$ be a continuous parametrized curve, with range the continuous curve $\mathbf{\Gamma} = \gamma([a, b])$.*[12] *Then, the following limits exist and*

$$L(\mathbf{\Gamma}) = V(\gamma) = \lim_{\tilde{\sigma}\supseteq\sigma} V_{\tilde{\sigma}}(\gamma) = \lim_{|\sigma|\to 0^+} V_{\sigma}(\gamma)\ (= \sup_{\sigma} V_{\sigma}(\gamma)), \tag{9.42}$$

[12] Note that the curve $\mathbf{\Gamma}$ is allowed to be either rectifiable or unrectifiable ($L(\mathbf{\Gamma}) < \infty$ or else $L(\mathbf{\Gamma}) = +\infty$); i.e., γ may be of *bounded* (finite) or *unbounded* (infinite) total variation.

where each of the limits, as well as the supremum, holds in $[0, +\infty]$ *and where the supremum is taken over all of the partitions* σ *of* $[a, b]$.[13] *The second limit in* (9.42) *means that for every* $\varepsilon > 0$, *there exists* $\eta > 0$ *such that for any (finite) partition* σ *of* $[a, b]$,

$$|\sigma| \leq \eta \Rightarrow 0 \leq V(\gamma) - V_\sigma(\gamma) \leq \varepsilon. \tag{9.44}$$

Hence, if $\mathbf{Pol}_\sigma(\mathbf{\Gamma})$ *denotes the polygonal approximation of* $\mathbf{\Gamma}$ *associated with the partition* σ, *then the following limits exist and are given by*

$$L(\mathbf{\Gamma}) = \lim_{\tilde{\sigma} \supseteq \sigma} L(\mathbf{Pol}_{\tilde{\sigma}}(\mathbf{\Gamma})) = \lim_{|\sigma| \to 0^+} L(\mathbf{Pol}_\sigma(\mathbf{\Gamma}))(= \sup_\sigma L(\mathbf{Pol}_\sigma(\mathbf{\Gamma})), \tag{9.45}$$

where the second (resp., first) limit in (9.45) *holds in* $[0, +\infty]$ *and is interpreted exactly as in the obvious counterpart of* (9.44) *(resp., of* (9.43) *in footnote* 13).

Proof. The proof is divided into the following three steps:

Step 1. *The first limit in* (9.42) *exists and equals* $V(\gamma)$.

Indeed, as we have already seen (and as follows at once from the definition of $V_\sigma(\gamma)$ given on the right-hand side of (9.4)), the map $\sigma \mapsto V_\sigma(\gamma)$ is a nondecreasing function, as σ increases relative to the partial ordering defined by the refinement of partitions. Namely, if $\tilde{\sigma} \supseteq \sigma$ (i.e., if $\tilde{\sigma}$ is a refinement of σ), then $V_{\tilde{\sigma}}(\gamma) \geq V_\sigma(\gamma)$.

Therefore, it follows from general principles that the first limit in (9.42) exists and equals $V(\gamma)$. More specifically (since the present situation may be more abstract than what some readers may be accustomed to), since, by definition, $V(\gamma) = \sup_\sigma V_\sigma(\gamma)$, given $\varepsilon > 0$, there exists a partition σ_0 such that[14] $0 \leq V(\gamma) - V_{\sigma_0}(\gamma) \leq \varepsilon$. Hence, for any refinement $\tilde{\sigma}$ of σ_0 (i.e., if $\tilde{\sigma} \supseteq \sigma_0$), we deduce that $0 (\leq V(\gamma) - V_{\sigma_0}(\gamma)) \leq V(\gamma) - V_{\tilde{\sigma}}(\gamma) \leq \varepsilon$, as required; see (9.43) in footnote 13.

If $V(\gamma) = +\infty$, we reason as follows. Given $A > 0$, there exists σ_0 such that $V_{\sigma_0}(\gamma) \geq A$. Consequently, for any refinement $\tilde{\sigma}$ of σ_0, we must have that $V_{\tilde{\sigma}}(\gamma) \geq V_{\sigma_0}(\gamma) \geq A$. This proves that the first limit in (9.42) exists and equals $+\infty$, as desired in this case. This concludes the proof of Step 1.

---

[13] Here, the first limit in (9.42) is taken along the natural filter of (finite) partitions of $[a, b]$ (partially ordered by refinement) and concretely means that the following property holds: Given $\varepsilon > 0$, there exists a partition σ_0 such that

$$\tilde{\sigma} \supseteq \sigma_0 \text{ (i.e., } \tilde{\sigma} \text{ is a refinement of } \sigma_0) \Rightarrow 0 \leq V(\gamma) - V_{\tilde{\sigma}}(\gamma) \leq \varepsilon. \tag{9.43}$$

[14] For simplicity, we assume implicitly here that $V(\gamma) < \infty$ and will indicate below the small changes needed in order to deal with the case when $V(\gamma) = +\infty$.

Step 2. *The second limit in* (9.42) *exists and equals* $V(\gamma)$.

The proof of Step 2 relies on somewhat different arguments than in Step 1, which should, however, be familiar to the reader acquainted with the rigorous theory of the Riemann integral. We first assume that $V(\gamma) < \infty$. (The case when $V(\gamma) = +\infty$ can be dealt with much as in Step 1.)

Given $\varepsilon > 0$ and since, by definition, $V(\gamma) = \sup_\sigma V_\sigma(\gamma)$, there exists a (finite) partition σ_0 of $[a, b]$ such that

$$0 \leq V(\gamma) - V_{\sigma_0}(\gamma) \leq \varepsilon/3, \tag{9.46}$$

say, $\sigma_0 = \{t_0 := a < t_1 < \cdots < t_{k_0} := b\}$.

Since γ is uniformly continuous on $[a, b]$ (because it is continuous on the compact interval $[a, b]$), there exists $\eta > 0$ such that

$$|x - y| \leq \eta \Rightarrow |\gamma(x) - \gamma(y)| \leq \varepsilon/3k_0; \tag{9.47}$$

clearly, we may also assume that $\eta < \min_{j=1,\ldots,k_0}(t_j - t_{j-1})$.

Next, consider an arbitrary (finite) partition σ of $[a, b]$ with mesh $|\sigma| \leq \eta$. By our choices of η and σ, we know that each open subinterval (t_{j-1}, t_j) of σ_0 (for $j = 1, \ldots, k_0$) must contain at least one point of σ. Now, for $j = 2, \ldots, k_0$, let us denote by t^*_{j-1} (resp., t^{**}_{j-1}) the greatest (resp., smallest) of the points τ of the partition σ such that $\tau < t_{j-1}$ (resp., $t_{j-1} < \tau$); furthermore, let

$$U_{j-1} := \gamma(t_{j-1}), \quad U^*_{j-1} := \gamma(t^*_{j-1}), \quad \text{and} \quad U^{**}_{j-1} = \gamma(t^{**}_{j-1}). \tag{9.48}$$

Denoting by $\sigma \cup \sigma_0$ the joint (or union) of σ and σ_0 (the partition consisting, by definition, precisely of the points belonging to either σ or σ_0, but without counting multiplicities),[15] we have that (see (9.48) and (9.47))

$$\begin{aligned} V_{\sigma\cup\sigma_0}(\gamma) &\leq \sum_j (|U^*_{j-1} - U_{j-1}| + |U^{**}_{j-1} - U_{j-1}| - |U^{**}_{j-1} - U^*_{j-1}|) \\ &\leq 2\,(\varepsilon/3k_0)\,k_0 = 2\varepsilon/3. \end{aligned} \tag{9.49}$$

Since $V_{\sigma_0} \leq V_{\sigma\cup\sigma_0}$ (because $\sigma \cup \sigma_0$ is a refinement of σ_0) and since (by (9.46) and (9.49)) $0 \leq V(\gamma) - V_{\sigma_0}(\gamma) \leq \varepsilon/3$ and $V_{\sigma\cup\sigma_0}(\gamma) \leq 2\varepsilon/3$, we obtain (by adding the corresponding three inequalities) that (see footnote 15)

$$\begin{aligned} 0 \leq V(\gamma) - V_\sigma(\gamma) &\leq (V(\gamma) - V_{\sigma_0}(\gamma)) + (V_{\sigma_0}(\gamma) - V_{\sigma\cup\sigma_0}(\gamma)) + V_{\sigma\cup\sigma_0}(\gamma) \\ &\leq (V(\gamma) - V_{\sigma_0}(\gamma)) + V_{\sigma\cup\sigma_0}(\gamma) \leq (2\varepsilon/3) + (\varepsilon/3) = \varepsilon, \end{aligned} \tag{9.50}$$

as desired. This yields (9.44) and concludes the proof of Step 2.

---

[15] Observe that, by construction, $\sigma \cup \sigma_0$ is a common refinement of both σ and σ_0; hence, $V_{\sigma_0}(\gamma) - V_{\sigma\cup\sigma_0}(\gamma) \leq 0$.

Step 3. *The exact counterparts of the statements of Step* 1 *and of Step* 2 (*i.e., the counterpart of* (9.42) *in the first part of Theorem* 9.20) *hold, but with total variations replaced by lengths; i.e.,* (9.45) *holds in the second part of Theorem* 9.20.

Indeed, in light of Steps 1 and 2, which show that (9.42) holds, Step 3 follows at once from the relevant definitions, according to which, in particular, $L(\mathbf{\Gamma}) = V(\gamma)$ and $L(\boldsymbol{Pol}_\sigma(\gamma)) = V_\sigma(\gamma)$, for any (finite) partition σ of $[a,b]$; see (9.6) and (9.4), respectively. Also, clearly, we may assume here that each polygonal approximation $\boldsymbol{Pol}_\sigma(\Gamma)$ of $\mathbf{\Gamma}$ is parametrized on the same interval $[a,b]$ as the parametrized curve $\gamma : [a,b] \to \mathbb{R}^m$.

This concludes the proof of Step 3—as well as of (9.45), (9.42) and hence also of the entire theorem (i.e., Theorem 9.20). □

Remark 9.21 (*Lack of Continuity of the Length and of the Total Variation*). We caution the reader that neither the length nor the total variation is continuous with respect to the Hausdorff metric d_H or with respect to uniform convergence $\|\cdot\|_\infty$. For example, one can find a sequence $\{\mathbf{\Gamma}_n\}_{n=1}^\infty$ of simple curves (associated with the parametrized curves $\gamma_n : [a,b] \to \mathbb{R}^m$) and a simple curve $\mathbf{\Gamma}$ (all of finite length) such that either

$$d_H(\mathbf{\Gamma}_n, \mathbf{\Gamma}) \to 0, \text{ as } n \to \infty, \tag{9.51}$$

or

$$\gamma_n \to \gamma, \text{ uniformly on } [a,b] \tag{9.52}$$

(i.e., $\|\gamma_n - \gamma\|_\infty := \sup_{t\in[a,b]} |\gamma_n(t) - \gamma(t)| \to 0$, as $n \to \infty$), but

$$L(\mathbf{\Gamma}_n) \nrightarrow L(\mathbf{\Gamma}), \text{ as } n \to \infty, \tag{9.53}$$

and hence (since $V(\gamma_n) = L(\mathbf{\Gamma}_n)$ and $V(\gamma) = \boldsymbol{L}(\mathbf{\Gamma})$, by (9.4) and (9.6), respectively),

$$V(\gamma_n) \nrightarrow V(\gamma), \text{ as } n \to \infty. \tag{9.54}$$

For a counterexample for which (9.51) and (9.53)–(9.54) hold; see Example 9.19 above (as discussed, e.g., in [**593**, Section 5.5]), while for a counterexample for which (9.51)–(9.54) hold, see Exercise 9.15 above (as discussed in [**81**, VI.3.§11]).

Theorem 9.22 (Semi-Continuity of the Length Function). *With the notation of Theorem* 9.13 (*introduced just before the statement of that theorem*), *we have that*

$$L(\mathbf{\Gamma}) \le \varliminf_{n\to\infty} L(\mathbf{\Gamma}_n), \tag{9.55}$$

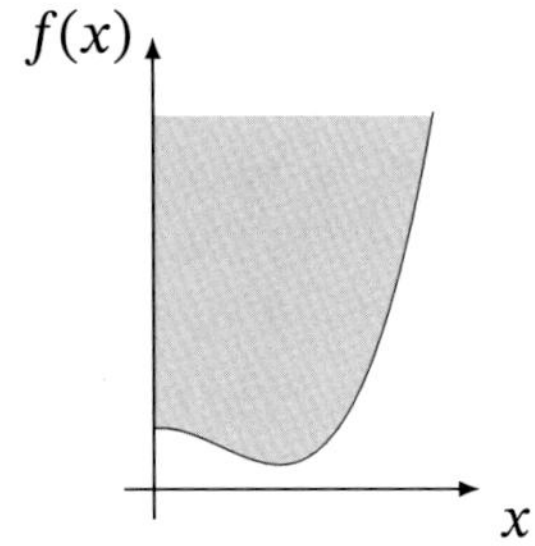

Figure 9.7. The epigraph $E(f)$ of a function $f : [a,b] \to \overline{\mathbb{R}}$, as defined in (9.57), is represented by the shaded area depicted here.

where $\{\mathbf{\Gamma}_n\}_{n=0}^{\infty}$ *is a sequence of* simple *curves* $\mathbf{\Gamma}_n$ *converging in the Hausdorff metric to the* simple *curve* $\mathbf{\Gamma}$*—or such that, as a special case (by Theorem* 9.13*), the sequence* $\{\gamma_n\}_{n=0}^{\infty}$ *of (simple) parametrized curves underlying the sequence* $\{\mathbf{\Gamma}_n\}_{n=0}^{\infty}$*, converges uniformly to* γ *on* $[a,b]$*, where* γ *is the (simple) parametrized curve underlying* $\mathbf{\Gamma}$.

Proof.

(a) We first prove the statement concerning the uniform convergence and leave the other statement as an exercise for the reader; see Exercise 9.23 below.

By definition,

$$V(\gamma) = \sup_{\sigma} V_{\sigma}(\gamma), \quad \text{i.e.,}\ L(\mathbf{\Gamma}) = \sup_{\sigma} L(\mathbf{\Gamma}_{\sigma}). \tag{9.56}$$

Now, it is easy to check that for each (finite) partition σ of $[a,b]$, $V_{\sigma}(\gamma)$ is a semi-continuous function. Hence, in light of the first equality of (9.56), $V(\gamma)$ is semi-continuous, as the supremum of a family of semi-continuous functions; see, e.g., [**81**, V1–3–§8]. Therefore, in light of the second equality of (9.56), the same is true of the length $L(\mathbf{\Gamma})$.

The simplest way to establish the above "supremum property" (according to which a supremum of lower semi-continuous functions is lower semi-continuous) is to use the following geometric characterization of lower semi-continuity;[16] let $f : [a,b] \to \overline{\mathbb{R}}$ be a map, where $\overline{\mathbb{R}} := [-\infty, +\infty] = \mathbb{R} \cup \{\pm\infty\}$ is the extended real line. Then, it is not difficult to show that f is lower semi-continuous iff its *epigraph*,

$$E(f) := \{(x,y) \in [a,b] \times \overline{\mathbb{R}} : y \geq f(x)\}, \tag{9.57}$$

is closed. Here, $E(f)$ is viewed as a subset of $[a,b] \times \overline{\mathbb{R}}$, equipped with the product topology; see the shaded area in Figure 9.7. One then uses the fact that $E_{\sup f_j} = \bigcap_j E_{f_j}$ and the intersection of any family of closed sets is closed.

[16]The reader who is not very comfortable with finer elements of point-set topology may wish to ignore the present comments completing the proof of the first part of the theorem.

Here, the topology on $\overline{\mathbb{R}}$ is obtained either by using the "interval topology" induced by the total (linear) order on $\overline{\mathbb{R}}$—according to which (i) for $x, y \in \mathbb{R}$, $x \leq y$ in $\overline{\mathbb{R}}$ iff $x \leq y$ in $\mathbb{R}$ and (ii) $-\infty \leq x \leq +\infty$, for all $x \in \mathbb{R}$—or, equivalently, by transporting the topology of the compact interval $[-\pi/2, \pi/2] \subseteq \mathbb{R}$ via the tangent map, $t \mapsto \tan(t)$, and letting, of course, $\tan\left(\pm\frac{\pi}{2}\right) := \lim_{t\to\pm\infty} \tan(t) = \pm\infty$. Then, this extended tangent map becomes a homeomorphism of the compact interval $[-\pi/2, \pi/2]$ onto $\overline{\mathbb{R}}$ and $\overline{\mathbb{R}}$ itself becomes a compact, Hausdorff, and metrizable space; furthermore, a metric $d_{\overline{\mathbb{R}}}$ on $\overline{\mathbb{R}}$ inducing the topology of $\overline{\mathbb{R}}$ is provided by $d_{\overline{\mathbb{R}}}(u, v) = |\tan^{-1}(u) - \tan^{-1}(v)|$, for all $u, v \in \overline{\mathbb{R}}$, with the obvious conventions. In fact, since $(-\pi/2, \pi/2)$ is clearly homeomorphic to $\mathbb{R}$ and, by construction, $[-\pi/2, \pi/2]$ is homeomorphic to $\overline{\mathbb{R}}$, the extended real line $\overline{\mathbb{R}}$ can be viewed as a two-point compactification of the ordinary real line $\mathbb{R} = (-\infty, +\infty)$, with the two added points being $\{-\infty, +\infty\}$.

As was mentioned earlier, in order to prove the supremum property of lower semi-continuous functions, we first observe that if $\{f_j\}_{j\in J}$ is any family of $\overline{\mathbb{R}}$-valued functions on $[a, b]$, then (in light of (9.57) and Figure 9.7) with $f := \sup_{j\in J} f_j$, we have that

$$E(f) = \bigcap_{j\in J} E(f_j). \tag{9.58}$$

According to (9.58), it then suffices to appeal to the above geometric characterization of lower semi-continuity combined with the fact that an arbitrary intersection of closed sets is also closed.

This finally concludes the proof of the first part of the theorem.

(b) In order to prove the second part of the theorem (i.e., the lower semi-continuity of the length with respect to the Hausdorff metric), we can proceed as follows (see [**593**, Section 5.5]). Let $\{\mathbf{\Gamma}_n\}_{n=1}^{\infty}$ be a sequence of curves such that $d_H(\mathbf{\Gamma}_n, \mathbf{\Gamma}) \to 0$, as $n \to \infty$. Then, we want to show that[17]

$$L(\mathbf{\Gamma}) \leq \varliminf_{n\to\infty} L(\mathbf{\Gamma}_n). \tag{9.60}$$

[17] If $\mathbf{\Gamma}$ is not rectifiable (i.e., if $L(\mathbf{\Gamma}) = +\infty$), then the inequality (9.60) is equivalent to the following equality:

$$\lim_{n\to\infty} L(\mathbf{\Gamma}_n) = +\infty. \tag{9.59}$$

Indeed, (9.60) implies that $\varlimsup_{n\to\infty} L(\mathbf{\Gamma}_n) \geq \varliminf_{n\to\infty} L(\mathbf{\Gamma}_n) = +\infty$, and thus, in shorthand notation (see Appendix A), the above limit exists and is given by $\lim = \overline{\lim} = \underline{\lim} = +\infty$. The converse is obvious.

Hence, given any $\varepsilon > 0$, we must find an integer $N_0 \geq 1$ such that

$$n \geq N_0 \Rightarrow L(\mathbf{\Gamma}_n) \geq L(\mathbf{\Gamma}) - \varepsilon, \tag{9.61}$$

assuming here for the simplicity of exposition that $L(\mathbf{\Gamma}) < \infty$ (i.e., $\mathbf{\Gamma}$ is rectifiable).[18]

Consider a polygonal approximation $\mathbf{\Gamma}_{Pol}$ of $\mathbf{\Gamma}$ with vertices

$$A =: P_0, P_1, \dots, P_{k-1}, P_k := B. \tag{9.62}$$

Note that, as before, the vertices of $\mathbf{\Gamma}_{Pol}$ belong to $\mathbf{\Gamma}$ and that the endpoints of $\mathbf{\Gamma}_{Pol}$ coincide with the endpoints of A and B of $\mathbf{\Gamma}$.

Let $2\delta_n := d_H(\mathbf{\Gamma}_n, \mathbf{\Gamma})$, for any $n \in \mathbb{N}$. Without loss of generality, clearly, we may assume—by taking N_0 and hence also $n \geq N_0$, sufficiently large—that, for $j = 0, 1, \dots, k$, the closed balls B_j (in $\mathbb{R}^m$) with centers P_j (as in (9.62) just above) and radius δ_n are disjoint. By construction, $d_H(\mathbf{\Gamma}_n, \mathbf{\Gamma}) < \delta_n$ and so, $\mathbf{\Gamma} \subseteq [\mathbf{\Gamma}_n]_{\delta_n}$ (the closed δ_n-neighborhood of $\mathbf{\Gamma}_n$), as well as $\mathbf{\Gamma}_n \subseteq [\mathbf{\Gamma}]_{\delta_n}$. Thus, each of these balls B_j contains at least one point $Q_{j,n}$ lying on the curve $\mathbf{\Gamma}_n$, for $j = 0, 1, \dots, k$.[19]

Now, for $j = 1, \dots, k$, the length of each polygonal line segment $\overline{Q_{j-1,n}Q_{j,n}}$ must satisfy the inequality

$$|Q_{j,n} - Q_{j-1,n}| \geq |P_j - P_{j-1}| - 2\delta_n. \tag{9.63}$$

Since there are k such segments and the length $L(\mathbf{\Gamma}_n)$ does not exceed the length of the polygonal curve $A =: Q_{0,n}, Q_{1,n}, \dots, Q_{k-1,n}, Q_{k,n} := B$ approximating $\mathbf{\Gamma}_n$, we deduce from (9.63) that

$$\begin{aligned} L(\mathbf{\Gamma}_n) &\geq \sum_{j=1}^{k} |Q_{j,n} - Q_{j-1,n}| \\ &\geq \sum_{j=1}^{k} |P_j - P_{j-1}| - 2k\delta_n = L(\mathbf{\Gamma}_{Pol}) - 2k\delta_n. \end{aligned} \tag{9.64}$$

Since, by definition of the length $L(\mathbf{\Gamma})$ of $\mathbf{\Gamma}$, the length $L(\mathbf{\Gamma}_{Pol})$ of the polygonal approximation $\mathbf{\Gamma}_{Pol}$ of $\mathbf{\Gamma}$ can be chosen arbitrarily close to $L(\mathbf{\Gamma})$ (but, of course, also with $L(\mathbf{\Gamma}) \geq L(\mathbf{\Gamma}_{Pol})$), we can choose $\mathbf{\Gamma}_{Pol}$ so that we have $0 \leq L(\mathbf{\Gamma}) - L(\mathbf{\Gamma}_{Pol}) \leq \varepsilon/2$. Then, for the corresponding value of $k = k(\varepsilon)$ fixed accordingly—and since, by hypothesis, $\delta_\ell \to 0$, as $\ell \to \infty$—we can find an integer $N_0 = N_0(\varepsilon) \geq 1$ so large that $N \geq N_0 \Rightarrow \delta_n \leq \varepsilon/2k$; i.e., $2k\delta_n \leq \varepsilon/2$.

[18] In light of footnote 17, if, instead, we assume that $L(\mathbf{\Gamma}) = +\infty$, then we must show that (9.59) in footnote 17 holds; i.e., given any $M > 0$, we must find an integer $N \geq 1$ such that the following implication holds: $n \geq N \Rightarrow L(\mathbf{\Gamma}_n) \geq M$. The proof of this implication follows the same line as that of (9.61) and is left as a simple exercise for the interested reader.

[19] Recall from an earlier comment that P_j belongs to $\mathbf{\Gamma}$, for each $j = 0, 1, \dots, k$. Also, clearly, we may assume that $Q_{0,n} := A$ and $Q_{k,n} := B$.

Consequently, by combining (9.64) and the latter two inequalities, we obtain that $n \geq N_0 \Rightarrow L(\mathbf{\Gamma}_n) \geq L(\mathbf{\Gamma}) - 2(\varepsilon/2) = L(\mathbf{\Gamma}) - \varepsilon$, which is the sought for implication (9.61).

This completes the proof of part (b) and hence of the entire theorem, i.e., of Theorem 9.22. □

Exercise 9.23.

(a) Give a detailed proof of the geometric characterization of lower semicontinuity, as well as of the identity (9.58) (with $f := \sup_{j \in J} f_j$), both provided towards the end of part (a) of the proof of Theorem 9.22.

(b) Describe in your own words the proof of part (b) of Theorem 9.22.

The next result—like the other main results stated in this section—will also be proved and used in [**116**]. To our knowledge, it has not been previously obtained in the literature. We refer to Remark 9.25 below for the advantages of Theorem 9.24, compared to Theorem 9.17, for example.

Theorem 9.24 (A Sufficient Condition for the Interchange of Lengths and Limit). *Let $\mathbf{\Gamma}$ be a continuous curve (not necessarily assumed to be simple), and let $\{\mathbf{\Gamma}_{n,Pol}\}_{n=0}^{\infty}$ be a sequence of polygonal approximations to $\mathbf{\Gamma}$ satisfying the following two hypotheses:*

(i) *$\{\mathbf{\Gamma}_{n,Pol}\}_{n=0}^{\infty}$ converges to $\mathbf{\Gamma}$ in the Hausdorff metric; i.e.,*

$$\mathbf{\Gamma}_{n,Pol} \to \mathbf{\Gamma}, \text{ in } (\mathbb{K}(\mathbb{R}^m), d_H). \tag{9.65}$$

(ii) *The sequence of lengths $\{L(\mathbf{\Gamma}_{n,Pol}\}_{n=0}^{\infty}$ is nondecreasing:*

$$L(\mathbf{\Gamma}_{n,Pol}) \leq L(\mathbf{\Gamma}_{n+1,Pol}), \text{ for all } n \geq 0. \tag{9.66}$$

Then, the following limit exists (in $[0, +\infty]$) and is given by[20]

$$L(\mathbf{\Gamma}) = \lim_{n \to \infty} L(\mathbf{\Gamma}_{n,Pol}) = \sup_{n \geq 0} L(\mathbf{\Gamma}_{n,Pol}). \tag{9.67}$$

Hence, the curve $\mathbf{\Gamma}$ can be rectifiable or not rectifiable, according to whether the limit (or, equivalently, the supremum in (9.67)) is finite or infinite, respectively.

Remark 9.25.

(a) A key advantage of Theorem 9.24 is that $\mathbf{\Gamma}$ is not assumed to be simple and that the hypotheses (i) and (ii) in Theorem 9.24 are very often satisfied and easy to verify in the case of fractal curves constructed via a self-similar (or else a self-affine) construction—or, more generally, where $\mathbf{\Gamma}$ is the attractor (or invariant set) of a finite family of (possibly nonlinear) maps; see Sections 4.2–4.4. We will repeatedly apply (implicitly or explicitly) Theorem 9.24 (as well as several of the other results stated in this section) later in this chapter.

[20]Note that this first assertion follows at once from the monotonicity of the sequence of nonnegative positive numbers $\{L(\mathbf{\Gamma}_{n,Pol})\}_{n=0}^{\infty}$. Furthermore, since this sequence is nondecreasing, its limit is also given by the supremum appearing on the right-hand side of (9.67).

(b) We close this discussion by referring the interested reader to [**593**, Section 13.7] for a useful criterion to show that a (planar) *self-similar curve* Γ—in the sense of [**593**, Section 13.2], which is more specific than that of Section 4.2 above[21] and obtained as the range of a continuous parametrized curve $\gamma : [a, b] \to \mathbb{R}^2$—is *simple.*

9.3. Interlude: An Introduction to Complex Dimensions and Fractal Zeta Functions

The main goal of this section is to provide a brief introduction to the theory of fractal complex dimensions and fractal zeta functions—as developed by the first author (Michel L. Lapidus) and his collaborators—including, especially, Machiel van Frankenhuijsen, in the research books (published between 2000 and 2013) [**370**], [**374**], and [**375**] (for the 1-dimensional theory, namely, the theory of fractal strings and their complex dimensions), as well as Goran Radunović and Darko Žubrinić, in the 2017 research monograph [**356**] (and a series of accompanying papers, [**357–365**] and [**355**]), for the more recent higher-dimensional theory of complex dimensions, fractal zeta functions, and relative fractal drums (RFDs), which, in particular, can be applied to any bounded subset of m-dimensional Euclidean space $\mathbb{R}^m$, with $m \in \mathbb{N}$ arbitrary.[22]

We refer the interested reader to the first author's survey article, [**329**], entitled *An overview of complex fractal dimensions: From fractal strings to fractal drums, and back*, for an introduction to both aspects of the theory—i.e., the 1-dimensional case of fractal strings and the higher-dimensional case of fractal drums (as well as, more generally, of relatively fractal drums).

Since our emphasis in this section is on the higher-dimensional theory, we refer to Section 8.2 above for a very brief introduction to fractal strings, their geometric zeta functions and their complex dimensions. The sequel to this book, [**119**], will be devoted to providing a much more thorough, but still accessible, introduction to the theory of complex dimensions of fractal strings and drums. See also [**116**] and [**117**] for a significant extension of the theory and for its applications to the study of fractal curves—based, in particular, on the work developed by Claire David and the first author in [**106–115**]; see, especially, [**106–114**].

---

[21] In particular, Γ is a self-similar set, i.e., the attractor (or invariant set) of an IFS consisting of finitely many similarities of $\mathbb{R}^2$.

[22] In fact, the theory developed in [**356**] can be extended (essentially without any modification) to a large class of metric spaces equipped with a measure compatible with the underlying metric—the so-called Ahlfors spaces or spaces of homogeneous type [**98**]; see Sean Watson's thesis, [**606**], along with the paper by the first author of this book and the latter author, [**379**].

9.3.1. Lightning Speed Review of Complex Analysis. We very quickly recall here a few basic definitions and results from complex analysis for functions of one complex variable. General, and quite accessible, references on complex analysis include the books by J. B. Conway [**93**], L. V. Ahlfors [**1**], W. Rudin [**523**], and B. Simon [**556**].

Let $f : U \to \mathbb{C}$ be a complex-valued function of a complex variable, where U is an open subset of $\mathbb{C}$. Then, f is said to be *holomorphic at* $s_0 \in U$ if its *difference quotient at* s_0 (namely, $(f(s) - f(s_0))/(s - s_0)$) converges in $\mathbb{C}$, as $s \to s_0$, with $s \in \mathbb{C}$ and $s \neq s_0$. The corresponding limit is denoted by $f'(s_0)$ and is called the *(complex) derivative of* f *at* s_0; note that, by definition, $f'(s_0) \in \mathbb{C}$. Furthermore, f is said to be *holomorphic on* U if it is holomorphic at every $s_0 \in U$. As is well known and is characteristic of the rigidity of complex analysis, it turns out that if f is holomorphic on U, then all of its (complex) higher-order derivatives $f^{(n)}$ (with $n \in \mathbb{N}_0$ and $f^{(0)} := f$) exist and are continuous on U.

Moreover, if the open set U is connected (i.e., if it is a *domain* of $\mathbb{C}$), a complex-valued function h on U is *meromorphic* if it is the quotient of two holomorphic functions f and g on U, where g is not the zero function on U. (Conversely, even if U is not connected, any such quotient is a meromorphic function on U.) Alternatively, a meromorphic function on U is a complex-valued function that is holomorphic, except on a (possibly empty) discrete (and hence also finite or countable) set $\mathcal{P} = \mathcal{P}(h)$ of points in U,[23] called the *poles* of h (in U), and more precisely defined just below.

Let $h : U\backslash\{\omega\} \to \mathbb{C}$ be holomorphic, where $\omega \in U$ and U is a domain of $\mathbb{C}$. Then, h can have three different types of *isolated singularities* (see, e.g., [**356**, Subsection 1.3.2] and the relevant reference therein):

(i) Either ω is a *removable singularity* of h; i.e., $\lim_{s\to\omega} h(s)$ exists in $\mathbb{C}$, in which case h can be (uniquely) extended to a holomorphic function on U.

(ii) Or ω is a *pole* of h; i.e., there exists a (necessarily unique) smallest positive integer l such that $\lim_{s\to\omega}(s - \omega)^l \cdot h(s)$ exists in $\mathbb{C}\backslash\{0\}$. Then, ω is called a *pole of multiplicity* (or *of order*) l of h.[24] Furthermore, if $l = 1$, ω is a *simple pole* of h and the *residue* of h at ω, denoted $\operatorname{res}(h, \omega)$, is given by the following

[23] By "discrete", we mean that every point of $\mathcal{P}$ is *isolated*; i.e., $\mathcal{P} \subseteq U$ and given any $s_0 \in \mathcal{P}$, there exists a small disk D centered at s_0 and contained in U, such that $D \cap \mathcal{P} = \{s_0\}$. It follows that $\mathcal{P}$ must be at most countable because, for any compact (and thus also bounded) subset K of $\mathbb{C}$, $\mathcal{P} \cap K$ is finite. It also follows that $\mathcal{P} = \mathcal{P}(h)$ is closed, and thus, $U\backslash\mathcal{P}$ is open in $\mathbb{C}$ (as well as in U).

[24] Consequently, we have that $\lim_{s\to\omega} |f(s)| = +\infty$; and thus, one often writes $h(\omega) := \infty$, in the literature, at any pole ω of h. Here, the symbol ∞ stands for the point at infinity in the Riemann sphere $\widetilde{\mathbb{C}} := \mathbb{C} \cup \{\infty\}$.

nonzero complex number:[25]

$$\text{res}(h, \omega) := \lim_{s \to \omega} (s - \omega) h(s). \tag{9.68}$$

If ω is not simple (i.e., if $l \geq 2$), it is called a *multiple pole* of h.

(iii) Or else ω is an *essential singularity* of h; i.e., it is an isolated singularity of h which is neither a removable singularity nor a pole of h. A classic example of this case is given by $U := \mathbb{C}$, $\omega := 0$ and $h(s) := e^{1/s}$, for all $s \in \mathbb{C}$. In order to see this, observe that $h(s)$ is given by an infinite (to the left) Laurent series; namely, we have that $h(s) = \sum_{n=0}^{\infty} (s^{-n} \backslash n!)$, for all nonzero $s \in \mathbb{C}$.

Accordingly, a meromorphic function cannot have essential singularities, and its set of poles, $\mathcal{P}$, is discrete—and thus is at most countable, as well as consists of isolated singularities. It is also closed in $\mathbb{C}$ (as well as in U).[26]

Observe that a holomorphic function f on U is just a meromorphic function on U which does not have any pole in U; i.e., f is a meromorphic function on U such that $\text{res}(f, \omega) = 0$, for all $\omega \in U$—or, equivalently, such that $\mathcal{P} := \mathcal{P}(f) = \emptyset$.

It is also noteworthy that—in light, in particular, of footnote 24—a meromorphic function on $U \subseteq \mathbb{C}$ can be viewed as a continuous function on U with values in the Riemann sphere $\widetilde{\mathbb{C}} := \mathbb{C} \cup \{\infty\}$, equipped with an appropriate metric (the so-called chordal metric); see, e.g., [**375**, Section C.1, p. 506] and [**356**], along with the relevant references therein.

Finally, we wish to recall a very important and useful consequence (for meromorphic functions) of the Principle of Analytic Continuation, a key feature of complex analysis and geometry. Let $h_1, h_2 : U \to \mathbb{C}$. Assume that $h_1 = h_2$ on a connected open set (i.e., a *domain*) U of $\mathbb{C}$ and that h_2 is meromorphic on some connected open set $V \supseteq U$ (i.e., U is a subdomain of V). Then, h_1 can be meromorphically continued (necessarily *uniquely*) to all of V and the identity $h_1 = h_2$ continues to hold in V. This principle—referred to here as the *Principle of Persistence of Identities* or, more briefly, as the *Identity Theorem*—will often be used, most of the time implicitly, in the remainder of this chapter.

---

[25]Henceforth, as well as in the literature, the notation $\text{res}(h, \omega)$ is often used even if the limit on the right-hand side of (9.68) is equal to 0. In this case, ω is *not* a pole of h—and hence, h is holomorphic at ω (provided it is assumed to be meromorphic). It then follows that $\text{res}(h, \omega) = 0$.

[26]In the terminology and with the notation to be introduced in Subsections 9.3.2–9.3.8, it follows that the exact same properties are satisfied by the set $\mathcal{D}_A(U)$ of (visible) complex dimensions of a bounded set $A \subseteq \mathbb{R}^m$. Indeed, we have that $\mathcal{D}_A(U) = \mathcal{P}(\zeta_A)$, where the meromorphic function ζ_A is the distance zeta function of A. (More generally, the same statement is true for any RFD (A, Ω) of $\mathbb{R}^m$.)

Recall that the *Principle of Analytic Continuation*—from which the just-mentioned Identity Theorem follows—can be stated exactly as above, but by replacing "meromorphic" by "holomorphic" throughout.[27] In particular, if a holomorphic function f vanishes identically on a strict subdomain U of the domain V, then f vanishes identically on all of V.

Of course, many other results from complex analysis (and other fields of mathematics, including number theory, harmonic analysis, functional analysis, and distribution theory)—such as Cauchy's Integral Formula, Weierstrass's Theorem for Series (or for Sequences) of Holomorphic Functions, and the Residue Theorem—are needed in order to prove the theorems stated or alluded to in Section 9.3. However, we limit ourselves here to what is needed in order to understand the statements (but not the proofs) of those theorems.

9.3.2. Distance Zeta Function. *From now on, in this section, unless explicitly stated otherwise, A denotes an arbitrary (nonempty) bounded subset of the m-dimensional Euclidean space $\mathbb{R}^m$, where $m \geq 1$ is any positive integer.* Furthermore, as before, given $\boldsymbol{x} \in \mathbb{R}^m$, $\operatorname{dist}(\boldsymbol{x}, A)$ denotes the Euclidean distance from $\boldsymbol{x}$ to A:[28]

$$\operatorname{dist}(\boldsymbol{x}, A) = \inf\{|\boldsymbol{x} - \boldsymbol{a}| \,:\, \boldsymbol{a} \in A\}. \tag{9.69}$$

Note that if A is closed (and thus, in particular, if A is compact), then the infimum in (9.69) is actually equal to zero iff $\boldsymbol{x} \in A$. Indeed, since A is closed, then $\overline{A} = A$ and hence, $\boldsymbol{x} \in A$ iff $\operatorname{dist}(\boldsymbol{x}, A) = 0$; see Exercise 6.8.

In 2009, the first author introduced the following *distance zeta function* of A, denoted by ζ_A; until then, this was the missing piece for developing a useful higher-dimensional theory of complex dimensions.

Fix $\delta > 0$. Then, for any $s \in \mathbb{C}$ with $\operatorname{Re}(s)$ sufficiently large,[29] the *distance zeta function of A* is given by

$$\zeta_A(s) := \int_{A_\delta} \operatorname{dist}(\boldsymbol{x}, A)^{s-m} \, d\boldsymbol{x}, \tag{9.70}$$

---

[27] It can be stated much more generally; for example, it would suffice to assume that the identity $h_1(s) = h_2(s)$ holds for all s in a subset of U having an accumulation point in U. However, we will not need this extension here.

[28] Here, $|\boldsymbol{x} - \boldsymbol{a}| = d_m(\boldsymbol{x}, \boldsymbol{a})$, the Euclidean distance from $\boldsymbol{x}$ to $\boldsymbol{a}$, as defined in (4.1).

[29] For notational simplicity, we let $\operatorname{dist}(\boldsymbol{x}, A)^z := (\operatorname{dist}(\boldsymbol{x}, A))^z$, for all $\boldsymbol{x} \in \mathbb{R}^m$ and all $z \in \mathbb{C}$. For any $z \in \mathbb{C}$ with $\operatorname{Re}(z) > 0$, it is natural to let $0^z := 0$; as is justified in [**356**, Chapter 2], we continue to use this same convention even if $\operatorname{Re}(z) \leq 0$. Consequently, we use the convention according to which $0^z := 0$, for any $z \in \mathbb{C}$. Alternatively, without loss of generality (see [**356**, Proposition 2.1.22, p. 66]), in (9.70), one could replace A_δ by $A_\delta \setminus \overline{A}$, the open subset of A_δ on which the function $\operatorname{dist}(\cdot, A)$ is strictly positive; indeed, $\overline{A} = \{\boldsymbol{x} \in \mathbb{R}^m \,:\, \operatorname{dist}(\boldsymbol{x}, A) = 0\}$.

where the integral on the right-hand side of (9.70) is interpreted as a Lebesgue integral (see Section 3.6 above) and where, as before,

$$A_\delta := \{x \in \mathbb{R}^m : \operatorname{dist}(x, A) < \delta\} \tag{9.71}$$

stands for the (open) δ-neighborhood of A.[30]

Notation 9.26. Here and thereafter, given $\alpha \in \mathbb{R} \cup [\pm\infty]$, we use the shorthand notation $\{\operatorname{Re}(s) > \alpha\}$ for the open right half-plane $\{s \in \mathbb{C} : \operatorname{Re}(s) > \alpha\}$. By convention, if $\alpha = -\infty$ (resp., if $\alpha = +\infty$), this "half-plane" stands for all of $\mathbb{C}$ (resp., for the empty set $\emptyset$).

We use a similar notation, $\{\operatorname{Re}(s) \geq \alpha\}$, for the closed right half-plane $\{s \in \mathbb{C} : \operatorname{Re}(s) \geq \alpha\}$, for example.

Remark 9.27.

(a) (*The Choice of δ is Unimportant*). The choice of $\delta > 0$ in the definition of $\zeta_A = \zeta_{A,\delta}$ in (9.70) is unimportant, from the point of view of the theory of complex dimensions (which are poles or singularities of ζ_A). Indeed, if we replace δ by $\delta^* > \delta$, say, then it is easy to see that $\zeta_{A,\delta^*} - \zeta_{A,\delta}$ is an entire function (i.e., it is holomorphic in all of $\mathbb{C}$). Hence, the poles of $\zeta_{A,\delta}$ (i.e., the complex dimensions of A) are independent of the choice of $\delta > 0$ and so are the residues (or, more generally, the principal parts, in the case of multiple poles) at those poles. In practical computations of ζ_A, however, it is often convenient to choose $\delta > 0$ appropriately, in order to take into account the geometry and topology of A.

An entirely similar comment can be made about the tube zeta function of A, denoted $\tilde{\zeta}_A$ and to be discussed in Subsection 9.3.3 below. We note that parts (b) and (c) just below also have an obvious counterpart for $\tilde{\zeta}_A$.

(b) (*Dirichlet-Type Integral*). Because the Lebesgue integral defining ζ_A in (9.70) is a so-called *Dirichlet-type integral* (see [**356**, Appendix A]), it admits an *abscissa of (absolute) convergence*, σ_A, defined by any of the following four equalities:

(9.72)

$$\begin{aligned}
\sigma_A &:= \inf\left\{\alpha \in \mathbb{R} : \int_{A_\delta} \operatorname{dist}(x, A)^{\alpha - m} dx < \infty\right\} \\
&= \inf\left\{\operatorname{Re}(s) : s \in \mathbb{C}, \int_{A_\delta} \operatorname{dist}(x, A)^{\operatorname{Re}(s) - m} dx < \infty\right\} \\
&= \inf\left\{\operatorname{Re}(s) : s \in \mathbb{C}, \int_{A_\delta} |\operatorname{dist}(x, A)^{s - m}| dx < \infty\right\} \\
&= \inf\{\operatorname{Re}(s) : s \in \mathbb{C}, x \mapsto \operatorname{dist}(x, A)^{s-m} \text{ is Lebesgue integrable on } A_\delta\}.
\end{aligned}$$

[30] In the present context of Euclidean space $\mathbb{R}^m$, it turns out that whether one takes an open (A_δ) or a closed ($[A_\delta]$) δ-neighborhood of A is unimportant.

It also follows that ζ_A is holomorphic (i.e., complex differentiable) for $\mathrm{Re}(s) > \sigma_A$. As we will soon see, σ_A coincides with the upper Minkowski dimension of A, denoted $\overline{D_A}$ and defined in Section 8.1.[31] In particular,[32] we have that

$$0 \leq \sigma_A \leq m. \tag{9.73}$$

Then, the open right half-plane $\{\mathrm{Re}(s) > \overline{D_A}\}$ is called the *half-plane of convergence of* ζ_A.

(c) ($\zeta_A = \zeta_{\overline{A}}$). Since $\mathrm{dist}(\boldsymbol{x}, A) = \mathrm{dist}(\boldsymbol{x}, \overline{A})$, for all $\boldsymbol{x} \in \mathbb{R}^m$ (see Exercise 9.28 below), we also have (by (9.70)) that $\zeta_A = \zeta_{\overline{A}}$. Hence, replacing A by its closure $\overline{A}$ does not change anything; in particular, *given any bounded set $A \subseteq \mathbb{R}^m$, A and its closure $\overline{A}$ have the same complex dimensions*: $\zeta_A = \zeta_{\overline{A}}$; hence, in the notation of Subsection 9.3.4 below, we also have that $\mathcal{D}_A(U) = \mathcal{D}_{\overline{A}}(U)$ (with the same multiplicities). Consequently, without loss of generality, we may assume here that $A \subseteq \mathbb{R}^m$ is compact, i.e., that A is closed and bounded in $\mathbb{R}^m$.

The following easy exercise is useful, in light of the above definition and comments.

Exercise 9.28. Let $A \subseteq \mathbb{R}^m$ be nonempty.

(a) Show that $\mathrm{dist}(\boldsymbol{x}, A) = \mathrm{dist}(\boldsymbol{x}, \overline{A})$, for all $\boldsymbol{x} \in \mathbb{R}^m$.

(b) Furthermore, show that the function $\phi : \mathbb{R}^m \to [0, +\infty]$ given by $\phi(\boldsymbol{x}) := \mathrm{dist}(\boldsymbol{x}, A)$, for all $\boldsymbol{x} \in \mathbb{R}^m$, is continuous—and hence is also Borel measurable (in the sense of Chapter 3).

(c) Finally, show that the identity $\zeta_A = \zeta_{\overline{A}}$, stated in part (c) of Remark 9.27 just above, holds.

The first main result concerning ζ_A is the following theorem, [**356**, Theorem 2.1.11, p. 57]; see also [**357**].

Theorem 9.29 (Minkowski Dimension, Abscissa of Convergence, and Holomorphicity; M. L. Lapidus, G. Radunović, and D. Žubrinić, [**356**] and [**357**], 2017). *Let A be any (nonempty) bounded subset of $\mathbb{R}^m$.* Then:

(i) *The upper Minkowski dimension of A, denoted $\overline{D_A} := \overline{\dim}_M(A)$, coincides with the abscissa of convergence, σ_A, of ζ_A (as defined in part* (b) *of Remark* 9.27; *see* (9.72)):

$$\overline{D_A} = \sigma_A. \tag{9.74}$$

It follows, in particular, that $0 \leq \overline{D_A} \leq m$.

[31] Most of the time, in Chapter 8, $\overline{D_A}$ is denoted by $\overline{\dim}_M(A)$.

[32] A priori, $\sigma_A \in [-\infty, +\infty]$, with the obvious conventions concerning $\sigma_A = -\infty$ and $\sigma_A = +\infty$, but it turns out that for the present case of a bounded set $A \subseteq \mathbb{R}^m$, the finite bounds in (9.73) hold.

(ii) *The distance zeta function* ζ_A *is holomorphic for* $\mathrm{Re}(s) > \sigma_A$ *and its* (complex) *derivative can be obtained in this same half-plane by differentiating under the integral sign on the right-hand side of* (9.70), *and similarly for the higher-order (complex) derivatives of* ζ_A.

(iii) *If the Minkowski dimension of A, denoted* $D_A := \dim_M(A)$, *exists,*[33] $\mathcal{M}_{*,A} := \mathcal{M}_*^{D_A} > 0$,[34] *and* $D_A < m$ *(i.e.,* $D_A \neq m$*),*[35] *then* $\zeta_A(s) \to +\infty$, *as* $s \to D_A$, *with* $s \in \mathbb{R}$ *and* $s > D_A$.

Hence, D_A *is a singularity of* ζ_A, *in this case—and so,* $\{\mathrm{Re}(s) > \sigma_A\} = \{\mathrm{Re}(s) > D_A\}$ *is the largest open half-plane to which* ζ_A *can be holomorphically continued; it is called the* half-plane of holomorphic continuation of ζ_A. *Thus, under the present hypotheses* (of part (iii)), the half-plane of convergence of ζ_A coincides with the half-plane of holomorphic continuation of ζ_A.

9.3.3. Tube Zeta Function and Functional Equation. The other main fractal zeta function is the *tube zeta function of A*, denoted $\tilde{\zeta}_A$ and initially defined for Re(s) sufficiently large (in fact, for $\mathrm{Re}(s) > \overline{D_A}$), as well as for any given $\delta > 0$, by[36]

$$\tilde{\zeta}_A(s) = \int_0^\delta t^{s-m-1} \cdot V_A(t)dt = \int_0^\delta t^{s-m} \cdot V_A(t)\frac{dt}{t}. \tag{9.75}$$

Here, the *tube function* of A, denoted $V_A = V_A(t)$, is defined for any $t > 0$ by the volume of the t-neighborhood of A:

$$V_A(t) := \mathcal{L}^m(A_t). \tag{9.76}$$

Note that, essentially, and in light of (9.75), $\tilde{\zeta}_A$ *is the (truncated, m-dimensional)* Mellin transform *of the tube function* V_A.

The following key property is also established in loc. cit.; see [**356**, equation (2.2.23), p. 119]. The distance and tube zeta functions, $\zeta_A = \zeta_{A,\delta}$ and $\tilde{\zeta}_A = \tilde{\zeta}_{A,\delta}$—defined for the same value of $\delta > 0$ in (9.70) and (9.75), respectively—are connected via the following *functional equation*:

$$\zeta_A(s) = (m - s)\tilde{\zeta}_A(s) + \delta^{s-m} \cdot V_A(\delta). \tag{9.77}$$

Furthermore, given a domain $U \subseteq \mathbb{C}$ containing the closed half-plane $\{\mathrm{Re}(s) \geq \overline{D_A}\}$, and provided $\overline{D_A} < m$ (i.e., $\overline{D_A} \neq m$), then ζ_A admits a (necessarily unique) meromorphic continuation to all of U iff $\tilde{\zeta}_A$ does. Moreover,

[33] According to Definition 8.3 in Section 8.1, it means that the lower and upper Minkowski dimensions of A coincide: in short, $\underline{D_A} = \overline{D_A}$.

[34] Here, $\mathcal{M}_*^{D_A}$ denotes the *lower Minkowski content* of A; see Section 8.1 above.

[35] These conditions are satisfied by all of the fractal curves which will be studied in this chapter, except, of course, by plane-filling curves—such as the Peano and Hilbert curves (see Subsection 1.4.3 and Remark 9.63), for which $D_A = m := 2$.

[36] Observe that dt/t is the natural translation-invariant measure (or Haar measure) of the abelian multiplicative group $(0, +\infty)$.

in this case, the identity (i.e., the functional equation) (9.77) continues to hold for all $s \in U$. In particular, with the terminology introduced in Subsection 9.3.4 below, *the visible complex dimensions of A (relative to the domain $U \subseteq \mathbb{C}$) are the same, whether they are defined by means of the distance zeta function ζ_A or by means of the tube zeta function $\tilde{\zeta}_A$.*

We should also point out that the exact counterpart of Theorem 9.29 above holds without change,[37] except for ζ_A replaced by $\tilde{\zeta}_A$ and σ_A replaced by $\tilde{\sigma}_A$, the *abscissa of convergence* of the Dirichlet-type integral $\tilde{\zeta}_A$ in (9.75).[38] (See [**356**, Theorem 2.2.11, p. 121].) In particular, we have that $\overline{D_A} = \tilde{\sigma}_A$ ($= \sigma_A$, as was stated in (9.74)).

Remark 9.30 (ζ_A *vs.* $\tilde{\zeta}_A$). The two *fractal zeta functions,* ζ_A and $\tilde{\zeta}_A$, contain essentially the same information (in particular, about the complex dimensions of A; see Subsection 9.3.4 below). For certain problems or classes of examples, however, it is more convenient to use one rather than the other—and then to deduce the other fractal zeta function by means of the above functional equation (9.77).[39] This comment will be illustrated in the relevant parts of Sections 9.4–9.7 below, about diverse examples of fractal curves.

9.3.4. Complex Dimensions. By definition, the complex dimensions of A are the poles of the meromorphic continuation of ζ_A (when it exists). More specifically, given a connected open set U (i.e., a domain of $\mathbb{C}$) containing the closed half-plane $\{\mathrm{Re}(s) \geq \overline{D_A}\}$ and such that ζ_A admits a meromorphic continuation (necessarily unique and still denoted ζ_A) to all of U,[40] the *visible complex dimensions* of A (in U) are the poles of ζ_A belonging to U. We denote by $\mathcal{D}_A(U)$ the resulting *set of visible complex dimensions.*[41]

Since ζ_A is holomorphic in $\{\mathrm{Re}(s) > \overline{D_A}\}$ (see parts (i) and (ii) of Theorem 9.29 in Subsection 9.3.2 above), we have the inclusion $\mathcal{D}_A(U) \subseteq \{\mathrm{Re}(s) \leq \overline{D_A}\}$.

If—as will be the case for all of the examples of fractal curves to be studied in Sections 9.4–9.7 below—ζ_A (and hence also $\tilde{\zeta}_A$) admits a meromorphic continuation to all of $\mathbb{C}$, then we can let $U := \mathbb{C}$ in the above definition. The resulting poles of ζ_A (in $\mathbb{C}$) are then simply called the *complex dimensions* of A and the resulting *set of complex dimensions* is denoted $\mathcal{D}_A := \mathcal{D}_A(\mathbb{C})$.

---

[37] A minor point is that in the analog (for $\tilde{\zeta}_A$) of part (iii) of Theorem 9.29, however, one no longer needs to assume that $D_A < m$.

[38] Note that $\tilde{\sigma}_A$ is defined exactly as in (9.72), except for the integral initially defining ζ_A in (9.70) replaced by that defining $\tilde{\zeta}_A$ in (9.75).

[39] This is the case, for instance, in order to obtain a fractal tube formula or to calculate the complex dimensions.

[40] The uniqueness of the meromorphic continuation of ζ_A to the domain U (when it exists) is a consequence of the principle of Analytic Continuation (for meromorphic functions)—or, equivalently, of the Identity Theorem; see the end of Subsection 9.3.1 above.

[41] See Subsection 9.3.1 above for a reminder of the definition of the poles of a meromorphic function.

In light of the functional equation (9.77) connecting ζ_A and $\tilde{\zeta}_A$, and provided $\overline{D_A} < m$ (i.e., $\overline{D_A} \neq m$), the distance and tube zeta functions have exactly the same poles in U (see Subsection 9.3.3 above) and hence define the same (visible) complex dimensions. Thus, if $\overline{D_A} < m$, the complex dimensions of A can be defined either via ζ_A or via $\tilde{\zeta}_A$. Also, ζ_A has a meromorphic continuation to U iff A does and then, the corresponding meromorphic continuations of ζ_A and of $\tilde{\zeta}_A$ are still connected by the functional equation (9.77). Furthermore, in this case (see [**356**, equation (12.2.55), p. 127]),[42] we have that

$$\text{res}(\zeta_A, \omega) = (m - \omega)\,\text{res}(\tilde{\zeta}_A, \omega). \tag{9.78}$$

If $\text{res}(\zeta_A, \omega) \neq 0$ (i.e., if ω is a simple pole of ζ_A), then (still assuming that $\overline{D_A} < m$), we have that $\text{res}(\tilde{\zeta}_A, \omega) \neq 0$ and so, ω is also a simple pole of $\tilde{\zeta}_A$. (The converse is also true, provided $\omega \neq m$.) We just say that ω is a *simple complex dimension* of A.[43]

We close this subsection by providing one more useful definition. Assume that $\overline{D_A} < m$ (i.e., $\overline{D_A} \neq m$) and that ζ_A—and hence also $\tilde{\zeta}_A$—admits a (necessarily unique) meromorphic continuation to a connected open set (i.e., a domain) $U \supseteq \{\text{Re}(s) \geq \overline{D_A}\}$, as above. Then, $\mathcal{D}_{A,pr}$, the *set of principal complex dimensions* of A, is the set of complex dimensions of A in U with real part $\overline{D_A}$:[44]

$$\begin{aligned} \mathcal{D}_{A,pr} &= \{\omega \in U : \zeta_A(\omega) = \infty,\ \text{Re}(\omega) = \overline{D_A}\} \\ &= \{\omega \in U : \tilde{\zeta}_A(\omega) = \infty,\ \text{Re}(\omega) = \overline{D_A}\}. \end{aligned} \tag{9.79}$$

As can be easily checked, $\mathcal{D}_{A,pr}$ is then independent of the choice of the domain U satisfying the same conditions as above.

Finally, we observe that, under the hypotheses of part (iii) of Theorem 9.29 above, if ζ_A (and hence also $\tilde{\zeta}_A$) can be meromorphically continued to a neighborhood of D_A $(= \overline{D_A})$ (in U), *then, D_A, the Minkowski dimension of A, is the (visible and necessarily unique) complex dimension of A with maximal real part, located on the real axis*:

$$D_A = \max\{\text{Re}(\omega) : \omega \in \mathcal{D}_A(U)\}. \tag{9.80}$$

9.3.5. Definition of Fractality and Geometric Interpretation. Within the theory of complex dimensions (see, e.g., [**370**], [**374**], and [**375**], for the $m = 1$ case, and [**356**], [**329**], and [**330**], for the general case when $m \geq 1$), one can give a very general and rather accurate *definition of fractality as the presence of* nonreal *complex dimensions*. More specifically, a bounded set $A \subseteq \mathbb{R}^m$ is

---

[42] Here and henceforth, as in Subsection 9.3.1, $\text{res}(\zeta_A, \omega)$ denotes the *residue* of ζ_A at $\omega \in U$. Recall that if ω is a simple pole of ζ_A, $\text{res}(\zeta_A, \omega) = \lim_{s\to\omega}(s-\omega)\zeta_A(s)$, with the variable s constrained to remain in U.

[43] For the simplicity of the exposition, we will not discuss here in detail the case of multiple poles (i.e., of *multiple complex dimensions*), except to mention that the principal parts of ζ_A and $\tilde{\zeta}_A$ are then related by a suitable analog of (9.78); see [**356**].

[44] We use here the notation introduced in footnote 24 of Subsection 9.3.1 above.

said to be *fractal* if it admits at least one *nonreal* complex dimension, i.e., if $\mathcal{D}_A(U) \cap (\mathbb{C}\backslash\mathbb{R}) \neq \emptyset$.[45]

Also, A is said to be *critically* (or *principally*) *fractal* if it admits at least one *nonreal* complex dimension *with real part* $\overline{D_A}$, i.e., if $\mathcal{D}_{A,pr} \cap (\mathbb{C}\backslash\mathbb{R}) \neq \emptyset$.

Finally, given $d \in \mathbb{R}$, A is said to be *fractal in dimension* d if it admits at least one nonreal complex dimension with real part equal to d.[46] Furthermore, it is said to be *subcritically fractal* if it is fractal but *not* critically fractal.

These definitions will be illustrated and motivated on multiple occasions in Sections 9.4–9.7 below. For example, we will see that the Koch curve, the Koch snowflake curve, the Sierpinski gasket, the Sierpinski carpet, and the Weierstrass curve are all critically (or principally) fractal. Furthermore, the Devil's staircase (i.e., the graph of the Cantor function) is fractal (in the above sense) whereas contrary to one's geometric intuition, it is not fractal according to Mandelbrot's definition; see Subsection 9.4.6 below. On the other hand, the Devil's staircase is subcritically fractal; indeed, it is *not* critically fractal, because it has Minkowski dimension 1 but its nonreal complex dimensions have real part $\log_3 2 < 1$.

In addition, we will see in Section 9.7 (especially in Subsection 9.7.3) that the Weierstrass curve is both critically (or principally) fractal and fractal in infinitely many (real) dimensions d_k, with $d_k \to -\infty$, as $k \to -\infty$, a fact which reflects the great geometric complexity of this type of fractal curve.

As for the Peano curve and other plane-filling curves, it is fractal (and, more precisely, subcritically fractal) in a rather subtle way, when viewed from a higher-dimensional perspective (say, in $\mathbb{R}^3$ instead of in $\mathbb{R}^2$, here); see Remark 9.63 at the end of Subsection 9.4.7 below. Once again, this fact perfectly fits our geometric intuition gained when looking at the construction of plane-filling curves. An entirely similar comment can be made about higher-dimensional space-filling curves.

The geometric interpretation of the complex dimensions is that they correspond in a very faithful way to the *intrinsic geometric oscillations* of fractal-like objects. One can imagine a *"geometric wave"* propagating through the fractal whose *amplitudes* are governed by *the real parts of the complex dimensions* and whose *natural frequencies* are governed by *the imaginary parts of the complex dimension*; *see* [**329**] and [**330**].

---

[45]Observe that *nonreal* complex dimensions necessarily come in complex conjugate pairs. Indeed, $\zeta_A(\bar{s}) = \overline{\zeta_A(s)}$, and similarly for $\tilde{\zeta}_A(s)$. These identities can easily be deduced from (9.70) and (9.75), respectively, followed by an application of the Identity Theorem stated towards the end of Subsection 9.3.2 above.

[46]Note that, here, $d \in \mathbb{R}$ is not assumed to be a (visible) complex dimension of A.

Accordingly, the above definition of fractality, given in terms of the presence of nonreal complex dimensions, can be interpreted as *the presence of intrinsic geometric oscillations or vibrations* within the underlying space of scales or within the fractal itself.

9.3.6. Residues and Ordinary or Average Minkowski Content. Let A be a (nonempty) bounded subset of $\mathbb{R}^m$ which is Minkowski nondegenerate (i.e., $0 < \mathcal{M}_* \leq \mathcal{M}^* < \infty$), where $\mathcal{M}_* := \mathcal{M}_*^{D_A}(A)$ and $\mathcal{M}^* := \mathcal{M}^{*,D_A}(A)$ denote, respectively, the lower Minkowski content and the upper Minkowski content of A. Recall from Section 8.1 that this hypothesis implies that the Minkowski dimension of A, $D_A = \dim_M(A)$, exists. Further assume that ζ_A admits a meromorphic extension to an open and connected neighborhood of D_A. Then, provided $D_A < m$ (i.e., $D_A \neq m$), we deduce that D_A is a simple pole of ζ_A and

$$(m - D_A)\mathcal{M}_* \leq \operatorname{res}(\zeta_A, D_A) \leq (m - D_A)\mathcal{M}^*. \tag{9.81}$$

Moreover, the value of $\operatorname{res}(\zeta_A, D_A)$, the residue of ζ_A at D_A, is independent of the choice of $\delta > 0$ (in the initial definition of ζ_A, given by (9.70)).

In particular, if A is Minkowski measurable, with Minkowski content $\mathcal{M}$ (i.e., if $\mathcal{M}_* = \mathcal{M}^* =: \mathcal{M}$, here), then

$$\operatorname{res}(\zeta_A, D_A) = (m - D_A)\mathcal{M}. \tag{9.82}$$

This is the content of [**356**, Theorem 2.2.3, p. 114].[47]

According to [**356**, Theorem 2.2.14, p. 123], and in light of the identity (9.78) above, the exact same result holds for $\tilde{\zeta}_A$ (instead of for ζ_A, but now without the need to assume that $D_A < m$), except with (9.81) and (9.82) respectively replaced by

$$\mathcal{M}_* \leq \operatorname{res}(\tilde{\zeta}_A, D_A) \leq \mathcal{M}^* \quad \text{and} \tag{9.83}$$

$$\operatorname{res}(\tilde{\zeta}_A, D_A) = \mathcal{M}. \tag{9.84}$$

We next consider a situation and a result that will be encountered and used several times in the remainder of this chapter, in the case when A is not Minkowski measurable; see, e.g., Subsections 9.6.3 and 9.7.3. The corresponding result is the content of [**356**, Theorem 2.3.25, p. 157].[48]

Theorem 9.31 (Residue and Average Minkowski Content: Minkowski Nonmeasurable Case). *Assume that there exist constants $D \geq 0$ and $\alpha > 0$, along with $G : \mathbb{R} \to (0, +\infty)$, a* nonconstant *$T$-periodic function (with minimal period $T > 0$), such that*

$$V_A(\varepsilon) := \mathcal{L}^m(A_\varepsilon) = \varepsilon^{m-D} \cdot (G(\log(\varepsilon^{-1})) + o(\varepsilon^\alpha)), \text{ as } \varepsilon \to 0^+. \tag{9.85}$$

[47] For notational simplicity, we write here $\mathcal{M}_*, \mathcal{M}^*$, and $\mathcal{M}$, as above, instead of $\mathcal{M}_*^{D_A}, \mathcal{M}^{*,D_A}$, and $\mathcal{M}^{D_A}$ (as in Section 8.1), respectively.

[48] By necessity of concision, we only give the part of the theorem that is most relevant to us.

Then, G is continuous, the Minkowski dimension of $A, D_A := \dim_M(A)$*, exists, and* $D_A = D$*. Furthermore, A is Minkowski nondegenerate but* not *Minkowski measurable, while its lower and upper Minkowski contents are respectively given by*

$$0 < \mathcal{M}_* = \min_{[0,T]} G \quad \text{and} \quad \mathcal{M}^* = \max_{[0,T]} G < \infty, \tag{9.86}$$

which implies that the range of G on $[0,T]$ *is equal to* $[\mathcal{M}_*, \mathcal{M}^*]$*. Moreover,* $\tilde{\zeta}_A$ *can be (necessarily uniquely) meromorphically continued to (at least) the open right half-plane* $\{\mathrm{Re}(s) > D_A - \alpha\}$*.*

Finally, the average Minkowski content $\widetilde{\mathcal{M}}$ *of A (as given in Definition* 9.32 *below) exists in* $(0,+\infty)$ *and coincides with the residue of* $\tilde{\zeta}_A$ *at* $D = D_A$*, as well as with the mean value of the T-periodic function G (over* $[0,T]$*):*

$$\mathrm{res}(\tilde{\zeta}_A, D_A) = \widetilde{\mathcal{M}} = \frac{1}{T}\int_0^T G(t)dt. \tag{9.87}$$

In particular, we have that $0 < \mathcal{M}_* < \widetilde{\mathcal{M}} < \mathcal{M}^* < \infty$*.*

We next provide the definition of the average Minkowski content, as used in (9.87) of Theorem 9.31 just above; see [**356**, Subsection 2.4.2] for more information.

Definition 9.32 (*Average Minkowski Content; see* [**356**, Definition 2.4.1, p. 178]). Assume that $A \subseteq \mathbb{R}^m$ is bounded and Minkowski nondegenerate. Fix $\delta > 0$. Then, the *average Minkowski content* (or *Cesàro logarithmic mean*) of A, denoted $\widetilde{\mathcal{M}} = \widetilde{\mathcal{M}}_A$, is said to *exist* (in $(0,+\infty)$) if the following limit exists (in $(0,+\infty)$):

$$\widetilde{\mathcal{M}} := \lim_{\tau \to +\infty} \frac{1}{\log \tau} \int_{1/\tau}^{\delta} t^{D_A - m} \cdot V_A(t) \frac{dt}{t}. \tag{9.88}$$

The set A is then said to be (*logarithmically*) *Cesàro Minkowski measurable.*

The value of $\widetilde{\mathcal{M}} = \widetilde{\mathcal{M}}_A$ is independent of the choice of $\delta > 0$, so that, without loss of generality, we may choose $\delta := 1$ in (9.88).

We close this subsection by mentioning that a *Minkowski measurability criterion* is obtained in [**356**], extending to higher dimensions the corresponding criterion obtained in [**375**, Section 8.3] for fractal strings (i.e., when $m = 1$). Roughly speaking, it can be stated as follows:

Under suitable hypotheses and given any fixed $m \in \mathbb{N}$, the bounded set $A \subseteq \mathbb{R}^m$ is Minkowski measurable iff it does *not* have any *nonreal* complex dimension with real part D_A, the Minkowski dimension of A, i.e., iff its set of principal complex dimensions, $\mathcal{D}_{A,pr}(U)$, is reduced to $\{D_A\}$: $\mathcal{D}_{A,pr}(U) = \{D_A\}$.

Intuitively—and in light of the fractal tube formulas to be discussed in Subsection 9.3.7 just below—it means that there are *no leading oscillations* (necessarily of order D_A) in the geometry of A. See [**356**, Theorems 5.4.20 and 5.4.25 in Subsection 5.4.3] for a precise statement of this criterion, along with [**362**] for further refinements and extensions.[49]

9.3.7. Fractal Tube Formulas. Building on the work on general fractal explicit formulas and fractal tube formulas for fractal strings (i.e., $m := 1$) developed by the first author and M. van Frankenhuijsen in [**370, 374, 375**] (see, especially, [**375**, Chapters 5 and 8]),[50] the first author, G. Radunović, and D. Žubrinić developed in [**356**] (and in [**361**]) an extensive higher-dimensional theory of fractal tube formulas, both for general bounded subsets $A \subseteq \mathbb{R}^m$ and for the more general case of relative fractal drums (RFDs), to be briefly discussed in Subsection 9.3.8. See, especially, [**356**, Chapter 5], along with [**361**].

These fractal tube formulas give the true geometric meaning of the complex dimensions and of the notion of fractality described in terms of geometric waves, multiplicative oscillations (or vibrations), as well as amplitudes and frequencies, towards the end of Subsection 9.3.5 above. They will be illustrated by many examples in [**119**], [**116, 117**], the sequels to this volume, as well as in the remainder of this chapter (especially Sections 9.4–9.6).

For now, by necessity of concision, we limit ourselves to a minimalist presentation of the theory, reduced to a statement of a few of the resulting general formulas, and mostly in the case of simple complex dimensions. We also limit ourselves to the case when $A \subseteq \mathbb{R}^m$ is bounded, although the corresponding statements would be identical in the more general case of RFDs discussed in Subsection 9.3.8 below.

Under suitable "languidity conditions" (roughly, under appropriate polynomial growth conditions on the fractal zeta function ζ_A or $\tilde{\zeta}_A$) and under some additional mild hypotheses, one obtains a (pointwise or distributional) exact or asymptotic *fractal tube formula* for the tube function of A (possibly viewed as a generalized function or distribution),

$$V_A = V_A(\varepsilon) := \mathcal{L}^m(A_\varepsilon), \text{ for } \varepsilon > 0, \tag{9.89}$$

enabling us to express it as a (typically countably infinite) sum over the set $\mathcal{D}_A(U)$ of all of the (visible) complex dimensions of A.

[49] See also [**356**, Theorems 5.4.2 and 5.4.15], along with [**362**], for, respectively, sufficient and necessary conditions for Minkowski measurability.

[50] See also the extension of the fractal tube formulas to certain higher-dimensional cases (the Koch curve, along with monophase self-similar sprays and tilings) obtained by the first author and E. P. J. Pearse in [**347–349**], as well as—in the case of general (multiphase) fractal sprays (in the sense of [**354**])—by the same authors and S. Winter in [**350, 351**].

For example, in the case of the tube zeta function $\tilde{\zeta}_A$, it takes the following form, for all $\varepsilon > 0$ sufficiently small (see [**356**, Sections 5.1 and 5.2]):[51]

$$V_A(\varepsilon) = \sum_{\omega \in \mathcal{D}_A(U)} \operatorname{res}(\varepsilon^{m-s} \cdot \tilde{\zeta}_A(s), \omega) + R(\varepsilon), \tag{9.90}$$

where the *error term* $R = R(\varepsilon)$ can be estimated in terms of a suitable power of ε; namely, $R(\varepsilon) = o(\varepsilon^{m-\overline{D}_A})$, as $\varepsilon \to 0^+$. When $\tilde{\zeta}_A$ is "strongly languid", then $R(\varepsilon) \equiv 0$ and the resulting fractal tube formula is said to be *exact*; in this case, we can take $U := \mathbb{C}$ and so, $\mathcal{D}_A(U) := \mathcal{D}_A(\mathbb{C}) = \mathcal{D}_A$, the set of all complex dimensions of A (see Subsection 9.3.4 above).

Then, in the special case of most interest to us (in the remainder of this chapter) of an exact, pointwise fractal tube formula, and when $\tilde{\zeta}_A$ has only *simple poles* (see [**356**, Theorem 5.1.16, p. 427]), the fractal tube formula (9.90) becomes[52]

$$V_A(\varepsilon) = \sum_{\omega \in \mathcal{D}_A} \operatorname{res}(\tilde{\zeta}_A, \omega) \cdot \varepsilon^{m-\omega}. \tag{9.91}$$

In this case, we assume, in particular, that $\tilde{\zeta}_A$ admits a meromorphic continuation to all of $\mathbb{C}$—and hence, we let $U := \mathbb{C}$; recall that $\mathcal{D}_A := \mathcal{D}_A(\mathbb{C})$.

Similarly, and under appropriate and slightly different assumptions, one can also obtain corresponding fractal tube formulas via the distance zeta function ζ_A (instead of via the tube zeta function $\tilde{\zeta}_A$); see [**356**, Section 5.3] or [**361**]. Then, the counterpart of (9.90) takes the following form:[53]

$$V_A(\varepsilon) = \sum_{\omega \in \mathcal{D}_A(U)} \operatorname{res}\left(\frac{\varepsilon^{m-s}}{m-s} \cdot \zeta_A(s), \omega\right) + R(\varepsilon), \tag{9.92}$$

while in the case when ζ_A has only simple poles and for the exact, pointwise fractal formula now expressed via ζ_A (instead of $\tilde{\zeta}_A$), the counterpart of (9.91) becomes (see [**356**, Theorem 5.3.16 and Corollary 5.3.17])[54]

$$V_A(\varepsilon) = \sum_{\omega \in \mathcal{D}_A(U)} \frac{\varepsilon^{m-\omega}}{m-\omega} \cdot \operatorname{res}(\zeta_A, \omega). \tag{9.93}$$

This concludes our brief discussion of fractal tube formulas.

---

[51] Recall from our discussion in Subsection 9.3.4 that $\mathcal{D}_A(U)$, being the set of poles of a meromorphic function, is a discrete subset of U (i.e., consists of isolated points in U)—and hence is at most countable.

[52] For a meromorphic function f, we typically write $\operatorname{res}(f, \omega)$ instead of $\operatorname{res}(f(s), \omega)$, when no ambiguity may arise.

[53] Here, the error term $R(\varepsilon)$ still satisfies the asymptotic estimate $R(\varepsilon) = o(\varepsilon^{m-\overline{D}_A})$, as $\varepsilon \to 0^+$—and, in particular, we have that $R(\varepsilon) \equiv 0$ and $U := \mathbb{C}$ (resulting in an *exact* fractal tube formula) if, in addition, ζ_A is strongly languid.

[54] Recall from the identity (9.78) in Subsection 9.3.4 that $\operatorname{res}(\tilde{\zeta}_A, \omega) = \operatorname{res}(\zeta_A, \omega)/(m - \omega)$, for any simple pole ω of ζ_A (or, equivalently, of $\tilde{\zeta}_A$). We assume here that $D_A < m$.

Remark 9.33 (*Analogy Between Fractal Tube Formulas and Steiner's Formula: Fractal Curvatures and Complex Dimensions*). There is an important analogy between the fractal tube formulas (such as those obtained in [**356, 370, 374, 375**]) and the celebrated Steiner formula for nonempty compact convex subsets K of Euclidean space $\mathbb{R}^m$, with $m \geq 2$. Indeed, the latter formula expresses the volume $V(\varepsilon)$ of the ε-neighborhoods of K as a polynomial of degree m in ε, with coefficients interpreted as suitable curvatures, while fractal tube formulas express $V(\varepsilon)$ (the tube function of a given fractal or of a given relative fractal drum in $\mathbb{R}^m$) as an appropriate fractal power series in ε, with exponents the associated complex codimensions (i.e., $m - \omega$, where ω runs through the set of possible complex dimensions—assumed to be simple here, for simplicity).

This analogy suggests, in particular, that the coefficients of the latter *fractal power series* (i.e., the residues at the complex dimensions ω, still in the case of simple poles) could be interpreted in terms of a suitable notion of *fractal curvatures associated with the underlying complex dimensions* of the fractal in question. We refer the interested reader to [**375**, Subsection 13.1.3] or the introduction of [**356**, Chapter 5], and to [**329**], [**350**], for a more detailed discussion of the aforementioned analogy between fractal tube formulas and the Steiner formula, as well as about the possible existence of suitable fractal curvatures indexed by the complex dimensions of the underlying fractal.

The interested reader can also find in loc. cit. a number of relevant references connected to Steiner's formula ([**566**], 1840) and its many generalizations in smooth differential geometry (beginning with H. Weyl ([**612**], 1939) and as later expanded and generalized to Riemannian geometry in A. Gray's 2004 book [**220**]; see also the 1988 book by M. Berger and B. Gostiaux [**39**]) and in geometric measure theory [**182**]—beginning with H. Federer's celebrated "curvature measures" introduced in his 1959 paper [**181**] and later pursued in many directions, including in the work of M. Zähle [**625–629**], D. Hug, G. Last, and W. Weil [**256**], S. Winter [**619**] (see also, e.g., [**620**]), S. Kombrink [**303,304**], M. Kesseböhmer and S. Kombrink [**290**]. For thorough information on this latter subject, we refer to the 2019 book by J. Rataj and M. Zähle [**503**]. For further general information on Steiner's formula, Minkowski's work on convex bodies ([**445**], 1911), and their many extensions, we also refer to R. Schneider's interesting book ([**540**], 2003) and, from a combinatorial as well as geometric probabilistic point of view, the enticing book by D. A. Klain and G.-C. Rota ([**296**], 1999).

Finally, we mention that fractal tube formulas and explicit formulas of different kinds—in the sense of (or in the spirit of) complex dimensions—are obtained (and/or used), for example, in [**369, 370, 374, 375**], [**329**], [**330**], [**347–351**], [**356, 360–362**], [**355**], [**335–337**], [**126–129**], and [**469, 470**], [**119**], as well as in [**106–118**], from the point of view of the extended higher-dimensional theory of complex dimensions developed in loc. cit. and to be briefly discussed in Section 9.7 below.

9.3.8. Relative Fractal Drums and Symmetries. Finally, we briefly discuss the conceptually and technically important concept of a relative fractal drum, (or RFD, in short) which extends both that of any bounded set in $\mathbb{R}^m$, as well as that of a fractal drum in $\mathbb{R}^m$, i.e., a drum with fractal boundary (see, e.g., [**441, 442**], [**40, 41**], [**59**], [**190**], [**318–320**], [**352, 353**], [**210–212**], [**322**], [**191**], [**354**], [**168**], [**122**], [**63**], [**598**], [**446**], [**235**], [**369**], [**370**], [**374**], [**375**], [**356–365**], [**329**], as well as the relevant references therein) and, in particular, that of a fractal string, corresponding to the case when $m = 1$ (see, e.g., [**352, 353**], [**339,340**], [**369,371**], [**370,374,375**], [**326**], [**367**], [**333–338**], [**332**], [**328**], [**356**], [**376**], [**377**], [**378**], along with [**329**] and [**240–242**]). For more information about relative fractal drums (RFDs), see [**356**, Chapters 4 and 5].

Let A be a (nonempty) bounded subset of $\mathbb{R}^m$, and let Ω be a (possibly unbounded) open subset of $\mathbb{R}^m$ with finite volume (i.e., $\mathcal{L}^m(\Omega) < \infty$) such that $\Omega \subseteq A_\delta$, for some $\delta > 0$. Then, the ordered pair (A, Ω) is called a *relative fractal drum* (or RFD, in short) in $\mathbb{R}^m$.

Here are three important special cases of relative fractal drums (RFDs):

(i) (*Fractal String; see Section* 8.2 *above*). When $m = 1$, Ω is a bounded open subset of $\mathbb{R}$, and $A := \partial\Omega$, the boundary of Ω.

(ii) (*Fractal Drum or Drum with Fractal Boundary*; see loc. cit.). Same as in (i), but with $\Omega \subseteq \mathbb{R}^m$, where $m \in \mathbb{N}$ is arbitrary.[55]

(iii) (*Arbitrary Bounded Set in* $\mathbb{R}^m$, *as in Subsections* 9.3.2–9.3.7 *above*). Still for any $m \in \mathbb{N}$, let $A \subseteq \mathbb{R}^m$ be bounded (or, more generally, of finite volume), and let $\Omega := A_\delta$, for some fixed, but arbitrary $\delta > 0$.

Then, all of the definitions of (lower, upper) Minkowski dimensions, $\underline{\dim}_M(A, \Omega)$ and $\overline{\dim}_M(A, \Omega)$, Minkowski dimension, $\dim_M(A, \Omega)$, (lower, upper) Minkowski contents, $\mathcal{M}_* = \mathcal{M}_{*,A,\Omega}$ and $\mathcal{M}^* = \mathcal{M}^*_{A,\Omega}$, Minkowski content, $\mathcal{M} = \mathcal{M}_{A,\Omega}$, and average Minkowski content, $\widetilde{\mathcal{M}} = \widetilde{\mathcal{M}}_{A,\Omega}$, as well as of (distance and tube) fractal zeta functions, $\zeta_{A,\Omega}$ and $\tilde{\zeta}_{A,\Omega}$, can be easily extended to the present, more general, context of an RFD (A, Ω) in $\mathbb{R}^m$ (rather than of a bounded set $A \subseteq \mathbb{R}^m$). Indeed, all of these notions have a natural *relative* counterpart (e.g., relative Minkowski dimension, content, fractal zeta function,

[55] More generally, Ω may be unbounded but have finite volume: $\mathcal{L}^m(\Omega) < \infty$. The same comment applies to case (i) just above, where $m = 1$.

etc.).[56] In short, it suffices to replace the ordinary tube function $V_A = V_A(\varepsilon)$ by the *relative tube function*, $V_{A,\Omega}(\varepsilon) := \mathcal{L}^m(A_\varepsilon \cap \Omega)$, in the corresponding definitions—or in the case of the *relative distance zeta function* (denoted by $\zeta_{A,\Omega}$), to integrate over $A_\delta \cap \Omega$ instead of over A_δ, in the counterpart of (9.70) in Subsection 9.3.2 above.

Moreover, all of the results stated in Subsections 9.3.2–9.3.7 have relative analogs valid for general RFDs (A, Ω) in $\mathbb{R}^m$, rather than for bounded subsets A of $\mathbb{R}^m$, with $m \geq 1$ arbitrary; see [**356**, Chapters 4 and 5], along with [**359**] and [**361**].

As was alluded to earlier, there is, however, one intriguing difference. Indeed—in contrast to the case of (nonempty) bounded sets $A \subseteq \mathbb{R}^m$, for which $0 \leq \sigma_A = \overline{D_A} \leq m$, and $\sigma_A = \tilde{\sigma}_A$ if $\overline{D_A} \neq m$ (see Theorem 9.29 and Subsection 9.3.3 above)—there exist RFDs (A, Ω) such that we have $\overline{D_{A,\Omega}} := \overline{\dim}_M(A, \Omega) < 0$ and even $\overline{D_{A,\Omega}} = -\infty$, even though, for any RFD (A, Ω), we still have that $\sigma_{A,\Omega} = \overline{D_{A,\Omega}} \leq m$ and (provided $\overline{D_{A,\Omega}} \neq m$) $\tilde{\sigma}_{A,\Omega} = \sigma_{A,\Omega}$.[57] See [**356**, Proposition 4.1.35 and Corollary 4.1.38 in Subsection 4.1.2] for such examples of RFDs for which, in addition, the Minkowski dimension $D_{A,\Omega}$ also exists and can take any prescribed value in $[-\infty, m]$. In the case of a curve A, for instance, these examples suggest an interesting connection between the existence of a *negative* Minkowski dimension and the local geometric features of A (e.g., the presence of a suitable cusp); see loc. cit.

The advantage of using this unifying and more flexible concept of RFD is that even for the specific example of a given bounded set $A \subseteq \mathbb{R}^m$, in order to compute its fractal zeta functions and its complex dimensions, or to obtain the associated fractal tube formula, it is often very convenient to break up A into finitely or countably many "sub-RFDs" (see, e.g., part (iii) of Theorem 9.34 below), for which the corresponding computation is much simpler. One then uses the underlying symmetry and scaling properties (see parts (i) and (ii) of Theorem 9.34 just below, along with Corollary 9.35) in order to obtain the sought for result.

We next give one important example of such properties, in the case of the relative distance zeta function $\zeta_{A,\Omega}$; see [**356**, Theorems 4.1.40 and 4.1.44, pp. 267 and 270], along with [**356**, Definition 4.1.43, p. 270].

[56] One important difference is that—unlike the Minkowski dimension D_A of a bounded set A, for which $D_A \in [0, m]$—we have, in general, that the *relative Minkowski dimension* $D_{A,\Omega}$ of an RFD (A, Ω) can now take any value in $[-\infty, m]$, including the value $-\infty$; see [**356**, Subsection 4.1.2].

[57] More specifically; for a suitable RFD (A, Ω), $D_{A,\Omega} := \dim_M(A, \Omega)$ (and $\overline{D_{A,\Omega}} = \overline{\dim}_M(A, \Omega)$) can take any given value in $[-\infty, m]$, *including the value* $-\infty$. This fact motivates our convention according to which $D_{\emptyset,\emptyset} = -\infty$ (rather than -1)—and, more generally, $\dim(\emptyset) = -\infty$ (rather than -1), for any choice of dimension $\dim \in \{\dim_T, \dim_H, \dim_B, \dim_M, \dim_P\}$ encountered in this book.

Theorem 9.34 (Symmetries and Scaling—Unions of RFDs; M. L. Lapidus, G. Radunović, and D. Žubrinić, [**356**], 2017). *Let* (A, Ω) *be an* RFD *in* $\mathbb{R}^m$. *Furthermore, let* U *be any domain of* $\mathbb{C}$ *containing the closed half-plane* $\{\mathrm{Re}(s) \geq \overline{D_{A,\Omega}}\}$ *and such that* $\zeta_{A,\Omega}$ *can be (necessarily uniquely) meromorphically continued to* U. *Then:*

(i) (Isometry Invariance). $\zeta_{A,\Omega}$ *is invariant under the action of any isometry* Ψ *(i.e., a rotation, a translation, or a reflection, or else any finite product of such transformations) of* $\mathbb{R}^m$. *More specifically, with* $\Psi \cdot (A, \Omega) := (\Psi(A), \Psi(\Omega))$, *we have that* $\overline{D_{\Psi\cdot(A,\Omega)}} = \overline{D_{(A,\Omega)}}$ *and, for all* $s \in \mathbb{C}$ *with* $\mathrm{Re}(s) > \overline{D_{A,\Omega}}$*—but also, in fact (upon analytic continuation), for all* $s \in U$*—we have that*

$$\zeta_{\Psi\cdot(A,\Omega)}(s) = \zeta_{A,\Omega}(s). \tag{9.94}$$

Hence, $\mathcal{D}_{\Psi\cdot(A,\Omega)}(U) = \mathcal{D}_{A,\Omega}(U)$.

(ii) (Scale Covariance). *For any* $\lambda > 0$ *and with* $\lambda \cdot (A, \Omega) := (\lambda A, \lambda\Omega)$, *we have that* $\overline{D_{\lambda\cdot(A,\Omega)}} = \overline{D_{A,\Omega}}$ *and, for all* $s \in \mathbb{C}$ *with* $\mathrm{Re}(s) > \overline{D_{A,\Omega}}$*—but also, in fact (upon meromorphic continuation), for all* $s \in U$*—we have that*

$$\zeta_{\lambda\cdot(A,\Omega)}(s) = \lambda^s \cdot \zeta_{A,\Omega}(s). \tag{9.95}$$

Hence, $\mathcal{D}_{\lambda\cdot(A,\Omega)}(U) = \mathcal{D}_{A,\Omega}(U)$, *where (for example)* $\mathcal{D}_{A,\Omega}(U)$ *denotes the set of visible complex dimensions of the RFD* (A, Ω). *Furthermore, for all* $\omega \in U$ *(and, in particular, for all* $\omega \in \mathcal{D}_{A,\Omega}(U)$*), we have that*

$$\mathrm{res}(\zeta_{\lambda\cdot(A,\Omega)}, \omega) = \lambda^\omega \cdot \mathrm{res}(\zeta_{A,\Omega}, \omega), \tag{9.96}$$

and similarly for the principal parts at ω, *in case* ω *is a multiple pole.*

(iii) (Additivity Under Suitable Unions). *If* $\{(A_j, \Omega_j)\}_{j\geq 1}$ *is a finite or countable family of* RFDs *in* $\mathbb{R}^m$ *and if* $(A, \Omega) = \bigcup_{j\geq 1}(A_j, \Omega_j)$ *is the* RFD *in* $\mathbb{R}^m$ *obtained as the union of those* RFDs,[58] *satisfying the condition*[59]

$$\mathrm{dist}(\boldsymbol{x}, A) = \mathrm{dist}(\boldsymbol{x}, A_j), \text{ for all } j \geq 1 \text{ and } \boldsymbol{x} \in \Omega_j, \tag{9.97}$$

then, for all $s \in \mathbb{C}$ *with* $\mathrm{Re}(s) > \overline{D_{A,\Omega}}$ *(and hence also upon analytic continuation, for all* $s \in U$*),*[60] *we have that*

$$\zeta_{A,\Omega}(s) = \sum_{j\geq 1} \zeta_{A_j,\Omega_j}(s). \tag{9.98}$$

Corollary 9.35 (Covariance Under the Action of Similitudes). *Let* S *be a similarity transformation of* $\mathbb{R}^m$ *(not necessarily a contraction), with scaling ratio* $r > 0$. *Then, with* $S \cdot (A, \Omega) := (S(A), S(\Omega))$, *and under the same hypotheses on*

---

[58]This means that the family of open sets $\{\Omega_j\}_{j=1}^\infty$ is (pairwise) *disjoint* and that $A := \bigcup_{j\geq 1} A_j$ and $\Omega := \bigcup_{j\geq 1} \Omega_j$. As for any RFD, we also assume that $\Omega \subseteq A_\delta$, for some $\delta > 0$.

[59]Condition (9.97) is satisfied, for example, if each Ω_j is a bounded open subset of $\mathbb{R}^m$ and $A_j = \partial\Omega_j$, for all $j \geq 1$.

[60]Note that we also have that $\overline{D_{A,\Omega}} \geq \sup_{j\geq 1} \overline{D_{A_j,\Omega_j}}$.

U and $\zeta_{A,\Omega}$ as in Theorem 9.34, *we have that* $\overline{D_{S\cdot(A,\Omega)}} = \overline{D_{(A,\Omega)}}$, $\mathcal{D}_{S\cdot(A,\Omega)}(U) = \mathcal{D}_{A,\Omega}(U)$, *and*

$$\zeta_{S\cdot(A,\Omega)}(s) = r^s \cdot \zeta_{A,\Omega}(s), \text{ for all } s \in U, \tag{9.99}$$

along with (for all $\omega \in \mathcal{D}_{A,\Omega}(U)$)

$$\operatorname{res}(\zeta_{S\cdot(A,\Omega)}, \omega) = r^\omega \cdot \operatorname{res}(\zeta_{A,\Omega}, \omega). \tag{9.100}$$

We close this subsection by very briefly discussing the enlightening example of a fractal string (see Section 8.2 above for more details).

Example 9.36 (*Fractal Strings, Revisited*). Let Ω be a bounded open subset of $\mathbb{R}$ and let $\mathcal{L} = \{\ell_j\}_{j\geq 1}$ be the associated sequence of the lengths of its connected components (which are bounded open intervals of $\mathbb{R}$, here). We may alternatively view the fractal string $\mathcal{L}$ as the RFD in $\mathbb{R}$ given by $(\partial\Omega, \Omega)$. (For example, $\Omega = [0,1]\setminus C$, where C is the ternary Cantor set and $\mathcal{L} = \{1/3, 1/9, 1/9, \ldots\}$ is the Cantor string, as in Example 8.17 of Section 8.2, while $(\partial\Omega, \Omega) = (C, \Omega)$ is the *Cantor string RFD*.) Then, it is shown in [**356**, Corollary 2.1.61, p. 92] that $\overline{D_{\mathcal{L}}} = \overline{D_{\partial\Omega,\Omega}}$ and that *the geometric zeta function $\zeta_{\mathcal{L}}$ of $\mathcal{L}$ and the relative distance zeta function $\zeta_{\partial\Omega,\Omega}$ of the* RFD $(\partial\Omega, \Omega)$ *are connected via the following functional equation*—first, for all $s \in \mathbb{C}$ with $\operatorname{Re}(s) > \overline{D_{\mathcal{L}}}$, and then, upon analytic continuation, for all $s \in U$:[61]

$$\zeta_{\partial\Omega,\Omega}(s) = \frac{2^{1-s}}{s} \cdot \zeta_{\mathcal{L}}(s). \tag{9.101}$$

Hence, $\mathcal{D}_{\partial\Omega,\Omega}(U) = \mathcal{D}_{\mathcal{L}}(U) \cup \{0\}$, if $0 \in U$ and $\zeta_{\mathcal{L}}(0) \neq 0$ (resp., $= \mathcal{D}_{\mathcal{L}}(U)$, otherwise).[62]

9.4. Devil's Staircase (or Cantor Graph): A Self-Affine and Rectifiable Fractal Curve

The *Devil's staircase* (or Cantor graph)—defined as the graph of the Cantor function f_{DS} (see Subsection 1.4.1 above)—along with its natural generalizations, is an ubiquitous fractal curve, which can be found in various disguises in many fields of mathematics and physics, for example, in number theory (Diophantine approximation), as well as in the theory of dynamical systems and chaos, where it plays a key role in the study of mode locking and the notion of a rotation number (see, e.g., [**96**], [**252**], and [**604**]), Fourier and harmonic analysis, real analysis (e.g., in the theory of the Lebesgue integral and of singular continuous functions and measures), as well as in the experimental and theoretical results concerning the fractional quantum Hall effect (see, e.g., [**520**]).

[61] Here, the connected open set $U \subseteq \mathbb{C}$ satisfies the same hypotheses as in Theorem 9.34.

[62] More specifically, if $0 \in U$ and $\zeta_{\mathcal{L}}(0) \neq 0$, the multiplicity of 0 as a (possible) pole of $\zeta_{\mathcal{L}}$ is increased by one, when viewed as a pole of $\zeta_{\partial\Omega,\Omega}$.

The Devil's staircase was introduced by Georg Cantor in his celebrated 1884 paper [**70**];[63] it was then further studied by Giuseppi Vitali (in [**602**], 1905)—who first observed that f_{DS} was not equal to the integral of its derivative—and by Henri Lebesgue in his seminal 1904 book on integration theory [**389**], in which he obtained a suitable analog (*not* satisfied by f_{DS}, in light of Vitali's result) of the Fundamental Theorem of Calculus within his theory of what is now known as the Lebesgue integral; see, e.g., [**5**], [**57**], [**91**], [**193**], [**521**], or [**523**].

Historically, the Devil's staircase (or Cantor graph) $\mathbf{\Gamma}_{DS}$ has provided a significant challenge for mathematicians because—besides yielding an obvious counterexample to a naive extension of the Fundamental Theorem of Calculus—it is clearly a "fractal curve", whatever meaning one wishes to give to the adjective "fractal". On the other hand, it is also rectifiable (i.e., has finite length) and self-affine, whereas it does not have a "fractional dimension". In fact, its topological, Minkowski (or box), and Hausdorff dimensions coincide and are all equal to 1. Hence, the Devil's staircase should not be "fractal", according to Mandelbrot's definition given in his classic book, [**419**]—and which would require that the Hausdorff dimension of $\mathbf{\Gamma}_{DS}$ be strictly greater than its topological dimension (which is also equal to 1).

This apparent paradox has long intrigued and bothered the first author (M. L. Lapidus) and was one of his main motivations for eventually developing (with several of his collaborators) a rigorous theory of complex fractal dimensions (see, e.g., [**370, 374, 375**], [**356**], [**329**], [**330**]), as well as for defining a *fractal* by the presence of at least one *nonreal* complex dimension. In the case of the Devil's staircase, it is shown in [**359**] and [**356**] that the *nonreal* complex dimensions of $\mathbf{\Gamma}_{DS} \subseteq \mathbb{R}^2$ are the same as those of the Cantor set (or string), namely,[64]

$$\log_3 2 + i\, n\frac{2\pi}{\log 3}, \quad \text{where } n \in \mathbb{Z}\ (n \neq 0) \text{ is arbitrary.} \tag{9.102}$$

As a result, $\mathbf{\Gamma}_{DS}$ is, indeed, fractal (in the sense of the theory of complex dimensions) and there is not any paradox anymore!

In fact—as is shown in the aforementioned theory of complex dimensions (see, especially, [**356**] and [**359, 361**])—the Devil's staircase has natural, *intrinsic geometric oscillations* directly connected to the presence of these nonreal complex dimensions (albeit of a lower order than the nonoscillatory leading term associated to the Minkowski dimension).

[63] See also, H. J. Smith's 1874–1875 paper [**560**] on the integration of discontinuous functions.

[64] We point out that $\log_3 2 = \log 2/\log 3$ is the Minkowski dimension of the Cantor set C, while the Minkowski dimension of the Devil's staircase $\mathbf{\Gamma}_{DS}$ is equal to 1, because (as will be shown in this section) $\mathbf{\Gamma}_{DS}$ is a *rectifiable* continuous curve. Note that if we let $n = 0$ on the left-hand side of (9.102), then we obtain $\log_3 2$, which is also a complex dimension of the Cantor graph but is real.

As a fractal curve, the Devil's staircase possesses many other beautiful and often baffling properties, as will be discussed in this section and, with more details and complete proofs, in the corresponding chapters of the sequels to this book, [**119**] and [**116**].

9.4.1. First Construction of the Devil's Staircase: via Monotonic or Continuous Extension. Let $f_{DS} : [0,1] \to \mathbb{R}$ denote the Cantor function—introduced in Chapter 1 (see Example 1.32, along with Exercises 1.33–1.35) and sometimes also called the Cantor–Lebesgue function in the literature.

The *Devil's staircase* (or *Cantor graph*), $\mathbf{\Gamma} = \mathbf{\Gamma}_{DS}$, is defined as the graph of the Cantor function f_{DS}. Accordingly, $\mathbf{\Gamma}_{DS}$ is a continuous planar curve, as will be further explained in this section. In particular (see Example 1.32 and Exercise 1.33 in Subsection 1.4.1), $\mathbf{\Gamma}_{DS} = \gamma_{DS}([0,1])$ is the image of the continuous map $\gamma_{DS} : [0,1] \to \mathbb{R}^2$ (viewed as a continuous parametrized curve) defined by $\gamma_{DS}(t) := (t, f_{DS}(t)) \in \mathbb{R}^2$, for all $t \in [0,1]$.

Recall that f_{DS} is piecewise (or locally) constant on the open set $\Omega \subseteq \mathbb{R}$—called the *Cantor string* in [**329, 370, 375**] (see Examples 1.28 and Section 8.2 above, especially Example 8.17)—consisting of the union of countably many disjoint intervals, namely, the deleted open intervals obtained at every stage of the construction of the (classic ternary) Cantor set C: $\Omega = [0,1]\backslash C$. Accordingly, $\Omega \subseteq [0,1]$ and its boundary $\partial\Omega$ coincides with the Cantor set: $\partial\Omega = C$.

Furthermore, the endpoints of the open intervals (or connected components) out of which Ω is built are the *ternary points* (i.e., the points of $[0,1]$ of the form $2^m/3^n$, with $m \in \mathbb{N}_0$, $n \in \mathbb{N}_0$, and $2^m \leq 3^n$). We denote by T the set of such ternary points. Clearly, $T \subseteq C$; in fact, recall that T is dense in C:

$$C = T \cup (C\backslash T) = \overline{T}, \tag{9.103}$$

where $C\backslash T$, the complement of T in C, coincides with the set of accumulation points of C.[65]

More specifically, the Cantor function f_{DS} is constant, with value $1/2$, on the first (open) "middle-third" interval, $I_{1,1} = (1/3, 2/3)$, which is deleted from $[0,1]$ at the first stage of the construction of the Cantor set C. Furthermore, f_{DS} is constant, with respective values $1/4$ and $3/4$, on each of the $2 = 2^1$ (open) middle-third intervals, $I_{1,2} = \left(\frac{1}{9}, \frac{2}{9}\right)$ and $I_{2,2} = \left(\frac{7}{9}, \frac{8}{9}\right)$, which are deleted at the second stage of the construction of C. Moreover, f_{DS} is constant, with respective values $1/8, 3/8, 5/8$, and $7/8$, on each of the $4 = 2^2$ (open) middle-third intervals which are deleted at the third stage of the construction of C, namely, $I_{1,3} = \left(\frac{1}{27}, \frac{2}{27}\right), I_{2,3} = \left(\frac{7}{27}, \frac{8}{27}\right), I_{3,3} = \left(\frac{19}{27}, \frac{20}{27}\right)$, and $I_{4,3} = \left(\frac{25}{27}, \frac{26}{27}\right)$; see Figure 9.8. And so on, ad infinitum.

[65] Here, an *accumulation point* of T is viewed as the limit of a (convergent) sequence of *distinct* points of T, by contrast to a *limit point* of T, which is the limit of a (convergent) sequence of (not necessarily distinct) points of T. Accordingly, the set of limit points of T is all of C.

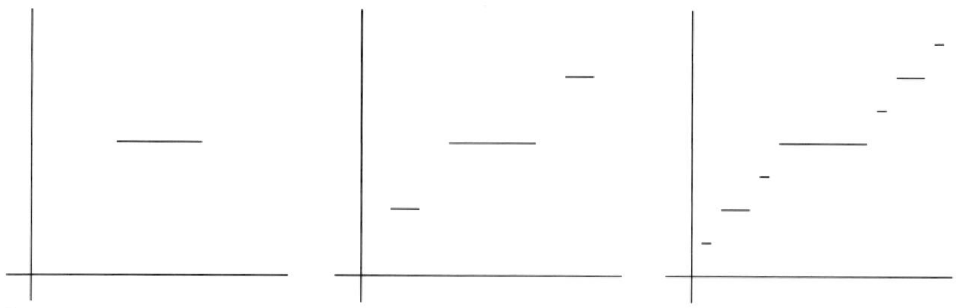

(a) (b) (c)

Figure 9.8. The first three stages of the construction of the Cantor function f_{CS}, as depicted in (a), (b), and (c), respectively.

The above discussion defines f_{DS} on the open set $\Omega \subseteq \mathbb{R}$, the complement of the Cantor set C in $[0,1]$, given by

$$\Omega = \bigcup_{n=1}^{\infty} \left(\bigcup_{k=1}^{2^{n-1}} I_{k,n} \right), \tag{9.104}$$

where the intervals $I_{k,n}$ (with $1 \leq k \leq 2^{n-1}$) are the 2^{n-1} disjoint (open) intervals deleted from $[0,1]$ at the n-th stage of the construction of C.

Note that, taken all together, the sequence of deleted open intervals,

$$\{I_{k,n} : n \in \mathbb{N}, 1 \leq k \leq 2^{n-1}\}, \tag{9.105}$$

constitutes the countably infinite family of (necessarily open) connected components of the open set $\Omega \subseteq \mathbb{R}$.

Next, one can uniquely extend f_{DS} from Ω to all of $[0,1]$ in either of two ways: (i) by monotonicity (i.e., more precisely, so as to become nondecreasing on $[0,1]$) or (ii) by continuity (i.e., so as to become continuous on all of $[0,1]$).

Note that in case (i), we must have

$$f_{DS}(t) = \inf_{\tau > t} f_{DS}(\tau) = \sup_{\tau < t} f_{DS}(\tau), \tag{9.106}$$

where the first (resp., second) equality in (9.106) holds for all $t \in [0,1)$ (resp., for all $t \in (0,1]$).

On the other hand, in case (ii), we must have

$$f(t) = f_{DS}(t^-) := \lim_{\tau \to t, \tau < t} f_{DS}(\tau) = f_{DS}(t^+) := \lim_{\tau \to t, \tau > t} f_{DS}(\tau), \tag{9.107}$$

where—much as for (9.106) just above—the first (resp., third) equality in (9.107) holds for all $t \in (0,1]$ (resp., for all $t \in [0,1)$).

Exercise 9.37 (*Justification of the First Construction of f_{DS}*). Proceed in three steps, as follows:

(a) First, prove that T, the set of ternary points of $[0,1]$, is dense in the Cantor set C; i.e., $\overline{T} = C$, where $\overline{T}$ denotes the closure of T in $[0,1]$ (or, equivalently, in $\mathbb{R}$). In particular, first verify that $T \subseteq C$ and T is dense in C, as was stated in the above discussion.

(b) Then, show that whether we extend f_{DS} by monotonicity (as in (i) above) or by continuity (as in (ii) above),[66] we necessarily have the following properties: $f_{DS}(0) = 0$, $f_{DS}(1) = 1$, and, more generally, for any $t \in T$, $f_{DS}(t)$ is equal to the constant value corresponding to the open interval $I_{k,n}$ (for some $n \in \mathbb{N}$ and $k \in \{1, \ldots, 2^{n-1}\}$) to which it belongs; for the notation used here, see (9.104) and the discussion surrounding it.

Deduce that f is constant on the closure, $\overline{I_{k,n}}$, of each interval $I_{k,n}$ appearing in (9.104)—and hence that f is piecewise constant on $\Omega \cup T$.

(c) Finally, conclude that f_{DS}—initially defined as a piecewise constant function on Ω—*can*, indeed, be extended to a *unique monotonic* (i.e., nondecreasing) *and continuous* (as well as uniformly continuous) *function on all of* $[0,1]$. In particular, the method of extension by monotonicity (as in case (i)) or by continuity (as in case (ii)) can be used to define the Cantor function f_{DS} without ambiguity (i.e., *uniquely*) on all of $[0,1]$.

The well-known results of the following easy exercise were used implicitly in Exercise 9.37 just above, in the case of the extension of f_{DS} by continuity.

Exercise 9.38 (*Unique Extension, by Density, of a Uniformly Continuous Function*). Let X, Y be metric spaces, with Y complete, and let $A \subseteq X$ be dense in X. Then, show that any uniformly continuous map $\theta : A \to Y$ can be uniquely extended to a uniformly continuous (and hence also continuous) function $\overline{\theta} : X \to Y$.

9.4.2. Second Construction of the Devil's Staircase: Uniform Approximation by Piecewise Linear Functions.

A second, very useful, construction of f_{DS} consists in defining it as the uniform limit of a sequence of piecewise linear and continuous functions, as was alluded to in Example 1.32, along with Exercises 1.34 and 1.35 of Subsection 1.4.1; see, in particular, Figure 1.8 in Subsection 1.4.1.

---

[66] In fact, in part (ii), one can extend f_{DS} by *uniform continuity*, which enables one to guarantee both the existence and uniqueness of the continuous extension of f_{DS}; see Exercise 9.38.

More specifically, we start, exactly as before, by defining f_{DS} as a piecewise constant function.[67] Then, at each stage of the construction (say, at the n-th stage, with $n \in \mathbb{N}$), we linearly interpolate between the values taken by f_{DS} at the endpoints of two consecutive intervals (say, $I_{k,n}$ and $I_{k+1,n}$, for some arbitrary $k \in \{1, \ldots, 2^{n-1}-1\}$), in the notation introduced in (9.104). See Figure 1.8 of Subsection 1.4.1 for the first three stages of this construction of f_{DS}, giving rise to the functions $f_{1,DS}$, $f_{2,DS}$, and $f_{3,DS}$.

At the end of the process, we obtain a sequence, $\{f_{n,DS}\}_{n=1}^{\infty}$, of piecewise linear and continuous functions on $[0,1]$. One must then show that $\{f_{n,DS}\}_{n=1}^{\infty}$ converges uniformly on $[0,1]$ to the same function f_{DS} as the one obtained above via the first construction. This is the object of the next exercise.

Exercise 9.39 (*Justification of the Second Construction of f_{DS}*)**.** Proceed in three steps, as follows:

(a) First, prove that the sequence of piecewise linear functions $\{f_{n,DS}\}_{n=1}^{\infty}$ defined just above is uniformly Cauchy on $[0,1]$; i.e., it is a Cauchy sequence in $C([0,1])$ (equipped with the metric δ_∞ of uniform convergence), in the notation of Section 9.2; see, especially, (9.19) and (9.20).

(b) Then, deduce from the completeness of $C([0,1])$ (see Corollary 9.10 of Theorem 9.9 in Section 9.2) that $\{f_{n,DS}\}_{n=1}^{\infty}$ converges uniformly on $[0,1]$ to a continuous function on $[0,1]$, still denoted by f_{DS}.

(c) Finally, show that the function f_{DS} obtained in part (b) coincides on all of $[0,1]$ with the Cantor function defined in the first construction.

Of course, this second construction shows naturally that the Cantor (or Devil's staircase) function f_{DS} is both continuous and nondecreasing, respectively as the uniform limit of a sequence of continuous functions on $[0,1]$ and as the pointwise limit of a sequence of nondecreasing functions on $[0,1]$. In particular, *it directly yields a (continuous) parametrization of the Cantor graph* (or Devil's staircase) $\boldsymbol{\Gamma}_{DS}$, the graph of f_{DS}. Namely, $\boldsymbol{\Gamma}_{DS} = \gamma_{DS}([0,1])$, where $\gamma_{DS} : [0,1] \to \mathbb{R}^2$ is the continuous, parametrized planar curve given by $\gamma_{DS}(t) = (t, f_{DS}(t))$, for all $t \in [0,1]$.

9.4.3. Rectifiability and Exact Length. Another advantage of the above second construction of $\boldsymbol{\Gamma}_{DS}$ (discussed in Subsection 9.4.2 just above) is that it not only enables us to show that the Devil's staircase is rectifiable (i.e., has finite length) but, in fact, to also explicitly calculate its precise length.

[67] We also decree that f_{DS} takes the same value at the endpoints of an arbitrary open interval $I_{k,n}$ (i.e., a connected component of Ω) as the constant value it takes on $I_{k,n}$; see the notation introduced in (9.104) and the text surrounding (9.104). It follows that f_{DS} is constant on each compact interval $J_{k,n} := \overline{I_{k,n}}$, for $k = 1, \ldots, 2^{n-1}$ and with $n \in \mathbb{N}$ arbitrary.

Proposition 9.40 (Rectifiability and Length of $\mathbf{\Gamma}_{DS}$). *The Devil's staircase $\mathbf{\Gamma}_{DS}$ is rectifiable. More specifically, it has length two*: $L(\mathbf{\Gamma}_{DS}) = 2$.

Proof. We have already shown in Example 9.7 of Section 9.1 that $L(\mathbf{\Gamma}_{DS}) \geq 2$; see (9.16) and the discussion surrounding it. One can then obtain the reverse inequality by invoking the semi-continuity of the length function with respect to uniform convergence (see Theorem 9.22 in Section 9.2).[68] Accordingly, since (in the notation of footnote 68) $\mathbf{\Gamma}_{n,DS} = \gamma_{n,DS}([0,1])$ and $\mathbf{\Gamma}_{DS} = \gamma_{DS}([0,1])$, we have—by the semi-continuity of the length function; see Theorem 9.22 in Section 9.2 above—that

$$L(\mathbf{\Gamma}_{DS}) \leq \liminf_{n\to\infty} L(\mathbf{\Gamma}_{n,DS}) = \lim_{n\to\infty} L(\mathbf{\Gamma}_{n,DS}), \tag{9.108}$$

where we have used the fact (proved in and around (9.15) and (9.16) of Example 9.7) that the following limit exists and is given by

$$\begin{aligned} \lim_{n\to\infty} L(\mathbf{\Gamma}_{n,DS}) &= \lim_{n\to\infty} \left(h_n + \frac{1}{3} \sum_{j=0}^{\infty} \left(\frac{2}{3}\right)^j \right) \\ &= \left(\lim_{n\to\infty} h_n \right) + 1 = 1 + 1 = 2, \end{aligned} \tag{9.109}$$

where h_n is the total length of the oblique (i.e., not horizontal) steps of $\mathbf{\Gamma}_{n,DS}$. Recall from the discussion following formula (9.15) that the oblique steps of $\mathbf{\Gamma}_{n,DS}$ are becoming increasingly vertical, as $n \to \infty$; in fact, their slopes tend to $+\infty$, as $n \to \infty$. Furthermore, the projections (or shadows) of these oblique steps onto the y-axis has total length 1, which explains why $\lim_{n\to\infty} h_n = 1$, in (9.109).

By combining (9.108) and (9.109), we deduce that $L(\mathbf{\Gamma}_{DS}) \leq 2$. Hence, since we both have that $L(\mathbf{\Gamma}_{DS}) \geq 2$ and $L(\mathbf{\Gamma}_{DS}) \leq 2$, we conclude that $L(\mathbf{\Gamma}_{DS}) = 2$, as claimed. In particular, $L(\mathbf{\Gamma}_{DS}) < \infty$ and so, $\mathbf{\Gamma}_{DS}$ is rectifiable, as desired. This completes the proof of the proposition. □

Remark 9.41 (*Shadow Cast by the Devil's Staircase*). Intuitively, the length of the Devil's staircase $\mathbf{\Gamma}_{DS} \subseteq [0,1]^2$ is equal to $2 = 1 + 1$ because the projection (or "shadow") of $\mathbf{\Gamma}_{DS}$ onto the x-axis (i.e., the t-axis, in our current notation) is equal to 1—since $\mathcal{L}^1(C) = 0$, $\mathcal{L}^1(T) = 0$, and so, $\mathcal{L}^1(\Omega \cup T) = 1$)—while, as was explained just above (as well as in Example 9.7), the projection (or "shadow") of $\mathbf{\Gamma}_{DS}$ onto the y-axis is also equal to 1.

As a result, the length $L(\mathbf{\Gamma}_{DS})$ of the Devil's staircase $\mathbf{\Gamma}_{DS}$—namely, $L(\mathbf{\Gamma}_{DS}) = 1 + 1 = 2$—is equal to the total length of the shadow cast by the Devil's staircase $\mathbf{\Gamma}_{DS}$ onto the horizontal x-axis (i.e., t-axis) and onto the vertical y-axis,

[68] Note that $\mathbf{\Gamma}_n$ (or $\mathbf{\Gamma}_{n,Pol}$) and $\mathbf{\Gamma}$ in Example 9.7 are now denoted $\mathbf{\Gamma}_{n,DS}$ and $\mathbf{\Gamma}_{DS}$, respectively. Furthermore, since $f_{n,DS}$ converges uniformly to f_{DS} on $[0,1]$, as $n \to \infty$, it follows at once that $\gamma_{n,DS}$ converges uniformly to γ_{DS} on $[0,1]$, as $n \to \infty$, where $\gamma_{n,DS}(t) := (t, f_{n,DS}(t))$ and $\gamma_{DS}(t) := (t, f_{DS}(t))$, for all $t \in [0,1]$.

as is explained just above. Hence, geometrically, $\boldsymbol{L}(\boldsymbol{\Gamma}_{DS}) = 2$ is just the length of the "half-square" $\overline{AO}$ and $\overline{OB}$, consisting of the vertical line segment $\overline{AO}$ and the horizontal line segment $\overline{OB}$ (in that order), each of length 1 and where $A := (0,1)$, $O := (0,0)$,[69] and $B := (1,0)$. Equivalently, of course, $\boldsymbol{\Gamma}_{DS} = 2$ is the length of the symmetric "half-square", $\overline{AD}$ and $\overline{DB}$, consisting of the horizontal line segment $\overline{AD}$ and of the vertical line segment $\overline{DB}$ (in that order and each of length 1), where $D := (1,1)$ is the rightmost point of the Cantor graph $\boldsymbol{\Gamma}_{DS}$.

Exercise 9.42. Investigate whether or not the results of Section 9.2 allowing us (under appropriate conditions) to interchange length and uniform or Hausdorff limits (see Theorems 9.13, 9.17, 9.20, and 9.24) can be used in the present context in order to show that the following limit exists and is given by $\lim_{n\to\infty} L(\boldsymbol{\Gamma}_{n,DS}) = L(\boldsymbol{\Gamma}_{DS})$.

9.4.4. Third Construction: via Explicit Ternary Expansions. There is a third construction—discussed in some detail in Example 1.32 and the associated Exercises 1.33–1.35 in Chapter 1—of the Cantor function f_{DS} and hence also of the Cantor graph (or Devil's staircase) $\boldsymbol{\Gamma}_{DS}$. It provides explicit values for $f_{DS}(t)$ at every point t of $[0,1]$ and, in particular, at every point of the Cantor set C. Consequently, *it also provides an explicit continuous parametrization,* $t \mapsto \gamma_{DS}(t) = (t, f_{DS}(t))$, for all $t \in [0,1]$, *for the Devil's staircase.* We refer the interested reader to the just-mentioned part of Chapter 1 for this third construction of f_{DS} and hence also of $\boldsymbol{\Gamma}_{DS}$. We further refer to Exercise 1.34 (resp., Exercise 1.35) for the equivalence of this third construction with the first (resp., second) construction of f_{DS} and of $\boldsymbol{\Gamma}_{DS}$.

Exercise 9.43 (*The Third Construction of f_{DS} and hence also of $\boldsymbol{\Gamma}_{DS}$*). Proceed in two steps, as follows:

(a) Without first looking at Example 1.32 and the associated exercises (Exercises 1.33–1.35), try to find an explicit way of writing down the ternary expansion $.y_1y_2\ldots_3$ of $y := f_{DS}(x)$,[70] given by

$$y = f_{DS}(x) = .y_1y_2\ldots_3 = \sum_{n=1}^{\infty} \frac{y_n}{3^n}, \tag{9.110}$$

with $y_n \in \{0,1,2\}$, for each $n \in \mathbb{N}$, where

$$x = .x_1x_2\ldots_3 = \sum_{n=1}^{\infty} \frac{x_n}{3^n} \tag{9.111}$$

[69] Note that O, the origin of $\mathbb{R}^2$, is the leftmost point of the Cantor graph $\boldsymbol{\Gamma}_{DS}$; indeed, $\gamma_{DS}(0) = 0$.

[70] We are using here the notation (for ternary and binary expansions, in particular) introduced in Example 1.31 of Subsection 1.4.1.

is the ternary expansion of an arbitrary $x \in [0,1]$, and with $x_n \in \{0,1,2\}$, for all $n \in \mathbb{N}$. Observe that $f_{DS} : [0,1] \to [0,1]$, so that $y = f_{DS}(x) \in [0,1]$, for any $x \in [0,1]$.[71]

(b) For each $x \in [0,1]$, rewrite in base 2 (rather than in base 3) the value $y = f_{DS}(x)$ given (in base 3) in identity (9.110) of part (a). Then, justify in detail all of the statements made in Example 1.32 and in the associated exercises, namely, Exercises 1.33–1.35.

9.4.5. Singularity of the Cantor–Lebesgue Function. As was already discussed in Subsection 1.4.1, the Cantor function f_{DS} is differentiable almost everywhere on $[0,1]$, with derivative zero,

$$f'_{DS}(t) = 0, \text{ for a.e. } t \in [0,1], \tag{9.112}$$

because it is piecewise constant on the open set $\Omega = [0,1]\backslash C$, the complement of the Cantor set in $[0,1]$, and since $\mathcal{L}^1(C) = 0$; i.e., C has 1-dimensional Lebesgue measure zero.

Nevertheless, $f_{DS} : [0,1] \to [0,1]$ is continuous and nondecreasing on $[0,1]$; also, $f_{DS}(0) = 0$ and $f_{DS}(1) = 1$. Consequently, by the Intermediate Value Theorem, along with Exercise 9.44 below, $f_{DS}([0,1]) = [0,1]$; i.e., f_{DS} takes on all of the values between 0 and 1. Of course, however, f_{DS} is *not* injective (because it is locally constant on Ω) and hence is certainly *not* a bijection—let alone a homeomorphism—of $[0,1]$ onto $[0,1]$.

Exercise 9.44. Show that $f_{DS}([0,1]) \subseteq [0,1]$.

Clearly—much as we have shown in Example 9.7 of Section 9.1 that the standard notion of velocity did not make sense in this context—the naïve counterpart of the Fundamental Theorem of Calculus does not apply to the Cantor function f_{DS}, also known in the literature as the Cantor–Lebesgue function. Indeed, otherwise, we would have that

$$f_{DS}(1) - f_{DS}(0) = \int_0^1 f'_{DS}(t)dt; \tag{9.113}$$

i.e., here, since $f_{DS}(0) = 0$, $f_{DS}(1) = 1$, and (9.112) holds, we must have that $1 = 0$, which is, of course, impossible. A fortiori, as was first noted by G. Vitali in 1905 in [**602**], f_{DS} cannot be equal to the integral of its (almost everywhere defined) derivative; i.e., we cannot have that

$$f_{DS}(t) = f_{DS}(0) + \int_0^t f'_{DS}(\tau)d\tau, \text{ for all } t \in [0,1]. \tag{9.114}$$

[71] *Caution*: Recall that a ternary expansion of a point of $[0,1]$ is not unique, in general. More specifically, it is unique for points of $[0,1]$ which are not ternary points. However, there is a slight ambiguity if the corresponding point of $[0,1]$ is a ternary point, in which case there are two possible choices of ternary expansions, one infinite and one finite. Hence, some care must be exercised in defining the respective ternary expansions of $x \in [0,1]$ and of $y := f_{DS}(x) \in [0,1]$ in this context.

Indeed, since $f_{DS}(0) = 0$ and, in light of (9.112), f_{DS} would have to be identically equal to zero on $[0,1]$, which is clearly not true, since, for example, $f_{DS}(1) = 1$.

Once again, we see that the standard notions of calculus and elementary kinematics—such as the position and the velocity (or momentum) of a particle moving along the Devil's staircase $\mathbf{\Gamma}_{DS}$, for example—do not make sense in the present context. One has to reinvent an entirely new, and much more general, type of calculus for the 21st century. Aspects of this research program—based on and motivated in part by the theory of complex dimensions (see, e.g., [**370**, **374**, **375**], [**356**] and [**329**], [**330**]), along with the work of C. David and the first author on nowhere differentiable functions and fractal curves (see, e.g., [**106**–**115**]) to be briefly discussed in Section 9.7 on the Weierstrass curve—should be the object of parts of two of the sequels to this book, [**117**] and, especially, [**116**].

Continuous functions—which, like the Cantor (or Cantor–Lebesgue) function, do not satisfy an analog of the Fundamental Theorem of Calculus—are called *singular* (or, rather, *singular continuous*).[72] Accordingly, and because the identity (9.112) holds, the Cantor–Lebesgue function is said to be a *singular function* (or, equivalently, the Lebesgue–Stieltjes measure $\mu(dx) := df_{DS}$) is said to be *singular* with respect to 1-dimensional Lebesgue measure $dx = \mathcal{L}^1$;[73] i.e., $\mathcal{L}^1(C) = 0$.[74] The example of f_{DS} (and of similar staircase functions) is precisely what led H. Lebesgue in the early 1900s to find a proper analog of the Fundamental Theorem of Calculus within the framework of his theory of integration; see [**389**]. Lebesgue's Theorem is valid for so-called absolutely continuous functions; see, e.g., [**5**], [**57**], [**91**], [**193**], [**521**], or [**523**]. Such functions are differentiable $\mathcal{L}^1$-almost everywhere and satisfy the exact counterpart of (9.113) (or of (9.114)), where the derivative inside the integral is interpreted as holding pointwise $\mathcal{L}^1$-almost everywhere. Lebesgue's interest in, and further study of, the Cantor function f_{DS} justifies the fact that f_{DS} is also called the Cantor–Lebesgue function in the literature.

Exercise 9.45. Show that f_{DS} is not differentiable at any point t of the Cantor set C. Furthermore, investigate whether or not, for any $t \in C$ (and, in particular, for any ternary point $t \in T$), f_{DS} has an infinite or a finite (possibly zero) right and/or left derivative.

[72]More precisely, they have a nontrivial singular component in their Jordan decomposition as the sum of an absolutely continuous function and of a singular continuous function; see, e.g., [**193**], [**521**], or [**523**].

[73]Accordingly, μ (also called the *Cantor–Lebesgue measure*) is the Borel measure on $\mathbb{R}$ (or, in fact, on $[0,1]$, here) associated with the nondecreasing function f_{DS}; see also Example B.25 in Section B.4 of Appendix B.

[74]This statement means that the measure μ is concentrated on a set of $\mathcal{L}^1$-measure zero. Here, such a set (i.e., the support of the measure $\mu = \mu_{DS}$) is the Cantor set C, for which $\mu(C) = 1$ and $\mathcal{L}^1(C) = 0$.

More information about the (right, left, or ordinary) derivatives, as well as about Dini derivatives—the latter being defined as upper (or lower) limits of difference quotients—and about Lebesgue derivatives (see below) can be found in Sections 7–9 of the survey article by O. Dovgoshey et al., [**145**].

The *Lebesgue derivative* (or first *symmetric derivative*) of a real-valued function g defined on an interval J is given by the following limit (provided it exists): $SD(g)(t) = \lim_{h\to 0^+} (f(t+h) - f(t-h))/2h$, for all $t \in J$, with the usual convention if t is a left or right endpoint of J. It is an elementary exercise to show that, if g is differentiable at $t \in J$, then its Lebesgue derivative at t exists and coincides with $g'(t)$, the ordinary derivative of g at t: $SD(g)(t) = g'(t)$,[75] and similarly at the (possible) endpoints of J.

The following simple result, due to J. Uher in [**596**], is geometrically intuitive—if one thinks of the Cantor graph as a true staircase, for example—and provides a much cleaner statement than with ordinary and Dini derivatives. We leave it as an exercise for the interested reader to establish it.

Proposition 9.46 (Lebesgue Derivative of f_{DS}; J. Uher [**596**], 1987). *The Cantor function f_{DS} has an infinite Lebesgue (or symmetric) derivative (i.e., $SD(f_{DS})(t) = +\infty$) at every point $t \in C\backslash\{0,1\}$, while (unsurprisingly) $SD(f_{DS})(t) = f'_{DS}(t) = 0$, for all $t \in \Omega$; here, as before, $\Omega := [0,1]\backslash C$.*

Exercise 9.47. Show that f_{DS} maps the Cantor set C onto $[0,1]$ (i.e., $f_{DS}(C) = [0,1]$) and interpret this fact geometrically. Is f_{DS} strictly increasing on C? Justify your answer and try to sketch (at least in your mind's eye) the graph of the restriction of f_{DS} to C, i.e., the part of the Devil's staircase $\mathbf{\Gamma}_{DS}$ which lies above C.

The interested reader may also try to prove the following result, which provides an interesting geometric interpretation of the Cantor (or Cantor–Lebesgue) function.

Theorem 9.48 (Cantor Function and Hausdorff Measure of the Cantor Set). *Let $\mathcal{H}^D$ denote the D-dimensional Hausdorff measure (as defined in Section 7.1), where $D := \log_3 2$ is the Hausdorff dimension (also equal to the Minkowski*

---

[75]Indeed, it suffices to note that each of the following expressions admits the indicated limit below, as $h \to 0^+$:

$$\frac{f(t+h)-f(t-h)}{2h} = \frac{1}{2}\left(\left(\frac{f(t+h)-f(t)}{h}\right) + \left(\frac{f(t)-f(t-h)}{h}\right)\right),$$

$$\frac{f(t+h)-f(t-h)}{2h} = \frac{1}{2}\left(\left(\frac{f(t+h)-f(t)}{h}\right) + \left(\frac{f(t-h)-f(t)}{-h}\right)\right)$$

$$\to \frac{1}{2}(f'(t)+f'(t)) = f'(t), \text{ as } h \to 0^+.$$

dimension) of the Cantor set C. Then,

$$f_{DS}(t) = \mathcal{H}^D([0,t] \cap C) = \mathcal{H}_C([0,t]), \tag{9.115}$$

for all $t \in [0,1]$, where $\mathcal{H}_C$ denotes the natural Hausdorff measure on C with total mass 1 (i.e., such that $\mathcal{H}_C(C) = 1$).[76]

Rephrased: The value $f_{DS}(t)$ of the Cantor function at an arbitrary point $t \in [0,1]$ is equal to $\mathcal{H}^D([0,t])$, the total mass (or "D-dimensional volume") of the portion of C contained in $[0,t]$—or, equivalently, in $(0,t)$, since $\mathcal{H}^D(\{x\}) = 0$, for any $x \in [0,1]$.

Exercise 9.49. Prove Theorem 9.48.

[*Hint*: (i) First, show that (9.115) holds for any $t \in \Omega$ and for every $t \in T$, by verifying that the right-hand side of (9.115) defines a piecewise constant function on Ω taking the same values as f_{DS} on the closure of each interval $I_{k,n}$, in the notation of (9.105) and the text preceding it, as well as of (9.104) and (9.105). (ii) Then, verify that (9.115) holds for all $t \in C$—or, equivalently, for all $t \in C \backslash T$, since the case of ternary points is taken care of by the first step (in (i)), because T consists of all the endpoints of the intervals $I_{k,n}$.]

The Cantor function f_{DS}, being continuous on the compact interval $[0,1]$, is uniformly continuous on $[0,1]$. In fact, significantly more is true, as advanced readers may attempt to show; see, e.g., [**145**, Proposition 10.1].

Theorem 9.50 (Optimal Hölder Continuity of the Cantor Function). *The Cantor function f_{DS} is Hölder continuous, with* optimal *Hölder exponent $D = \log_3 2$,*[77] *the common value of the Minkowski and Hausdorff dimensions of the Cantor set C.*

9.4.6. Self-Affinity, Symmetries, and Functional Equations. We begin this subsection by discussing several interesting symmetries and "scaling relations" (in physicists' terminology) satisfied by the Cantor function f_{DS}—and, in some cases, by the Devil's staircase $\mathbf{\Gamma}_{DS}$ as well.

In the sequel, it may be useful to recall from Example 4.23 in Section 4.3 that the standard IFS $\mathbf{\Phi}_C = \{\varphi_1, \varphi_2\}$ defining the Cantor set C as a self-similar set is given by $\varphi_1(t) := \frac{1}{3}t$ and $\varphi_2(t) := \frac{1}{3}t + \frac{2}{3}$, for all $t \in [0,1]$.

---

[76] It can be shown that the natural Hausdorff measure of the Cantor set C—viewed as a probability measure—is given, for any Borel set A of $\mathbb{R}$ by $\mathcal{H}_C(A) := \mathcal{H}^D(A \cap C)$. Also, $\mathcal{H}_C$ is the unique self-similar measure (in the sense of Definition 4.84 and Theorem 4.82) with equal weights $(1/2, 1/2)$ and associated with the natural IFS $\mathbf{\Phi} = \{\varphi_1, \varphi_2\}$ for the Cantor set C; see Example 4.23 in Section 4.3.

[77] This statement means that f_{DS} is Hölder continuous of order D, but that it is not Hölder continuous of any order (i.e., of Hölder exponent) $\alpha \in (D, 1]$.

Proposition 9.51 (Symmetries and Scaling Relations for f_{DS}). *The Cantor function f_{DS} satisfies the following symmetries and scaling relations or functional equations, for all $t \in [0,1]$:*[78]

(i) $f_{DS}(1-t) = 1 - f_{DS}(t)$;

(ii) $f_{DS}(\frac{t}{3})(= f_{DS}(\varphi_1(t))) = \frac{1}{2} f_{DS}(t)$;

(iii) $f_{DS}(\frac{t+2}{3})(= f_{DS}(\varphi_2(t))) = \frac{1+f_{DS}(t)}{2}$;

(iv) $f_{DS}(1-\frac{t}{3})(= f_{DS}(1-\varphi_1(t))) = 1 - \frac{f_{DS}(t)}{2}$; and

(v) $f_{DS}(\frac{1+t}{3}) = f_{DS}(\frac{1}{3} + \varphi_1(t)) = \frac{1}{2}$.

Remark 9.52. Let $\widetilde{\varphi}_1, \widetilde{\varphi}_2, S : [0,1] \to [0,1]$ be given by $\widetilde{\varphi}_1(t) = \frac{t}{2}$, $\widetilde{\varphi}_2(t) := \frac{t}{2} + \frac{1}{2}$, and $S(t) := 1 - t$, for all $t \in [0,1]$.[79] Then, the symmetries (i)–(iv) in Proposition 9.51 can be restated as follows in terms of composite maps:

(i) $f_{DS} \circ S = S \circ f_{DS}$;

(ii) $f_{DS} \circ \varphi_1 = \widetilde{\varphi}_1 \circ f_{DS}$;

(iii) $f_{DS} \circ \varphi_2 = \widetilde{\varphi}_2 \circ f_{DS}$; and

(iv) $f_{DS} \circ S \circ \varphi_1 = S \circ \widetilde{\varphi}_1 \circ f_{DS}$.

Exercise 9.53. Prove Proposition 9.51. (See Exercise 9.49 above.)

As it turns out, the symmetries (or functional equations) in parts (i) and (ii) of Proposition 9.51 (or, equivalently, of Remark 9.52) uniquely characterize f_{DS} among all nondecreasing functions on $[0,1]$, an elegant and somewhat surprising result, due to D. R. Chalice in [**80**].

Proposition 9.54 (First Functional Equation Characterization of f_{DS}; D. R. Chalice, [**80**], 1991). *The Cantor function f_{DS} is the unique real-valued, nondecreasing function $g : [0,1] \to \mathbb{R}$ satisfying the following functional equations—the same ones as in parts* (i) *and* (ii) *of Proposition* 9.51 (*or, equivalently, of Remark* 9.52 *above*):

$$g(1-t) = 1 - g(t), \text{ for all } \in [0,1], \text{ and} \tag{9.116}$$

$$g\left(\frac{t}{3}\right) = \frac{1}{2} g(t), \text{ for all } t \in [0,1]. \tag{9.117}$$

Proof. We already know from parts (i)–(ii) of Proposition 9.51 and earlier results that f_{DS} is a nondecreasing function satisfying the functional equations (9.116) and (9.117).

[78] Recall that f_{DS} maps $[0,1]$ onto all of $[0,1]$ (i.e., $f_{DS}([0,1] = [0,1])$, so that, for example, the property stated in part (i) makes sense. Also, note that (v) obviously holds since $f_{DS} = 1/2$ on $[1/3, 2/3]$ and $(1+t)/3 \in [1/3, 2/3]$, for all $t \in [0,1]$.

[79] Note that, clearly, S^2 is the identity map of $[0,1]$.

Next, assume that the function g satisfies the properties stated in the proposition. We then claim that $g = f_{DS}$. The proof of this identity is divided into two steps:

Step 1. *$g = f_{DS}$ on $\Omega \cup T = \bigcup_{n=1}^{\infty} \bigcup_{k=1}^{2^{n-1}} J_{k,n}$, where $J_{k,n} := \overline{I_{k,n}}$, the closure of $I_{k,n}$, for each $n \in \mathbb{N}$ and for $k = 1, \ldots, 2^{n-1}$;* see (9.104) and the text surrounding it.[80]

Indeed, observe that by (ii) (i.e., by (9.117) applied to $t := 0$, here)), $g(0) = g(0)/2$ and so, $g(0) = 0$. Thus, by (i) (i.e., by (9.116) also applied to $t := 0$, here), $g(1) = 1 - g(0) = 1 - 0 = 1$. Now, by (ii) (i.e., by (9.117) applied to $t := 1$, here),

$$g\left(\frac{1}{3}\right) = g\left(\frac{1}{3}(1)\right) = \frac{1}{2}g(1) = \frac{1}{2}(1) = \frac{1}{2};$$

and so, by (i) (or (9.116)), again applied to $t := 1/3$, we have that

$$g\left(\frac{2}{3}\right) = 1 - g\left(\frac{1}{3}\right) = 1 - \frac{1}{2} = \frac{1}{2}.$$

But, since g is assumed to be nondecreasing, and since $g(1/3) = g(2/3) = 1/2$, we deduce that g is constant (with value 1/2) on the closed interval $J_{11} := \overline{I_{11}} = [1/3, 2/3]$, as desired. Thus, $g = f_{DS}$ on J_{11}.

We continue in like fashion at the second step of the construction of the Cantor set C. Hence, we want to show that the following identities hold:

$$g = f_{DS} = \frac{1}{4} \text{ on } J_{1,2} := \left[\frac{1}{9}, \frac{2}{9}\right] \quad \text{and} \quad g = f_{DS} = \frac{3}{4} \text{ on } J_{2,2} := \left[\frac{7}{9}, \frac{8}{9}\right]. \tag{9.118}$$

In order to prove this, note that by (ii) (or (9.117)), applied to $t := 1/3$,

$$g\left(\frac{1}{9}\right) = g\left(\frac{1}{3}\left(\frac{1}{3}\right)\right) = \frac{1}{2}g\left(\frac{1}{3}\right) = \frac{1}{2}\left(\frac{1}{2}\right) = \frac{1}{4} \tag{9.119}$$

and (since $g(2/3) = 1/2$, as we saw above) by (ii), applied to $t := 2/3$,

$$g\left(\frac{2}{9}\right) = g\left(\frac{1}{3}\left(\frac{2}{3}\right)\right) = \frac{1}{2}g\left(\frac{2}{3}\right) = \frac{1}{2}\frac{1}{2} = \frac{1}{4}, \tag{9.120}$$

while by (i), applied to $t := 2/9$, and in light of (9.120), we obtain that

$$g\left(\frac{7}{9}\right) = 1 - g\left(\frac{2}{9}\right) = 1 - \frac{1}{4} = \frac{3}{4}. \tag{9.121}$$

Similarly, by (i) (applied to $t := 1/9$) and by (9.119),

$$g\left(\frac{8}{9}\right) = 1 - g\left(\frac{1}{9}\right) = 1 - \frac{1}{4} = \frac{3}{4}. \tag{9.122}$$

In light of (9.119) and (9.120) (resp., (9.121) and (9.122)), and since $g(1/9) = g(2/9) = 1/4$ (resp., $g(7/9) = g(8/9) = 3/4$), it follows much as above from the

[80]Note that $J_{k,n}$ is obtained by adjoining to $I_{k,n}$ its two endpoints, which are necessarily ternary points (i.e., elements of T). Also, recall from (9.104) that $\Omega := [0,1]\backslash C = \bigcup_{n=1}^{\infty} \bigcup_{k=1}^{2^{n-1}} I_{k,n}$.

monotonicity of g that g is constant, with constant value 1/4 (resp., 3/4), on the closed interval $J_{1,2} := \overline{I_{1,2}} = [\frac{1}{9}, \frac{2}{9}]$ (resp., $J_{2,2} := \overline{I_{2,2}} = [\frac{7}{9}, \frac{8}{9}]$). Hence, (9.118) holds, as desired.

Exactly in the same way as just above, we prove by induction on $n \in \mathbb{N}$ that $g = f_{DS}$ on $\bigcup_{l=1}^{n} \bigcup_{k=1}^{2^{l-1}} J_{k,l}$, for all $n \in \mathbb{N}$, from which it follows that, as desired, we have $g = f_{DS}$ on $\bigcup_{l=1}^{\infty} \bigcup_{k=1}^{2^{l-1}} J_{k,l} = \Omega \cup T$ (with $J_{k,\ell} := \overline{I_{k,\ell}}$). This concludes the proof of Step 1.

Step 2. $g = f_{DS}$ *on* $[0,1]$.

Indeed, in light of Step 1, this follows from the uniqueness of a monotonic extension of f_{DS} from Ω (and hence also from $\Omega \cup T$) to all of $[0,1]$, as was shown in Subsection 9.4.1

Finally, since $[0,1] = \Omega \cup T \cup (C\backslash T)$, we deduce at once from Steps 1 and 2 that $g = f_{DS}$, as required. This completes the proof of the proposition. □

The following result is essentially due to W. A. Coppel, [**94**]; see the survey article by O. Dovgoshey, O. Martio, V. Ryazanov, and M. Vuorinen, [**145**], pages 9–10.

Proposition 9.55 (Second Functional Equation: Characterization of f_{DS}; W. A. Coppel, [**94**], 1983). *The Cantor function* f_{DS} *is the unique bounded and real-valued function* g *on* $[0,1]$ *satisfying the following functional equation, for all* $t \in [0,1]$:

$$g(t) = \begin{cases} \frac{1}{2}g(3t), & \text{if } 0 \le t \le \frac{1}{3}, \\ \frac{1}{2}, & \text{if } \frac{1}{3} < t < \frac{2}{3}, \\ \frac{1}{2} + \frac{1}{2}g(3t-2), & \text{if } \frac{2}{3} \le t \le 1. \end{cases} \tag{9.123}$$

Proof. The proof of this result follows from the Banach Fixed Point Theorem (or Contraction Mapping Principle, Theorem 2.58 in Section 2.3, applied to the contraction mapping $\mathcal{T} : \mathcal{E} \to \mathcal{E}$, where the real vector space $\mathcal{E}$ is the complete metric space [81] of bounded real-valued functions on $[0,1]$, equipped with the metric δ_∞ of uniform convergence on $[0,1]$. Namely, δ_∞ is given by $\delta_\infty(\psi_1, \psi_2) := \sup_{0\le t\le 1} |\psi_1(t) - \psi_2(t)|$, and $\mathcal{T} : \mathcal{E} \to \mathcal{E}$ is given (for every $\psi \in \mathcal{E}$ and for all $t \in [0,1]$) by

$$\mathcal{T}(\psi)(t) := \begin{cases} \frac{1}{2}\psi(3t), & \text{if } 0 \le t \le \frac{1}{3}, \\ \frac{1}{2}, & \text{if } \frac{1}{3} < t < \frac{2}{3}, \\ \frac{1}{2} + \frac{1}{2}\psi(3t-2), & \text{if } \frac{2}{3} \le t \le 1. \end{cases} \tag{9.124}$$

[81] Really, here, $\mathcal{E}$ is a complete normed linear space; i.e., it is a Banach space.

Indeed, it follows from the aforementioned Contraction Mapping Principle (Theorem 2.58) that $\mathcal{T}$ has a unique fixed point $\psi^* \in \mathcal{E}$; i.e., $\mathcal{T}(\psi^*) = \psi^*$. According to our choice of $\mathcal{T}$ (in (9.124)) and of $\mathcal{E}$, this precisely means that the functional equation (9.123) possesses a real-valued and bounded solution (namely, ψ^*) and that this solution is unique, as stated in the proposition.

Also, it can be easily verified that the Cantor function f_{DS} satisfies the functional equation (9.123) and that f_{DS} belongs to $\mathcal{E}$ (since $f_{DS}([0,1]) \subseteq [0,1]$). Hence, $\mathcal{T}(f_{DS}) = f_{DS}$. By uniqueness of the fixed point of $\mathcal{T}$, we conclude that $f_{DS} = \psi^*$, as needed to be shown.

It now suffices to check that the linear map $\mathcal{T}$ is a contraction (with Lipschitz constant $1/2 < 1$, here), which is deduced at once from the following easily verified inequality (in light of (9.123) and (9.124)): For all $\psi_1, \psi_2 \in \mathcal{E}$,

(9.125)
$$\delta_\infty(\mathcal{T}(\psi_1), \mathcal{T}(\psi_2)) = \delta_\infty(\mathcal{T}(\psi_1 - \psi_2), 0) \leq \frac{1}{2}\delta_\infty(\psi_1 - \psi_2, 0) = \frac{1}{2}\delta_\infty(\psi_1, \psi_2).$$

This completes the proof of the proposition. □

Corollary 9.56. *The Cantor function f_{DS} is the unique bounded and real-valued function g on* $[0,1]$ *satisfying the functional equations* (ii), (iv), *and* (v) *in the statement of Proposition* 9.51, *except with f_{DS} replaced by g.*

Remark 9.57 (*Self-Affinity and Scaling Relations*). In light of the symmetry $f_{DS}(1-t) = 1 - f_{DS}(t)$—expressed in part (i) of Proposition 9.51 (or in the functional equation (9.116) of Proposition 9.54)—the scaling relations stated in parts (ii)–(iv) of Proposition 9.51) are concrete manifestations of the self-affinity of the Cantor graph (i.e., of the Devil's staircase) $\mathbf{\Gamma}_{DS}$; see Remark 9.58 just below.

Remark 9.58 (*Inhomogeneous Self-Affinity of the Devil's Staircase $\mathbf{\Gamma}_{DS}$*). We refer to Example 4.93 in Section 4.10 above for a detailed discussion of the Devil's staircase (or Cantor graph) $\mathbf{\Gamma}_{DS}$, viewed as an inhomogeneous self-affine set.

9.4.7. Complex Dimensions and Fractality of the Devil's Staircase. The complex dimensions of the Devil's staircase (or Cantor graph) $\mathbf{\Gamma}_{DS}$ express the oscillations that are intrinsic to the geometry of $\mathbf{\Gamma}_{DS}$. They have first been semi-heuristically determined by the first author and M. van Frankenhuijsen in [**370**, Section 10.3] and [**374**], as well as in [**375**, Section 12.1], via an approximate computation of $V_{DS}(\varepsilon)$, the volume (area, here) of the ε-neighborhood of $\mathbf{\Gamma}_{DS}$. At the time of [**370**], [**374**], and [**375**], there was not yet a comprehensive higher-dimensional theory of complex fractal dimensions. We only had at our disposal the 1-dimensional theory of complex dimensions, i.e., that of fractal strings (and of fractal sprays, [**354**], [**319**], [**320**], [**348–350**]).

The situation changed radically, however, when the first author (M. L. Lapidus) discovered (in 2009) the so-called distance zeta function, out of which (along with its close relative, the tube zeta function) the higher-dimensional theory of complex dimensions was developed by M. L. Lapidus, G. Radunović, and D. Žubrinić, in the book [**356**] and a series of accompanying papers, [**357–365, 367**], [**355**]. See Section 9.3 above, especially Subsection 9.3.8. This enabled the authors of the just-mentioned references to verify and establish the predictions of [**370**], [**374**], [**375**] regarding the complex dimensions of the Devil's staircase $\mathbf{\Gamma}_{DS}$ (viewed as a compact subset of $\mathbb{R}^2$) and to thereby show that the set $\mathcal{D}_{DS,pos}$ of (possible) complex dimensions of $\mathbf{\Gamma}_{DS}$ is given by

$$\begin{aligned}\mathcal{D}_{DS,pos} &= \{0,1\} \cup \left\{ \log_3 2 + in\frac{2\pi}{\log 3} \ : \ n \in \mathbb{Z} \right\} \\ &= \{0\} \cup \left\{ \frac{\log 2}{\log 3} + in\frac{2\pi}{\log 3} \ : \ n \in \mathbb{Z} \right\} \cup \{1\},\end{aligned} \tag{9.126}$$

where, in the second equality of (9.126), we have ordered the (possible) complex dimensions of $\mathbf{\Gamma}_{DS}$ according to their increasing real parts, $0, \log_3 2$, and 1.

Note that the set of (possible) complex dimensions of $\mathbf{\Gamma}_{DS}$ with real part $\log_3 2$ coincides with the set $\mathcal{D}_{CS}$ of (exact) complex dimensions of the Cantor string $\mathcal{L}_{CS}$ (or, equivalently, here, of the Cantor set C, viewed as a compact subset of $\mathbb{R}$), as given by

$$\mathcal{D}_{CS} = \left\{ \log_3 2 + in\frac{2\pi}{\log 3} \ : \ n \in \mathbb{Z} \right\} = \{ D_{CS} + in\mathbf{p} \ : \ n \in \mathbb{Z} \}, \tag{9.127}$$

where

$$D_{CS} = \dim_M(\mathcal{L}_{CS}) = \dim_M(C) = \log_3 2 \tag{9.128}$$

is the Minkowski dimension of the Cantor string (or, equivalently, of the Cantor set), while

$$\mathbf{p} = \mathbf{p}_{DS} = \mathbf{p}_{CS} := \frac{2\pi}{\log 3} \tag{9.129}$$

is the *oscillatory period* of the Cantor string (as well as of the Cantor set).[82]

Accordingly, since it is shown in loc. cit. (see [**359**], [**356**, Subsection 5.5.4]) that the complex dimensions of $\mathbf{\Gamma}_{DS}$ must be simple, the *fractal tube formula* for the Devil's staircase (or Cantor graph) $\mathbf{\Gamma}_{DS}$ must be of the following form

[82] See formula (8.56) of Example 8.17 in Section 8.2. Note that in Example 8.17, $\mathcal{L}_{CS}, \mathcal{D}_{CS}, D_{CS}$, and $\mathbf{p}_{CS}$ are denoted $\mathcal{L}_{can}, \mathcal{D}_{can}, D_{can}$, and $\mathbf{p}_{can}$, respectively.

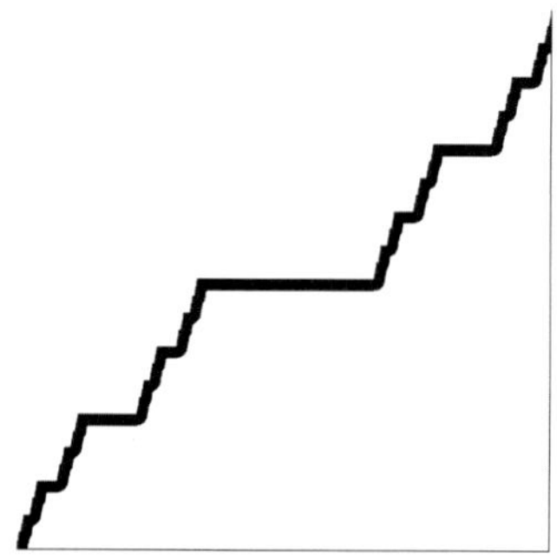

Figure 9.9. A tubular neighborhood (or ε-neighborhood) of the Devil's staircase (with respect to the Euclidean metric). The tubular volume $V_{DS}(\varepsilon)$ is the volume (area, here) of this ε-neighborhood.

(see Subsection 9.3.7 above and [**356**, Chapter 5]):[83] For all $\varepsilon > 0$ small enough,

$$(9.130)\quad V_{DS}(\varepsilon) = \sum_{\omega \in \mathcal{D}_{DS,pos}} \operatorname{res}(\tilde{\zeta}_{DS}, \omega) \cdot \varepsilon^{2-\omega} = \sum_{\omega \in \mathcal{D}_{DS,pos}} \frac{\operatorname{res}(\zeta_{DS}, \omega) \cdot \varepsilon^{2-\omega}}{2-\omega},$$

where $V_{DS}(\varepsilon)$ denotes the volume (i.e., the area or 2-dimensional Lebesgue measure, here) of the ε–neighborhood (or tubular neighborhood; see Figure 9.9) of $\mathbf{\Gamma}_{DS}$, while $\tilde{\zeta}_{DS}$ and ζ_{DS} are, respectively, the tube and distance zeta functions of the Devil's staircase $\mathbf{\Gamma}_{DS}$.

The *fractal power series* appearing in (9.130) can be rewritten as follows, with the coefficients c_ω in the first two equalities of (9.131) just below given in footnote 83, for every $\omega \in \mathcal{D}_{DS,pos}$:[84]

(9.131)

$$V_{DS}(\varepsilon) = \sum_{\omega \in \mathcal{D}_{DS,pos}} c_\omega \cdot \varepsilon^{2-\omega} = c_0 \cdot \varepsilon^2 + \sum_{\operatorname{Re}(\omega)=\log_3 2} c_\omega \cdot \varepsilon^{2-\omega} + c_2 \cdot \varepsilon^{2-1}$$

$$= c_0 \cdot \varepsilon^2 + \varepsilon^{2-\log_3 2} \cdot G_{DS}(\log_3(\varepsilon^{-1})) + c_2 \cdot \varepsilon,$$

where $c_2 > 0$ and $G_{DS} = G_{DS}(x)$ is a continuous, bounded, and 1-periodic function on $\mathbb{R}$, whose Fourier coefficients can easily be expressed in terms of the coefficients c_ω, with $\operatorname{Re}(\omega) = \log_2 3$, i.e., with $\omega = D_{CS} + in\mathbf{p}$, where $n \in \mathbb{Z}$.[85] Equivalently, still with $x := \log_3(\varepsilon^{-1})$ (for $0 < \varepsilon < 1$), $g_{DS}(\varepsilon) := G_{DS}(x)$ is a

[83] It is a general fact (see [**356**, Corollary 2.2.20, equation (2.2.55)]) that the residues of the distance and tube zeta functions at any (simple) $\omega \in \mathcal{D}_{DS,pos}$ are related by the following relation: $c_\omega := \operatorname{res}(\tilde{\zeta}_{DS}, \omega) = \operatorname{res}(\zeta_{DS}, \omega)/(2-\omega)$; see identity (9.78) above.

[84] It is possible that some of the coefficients c_ω in (9.131) (or, equivalently, of the residues in (9.130)) be equal to zero, in case the corresponding possible complex dimension ω is not an actual complex dimension of $\mathbf{\Gamma}_{DS}$; i.e., it is not an actual pole of $\tilde{\zeta}_{DS}$ (or, equivalently, of ζ_{DS}), which conjecturally, should never happen—or, at least, happen only exceptionally.

[85] Note that (with $x := \log_3(\varepsilon^{-1}) > 0$, for $0 < \varepsilon < 1$), $\varepsilon^{-in2\pi/\log 3} = e^{in2\pi \log_3(\varepsilon^{-1})}$; and so, $\varepsilon^{2-(\log_3 2 + in\mathbf{p}_{CS})} = \varepsilon^{2-\log_3 2} \cdot e^{in2\pi x}$.

continuous, bounded, and multiplicatively periodic (or "log periodic") function on $(0,+\infty)$, the "space of scales".

Since ε is small (and, in particular, $\varepsilon \in (0,1)$), the *dominant term* (or *leading term*) in (9.131) is $c_2 \cdot \varepsilon$, which corresponds to $\omega = 1$, the Minkowski dimension of $\mathbf{\Gamma}_{DS}$ (i.e., $\dim_M(\mathbf{\Gamma}_{DS}) = 1$),[86] the term involving the periodic function G_{DS} in the last equality of (9.131) is the *subdominant term* and corresponds to the possible complex dimensions of $\mathbf{\Gamma}_{DS}$ with real part $\log_3 2$, with $0 < \log_3 2 = D_{CS} < 1$ (i.e., to the complex dimensions of the Cantor string or, equivalently, of the Cantor set, given by (9.127)), while the first term, $c_0 \cdot \varepsilon^2$, in (9.131), is the smallest order term (as $\varepsilon \to 0^+$) and corresponds to the (possible) complex dimension $\omega = 0$ of $\mathbf{\Gamma}_{DS}$.

Consequently, it is natural to rewrite (9.131) in the order of decreasing (rather than increasing) real parts of the underlying complex dimensions. Accordingly, we have that

$$V_{DS}(\varepsilon) = c_2 \cdot \varepsilon + \varepsilon^{2-\log_3 2} \cdot G_{DS}(\log_3(\varepsilon^{-1})) + c_0 \cdot \varepsilon^2, \tag{9.132}$$

as $\varepsilon \to 0^+$, from which it follows immediately that the Minkowski dimension D_{DS} of $\mathbf{\Gamma}_{DS}$ exists and equals 1 and that $\mathbf{\Gamma}_{DS}$ is Minkowski measurable, with Minkowski content $\mathcal{M}_{DS}$ given by

$$\begin{aligned} \mathcal{M}_{DS} &= \lim_{\varepsilon \to 0^+} \frac{V_{DS}(\varepsilon)}{\varepsilon^{2-1}} = c_2 = \operatorname{res}(\tilde{\zeta}_{DS}, 1) \\ &= \frac{\operatorname{res}(\zeta_{DS}, 1)}{2-1} = \operatorname{res}(\zeta_{DS}, 1); \end{aligned} \tag{9.133}$$

see footnote 83 and Subsection 9.3.6 above, along with [**356**].

Exercise 9.59 (*Length and Minkowski Content of* $\mathbf{\Gamma}_{DS}$).

(a) Try to relate the Minkowski content $\mathcal{M}_{DS}$ and the (finite) length, $L_{DS} = 2$, of the Devil's staircase $\mathbf{\Gamma}_{DS}$; see Proposition 9.40 in Subsection 9.4.3 above.

(b) Answer perhaps a simpler question, namely, the exact same question as in part (a) except for the fact that the Devil's staircase $\mathbf{\Gamma}_{DS}$ is now replaced by its RFD counterpart, $(\mathbf{\Gamma}_{DS}, \Omega_{DS})$, discussed just below. (In fact, we will give the answer in (9.139) below.)

---

[86] The fact that $\dim_M(\mathbf{\Gamma}_{DS}) = 1$ follows from the general theory of complex dimensions, according to which the abscissa of convergence of ζ_{DS} and of $\tilde{\zeta}_{DS}$ (necessarily equal to 1, here, since $c_2 \neq 0$) coincides with the Minkowski dimension of $\mathbf{\Gamma}_{DS}$, $\overline{\dim}_M(\mathbf{\Gamma}_{DS}) = 1$; see part (a) of Theorem 2.1.11 in [**356**] or Subsection 9.4.3 above. Although it is not really needed for this discussion, we point out that it can be shown that $\dim_M(\mathbf{\Gamma}_{DS})$ exists and hence is given by $\dim_M(\mathbf{\Gamma}_{DS}) = 1$. Note that the fact that $\dim_M(\mathbf{\Gamma}_{DS}) = 1$ can also be deduced from the fact that the Devil's staircase $\mathbf{\Gamma}_{DS}$ is a rectifiable and continuous curve; see [**182**].

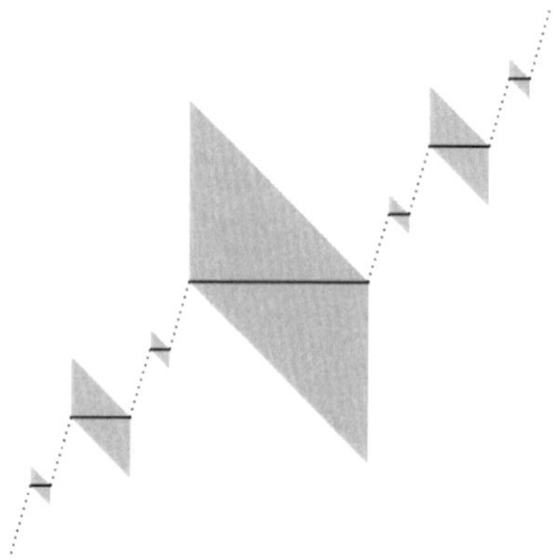

Figure 9.10. The Devil's staircase (or Cantor graph) RFD, $(\mathbf{\Gamma}_{DS}, \Omega_{DS})$, where Ω_{DS} roughly (but not exactly) corresponds to a suitable ε-neighborhood of the usual Devil's staircase $\mathbf{\Gamma}_{DS}$ (but in a different metric than the Euclidean metric on $\mathbb{R}^2$).

In [**356**, Example 5.5.14, Subsection 5.5.4] (see also [**359**] and [**361**]) is introduced a relative fractal drum (or RFD), $(\mathbf{\Gamma}_{DS}, \Omega_{DS})$, of $\mathbb{R}^2$, called the *Cantor graph* (or *Devil's staircase*) RFD.[87] This RFD is depicted in Figure 9.10 and briefly described in the caption of that figure. Roughly speaking, Ω_{DS} corresponds to (but does not quite coincide with) a suitable ε-neighborhood of the Devil's staircase $\mathbf{\Gamma}_{DS}$, but in the ordinary "sup metric" $d_{2,\infty}$ instead of the Euclidean metric in $\mathbb{R}^2$; recall that in this context, and for any $\mathbf{x} = (x_1, x_2)$, $\mathbf{y} = (y_1, y_2)$ in $\mathbb{R}^2$,

$$d_{2,\infty}((x_1, y_2), (x_2, y_2)) := \max(|x_2 - x_1|, |y_2 - y_1|). \tag{9.134}$$

It turns out that, for the Devil's staircase RFD, $(\mathbf{\Gamma}_{DS}, \Omega_{DS})$, the set $\mathcal{D}_{DS,F} = \mathcal{D}_{\mathbf{\Gamma}_{DS},\Omega_{DS}}$ of complex dimensions is exactly equal to the set of (its) possible complex dimensions, as well as coincides with $\mathcal{D}_{DS,pos}$, the set of possible complex dimensions of $\mathbf{\Gamma}_{DS}$ (see [**356**, Example 5.5.14], [**359**], or [**361**]):

$$\begin{aligned} \mathcal{D}_{DS} \subseteq \mathcal{D}_{DS,F} &= \mathcal{D}_{\mathbf{\Gamma}_{DS},\Omega_{DS}} (= \mathcal{D}_{\mathbf{\Gamma}_{DS},\Omega_{DS,pos}}) = \mathcal{D}_{DS,pos} \\ &= \{0,1\} \cup \{D_{DS} + in\mathbf{p}_{DS} : n \in \mathbb{Z}\} = \{0,1\} \cup \mathcal{D}_{CS}, \end{aligned} \tag{9.135}$$

where $\mathcal{D}_{DS,pos}$ and $\mathcal{D}_{CS}$ are given, respectively, by (9.126) and (9.127), and with $D_{CS} = \log_3 2$ and $\mathbf{p}_{CS} = 2\pi/\log_3$, as before in (9.128) and (9.129), respectively; see Figure 9.11.

[87] See Subsection 9.3.8 above for a brief discussion of the notion of a relative fractal drum, which—within the context of the higher-dimensional theory of complex dimensions [**356**]—extends that of a bounded set in $\mathbb{R}^m$; here, $m = 2$.

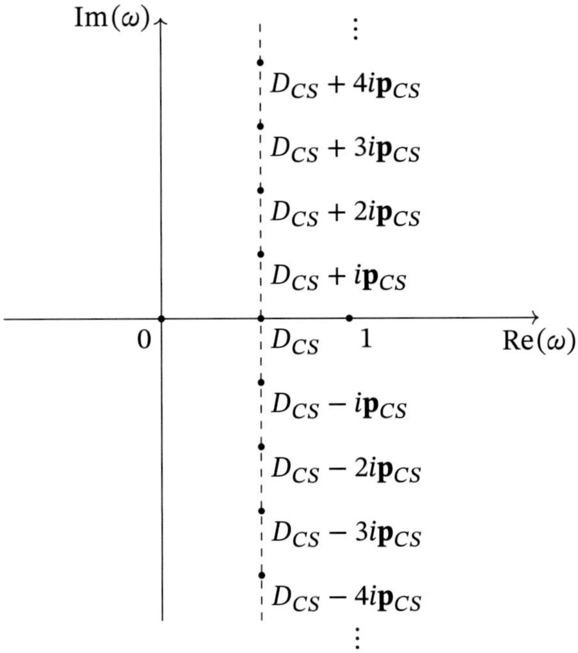

Figure 9.11. The set of possible complex dimensions of the Devil's staircase (or Cantor graph) RFD, $(\mathbf{\Gamma}_{DS}, \Omega_{DS})$: $\mathcal{D}_{DS,pos} = \mathcal{D}_{DS,F} = \{0,1\} \cup \{D_{CS} + in\mathbf{p}_{CS} : n \in \mathbb{Z}\}$, where $D_{CS} = \log_3 2$ is the Minkowski dimension of the Cantor string (or set) and $\mathbf{p}_{CS} = 2\pi/\log 3$ is the underlying oscillatory period.

In addition, by using the scaling and symmetry invariance properties of distance zeta functions briefly recalled in Subsection 9.3.8 above, it is shown in [**356**, Subsection 5.5.4] that the distance zeta function of $(\mathbf{\Gamma}_{DS}, \Omega_{DS})$, denoted $\zeta_{\mathbf{\Gamma}_{DS},\Omega_{DS}} = \zeta_{DS,F}$, is a meromorphic function in all of $\mathbb{C}$ given by

$$\zeta_{DS,F}(s) = \frac{2}{s(3^s - 2)(s-1)}, \quad \text{for all } s \in \mathbb{C}, \tag{9.136}$$

from which one deduces that the set $\mathcal{D}_{\mathbf{\Gamma}_{DS},\Omega_{DS}} = \mathcal{D}_{DS,F}$ of (exact) complex dimensions of $(\mathbf{\Gamma}_{DS}, \Omega_{DS})$ is, indeed, given by (9.135) and solely consists of simple complex dimensions, i.e., of simple poles of $\zeta_{\mathbf{\Gamma}_{DS},\Omega_{DS}}$ (or, equivalently, of the tube zeta function $\tilde{\zeta}_{\mathbf{\Gamma}_{DS},\Omega_{DS}}$).

Moreover, it is shown in loc. cit. (or in [**361**]) that the (exact, pointwise), fractal tube formula for the Devil's staircase RFD $(\mathbf{\Gamma}_{DS}, \Omega_{DS})$ takes the same form as the expected one for the Devil's staircase $\mathbf{\Gamma}_{DS}$ (compare (9.131) or (9.132) above) and can be precisely calculated as follows, for all $\varepsilon \in (0,1)$:

$$\begin{aligned} V_{DS,F}(\varepsilon) = V_{\mathbf{\Gamma}_{DS},\Omega_{DS}}(\varepsilon) &= 2\varepsilon^{2-D_{DS}} + \varepsilon^{2-D_{CS}} \cdot G_{DS,F}(\log_3(\varepsilon^{-1})) + \varepsilon^2 \\ &= 2\varepsilon + \varepsilon^{2-D_{CS}} \cdot G_{DS,F}(\log_3(\varepsilon^{-1})) + \varepsilon^2, \end{aligned} \tag{9.137}$$

where $D_{DS} = D_{DS,F} = 1$ is the Minkowski dimension of $(\mathbf{\Gamma}_{DS}, \Omega_{DS})$. Furthermore, in (9.137), $G_{DS,F} = G_{DS,F}(x)$ is a bounded and continuous 1-periodic function on $\mathbb{R}$ which is *nonconstant* and is given by the following explicit and convergent (because absolutely convergent) Fourier series (*with only nonzero Fourier coefficients*), where $\omega_n := D_{CS} + in\mathbf{p}_{CS}$, for all $n \in \mathbb{Z}$:[88]

$$G_{DS,F}(x) = \frac{1}{\log 3} \cdot \sum_{n \in \mathbb{Z}} \frac{e^{2\pi i n x}}{(2 - \omega_n)(\omega_n - 1)\omega_n}, \text{ for all } x \in \mathbb{R}, \tag{9.138}$$

It also follows from the fractal tube formula (9.137) that the RFD $(\mathbf{\Gamma}_{DS}, \Omega_{DS})$ is Minkowski measurable and that its Minkowski content $\mathcal{M}_{DS,F}$ is equal to the length of the Devil's staircase; namely,

$$\mathcal{M}_{DS,F} = 2 = L_{DS,F} := L(\mathbf{\Gamma}_{DS}). \tag{9.139}$$

Now, why is the example of the Devil's staircase $\mathbf{\Gamma}_{DS}$ or of its RFD counterpart $(\mathbf{\Gamma}_{DS}, \Omega_{DS})$ so interesting and important in fractal geometry, as well as for the theory of complex dimensions? Because, as was alluded to in the introduction of the present section (i.e., of Section 9.4), anyone with a good geometric intuition would not hesitate calling $\mathbf{\Gamma}_{DS}$ or $(\mathbf{\Gamma}_{DS}, \Omega_{DS})$ a "fractal", whatever definition of fractality she or he would like to adopt.

However, Benoît Mandelbrot, in his impactful and fascinating book, *The Fractal Geometry of Nature* [**419**], introduced (admittedly reluctantly) the following definition of "fractality":

A geometric object is *fractal* if its Hausdorff dimension H is strictly greater than its topological dimension T; namely, $T < H$. (Recall from Theorem 8.50 in Subsection 8.6.2 that the topological dimension never exceeds the Hausdorff dimension; i.e., $T \leq H$.)

Well, here, whether for $\mathbf{\Gamma}_{DS}$ or for its RFD counterpart, $(\mathbf{\Gamma}_{DS}, \Omega_{DS})$,[89] this definition is definitely not satisfied. Indeed, for the Devil's staircase, for example, and denoting by $M = D_{DS}$ the Minkowski dimension of $\mathbf{\Gamma}_{DS}$, we have that[90]

$$T = 1, \quad H = 1, \quad M\ (= D_{DS}) = 1. \tag{9.140}$$

---

[88]In fact, it follows from (9.136) that, for all $n \in \mathbb{Z}$, $\operatorname{res}(\zeta_{DS,F}, \omega_n) = 1/\log(3(\omega_n - 1)\omega_n)$, and from the general higher-dimensional theory of complex dimension developed in [**356**] that $\operatorname{res}(\tilde{\zeta}_{DS,F}, \omega_n) = \operatorname{res}(\zeta_{DS}, \omega_n)/(2 - \omega_n)$; see footnote 83 on page 434 above.

[89]Assuming, in the latter case, that one can appropriately extend to RFDs the notions of topological and Hausdorff dimensions, or that, alternatively, one replaces the Hausdorff dimension by the relative Minkowski dimension of the RFD, as defined and studied in [**356**, Chapter 4].

[90]Let us briefly explain why (9.140) holds: (i) $T = 1$ because it is the topological dimension of the graph of a continuous parametrized curve (since $\mathbf{\Gamma}_{DS}$ is the graph of the continuous Cantor function). (ii) $M = 1$ because it is the Minkowski dimension of a continuous *rectifiable* curve (see [**182**]), as can be shown directly or follows from the aforementioned results of [**356**] or of [**359**]; see (9.132) (or (9.131)) and the text following it. (iii) Finally (in light of Proposition 8.39 and Theorem 8.50), since $T \leq H \leq M$, with $T = M = 1$, it follows that $T = H = M = 1$, as claimed in (9.140).

It follows, in particular, that $\mathbf{\Gamma}_{DS}$ is *not* fractal, according to Mandelbrot's definition, contrary to all expectations and one's geometric intuition.

This paradox can now be resolved by means of the theory of complex fractal dimensions—according to which, as we recall, a geometric object is "fractal" if it admits at least one *nonreal* complex dimension (thereby giving rise to geometric oscillations, for instance, via the fractal tube formulas). Here, for the Devil's staircase RFD, we even have infinitely many complex conjugate pairs of nonreal complex dimensions—namely, the elements of

$$\mathcal{D}_{CS}\backslash\{0\} = \{\omega_n := D_{CS} + in\mathbf{p}_{CS} \, : \, n \in \mathbb{Z}, n \neq 0\}, \tag{9.141}$$

with $D_{CS} = \log_3 2$ and $\mathbf{p}_{CS} = 2\pi/\log 3$—which give rise, via the fractal tube formula (9.137) above, to *subcritical* (or *subdominant*) *oscillations* in the volume of tubular neighborhoods of $(\mathbf{\Gamma}_{DS}, \Omega_{DS})$.[91] Therefore, we conclude that $(\mathbf{\Gamma}_{DS}, \Omega_{DS})$, the Devil's staircase (or Cantor graph) RFD, is fractal, as claimed. (See also Remark 9.62 below.)

Exactly the same conclusions could be drawn for the Devil's staircase itself (or Cantor graph) $\mathbf{\Gamma}_{DS}$, provided—as expected and stated in [**356**, Example 5.5.14]—the following conjecture can be proved.[92]

Conjecture 9.60 (Fractality of the Devil's Staircase). *The possible* nonreal complex *dimensions* $\omega_n = D_{CS} + in\mathbf{p}_{DS}$ *(with* $n \in \mathbb{Z}\backslash\{0\}$ *arbitrary) of the Devil's staircase* $\mathbf{\Gamma}_{DS}$*—or at least one of them,* ω_n *(and hence also, necessarily, its complex conjugate* $\overline{\omega}_n = \omega_{-n}$*), which would suffice here—are actual complex dimensions of* $\mathbf{\Gamma}_{DS}$ *(i.e., are poles of* ζ_{DS} *or, equivalently, of* $\tilde{\zeta}_{DS}$*).*[93]

We point out that, according to the aforementioned results of [**356**] *and* [**361**]*, its RFD counterpart,* $(\mathbf{\Gamma}_{DS}, \Omega_{DS})$*, is* fractal*, in the sense of the theory of complex dimensions, in agreement with geometric intuition.*[94]

---

[91]We use here the phrase "subcritical" (or *subdominant*) *oscillations* because in the fractal tube formula (9.137), the *dominant term* (as $\varepsilon \to 0^+$) is 2ε, while the *subdominant term* is the next term, which is bounded by a finite and positive constant times $\varepsilon^{2-D_{CS}}$ (with $2 - D_{CS} = 2 - \log_3 2 \approx 1.3691 > 1$) and corresponds to the nonreal complex dimensions (all of real part $D_{CS} = \log_3 2$), as well as to D_{DS} itself.

[92]This conjecture was essentially (but certainly not fully) proved in [**370**, Section 10.2] (see also [**374**, Section 12.2] or [**375**, Section 12.1]) via an approximate computation, which, however, was not published in detail in the corresponding books and preceded by many years the development of the higher-dimensional theory of complex fractal dimensions in [**356**] and a series of accompanying papers. According to loc. cit., not only 1 belongs to $\mathcal{D}_{DS}$, but also $\mathcal{D}_{CS} = \{D_{DS} + in\mathbf{p} \, : \, n \in \mathbb{Z}\} \subset \mathcal{D}_{DS}$. We would also like to obtain an exact fractal formula, much as in (9.131) or (9.132), but now, with explicit coefficients, if at all possible.

[93]That is, $\operatorname{res}(\zeta_{DS}, \omega_n) \neq 0$ or, equivalently, in light of footnote 83 on page 434, we have that $\operatorname{res}(\tilde{\zeta}_{DS}, \omega_n) \neq 0$, for some $n > 0$ but also, most likely, for infinitely many $n > 0$ (and hence for infinitely many pairs $(n, -n)$, with $n \in \mathbb{Z}\backslash\{0\}$).

[94]Indeed, in loc. cit., this fact is verified numerically by checking that the residue of $\tilde{\zeta}_{\mathbf{\Gamma}_{DS}, \Omega_{DS}}$ at ω_n (and hence also at $\omega_{-n} = \overline{\omega}_n$) is nonzero, for finitely many values of $n \in \mathbb{N}$.

We close this discussion by quoting an excerpt of Mandelbrot's discussion of the Devil's staircase $\mathbf{\Gamma}_{DS}$, in his celebrated book *The Fractal Geometry of Nature* (see [**419**], page 83):

> *One would love to call the present curve* [the Devil's staircase $\mathbf{\Gamma}_{DS}$] *a fractal, but to achieve this goal we would have to define fractals less stringently, on the basis of notions other than* [the Hausdorff dimension] *H alone.*[95]

Our contention is that the missing notion was that of complex dimensions, which enables one to capture the essence of fractality via the presence of intrinsic geometric oscillations—themselves associated with the presence of *nonreal* complex dimensions.

The aforementioned results from [**356**] in (9.137), (9.138), and (9.139), concerning the Devil's staircase RFD $(\mathbf{\Gamma}_{DS}, \Omega_{DS})$, along with the earlier results from [**370**], [**374**], and [**375**] mentioned in footnote 92, provide strong corroborative evidence for Conjecture 9.60. We invite the interested reader to attempt completely solving this challenging open problem.

Remark 9.61. Thanks to the new approach to—and significant refinement as well as extension of—the higher-dimensional theory of complex dimensions (in [**356**]) developed by C. David and the first author in [**106–115**]—by means, in particular, of prefractal approximations, iterated fractal drums (IFDs), effective fractal zeta functions, and polyhedral measures[96]—Conjecture 9.60, suitably interpreted in terms of the Devil's staircase iterated fractal drum (IFD) naturally associated with the Cantor graph, would now seem to be within reach. We also expect the corresponding exact (and not just the possible) complex dimensions can be precisely determined, much as is done for the Weierstrass curve (or IFD) in [**110**] (announced in [**112**] and [**108, 109**]) and, for the Koch curve, in [**111**].

Remark 9.62 (*Subcriticality Index of Fractality*)**.** Given any (nonempty) bounded subset A of $\mathbb{R}^m$—or, more generally, any RFD (A, Ω) of $\mathbb{R}^m$—with $m \geq 1$, one can define (as in [**356**], equation (5.5.65), page 500) the following *subcriticality index* (or *index of fractality*):

$$\alpha_A = \sup\{d \in \mathbb{R} : A \text{ is fractal in dimension } d\}, \tag{9.142}$$

and entirely similarly for $\alpha_{A,\Omega}$, with A replaced by (A, Ω). (By convention, $\sup(\emptyset) = -\infty$.) Here, A (or, more generally, (A, Ω)) is said to be *fractal in dimension d* if it has at least one *nonreal* complex dimension ω with real part d : $\mathrm{Re}(\omega) = d$, $\omega \neq d$—and hence, $\overline{\omega}$, the complex conjugate of ω, is also a complex dimension satisfying $\mathrm{Re}(\overline{\omega}) = d$, $\overline{\omega} \neq d$.

---

[95] In Mandelbrot's original text, H was denoted by D.

[96] See Section 9.7, especially Subsections 9.7.2 and 9.7.3, for a very brief introduction to this extended theory.

Then, according to the aforementioned results, for the Devil's staircase RFD $(\mathbf{\Gamma}_{DS}, \Omega_{DS})$, the subcriticality index (denoted by $\alpha_{DS,F}$, for simplicity) is given by

$$\alpha_{DS,F} = D_{CS} = \log_3 2 < 1 = D_{DS,F}, \tag{9.143}$$

while, according to Conjecture 9.60 just above, exactly the same relation should hold for the Devil's staircase $\mathbf{\Gamma}_{DS}$ itself; namely, the subcriticality index of $\mathbf{\Gamma}_{DS}$ (also denoted by α_{DS}, for simplicity) should be given by

$$\alpha_{DS} = \alpha_{DS,F} = D_{CS} = \log_3 2 < 1 = D_{DS}. \tag{9.144}$$

Note that, according to footnote 94 on page 439, the inequality stated in (9.144) —and hence also the fractality of the Devil's staircase $\mathbf{\Gamma}_{DS}$—has been verified numerically in [**356**] and [**361**].

This fact justifies, a posteriori, the use (without any accompanying definition) of the phrase *borderline fractal* in several texts on fractal geometry, such as [**483**]. Now, however, thanks to the theory of complex dimensions, we have a precise and mathematical way of defining both fractality (or *strong fractality*, if α_A or $\alpha_{A,\Omega}$ coincides with the Minkowski dimension of A or of (A, Ω), respectively) and *subfractality* (or *borderline fractality*), otherwise, i.e., if $\alpha_A < D_A = \dim_M(A)$ or $\alpha_{A,\Omega} < D_{A,\Omega} = \dim_M(A, \Omega)$, respectively.

Remark 9.63 (*Fractality of the Peano and the Hilbert Plane-Filling Curves*). Just like the Devil's staircase, the Peano curve ([**478, 479**]; see Figure 1.10 on page 33 in Subsection 1.4.3)—as well as other plane-filling curves, such as the Hilbert curve ([**247**]; see Figure 9.12 on page 442 in this section) and their higher-dimensional analogs, called space-filling curves—are clearly "fractal", whatever one may wish to mean by this term. However, just like the Devil's staircase, plane-filling curves $\mathbf{\Gamma}_2$—or, more generally, space-filling curves $\mathbf{\Gamma}_m$ (which fill the unit cube in $\mathbb{R}^m$, $Q_0 := [0,1]^m$, with $m \in \mathbb{N}$ and $m \geq 2$)—are not fractal according to Mandelbrot's definition since, obviously, $\mathbf{\Gamma}_m = Q_0$ (with $m := 2$ in the case of a plane-filling curve), and hence,

$$\dim_T(\mathbf{\Gamma}_m) = \dim_H(\mathbf{\Gamma}_m)\ (= \dim_M(\mathbf{\Gamma}_m) = \dim_B(\mathbf{\Gamma}_m)) = m.$$

As a result, the first author has long conjectured (since the early-to-mid 2000s) that all space- (or plane)-filling curves are fractal, in the sense of the theory of complex fractal dimensions (now developed in [**370, 374, 375**] and in [**356**], for example). He has asked his Ph.D. student, Adam Richardson, to investigate this question and suggested to him to use for this purpose a suitable relative fractal drum (RFD) in $\mathbb{R}^3$ (when $m = 2$)—or, more generally, in $\mathbb{R}^{m+1}$, when $m \geq 2$ is arbitrary. It turns out that, indeed as is shown by A. Richardson in his Ph.D. Dissertation [**508**], the Peano curve and the Hilbert curve—as

Figure 9.12. The Hilbert curve (and its prefractal approximations), another celebrated example of a plane-filling curve.

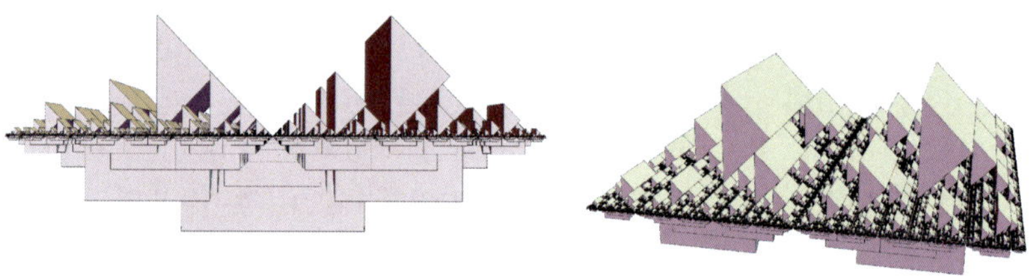

Figure 9.13. The Hilbert RFD: (a) front view and (b) side view.

well as, more generally, a large class of plane-filling curves—are fractal, in the sense of the higher-dimensional theory of complex dimensions [**356**], because the associated RFD constructed in loc. cit. has (countably many) nonreal complex dimensions with real part zero. (See Figure 9.13 in this subsection for two different views of the Peano RFD.) This construction and result also seem to extend to the higher-dimensional case, for any integer $m \geq 3$.

It is natural to expect that the refined higher-dimensional theory of complex dimensions developed by C. David and the first author in [**106–115**][97]—as is briefly discussed in Section 9.7 below—will enable one to obtain significantly more precise results about the complex dimensions of plane- (and space)-filling curves, without having recourse to RFDs but by working instead with the prefractal approximations to $\mathbf{\Gamma}_m$ and their limit, along with the associated (effective) local and global fractal zeta functions of the corresponding iterated fractal drums (IFDs); see especially [**106–113**].

[97]See also [**116**] and [**117**], the second and third sequels to this book, along with [**115**].

9.5. Koch Curve: A Self-Similar and Unrectifiable Curve

The Koch curve (or von Koch curve) $\mathbf{\Gamma}_{KC}$ is a classic example of a self-similar and fractal curve which is continuous but is nowhere differentiable; i.e., it does not admit a tangent line at any point. This fractal curve is simple; it is a Jordan arc—but it is not closed, unless we identify its endpoints, as will be done just below.

A natural closed curve analog of the Koch curve $\mathbf{\Gamma}_{KC}$ is the Koch snowflake curve (also called snowflake curve) $\mathbf{\Gamma}_{KS}$, a Jordan curve which is obtained by suitably putting together three (isometric) copies of the Koch curve $\mathbf{\Gamma}_{KC}$—so that any two adjacent copies of $\mathbf{\Gamma}_{KC}$ have a single point in common.

The Koch curve $\mathbf{\Gamma}_{KC}$ and the Koch snowflake curve $\mathbf{\Gamma}_{KS}$ (unlike the Devil's staircase $\mathbf{\Gamma}_{DS}$ studied in Section 9.4 above) have very similar geometric properties. In particular, they both have infinite length, are simple continuous and nowhere differentiable curves, are self-similar, have the same fractal (i.e., here, Minkowski and Hausdorff) dimensions ($D_{KC} = H_{KC} = D_{KS} = H_{KS} = \log_3 4$), as well as the same infinite set of complex dimensions, to be discussed further on and most of which are nonreal, very much as is the case for the Devil's staircase $\mathbf{\Gamma}_{DS}$ and its RFD counterpart; see Subsection 9.4.7 above. Accordingly, they both are fractal curves, but this time, they are already fractal in the largest possible real dimension, namely, the Minkowski dimension $D_{KC} = \log_3 4$, unlike the Devil's staircase $\mathbf{\Gamma}_{DS}$ which was fractal in dimension $d := D_{CS} = \log_3 2 < D_{DS} = 1$ and hence, could be called subcritically fractal or "subfractal"; see Remark 9.62 above.

Since the Koch curve $\mathbf{\Gamma}_{KC}$ and the Koch snowflake curve $\mathbf{\Gamma}_{KS}$ share essentially all of the same geometric properties (except for closedness), we will mostly focus on $\mathbf{\Gamma}_{KC}$. We point out, however, that the Koch snowflake curve is the boundary of a bounded and simply connected domain—that is, a bounded open set, Ω_{KS}, which is connected and cannot be continuously deformed to a point (i.e., it is simply connected); see Figure 9.14. It serves as an important model for many physically and technologically motivated problems, such as the study of the vibrations of a fractal drum—the so-called Koch drum or Koch snowflake drum (see, e.g., [**40, 41**], [**318–320**], [**322**], [**210–212**], [**235**], [**341–351**], [**370**], [**374**], [**375**], [**329**], [**356**], [**111**], as well as the relevant references therein), used, in particular, to understand the origins of fractality in nature (see, e.g., [**217**], [**534**], [**342**], and [**346**]), as well as the orbits of a fractal billiard—in this case, the Koch billiard or the Koch snowflake billiard, having $\overline{\Omega_{KS}} = \Omega_{KS} \cup \mathbf{\Gamma}_{KS}$ as its billiard table (see, e.g., [**341**], [**343–345**], [**375**, Section 12.5])—and for many engineering, computer graphics, or high-technology applications, including fractal antennae [**89**] and improved cell phone technology, based on either the Koch curve or the snowflake curve, and variants thereof (such as the square snowflake); see also the relevant references cited just above.

Figure 9.14. The Koch snowflake curve $\mathbf{\Gamma}_{KS}$, viewed as the boundary $K_S = \partial\Omega_{KS}$ of the Koch snowflake domain Ω_{KS} (also called Koch snowflake drum or Koch billiard).

The (deterministic) Koch curve—along with random versions of it, such as statistically self-similar Koch curves—has also often been used as a model of coastline, such as Brittany's coastline in Benoît Mandelbrot's celebrated 1967 article entitled *How long is the coast of Britain? Statistical self-similarity and fractional dimension* [**415**], which was one of the first papers on "fractals" that truly caught the attention of the scientific community and the broader scientifically inclined public. See also the relevant parts of Mandelbrot's books, [**418**], [**419**], and [**420**].[98]

[98]The 20th-century British mathematician, physicist, and meteorologist Lewis Fry Richardson, however, deserves the credit for having discovered—through his multifaceted work on weather forecast, turbulence and the prediction of wars (or conflicts), in particular—the importance of *power laws* in nonlinear phenomena, along with the related fact that the length of a coastline (or of the common border between two potential belligerent countries) grows without bound as the unit of measurement (i.e., the scale) becomes smaller and smaller; see [**510–512**], along with [**146**]. As such, Lewis Fry Richardson also deserves to be recognized as one of the fathers of the notion of fractality in Nature; see also the frequent references to his work in Mandelbrot's treatise [**419**].

Finally, we close these comments by mentioning that the von Koch curve was introduced by the Swedish mathematician Helge von Koch in 1904 in the paper [**603**]. Von Koch's motivation was to provide an "elementary geometric construction" for an example of a continuous curve which was nowhere differentiable and hence did not have a tangent line at any point.

As we will see in Section 9.7, thirty years earlier, Karl Weierstrass [**609**] had already introduced his famous (and much lauded, as well as decried) example of a continuous but nowhere differentiable curve, $\mathbf{\Gamma}_W$, now universally known as the Weierstrass curve. This example, however, was significantly more complicated than that of the Koch curve—as will be clear both geometrically and topologically, as well as by comparing the set $\mathcal{D}_{KC} = \mathcal{D}_{KS}$ of complex dimensions of the Koch curve $\mathbf{\Gamma}_{KC}$ or, equivalently, of the Koch snowflake $\mathbf{\Gamma}_{KS}$—as obtained in [**347**]—and the set $\mathcal{D}_W$ of complex dimensions of $\mathbf{\Gamma}_W$—as obtained in [**106**], which has a richer structure than that of $\mathcal{D}_{KC}$; compare Subsections 9.5.3 and 9.7.3 below. See [**106**] (announced in [**108**]).[99]

9.5.1. Polygonal Approximations, Unrectifiability, and Nowhere Differentiability.

The Koch curve has been defined in Section 4.3 (Example 4.31) as a self-similar set K_C associated with a certain self-similar system $\mathbf{\Phi}_{KC} = \{\varphi_j\}_{j=1}^4$. Later in this section (see Theorem 9.70 in Subsection 9.5.2 below), we will show that K_C coincides with the curve $\mathbf{\Gamma}_{KC}$, as introduced next. The Koch curve—denoted, henceforth, $\mathbf{\Gamma}_{KC}$—can be directly constructed via a sequence of polygonal curves $\{\mathbf{\Gamma}_{n,KC}\}_{n=0}^{\infty}$, as follows (see Figure 4.4 in Section 4.3, along with Figure 9.15 in the present section):

$(\mathrm{i})_0$ At the 0-th step (i.e., for $n = 0$), start with the unit horizontal line segment $\mathbf{\Gamma}_{0,KC} = \overline{AB}$ (parametrized by the unit segment $[0,1]$). The line segment $\Gamma_{0,KC} = \overline{AB}$ is called the *initiator* of the Koch curve; see part (a) of Figure 9.16.

$(\mathrm{i})_1$ At the first step (i.e., for $n = 1$), open up the middle-third interval (or line segment) in the initiator $\mathbf{\Gamma}_{0,KC} = \overline{AB}$ and replace it by an equilateral triangle of side lengths $1/3$; see part (a) of Figure 9.15. We call the resulting polygonal curve $\mathbf{\Gamma}_{1,KC}$ the *generator* of the Koch curve.

$(\mathrm{i})_2$ At the second step (i.e., for $n = 2$), repeat the first step—but this time, with $\mathbf{\Gamma}_{0,KC}$ replaced by $\mathbf{\Gamma}_{1,KC}$—thereby obtaining the polygonal curve, $\mathbf{\Gamma}_{2,KC}$, also with endpoints A and B; see part (b) of Figure 9.15.

And so on, ad infinitum. More specifically, for each $n \in \mathbb{N}$, we then repeat the process described in the first step (see $(\mathrm{i})_1$ above)—except for the fact that, at the n-th stage of the approximation (with $n \in \mathbb{N}$), $\mathbf{\Gamma}_{0,KC}$ must be replaced

[99] See also [**110**] (announced in [**112**]), along with [**111**], where the actual (or exact) complex dimensions of the Weierstrass curve $\mathbf{\Gamma}_W$—along with those of the Koch curve $\mathbf{\Gamma}_{KC}$, in the case of [**111**]—are obtained, respectively, from a new point of view.

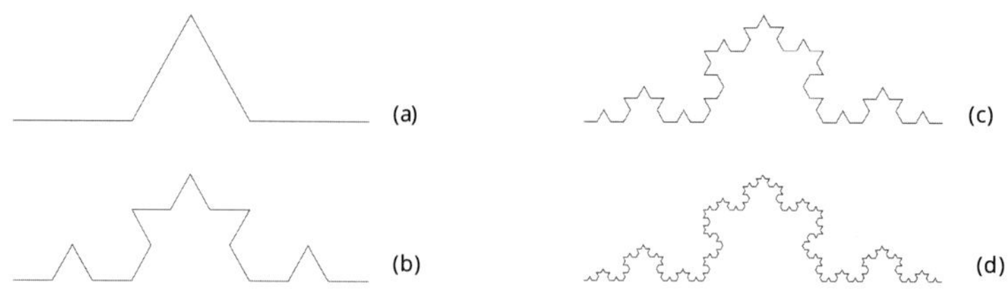

Figure 9.15. The first four stages of the construction (via natural polygonal or graph prefractal approximations) of the Koch curve $\mathbf{\Gamma}_{KC}$, as depicted in (a), (b), (c), and (d), respectively.

by $\mathbf{\Gamma}_{n-1,KC}$. The resulting polygonal curve (still with endpoints A and B) is denoted by $\mathbf{\Gamma}_{n,KC}$ and is called the n-th *natural polygonal approximation* to the Koch curve, a terminology that will be justified shortly.[100] It is sometimes simply called the n-th *polygonal approximation* to $\mathbf{\Gamma}_{KC}$.

It can be easily shown by induction on $n \in \mathbb{N}_0$ that the n-th natural polygonal approximation, $\mathbf{\Gamma}_{n,KC}$, is a polygonal curve consisting of 4^n line segments of length $3^{-n} = 1/3^n$. Hence,

$$L(\mathbf{\Gamma}_{n,KC}) = 4^n \cdot \frac{1}{3^n} = \left(\frac{4}{3}\right)^n, \text{ for all } n \in \mathbb{N}_0. \tag{9.145}$$

At this point, several questions arise naturally:

(1) How do we know that the process converges and yields, in the limit, a continuous curve, called the Koch curve and denoted by $\mathbf{\Gamma}_{KC}$?

(2) Why are the polygonal curves $\mathbf{\Gamma}_{n,KC}$ actual polygonal approximations to $\mathbf{\Gamma}_{KC}$, in the sense of Section 9.1 (see the discussion surrounding formula (9.2) in Section 9.1)?

(3) In which precise sense does the sequence of natural polygonal approximations $\{\mathbf{\Gamma}_{n,KC}\}_{n=0}^{\infty}$ converge to $\mathbf{\Gamma}_{KC}$? Is it in the sense of the Hausdorff metric in $(\mathbb{K}(\mathbb{R}^2), d_H)$? Is it also (or instead) in the sense of the uniform convergence (on $[0,1]$, say) of the underlying parametrized curves $\{\gamma_{n,KC}\}_{n=0}^{\infty}$ (associated with $\{\mathbf{\Gamma}_{n,KC}\}_{n=0}^{\infty}$) to γ_{KC}, the parametrized curve associated with $\mathbf{\Gamma}_{KC}$? (The answer to both questions is "yes".)[101] And, if so, how do we know that $\mathbf{\Gamma}_{KC}$ is a continuous curve and hence has a continuous parametrization $\gamma_{KC} : [0,1] \to \mathbb{R}^2$, in the first place?

[100] The cases when $n = 0$ and when $n = 1$ correspond, respectively, to the initiator $\mathbf{\Gamma}_{0,KC} = \overline{AB}$ and to the generator $\mathbf{\Gamma}_{1,KC}$ of the Koch curve; see $(\mathrm{i})_0$ and $(\mathrm{i})_1$.

[101] In fact, in light of the first part of Theorem 9.17 in Section 9.2 above and since, here, each line segment of $\mathbf{\Gamma}_{n,KC}$ has the same length—namely, 3^{-n}, so that $\mathrm{mesh}(\mathbf{\Gamma}_{n,KC}) = 3^{-n} \to 0$, as $n \to \infty$—it suffices to observe that $\mathbf{\Gamma}_{KC} := \gamma_{KC}([0,1])$ is a *simple* curve. More specifically, one therefore first shows that $\{\gamma_{n,KC}\}_{n=0}^{\infty}$ converges uniformly to some $\gamma_{KC} \in C([0,1])$, then verify that $\mathbf{\Gamma}_{KC}$ is simple (i.e., γ_{KC} is injective), and finally apply the aforementioned Theorem 9.17.

We point out that the answer to the latter question appeals—as might be expected by the attentive reader—to the use of Cauchy sequences: One simply shows that $\{\gamma_{n,KC}\}_{n=0}^{\infty}$ is a Cauchy sequence in $C([0,1]) = C([0,1],\mathbb{R}^2)$, by using Corollary 9.10 to Theorem 9.9 (about the completeness of the space $(C([0,1]),\delta_\infty)$)[102]—and hence that it converges uniformly on $[0,1]$ to an element, γ_{KC}, of $C([0,1])$. By construction, then, γ_{KC} is a continuous parametrization of the Koch curve $\mathbf{\Gamma}_{KC}$ and so, $\mathbf{\Gamma}_{KC}$ is a well-defined continuous curve—with underlying continuous parametrization γ_{KC}, itself the uniform limit of the sequence $\{\gamma_{n,KC}\}_{n=0}^{\infty}$ of elements of $C([0,1])$. This is, by far, the simplest and most efficient way to deal with this question (without having to provide an *explicit*, and rather cumbersome, continuous parametrization of $\mathbf{\Gamma}_{KC}$) and thereby, to answer all of the previous questions, as well as the next one, namely, (4) just below, via one of the length approximation theorems discussed in Section 9.2 above, namely, either Theorem 9.24 or Theorem 9.17.

(4) Once the above question has been answered, how do we know that the length $L(\mathbf{\Gamma}_{KC})$ of the Koch curve $\mathbf{\Gamma}_{KC}$ is the limit as $n \to \infty$ of the lengths $L(\mathbf{\Gamma}_{n,KC})$ of its (natural) polygonal approximations $\mathbf{\Gamma}_{n,KC}$? As was alluded to just above, one can simply appeal to either Theorem 9.24 or to the second part of Theorem 9.17 of Section 9.2 in order to conclude that (in light of (9.145), for the third equality in (9.146) just below)

$$L(\mathbf{\Gamma}_{KC}) = \lim_{n\to\infty} L(\mathbf{\Gamma}_{n,KC}) = \lim_{n\to\infty} \left(\frac{4}{3}\right)^n = +\infty. \tag{9.146}$$

Consequently, the Koch curve $\mathbf{\Gamma}_{KC}$ is *unrectifiable*; i.e., it has infinite length: $L(\mathbf{\Gamma}_{KC}) = +\infty$. Note that this is in sharp contrast with the Devil's staircase $\mathbf{\Gamma}_{DS}$—which, as we have seen in Subsection 9.4.3, has finite length (equal to two) and hence is rectifiable.

In summary, we proceed as follows. We first parametrize in a natural way the polygonal curves $\{\mathbf{\Gamma}_{n,KC}\}_{n=0}^{\infty}$ constructed inductively above. Then, we show that the corresponding sequence of continuous parametrized curves $\{\gamma_{n,KC}\}_{n=0}^{\infty}$ —where for each $n \in \mathbb{N}_0$, $\gamma_{n,KC} : [0,1] \to \mathbb{R}^2$ is a (piecewise linear) continuous map on $[0,1]$—is a Cauchy sequence in $C([0,1],\delta_\infty)$, where $C([0,1]) := C([0,1],\mathbb{R}^2)$. We then apply Corollary 9.10 to Theorem 9.9 from Section 9.2 in order to deduce that $\gamma_{n,KC}$ converges uniformly on $[0,1]$ to some unique $\gamma_{KC} \in C([0,1])$. The resulting continuous map γ_{KC} from $[0,1]$ to $\mathbb{R}^2$ is then used to *define* the Koch curve as follows: $\mathbf{\Gamma}_{KC} = \gamma_{KC}([0,1])$. One further verifies that $\mathbf{\Gamma}_{KC}$ so defined is a simple curve (i.e., that γ_{KC} is injective) and that each polygonal curve $\mathbf{\Gamma}_{n,KC}$ is, indeed, a polygonal approximation to $\mathbf{\Gamma}_{KC}$ (in the sense of Section 9.1), which is not too difficult to check, at this stage.

[102]Here, δ_∞ denotes the uniform metric (or "sup metric") on $C([0,1]) = C([0,1],\mathbb{R}^2)$, as given in formula (9.20) of Section 9.2.

Moreover, one applies Theorem 9.17 (from Section 9.2) in order to conclude that $\{\mathbf{\Gamma}_{n,KC}\}_{n=0}^{\infty}$ also converges to $\mathbf{\Gamma}_{KC}$ in the Hausdorff metric on $\mathbb{R}^2$, i.e.,

$$d_H(\mathbf{\Gamma}_{n,KC}, \mathbf{\Gamma}_{KC}) \to 0, \text{ as } n \to \infty, \tag{9.147}$$

and that (9.146) holds (i.e., $L(\mathbf{\Gamma}_{KC}) = \lim_{n\to\infty} L(\mathbf{\Gamma}_{n,KC}) = +\infty$) and hence that $\mathbf{\Gamma}_{KC}$ is unrectifiable. Alternatively, in order to verify that (9.146) holds, one could simply apply Theorem 9.24 (from Section 9.2) since the sequence of lengths $\{L(\mathbf{\Gamma}_{n,KC})\}_{n=0}^{\infty}$ of the polygonal approximations $\{\mathbf{\Gamma}_{n,KC}\}_{n=0}^{\infty}$ to the Koch curve $\mathbf{\Gamma}_{KC}$ is obviously increasing (because $L(\mathbf{\Gamma}_{n,KC}) = (4/3)^n$, for every $n \geq 0$, and since $4/3 > 1$), which presents the advantage of not having to show a priori that $\mathbf{\Gamma}_{KC}$ is a simple curve; see part (a) of Remark 9.25.

The next three exercises (Exercises 9.64–9.66) provide a step-by-step guideline as to how to go about proving the facts stated in the above discussion, as well as being summarized in Theorem 9.67 below. The previous discussion also gives appropriate hints as to how to solve them.

Exercise 9.64 (*Uniform Convergence and Continuity*).

(a) Prove that the sequence of polygonal (and hence also continuous) parametrized curves $\{\gamma_{n,KC}\}_{n=0}^{\infty}$—associated with the sequence of polygonal curves $\{\mathbf{\Gamma}_{n,KC}\}_{n=0}^{\infty}$—is a Cauchy sequence in $(C([0,1]), \delta_\infty)$.

(b) Deduce from part (a) that $\{\gamma_{n,KC}\}_{n=0}^{\infty}$ converges uniformly on $[0,1]$ to a unique continuous parametrized curve γ_{KC}, with $\gamma_{KC} : [0,1] \to \mathbb{R}^2$. Then, define the Koch curve by $\mathbf{\Gamma}_{KC} := \gamma_{KC}([0,1])$, as above.

Exercise 9.65 (*Polygonal Approximations*). Show that the sequence of polygonal curves $\{\mathbf{\Gamma}_{n,KC}\}_{n=0}^{\infty}$ is, indeed, a sequence of *polygonal approximations* to the Koch curve $\mathbf{\Gamma}_{KC}$, in the sense of Section 9.1.

[*Hint*: When one goes from one step of the construction of $\mathbf{\Gamma}_{KC}$ outlined at the beginning of Subsection 9.5.1 to the next step—say, specifically, from $\mathbf{\Gamma}_{n,KC}$ to $\mathbf{\Gamma}_{n+1,KC}$—one can see that the vertices of $\mathbf{\Gamma}_{n,KC}$ remain vertices of $\mathbf{\Gamma}_{n+1,KC}$, while new vertices are formed. Hence, once a point is a vertex of $\mathbf{\Gamma}_{n,KC}$, for some $n \in \mathbb{N}_0$, it remains a vertex of $\mathbf{\Gamma}_{l,KC}$, for every $l \geq n$—and thus also, by a limiting argument, it becomes a point of $\mathbf{\Gamma}_{KC}$.]

Exercise 9.66 (*Simplicity and Unrectifiability*).

(a) Show that $\mathbf{\Gamma}_{KC}$ is a simple curve; i.e., it is a Jordan arc—or else a Jordan curve, if its endpoints are identified. Furthermore, show that the sequence of polygonal paths $\{\mathbf{\Gamma}_{n,KC}\}_{n=0}^{\infty}$ also converges to $\mathbf{\Gamma}_{KC}$, in the Hausdorff metric of $\mathbb{R}^2$; i.e., (9.147) holds.

(b) Show that the sequence of lengths $\{L(\mathbf{\Gamma}_{n,KC})\}_{n=0}^{\infty}$ converges to $L(\mathbf{\Gamma}_{KC})$ in $[0,+\infty]$. Deduce that $L(\mathbf{\Gamma}_{KC}) = +\infty$ and hence that the Koch curve $\mathbf{\Gamma}_{KC}$ is unrectifiable.

We now conclude this subsection by summarizing our previous discussion in the following theorem.

Theorem 9.67 (Continuous Parametrization of, and Natural Polygonal Approximations to, the Koch Curve $\mathbf{\Gamma}_{KC}$).

(i) *With the above notation, the sequence* $\{\gamma_{n,KC}\}_{n=0}^{\infty}$ *converges uniformly (i.e., in* $C([0,1]) = C([0,1],\mathbb{R}^2))$ *to a unique* $\gamma \in C([0,1])$*, as* $n \to \infty$*. Therefore,* $\mathbf{\Gamma}_{KC} := \gamma_{KC}([0,1])$ *is a well-defined continuous curve, with continuous parametrization* γ_{KC}.

(ii) *Furthermore,* $\mathbf{\Gamma}_{KC}$ *is a simple curve and* $\{\mathbf{\Gamma}_{n,KC}\}_{n=0}^{\infty}$ *is a sequence of polygonal approximations to the Koch curve* $\mathbf{\Gamma}_{KC}$*. It is called* the sequence of natural polygonal approximations to $\mathbf{\Gamma}_{KC}$.

(iii) *Moreover,* $\{\mathbf{\Gamma}_{n,KC}\}_{n=0}^{\infty}$ *converges (as* $n \to \infty$*) to* $\mathbf{\Gamma}_{KC}$*, in the sense of the Hausdorff metric (i.e., in* $(\mathbb{K}(\mathbb{R}^2), d_H)$*), which means that* (9.147) *holds.*

(iv) *Finally, the lengths of the natural polygonal approximations to* $\mathbf{\Gamma}_{KC}$ *tend to the length of* $\mathbf{\Gamma}_{KC}$: $L(\mathbf{\Gamma}_{n,KC}) \to L(\mathbf{\Gamma}_{KC})$*, as* $n \to \infty$*. Hence,* $\mathbf{\Gamma}_{KC}$ *has infinite length; i.e., the Koch curve* $\mathbf{\Gamma}_{KC}$ *is unrectifiable.*

Remark 9.68 (*Sketch of the Proof of the Simplicity of* $\mathbf{\Gamma}_{KC}$). In order to prove the simplicity of $\mathbf{\Gamma}_{KC}$ in part (ii) of Theorem 9.67, one can argue as follows. Reason by contradiction and assume that $\gamma_{KC}(t) = \gamma_{KC}(\tilde{t})$, for some $t, \tilde{t} \in [0,1]$, with $t \neq \tilde{t}$. Then, fix $\varepsilon > 0$ and use the uniform convergence of $\{\gamma_{k,KC}\}_{k=0}^{\infty}$ to γ_{KC} (see part (i) of Theorem 9.67) in order to deduce the existence of $n \in \mathbb{N}_0$ such that $|\gamma_{k,KC}(\tau) - \gamma_{KC}(\tau)| \leq \varepsilon/2$, for all $k \geq n$ and $\tau \in [0,1]$. Hence, for $k \geq n$ fixed, and with either $\tau := t$ or $\tau := \tilde{t}$, we deduce from the Triangle Inequality that (since $\gamma_{KC}(t) = \gamma_{KC}(\tilde{t})$, by assumption)

$$
\begin{aligned}
&|\gamma_{k,KC}(t) - \gamma_{k,KC}(\tilde{t})| \\
&\leq |\gamma_{k,KC}(t) - \gamma_{KC}(t)| + |\gamma_{KC}(t) - \gamma_{KC}(\tilde{t})| + |\gamma_{KC}(\tilde{t}) - \gamma_{k,KC}(\tilde{t})| \\
&\leq \frac{\varepsilon}{2} + 0 + \frac{\varepsilon}{2} = \varepsilon.
\end{aligned}
\tag{9.148}
$$

Since $\varepsilon > 0$ is arbitrary, it follows from (9.148) that $\gamma_{k,KC}(t) = \gamma_{k,KC}(\tilde{t})$, for all integers $k \geq n$. A moment's reflection, along with the visualization of the corresponding figure, shows that this is impossible. We therefore conclude that the Koch curve $\mathbf{\Gamma}_{KC}$ is simple, as desired.

We invite the interested reader to complete the omitted details in the above sketch of the simplicity of $\mathbf{\Gamma}_{KC}$.

9.5.2. Self-Similarity and Finite Graph Approximations.

As we have seen in Example 4.31 of Section 4.3, the Koch curve $\mathbf{\Gamma}_{KC}$ is a self-similar set, associated with the self-similar system $\mathbf{\Phi}_{KC} = \{\varphi_j\}_{j=1}^{4}$ consisting of four contractive similarities of $\mathbb{R}^2$ with the same scaling ratio $r_j = 1/3$, for $j = 1, \ldots, 4$. We refer to the aforementioned example for a precise description of the maps $\varphi_1, \varphi_2, \varphi_3$, and φ_4. (Recall that $\mathbf{\Gamma}_{KC}$ is denoted K_C in Example 4.31.) See also Figure 4.4 in Section 4.3. Since, as a similar set, associated with $\mathbf{\Phi}_{KC}$, the Koch curve $\mathbf{\Gamma}_{KC}$ satisfies the Open Set Condition, we know from Theorem 5.40 (Moran's Theorem) that

$$
\begin{aligned}
D_{KC} &:= \dim_M(\mathbf{\Gamma}_{KC}) = \dim_B(\mathbf{\Gamma}_{KC}) = \dim_H(\mathbf{\Gamma}_{KC}) \\
&=: H_{KC} = \frac{\log 4}{\log 3} = \sigma_{KC},
\end{aligned}
\tag{9.149}
$$

the similarity dimension of the self-similar set $\mathbf{\Gamma}_{KC}$.[103] We will see in Subsection 9.5.3 that, at least as far as the Minkowski dimension $D_{KC} = \dim_M(\mathbf{\Gamma}_{KC})$ is concerned, there is another way—based on a fractal tube formula from [**347**] and the theory of complex fractal dimensions developed in [**375**] and [**356**]—of showing that the Minkowski dimension of $\mathbf{\Gamma}_{KC}$ exists (in the sense of Definition 8.3) and equals $\log 4/\log 3$.

By using the self-similarity of $\mathbf{\Gamma}_{KC}$, much as in the general theory of symbolic dynamics for IFSs described in Section 4.5, one can give an address to every vertex of each of the natural polygonal approximations $\mathbf{\Gamma}_{n,KC}$ to $\mathbf{\Gamma}_{KC}$ introduced in Subsection 9.5.1 above.[104] Such points, which necessarily belong to the Koch curve $\mathbf{\Gamma}_{KC}$ (see Exercise 9.65 and its hint), are called either *vertices* or else *junction points* of $\mathbf{\Gamma}_{KC}$.

Of course, according to part (vii) of Theorem 4.48 in Section 4.5, one could also give an (infinite) address to every point of $\mathbf{\Gamma}_{KC}$, namely, not only to the vertices of $\mathbf{\Gamma}_{KC}$ but also to the limit points of the set of these vertices.

Here and henceforth, it is convenient to identify the endpoints A and B of the initiator $\mathbf{\Gamma}_{0,KC} := \overline{AB}$ of the Koch curve.

Exercise 9.69 (*Addressing System for the Vertices of* $\mathbf{\Gamma}_{KC}$).

(a) Show that for each $j = 1, \ldots, 4$, the unique fixed point of the contractive similitude φ_j is the point P_j, defined as the j-th vertex of the generator $\mathbf{\Gamma}_{1,KC}$ of the Koch curve $\mathbf{\Gamma}_{KC}$, oriented from left to right; see Figure 9.16. Here, we let $A := P_1$ and, according to the aforementioned convention, we also have $B := P_5 = P_1 = A$.

(b) Extend part (a) by determining the unique fixed point of the map φ_J, as defined by (4.26) in Section 4.5, for any finite word (of length n, with $n \in \mathbb{N}$) $J \in \mathfrak{J}_n$ based on the alphabet $S := \{1, \ldots, 4\}$; see Section 4.5.

---

[103]Since $\mathbf{\Phi}_{KC}$ consists of 4 similarities of scaling ratio 1/3, the similarity dimension $\sigma = \sigma_{KC}$ of $\mathbf{\Gamma}_{KC}$ is uniquely determined by the equation $\sum_{j=1}^{4} r_j^{\sigma} = 1$, with $\sigma > 0$; i.e., $4(1/3)^{\sigma} = 1$ or, equivalently, $3^{\sigma} = 4$, which, indeed, yields $\sigma = \sigma_{KC} = \log 4/\log 3$.

[104]By necessity of concision, we will not do this explicitly here but instead refer the interested reader to the exploratory Exercise 9.69.

(c) Show that one can associate an explicit (finite or infinite) address (in the sense of Section 4.5) to every vertex (i.e., junction point) of the Koch curve $\mathbf{\Gamma}_{KC}$; see Figure 9.16. Is this address unique? If not, determine how many finite and infinite addresses can be associated to a given vertex; if necessary, limit yourself to providing concrete examples of nonuniqueness.

(d) Let $\mathcal{V}_\infty$ denote the set of all vertices of $\mathbf{\Gamma}_{KC}$. Show that every point of $\mathbf{\Gamma}_{KC} \backslash \mathcal{V}_\infty$ can be given an infinite address, i.e., an address associated with an infinite word; determine whether or not this address is unique. Also, show that in contrast with the points of $\mathcal{V}_\infty$, no point of $\mathbf{\Gamma}_{KC} \backslash \mathcal{V}_\infty$ can be given a finite address. Moreover, try to imagine in your mind's eye the resulting addressing system.

(e) Find a different addressing system for the points of the Koch snowflake curve which is a closed curve. (See the addressing system used in [**346**].)

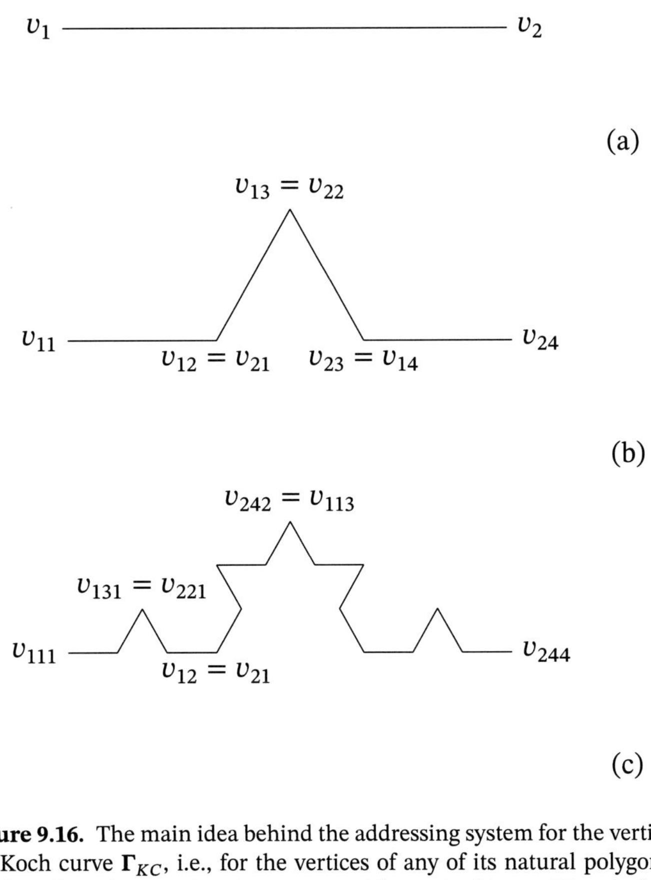

Figure 9.16. The main idea behind the addressing system for the vertices of the Koch curve $\mathbf{\Gamma}_{KC}$, i.e., for the vertices of any of its natural polygonal or graph prefractal approximations $\mathbf{\Gamma}_{n,KC}$, with $n \geq 0$ arbitrary. Here, in (a), (b), and (c), respectively, we have $n = 0, 1, 2$. (b) $|J| = 1$: Index each vertex of the generator $\mathbf{\Gamma}_{1,KC}$ of $\mathbf{\Gamma}_{KC}$ by a word of length 1, as indicated. (c) $|J| = 2$: Index each vertex of $\mathbf{\Gamma}_{2,KC}$ by an address or a word of length 2.

A natural question arises in this context; it was purposely left aside up to now, but it can no longer be cast away. Namely, how do we know that K_C—the self-similar set (associated with the self-similar system or IFS $\mathbf{\Phi}_{KC} = \{\varphi_j\}_{j=1}^4$) described in Example 4.31 of Section 4.3—and the Koch curve $\mathbf{\Gamma}_{KC}$—defined and studied in Subsection 9.5.1 (and up to now, in Subsection 9.5.2) as a suitable limit of the sequence $\{\mathbf{\Gamma}_{n,KC}\}_{n=0}^{\infty}$ of its natural polygonal approximations—coincide? In short, is it true that $K_C = \mathbf{\Gamma}_{KC}$, where K_C and $\mathbf{\Gamma}_{KC}$ are the nonempty compact subsets of $\mathbb{R}^2$ (i.e., the elements of $\mathbb{K}(\mathbb{R}^2)$)—respectively defined in Example 4.31 (of Section 4.3) and in Subsection 9.5.1 just above? The answer is "yes", of course, as we now explain.

For each $n \in \mathbb{N}_0$ and for any $j \in \{1, \ldots, 4^n\}$, let us denote by

$$\boldsymbol{L}_{j,n} := \overline{P_{j-1,n}P_{j,n}} \tag{9.150}$$

the line segment connecting the two adjacent vertices $P_{j-1,n}$ and $P_{j,n}$ of the n-th natural polygonal approximation $\mathbf{\Gamma}_{n,KC}$ of the Koch curve $\mathbf{\Gamma}_{KC}$ defined in Subsection 9.5.1. Here, $\{P_{j,n}\}_{j=0}^{4^n}$ stands for the finite sequence of the (successive) vertices of the polygonal curve $\mathbf{\Gamma}_{n,KC}$, ordered from left to right. In particular, $P_{0,n} := A$ and $P_{4^n,n} = B$, for each $n \in \mathbb{N}_0$—provided, for $n = 0$, we let $\mathbf{\Gamma}_{0,KC} = \overline{AB} =: \boldsymbol{L}_{1,0}$, with $A =: P_{0,0}$ and $B =: P_{1,0}$ identified, for convenience, so that $\mathbf{\Gamma}_{n,KC}$ consists of exactly 4^n (instead of $4^n + 1$) distinct vertices.

Accordingly, we have that for each $n \in \mathbb{N}_0$,

$$\mathbf{\Gamma}_{n,KC} = \bigcup_{j=1}^{4^n} \boldsymbol{L}_{j,n}. \tag{9.151}$$

A moment's reflection shows that, with the notation used for the contraction map $\mathbf{\Phi} : \mathbb{K}(\mathbb{R}^2) \to \mathbb{K}(\mathbb{R}^2)$ (introduced in formula (4.8) of Section 4.2; see

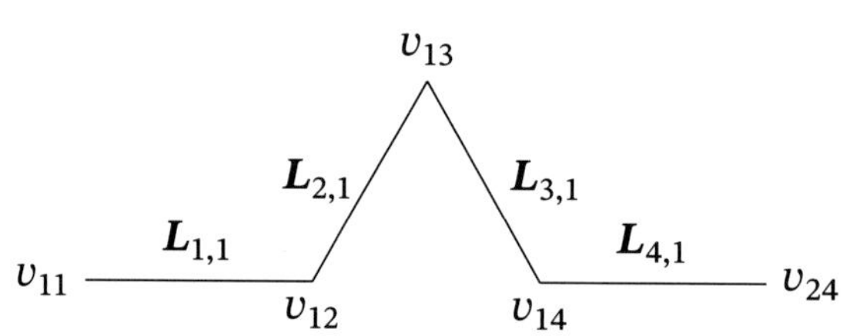

Figure 9.17. The labeling of the vertices and of the line segments (or edges) for the first polygonal or graph prefractal approximation $\mathbf{\Gamma}_{1,KC}$ of $\mathbf{\Gamma}_{KC}$. Essentially (see also Figure 9.18), by construction (and with $\mathbf{\Gamma}_{0,KC} := \overline{AB}$), we have that $\mathbf{\Gamma}_{1,KC} = \bigcup_{l=1}^4 \boldsymbol{L}_{l,1} = \bigcup_{l=1}^4 \varphi_l(\boldsymbol{L}_{1,0}) = \mathbf{\Phi}(\mathbf{\Gamma}_{0,KC})$. As a simple exercise, the interested reader may wish to depict similarly $\mathbf{\Gamma}_{2,KC}$, the second polygonal approximation of $\mathbf{\Gamma}_{KC}$, and verify that the identity $\mathbf{\Gamma}_{2,KC} = \mathbf{\Phi}(\mathbf{\Gamma}_{1,KC})$ holds.

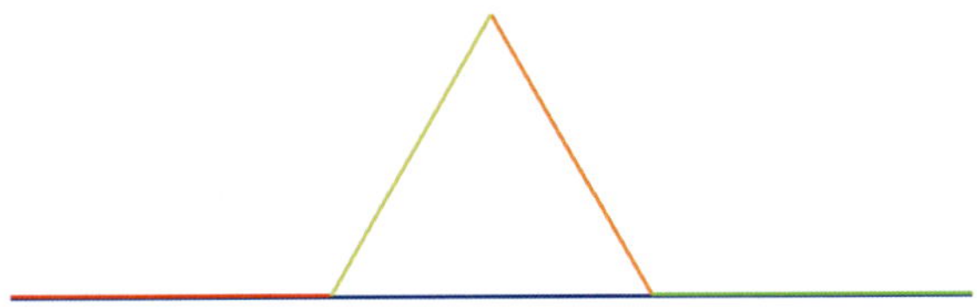

Figure 9.18. An illustration, in the case when $n = 0$, of the identity (9.156); namely, $\mathbf{\Phi}(\mathbf{\Gamma}_{n,KC}) = \mathbf{\Gamma}_{n+1,KC}$. Here, $\mathbf{\Gamma}_{0,KC} := \overline{AB} = \boldsymbol{L}_{1,0}$, the initial horizontal line segment depicted in blue color, while, its images by φ_j, for $j = 1, \ldots, 4$, are depicted in red, yellow, orange, and green, respectively. See also Figure 9.17.

also Section 4.5), given by

$$\mathbf{\Phi}(K) := \bigcup_{l=1}^{4} \varphi_l(K), \text{ for all } K \in \mathbb{K}(\mathbb{R}^2), \tag{9.152}$$

where $(\mathbb{K}(\mathbb{R}^2), d_H)$ is the complete metric space $\mathbb{K}(\mathbb{R}^2)$ of nonempty compact subsets of $\mathbb{R}^2$, equipped with the Hausdorff metric d_H (see Section 4.1), we have successively, in light of (9.151), (9.150), and (9.152):

$$\begin{aligned} \mathbf{\Phi}(\mathbf{\Gamma}_{n,KC}) &= \bigcup_{l=1}^{4} \varphi_l(\mathbf{\Gamma}_{n,KC}) = \bigcup_{l=1}^{4} \varphi_l\left(\bigcup_{j=1}^{4^n} \boldsymbol{L}_{j,n}\right) \\ &= \bigcup_{l=1}^{4}\bigcup_{j=1}^{4^n} \varphi_l(\boldsymbol{L}_{j,n}) = \bigcup_{l=1}^{4}\bigcup_{j=1}^{4^n} \boldsymbol{L}_{4(j-1)+l,n+1} = \mathbf{\Gamma}_{n+1,KC}; \end{aligned} \tag{9.153}$$

see Figures 9.17 and 9.18. Here, in the fourth equality of (9.153), we have used the following identity, valid for all $l \in \{1, \ldots, 4\}$ and $j \in \{1, \ldots, 4^n\}$:

$$\varphi_1(\boldsymbol{L}_{j,n}) = \boldsymbol{L}_{4(j-1)+l,n+1}. \tag{9.154}$$

Furthermore, note that in order to obtain the fifth (i.e., last) equality of (9.153), we have implicitly partitioned the interval $[1, 4^{n+1}]_{\mathbb{N}_0}$ in the following manner:[105]

$$[1, 4^{n+1}]_{\mathbb{N}_0} = [1, 4]_{\mathbb{N}_0} \cup [5, 8]_{\mathbb{N}_0} \cup \cdots \cup [4(4^n - 1) + 1, 4^{n+1}]_{\mathbb{N}_0}. \tag{9.155}$$

In conclusion, we have shown in (9.153) above that

$$\mathbf{\Phi}(\mathbf{\Gamma}_{n,KC}) = \mathbf{\Gamma}_{n+1,KC}, \text{ for all } n \in \mathbb{N}_0. \tag{9.156}$$

(See Figure 9.18 for the $n = 0$ case of this identity.) Now, since $\mathbf{\Phi}$ is a contraction on the complete metric space $(\mathbb{K}(\mathbb{R}^2), d_H)$, it is Lipschitz and hence also

[105] Here, the notation $[a, b]_{\mathbb{N}_0}$, with $a, b \in \mathbb{N}_0$, means that

$$[a, b]_{\mathbb{N}_0} := [a, b] \cap \mathbb{N}_0 = \{k \in \mathbb{N}_0 \,:\, a \le k \le b\}.$$

continuous, still with respect to the Hausdorff metric d_H on $\mathbb{K}(\mathbb{R}^2)$. Moreover, as we saw in part (iii) of Theorem 9.67,

$$\mathbf{\Gamma}_{n,KC} \to \mathbf{\Gamma}_{KC}, \text{ as } n \to \infty, \text{ in } (\mathbb{K}(\mathbb{R}^2), d_H). \tag{9.157}$$

Therefore, by passing to the limit as $n \to \infty$ in (9.156), we obtain that

$$\mathbf{\Phi}(\mathbf{\Gamma}_{KC}) = \mathbf{\Gamma}_{KC}; \tag{9.158}$$

i.e., $\mathbf{\Gamma}_{KC}$ is a *fixed point* of the contraction mapping $\mathbf{\Phi}$ of the complete metric space $(\mathbb{K}(\mathbb{R}^2), d_H)$. Since we know that *such a fixed point is unique* (see the Contraction Mapping Principle, Theorem 2.58 in Section 2.3, or else Theorem 4.17 in Section 4.2), it follows from (9.158) that $\mathbf{\Gamma}_{KC}$ must be the attractor of the IFS $\mathbf{\Phi}_{KC} = \{\varphi_j\}_{j=1}^4$—or, equivalently, that it must coincide with K_C, the unique self-similar set associated with the self-similar system $\mathbf{\Phi}_{KC} = \{\varphi_j\}_{j=1}^4$ (as described in Example 4.31): $\mathbf{\Gamma}_{KC} = K_C$, as desired.

We summarize the above discussion in the following theorem.

Theorem 9.70 (Self-Similarity and Polygonal Approximations). *The self-similar set K_C (defined in Example* 4.31 *of Section* 4.3) *and the Koch curve $\mathbf{\Gamma}_{KC}$ (defined via the sequence $\{\Gamma_{n,KC}\}_{n=0}^\infty$ of its natural polygonal approximations in Subsection* 9.5.1*; see Theorem* 9.67) *coincide*:

$$K_C = \mathbf{\Gamma}_{KC}. \tag{9.159}$$

Indeed, $\mathbf{\Gamma}_{KC}$ is the unique solution of the fixed point equation $\mathbf{\Phi}(K) = K$, with $K \in \mathbb{K}(\mathbb{R}^2)$, as follows from the "Almost Self-Similar Identity"

$$\mathbf{\Phi}(\mathbf{\Gamma}_{n,KC}) = \mathbf{\Gamma}_{n+1,KC}, \text{ for all } n \in \mathbb{N}_0, \tag{9.160}$$

where $\mathbf{\Phi}_{KC} = \{\varphi_l\}_{l=1}^4$ is the self-similar system defining K_C (= $\mathbf{\Gamma}_{KC}$) as a self-similar set of $\mathbb{R}^2$ and $\mathbf{\Phi}$ is the associated contraction map of $(\mathbb{K}(\mathbb{R}^2), d_H)$.

As a result, by part (iv) *of Theorem* 9.67, *$\mathbf{\Gamma}_{KC} = K_C$ is a fractal,*[106] *continuous, unrectifiable, and self-similar curve, with* initiator $\mathbf{\Gamma}_{0,KC} = \overline{AB} = \mathbf{L}_{1,0}$ *and* generator $\mathbf{\Gamma}_{1,KC}$ *given by* (*in the notation of* (9.150) *and* (9.151))

$$\mathbf{\Gamma}_{1,KC} = \mathbf{\Phi}(\mathbf{\Gamma}_{0,KC}) = \bigcup_{l=1}^{4} \varphi_l(\mathbf{L}_{1,0}) = \bigcup_{l=1}^{4} \mathbf{L}_{l,1}. \tag{9.161}$$

We close this subsection by briefly discussing the nowhere differentiability of the continuous curve $\mathbf{\Gamma}_{KC}$ and, especially, by explaining how to view the Koch curve, $\mathbf{\Gamma}_{KC}$, as the Hausdorff limit of a sequence of finite graphs, an important fact which is common to many classic fractal curves, such as the Sierpinski gasket—which is a key model for developing various aspects of analysis and geometric analysis on suitable self-similar fractals (see, e.g., [**72**], [**84**],

[106] The use of the adjective "fractal" in this context will be justified in Subsection 9.5.3 below, by means of the theory of complex dimensions.

[**292–295**], [**311, 312**], [**317**], [**320**], [**323–325**], [**368**], [**372**], [**500, 501**], [**528–530**], [**574**], [**586, 587**] and the relevant references therein, along with Section 9.6 below)—and the Weierstrass curve, in which the underlying IFS consists of *nonlinear* and *noncontractive* maps (see, e.g., [**103–105**], [**106–115**] and the relevant references therein, along with Section 9.7 below).

For any $n \in \mathbb{N}_0$, let $\mathcal{V}_n$ denote the set of *vertices*—also called, indifferently, *nodal points* or *junction points*—of $\mathbf{\Gamma}_{n,KC}$, the n-th (natural) polygonal approximation of the Koch curve $\mathbf{\Gamma}_{KC}$, as discussed in Subsection 9.5.1 and earlier in the present subsection. Recall from (9.150), (9.151), and the surrounding discussion that $\mathcal{V}_n = \{P_{j,n}\}_{j=0}^{4^n}$ consists of 4^n vertices—provided, as we continue to assume, that $P_{0,n} := A$ and $P_{4^n,n} := B$ are identified—and that, for all integers $n \geq 0$ and $j \in \{1, \dots, 4^n\}$,

$$(9.162) \qquad \mathbf{\Gamma}_{n,KC} = \bigcup_{j=1}^{4^n} \boldsymbol{L}_{j,n}, \quad \text{with } \boldsymbol{L}_{j,n} := \overline{P_{j-1,n}P_{j,n}}.$$

One can therefore think of $\boldsymbol{G}_n := (\mathcal{V}_n, E_n)$ as a (directed) finite graph, where E_n consists of the 4^n *edges* (i.e., the line segments) $\boldsymbol{L}_{j,n} = \overline{P_{j-1,n}P_{j,n}}$ connecting any two adjacent vertices $P_{j-1,n}$ and $P_{j,n}$, as in (9.150); see also Figures 9.17 and 9.18. This graph, $\boldsymbol{G}_n$, is the n-th (*natural*) *approximating finite graph* (to the Koch curve $\mathbf{\Gamma}_{KC}$) alluded to in the title of this subsection, a terminology which will be justified shortly.

Observe that, by construction,

$$(9.163) \qquad \mathcal{V}_n \subseteq \mathcal{V}_{n+1} \subseteq \mathbf{\Gamma}_{KC}, \quad \text{for all } n \in \mathbb{N}_0,$$

as we noted earlier. Indeed, once a vertex is formed (say, at stage $n \geq 0$) in the construction of $\mathbf{\Gamma}_{KC}$ described in Subsection 9.5.1, it remains a vertex of all later (natural) polygonal approximations $\mathbf{\Gamma}_{k,KC}$, with $k \geq n$—and hence, upon a passage to the limit in $(\mathbb{K}(\mathbb{R}^2), d_H)$, it becomes a point (also called a *vertex*) of the Koch curve $\mathbf{\Gamma}_{KC}$ itself.[107]

Next, let $\mathcal{V}_\infty$ denote the countable union (or set-theoretic limit) of all of the vertex sets $\mathcal{V}_n$ at each finite stage $n \geq 0$:

$$(9.164) \qquad \mathcal{V}_\infty := \bigcup_{n=0}^{\infty} \mathcal{V}_n \subseteq \mathbf{\Gamma}_{KC}.$$

The set $\mathcal{V}_\infty$ is called the *vertex set* or *set of vertices* (as well as, indifferently, of *nodal points* or *junction points*) of the Koch curve $\mathbf{\Gamma}_{KC}$. Observe that since, for each $n \geq 0$, the set $\mathcal{V}_n$ is finite, and a countable family of (nonempty) finite (or even countable) sets is countable, $\mathcal{V}_\infty$ is a *countably infinite* subset of $\mathbf{\Gamma}_{KC}$.

[107] Observe, however, that two adjacent vertices in $\mathbf{\Gamma}_{n,KC}$ are often no longer adjacent in $\mathbf{\Gamma}_{n+1,KC}$.

However, $\mathscr{V}_\infty$ is far from discrete (in the topological sense, [**81**], [**82**], [**456**]), as we next explain.[108]

A key property—which can be proved by using the Hausdorff convergence of $\mathbf{\Gamma}_{n,KC}$ to $\mathbf{\Gamma}_{KC}$ (or, correspondingly, the uniform convergence of $\gamma_{n,KC}$ to γ_{KC}) as $n \to \infty$ (see Theorem 9.67 in Subsection 9.5.1 above)—is that $\mathscr{V}_\infty$ is dense in $\mathbf{\Gamma}_{KC}$ (i.e., $\overline{\mathscr{V}_\infty} = \mathbf{\Gamma}_{KC}$) and that, furthermore, $\mathbf{\Gamma}_{KC}$ is the set of accumulation points of $\mathscr{V}_\infty$ (in $\mathbb{R}^2$).[109] Finally, $\{\mathscr{V}_n\}_{n=0}^\infty$ converges to $\mathbf{\Gamma}_{KC}$ in $(\mathbb{K}(\mathbb{R}^2), d_H)$,[110] and (as was seen in part (iii) of Theorem 9.67 above) so does the sequence of (natural) polygonal approximations $\{\mathbf{\Gamma}_{n,KC}\}_{n=0}^\infty$—or, equivalently, the sequence $\{E_n\}_{n=0}^\infty$ of n-th edge sets of the finite graphs $\{\mathbf{G}_n\}_{n=0}^\infty$.

We can now summarize the above information in the following theorem.

Theorem 9.71 (Finite Graph Approximations to the Koch Curve).

(i) *The sequence of n-th vertex sets $\{\mathscr{V}_n\}_{n=0}^\infty$ is monotonically increasing to $\mathscr{V}_\infty$, the countably infinite vertex set of the Koch curve $\mathbf{\Gamma}_{KC}$:*

$$\mathscr{V}_n \subseteq \mathscr{V}_{n+1} \subseteq \mathscr{V}_\infty := \bigcup_{k=0}^\infty \mathscr{V}_k \subseteq \mathbf{\Gamma}_{KC}, \text{ for all } n \in \mathbb{N}_0. \tag{9.165}$$

(ii) *Furthermore, $\mathscr{V}_\infty$ is dense in $\mathbf{\Gamma}_{KC}$: $\overline{\mathscr{V}_\infty} = \mathbf{\Gamma}_{KC}$. In fact, $\mathbf{\Gamma}_{KC}$ is the set of accumulation points of $\mathscr{V}_\infty$.*

(iii) *Moreover, $\{\mathscr{V}_n\}_{n=0}^\infty$, the sequence of n-th* vertex sets of $\{\mathbf{G}_n\}_{n=0}^\infty$, *converges (as $n \to \infty$) to $\mathbf{\Gamma}_{KC}$ in the Hausdorff metric of $\mathbb{R}^2$ (that is, in $(\mathbb{K}(\mathbb{R}^2), d_H)$) —and hence so does $\{E_n\}_{n=0}^\infty$, the sequence of all n-th edges of $\{\mathbf{G}_n\}_{n=0}^\infty$.*[111]

(iv) *Consequently, in light of* (i)–(iii), *we are justified in calling the sequence of finite graphs $\{\mathbf{G}_n := (\mathscr{V}_n, E_n)\}_{n=0}^\infty$* the sequence of (natural) finite graph approximations to the Koch curve $\mathbf{\Gamma}_{KC}$ *and to state that it converges (in an appropriate sense) to $\mathbf{\Gamma}_{KC}$, as $n \to \infty$.*

We just note here, with an eye towards proving part (iii), that, by using the easily extended notion of Hausdorff metric—valid for nonempty bounded (and not just compact) subsets of $\mathbb{R}^2$, and according to which $d_H(A, B) = d_H(\overline{A}, \overline{B})$,

[108] A subset A of $\mathbf{\Gamma}_{KC}$ is said to be *discrete* if, given any point $\boldsymbol{x} \in A$, there exists a small open neighborhood of $\mathbf{\Gamma}_{KC}$ containing $\boldsymbol{x}$ (here, say, a small open arc of $\mathbf{\Gamma}_{KC}$ centered at $\boldsymbol{x}$) and intersecting A only at $\boldsymbol{x}$: $A \cap V = \{\boldsymbol{x}\}$. In other words, every point of A is isolated in $\mathbf{\Gamma}_{KC}$.

[109] Recall that to say that $\boldsymbol{x} \in \mathbb{R}^2$ is an *accumulation point* of $\mathscr{V}_\infty$ means that $\boldsymbol{x}$ is the limit of a sequence of *distinct* points of $\mathscr{V}_\infty$. By contrast, to say that $\boldsymbol{x}$ is in the closure $\overline{\mathscr{V}_\infty}$ of $\mathscr{V}_\infty$—or , equivalently, that $\boldsymbol{x}$ is a *limit point* of $\mathscr{V}_\infty$—would allow for the above sequence to be eventually constant, instead of being comprised of distinct points.

[110] Note that each $\mathscr{V}_n$ (with $n \geq 0$) is compact (and nonempty) since it is a finite (and nonempty) subset of $\mathbb{R}^2$; hence, $\mathscr{V}_n \in \mathbb{K}(\mathbb{R}^2)$.

[111] Here, E_n is identified with $\mathbf{\Gamma}_{n,KC}$, viewed as the ordered list of its line segments (or edges) of level n; see (9.162).

for A, B arbitrary bounded (and nonempty) subsets of $\mathbb{R}^2$, we have that

$$d_H(\mathcal{V}_n, \mathbf{\Gamma}_{KC}) = d_H(\overline{\mathcal{V}_n}, \overline{\mathcal{V}_\infty}) = d_H(\mathcal{V}_n, \mathcal{V}_\infty) \to 0, \text{ as } n \to \infty, \tag{9.166}$$

as follows at once from part (i) of Theorem 9.71. Note that, for all $n \geq 0$, $\mathcal{V}_n = \overline{\mathcal{V}_n}$ because $\mathcal{V}_n$ is finite, and, by part (ii) of Theorem 9.71, $\overline{\mathcal{V}_\infty} = \mathbf{\Gamma}_{KC}$.

Exercise 9.72. Attempt proving parts or all of Theorem 9.71 just above.

[*Hint*: For part (iii), use Lemma 4.7 and the criterion in (4.7).]

Remark 9.73. An important application of the density of $\mathcal{V}_\infty$ in $\mathbf{\Gamma}_{KC}$ is that any continuous function on $\mathbf{\Gamma}_{KC}$—or, equivalently, any uniformly continuous function on $\mathbf{\Gamma}_{KC}$, since the metric space $\mathbf{\Gamma}_{KC}$ is compact—is uniquely determined by its restriction to $\mathcal{V}_\infty$.

Conversely, also since $\mathcal{V}_\infty$ is dense in $\mathbf{\Gamma}_{KC}$, any uniformly continuous function on $\mathcal{V}_\infty$ can be uniquely extended to a uniformly continuous—and hence also continuous—function on all of $\mathbf{\Gamma}_{KC}$.

We conclude this subsection by asking the reader to come up with both an informal (or intuitive) argument and a rigorous proof for the nowhere differentiability of the Koch curve $\mathbf{\Gamma}_{KC}$. We first state the corresponding result.

Theorem 9.74 (Continuity but Nowhere Differentiability of $\mathbf{\Gamma}_{KC}$). *The Koch curve $\mathbf{\Gamma}_{KC}$ is a self-similar curve* (*as we already know*) *which is* nowhere differentiable; *i.e., its underlying continuous parametrization $\gamma_{KC} : [0,1] \to \mathbb{R}$* (*as obtained in part* (i) *of Theorem* 9.67) *is* not differentiable *at any point of* $[0,1]$.

Geometrically, this means that the Koch curve $\mathbf{\Gamma}_{KC}$ does not *admit a tangent line at any of its points.*

Exercise 9.75. Provide an intuitive or semi-heuristic analytic or geometric argument for establishing Theorem 9.74. Then, try to transform your argument into a rigorous proof of this theorem.

[*Hint*: If P is an arbitrary point of $\mathbf{\Gamma}_{KC}$, with $Q \neq P$, and Q is another, nearby, point of $\mathbf{\Gamma}_{KC}$, then the direction of the secant connecting P and Q (i.e, the line segment $\overline{QP}$) keeps oscillating as Q tends to P.]

From a more advanced point of view, the issue of nowhere differentiability will be revisited in Section 9.7—based on the work from [**106**], in the much more complicated case of the Weierstrass curve—as well as, in more detail, in [**110, 116**], in terms of optimal Hölder and (discrete) reverse Hölder continuity. A similar (but simpler) argument could be used here as well.

9.5.3. Fractal Tube Formulas and Complex Dimensions.

Our aim in this subsection is to describe the main results obtained by the first author and Erin P. J. Pearse [**347**] (see also [**348**] and [**349**]) concerning a fractal tube formula

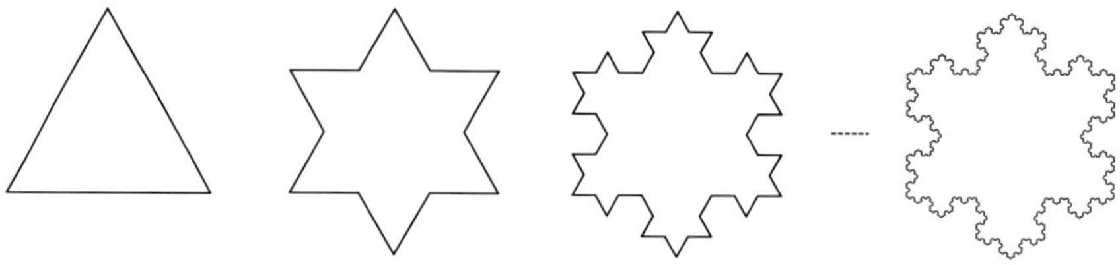

Figure 9.19. The Koch snowflake drum, Ω_{KS}, and its boundary, $\mathbf{\Gamma}_{KS} = \partial\Omega_{KS}$, the Koch snowflake curve.

for the Koch curve $\mathbf{\Gamma}_{KC}$ and the complex dimensions of $\mathbf{\Gamma}_{KC}$—or, equivalently, of the Koch snowflake curve $\mathbf{\Gamma}_{KS}$. We follow here, in part, the exposition of these results given in [**375**, Subsection 12.2.1] as well as take into account the additional complements of information provided therein.

Let $\mathbf{\Gamma}_{KS}$ denote the *Koch snowflake curve* (or *snowflake curve*, in short), sometimes also called the *triangular snowflake curve* (in order to distinguish it from other variants, such as the square snowflake curve) or else *von Koch snowflake.* Recall that it is the boundary $\partial\Omega_{KS}$ of the *Koch snowflake drum* (or *snowflake drum*, in short): $\mathbf{\Gamma}_{KS} = \partial\Omega_{KS}$, where Ω_{KS} is the bounded open set of $\mathbb{R}^2$ obtained as the unique bounded (and necessarily open) connected component of the complement, $\mathbb{R}^2\backslash\mathbf{\Gamma}_{KS}$, of the snowflake curve $\mathbf{\Gamma}_{KS}$ in $\mathbb{R}^2$, where $\mathbf{\Gamma}_{KS}$ is viewed as a Jordan curve and hence, in particular, as a closed curve; see Figure 9.14.[112] Also recall that $\mathbf{\Gamma}_{KS}$ is obtained by putting together three congruent (and thus, isometric) copies of the Koch curve $\mathbf{\Gamma}_{KC}$. Just as is the case for the Koch curve $\mathbf{\Gamma}_{KC}$, the Koch snowflake curve $\mathbf{\Gamma}_{KS}$ can also be seen as a self-similar set and hence also as a *self-similar (Jordan) curve*; see Figures 9.14 and 9.19. Furthermore, it is approximated by a sequence $\{\mathbf{\Gamma}_{n,KS}\}_{n=0}^{\infty}$ of (natural) polygonal approximations—which, unlike for $\mathbf{\Gamma}_{KC}$ and the corresponding sequence $\{\mathbf{\Gamma}_{n,KC}\}_{n=0}^{\infty}$ of polygonal approximations of $\mathbf{\Gamma}_{KC}$, consists of *closed* polygonal curves.

Exercise 9.76. Justify the above statements concerning the self-similarity of $\mathbf{\Gamma}_{KS}$ and the sequence $\{\mathbf{\Gamma}_{n,KS}\}_{n=0}^{\infty}$ of (natural) polygonal approximations of $\mathbf{\Gamma}_{KS}$. In particular, determine a self-similar system having $\mathbf{\Gamma}_{KS}$ as attractor, as well as specify and depict graphically the corresponding initiator $\mathbf{\Gamma}_{0,KS}$ and the generator $\mathbf{\Gamma}_{1,KS}$; compare with Example 4.31.

Can you find *outer* (rather than inner) natural (closed) polygonal approximations of $\mathbf{\Gamma}_{KS}$?

By symmetry, and since Lebesgue measure $\mathcal{L}^m$ on $\mathbb{R}^m$ is invariant under rotations and translations ($m = 2$, in the present case), it is clear that, with the

[112]Since $\mathbf{\Gamma}_{KS}$ is compact—as the union of three compact sets, isometric and hence also homeomorphic to the compact set $\mathbf{\Gamma}_{KC}$—it is also a closed subset of $\mathbb{R}^2$, so that $\mathbb{R}^2\backslash\mathbf{\Gamma}_{KS}$ is, indeed, open in $\mathbb{R}^2$.

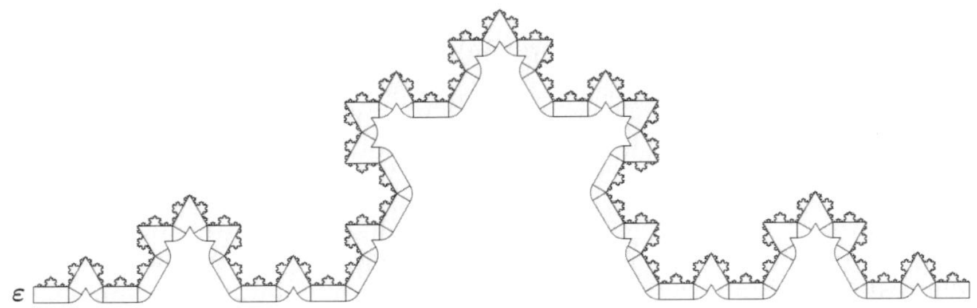

Figure 9.20. The inner (or one-sided) ε-neighborhood of the Koch curve $\mathbf{\Gamma}_{KC}$, viewed as the relevant part of the boundary of the Koch snowflake drum Ω_{KS}.

obvious notation,[113] $V_{KS} = 3 \cdot V_{KC}$. Hence, via Mellin transform, and upon analytic continuation (see Theorem 9.79 below and its proof, along with the discussion preceding it), we deduce that

$$\tilde{\zeta}_{KS} = 3 \cdot \tilde{\zeta}_{KC}, \text{ which implies that } \zeta_{KS} = 3 \cdot \zeta_{KC}, \tag{9.167}$$

via the functional equation (9.77) connecting the tube and distance zeta functions, respectively, $\tilde{\zeta}_A$ and ζ_A, of the same bounded set $A \subseteq \mathbb{R}^2$. Here, $A := \mathbf{\Gamma}_{KS} = \partial\Omega_{KS}$ or $A := \mathbf{\Gamma}_{KC} = \partial\Omega_{KC}$, respectively; also, observe that, with the notation of (9.76) and (9.77), $V_{KS}(\delta) = 3 \cdot V_{KC}(\delta)$.

Also, for the same reasons as above, we have that

$$\begin{aligned} 2 > \log_3 4 = D_{KC} &:= \dim_M(\mathbf{\Gamma}_{KC}) = \dim_M(\mathbf{\Gamma}_{KS}) =: D_{KS} \\ &= H_{KC} := \dim_H(\mathbf{\Gamma}_{KC}) = \dim_H(\mathbf{\Gamma}_{KS}) =: H_{KS}. \end{aligned} \tag{9.168}$$

We note that a direct proof of the equalities concerning the Minkowski dimension in (9.168) follows at once from the fractal tube formula in Theorem 9.77 below and from the theory of complex dimensions.[114]

It follows from the above discussion that the set of complex dimensions, as well as that of possible complex dimensions, of $\mathbf{\Gamma}_{KS}$ and of $\mathbf{\Gamma}_{KC}$ coincide:

$$\mathcal{D}_{KS} = \mathcal{D}_{KC} \quad \text{and} \quad \mathcal{D}_{KS,pos} = \mathcal{D}_{KC,pos}. \tag{9.169}$$

[113]Note that, here, $V_{KS}(\varepsilon)$ stands for the volume of the one-sided ε-neighborhood (or tubular neighborhood of size $\varepsilon > 0$), relative to Ω_{KS}, and similarly for V_{KC}. Strictly speaking, since we are working here with one-sided (rather than two-sided) tubular neighborhoods of $\mathbf{\Gamma}_{KS}$ and $\mathbf{\Gamma}_{KC}$—namely, $[\mathbf{\Gamma}_{KS}]_\varepsilon \cap \Omega_{KS}$ and $[\mathbf{\Gamma}_{KC}]_\varepsilon \cap \Omega_{KC}$, respectively (see Figure 9.20, for the case of $\mathbf{\Gamma}_{KC}$), where Ω_{KC} is the part of the bounded (and necessarily open) connected component of $\mathbb{R}^2 \setminus \mathbf{\Gamma}_{KC}$ lying above the x-axis—we should really use the notation for RFDs (in the sense of Subsection 9.3.8 above), namely, $(\mathbf{\Gamma}_{KS}, \Omega_{KS})$ and $(\mathbf{\Gamma}_{KC}, \Omega_{KC})$, respectively, and similarly for the corresponding fractal zeta functions, etc. For notational simplicity, however, we will not do so in this subsection.

[114]An alternative proof of all of the equalities in (9.168) relies on Moran's Theorem (Theorem 5.40 in Section 5.4 above) combined with the fact that $\mathbf{\Gamma}_{KC}$ and $\mathbf{\Gamma}_{KS}$ are self-similar, with the same scaling ratios, as well as satisfy the Open Set Condition.

We next proceed to stating (and completing) the main result of [**347**], providing an explicit formula (in the sense of [**370, 374**] and [**375**, Chapters 5 and 8]) for the (inner) tube function of $\mathbf{\Gamma}_{KC}$—as well as, equivalently, of $\mathbf{\Gamma}_{KS}$. The corresponding results concerning the (possible and actual) complex fractal dimensions of $\mathbf{\Gamma}_{KC}$ and of $\mathbf{\Gamma}_{KS}$ will follow at once, by means of a simple Mellin transformation followed by meromorphic continuation to all of $\mathbb{C}$, in light of the higher-dimensional theory of complex dimensions developed since then (see, e.g., [**356**]).

In view of (9.75), this would first yield $\tilde{\zeta}_{KS}$—and then, ζ_{KS}, via the functional equation (9.77)—from which, by (9.167), the corresponding expressions for $\tilde{\zeta}_{KC}$ and ζ_{KC} would follow at once.

Theorem 9.77 (Fractal Tube Formula for the Koch Snowflake Curve; M. L. Lapidus and E. P. J. Pearse, [**347**], 2006). *The volume V_{KS} (resp., V_{KC}) of the inner ε-neighborhood of the snowflake curve $\mathbf{\Gamma}_{KS}$ (resp., of the Koch curve $\mathbf{\Gamma}_{KC}$) is given by the following pointwise, fractal tube formula, for all positive ε sufficiently small*:

$$V_{KS}(\varepsilon) = G_{KS,pr}(\varepsilon) \cdot \varepsilon^{2-D_{KS}} + G_{KS,0}(\varepsilon) \cdot \varepsilon^2 \tag{9.170}$$

(resp., $V_{KC}(\varepsilon) = V_{KS}(\varepsilon)/3$, with $V_{KS}(\varepsilon)$ as in (9.170) *just above.*[115] *Here, the first term on the right-hand side of* (9.170) *yields the leading asymptotics of $V_{KS}(\varepsilon)$ with multiplicatively periodic (or "log periodic") oscillations, corresponding here to the set $\mathcal{D}^{pos}_{KS,pr} = \mathcal{D}^{pos}_{KC,pr}$ of (possible) principal complex dimensions of $\mathbf{\Gamma}_{KS}$ and $\mathbf{\Gamma}_{KC}$ (i.e., the complex dimensions with maximal real part $D_{KS} = D_{KC} =$ $\log_3 4 \in (1,2)$, while the second term on the right-hand side of* (9.170) *yields (potential) lower-order oscillations associated with the (possible) complex dimensions of real part* 0.[116]

Furthermore, $D_{KS} = D_{KC} = \log_3 4$ (resp., $\mathbf{p}_{KS} = \mathbf{p}_{KC} = 2\pi/\log 3$) is the common Minkowski dimension (resp., the common oscillatory period) of $\mathbf{\Gamma}_{KS}$ and $\mathbf{\Gamma}_{KC}$. Moreover, $G_{KS,pr}$ and $G_{KS,0}$ are continuous, real-valued, and multiplicatively periodic functions (of multiplicative period 3*) of the multiplicative variable ε*[117] *or, equivalently,* 1*-periodic functions of the additive variable $x :=$ $\log_3(\varepsilon^{-1}) \in \mathbb{R}$.*

Remark 9.78 (*Nonlinear Devil's Staircase and Fourier Coefficients*). By necessity of concision, we will not give here the rather "explicit" expressions—in terms of the Fourier coefficients of a certain nonlinear and periodic analog of the Cantor staircase function having a natural geometric interpretation in this context—for the (complex) Fourier coefficients A_n and B_n (with $n \in \mathbb{Z}$) of the

[115] So that $V_{KC}(\varepsilon) = V_{KS}(\varepsilon)/3$ is given by the exact counterpart of (9.170), except with $G_{KC,pr}(\varepsilon) = G_{KS,pr}(\varepsilon)/3$ (resp., $G_{KC,0}(\varepsilon) = G_{KS,0}(\varepsilon)/3$) instead of $G_{KS,pr}(\varepsilon)$ (resp., $G_{KS,0}(\varepsilon)$).

[116] See Theorem 9.79 and the discussion preceding it, along with Remark 9.78 and part (a) of Remark 9.80.

[117] That is, $G_{KS,pr}(\varepsilon) = G_{KS,pr}(\varepsilon/3)$, for all $\varepsilon > 0$, and similarly for $G_{KS,0}$.

periodic functions $G_{KS,pr}$ and $G_{KS,0}$, which we simply write in the following form, with $x := \log_3(\varepsilon^{-1}) \in \mathbb{R}$, now viewed as an additive variable (as opposed to the multiplicative variable $\varepsilon > 0$):[118]

$$G_{KS,pr}(\varepsilon) = \sum_{n\in\mathbb{Z}} A_n \cdot \varepsilon^{-in\mathbf{p}_{KS}} = \sum_{n\in\mathbb{Z}} A_n \cdot e^{in2\pi x} = G_{KS,pr}(x) \tag{9.171}$$

and, similarly,

$$G_{KS,0}(\varepsilon) = \sum_{n\in\mathbb{Z}} B_n \cdot \varepsilon^{-in\mathbf{p}_{KS}} = \sum_{n\in\mathbb{Z}} B_n \cdot \varepsilon^{in2\pi x} = G_{KS,0}(x). \tag{9.172}$$

Accordingly, $G_{KS} = G_{KS}(x)$ (resp., $G_{KS,0} = G_{KS,0}(x)$) is a continuous, 1-periodic function with complex Fourier coefficients A_n (resp., B_n), for all $n \in \mathbb{Z}$. Furthermore, for every $n \in \mathbb{Z}$, $A_{-n} = \overline{A_n}$ (the complex conjugate of A_n) and, similarly, $B_{-n} = \overline{B_n}$, so that $G_{KS,pr}$ and $G_{KS,0}$ are real-valued—and hence, $V_{KS}(\varepsilon)$ in formula (9.170) of Theorem 9.77 is a real number, as should be the case since it is a volume (i.e., an area, here, since $m = 2$).[119]

Moreover, the Fourier series of $G_{KS,pr}$ and $G_{KS,0}$ in (9.171) and (9.172), respectively, are absolutely convergent (and hence also convergent) on $\mathbb{R}$.

In addition, it can be shown (see [**347**]) that $A_0 \neq 0$, as well as verified numerically that at least some of the coefficients A_n (with $|n| \geq 1$) are also nonzero, so that $G_{KS,pr}$ is *nonconstant* and its Fourier series in (9.171) truly contains *oscillatory terms*. In fact, we conjecture that infinitely many (and possibly all) coefficients A_n (with $n \in \mathbb{Z}$) are nonzero, but we will not be using this conjecture in the sequel.

We also conjecture that $G_{KS,pr}$ is positive and is, in fact, bounded away from zero and infinity, a statement which will imply the Minkowski nondegeneracy of $\mathbf{\Gamma}_{KS}$ and $\mathbf{\Gamma}_{KC}$ and does not follow from the results obtained in [**347**] (or from their refinement stated in [**375**, Subsection 12.2.1]), although it could also probably be verified numerically.

We next set the stage for stating the new results concerning the (possible) complex dimensions of $\mathbf{\Gamma}_{KS}$ and $\mathbf{\Gamma}_{KC}$, as well as their geometric consequences, while explaining how they can be derived rigorously, by using the higher-dimensional theory of complex dimensions developed in [**356**] and, in particular, the definition of (and the results pertaining to) the fractal zeta functions (see Subsections 9.3.2 and 9.3.3 above)—especially, here, the tube zeta function $\tilde{\zeta}_{KS} = \tilde{\zeta}_{KS}(s)$ of $\mathbf{\Gamma}_{KS}$, given initially (for a fixed arbitrary $\delta > 0$) by the

[118]We refer the interested reader to the results of [**347**] and to the refined and even more specific exposition of those results given in [**375**, Subsection 12.2.1, pp. 346–353] for a more precise expression of the coefficients A_n and B_n, as well as for a discussion of the aforementioned *nonlinear* (and *periodic*) Devil's staircase, depicted (multiplicatively) in [**347**] and in [**375**, Figure 12.10, p. 352].

[119]This is called the *Reality Principle* in [**375**].

following (truncated) Mellin transform (since $\mathbf{\Gamma}_{KS} \subseteq \mathbb{R}^2$):[120]

$$\text{(9.173)} \quad \tilde{\zeta}_{KS}(s) := \int_0^\delta t^{s-3} \cdot V_{KS}(t)dt = \int_0^\delta t^{s-2} \cdot V_{KS}(t)\frac{dt}{t}, \text{ for } \mathrm{Re}(s) > D_{KS}.$$

We then apply the fractal tube formula (9.170) in Theorem 9.77 and deduce—upon substituting the resulting expression for $V_{KS} = V_{KS}(t)$ in the last integral of (9.173), as well as by taking (9.169) and (9.168) into account, and then interchanging integral and sum (which can be justified here)—that $\tilde{\zeta}_{KS}$ is given, initially only for $\mathrm{Re}(s) > D_{KS}$, by

(9.174)

$$\tilde{\zeta}_{KS}(s) = \sum_{n\in\mathbb{Z}} \frac{A_n}{s-(D_{KS}+in\mathbf{p}_{KS})}\delta^{s-(D_K+in\mathbf{p}_{KS})} + \sum_{n\in\mathbb{Z}} \frac{B_n}{s-in\mathbf{p}_{KS}}\delta^{s-in\mathbf{p}_{KS}},$$

with $A_n = O(1/n)$ and $B_n = O(1/n)$, as $|n| \to \infty$, as is shown in [**347**]. Finally, it next suffices to apply the Principle of Analytic Continuation in order to conclude that ζ_{KS} admits a (necessarily unique) meromorphic extension to all of $\mathbb{C}$, which is still given by (9.174), but is now valid for all $s \in \mathbb{C}$.

We can also read off the (possible) singularities (i.e., simple poles, here) of $\tilde{\zeta}_{KS}$ in (9.174) and conclude that the (possible) complex dimensions of $\mathbf{\Gamma}_{KS}$—and hence also of $\mathbf{\Gamma}_{KC}$, in light of (9.167)—are all simple and given by

$$\text{(9.175)} \quad \mathcal{D}_{KS,pos} = \mathcal{D}_{KC,pos} = \{D_{KS} + in\mathbf{p}_{KS} : n \in \mathbb{Z}\} \cup \{in\mathbf{p}_{KS} : n \in \mathbb{Z}\},$$

with the sets of (actual) complex dimensions of $\mathbf{\Gamma}_{KS}$ and $\mathbf{\Gamma}_{KC}$ satisfying

$$\text{(9.176)} \quad \mathcal{D}_{KS} = \mathcal{D}_{KC} \subseteq \mathcal{D}_{KS,pos} = \mathcal{D}_{KC,pos},$$

where (in light of Remark 9.78) $0 \in \mathcal{D}_{KS}, D_{KS} = \log_3 4 \in \mathcal{D}_{KS}$, and where $D_{KS} \pm in\mathbf{p}_{KS} \neq 0$, for some $n \in \mathbb{N}$, are *actual, nonreal* complex dimensions of $\mathbf{\Gamma}_{KC}$ (and of $\mathbf{\Gamma}_{KS}$);[121] see Figure 9.21.[122]

We can now summarize the above discussion in the following theorems. Note that for the results concerning ζ_{KS}, we simply apply the functional equation (9.77) connecting $\tilde{\zeta}_{KS}$ and ζ_{KS}; see Subsection 9.3.3 above.

Theorem 9.79 (Fractal Zeta Functions, Complex Dimensions, and Fractality of $\mathbf{\Gamma}_{KS}$ and $\mathbf{\Gamma}_{KC}$). *The tube and distance zeta functions (with parameter $\delta > 0$), $\tilde{\zeta}_{KS}$*

---

[120]We note that in [**347**], as well as in the exposition given in [**375**, Subsection 12.2.1]—and much as was done in [**370**, Sections 10.2 and 10.3] and [**374**, Subsections 12.2.2 and 10.2.3] via an approximate tube formula for the Devil's staircase and the Koch curve—the complex dimensions were "derived" by analogy with the 1-dimensional theory of complex dimensions (i.e., for fractal strings) and the associated fractal tube formulas (as obtained in [**370, 374**] or [**375**, Chapter 8]). Indeed, appropriate fractal zeta functions (and, therefore, a suitable definition of complex dimensions in $\mathbb{R}^m$, with $m \geq 2$) were not yet available at the time.

[121]Note that, for any $n \in \mathbb{Z}$, $\omega_{n,1} := D_{KS} + in\mathbf{p}_{KS}$ (resp., $\omega_n = in\mathbf{p}_{KS}$) belongs to $\mathcal{D}_{KS}$ iff $A_n \neq 0$ (resp., $B_n \neq 0$), and similarly for $\mathcal{D}_{KC}$.

[122]Observe that since $D_{KS} = \log_3 4 < 2$, the complex dimensions of $\mathbf{\Gamma}_{KS}$ can be indifferently defined via either $\tilde{\zeta}_{KS}$ or ζ_{KS}, and similarly for $\mathbf{\Gamma}_{KC}$.

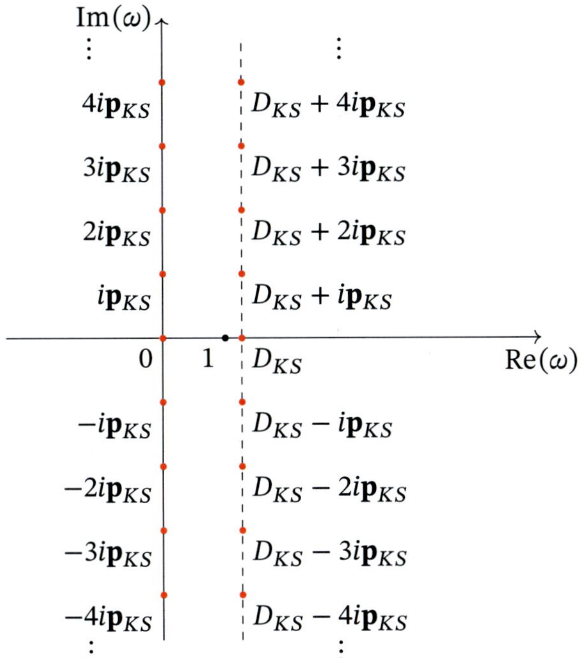

Figure 9.21. The possible complex dimensions of the Koch snowflake curve $\mathbf{\Gamma}_{KS} = \partial\Omega_{KS}$ (as well as of the Koch curve $\mathbf{\Gamma}_{KC}$), relative to the snowflake domain Ω_{KS}, the bounded region enclosed by $\mathbf{\Gamma}_{KC}$. Here, $D_{KS} = D_{KC} = \log_3 4 \in (1, 2)$ is the common Minkowski dimension of $\mathbf{\Gamma}_{KS}$ and $\mathbf{\Gamma}_{KC}$, while $\mathbf{p} := \mathbf{p}_{KS} = \mathbf{p}_{KC} = 2\pi/\log 3$ is their common oscillatory period.

and ζ_{KS}*, respectively, admit a (necessarily unique)* meromorphic continuation, *given for all* $s \in \mathbb{C}$ *by* (9.174) *above (for* $\tilde{\zeta}_{KS}$*) and by* (9.177) *just below (for* ζ_{KS}*):*

$$\zeta_{KS}(s) = (2-s)\tilde{\zeta}_{KS}(s) + \delta^{s-2} \cdot V_{KS}(\delta), \tag{9.177}$$

with $V_{KS}(\delta)$ *given by* (9.170) *(with* ε *replaced by* δ*) in which the expressions for* $G_{KS,pr}$ *and* $G_{KS,0}$ *are given as in* (9.171) *and* (9.172) *of Remark* 9.78.

Consequently, $\tilde{\zeta}_{KC} = \tilde{\zeta}_{KS}/3$ *and* $\zeta_{KC} = \zeta_{KS}/3$*, with* $\tilde{\zeta}_{KS}$ *and* ζ_{KS} *respectively given by* (9.174) *and* (9.177)*, for all* $s \in \mathbb{C}$*. In particular,* $\tilde{\zeta}_{KC}$ *and* ζ_{KC} *are also meromorphic in all of* $\mathbb{C}$*; furthermore, they have the same possible (respectively, exact) poles in* $\mathbb{C}$ *(or in any domain of* $\mathbb{C}$*).*

It follows that the sets of possible (as well as of actual) complex dimensions of $\mathbf{\Gamma}_{KS}$ *and* $\mathbf{\Gamma}_{KC}$ *coincide and are given by* (9.175) *and* (9.176)*, respectively, with* $\mathcal{D}_{KS} = \mathcal{D}_{KC}$ *containing* 0*,* $D_{KS} = D_{KC} = \log_3 4$ *(which must necessarily be the common Minkowski dimension of* $\mathbf{\Gamma}_{KS}$*, and with* $\mathbf{\Gamma}_{KC}$*), and with at least one complex conjugate pair of nonreal principal complex dimensions,* $D_{KC} \pm in\mathbf{p}_{KS}$*, for some* $n \in \mathbb{N}$ *(hence,* $n \neq 0$*).*

In particular, as expected, since they have (actual) nonreal complex dimensions, the Koch snowflake curve $\mathbf{\Gamma}_{KS}$ *and the Koch curve* $\mathbf{\Gamma}_{KC}$ *are* fractal, *in the sense of the theory of complex dimensions.*

Remark 9.80.

(a) The leading oscillations in the geometry of $\mathbf{\Gamma}_{KS}$ and $\mathbf{\Gamma}_{KC}$ in the fractal tube formula[123] (9.170) of Theorem 9.77 correspond to the common set of (possible) principal complex dimensions

$$\mathcal{D}^{pos}_{KS,pr} = \mathcal{D}^{pos}_{KC,pr} = \{D_{KS} + in\mathbf{p}_{KS} \,:\, n \in \mathbb{Z}\}, \tag{9.178}$$

which we conjecture to also coincide with $\mathcal{D}_{KS,pr} = \mathcal{D}_{KC,pr}$.

(b) Let us assume that $\delta = 1$, for notational simplicity. It then follows from (9.174) that (with $\omega_{n,1} := D_{KS} + in\mathbf{p}_{KS}$ and $\omega_n := in\mathbf{p}_{KS}$)

$$\operatorname{res}(\tilde{\zeta}_{KS}, \omega_{n,1}) = A_n \quad \text{and} \quad \operatorname{res}(\tilde{\zeta}_{KS}, \omega_{n,2}) = B_n, \tag{9.179}$$

for all $n \in \mathbb{Z}$.[124] We thus deduce that the fractal tube formula (9.170) is compatible with the general results about fractal tube formulas in the higher-dimensional theory of complex dimensions; see [**356**] or [**361**], along with Subsection 9.3.7 above. Indeed, according to [**356**, Theorem 5.1.16, p. 427], in the case of simple poles (which is our present case), we have, under suitable (polynomial) growth assumptions on $\tilde{\zeta}_{KS}$ (the so-called "strong languidity conditions"), that we obtain the same expression as what follows by combining (9.170), (9.169), (9.168), and (9.179); namely,

$$\begin{aligned} V_{KS}(\varepsilon) &= \sum_{\omega \in \mathcal{D}_{KS}} \operatorname{res}(\tilde{\zeta}_{KS}, \omega) \cdot \varepsilon^{2-\omega} \\ &= \sum_{n\in\mathbb{Z}} A_n \cdot \varepsilon^{2-(D_{KS}+in\mathbf{p}_{KS})} + \sum_{n\in\mathbb{Z}} B_n \cdot \varepsilon^{2-in\mathbf{p}_{KS}}. \end{aligned} \tag{9.180}$$

Naturally, exactly the same comment can be made about ζ_{KS} instead of about $\tilde{\zeta}_{KS}$.[125]

We next discuss the geometric consequences of the above result; see, especially, Theorems 9.77 and 9.79, along with Remarks 9.78 and 9.80. Note that Theorem 9.81 below was not stated in [**347**] (or in [**375**, Subsection 12.2.1]). Indeed, besides from Theorem 9.77 (the fractal tube formula, from [**347**]) and the end of Remark 9.78, it follows from [**356**, Theorem 2.3.25, p. 157] obtained in the higher-dimensional theory of complex dimensions; see Theorem 9.31 in Subsection 9.3.6 above. (We invite the reader to briefly review Definitions 8.2, 8.3, and 8.6 in Section 8.1, about the Minkowski dimension and content.)

---

[123] See the first term on the right-hand side of (9.170), including the 1-periodic function $G_{KS,pr}$, given by formula (9.171) in Remark 9.78.

[124] Also, for all $n \in \mathbb{Z}$, $\operatorname{res}(\zeta_{KS}, \omega_{n,1}) = (2-\omega_{n,1})A_n$ and $\operatorname{res}(\zeta_{KS}, \omega_{n,2}) = (2-\omega_{n,2})B_n$.

[125] In the case of $\tilde{\zeta}_{KS}$, we must replace $\operatorname{res}(\tilde{\zeta}_{KS}, \omega)$ by $\operatorname{res}(\zeta_{KS}, \omega)/(2-\omega)$.

Theorem 9.81 (Minkowski Nonmeasurability, Minkowski Nondegeneracy, and Average Minkowski Content of $\mathbf{\Gamma}_{KS}$ and $\mathbf{\Gamma}_{KC}$). *The Koch snowflake curve $\mathbf{\Gamma}_{KS}$ and the Koch curve $\mathbf{\Gamma}_{KC}$ are not Minkowski measurable (since the periodic function is not constant). Furthermore, their average Minkowski contents, $\widetilde{\mathcal{M}}_{KS}$ and $\widetilde{\mathcal{M}}_{KC}$, exist and are given, respectively (see Definition* 9.82 *below), by*

$$\widetilde{\mathcal{M}}_{KS} = \operatorname{res}(\tilde{\zeta}_{KS}, D_{KS}) = \int_0^1 G_{KS,pr}(t)dt, \tag{9.181}$$

the mean value of the 1*-periodic function $G_{KS,pr}$, and similarly for $\widetilde{\mathcal{M}}_{KC} = \widetilde{\mathcal{M}}_{KS}/3$. Note that $\mathcal{M}_{*,KS} < \widetilde{\mathcal{M}}_{KS} < \mathcal{M}^*_{KS}$, and similarly for $\widetilde{\mathcal{M}}_{KC}$.*

Moreover, since (according to the end of Remark 9.78 *above) the* 1*-periodic, nonconstant, and continuous function $G_{KS,pr}$ is positive—and is, in fact, bounded away from zero and infinity—$\mathbf{\Gamma}_{KS}$ is Minkowski nondegenerate (i.e., here, $0 < \mathcal{M}_{*,KS} < \mathcal{M}^*_{KS} < \infty$), $\widetilde{\mathcal{M}}_{KS} \in (0,+\infty)$, and the lower and upper Minkowski contents, $\mathcal{M}_{*,KS}$ and $\mathcal{M}^*_{KS}$, of $\mathbf{\Gamma}_{KS}$ are given by*

$$\mathcal{M}_{*,KS} = \min_{[0,1]} G_{KS,pr} \quad \text{and} \quad \mathcal{M}^*_{KS} = \max_{[0,1]} G_{KS,pr}. \tag{9.182}$$

It follows, in particular, that $0 < \mathcal{M}_{*,KS} < \widetilde{\mathcal{M}}_{KS} < \mathcal{M}^*_{KS}$.

Entirely similar statements to those in and around (9.181) *and* (9.182) *above hold for $\mathbf{\Gamma}_{KC}$ (instead of for $\mathbf{\Gamma}_{KS}$), except for $G_{KS,pr}$ replaced by $G_{KC,pr} = G_{KS,pr}/3$.*

We close this discussion (and this section) by recalling a definition and providing several comments related to Theorem 9.81 and some of the previous results stated in the present subsection.

Definition 9.82 (*Average Minkowski Content; see* [**356**, Definition 2.4.1, p. 178]). The *average Minkowski content* of $\mathbf{\Gamma}_{KS}$, denoted $\widetilde{\mathcal{M}}_{KS}$, is said to exist if it is given by the following Cesàro logarithmic mean (or average), provided the corresponding limit exists:

$$\widetilde{\mathcal{M}}_{KS} = \lim_{t\to+\infty} \frac{1}{\log t} \int_{1/t}^{\delta} \tau^{D-2} \cdot V_{KS}(\tau)\frac{d\tau}{\tau}. \tag{9.183}$$

The value of $\widetilde{\mathcal{M}}_{KS}$ is then independent of the choice of $\delta > 0$, so that, without loss of generality, we may choose $\delta := 1$.[126]

[126] Naturally, with the obvious change of notation, the corresponding definition in [**356**] is valid for any bounded set A (and, more generally, for any RFD (A,Ω) in $\mathbb{R}^m$. Here, $m = 2$ but in the general case, one must replace 2 by m in (9.183).

Remark 9.83 (*Rigorous vs. Numerical Verification*).

(a) Recall from the discussion towards the end of Remark 9.78 that the new geometric (and analytic) methods of C. David and the first author in [**106**] and [**110**] (regarding the Weierstrass curve)—and extended to the present case of the Koch curve or of the snowflake curve—enable us to prove rigorously (and not just verify numerically or graphically) that the appropriate counterpart of $G_{KS,pr}$ is nonconstant as well as bounded away from zero and infinity. In fact, as will be briefly discussed in part (c) below, much more is true: the authors of loc. cit. determine the *exact* complex dimensions (defined in a different manner) of the Koch curve as well as obtain exact expressions for the associated (local and global, polyhedral, effective) fractal zeta functions; see [**111**], along with Subsection 9.7.3 below. (See also part (d) below.)

(b) (*Sharp Hölder Continuity*). It is well known—and should also follow from the aforementioned methods and results of [**111**], [**108**], [**110**], and especially [**106**]—that the Koch curve, viewed as a parametrized curve $\gamma_{KC} : [0,1] \to \mathbb{R}^2$, is Hölder continuous with *optimal* Hölder exponent $\alpha_{KC} := 2 - D_{KC} = 2 - \log_3 4 \in (0,1)$; and entirely similarly for the Koch snowflake curve, viewed as a (closed) parametrized curve γ_{KS}.

It follows from the sharpness of this result (i.e., from the optimality of α_{KC}) that the Koch curve γ_{KC} and the Koch snowflake curve γ_{KS} are continuous but *nowhere* differentiable functions—and hence, equivalently, that the curves $\mathbf{\Gamma}_{KC}$ and $\mathbf{\Gamma}_{KS}$ do not admit a tangent line at any point.

Indeed, if γ_{KC}, say, were differentiable at some point $t_0 \in [0,1]$, then it would have to be (locally) Lipschitz at t_0, which would contradict the fact that $\alpha_{KC} < 1$ is the best possible Hölder exponent for γ_{KC}.

(c) (*Comparison with Earlier Approximate Results*) In [**370**, Subsection 10.3.4], an approximate (fractal) tube formula for $\mathbf{\Gamma}_{KS}$ (and $\mathbf{\Gamma}_{KC}$) was obtained that enabled the authors of loc. cit. to "guess" (by analogy with the fractal tube formulas obtained for fractal strings in [**370**, Chapter 6]) the values of the (possible) complex dimensions of $\mathbf{\Gamma}_{KS}$ (and $\mathbf{\Gamma}_{KC}$), as provided in formula (9.175) and Theorem 9.79 above and later obtained, still by using the same analogy, in [**347**]. Of course, now, thanks to the higher-dimensional theory of complex dimensions developed in [**356**], we can fully justify the aforementioned results of Theorem 9.79 and formula (9.175) concerning the complex dimensions of $\mathbf{\Gamma}_{KS}$ and $\mathbf{\Gamma}_{KC}$.

(d) (*Exact Intrinsic Complex Dimensions of the Koch Curve.*) Using so-called polyhedral neighborhoods rather than Euclidean tubular neighborhoods—as well as building on the new methods introduced in their earlier work in ([**106**], [**108–110**], [**112**])—C. David and the first author have rigorously determined in [**111**] the *exact* intrinsic complex dimensions of the Koch curve $\mathbf{\Gamma}_{KC}$ (or of the Koch iterated fractal drum)—and hence also, equivalently, of the Koch snowflake curve $\mathbf{\Gamma}_{KS}$. In addition to being exact, these (intrinsic) complex dimensions are all simple and are given as follows (with the obvious notation):

$$\widetilde{\mathcal{D}}_{KC} = \{D_{KC} - k(2 - D_{KC}) \pm ik\widetilde{\mathbf{p}}_{KC} \,:\, k \in \mathbb{N}_0 \text{ arbitrary}\},$$
$$\widetilde{\mathcal{D}}_{KS} = \{D_{KS} - k(2 - D_{KS}) \pm ik\widetilde{\mathbf{p}}_{KS} \,:\, k \in \mathbb{N}_0 \text{ arbitrary}\}, \tag{9.184}$$

where $D_{KC} = D_{KS} = \log 4/\log 3$ is the Minkowski dimension of $\mathbf{\Gamma}_{KC}$ and of $\mathbf{\Gamma}_{KS}$, while $\widetilde{\mathbf{p}}_{KC} = \widetilde{\mathbf{p}}_{KS} = \pi/3\log 3$ is the corresponding (intrinsic) oscillatory period of $\mathbf{\Gamma}_{KC}$ and of $\mathbf{\Gamma}_{KS}$. Note that the latter oscillatory period is different from the one introduced in [**347**] and used in this subsection; namely, $\mathbf{p}_{KC} = \mathbf{p}_{KS} = 2\pi/\log 3$.

9.6. Sierpinski Gasket and Carpet: Universality and Beyond

The Sierpinski gasket $\mathbf{\Gamma}_{SG}$, often denoted **SG**, is one of the most popular models of fractals: it is a self-similar curve, is finitely ramified, as well as being naturally approximated by polygonal curves and by finite graphs. For these reasons, it is one of the key fractal objects on which to first develop a suitable counterpart of geometric and harmonic analysis.

The Sierpinski carpet $\mathbf{\Gamma}_{SC}$ (henceforth, also often denoted by **SC**) satisfies many of the aforementioned key properties of the Sierpinski gasket, except for one: the finite ramification. Indeed, it is infinitely ramified, which makes it significantly harder to work with, from the point of view of analysis on fractals.[127] This is especially true for its higher-dimensional analogs, such as the Menger sponge—a space, rather than a planar curve—denoted by $\mathbf{\Gamma}_{MS}$ or simply **MS**. A key consequence of the infinite ramification of **SC** (resp., **MS**) is that it is universal among all planar (resp., space) curves, in a sense to be specified in Subsection 9.6.2. By contrast, the Sierpinski gasket, along with its higher-dimensional analogs, is not universal among planar (or higher-dimensional) curves.

In closing this introduction, we mention that the Sierpinski gasket (or curve) **SG** and the Sierpinski carpet **SC** were both introduced and studied by Waclaw Sierpinski in [**551**], [**554**], and [**552**] between 1915 and 1917; see also his collected works in [**555**]. We note that, at about the same time, Sierpinski also introduced a new type of plane-filling curve, in [**550**], as well as an analog of the Cantor–Lebesgue function, in [**553**]. Moreover, the Menger sponge was introduced by Karl Menger in 1926, in [**439**], as well as further studied in his 1928 book on *Dimension Theory*, [**440**].

[127]See, e.g., the books [**293**], [**574**], [**72**], [**372**], [**370**], [**374**], [**375**], [**356**], [**7**] and [**27**], [**84**], [**103–115**], [**207**], [**221**], [**225**], [**292**], [**294, 295**], [**311, 312**], [**317**], [**320**], [**323, 324**], [**325**], [**368**], [**414**], [**500**], [**501**], [**528–530**], [**586, 587**], along with the relevant references therein.

9.6.1. Finite vs. Infinite Ramification. As we have seen earlier (see Example 4.33 in Section 4.3), the Sierpinski gasket **SG** (or $\mathbf{\Gamma}_{SG}$) is a self-similar curve which can be defined as the (unique) attractor of the following self-similar system, $\mathbf{\Phi}_{SG}$. Consider an equilateral triangle (with side lengths 1), $T_0 = v_1 v_2 v_3$, with vertices v_1, v_2, and v_3. Then, $\mathbf{\Phi}_{SG} := \{\varphi_1, \varphi_2, \varphi_3\}$, the self-similar system defining **SG** as a self-similar set on $\mathbb{R}^2$, can be given as follows:

$$\varphi_j(x) = \frac{1}{2}(\boldsymbol{x} - v_j) + v_j, \text{ for } j = 1, 2, 3 \text{ and all } \boldsymbol{x} \in \mathbb{R}^2. \tag{9.185}$$

Each φ_j, for $j = 1, 2, 3$, is a similarity transformation of $\mathbb{R}^2$ of scaling ratios 1/2 (and with unique fixed point v_j). Since **SG** satisfies the Open Set Condition, it follows from Moran's Theorem (Theorem 5.40) that **SG** has Hausdorff and Minkowski (or box) dimensions H_{SG} and D_{SG} coinciding with its similarity dimension σ_{SG}. Namely, since $\mathbf{\Phi}_{SG}$ consists of three similarities, each having the same scaling ratio 1/2, we have, in agreement with part (vii) of Example 1.19, that

$$H_{SG} = D_{SG} = \sigma_{SG} = \frac{\log 3}{\log 2} = \log_2 3 \approx 1.5850 \in (1, 2). \tag{9.186}$$

Observe that it clearly follows from the above comments that the Sierpinski gasket **SG** is a *lattice* self-similar set, in the sense of Definition 4.53 in Section 4.6.

There is another way to define **SG**, which is very reminiscent of the classic construction of the (ternary) Cantor set. Start with $\mathbf{SG}_0 := T_0$, the (closed) initial equilateral triangle, of side lengths 1. Next, delete from $\mathbf{SG}_0$ the open middle equilateral triangle, of side lengths 1/2 and with vertices the midpoints of the sides of $\mathbf{SG}_0 = T_0$. Then, let $\mathbf{SG}_1$ denote the remaining subset of $\mathbf{SG}_0$: it consists of 3 closed equilateral triangles of side lengths 1/2. And repeat the process, ad infinitum. What remains after infinitely many steps also coincides with the Sierpinski gasket $\mathbf{SG} : \bigcap_{n=0}^{\infty} \mathbf{SG}_n = \mathbf{SG}$.

More specifically, at the n-th step of the above construction, $\mathbf{SG}_n$ consists of 3^n equilateral triangles of side lengths $1/2^n$. Furthermore, $\{\mathbf{SG}_n\}_{n=1}^{\infty}$ is a decreasing sequence of nonempty compact subsets of $\mathbb{R}^2$. Hence, $I_{SG} := \bigcap_{n=0}^{\infty} \mathbf{SG}_n$ is both nonempty and compact; i.e., $I_{SG} \in \mathbb{K}(\mathbb{R}^2)$. It is now an easy task to check—much as we did for the Koch curve in Subsection 9.5.2 above—that I_{SG} satisfies the fixed point equation, $\mathbf{\Phi}_{SG}(I_{SG}) = I_{SG}$, from which it follows that $I_{SG} = \mathbf{SG}$, the unique self-similar set defined by the self-similar system $\mathbf{\Phi}_{SG}$; see Figure 9.22, along with Sections 4.1–4.3 above.

Figure 9.22. The second construction of the Sierpinski gasket, via repeated deletions. The shadowed area corresponds to the deleted open equilateral triangles, at each step of the construction. What is left in the end, $I_{SG} = \bigcap_{n=0}^{\infty} \mathbf{SG}_n$, is clearly self-similar (it is *the* attractor of the self-similar system $\mathbf{\Phi}_{SG}$) and hence coincides with **SG**: $I_{SG} = \mathbf{SG}$.

Exercise 9.84.

(a) Show that I_{SG} does coincide with **SG**, as claimed just above and by following the indicated method. Furthermore, illustrate graphically the self-similarity (at all scales) of **SG**, as was discussed in Example 1.9 and, especially, in Example 4.33.

(b) Determine whether or not $\mathbf{\Gamma}_{SG} = \mathbf{SG}$ is a continuous curve, i.e., the image of a continuous parametrized curve. Note that it is clearly a compact, connected, and nonempty subset of $\mathbb{R}^2$, what is often simply called a (planar) curve, in topology.

(c) Show that $\mathbf{\Gamma}_{SG} = \mathbf{SG}$ is a *simple* curve (i.e., a Jordan arc), and show directly that it is unrectifiable (i.e., $L(\mathbf{\Gamma}_{SG}) = +\infty$). (See Section 9.2.)

Just as the Koch curve can be approximated by a sequence of finite graphs (as was seen in Subsection 9.5.2), the Sierpinski gasket is the limit of an increasing sequence of finite graphs. More specifically, if for each $n \in \mathbb{N}_0$, $\boldsymbol{G}_n := (\mathcal{V}_n, E_n)$, where $\mathcal{V}_n$ is the finite set of *vertices* of $\boldsymbol{G}_n$ (also called *junction points* or *nodal points*) and E_n is the corresponding set of 4^n edges (or line segments) connecting any two adjacent vertices (see Figure 9.23), then $\mathcal{V}_n \subseteq \mathcal{V}_{n+1}$, for each $n \geq 0$, and

$$d_H(\mathcal{V}_n, \mathbf{SG}) \to 0, \text{ as } n \to \infty; \tag{9.187}$$

i.e., **SG** is the Hausdorff limit (in $\mathbb{K}(\mathbb{R}^2)$) of $\{\mathcal{V}_n\}_{n=0}^{\infty}$, in the sense of Section 4.1.

In addition, if $\mathcal{V}_\infty := \bigcup_{n=0}^{\infty} \mathcal{V}_n$ denotes the countably infinite set of all vertices of $\boldsymbol{SG}$ at every stage and is called the set of *vertices* (or of *junction* points) of **SG**, then $\mathcal{V}_\infty$ is dense in **SG** (i.e., $\overline{\mathcal{V}_\infty} = \mathbf{SG}$), an important fact which we leave for the interested reader to verify in the following exercise.

Exercise 9.85.

(a) Prove that (9.187) holds and that $\overline{\mathcal{V}_\infty} = \mathbf{SG}$. Also, calculate $\#\mathcal{V}_n$, the number of vertices of the graph $\boldsymbol{G}_n$ (i.e., in short, of $\mathcal{V}_n$), for each $n \in \mathbb{N}_0$.

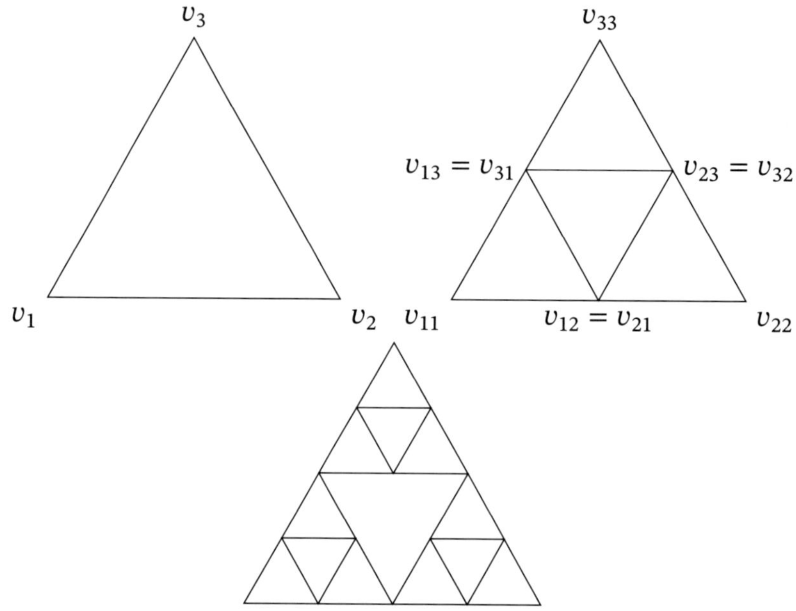

Figure 9.23. The increasing sequence of finite graphs tending to **SG** in the Hausdorff metric, i.e., in $(\mathbb{K}(\mathbb{R}^2), d_H)$, in the notation of Section 4.1. Here, the vertex sets $\mathcal{V}_0, \mathcal{V}_1$, and $\mathcal{V}_2$ are depicted; clearly, $\mathcal{V}_0 \subseteq \mathcal{V}_1 \subseteq \mathcal{V}_2$.

(b) In addition, show that, for any $n \in \mathbb{N}$, each point in $\mathcal{V}_n \backslash \mathcal{V}_0$ has exactly 3 neighbors in $\mathcal{V}_n$, whereas each point of $\mathcal{V}_0$ has precisely only 2 neighbors in $\mathcal{V}_0$.

(c) Consider the addressing system introduced in Section 4.5 for general self-similar sets, in particular, but now applied to **SG** and which the reader may wish to first illustrate graphically, by means of examples (see also Figure 9.23). Here, in light of (9.185) and the discussion surrounding it, the underlying alphabet is given by $S := \{1, 2, 3\}$. *True or false:*

(i) Every point of $\mathcal{V}_\infty$ has exactly one finite address and one infinite address.

(ii) Every point of $\mathbf{SG} \setminus \mathcal{V}_\infty$ has exactly a single (necessarily infinite) address. Accordingly, since $\mathcal{V}_\infty$ is countable and $\mathbf{SG} = \mathbf{\Gamma}_{SG}$ is in bijection with $[0, 1]$, there is a continuum of points of **SG** admitting a unique address. More precisely, all but countably many points of **SG** have a unique (and necessarily infinite) address.

(iii) (A little more difficult). Determine whether or not almost every point (with respect to the natural Hausdorff measure of **SG**) has a single (necessarily infinite) address.

The fact that **SG** can be approximated by a sequence of finite graphs is crucial for being able to develop relatively easily harmonic analysis on the Sierpinski gasket.[128] Actually, it is intimately connected with the fact that **SG** is *finitely ramified*, a key topological property of the Sierpinski gasket. Roughly speaking, it suffices to remove finitely many points of **SG** in order to disconnect **SG**.

More specifically, a point M of the continuous curve $\mathbf{SG} = \mathbf{\Gamma}_{SG}$ is said to be a *point of ramification* (of **SG**) provided one can find three continua (i.e., here, nonempty compact and connected subsets of $\mathbb{R}^2$) A_1, A_2, and A_3 such that $A_j \cap A_k = \{M\}$ for each $j \neq k$, with $j, k \in \{1, 2, 3\}$.

Then, Sierpinski showed in [**551,554**] that every point of $\mathbf{SG} = \mathbf{\Gamma}_{SG}$—other than the points of $\mathcal{V}_0 = \{v_1, v_2, v_3\}$—is a point of ramification of **SG**, whereas none of the points of $\mathcal{V}_0$ is a point of ramification of **SG**. Observe that this rather intuitive observation does require a proof because the Sierpinski curve $\mathbf{SG} = \mathbf{\Gamma}_{SG}$ is the union of $\mathcal{V}_\infty$ *and of its accumulation points* (i.e., the limits of convergent sequences of *distinct* points of $\mathcal{V}_\infty$). Note that the stated property is relatively easy to verify for the elements of $\mathcal{V}_\infty \setminus \mathcal{V}_0$.

We next discuss the Sierpinski carpet $\mathbf{SC} = \mathbf{\Gamma}_{SC}$. Recall that **SC** is a self-similar set, as described in Example 4.34 of Section 4.3.

Furthermore, much like the Sierpinski gasket, the Sierpinski carpet **SC** can also be obtained by deleting (open) middle squares (instead of triangles) at each stage of its construction; see Example 1.10 in Section 1.2, along with Figure 9.24.

In this respect, the interested reader may wish to solve the exact counterpart of Exercise 9.84 above, as well as to fully justify the following analog of formula (9.186). Namely, with the obvious notation (and, as we have seen in part (viii) of Example 1.19),

$$H_{SC} = D_{SC} = \sigma_{SC} = \frac{\log 8}{\log 3} = \log_3 8 \approx 1.8928 \in (1, 2). \tag{9.188}$$

Indeed, the self-similar system $\mathbf{\Phi}_{SC}$ defining **SC** as a self-similar set satisfies the Open Set Condition (OSC); see Definition 4.65. Hence, we can apply Moran's Theorem (Theorem 5.40) in order to conclude that (9.188) holds. Recall from Example 4.34 of Section 4.3 that $\mathbf{\Phi}_{SC}$ consist of 8 similitudes of $\mathbb{R}^2$, each with the same scaling ratio $1/3$.

The Sierpinski carpet **SC**, like the Sierpinski gasket **SG**, is a 1-dimensional continuum, i.e., a compact and connected topological space of *topological dimension one*. However, as we will see in more detail in the next subsection,

[128] See, e.g., [**292–295, 574**] and the relevant references therein. In fact, a more refined version of finite ramification—called p.c.f. (in short) or "post-critically finiteness" and of a dynamical, rather than of a topological, nature—is needed for this purpose; see loc. cit.

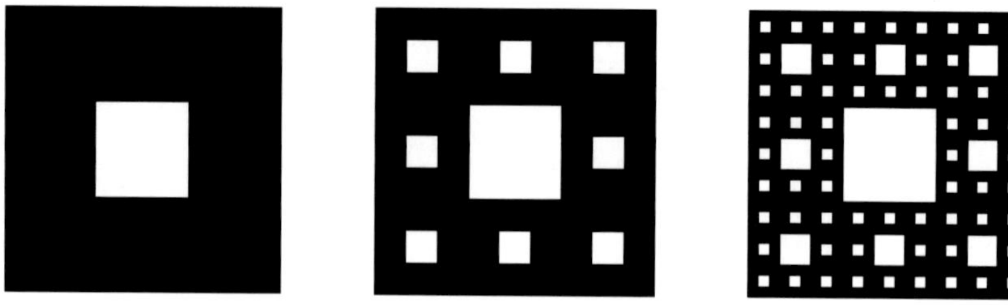

Figure 9.24. The Sierpinski carpet $\mathbf{SC} = \boldsymbol{\Gamma}_{SC}$ can be constructed by consecutively deleting open squares from the initial unit square. Only the first three generations of deleted squares are indicated here.

regarding the universality theorems, the Sierpinski carpet **SC** is a much richer and more complicated fractal curve than the Sierpinski gasket **SG**. In particular, it is *infinitely ramified* (i.e., it is *not* finitely ramified), as can be shown, for example, by observing that the horizontal, vertical, and diagonal cross-sections of **SC** are homeomorphic to the Cantor set.

Exercise 9.86 (*Higher-Dimensional Analogs of the Sierpinski Gasket and Carpet*). This exercise explores the counterpart of **SG** and **SC** in $\mathbb{R}^m$.

(a) Construct in two different, but equivalent, ways—via self-similarity and via successive deletions—the higher-dimensional (i.e., m-dimensional, with $m \geq 2$) analogs of **SG** and **SC**, denoted, respectively, $\mathbf{SG}^{(m)}$ and $\mathbf{SC}^{(m)}$, so that $\mathbf{SG}^{(2)} = \mathbf{SG}$ and $\mathbf{SC}^{(2)} = \mathbf{SC}$. The 3-dimensional analogs, $\mathbf{SG}^{(3)}$ and $\mathbf{SC}^{(3)} = \mathbf{MS}$, of **SG** and **SC**, are called, respectively, the *Sierpinski tetrahedron* and the *Menger sponge*; see Figure 9.25 and Figure 9.26, respectively.[129] (Also, observe that $\boldsymbol{SC}^{(1)} = C$, the ternary Cantor set.)

(b) Obtain the analog for $\mathbf{SG}^{(m)}$ and $\mathbf{SC}^{(m)}$ of formulas (9.186) and (9.188), respectively. Also, determine the topological dimension of $\mathbf{SG}^{(m)}$ and $\mathbf{SC}^{(m)}$, and show that $\mathbf{SG}^{(m)}$ and $\mathbf{SC}^{(m)}$ are continua (i.e., nonempty compact and connected subscts) of $\mathbb{R}^m$.

(c) Show that $\mathbf{SG}^{(m)}$ and $\mathbf{SC}^{(m)}$ are (lattice) self-similar sets, and determine associated IFSs—or, more precisely, self-similar systems—in $\mathbb{R}^m$.

9.6.2. Sierpinski Carpet and Menger Sponge: Universality Theorems.

We discuss here the main consequences of the infinite ramification—and hence of the great topological complexity—of the Sierpinski carpet **SC**. Namely, the Sierpinski carpet is universal among all planar, simple curves (i.e., all Jordan

[129] Recall that in Example 4.35 of Section 4.3, the Menger sponge has been defined as a self-similar set in $\mathbb{R}^3$.

Figure 9.25. The Sierpinski tetrahedron $\mathbf{SG}^{(3)}$, a 3-dimensional analog of the Sierpinski gasket.

Figure 9.26. The Menger sponge $\mathbf{MS} = \mathbf{SC}^{(3)}$, a 3-dimensional analog of the Sierpinski carpet.

arcs), as was originally established by Waclaw Sierpinski in 1916 in [**552**]. Actually, we will formulate more generally this universality property, by using the topological notion of a planar curve (or a *curve in* $\mathbb{R}^2$), as introduced below.

We will then discuss the extension by Karl Menger in the late 1920s in [**439**, **440**] to the higher-dimensional analogs $\mathbf{SC}^{(m)}$ of $\mathbf{SC} = \mathbf{SC}^{(2)}$, including the Menger sponge $\mathbf{MS} = \mathbf{SC}^{(3)}$; see Exercise 9.86 just above.

Following the terminology introduced in point-set topology during the first half of the 20th century, given any integer $m \geq 2$, we define a *curve in* $\mathbb{R}^m$ as being a continuum of $\mathbb{R}^m$ (i.e., a nonempty, compact, and connected subset of $\mathbb{R}^m$) of topological dimension one. If $m = 2$ (resp., if $m = 3$), then a curve in $\mathbb{R}^m$ is also called a *planar curve* (resp., a *space curve*).

Observe that every simple, continuous curve $\mathbf{\Gamma}$ in $\mathbb{R}^m$ (i.e., every Jordan arc in $\mathbb{R}^m$)—defined, as in Section 9.1, as the image $\mathbf{\Gamma} = \gamma([a, b])$ of a continuous and injective (i.e., simple) parametrized curve in $\mathbb{R}^m$, $\gamma : [a, b] \to \mathbb{R}^m$, with $a < b$—is a curve in $\mathbb{R}^m$, in the above more general sense. Indeed, because γ is a continuous bijection from $[a, b]$ onto $\mathbf{\Gamma}$, it follows that $\mathbf{\Gamma}$ is homeomorphic to the nonempty compact interval $[a, b]$;[130] hence, it is both compact, connected, and of topological dimension one because $[a, b]$ itself has these same topological properties.

We can now state the modern version of Sierpinski's Universality Theorem, a remarkable result, especially if one places it within its historical context.

Theorem 9.87 (Sierpinski's Universality Theorem; W. Sierpinski, [**552**], 1916). *The Sierpinski carpet* $\mathbf{SC} = \mathbf{SC}^{(2)}$ *is a* universal planar curve; *i.e., it is a planar curve and is universal among all planar curves (understood in the above general sense)—and hence, in particular, among all continuous and planar Jordan arcs, i.e., among all simple and continuous planar curves.*

This means that given any planar curve K (i.e., a curve in $\mathbb{R}^2$, in the above sense), there exists a continuous embedding (and hence also a homeomorphism onto its image, $h(K)$) $h : K \to$ $\mathbf{SC}$ *of K into* $\mathbf{SC}$.

Menger's generalization can next be easily stated as follows. Recall from Exercise 9.86 in Subsection 9.6.1 above that for any integer $m \geq 2$, we denote by $\mathbf{SC}^{(m)}$ the m-dimensional analog of the Sierpinski carpet $\mathbf{SC} = \mathbf{SC}^{(2)}$. It can be called the *(m-dimensional) Sierpinski cube*—or, more fairly, the *Menger curve* (in $\mathbb{R}^m$)—and is constructed either as a self-similar set (see Example 4.35 of Section 4.3 for the $m = 3$ case, corresponding to *Menger's sponge* $\mathbf{MS} = \mathbf{SC}^{(3)}$) or else by successive deletions of "middle" (m-dimensional) cubes from the initial (m-dimensional) unit cube $[0, 1]^m$. (See Figure 9.26 for a picture of Menger's sponge $\mathbf{MS} = \mathbf{SC}^{(3)}$.) By necessity of concision, we state Menger's Universality Theorem in an abbreviated form. Of course, the $m = 2$ case of

[130] In order to see this, note that the inverse map $\gamma^{-1} : \mathbf{\Gamma} \to [a, b]$ is continuous since the inverse image under γ^{-1} of any closed (and hence also compact) subset F of the compact set $\mathbf{\Gamma}$ is compact: $(\gamma^{-1})^{-1}(F) = \gamma(F)$ is compact, as the continuous image of a compact set.

the theorem reduces to Sierpinski's Universality Theorem (Theorem 9.87 just above).

Theorem 9.88 (Menger's Universality Theorem; K. Menger, [**439**], 1926, and [**440**], 1928). *The m-dimensional Sierpinski cube (or Menger curve)* $\mathbf{SC}^{(m)}$ *is a* universal curve *among all curves in* $\mathbb{R}^m$*—and, in particular, among all Jordan arcs in* $\mathbb{R}^m$*.*[131]

This means exactly the same thing as for Theorem 9.87 *above, except for the fact that* $\mathbb{R}^2$ *is replaced by* $\mathbb{R}^m$ *and* **SC** *by* $\mathbf{SC}^{(m)}$ *in the statement of the latter part of that theorem.*

A complete and rigorous proof of Theorem 9.88—and hence also of Theorem 9.87—will be given in [**116**]. In essence, it is based on a beautiful blend of topological (especially of deformation) arguments, as well as of geometric and analytic insights.[132]

We point out the following striking corollary of Theorems 9.87 (the $m = 2$ case) and 9.88 (the $m = 3$ and arbitrary $m \geq 2$ cases). In essence, this corollary follows from the fact that each of the cited curves is a Jordan arc (i.e., is simple)—and hence is also a curve, in the aforementioned extended sense.

Corollary 9.89. *Each of the Sierpinski gasket* $\mathbf{SG} = \mathbf{\Gamma}_{SG}$ *(Section* 9.6*), the Koch curve* $\mathbf{KC} = \mathbf{\Gamma}_{KC}$*, and the Koch snowflake curve* $\mathbf{KS} = \mathbf{\Gamma}_{KS}$ *(Section* 9.5*), along with the Devil's staircase* $\mathbf{\Gamma}_{DS}$ *(Section* 9.4*) and the Weierstrass curve* $\mathbf{\Gamma}_{W}$ *(Section* 9.7 *below), can be continuously—and hence also homeomorphically—embedded into the Sierpinski carpet.*

Similarly, the Sierpinski tetrahedron $\mathbf{SG}^{(3)}$ *(resp., more generally, the m-dimensional Sierpinski simplex or gasket* $\mathbf{SG}^{(m)}$*, the m-dimensional analog of* **SG**),[133] *with* $m \geq 3$ *arbitrary, can be continuously—and hence also homeomorphically—embedded into Menger's sponge* $\mathbf{MS} = \mathbf{SC}^{(3)}$ *(resp., into the m-dimensional Sierpinski cube* $\mathbf{SC}^{(m)}$*).*

We warn the reader that the following exercise is not a straightforward one.

Exercise 9.90. Try to directly prove Corollary 9.89 in the special case of the Sierpinski gasket **SG** (resp., of Sierpinski's tetrahedron $\mathbf{SG}^{(3)}$) by exhibiting a *concrete* continuous embedding of $\mathbf{SG} = \mathbf{SG}^{(2)}$ into the Sierpinski carpet $\mathbf{SC} = \mathbf{SC}^{(2)}$ (resp., of $\mathbf{SG}^{(3)}$ into the Menger sponge $\mathbf{MS} = \mathbf{SC}^{(3)}$).

---

[131]Therefore, Menger's sponge $\mathbf{MS} = \mathbf{SC}^{(3)}$ is universal among all space curves—and, in particular, among all Jordan arcs in $\mathbb{R}^3$, i.e., among all simple and continuous space curves.

[132]*Caution*: As is often the case, analysis is also necessary in order to make sure that one's geometric intuition is not misleading.

[133]See Exercise 9.86 above and, e.g., Example 2 in Section 3 of [**294**].

We complete this discussion of the Sierpinski carpet (and of its higher-dimensional analogs) by briefly mentioning two remarkable topological (and metric) characterizations of the Sierpinski carpet obtained by G. T. Whyburn in 1958 in [**614**].

The first topological characterization of the Sierpinski carpet can be stated as follows. A metrizable, topological space X is homeomorphic to the Sierpinski carpet iff it is a planar continuum of topological dimension 1 which is locally connected and has no local cut point, in other words, iff X is a planar curve (in the general topological sense discussed above) which is locally connected and does not have any local cut point.[134]

Here, a *local cut point* is a point q of X for which there exists a connected neighborhood V of q such that $V \setminus \{q\}$ is *not* connected.[135]

The second characterization of the Sierpinski carpet—also due to Whyburn in [**614**] and later generalized to any dimension $m \neq 3$ (with **SC** replaced by $\mathbf{SC}^{(m)}$) by J. W. Cannon in 1973 in [**66**]—is of the following form:

Let X be a continuum (i.e., a compact, connected, and nonempty metrizable topological space) embedded in the plane. Assume that the complement $\mathbb{R}^2 \backslash X$ of X in $\mathbb{R}^2$ has countably many (necessarily open) connected components $\{D_j\}_{j=1}^{\infty}$ satisfying the following properties:[136]

(i) $\operatorname{diam}(D_j) \to 0$, as $j \to \infty$;

(ii) $\partial D_j \cap \partial D_k = \emptyset$, for any $j \neq k$;

(iii) ∂D_j is a Jordan curve (or arc), for any $j \in \mathbb{N}$; and

(iv) $\bigcup_{j=1}^{\infty} \partial D_j$ is dense in X.

Then, X is homeomorphic to the Sierpinski carpet $\mathbf{SC} = \mathbf{SC}^{(2)}$.

Exercise 9.91. Verify that the Sierpinski carpet itself satisfies the properties stated in each of the above two topological (or metric) characterizations stated above.

Remark 9.92. The study of Sierpinski spaces, from various points of view, has contributed to the development of topology (and of metric topology, in particular), as well as of analysis and fractal geometry. Moreover, it continues to be a source of inspiration and an object of further investigation; see, e.g., M. Bonk's survey article [**54**] and the relevant references therein.

Early references on this subject include (in chronological order, starting in 1879) those by E. Netto [**461**], L. E. J. Brouwer [**60**], K. Borsuk [**55**], P. Urysohn

[134] Recall that a topological space X is said to be *locally connected* if every point $\boldsymbol{x} \in X$ has a neighborhood base consisting of *connected* open sets (of X) containing $\{\boldsymbol{x}\}$. Note that X itself need not be connected.

[135] For example, clearly, any point of a circle X is a local cut point of X, whereas no point of an open disk X is a local cut point of X.

[136] As usual, ∂D_j denotes the boundary of D_j (in $\mathbb{R}^2$, here).

[**597**], S. Lefschetz [**392**], G. Nöbeling [**464, 465**], followed by further investigations into the "universality" of Sierpinski and Menger spaces, along with their natural generalizations, by, notably, J. Nagata [**457–459**], K. Morita [**453**], H. J. Kowalsky [**306**], R. D. Anderson [**8, 9**], M. A. Štan'ko [**563**], S. L. Lipscomb [**404–409**], V. V. Fedorchuk [**184**], M. Bestvina [**48**], J. C. Perry [**489**], I. Ivanšić and U. Milutinović [**263**], J. C. Perry and S. L. Lipscomb [**490**], along with R. Miculescu and A. Mihail [**443**]. An in-depth exposition of this subject can be found in the book by S. L. Lipscomb, entitled *Fractals and Universal Spaces in Dimension Theory* [**408**]; see also the short (but informative) surveys by I. Stewart [**569**] and by S. L. Lipscomb [**409**], along with the book review by G. A. Edgar [**154**].

This subject—especially the universality of the Sierpinski carpet and its higher-dimensional analogs or infinite-dimensional generalizations—will be discussed in more detail in the second sequel to the present book, [**116**].

9.6.3. Complex Dimensions. We focus here on the Sierpinski gasket **SG**, its distance zeta function, and the corresponding fractal tube formula. We also give a good indication of how to carry out the associated computation—based on the results and techniques obtained and developed in [**356**] and the accompanying papers, especially [**357–359**], along with [**361**]. By using analogous techniques, entirely similar results could be obtained for the Sierpinski carpet **SC**—or, with significantly more effort, for the higher-dimensional counterparts, $\mathbf{SG}^{(m)}$ and $\mathbf{SC}^{(m)}$ of $\mathbf{SG} = \mathbf{SG}^{(2)}$ and of $\mathbf{SC} = \mathbf{SC}^{(2)}$, respectively, with $m \geq 3$ arbitrary—including the Sierpinski tetrahedron $\mathbf{SG}^{(3)}$ and Menger's sponge $\mathbf{MS} = \mathbf{SC}^{(3)}$, as discussed in the previous subsections; see, especially, Exercise 9.86 in Subsection 9.6.1 above.

We next state the corresponding result for the Sierpinski gasket **SG** and then provide a detailed sketch of the proof.

Theorem 9.93 (Sierpinski Gasket: Distance Zeta Function, Complex Dimensions, and Fractal Tube Formula; M. L. Lapidus, G. Radunović, and D. Žubrinić, [**356**], 2017). [137]

(i) (Distance Zeta Function). *The distance zeta function ζ_{SG} of* **SG** *admits a (necessarily unique) meromorphic extension to all of $\mathbb{C}$, given for all $s \in \mathbb{C}$ by the following expression:*[138]

$$\zeta_{SG}(s) = \frac{6(\sqrt{3})^{1-s} \cdot 2^{-s}}{s(s-1)(2^s-3)} + 2\pi\frac{\delta^s}{s} + 2\frac{\delta^{s-1}}{s-1}. \tag{9.189}$$

[137] See [**356**], Proposition 3.2.3, page 208.

[138] Here, $\delta > 0$ is the underlying parameter defining ζ_{SG} and we take $\delta > 1/4\sqrt{3}$ so that the corresponding δ-neighborhood of **SG** is simply connected.

(ii) (Complex Dimensions). *It follows from part* (i) *that the set* $\mathcal{D}_{SG}$ *of actual (i.e., exact) complex dimensions of* **SG** *is given by*[139]

$$\mathcal{D}_{SG} = \{0\} \cup \left\{\log_2 3 + ik\frac{2\pi}{\log 2} : k \in \mathbb{Z}\right\}, \tag{9.190}$$

where $D_{SG} = \log_2 3$ *and* $\mathbf{p}_{SG} = 2\pi/\log 2$ *are, respectively, the Minkowski dimension and oscillatory period of* **SG**; *each complex dimension in* (9.190) *is simple (i.e., is a simple pole of* ζ_{SG} *or, equivalently, of* $\tilde{\zeta}_{SG}$*). Hence, the set of* principal *complex dimensions of* **SG** *(i.e., the set of complex dimensions of* **SG** *with real part* $D_{SG} = \log_2 3$*, denoted by* $\mathcal{D}_{SG,pr}$*) is given by the last expression in* (9.190)*; namely,* $\mathcal{D}_{SG,pr} = \log_2 3 + i(2\pi/\log 2)\mathbb{Z}$.

(iii) (Fractal Tube Formula). *The tube function* $V_{\mathbf{SG}} = V_{\mathbf{SG}}(\varepsilon)$ *for* **SG** *is given by the following pointwise and exact fractal tube formula for all* $\varepsilon \in (0, 1/2\sqrt{3})$ *and with* $\delta > 1/4\sqrt{3}$:

$$\begin{aligned} V_{SG}(\varepsilon) &= \sum_{\omega \in \mathcal{D}_{SG}} \operatorname{res}(\zeta_{SG}, \omega) \cdot \frac{\varepsilon^{2-\omega}}{2-\omega} \\ &= \varepsilon^{2-D_{\mathrm{SG}}} \cdot G_{SG,pr}(\log_2(\varepsilon^{-1})) + \left(\frac{3\sqrt{3}}{2} + \pi\right)\varepsilon^2, \end{aligned} \tag{9.191}$$

where $G_{SG,pr}$ *is a positive, nonconstant, and* 1*-periodic function on* $\mathbb{R}$*, which is bounded away from zero and from infinity. It is given explicitly by the following (absolutely convergent and thus convergent) Fourier series (where* $\omega_k := D_{SG} + ik\mathbf{p}_{SG}$*, for all* $k \in \mathbb{Z}$*):*

$$G_{SG,pr}(x) = \frac{6\sqrt{3}}{\log 2} \cdot \sum_{k \in \mathbb{Z}} \frac{(4\sqrt{3})^{-\omega_k} \cdot e^{2\pi i k x}}{(2-\omega_k)(\omega_k - 1)\omega_k}, \quad \text{for all } x \in \mathbb{R}. \tag{9.192}$$

(iv) (Minkowski Nondegeneracy and Average Minkowski Content). *It follows from part* (iii) *that the Sierpinski gasket* **SG** *is Minkowski nondegenerate but is not Minkowski measurable (as was already pointed out in* [**320**]*), while its average Minkowski content* $\widetilde{\mathcal{M}}_{SG}$ *exists, is positive and finite, and is given by*

$$\begin{aligned} \widetilde{\mathcal{M}}_{SG} &= (2 - D_{SG}) \operatorname{res}(\zeta_{SG}, D_{SG}) = \frac{2}{3}(2 - D_{SG}) \frac{(\sqrt{3})^{1-D_{SG}}}{(D_{SG} - 1)\log 3} \\ &= \int_0^1 G_{SG,pr}(t)\,dt, \end{aligned} \tag{9.193}$$

the mean value of the 1*-periodic function* $G_{SG,pr}$ *(giving rise to the leading oscillatory term in* (9.191) *just above). Also, the lower and upper Minkowski contents*

[139]It turns out that 1 is not a pole of ζ_{SG} (and thus *not* a complex dimension of **SG**) because, in light of (9.189), a simple computation yields that $\operatorname{res}(\zeta_{SG}, 1) = -3 + 3 = 0$.

of **SG** *are, respectively, given by*

$$\mathcal{M}_{*,SG} = \min_{[0,1]} G_{SG,pr} \quad \text{and} \quad \mathcal{M}^*_{SG} = \max_{[0,1]} G_{SG,pr}. \tag{9.194}$$

Proof.

(i) *and* (ii): The main idea behind the proof of part (i) is to remember the second construction of **SG** (analogous to that of the Cantor set $\boldsymbol{C}$), by successive deletions of the (open) "middle" equilateral triangles; see Subsection 9.6.1 above.

For each $n \in \mathbb{N}$, let Ω_n denote the disjoint union of the 3^{n-1} deleted (open) equilateral triangles (of side lengths 2^{-n}), at the n-th stage of the construction. Furthermore, let $\Omega = \bigcup_{n=1}^{\infty} \Omega_n$, the disjoint union of all the deleted triangles, at any stage of the construction, so that $\mathbf{SG} = \partial\Omega$, the boundary of Ω, and Ω is a bounded open subset of $\mathbb{R}^2$; also, by construction, $\Omega \cup \mathbf{SG} = T$, the closed initial equilateral triangle (with unit side lengths). Consider the RFD (relative fractal drum) in $\mathbb{R}^2$ given by $(\partial\Omega, \Omega)$; see Subsection 9.3.8 above.

Then, a moment's reflection shows that

$$\zeta_{SG}(s) = \zeta_{\partial\Omega,\Omega}(s) + 2\pi\frac{\delta^s}{s} + 3\frac{\delta^{s-1}}{s-1}. \tag{9.195}$$

We next claim that $\zeta_{\partial\Omega,\Omega}(s)$ is given by the first term on the right-hand side of (9.189), thereby yielding (9.189) (thanks to (9.195)), first for $\mathrm{Re}(s) > \log_2 3 = D_{SG}$ and then, upon analytic (i.e., meromorphic, here) continuation, for all $s \in \mathbb{C}$. This last claim (about $\zeta_{\partial\Omega,\Omega}$) can be proved as follows. First, use the basic properties of the distance zeta function under scaling, isometry, and (countable) disjoint unions[140] in order to deduce that

$$\zeta_{\partial\Omega,\Omega}(s) = \sum_{n=1}^{\infty} \zeta_{\partial\Omega_n,\Omega_n}(s) = \zeta_{\mathcal{L}_{SG}}(s) \cdot \zeta_{\partial\Omega_1,\Omega_1}(s), \tag{9.196}$$

where $\zeta_{\mathcal{L}_{SG}}$ (the "scaling zeta function") is given by

$$\zeta_{\mathcal{L}_{SG}}(s) = \sum_{n=1}^{\infty} 3^{n-1} \cdot (2^{-n})^s = 2^{-s} \cdot \sum_{n=0}^{\infty} (3 \cdot 2^{-s})^n = \frac{1}{2^s - 3}, \tag{9.197}$$

and $\zeta_{\partial\Omega_1,\Omega_1}$, the distance zeta function of the basic RFD $(\partial\Omega_1, \Omega_1)$, is given, for all $s \in \mathbb{C}$, by

$$\zeta_{\partial\Omega_1,\Omega_1} = \frac{6(\sqrt{3})^{1-s} \cdot 2^{-s}}{s(s-1)}. \tag{9.198}$$

This last expression can be obtained by subdividing the first (open) deleted equilateral triangle Ω (of side lengths 1/2) into 6 congruent right triangles (as

[140] See Subsection 9.3.8 for a summary of those symmetry and decomposition properties, along with [**356**] for more details and complete proofs.

indicated in the left-hand side of Figure 9.27) and then calculating the distance zeta function of the RFD (A', Ω') associated with one of these right triangles, Ω' (with boundary $A' = \partial\Omega'$, say; see the right-hand side of Figure 9.27) as follows, by symmetry:

$$\begin{aligned}\zeta_{\partial\Omega_1,\Omega_1}(s) &= 6 \cdot \zeta_{A',\Omega'}(s) = 6 \int_{\Omega'} \operatorname{dist}((x,y),A')^{s-2}dxdy \\ &= 6 \int_0^{1/4} dx \int_0^{x\sqrt{3}} y^{s-2}dy = \frac{6(\sqrt{3})^{1-s} \cdot 2^{-s}}{s(s-1)},\end{aligned} \tag{9.199}$$

as claimed. A priori, the computation in (9.196)–(9.197) and in (9.198)–(9.199) is only valid for $\operatorname{Re}(s) > \log_2 3 = D_{SG}$ and for $\operatorname{Re}(s) > 1$, respectively. Upon meromorphic continuation, however, the conclusion of each of these computations remains valid for all $s \in \mathbb{C}$. Hence, by combining (9.196)–(9.199), we deduce that $\zeta_{SG} = \zeta_{SG}(s)$ is meromorphic in all of $\mathbb{C}$ and is given, for all $s \in \mathbb{C}$, by the first term on the right-hand side of (9.189), as desired.

This concludes the proof of part (i)—and hence also of part (ii) since, as we already noted in footnote 139, $\operatorname{res}(\zeta_{SG}, 1) = 0$ and so, $1 \notin \mathcal{D}_{SG}$.

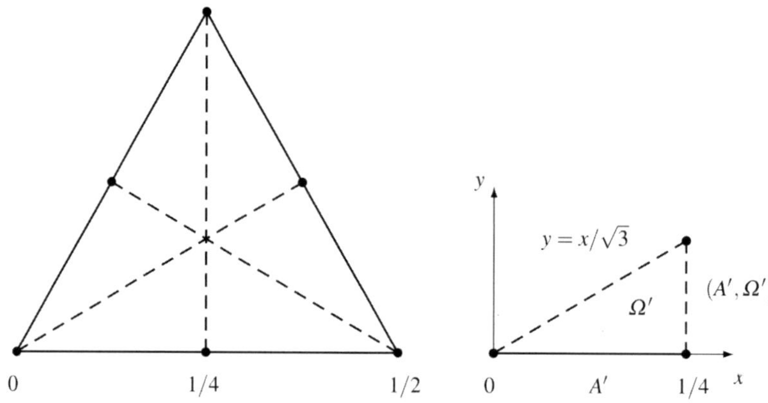

Figure 9.27. On the left is depicted the decomposition of the (open) equilateral triangle Ω_1 into six (open) congruent right triangles, one of which is Ω', with boundary $A' = \partial\Omega'$. On the right, the RFD (A', Ω') is depicted.

(iii) *and* (iv): The fractal tube formula in (9.191) of part (iii) can be obtained by simply applying to **SG** the general fractal tube formula for RFDs obtained in [**356**, Theorem 5.3.16, p. 448] (see also [**361**]), for the distance zeta function and in the present case of simple poles, as was briefly discussed in Subsection 9.3.7 above. We also observe that (still with $\omega_k := D_{SG} + ik\mathbf{p}_{SG}$, for each $k \in \mathbb{Z}$)

$$\operatorname{res}(\zeta_{SG}, \omega_k) = \frac{6(\sqrt{3})^{1-\omega_k}}{4^{\omega_k} \cdot (\log 2)\omega_k(\omega_k - 1)}, \text{ for all } k \in \mathbb{Z}, \text{ and} \tag{9.200}$$
$$\operatorname{res}(\zeta_{SG}, 0) = 3\sqrt{3} + 2\pi.$$

Finally, part (iv) now easily follows from part (iii) and the relevant results in [**356**] briefly recalled in Section 9.3 above, especially in Subsection 9.3.6. This concludes the proof of the theorem. □

Remark 9.94. Since $\mathcal{D}_{SG} = \log_2 3 < 2$, the complex dimensions of **SG** given in (9.190) could have equivalently been obtained as the (necessarily simple, here) poles of the tube zeta function $\tilde{\zeta}_{SG}$ of **SG**. The latter zeta function is meromorphic in all of $\mathbb{C}$ and the expression of $\tilde{\zeta}_{SG}(s)$, for all $s \in \mathbb{C}$, can be at once deduced from (9.190), via the functional equation connecting ζ_{SG} and $\tilde{\zeta}_{SG}$ (see Subsection 9.3.3), as we did in Sections 9.4 and 9.5 in the case of $\mathbf{\Gamma}_{DS}$ and $\mathbf{\Gamma}_{KS}$ (or of $\mathbf{\Gamma}_{KC}$), respectively. Since $D_{SC} < 2$, an entirely analogous comment applies to $\tilde{\zeta}_{SC}$, the tube zeta function of **SC**.

As was alluded to earlier, an entirely similar—and even easier to implement, geometrically—method can be applied to obtain the exact counterpart of Theorem 9.93 above, for the Sierpinski carpet $\mathbf{SC} \subseteq \mathbb{R}^2$, by using the results of [**356**] briefly discussed in Subsection 9.3.6 above; see [**356**, Proposition 3.2.11, p. 204] and [**361**]. We simply state the corresponding result for ζ_{SC} and $\mathcal{D}_{SC}$. Namely, ζ_{SC} admits a meromorphic continuation to all of $\mathbb{C}$, given for all $s \in \mathbb{C}$ (and for any value of the parameter $\delta > 1/6$) by

$$\zeta_{SC}(s) = \frac{8}{2^s \cdot s(s-1)(3^s - 8)} + 2\pi\frac{\delta^s}{s} + 4\frac{\delta^{s-1}}{s-1}. \tag{9.201}$$

Hence, the (actual) complex dimensions of **SC** are all simple and given by

$$\mathcal{D}_{SC} = \{0, 1\} \cup \mathcal{D}_{SC,pr}, \quad \text{with } \mathcal{D}_{SC,pr} = D_{SC,pr} + i\mathbf{p}_{SC} \cdot \mathbb{Z}, \tag{9.202}$$

the set of (actual) principal complex dimensions of **SC** (i.e., the complex dimensions with real part D_{SC}). Here, $D_{SC} = \log_3 8$ and $\mathbf{p}_{SC} = 2\pi/\log 3$ denote, respectively, the Minkowski dimension and the oscillatory period of **SC**. Furthermore, the exact analog of (9.200) holds (now with $\omega_k := D_{SC} + ik\mathbf{p}_{SC}$, for each $k \in \mathbb{Z}$); namely,

$$\operatorname{res}(\zeta_{SC}, \omega_k) = \frac{2^{-\omega_k}}{\omega_k(\omega_k - 1)(3^s - 8)}, \text{ for all } k \in \mathbb{Z}, \text{ and} \tag{9.203}$$
$$\operatorname{res}(\zeta_{SC}, 0) = 2\pi + \frac{8}{7}, \quad \operatorname{res}(\zeta_{SC}, 1) = \frac{16}{5}.$$

It follows that the (exact and pointwise) fractal tube formula for **SC** is given by the exact counterpart of (9.191) in Theorem 9.93, but with **SG** replaced by **SC** and now taking into account (9.202) and (9.203) instead of their analogs for **SC** (namely, (9.190) and (9.200)). We leave it to the interested reader to write down the resulting expression for $V_{SC}(\varepsilon)$ and the Fourier series for the 1-periodic function $G_{SC} = G_{SC}(\log_3(\varepsilon^{-1}))$, by using the results from [**356, 359, 361**] briefly recalled in Subsection 9.3.7. We also leave it to the reader to obtain the exact analog (for **SC** instead of for **SG**) of part (iv) of Theorem 9.93, by using the results from [**356, 359**] briefly discussed in Subsection 9.3.6 above.

We caution the reader that our last "exercise" in this section is rather challenging, although the methods needed in order to solve it are entirely similar to those outlined in the proof of Theorem 9.93 above. The underlying geometry, however, is more complicated in higher dimensions.

Exercise 9.95 (*Sierpinski's Tetrahedron, Menger's Sponge, and Beyond*). Try to obtain (all or part of) the counterpart of Theorem 9.93 above for the Sierpinski tetrahedron $\mathbf{SG}^{(3)} \subseteq \mathbb{R}^3$ and for Menger's sponge $\mathbf{MS} = \mathbf{SC}^{(3)} \subseteq \mathbb{R}^3$, as well as for their higher-dimensional analogs $\mathbf{SG}^{(m)} \subseteq \mathbb{R}^m$ and $\mathbf{SC}^{(m)} \subseteq \mathbb{R}^m$, for any integer $m \geq 4$; see Exercise 9.86 above.

9.7. Weierstrass Curve: A Continuous but Nowhere Differentiable Curve

The Weierstrass function $\mathcal{W}$ and the associated curve $\mathbf{\Gamma}_{\mathcal{W}}$ (i.e., the graph of $\mathcal{W}$) have played and continue to play a special role in the history of mathematics, as we next briefly discuss in Subsection 9.7.1. In Subsection 9.7.2, we will then describe several of the key symmetries and geometric properties of $\mathbf{\Gamma}_{\mathcal{W}}$, along with the nonlinear (and noncontractive) IFS of which $\mathbf{\Gamma}_{\mathcal{W}}$ is the attractor. Finally, in Subsection 9.7.3, we give the corresponding fractal tube formula and the complex dimensions, along with the average Minkowski content of the iterated fractal drum (IFD) associated with $\mathbf{\Gamma}_{\mathcal{W}}$ (all based on the work of C. David and M. L. Lapidus in [**106–115**]; see, especially, [**106–113**], building on [**370, 374, 375**], [**356**], and [**103–105**]) from which we deduce (as in loc. cit.), in particular, the Hölder continuity and the (discrete) reverse Hölder continuity of $\mathcal{W}$ with the optimal (i.e., best possible) Hölder exponent—and hence also the nowhere differentiability of $\mathcal{W}$, as well as of $\mathbf{\Gamma}_{\mathcal{W}}$.

The Weierstrass curve and function will be discussed in much greater detail in the second sequel to this book, [**116**].

9.7.1. Introduction: History and Complexity. Historically, the Weierstrass function $\mathcal{W}$ was the first published example of a continuous but nowhere differentiable function. Consequently, its graph, the Weierstrass curve, $\mathbf{\Gamma}_{\mathcal{W}} \subseteq \mathbb{R}^2$, does not admit a tangent line at any point. Karl Weierstrass introduced it in 1875 in [**609**] and a few years earlier, in a short communication in [**607**]; see also

[**608–610**]. His announcement and proof of the existence of such a function—and therefore also of such a continuous and nowhere differentiable curve—was greeted with astonishment and awe, by many, as well as with horror and fright, by quite a few other mathematicians; see, e.g., the relevant quotes heading this chapter. In fact, by a strange twist of history, it turns out—as was shown in the work of Norbert Wiener in the 1920s (see [**615, 616, 618**] and [**617**]) on a mathematical version of physical Brownian motion ([**159**] and [**488**]) and what came to be known as Wiener's probability measure (see, e.g., Chapters 2–4 in the book by G. W. Johnson and the first author, [**274**])—that almost every continuous function is nowhere differentiable; once again, this realization came as a shock to many mathematicians.

More specifically, *almost surely* (i.e., with probability 1 with respect to Wiener's probability measure) a real-valued continuous function on $[0, +\infty)$ (or on any given interval $[0, T]$, with $T > 0$), selected at random, is nowhere differentiable; see, e.g., [**274**, Chapter 3].[141] Perhaps less surprisingly, nowhere differentiable functions are generic, in the topological sense of Baire Category —and are dense, in particular—among all continuous functions; see [**18**].

The Weierstrass function, $\mathcal{W} = \mathcal{W}_{\lambda,b} : \mathbb{R} \to \mathbb{R}$, is defined by

$$\mathcal{W}(t) = \sum_{n=0}^{\infty} \lambda^n \cdot \cos(2\pi b^n \cdot t), \text{ for all } t \in \mathbb{R}, \tag{9.204}$$

where the real parameters λ and b satisfy[142]

$$0 < \lambda < 1 \quad \text{and} \quad \lambda b > 1. \tag{9.205}$$

See Figure 1.9 in Subsection 1.4.2 for a depiction of the Weierstrass curve $\mathbf{\Gamma}_{\mathcal{W}}$ (defined as the graph of $\mathcal{W}$), along with Figure 9.30 in Subsection 9.7.2 below for a picture of some of its polygonal (prefractal) approximations.

The Weierstrass function $\mathcal{W}$ and the Weierstrass curve $\mathbf{\Gamma}_{\mathcal{W}}$ are significantly more complicated than the other fractal functions and curves encountered thus far in this chapter, such as the Cantor function f_{DS} or the associated graph, the Devil's staircase $\mathbf{\Gamma}_{DS}$ (see Section 9.4) or the Koch curve $\mathbf{\Gamma}_{KC}$ and the Koch snowflake curve $\mathbf{\Gamma}_{KS}$ (see Section 9.5)—along with the corresponding parametrized curves γ_{KC} and γ_{KS}, viewed as maps from $[0, 1]$ to $\mathbb{R}^2$—or else the Sierpinski gasket **SG** and carpet **SC** (see Section 9.6). If one has a computer at his disposal, this claim is apparent by comparing a picture of $\mathbf{\Gamma}_{\mathcal{W}}$ (as in Figure 1.9 on page 32 of Chapter 1) and that of $\mathbf{\Gamma}_{DS}, \mathbf{\Gamma}_{KC}$, or $\mathbf{\Gamma}_{KS}$, or even of **SG** and **SC**, for example. An even simpler way to justify this statement is to compare the complex dimensions of $\mathbf{\Gamma}_{\mathcal{W}}$ to those of $\mathbf{\Gamma}_{KS}$, or of **SG** and **SC**, as described, respectively, in Subsections 9.4.1 and 9.6.3 above.

---

[141] See also [**257**] and [**474**] for a refined reformulation, using the notion of "prevalence".

[142] Because $|\lambda^n \cdot \cos(2\pi b^n \cdot t)| \leq \lambda^n$, for all $t \in \mathbb{R}$ and $n \in \mathbb{N}_0$, and $0 < \lambda < 1$, it is elementary to check that $\mathcal{W}$ is well defined and continuous on $\mathbb{R}$; see part (a) of Exercise 9.96 at the end of this subsection.

Indeed, as we will discuss in Subsection 9.7.2 below and as was shown by C. David and the first author in [**106**] (see also [**108**])—in contrast to all of the other aforementioned fractal curves, which have a single or at most finitely many (vertical, discrete) lines of complex dimensions—the Weierstrass curve $\mathbf{\Gamma}_{\mathcal{W}}$ has infinitely many (vertical, discrete) lines of actual (for (9.209)) or possible (for (9.210)) complex dimensions (all periodically distributed with the same period $\mathbf{p}_{\mathcal{W}}$), given in (9.209) and (9.210) below, including one such line consisting of the principal complex dimensions of $\mathbf{\Gamma}_{\mathcal{W}}$ (i.e., the complex dimensions with maximal real part, $D_{\mathcal{W}}$). Namely,

$$\mathcal{D}_{\mathcal{W},pr} = D_{\mathcal{W}} + i\mathbf{p}_{\mathcal{W}} \cdot \mathbb{Z} =: L_{0,D_{\mathcal{W}}}, \text{ where} \tag{9.206}$$

$$D_{\mathcal{W}} = 2 - \frac{\log(1/\lambda)}{\log b} = 2 - \log_b(1/\lambda) \in (1,2) \tag{9.207}$$

is the Minkowski dimension (or box-counting dimension) of $\mathbf{\Gamma}_{\mathcal{W}}$ and

$$\mathbf{p}_{\mathcal{W}} = \frac{2\pi}{\log b} \tag{9.208}$$

is the *oscillatory period* of $\mathbf{\Gamma}_{\mathcal{W}}$.[143] The countably many (discrete) vertical lines $L_{k,D_{\mathcal{W}}}$ of actual complex dimensions associated with $D_{\mathcal{W}}$ are given by

$$L_{k,D_{\mathcal{W}}} := D_{\mathcal{W}} - k(2 - D_{\mathcal{W}}) + i\mathbf{p}_{\mathcal{W}} \cdot \mathbb{Z}, \text{ with } k \in \mathbb{N}_0 \text{ arbitrary,} \tag{9.209}$$

so that, indeed, $L_{0,D_{\mathcal{W}}} = \mathcal{D}_{\mathcal{W},pr}$, as stated in (9.206). There is another countable infinity of (possible, this time) discrete, vertical lines, of the form

$$L_{k,1} := 1 - 2k + i\mathbf{p}_{\mathcal{W}} \cdot \mathbb{Z}, \text{ with } k \in \mathbb{N}_0 \text{ arbitrary.} \tag{9.210}$$

We also know that 0 and -2 are exact complex dimensions of $\mathbf{\Gamma}_{\mathcal{W}}$.[144]

It follows from the geometric proof of the above results in [**106, 108**] that the Weierstrass function is Hölder continuous, with optimal (i.e., best possible) Hölder exponent given by

$$\alpha_{\mathcal{W}} := 2 - D_{\mathcal{W}} = \frac{\log(1/\lambda)}{\log b} \in (1,2). \tag{9.211}$$

Therefore, since $\alpha_{\mathcal{W}} < 1$, we deduce that $\mathcal{W}$ is nowhere differentiable (even though it is continuous)—and hence also, equivalently, that $\mathbf{\Gamma}_{\mathcal{W}}$ does not have a

[143] Note that in light of (9.205), $1/\lambda > 1$ and $b > 1$ (since $b > 1/\lambda$). Also, observe that $0 < \alpha_{\mathcal{W}} := \log(1/\lambda)/\log b < 1$ and $D_{\mathcal{W}} = 2 - \alpha_{\mathcal{W}} \in (1,2)$.

[144] We point out that if—as is now rigorously established in [**110**] (announced in [**112**])—polyhedral (instead of Euclidean tubular) neighborhoods are used to define the corresponding fractal zeta functions and complex dimensions—then -2, 0 and the elements of the discrete lines $L_{k,1}$ given by (9.210) (for any $k \in \mathbb{N}_0$) are no longer complex dimensions; furthermore, the possible complex dimensions described by all of the discrete lines $L_{k,D_{\mathcal{W}}}$ given by (9.209) (with $k \in \mathbb{N}_0$ arbitrary) are the *actual* (or *exact*) complex dimensions of $\mathbf{\Gamma}_{\mathcal{W}}$, and they are all simple (i.e., of multiplicity one). These are called in [**110, 112**] the *intrinsic* complex dimensions of the Weierstrass curve $\mathbf{\Gamma}_{\mathcal{W}}$. See also Subsection 9.7.3 below, especially Theorem 9.110.

tangent line at any point—a well-known fact initially proved in 1875 by Weierstrass in [**607–610**] under the additional assumption that b is an integer and $\lambda b > 1 + (3\pi/2)$.

Weierstrass's original (and analytical) proof, to be provided in detail in [**116**], was remarkably clever and intricate but cannot be extended beyond the above range of the parameters λ and b.

It took another forty years before Godfrey H. Hardy established, for all admissible values of the parameters λ and b (see (9.205) above), the optimal (or sharp) Hölder continuity of $\mathcal{W}$—and hence also its nowhere differentiability—in his beautiful, but difficult, 1916 article [**228**].

Hardy's equally clever methods in [**228**] are based on harmonic analysis [**287**], including Fourier analysis [**565**], as well as the use of Green's functions and associated linear integral operators. They are flawless but not very intuitive, for most readers. By contrast, the methods used in [**106**] are essentially geometric (and, even, fractal geometric), based in part on a suitable extension (introduced and applied in this context by Claire David (see [**103–105**])—and, jointly with the first author, in [**106–115**])—of the use of (nonlinear, here) iterated function systems (IFSs), beyond the traditional setting of contraction mappings with respect to the Hausdorff metric (studied in Chapter 4 above)—combined with (and motivated by) the theory of complex fractal dimensions and fractal tube formulas developed in [**370, 374, 375**] and [**356**], along with [**347**], [**348**], and [**350**]; see also, e.g., [**326**], [**238**], [**87**], [**361**], and [**329**]. They also give more precise estimates regarding the (sharp) Hölder continuity and provide precise new information concerning an optimal type of (discrete) local reverse Hölder continuity of $\mathcal{W}$.

In fact, because of the great complexity of the fractal geometries involved, significant modifications and enhancements to the usual approach (in [**375**], [**356**], [**357**], [**359**], [**361**], [**329**], for example) are necessary in order to tackle this difficult problem, as will be expounded upon in the second sequel to this book, [**116**]. We will briefly discuss this issue later in this section; see Subsection 9.7.3.

Weierstrass's work was motivated, in large part, by a claim purportedly made by B. Riemann to his students during a lecture in the late 1850s (see also his thesis, published in 1868, [**513**] and [**514**])[145] that the following function now known as *Riemann's* (multifractal) *function* (and not to be mistaken with the celebrated Riemann zeta function)

$$\mathcal{R}(t) = \sum_{n=1}^{\infty} \frac{1}{n^2} \cdot \sin(n^2 \cdot t), \text{ for all } t \in \mathbb{R}, \tag{9.212}$$

[145] For an interesting history of this general topic, see M. Kline's book, [**298**].

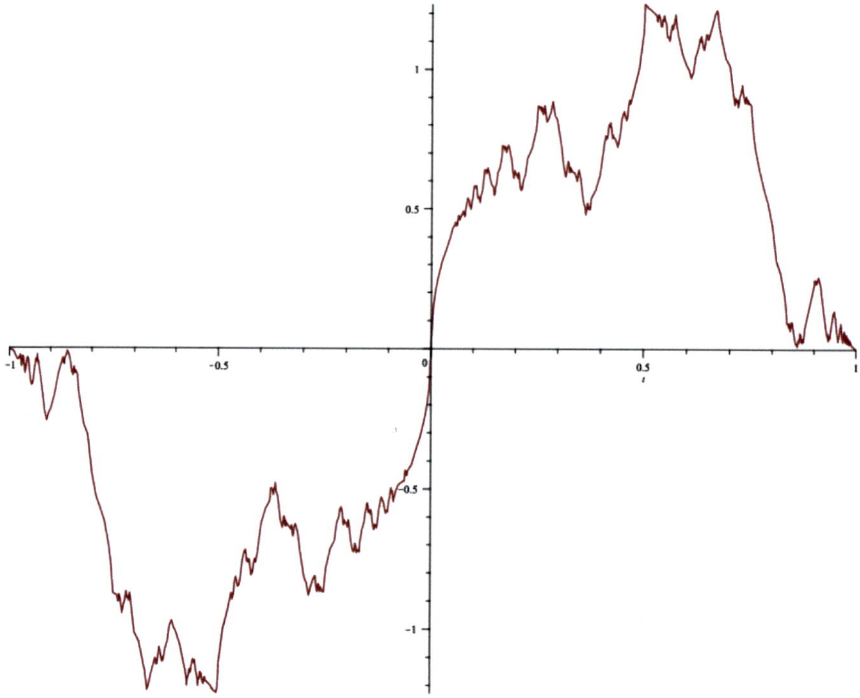

Figure 9.28. The graph of Riemann's multifractal function.

was nowhere differentiable. This claim had long puzzled Karl Weierstrass—prompting him, in particular, to come up with his own example of "pathological" function, namely, $\mathcal{W}$—and was eventually wrongly established in 1914 by Godfrey H. Hardy and John E. Littlewood in [**229**] (see also [**228**]), while much later, in 1971, Joseph Gerver [**214**], building on his earlier work in [**213**], precisely determined the points of differentiability of Riemann's ("nondifferentiable" but clearly continuous) function $\mathcal{R}$. See Figure 9.28 for a depiction of the graph of Riemann's function.

Later, Riemann's function became the object of further investigation—especially via the use of wavelet analysis and the multifractal thermodynamics formalism (motivated by statistical physics)[146]—in mathematics, physics, and engineering—and, along with $\mathcal{W}$ and other Weierstrass-like functions such as

[146]See, e.g., the work of B. Duplantier and of his collaborators (S. Sheffield, R. Rhodes, and V. Vargas) in [**149**], [**150**], and [**151**], with the relevant references therein—including on the so-called Liouville approach to quantum gravity.

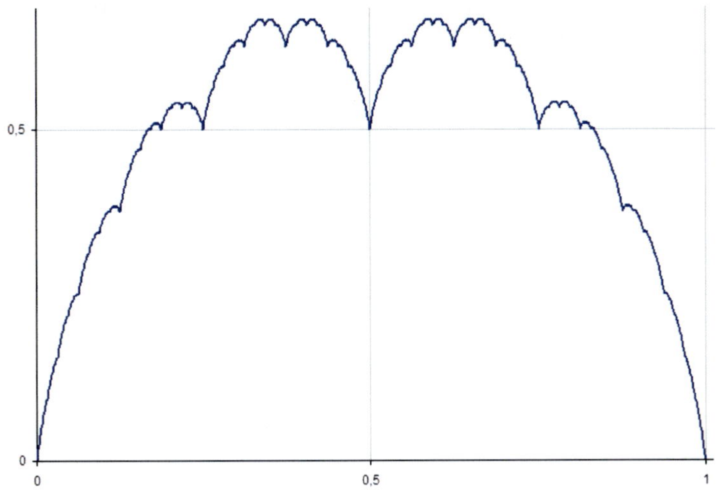

Figure 9.29. The graph $\mathbf{\Gamma}_{\mathcal{T}}$ of the original Takagi function [**582**], $\mathcal{T}(t) := \sum_{n=0}^{\infty}(1/2^n)\theta(2^n \cdot t)$, for all $t \in [0,1]$, where $\theta(u)$ denotes the distance to the nearest integer to $u \in \mathbb{R}$.

Takagi's function ([**6**], [**582**], [**583**]),[147] is now quite relevant to the study of turbulence, nonsmooth viscous solutions of the Navier–Stokes equation (see, e.g., U. Frisch and G. Parisi [**203**], M. Hata and M. Yamaguchi [**233**], and Y. Fujita, N. Hamamuki, A. Siconolfi and N. Yamaguchi [**206**]) and multifractality.[148] For the latter topic, see, e.g., A. Arneodo, E. Bacry, and J.-F. Muzy [**11**], M. Holschneider and P. Tchamitchian [**253**], S. Jaffard ([**265–268**] and the survey article [**269**]), S. Jaffard and Y. Meyer [**270**], along with the relevant references therein.

But let us next return to Weierstrass's extraordinary function. A long-time challenge had been to determine the fractal dimension—either the Hausdorff, packing, or else the Minkowski (or box-counting) dimension. This problem was taken up by a number of researchers, including at first A. S. Besicovitch

[147]In the notation of (9.241) of part (b) of Remark 9.115 below, the original Takagi function ([**582**], 1903) corresponds to the choice when $\Theta(t) := \theta(t)$ denotes the distance from $t \in [0,1]$ to the nearest integer, and $b \in \mathbb{N}\backslash\{1\}$ is given by $b := 2$ (and $\lambda := 2$); see the caption of Figure 9.29. On the other hand, the van der Waerden function is obtained by letting $b := 10$. These functions were later rediscovered several times, including by T. H. Hildebrandt in [**248**] and by G. de Rham in the mid-1950s in ([**132–134**]). See the interesting survey articles on the Takagi function by J. C. Lagarias [**310**] as well as by P. C. Allaart and K. Kawamura [**6**]. We refer to Figure 9.29 for a depiction of the graph of the Takagi function.

[148]Roughly speaking—unlike a "monofractal", which is homogeneous and thus admits the same *single* fractal (Hausdorff, packing, or Minkowski) dimension at every point—a "multifractal" is heterogeneous at all scales and hence typically admits a continuum of (real) fractal dimensions; see loc. cit.; see also, e.g., the work of J. A. Rock [**516**], M. L. Lapidus, J. A. Rock, and their collaborators ([**366**], [**332**], [**398**], [**135**], [**136**], [**160**]), as well as L. Olsen [**469, 470**], for various extensions of the theory of complex fractal dimensions to multifractals. Furthermore, we refer, e.g., to [**7**], [**61**], [**187**], [**321**], [**372**, Part 2], [**382**], [**383–385**], [**386–388**], [**399**], [**400**], [**402**], [**476**], and [**471, 472**], for a sample of additional references related to multifractals.

and H. D. Ursell [**47**], as well as B. B. Mandelbrot. The latter, besides attracting attention to the many interesting fractal properties of $\mathcal{W}$, conjectured in 1977 in [**419**] (see also [**418**]) that the Weierstrass curve $\mathbf{\Gamma}_{\mathcal{W}}$ has Hausdorff dimension $H_{\mathcal{W}}$ given by the right-hand side of (9.207):[149]

$$H_{\mathcal{W}} = 2 - \log_b(1/\lambda). \tag{9.213}$$

See also Falconer's 1986 book [**171**] for a discussion of this conjecture.

In the present case, it is also natural to expect that (9.207) holds, namely, that $D_{\mathcal{W}}$ (the Minkowski dimension of $\mathbf{\Gamma}_{\mathcal{W}}$) exists and equals $H_{\mathcal{W}}$, as given by (9.213).[150]

The fact that (9.207) holds was first proved (except for one geometric step), via analytical and dynamical systems arguments, by James L. Kaplan, John Mallet-Paret, and James A. Yorke in 1984 in [**286**].[151] Note that (9.207) is not implied by (9.213) since, in general, we only have that $H_{\mathcal{W}} \le D_{\mathcal{W}}$. Further results in this direction were obtained by Feliks Przytycki and Mariusz Urbánski in 1989 in [**497**], Tian-You Hu and Ka-Sing Lau in 1993 in [**254**] (where the packing dimension was shown to be equal to the right-hand side of (9.207)), as well as by Brian Hunt in 1998 in [**258**].

Moreover, building on seminal dynamical systems-based work of François Ledrappier in 1992 in [**390**], as well as on earlier work of the same author and Lai-Sang Young in 1985 in [**391**]—which established, in particular, further interesting connections between Lyapunov exponents, metric entropy, and fractal dimensions—Krzysztof Barański, Balázs Bárány, and Julia Romanowska obtained in 2014 in [**26**] partial results concerning Mandelbrot's Conjecture in (9.213); see [**25**] for a survey, as well as, e.g., [**21–24**] for earlier related work. (See also Gerhard Keller's 2017 paper [**289**] giving, in particular, a nice and simpler proof of the latter main result in [**26**].)

Finally, in 2018, Weixiao Shen [**546**] established Mandelbrot's Conjecture (in the case of the Hausdorff dimension), stated in (9.213), for every admissible value of the parameters λ and b (i.e., under condition (9.205)). His proof consists in examining every possible case and, as a result, is not very transparent. A challenge for future research in this area is to obtain a compelling and more conceptual proof of (9.213), still in the case of the Hausdorff dimension.

We close this introduction by mentioning that by using the nonlinear IFS to be discussed in the next subsection, Claire David proposed in [**103**] a new

[149] Actually, in his conjecture formulated in [**419**], Mandelbrot only used the phrase "fractal dimension", which could be interpreted, for example, as either Hausdorff or Minkowski dimension.

[150] Since the packing dimension of $\mathbf{\Gamma}_{\mathcal{W}}$ lies between $H_{\mathcal{W}}$ and $D_{\mathcal{W}}$ (with $H_{\mathcal{W}} \le D_{\mathcal{W}}$), this would also imply that the packing, Hausdorff, and Minkowski dimensions of $\mathbf{\Gamma}_{\mathcal{W}}$ coincide and are equal to $2 - \log_b(1/\lambda)$.

[151] The missing mathematical step in the proof—which was replaced by a one-line physical argument—corresponds, essentially, to a new version of Theorem 8.10 in Section 8.1 (and its proof) now obtained in [**110**]. The results of [**286**] show, however, that the Lyapunov dimension of an equivalent dynamical system is given by the right-hand side of (9.207)

method for obtaining a direct geometric proof of (9.207), while the latter author and Michel L. Lapidus [**110**] (see also [**112**]) obtained what is probably the most complete, rigorous, and transparent proof of (9.207) (i.e., of Mandelbrot's Conjecture interpreted in terms of the Minkowski dimension), by building on loc. cit. and on a new approach to the higher-dimensional theory of complex fractal dimensions, based on so-called local and global effective polyhedral, tube zeta functions. Accordingly, it is shown in [**110**] (announced in [**112**]) that the Minkowski dimension $D_{\mathcal{W}}$ of $\mathbf{\Gamma}_{\mathcal{W}}$ exists and coincides with the abscissa of convergence of the global (effective, polyhedral) tube zeta function of $\mathbf{\Gamma}_{\mathcal{W}}$, which happens to be equal to the right-hand side of (9.207), namely, $2 - \log_b(1/\lambda)$, as desired; see Subsection 9.7.3 below.[152]

Exercise 9.96.

(a) Show that (under condition (9.205)) the Weierstrass function $\mathcal{W}$, as given by (9.204), is well defined and continuous on $\mathbb{R}$.

(b) Show that if, in condition (9.205), we replace the inequality $\lambda b > 1$ by $\lambda b < 1$, then $\mathcal{W}$ is of class C^1 (i.e., is continuously differentiable) and its derivative can be obtained by differentiating term by term in the power series defining $\mathcal{W}$. Hence, by Theorem 9.3 in Section 9.1, $\mathbf{\Gamma}_{\mathcal{W}}$ is rectifiable in this case. Under which conditions is $\mathcal{W}$ of class C^n, for any given integer $n \geq 2$?

Remark 9.97 (*Nowhere Differentiable Functions*). The first example of a continuous nowhere differentiable function seems to be due to B. Bolzano in 1831 (unpublished at the time); this example was published by M. Jašek in 1922 in [**272**]; see also the later (1923) article [**307**]. Another early example is due to C. Cellérier in about 1860 and was also published posthumously in 1890 in [**76**]. Leaving aside the aforementioned work of B. Riemann (which, as was later discovered [**213, 214**], did not give rise to a fully nowhere differentiable function) and of K. Weierstrass [**609, 610**], other examples of nowhere differentiable functions are due, in chronological order, to G. Darboux in 1875–1879 [**99, 100**], U. Dini in 1877–1878 [**140, 141**], G. Peano in 1890 [**479**], D. Hilbert in 1891 [**247**], T. Takagi in 1903 [**582**], H. von Koch in 1904 [**603**], W. Sierpinski in 1912–1916 [**550–552**], K. Knopp in 1918 [**300**] (see also [**301**], 1928), A. S. Besicovitch in 1924 [**43**], S. Mazurkiewicz in 1931 [**433**], T. H. Hildebrandt in 1933

[152]For a sample of contemporary work (i.e., after 1950) on—or related to—the Weierstrass function (or curve) and its nowhere (or almost surely nowhere) differentiable generalizations, see, e.g., [**11**], [**21–25**], [**26**], [**42**], [**50**], [**103–105**], [**106–117**], [**118**], [**132–134**], [**171**], [**203**], [**206**], [**213, 214**], [**219**], [**232**], [**233**], [**245, 246**], [**253**], [**257, 258**], [**259**], [**265–269**], [**270**], [**271**], [**274**], [**281**], [**282**], [**286**], [**289**], [**298**], [**310**], [**390**], [**391**], [**408**], [**418–420**], [**431**], [**454**], [**474**], [**497**],[**593**], [**585**], [**595**], [**633**], along with the many references therein. For early references on this subject (i.e., essentially prior to 1950), see Remark 9.97, along with, e.g., [**488**] and [**615–618**].

[**248**], B. L. van der Waerden in 1930 [**599**], I. J. Schoenberg in 1938 [**541**], and G. de Rham in 1956–1957 [**132–134**], among many others.[153]

9.7.2. Geometric Properties and Nonlinear IFS. The Weierstrass function $\mathcal{W}$ has several interesting symmetry and scaling properties. By necessity of concision, we only mention the following ones. *We assume from now on that b is an integer* (i.e., $b \in \mathbb{N}\backslash\{1\}$, here), as was the case in Weierstrass's original paper [**609**]. Note that this hypothesis implies that $\mathcal{W}$ is 1-periodic. (See part (a) of Remark 9.115 below for the general case when $b \in \mathbb{R}$, with $b > 1$.) Moreover, we identify the points $A := (0, \mathcal{W}(0))$ and $B := (1, \mathcal{W}(1))$ of $\boldsymbol{\Gamma}_{\mathcal{W}}$. In the sequel, in light of the 1-periodicity of $\mathcal{W}$, we also define the Weierstrass curve $\boldsymbol{\Gamma}_{\mathcal{W}}$ to be the graph of the restriction of $\mathcal{W}$ to the compact unit interval $[0, 1]$. Because we identify its endpoints A and B, it is a closed curve. In fact, it is a Jordan curve and is unrectifiable.

Clearly, in light of the definition of $\mathcal{W}$ in (9.204) and since $b \in \mathbb{N}$,

$$\mathcal{W}(t) = \mathcal{W}(1-t), \text{ for all } t \in \mathbb{R}. \tag{9.214}$$

Hence, the *Weierstrass curve* $\boldsymbol{\Gamma}_{\mathcal{W}}$ (i.e., the graph of $\mathcal{W}$) *is symmetric with respect to the vertical line of abscissa* $t = 1/2$, as is noted and used repeatedly in [**106–114**].

In addition, $\mathcal{W}$ satisfies the following *scaling property*, which also easily follows from (9.204):

$$\lambda \cdot \mathcal{W}(bt) = \mathcal{W}(t) - \cos(2\pi t), \text{ for all } t \in \mathbb{R}. \tag{9.215}$$

As is also observed and used in [**106**], one sees by induction on $\ell \in \mathbb{N}^*$ that the scaling relation (9.215) has the following important consequence: For all $\ell \in \mathbb{N}^*$ and all $j \in \{0, \dots, \#\mathcal{V}_\ell - 1\}$, where $\#\mathcal{V}_\ell - 1 = (b-1)b^\ell$ (according to (9.224) below), we have that

$$\mathcal{W}\left(\frac{j}{(b-1)b^l}\right) = \lambda^\ell \cdot \mathcal{W}\left(\frac{j}{b-1}\right) + \sum_{k=0}^{\ell-1} \lambda^k \cdot \cos\left(\frac{2\pi b^k \cdot j}{(b-1)b^\ell}\right). \tag{9.216}$$

We next introduce the aforementioned *nonlinear* (and *noncontractive*) IFS, $\boldsymbol{\Phi}_{\mathcal{W}}$, based on the work by C. David in [**103**] (see also [**104**] and [**105**]):

$$\boldsymbol{\Phi}_{\mathcal{W}} := \{\varphi_0, \varphi_1, \dots, \varphi_{b-1}\}, \tag{9.217}$$

where, for any integer $j \in \{0, 1, \dots, b-1\}$ and any $(t, u) \in \mathbb{R}^2$,

$$\varphi_j(t, u) := \left(\frac{t+j}{b}, \lambda u + \cos\left(2\pi\left(\frac{t+j}{b}\right)\right)\right). \tag{9.218}$$

These nonlinear self-maps of $\mathbb{R}^2$ are not contractions—in the strict sense of the term, even though they do contract the area of a suitable sequence of

[153]See also, e.g., [**219**] for further historical and mathematical information about nowhere differentiable functions.

rectangles covering $\mathbf{\Gamma}_W$ (see [**104**]), due to a kind of hyperbolicity property similar to that of the chaotic dynamical system defining Smale's horseshoe (see, e.g., [**138**]). Therefore, one cannot apply the standard theory of IFSs in J. E. Hutchinson's 1981 work [**261**] or even in the 1985 paper by M. F. Barnsley and S. G. Demko [**30**]), as described in its basic form in Chapter 4 above (especially in Sections 4.2 and 4.5). Nevertheless, the Weierstrass curve $\mathbf{\Gamma}_W$ is the unique *attractor* (or *invariant set*) of the IFS $\mathbf{\Phi}_W$; i.e., as in the standard case (see Definition 4.18 and Theorem 4.17), it is the unique nonempty compact subset of $\mathbb{R}^2$ which satisfies the following fixed point equation:

$$\mathbf{\Phi}_W(\mathbf{\Gamma}_W) := \bigcup_{j=0}^{b-1} \varphi_j(\mathbf{\Gamma}_W) = \mathbf{\Gamma}_W. \tag{9.219}$$

Furthermore, each map φ_j (for $j = 0, \dots, b-1$) has a unique fixed point, say, P_j, which is given here by

$$P_j = (t_j, u_j) := \left(\frac{j}{b-1}, \frac{1}{1-\lambda}\cos\left(\frac{2\pi j}{b-1}\right)\right), \text{ for } j = 0, \dots, b-1. \tag{9.220}$$

We let

$$\mathcal{V}_0 := \{P_0, P_1, \dots, P_{b-1}\}, \text{ viewed as an ordered set.} \tag{9.221}$$

We can now approximate $\mathbf{\Gamma}_W$ by a sequence of directed *finite graphs*, $\{G_\ell := (\mathcal{V}_\ell, E_\ell)\}_{\ell=0}^{\infty}$, called the sequence of *prefractal graphs*—as was done in [**103**, **104**] and significantly used in [**120**] and [**106–115**]. Here, $\mathcal{V}_0$ is given by (9.221) just above and $\mathcal{V}_\ell$ is given inductively for every $\ell \in \mathbb{N}$ by

$$\mathcal{V}_\ell = \mathbf{\Phi}_W(\mathcal{V}_{\ell-1}) := \bigcup_{j=0}^{b-1} \varphi_j(\mathcal{V}_{\ell-1}), \tag{9.222}$$

much as was seen in Subsection 9.5.2 for the Koch curve $\mathbf{\Gamma}_{KC}$, for example. For each $\ell \in \mathbb{N}_0$, the finite set of *vertices*, $\mathcal{V}_\ell$, of the graph G_ℓ is viewed as an ordered set and is ordered according to increasing abscissae. It is called the *ℓ-th graph prefractal approximation* of $\mathbf{\Gamma}_W$. Two consecutive points are linked by an edge of the graph (i.e., an element of E_ℓ), and these are the only edges of G_ℓ. Clearly, we have the following properties, for each $\ell \in \mathbb{N}_0$:

$$\mathcal{V}_\ell \subseteq \mathcal{V}_{\ell+1}, \tag{9.223}$$

$$\#\mathcal{V}_\ell = (b-1)b^\ell + 1 \quad \text{and} \quad \#E_\ell = (b-1)b^\ell. \tag{9.224}$$

Note that, in light of (9.204) and (9.220), we have that $P_0 = P_{b-1} = A = B$, the (identified) endpoints of $\mathbf{\Gamma}$. Similarly, in light of the construction of $\mathcal{V}_\ell$ and of (9.222), the graph $G_\ell = (\mathcal{V}_\ell, E_\ell)$ naturally yields a closed polygonal curve $\mathbf{\Gamma}_{\ell,W}$, with (identified) endpoints $A = B$, and called the ℓ-th *natural polygonal approximation* of $\mathbf{\Gamma}_W$; see Figure 9.30.

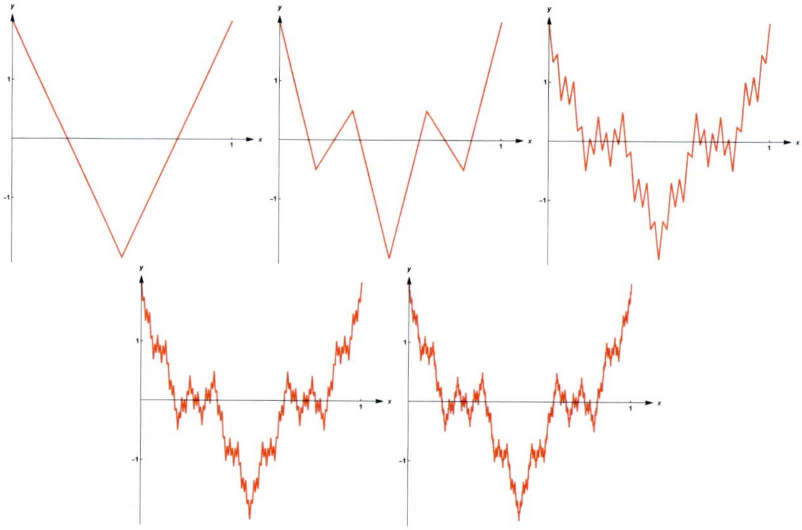

Figure 9.30. The first few natural polygonal (or finite graph) prefractal approximations to the Weierstrass curve $\mathbf{\Gamma}_{\mathcal{W}}$, for the values of the parameters $\lambda := 1/2$ and $b := 3$.

Moreover, it can be shown that $\{\Gamma_{\ell,\mathcal{W}}\}_{\ell=1}^{\infty}$ tends to $\mathbf{\Gamma}_{\mathcal{W}}$, as $\ell \to \infty$, in $(\mathbb{K}(\mathbb{R}^2), d_H)$; see part (a) of Exercise 9.100 below.

Another useful geometric property (established in [**107**]) is that the unique fixed point of the mapping φ_j (for $j = 0, \ldots, b-1$), namely, $P_j = (j/(b-1), \cos((2\pi j)/(b-1)))$, belongs to the Weierstrass curve $\mathbf{\Gamma}_{\mathcal{W}}$:

$$\mathcal{W}\left(\frac{j}{b-1}\right) = \frac{\cos(2\pi j)}{b-1}, \text{ for all } j \in \{0, \ldots, b-1\}. \tag{9.225}$$

Similarly, the coordinates of the vertices (or junction points) of each ℓ-th prefractal approximation $\mathbf{\Gamma}_{\ell,\mathcal{W}}$ of $\mathbf{\Gamma}_{\mathcal{W}}$ (with $\ell \in \mathbb{N}_0$) are calculated in [**106**], a very useful fact for obtaining the results stated in Subsection 9.7.3 below, including the fractal tube formula for $\mathbf{\Gamma}_{\mathcal{W}}$. Also, $\mathcal{V}_\infty := \bigcup_{\ell=0}^{\infty} \mathcal{V}_\ell$, the countably infinite set of all vertices of $\mathbf{\Gamma}_{\mathcal{W}}$, is dense in $\mathbf{\Gamma}_{\mathcal{W}}$; see [**106–110**].

A number of other geometric properties—regarding the so-called "elementary heights", the "elementary angles", and the "geometric angles"—are obtained in loc. cit. Other new properties regarding the polygons associated with the graph prefractal approximations to $\mathbf{\Gamma}_{\mathcal{W}}$ (as initially described in this context in [**103, 104**]) are obtained in [**107**], [**109**], and [**110**], along with several other geometric and analytic properties, including the following refined and sharp estimates, which are key to obtaining several of the main results in loc. cit., including the optimal Hölder continuity and discrete reverse Hölder continuity for the Weierstrass function $\mathcal{W}$ (see Theorem 9.114 in Subsection 9.7.3 below).

Theorem 9.98 (Explicit Upper and Lower Bounds for the Elementary Heights; [**106**], 2022). *For all $\ell \in \mathbb{N}$ and all integers j in $\{0, \ldots, (b-1)b^\ell = (\#\mathcal{V}_\ell)-1\}$, let $L_\ell = 1/((b-1)b^\ell)$ denote the corresponding* elementary horizontal length. *Then, the associated* elementary height *satisfies the following explicit inequalities, for the above values of ℓ and j:*

$$C_{\inf} \cdot L_\ell^{2-D_W} \leq \left| \mathcal{W}((j+1)L_\ell) - \mathcal{W}(jL_\ell) \right| \leq C_{\sup} \cdot L_\ell^{2-D_W}, \tag{9.226}$$

where the finite and positive constants $C_{\inf}$ and $C_{\sup}$ are independent of $\ell \in \mathbb{N}$ and given by (with $\mathcal{W}(q/(b-1))$ as in (9.225), for $q = 0, \ldots, b-1$),

$$C_{\inf} := (b-1)^{2-D_W} \cdot \min \left| \mathcal{W}\left(\frac{j+1}{b-1}\right) - \mathcal{W}\left(\frac{j}{b-1}\right) \right|,$$

where the minimum is taken over all integers j such that

$$0 \leq j \leq b-1, \text{ with } \mathcal{W}((j+1)/(b-1)) \neq \mathcal{W}(j/(b-1)) \text{ and}$$

$$C_{\sup} := (b-1)^{2-D_W} \cdot \left(\max \left| \mathcal{W}\left(\frac{j+1}{b-1}\right) - \mathcal{W}\left(\frac{j}{b-1}\right) \right| + \frac{2\pi}{(b-1)(\lambda b - 1)} \right),$$

where the maximum is taken over all integers j such that $0 \leq j \leq b-1$.

The next useful result (used, in particular, to prove some of the results of Subsection 9.7.3) is also obtained in [**106**], where it is given a more precise formulation in terms of the extreme vertices of polygons naturally associated with $\mathbf{\Gamma}_W$.

Theorem 9.99 (Local Extrema of $\mathcal{W}$; [**106**], 2022). *The set of local extrema of $\mathcal{W}$ is dense in the Weierstrass curve $\mathbf{\Gamma}_W$. Moreover, the set of local maxima of the restriction of $\mathcal{W}$ to $[0,1]$ is given by*

$$\left\{ \left(\frac{(b-1)k}{b^\ell}, \mathcal{W}\left(\frac{(b-1)k}{b^\ell} \right) \right) : 0 \leq k \leq b-1, \ell \in \mathbb{N}_0 \right\}, \tag{9.227}$$

while, when b is odd, the corresponding set of local minima is given by

$$\left(\left(\frac{(b-1)k + ((b-1)/2))}{(b-1)b^\ell} \right), \mathcal{W}\left(\frac{(b-1)k + ((b-1)/2)}{(b-1)b^\ell} \right) \right), \tag{9.228}$$

where $0 \leq k \leq b-1$ and $\ell \in \mathbb{N}_0$ are arbitrary.

In [**106**] are also obtained *sharp local discrete reverse Hölder inequalities* for $\mathcal{W}$ (of optimal exponent $2 - D_W = \log_b(1/\lambda)$), a property that was never established before and which plays a key role in obtaining the results of [**106**] and [**110**] concerning complex dimensions and (effective) fractal zeta functions discussed in Subsection 9.7.3. Because the corresponding precise statement is a little complicated and lengthy, we omit it here and refer the interested reader to loc. cit.; see Theorem 9.114 in Subsection 9.7.3 below for a very brief version.

Exercise 9.100.

(a) Show that $\{\mathbf{\Gamma}_{\ell,W}\}_{\ell=0}^{\infty}$, the sequence of natural polygonal approximations to $\mathbf{\Gamma}_W$ (as defined just after (9.224)), tends (as $\ell \to \infty$) to $\mathbf{\Gamma}_W$, in the Hausdorff metric of $\mathbb{R}^2$. State and prove a similar result for the sequence $\{\mathcal{V}_\ell\}_{\ell=0}^{\infty}$ of prefractal graphs.

(b) Examine the uniform convergence of $\{\gamma_{\ell,W}\}_{\ell=0}^{\infty}$ to γ_W, where γ_W (resp., $\gamma_{\ell,W}$, for each $\ell \in \mathbb{N}_0$) is the parametrized curve associated with $\mathbf{\Gamma}_W$ (resp., with $\mathbf{\Gamma}_{\ell,W}$), viewed as a continuous map from $[0,1]$ to $\mathbb{R}^2$.

(c) Show (as is proved in [**107**]) that $\mathcal{V}_\infty = \bigcup_{\ell=0}^{\infty} \mathcal{V}_\ell$, the set of all vertices (or junction points) of $\mathbf{\Gamma}_W$, is dense in the Weierstrass curve $\mathbf{\Gamma}_W$.

Exercise 9.101. Show that the Weierstrass curve $\mathbf{\Gamma}_W$ is simple (and hence is a Jordan curve) and, still under the assumption (9.205), has infinite length (i..e., is unrectifiable), whereas when $\lambda b < 1$, it is rectifiable (see part (b) of Exercise 9.96 above). Also, show that $\mathcal{W}$ is not injective, even when it is restricted to $[0, 1/2]$ (or, equivalently, to $[1/2, 1]$, by (9.214)).

[*Hint*: Use the appropriate result in Section 9.2.]

9.7.3. Complex Dimensions, Fractal Tube Formula, and Optimal Hölder Continuity. Unless explicitly mentioned otherwise, all of the results stated in this subsection have been obtained in the paper [**106**] (announced in [**108**]) by Claire David and Michel L. Lapidus, where they are stated with a lot more details and provided with complete proofs; see [**106–110**].[154]

The first result—which, like Theorem 9.108 below, is really obtained in [**106**] as a consequence of the prefractal tube formulas stated in Theorem 9.105 below—provides the complex dimensions of the Weierstrass curve $\mathbf{\Gamma}_W$.

We note that the statement concerning the Minkowski dimension of $\mathbf{\Gamma}_W$ in part (i) of Theorem 9.102 just below is actually proved by the same authors in [**110**], itself extending and adapting to the polyhedral realm some of the new results and methods of [**106**], as well as by using (via appropriate optimal covers) the notions of polyhedral neighborhoods and of a polyhedral measure introduced and studied in [**109**].

Theorem 9.102 (Minkowski Dimension and Complex Dimensions; C. David and M. L. Lapidus, [**106**], 2022, and [**110**], 2023).

(i) (Minkowski Dimension; [**110**]). *The Minkowski dimension D_W of $\mathbf{\Gamma}_W$ exists and is given by*

$$D_W = 2 - \frac{\log(1/\lambda)}{\log b} = 2 - \log_b(1/\lambda) \in (1,2). \tag{9.229}$$

[154]One such exception is part (i) of Theorem 9.102, part of which (namely, identity (9.229)) is proved in [**110**], in the spirit of the extended theory of complex dimensions, but by now using polyhedral (i.e., polygonal, since $m = 2$ here)—rather than tubular—neighborhoods.

It is a simple (and exact) complex dimension of $\mathbf{\Gamma}_W$, is real, and is the only complex dimension of $\mathbf{\Gamma}_W$ located on the real axis and with maximal real part. It also coincides with the abscissa of convergence of both the local effective tube and distance zeta functions, $\tilde{\zeta}_\ell^e$ and ζ_ℓ^e, for all $\ell \in \mathbb{N}$ sufficiently large (i.e., for all integers $\ell \geq \ell_0$, with $\ell_0 \in \mathbb{N}$ optimal).

(ii) (Complex Dimensions; [**106**]).[155] *Let*

$$\alpha_W := 2 - D_W = \frac{\log(1/\lambda)}{\log b} = \log_b(1/\lambda) \in (0,1) \tag{9.230}$$

denote the sharp Hölder exponent of $\mathcal{W}$—or, equivalently, here, the (real) Minkowski complex codimension *of $\mathbf{\Gamma}_W$; see Theorem* 9.114 *below. Then, apart from* 0 *and* -2, *which are (actual) complex dimensions of $\mathbf{\Gamma}_W$, the complex dimensions of the Weierstrass curve $\mathbf{\Gamma}_W$ are periodically distributed—with the same* oscillatory period $\mathbf{p}_W = 2\pi/\log b$—*along (countably) infinitely many vertical lines, of abscissae*[156]

$$D_W - k\alpha_W \text{ and } 1 - 2k, \text{ with } k \in \mathbb{N}_0 \text{ arbitrary.}$$

More specifically, the set $\mathcal{D}_{W,pos}$ of possible complex dimensions *of $\mathbf{\Gamma}_W$ is given by*[157]

$$\begin{aligned} \mathcal{D}_{W,pos} = &\{D_W - k\alpha_W + in\mathbf{p}_W \,:\, k \in \mathbb{N}_0, n \in \mathbb{Z}\} \\ &\cup \{1 - 2k + in\mathbf{p}_W \,:\, k \in \mathbb{N}_0, n \in \mathbb{Z}\} \cup \{-2, 0\}, \end{aligned} \tag{9.231}$$

while the set $\mathcal{D}_W$ of (actual) complex dimensions *of $\mathbf{\Gamma}_W$ satisfies*

$$\mathcal{D}_{W,pos} \supseteq \mathcal{D}_W \supseteq \{D_W - k\alpha_W + in\mathbf{p}_W \,:\, k \in \mathbb{N}_0, n \in \mathbb{Z}\} \cup \{-2, 0\}, \tag{9.232}$$

with the elements of $\mathcal{D}_W$ being simple *complex dimensions (i.e., simple poles of $\tilde{\zeta}_\ell^e$ (or, equivalently, of ζ_ℓ^e), for all sufficiently large $\ell \in \mathbb{N}$ (i.e., for all $\ell \geq \ell_0$).*[158]

Because of the great complexity of fractal curves such as the Weierstrass curve, we must introduce new definitions and techniques that may, at first, be somewhat surprising but which, upon reflection, become quite natural and are also perhaps, unavoidable. In particular, there is a natural length scale which intimately depends on the geometry of the underlying fractal curve (here, the

---

[155] See Remark 9.109 and the text preceding it for the definition of the notion of complex dimensions used in this theorem, defined as the poles of the effective local tube zeta functions $\tilde{\zeta}_\ell^e$, for all sufficiently large $\ell \in \mathbb{N}$.

[156] Of course, there are exceptional cases for special values of the parameters λ and b. For the corresponding somewhat lengthy discussion, we refer to [**106**].

[157] See Figure 9.31, on page 501 below, for a depiction of the exact complex dimensions of $\mathbf{\Gamma}_W$, given by $\omega_{k,n} := D_W - k(2 - D_W) + in\mathbf{p}$, where $k \in \mathbb{N}_0$ and $n \in \mathbb{Z}$ are arbitrary and which are associated with the integer scaled translates of the Minkowski dimension D_W.

[158] For the definition of the local fractal zeta functions $\tilde{\zeta}_\ell^e$ and ζ_ℓ^e, see Definition 9.107 below and the discussion preceding it.

Weierstrass curve $\mathbf{\Gamma}_W$).[159] The resulting notion was discovered in the process of constructing and characterizing the fractal cohomology of the Weierstrass curve (or IFD) in [**107**] (see also [**108, 113**].

Definition 9.103 (*ℓ-th Cohomology Infinitesimal*; [**106–108**]). We call *cohomology infinitesimal* of $\mathbf{\Gamma}_W$ the sequence $\{\varepsilon_\ell^\ell := (\varepsilon_\ell)^\ell\}_{\ell=0}^\infty$ of positive numbers given by

$$\varepsilon_\ell^\ell = (\varepsilon_\ell)^\ell = 1/((b-1)\cdot b^\ell) = L_\ell,$$

the ℓ-th elementary horizontal length introduced in Theorem 9.98 above.

Observe that both ε_ℓ and ε_ℓ^ℓ depend on $\ell \in \mathbb{N}_0$. Furthermore, we have that $\varepsilon_\ell^\ell \to 0$, as $\ell \to \infty$ (since $b > 1$), whereas $\varepsilon_\ell \to 1/b$, as $\ell \to \infty$. In fact, $\{\varepsilon_\ell\}_{\ell=0}^\infty$ decreases to $1/b > 0$.

For later purposes, we also introduce, as in [**110**], the *intrinsic cohomology infinitesimal*, given by the sequence $\{\varepsilon^\ell = (1/b)^\ell\}_{\ell=0}^\infty$, where $\varepsilon := 1/b$ is called the *intrinsic scale* of $\mathbf{\Gamma}_W$. Note that $\varepsilon_\ell \sim \varepsilon$, as $\ell \to \infty$.

We can now introduce the proper notion of tubular neighborhood and tubular volume we want to consider here, along with the key new notion of an iterated fractal drum (IFD).

Definition 9.104 (*$(\ell, \varepsilon_\ell^\ell)$-Neighborhood, ℓ-th Tubular Volume, and Weierstrass IFD*; [**106–110**]). Let $\ell \in \mathbb{N}_0$ be arbitrary, with ε_ℓ^ℓ as in Definition 9.103 just above. We call the *$(\ell, \varepsilon_\ell^\ell)$-tubular neighborhood* of $\mathbf{\Gamma}_W$ the (two-sided) ε_ℓ^ℓ-neighborhood of $\mathbf{\Gamma}_\ell$, the ℓ-th prefractal approximation to $\mathbf{\Gamma}_W$: $\mathcal{N}_\ell := [\mathbf{\Gamma}_\ell]_{\varepsilon_\ell^\ell}$.

Furthermore, we call the *ℓ-th tubular volume* (of $\mathbf{\Gamma}_W$) and denote by $V_\ell(\varepsilon_\ell^\ell) = \text{vol}_2(\mathcal{N}_\ell)$ the volume (i.e., the 2-dimensional Lebesgue measure) of the $(\ell, \varepsilon_\ell^\ell)$-tubular neighborhood just defined.

Moreover, we call the *Weierstrass iterated fractal drum* (or *Weierstrass IFD*, in short) the sequence $\{\mathcal{N}_\ell := [\mathbf{\Gamma}_\ell]_{\varepsilon_\ell^\ell}\}_{\ell\in\mathbb{N}_0}$ of $(\ell, \varepsilon_\ell^\ell)$-neighborhoods. Then, $\{\mathcal{N}_\ell\}_{\ell\in\mathbb{N}_0}$ tends to the Weierstrass curve $\mathbf{\Gamma}_W$, in the Hausdorff metric on $\mathbb{K}(\mathbb{R}^2)$, as is observed in [**107–109**]—and so, $V_\ell(\varepsilon_\ell^\ell) \to 0$, as $\ell \to \infty$.

We can now finally state the corresponding "prefractal formula", bearing in mind the notation introduced in Definitions 9.103 and 9.104 above.

[159] Intuitively, above the indicated critical length scale, the curve ceases to be "fractal" and becomes "classical" (or Euclidean), in the spirit of [**468**].

Theorem 9.105 (Prefractal Tube Formulas; [**106**], 2022). *Then, for all $\ell \in \mathbb{N}$ sufficiently large* (so that ε_ℓ^ℓ is sufficiently small)*—say, for all $\ell \geq \ell_0$ (with $\ell_0 \in \mathbb{N}$ minimal)—and with $\alpha_W := 2 - D_W = \log_b(1/\lambda)$, the Minkowski codimension of $\mathbf{\Gamma}_W$, as well as with $x_\ell^\ell := \log_b(1/\varepsilon_\ell^\ell)$, we have that*

$$V_\ell(\varepsilon_\ell^\ell) = \sum_{k=0}^{\infty} (\varepsilon_\ell^\ell)^{2-(D_W - k\alpha_W)} \cdot G_{k,D_W}(x_\ell^\ell) + \sum_{k=0}^{\infty} (\varepsilon_\ell^\ell)^{2-(1-2k)} \cdot G_{k,1}(x_\ell^\ell) + \pi(\varepsilon_\ell^\ell)^2 - \frac{\pi}{2}(\varepsilon_\ell^\ell)^4, \tag{9.233}$$

where for each $k \in \mathbb{N}_0$, $G_{k,D_W} = G_{k,D_W}(x)$ and $G_{k,1} = G_{k,1}(x)$ are (additive) 1-periodic functions respectively associated to the complex dimensions with real parts $D_W - k\alpha_W$ and $1-2k$ (see part (ii) *of Theorem* 9.102 *above).*[160] *Furthermore, all of the Fourier coefficients of G_{k,D_W} (with $k \in \mathbb{N}_0$) are nonzero; in particular, these functions are nonconstant. Moreover, the function G_{0,D_W} is positive and bounded away from zero and infinity.*

Remark 9.106 (*Analogy with the Higher-Dimensional Theory of Fractal Tube Formulas and Complex Dimensions from* [**356**]). By expanding into a Fourier series each 1-periodic function—and taking into account the fractal tube formula (9.233) in Theorem 9.105 above, along with Theorem 9.108 below about fractal zeta functions, while remembering that $x_\ell^\ell = \log_b(1/\varepsilon_\ell^\ell)$—we obtain that, for all $\ell \in \mathbb{N}$ sufficiently large (i.e., for all $\ell \geq \ell_0$),

(9.234)

$$V_\ell(\varepsilon_\ell^\ell) = \sum_{k=0}^{\infty} \sum_{n\in\mathbb{Z}} c_{k,D_W;n} \cdot (\varepsilon_\ell^\ell)^{2-((D_W - k\alpha_W)+in\mathbf{p}_W)} + \sum_{k=0}^{\infty} \sum_{n\in\mathbb{Z}} c_{k,1;n} \cdot (\varepsilon_\ell^\ell)^{2-((1-2k\alpha_W)+in\mathbf{p}_W)} + \pi \cdot (\varepsilon_\ell^\ell)^{2-0} - \frac{\pi}{2} \cdot (\varepsilon_\ell^\ell)^{2-(-2)}.$$

We note that all of the coefficients in the above *fractal power series* in (9.234) also depend on $\ell \geq \ell_0$—and, hence, so do the periodic functions G_{k,D_W} and $G_{k,1}$ in (9.233)—but, for notational simplicity, we do not indicate it explicitly. Furthermore, the coefficients $c_{k,D_W;n}$ (for all $k \in \mathbb{N}_0$ and $n \in \mathbb{Z}$) are nonzero.

We continue our brief exposition of the modifications to the higher-dimensional theory of complex dimensions [**356**] needed in [**106**] (and its sequels, especially in [**110**] and [**111**], but then, in the polyhedral case) in order to deal with fractal geometries of great complexity, such as the Weierstrass curve (in

[160] As functions of $\varepsilon = b^{-x} > 0$, these same functions are multiplicatively periodic (or log-periodic) functions, with multiplicative period $1/b$.

[**106**, **107**, **110**]) and the Koch curve [**111**]. Indeed, in loc. cit., both the tube function $V_\ell = V_\ell(t)$ and the usual fractal (i.e., tube or distance) zeta functions are replaced by their *effective* counterparts, to be defined in Definition 9.107 below. Furthermore, as may now be clear to the reader (see, especially, Theorem 9.102 above), one does not work directly with the fractal curve $\mathbf{\Gamma}_W$ but one instead first works with its prefractal polygonal (or graph) approximations $\mathbf{\Gamma}_\ell := \mathbf{\Gamma}_{W_\ell}$—i.e., with the associated (tubular) Weierstrass iterated fractal drum (IFD).

Definition 9.107 (*Effective Local Volumes and Effective Local Tubular Zeta Functions*). For each $\ell \in \mathbb{N}$ sufficiently large (i.e., as in the prefractal tube formula (9.234) or as in its equivalent form (9.233) given in Theorem 9.105), we introduce the ℓ-th *effective local tube function*—also called the ℓ-th *effective tubular volume* (of the Weierstrass IFD)—given by the continuous function $t \mapsto \widetilde{V}_\ell(t)$, defined for all $t \in [0, 1)$ and obtained by substituting t for ε_ℓ^ℓ on the right-hand side of (9.234) (or, equivalently, of (9.233)).[161]

Still for all ℓ large enough (i.e., for all $\ell \geq \ell_0$, with $\ell_0 \in \mathbb{N}$ optimal) and for $\text{Re}(s) > D_W$, we then introduce the *ℓ-th effective local tube zeta function* $\tilde{\zeta}_\ell^e = \tilde{\zeta}_\ell^e(s)$ (of the Weierstrass IFD)[162] via the following truncated Mellin transform of $\widetilde{V}_\ell$:[163]

$$\tilde{\zeta}_\ell^e(s) = \int_0^\varepsilon t^{s-3} \cdot \widetilde{V}_\ell(t)dt = \int_0^\varepsilon t^{s-2} \cdot \widetilde{V}_\ell(t)\frac{dt}{t}. \tag{9.235}$$

Then, guided by the functional equation (9.77) connecting the usual tube and distance zeta functions in [**356**] (see Subsection 9.3.3 above), we introduce the ℓ-th *effective local distance zeta function* $\zeta_\ell^e = \zeta_\ell^e(s)$ (of the Weierstrass IFD) via the following functional equation (still with $\varepsilon := 1/b$):

$$\zeta_\ell^e(s) = \varepsilon^{s-2} \cdot \widetilde{V}_\ell(\varepsilon) + (2 - s)\tilde{\zeta}_\ell^e(s). \tag{9.236}$$

In light of Definition 9.107 and of the functional equation (9.236), the next result is a simple consequence of Theorem 9.105 above.

---

[161] It follows from the results of [**106**] that (once $t \in [0, 1)$ has been substituted for ε_ℓ^ℓ) each of the series on the right-hand side of (9.234) is absolutely convergent—as well as locally uniformly (and normally) convergent in $t \in [0, 1)$. Thus, $\widetilde{V}_\ell$ is well defined and continuous on $[0, 1) \supseteq [0, \varepsilon]$. In addition, we stress that $V_\ell(t)$ *cannot* be equal to $\widetilde{V}_\ell(t)$, for all $t \in [0, \varepsilon_\ell^\ell]$. Indeed, otherwise, it would follow (from Theorem 9.108 from [**110**], and from [**356**]) that $D_W = 1$, the Minkowski dimension of the polygonal curve $\mathbf{\Gamma}_\ell$, which is impossible since $D_W > 1$.

[162] For notational simplicity, we write, for example, $\tilde{\zeta}_\ell^e$ instead of $\tilde{\zeta}_{\ell,W_\ell}^e$.

[163] Observe the analogies and the differences with the usual definition of a tube zeta function; compare with the corresponding definition in Subsection 9.3.3. Furthermore, it follows from Theorem 9.108 that $\tilde{\zeta}_\ell^e$ (and hence also ζ_ℓ^e) exists and is meromorphic in all of $\mathbb{C}$.

Theorem 9.108 (Local Fractal Zeta Functions; [**106**], 2022). *For every $\ell \in \mathbb{N}$ sufficiently large (i.e., for all integers $\ell \geq \ell_0$, for some optimal $\ell_0 \in \mathbb{N}^*$), the ℓ-th effective local tube zeta function $\tilde{\zeta}_\ell^e$ (of the Weierstrass IFD) admits a (necessarily unique) meromorphic continuation to all of $\mathbb{C}$, given for all $s \in \mathbb{C}$ by the following expression:*

$$\begin{aligned}(9.237)\quad \tilde{\zeta}_\ell^e(s) = &\sum_{k=0}^{\infty}\sum_{n\in\mathbb{Z}} d_{k,D_W;n}\cdot\frac{\varepsilon^{s-(D_W-k\alpha_W+in\mathbf{p}_W)}}{s-(D_W-k\alpha_W+in\mathbf{p}_W)}\\ &+\sum_{k=0}^{\infty}\sum_{n\in\mathbb{Z}} d_{k,1;n}\cdot\frac{\varepsilon^{s-(1-2k+in\mathbf{p}_W)}}{s-(1-2k+in\mathbf{p}_W)}+\pi\frac{\varepsilon^{s-0}}{s-0}-\frac{\pi}{4}\frac{\varepsilon^{(s-(-2))}}{s-(-2)}.\end{aligned}$$

Furthermore, $\tilde{\zeta}_\ell^e$ and ζ_ℓ^e are connected by the functional equation (9.236), *for all $s \in \mathbb{C}$, where $\tilde{\zeta}_\ell^e(s)$ and $\widetilde{V}_\ell(\varepsilon)$ are respectively given by* (9.237) *and the counterpart of* (9.234) *(or, equivalently, of* (9.233) *in Theorem* 9.105*). Moreover, the coefficients $d_{k,D_W;n}$ and $d_{k,1;n}$ of the relevant series in* (9.237) *also depend on $\ell \geq \ell_0$ (which we do not indicate explicitly, for notational simplicity) and are uniformly bounded in $k \in \mathbb{N}_0$ and $n \in \mathbb{Z}$, as well as in $\ell \geq \ell_0$. Finally, $d_{k,D_W;n} \neq 0$, for all $k \in \mathbb{N}_0$ and $n \in \mathbb{Z}$.*

Remark 9.109 (*Possible Complex Dimension in Theorem* 9.102). Theorem 9.108 implies at once part (ii) of Theorem 9.102, provided in the latter theorem, we interpret the phrase "possible complex dimension" of $\mathbf{\Gamma}_W$ (or of the Weierstrass IFD) as a possible pole of the local effective tube zeta functions $\tilde{\zeta}_\ell^e$ (or, equivalently, of ζ_ℓ^e), for all integers ℓ sufficiently large (i.e., for all $\ell \geq \ell_0$), which is naturally the definition we adopt in Theorem 9.102. Clearly, in light of (9.237), 0 and -2 are actually *exact* complex dimensions of $\mathbf{\Gamma}_W$ (or of the associated polyhedral Weierstrass IFD); i.e., they are actual poles of $\tilde{\zeta}_\ell^e$ (or, equivalently, of ζ_ℓ^e), for all $\ell \geq \ell_0$.

We next very briefly introduce the notion of *intrinsic* complex dimensions (directly inspired by the work in [**110**] and [**107**]) and the associated notion of a *global* effective *polyhedral* zeta function. For this purpose, we mention that if we follow a procedure similar (but not identical) to the above one—in which we replace ε_ℓ^ℓ by ε^ℓ (for each $\ell \in \mathbb{N}_0$), where $\varepsilon := 1/b$ is the intrinsic scale of $\mathbf{\Gamma}_W$, and the sequence of $(\ell, \varepsilon_\ell^\ell)$-tubular neighborhoods of $\mathbf{\Gamma}_W$ (in Definition 9.104) by an appropriate sequence of (ℓ, ε^ℓ)-polygonal (or polyhedral) neighborhoods of $\mathbf{\Gamma}_W$, where the polygons involved are naturally associated with the ℓ-th prefractal approximations $\mathbf{\Gamma}_\ell$ (or with the corresponding ℓ-th graph prefractal approximations $G_\ell = (\mathcal{V}_\ell, E_\ell)$)—one can define (upon a suitable rescaling of the variables involved) a sequence $\{\tilde{\zeta}_\ell^{e,\mathcal{P}}\}_{\ell\in\mathbb{N}_0}$ of *local* effective *polyhedral* zeta functions converging, in an appropriate sense, to a so-called *global* effective *polyhedral* zeta function, denoted $\tilde{\zeta}_W^e$ and whose exact poles (i.e., the *intrinsic* complex dimensions of $\mathbf{\Gamma}_W$) can be explicitly determined; see [**110**] for the details of the construction.[164]

We can now state the following key theorem (and definitions), which is established in [**110**] (and announced in [**108, 112**]).

Theorem 9.110 (Global Zeta Function and Intrinsic Complex Dimensions; [**110**], 2023). *There exists an optimal* $\ell_0 \in \mathbb{N}^*$ *such that the sequence of* local effective polyhedral zeta functions $\{\tilde{\zeta}_\ell^{e,\mathcal{P}}\}_{\ell \geq \ell_0}$ *converges locally uniformly (and hence also pointwise)*[165] *on all of* $\mathbb{C}$ *to a meromorphic function on* $\mathbb{C}$, *denoted* $\tilde{\zeta}_W^{e,\mathcal{P}}$ *and called the* global effective polyhedral zeta function *of the Weierstrass curve* $\mathbf{\Gamma}_W$ *(or of the* polyhedral Weierstrass IFD).[166]

Moreover, the intrinsic complex dimensions *of* $\mathbf{\Gamma}_W$ *(or of the associated IFD) —defined as the poles of* $\tilde{\zeta}_W^{e,\mathcal{P}}$ *—are all exact (i.e., actual) and simple; they are given by*[167]

$$\omega_{j_{k,\ell},k} := D_W - k(2 - D_W) \pm i n_{j_{k,\ell},k} \cdot \mathbf{p}_W, \tag{9.238}$$

where $\ell \in \mathbb{N}$ *(with* $\ell \geq \ell_0$*) is arbitrary,* $0 \leq k \leq \ell$, *and for suitable choices of* $n_{j_{k,\ell},k} \in \mathbb{N}_0$ *(corresponding to the so-called* 'cohomological integers' introduced *in* [**107**], [**110**], *and* [**114**]) *and, as before,* $\mathbf{p}_W = 2\pi/\log b$ *is the oscillatory period of* $\mathbf{\Gamma}_W$; *see Remark* 9.111 *just below and the end of the caption of Figure* 9.31.

Remark 9.111. As is explained in [**110**], the integers $n_{j_{\ell,\ell},k}$ (which necessarily depend on $k \in \mathbb{N}_0$ and $\ell \in \mathbb{N}_0$, with $\ell \geq \ell_0$) have a specific and very natural geometric and dynamical interpretation associated with an appropriate labeling of the vertices of the ℓ-th prefractal graph approximation of $\mathbf{\Gamma}_W$, as well as with the motion of points (i.e., vertices) across scales. Furthermore, for each integer $k \geq 0$, we have that $n_{j_{k,\ell},k} \geq 0$—and hence, in light of (9.238) above, $n_{j_{k,\ell},k} = 0$ iff $\omega_{j_{k,\ell},k} \in \mathbb{R}$ (i.e., iff $\omega_{j_{k,\ell},k} = D_W - k(2 - D_W)$). It follows that all *nonreal* intrinsic complex dimensions appear in complex conjugate pairs, as should be the case, geometrically; also, on each vertical line with abscissa

[164] Naturally, in order to define $\tilde{\zeta}_\ell^{e,\mathcal{P}}$ (for all $\ell \in \mathbb{N}^*$ sufficiently large), we must define and compute an appropriate *effective polyhedral volume function*. We note that, unlike in the tubular case studied in [**106**], the coefficients of the fractal power series for the ℓ-th effective polyhedral volume $\widetilde{V}_\ell$ in (9.233) and of the convergent series for the local effective polyhedral zeta function $\tilde{\zeta}_\ell^{e,\mathcal{P}}$ in the counterpart of (9.237) can be explicitly computed, for all $\ell \geq \ell_0$. Also, $\tilde{\zeta}_W^{e,\mathcal{P}}$ can be obtained as a limit of suitable Riemann sums; see [**110**].

[165] More precisely, one must first subtract from $\tilde{\zeta}_\ell^{e,\mathcal{P}}(s)$ each of the pole parts corresponding to the complex dimensions 0 and -2; see the end of (9.237).

[166] In this statement, the meromorphic functions $\tilde{\zeta}_\ell^{e,\mathcal{P}}$ (for all $\ell \geq \ell_0$) and $\tilde{\zeta}_W^{e,\mathcal{P}}$ are viewed as continuous functions with values in the Riemann sphere, $\widetilde{\mathbb{C}} := \mathbb{C} \cup \{\infty\}$, equipped with the chordal metric, as in [**370, 374, 375**], [**356**], and [**238**].

[167] For a fixed $\ell \in \mathbb{N}^*$, and still with $\ell \geq \ell_0$, the complex dimensions in (9.238)—and in the text following it, where $0 \leq k \leq \ell$—are just the poles of $\tilde{\zeta}_\ell^{e,\mathcal{P}}$ (or, equivalently, of $\zeta_\ell^{e,\mathcal{P}}$); see also [**110**].

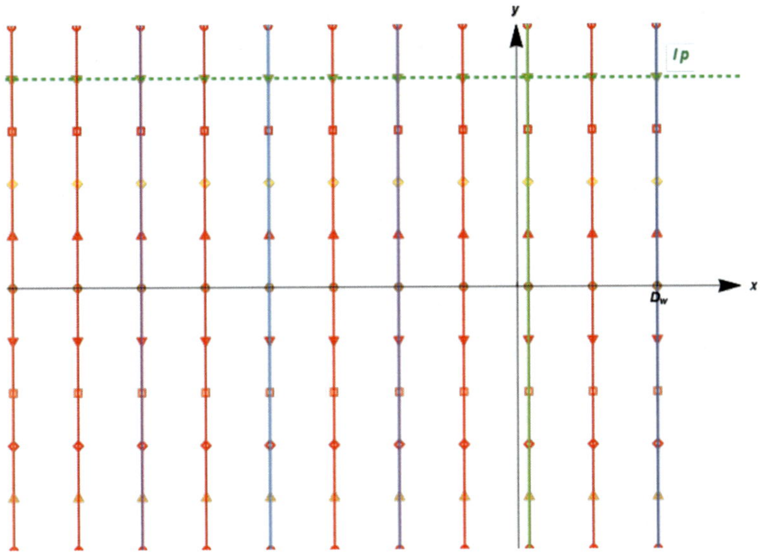

Figure 9.31. The exact complex dimensions, $\omega_{k,n} = D_W - k(2 - D_W) + in\mathbf{p}$, with $k \in \mathbb{N}_0$ and $n \in \mathbb{Z}$ arbitrary (attached to D_W and its integer scaled translates) of the Weierstrass curve $\mathbf{\Gamma}_W$ (see part (ii) of Theorem 9.102)—as given (in a different notation) by formula (9.238) in Theorem 9.110 and for the values of the parameters $\lambda = 1/2$ and $b = 3$: They are periodically distributed along countably many vertical lines, with abscissae $D_W - k(2 - D_W)$, where $k \in \mathbb{N}_0$ is arbitrary, and with vertical period $i\mathbf{p}_W = i(2\pi/\log b)$. Here, a different color is used for each vertical line, and a specific symbol is used to plot the imaginary parts of the complex dimensions associated with a given vertical line. For the sake of simplicity, we did not represent the complex dimensions -2 and 0, as well as the possible complex dimensions $1 + in\mathbf{p}_W$, with $n \in \mathbb{Z}$ arbitrary; see part (ii) of Theorem 9.102. Note that, in light of Theorem 9.110, the set of *intrinsic* complex dimensions (as given by (9.238)) is an appropriate subset of the set of complex dimensions described in part (ii) of Theorem 9.102: They appear on each vertical line—which they necessarily intersect at countably many points, both on the real axis, at a single point, and at countably many nonreal complex conjugate pairs.

$D_W - k(2 - D_W)$, with $k \in \mathbb{N}_0$, there are (infinitely) countably many intrinsic complex dimensions, including a single real one (namely, $D_W - k(2 - D_W)$ itself) and the others, consisting of countably (nonreal) complex conjugate pairs.[168]

It follows from the corresponding discussion in [**110**] that the oscillations (or vibrations) associated with the complex dimensions (see, e.g., [**329, 330, 356,**

[168] (*Added in Proof.*) It turns out that in the final version of [**106**] (not discussed here), the results of [**110**] and [**106**] are remarkably similar: In particular, the set of the ℓ-th prefractal complex dimensions is finite and increases with $\ell \geq \ell_0$. Furthermore, the (restricted) global tubular and polyhedral zeta functions exist and coincide. Consequently, the sets of all tubular and polyhedral (exact) complex dimensions coincide.

370, 374, 375]) now also have a natural geometric and dynamical origin and interpretation within the context of the extended theory of complex dimensions developed in [**106–115**]. In particular, the underlying geometric oscillations correspond to the subdivision process (into $b = \varepsilon^{-1}$ subparts) enabling one to go from the ℓ-th to the $(\ell + 1)$-th graph prefractal approximation of $\mathbf{\Gamma}_W$; note that here ε^{-1} is the reciprocal intrinsic scale.

The following result is also established in [**110**] (announced in [**108, 112**]) by using the polyhedral methods introduced in [**109**], as well as used in a key manner in [**110**] and [**111**]. It is proved by first showing that $\dim_B(\mathbf{\Gamma}_W)$ is given by the expected formula, via a reduction to polyhedral (i.e., polygonal, here) neighborhoods of $\mathbf{\Gamma}_W$, and then appealing to an appropriate extension of the proof of the well-known result according to which $\dim_B$ and $\dim_M$ coincide on the bounded subsets of Euclidean space $\mathbb{R}^m$ (here, $m = 2$); see Theorem 8.29 in Section 8.4. The aforementioned polyhedral method seems to provide the most direct and complete (as well as geometric) proof of Mandelbrot's Conjecture (interpreted in terms of the Minkowski dimension).[169]

Theorem 9.112 (Minkowski and Box Dimensions; [**110**], 2023). *The Minkowski and box-counting dimensions of the Weierstrass curve $\mathbf{\Gamma}_W$, defined exactly as usual (see Chapter* 8, *Definition* 8.3 *and Definition* 8.22, *respectively) exist, coincide, and are given by the following formula*:

(9.239)
$$\dim_M(\mathbf{\Gamma}_W) = \dim_B(\mathbf{\Gamma}_W) = D_W = 2 - \alpha_W, \text{ with } \alpha_W := 2 - D_W = \log_b(1/\lambda).$$

Here, α_W is also the optimal Hölder exponent of the Weierstrass function W; see Theorem 9.114 *below. Observe that* (9.239) *is in agreement with Mandelbrot's Conjecture in* [**418–420**] *and, in fact, proves it in the case of the Minkowski (or box-counting) dimension.*

Furthermore, we suitably adapt the standard definitions of Minkowski nondegeneracy and of Minkowski measurability, as well as of average Minkowski content, given in Section 8.1 *and Subsection* 9.3.6, *by taking into account the natural length scales powers ε^ℓ and the ℓ-th effective polyhedral volume (for all $\ell \geq \ell_0$). Then, it turns out that the Weierstrass polyhedral IFD—along with each of the prefractal approximations $\mathbf{\Gamma}_\ell$, for every $\ell \geq \ell_0$—is (effectively) Minkowski nondegenerate but is not (effectively) Minkowski measurable; moreover, it admits a positive and finite (effective) average Minkowski content. We refer the interested reader to* [**110**], [**108, 112**] *for the precise statements and definitions.*

[169] See also the analytical and dynamical systems proof provided in [**286**]—in which, however, a geometric step seems to be missing.

The following result is an immediate consequence of Theorems 9.102, 9.108, and 9.112, along with (for the intrinsic version) Theorem 9.110 and its local counterpart obtained in [**110**].[170]

Theorem 9.113 (Fractality of the Weierstrass IFD and of Its Prefractal Approximations; [**110**], 2023). *The Weierstrass IFD (or curve $\mathbf{\Gamma}_W$), along with its prefractal approximations $\mathbf{\Gamma}_\ell$, for any $\ell \geq \ell_0$, is fractal. It is even principally or critically fractal (i.e., it is fractal in the maximum possible dimension) $d := D_W = \sigma_W$ (for the Weierstrass IFD) and $d := \sigma_{\ell,W} = D_W$ (for the ℓ-th prefractal approximation, with $\ell \geq \ell_0$).*[171]

In the polyhedral case, there is, however, one important difference between the Weierstrass curve (or IFD) and its prefractal approximations. Indeed, $\mathbf{\Gamma}_W$ (or the associated Weierstrass IFD) is fractal in countably infinitely *many dimensions (namely, $d = d_k := D_W - k(2 - D_W)$, for every $k \in \mathbb{N}_0$), whereas its ℓ-th prefractal approximations $\mathbf{\Gamma}_\ell$ (for every $\ell \geq \ell_0$) are only fractal in* finitely *many dimensions (namely, $d = d_k := D_W - k(2 - D_W)$, with $k = 0, \ldots, \ell$).*[172]

Finally, we state the result concerning the sharp Hölder continuity and the sharp discrete reverse Hölder continuity of $\mathcal{W}$, which follow, in particular, from the sharp geometric estimates stated in Theorem 9.98; see estimate (9.226). It is also obtained in [**106**].[173]

Theorem 9.114 (Sharp Hölder Continuity and Discrete Reverse Hölder Continuity; [**106**], 2022). *The Weierstrass function $\mathcal{W}$ is Hölder continuous, with optimal Hölder exponent, $\alpha_W := 2 - D_W = \log_b(1/\lambda) \in (0, 1)$, the Minkowski codimension of $\mathbf{\Gamma}_W$ (and with $D_W = 2 - \log_b(1/\lambda) \in (1, 2)$, as before).*[174] *Hence, in particular, since $\alpha_W < 1$, it follows that $\mathcal{W}$ is nowhere differentiable*[175]

---

[170] See Subsection 9.4.7 and the beginning of Subsection 9.3.5 for the definitions of fractality, fractality in dimension $d \in \mathbb{R}$, and principal (or critical) fractality. In the latter case, we are using here a slight adaptation to our present situation of the definition of principal (or critical) fractality because we are replacing in that definition the Minkowski dimension by the abscissa of convergence, $\sigma_{\ell,W}$ or σ_W, of the corresponding local or global (effective) fractal zeta function.

[171] Note that D_W is equal to the Minkowski dimension of $\mathbf{\Gamma}_W$ and, for any $\ell \geq \ell_0$, is also equal to both $\sigma_{\ell,W}$ and σ_W, the abscissae of convergence of $\tilde{\zeta}^e_\ell$ and of $\tilde{\zeta}^e_W$, but it is clearly *not* equal to the Minkowski dimension of $\mathbf{\Gamma}_\ell$, which is equal to 1.

[172] In contrast, in the tubular case (described in Theorem 9.102), both $\mathbf{\Gamma}_W$ and $\mathbf{\Gamma}_\ell$ (for any $\ell \geq \ell_0$) are fractal in countably many dimensions; namely, $d = d_k := D_W - k(2 - D_W)$, for all $k \in \mathbb{N}_0$; see, however, footnote 168 above.

[173] Recall from our earlier discussion that G. H. Hardy had obtained in his 1916 paper [**228**] the optimal Hölder exponent, but that he had not obtained the sharp discrete reverse Hölder estimates established in [**106**] and implicit in Theorem 9.114.

[174] This means that for any given $t \in \mathbb{R}$, $\mathcal{W}(t + h) - \mathcal{W}(t) = O(h^{\alpha_W})$, as $h \to 0$, where the exponent α_W is the best possible. The latter fact comes from the optimal discrete *reverse* local Hölder continuity also obtained in [**106**] and stated (at the first stage of the induction) in the first inequality of (9.226) in Theorem 9.98: $|\mathcal{W}(t) - \mathcal{W}(\tilde{t})| \geq C \cdot |t - \tilde{t}|^{2-D_W}$, for some positive and finite constant C and for the abscissae t and $\tilde{t}$ of any two neighboring vertices of $\mathbf{\Gamma}_\ell$, for any $\ell \in \mathbb{N}$.

[175] Indeed, otherwise, if $\mathcal{W}$ happened to be differentiable at some point $t_0 \in \mathbb{R}$, then it would have to be (locally) Lipschitz around t_0, which is impossible since $\alpha_W < 1$ and α_W is optimal as a Hölder exponent of $\mathcal{W}$.

—or, equivalently, that the Weierstrass curve $\mathbf{\Gamma}_W$ *does not admit a tangent line at any point.*

Moreover, as is shown in [**106**], W *satisfies* sharp, discrete, reverse Hölder inequalities, *also with optimal Hölder exponent* α_W.

We conclude this subsection by providing several comments about further extensions of the results presented above.

Remark 9.115 (*Further Extensions*).

(a) (*The Non-Integer Case*: $b \in \mathbb{R} \backslash \mathbb{N}$ and the smooth case when $\lambda b < 1$). The above results should be naturally extended to the case when the parameter b is no longer an integer; i.e., (9.205) still holds ($0 < \lambda < 1$ and $\lambda b > 1$) but now, $b \in \mathbb{R} \backslash \mathbb{N}$, with $b > 1$. One significant difference with the case discussed above when $b \in \mathbb{N}$ is that the nonlinear (and noncontractive) IFS $\mathbf{\Phi}_W$ is no longer finite. In fact, $\mathbf{\Phi}_W$ is *countably infinite* and given by $\Phi_W = \{\varphi_j\}_{j \in \mathbb{Z}}$, where each φ_j (with $j \in \mathbb{Z}$) is a self-map of $\mathbb{R}^2$ given by (9.218) in Subsection 9.7.2. We then expect the possible complex dimensions to be given as in Theorem 9.102 (resp., Theorem 9.110), in the tubular (resp., polyhedral) case—along with the other results from [**106**], [**107**], [**108**], and [**110**] briefly described in the present subsection—to have natural counterparts in this more general context; see the corresponding discussion in [**106**].

It is also expected that the case when $\lambda b < 1$ (instead of when $\lambda b > 1$)—and hence, W is at least C^1 smooth on $\mathbb{R}$ rather than nowhere differentiable—can be handled in a similar way and with an analogous (but not identical) outcome. Remarkably, then, one obtains a C^1 curve $\mathbf{\Gamma}_W$ which is fractal (in the sense of the theory of complex dimensions) since it has infinitely many nonreal complex dimensions (located along countably many vertical lines), including the nonreal principal complex dimensions of $\mathbf{\Gamma}_W$. It is also shown in [**114**] that any nowhere differentiable Weierstrass curve $\mathbf{\Gamma}_W$ (with parameters (λ, b) such that $\lambda b > 1$) has a *dual* which is C^1 smooth but is necessarily fractal, namely, the Weierstrass curve $\mathbf{\Gamma}_{W^*}$ (*with parameters* (λ^*, b^*) such that $b^* := b$ and $\lambda^* := 1/\lambda b^2$) satisfying the following remarkable property: The global (effective) polyhedral zeta functions $\tilde{\zeta}_W^{e,\mathcal{P}}$ and $\tilde{\zeta}_{W^*}^{e,\mathcal{P}}$ (of $\mathbf{\Gamma}_W$ and $\mathbf{\Gamma}_{W^*}$, respectively) are connected via the *functional equation*[176]

$$\tilde{\zeta}_{W^*}^{e,\mathcal{P}}(s) = \tilde{\zeta}_W^{e,\mathcal{P}}(2 - s), \text{ for all } s \in \mathbb{C}. \tag{9.240}$$

When $\lambda b = 1$, then $W = W^*$ (i.e., $\mathbf{\Gamma}_W = \mathbf{\Gamma}_{W^*}$)—and hence the corresponding *critical* Weierstrass curve $\mathbf{\Gamma}_W$, along with its global (polyhedral) zeta function $\tilde{\zeta}_W^{e,\mathcal{P}}$ and the associated functional equation (9.240) (with $\tilde{\zeta}_{W^*}^{e,\mathcal{P}} = \tilde{\zeta}_W^{e,\mathcal{P}}$)—is *self-dual*. This result of [**114**] thereby establishes in the present context of Weierstrass IFDs a suitable counterpart of a conjecture made in [**326**] in connection

[176] An entirely analogous functional equation holds in the case of the ℓ-th polygonal approximation of $\mathbf{\Gamma}_W$ and $\mathbf{\Gamma}_{W^*}$, for any fixed $\ell \geq \ell_0$; see also [**114**].

with fractal flows and the zeta functions (or partition functions) of fractal membranes.

(b) (*Extension to Other Weierstrass-Type Functions*). In light of [**104**], the results of [**106**] (and [**110**]) described in this subsection can also be extended to a large class of Weierstrass-type nowhere differentiable functions and curves, including under suitable assumptions on the 1-periodic function and piecewise C^1 function Θ,[177] Knopp-type functions [**300, 301**], [**593**] (and the associated curve $\mathbf{\Gamma}_{\mathcal{K}}$, the graph of $\mathcal{K}$)

$$\mathcal{K}(t) = \sum_{n=0}^{\infty} \lambda^n \cdot \Theta(b^n \cdot t), \text{ for all } t \in \mathbb{R}; \tag{9.241}$$

see [**115**]. We note that this class of curves includes the Weierstrass curve, the Koch curve (enabling us to significantly improve the results of [**347**]), dealt with from the present perspective in [**111**], and the well-known Takagi function ([**582**], [**6**], [**310**]), which is often used in the applications (for example, in the study of turbulence and other subjects directly relevant to applied mathematics and physics; see, e.g., [**233**] and [**206**]).

(c) (*Fractal Cohomology of the Weierstrass Curve*). In [**107**] (announced in [**108**] and [**113**]), C. David and M. L. Lapidus—building on earlier suggestions and work in [**370, 374, 375**], [**326**], [**87**], [**329**], [**330**], and [**120**]—precisely defined and calculated the "fractal cohomology" of the Weierstrass curve $\mathbf{\Gamma}_{\mathcal{W}}$. It makes use of some of the results and methods of [**106**] (especially, Theorems 9.98, 9.99, and 9.114 above), as well as establishes intriguing and direct connections with appropriate modifications of the generalized (fractal) explicit formulas obtained in [**370, 374, 375**] (see [**375**, Chapters 5 and 8]) and in [**356**, Chapter 5]. Remarkably, the set of possible cohomological complex dimensions obtained in [**107**] is closely related to the set of possible complex dimensions obtained in Theorem 9.102 above (from [**106**])—and, especially, to the set of (exact) *intrinsic* complex dimensions obtained in Theorem 9.110 (from [**110**]). In this case too, we expect that these results—aiming at capturing the essence of the "topological shape" of the Weierstrass curve—should extend to other continuous and nowhere differentiable curves (as in part (b)), as well as to the non-integer case when $b \in \mathbb{R} \backslash \mathbb{N}$, with $b > 1$ (as in part (a) above), and, especially, to higher-dimensional "fractal manifolds".

Finally, we mention that the work in [**106**] and [**108**] is pursued in a variety of directions—including fractal Morse theory, fractal curvatures, polyhedral neighborhoods, and fractal power series expansions, as well as fractal manifolds, fractal flows, and analogies with number theory, differential geometry, algebraic topology, algebraic and arithmetic geometries—in [**109**], [**110**], [**111**], [**115**], [**116**], [**117**], [**118**], along with [**107**] and [**114**].

[177] More generally, one can consider the case when θ is almost periodic, in the sense of [**52, 287**] (as in [**286**] and/or [**254**]) and Θ is Hölder continuous of order $\beta \in (\alpha_{\mathcal{W}}, 1]$.

Part 4

Appendices

Appendix A

Upper and Lower Limits

The well-known notions of upper and lower limits, along with their basic properties, play an important role when we study the notions of upper and lower Minkowski contents, as well as the associated notions of Minkowski measurability and of (upper, lower) Minkowski dimensions; see Chapter 8. They are also used, in particular, when studying the notions of upper and lower box (or box-counting) dimensions; see Sections 8.3 and 8.8. These typically correspond to the upper and lower limits of suitable functions, as defined in Section A.2.

The case when the upper and lower limits are equal (and hence, the limit exists, in an extended sense) is particularly important. For example, the key notion of Minkowski measurability corresponds to the equality of the upper and lower Minkowski contents (in $(0, +\infty)$), that is, to the case when a certain limit (involving the volume of ε-neighborhoods) exists and is finite and nonzero. Moreover, the Minkowski (resp., box) dimension is said to exist when the upper and lower box dimensions coincide, where the latter notions are defined in terms of the upper and lower limits of a certain function (resp., involving the box-counting function). Similarly, the aforementioned upper and lower Minkowski dimensions are defined in terms of the upper and lower limits of a certain function (involving the tube function, viewed as the volume of ε-neighborhoods); see Chapter 8, especially Section 8.1.

We also use the notions of upper and lower limits of sequences (rather than of functions), as defined in Section A.1, when studying certain properties of fractal strings in Chapter 8, especially in Section 8.2.)

A.1. Upper and Lower Limits of Sequences

The *extended real line* $\overline{\mathbb{R}} = [-\infty, +\infty]$ consists of the ordinary real line $\mathbb{R} = (-\infty, +\infty)$ to which we have adjoined two distinct points $-\infty$ and $+\infty$ such that $-\infty < r < +\infty$, for all $r \in \mathbb{R}$. By convention, for every $r \in \mathbb{R}$ and every $a > 0$ (with $a \in \mathbb{R}$),

$$\begin{aligned} r + (+\infty) &= +\infty, & r + (-\infty) &= -\infty, \\ a \cdot (+\infty) &= +\infty, & a \cdot (-\infty) &= -\infty, \\ (-a) \cdot (+\infty) &= -\infty, & (-a) \cdot (-\infty) &= +\infty, \end{aligned}$$

$(+\infty) \cdot (\pm\infty) = \pm\infty$, but $0 \cdot (\pm\infty)$ and $(+\infty) + (-\infty)$ are not well defined.[1]

Definition A.1. Let $\{y_j\}_{j=1}^{\infty}$ be a sequence of real numbers, and let y be a real number. We say that y is a *limit point* of the sequence $\{y_j\}_{j=1}^{\infty}$ if for every $\varepsilon > 0$, the set $\{j \geq 1 : |y_j - y| \leq \varepsilon\}$ is infinite.

One can easily generalize this definition to the case when $y = +\infty$ or $y = -\infty$. For example, we say that $+\infty$ is a *limit point* of $\{y_j\}_{j=1}^{\infty}$ if for every $A > 0$, the set $\{j \geq 1 : y_j > A\}$ is infinite.

It is not difficult to prove the following result, as is requested in Exercise A.3 below.

Proposition A.2. *With the above notation, $y \in \overline{\mathbb{R}}$ is a limit point of $\{y_j\}_{j=1}^{\infty}$ iff y is the limit (in $\overline{\mathbb{R}}$) of a convergent subsequence $\{y'_j\}_{j=1}^{\infty}$ (that is, of a sequence of the form $\{y_{j_k}\}_{k=1}^{\infty}$, with $1 \leq j_1 < j_2 < \cdots$).*

Exercise A.3. Prove Proposition A.2.

[*Hint*: By symmetry, you may assume that $y \in \overline{\mathbb{R}}_+ := [0, +\infty]$; then, distinguish between the cases when $y \in \mathbb{R}_+ := [0, +\infty)$ and when $y = +\infty$.]

The "only if" part of the next exercise may be easier to solve when we have considered the notions of upper and lower limits in Definition A.6 below.

Exercise A.4.

(i) Show that a bounded sequence of real numbers $\{y_j\}_{j=1}^{\infty}$ is convergent (in $\mathbb{R}$) iff it has a single limit point $y \in \mathbb{R}$. In this case, we have that $\lim_{y \to \infty} y_j = y$.

(ii) Extend part (i) by removing the assumption that $\{y_j\}_{j=1}^{\infty}$ is bounded and, therefore, by allowing $y \in \overline{\mathbb{R}} = [-\infty, +\infty]$.

Since the upper and lower limits of a sequence will be defined in terms of the limit of a monotonic sequence, the following well-known and easy result will be very handy; see Definition A.6.

[1] Of course, all of these operations, when they are well defined, are commutative.

Proposition A.5. *If $\{y_j\}_{j=1}^{\infty}$ is a monotonic sequence (i.e., nondecreasing or nonincreasing) of elements of $\overline{\mathbb{R}}$, then it has a single limit point (in $\overline{\mathbb{R}}$) and is therefore convergent (also in $\overline{\mathbb{R}}$).*

More specifically, if $\{y_j\}_{j=1}^{\infty}$ is nondecreasing (resp., nonincreasing), then

$$\lim_{j\to\infty} y_j = \sup_{j\geq 1} y_j \quad \text{(resp.,} \quad \lim_{j\to\infty} y_j = \inf_{j\geq 1} y_j). \tag{A.1}$$

Here, in (A.1)*, the supremum and infimum are taken in the extended real line $\overline{\mathbb{R}} = [-\infty, +\infty]$. Moreover, if, in addition, the sequence $\{y_j\}_{j=1}^{\infty}$ is bounded, then its upper and lower limits are real numbers.*

In the following definition, we shall explicitly use Proposition A.5 in a key manner. Observe that unlike the limit of a sequence, the upper and lower limits of a sequence always exist—at least in the extended sense of being elements of $\overline{\mathbb{R}}$.

Definition A.6 (*Upper and Lower Limits of a Sequence*)**.** The *upper limit* (or lim sup) of a sequence $\{y_j\}_{j=1}^{\infty}$ of real numbers is the limit (in $\overline{\mathbb{R}}$) of the nonincreasing sequence $(Y_j := \sup_{k\geq j} y_k)_{j=1}^{\infty}$ of (possibly extended) real numbers:

$$\limsup_{j\to\infty} y_j := \lim_{j\to\infty} \sup_{k\geq j} y_k = \inf_{j\geq 1} \sup_{k\geq j} y_k. \tag{A.2}$$

Similarly, since the sequence $\{Z_j := \inf_{k\geq j} y_k\}_{j=1}^{\infty}$ is nondecreasing, then the *lower limit* (or lim inf) of $\{y_j\}_{j=1}^{\infty}$ is given by

$$\liminf_{j\to\infty} y_j := \lim_{j\to\infty} \inf_{k\geq j} y_k = \sup_{j\geq 1} \inf_{k\geq j} y_k. \tag{A.3}$$

Notation. We shall write indifferently

$$\limsup_{j\to\infty} y_j \quad \text{or} \quad \varlimsup_{j\to\infty} y_j \quad \text{and} \quad \liminf_{j\to\infty} y_j \quad \text{or} \quad \varliminf_{j\to\infty} y_j.$$

Sometimes, when no confusion may arise, we may even write $\limsup y_j$ or $\varlimsup y_j$, for example, as well as $\{y_j\}$ instead of $\{y_j\}_{j=1}^{\infty}$.

Observe that one can go from the upper limit to the lower limit (and conversely) simply by replacing y_j by $-y_j$, for each $j \geq 1$. Indeed, the reader will easily verify that

$$\limsup_{j\to\infty} (-y_j) = -\liminf_{j\to\infty} y_j \tag{A.4}$$

(i.e., $\varlimsup_{j\to\infty}(-y_j) = -\varliminf_{j\to\infty} y_j$), and hence, analogously, with the roles of the lower and upper limits interchanged. More generally, we have the following property, stated in the next exercise.

Exercise A.7.

(i) Show that for any $r \in \mathbb{R}$ and any $a > 0$, we have

$$\overline{\lim_{j\to\infty}}(y_j + r) = r + \overline{\lim_{j\to\infty}}\, y_j \quad \text{and} \quad \overline{\lim_{j\to\infty}}(a y_j) = a\, \overline{\lim_{j\to\infty}}\, y_j, \tag{A.5}$$

as well as exactly analogously, with the upper limits replaced by lower limits.

(ii) Deduce from (i) combined with (A.4) that, for example, $\overline{\lim}_{j\to\infty}(a y_j) = a\, \underline{\lim}_{j\to\infty}\, y_j$, for any $a < 0$ and analogously, with the roles of the upper and lower limits interchanged.

(iii) Are the conclusions of part (i) and of part (ii) still valid when $a = 0$?

The conclusion of the following important theorem could have been taken as the definition of the upper limit and of the lower limit. Note that the fact that the set of limit points has a maximum and minimum in $\overline{\mathbb{R}}$ follows from Exercise A.14 below.

Theorem A.8. *The upper limit (resp., the lower limit) of a sequence $\{y_j\}_{j=1}^{\infty}$ is the largest (resp., smallest) element of the set of limit points (in $\overline{\mathbb{R}}$) of $\{y_j\}_{j=1}^{\infty}$. In particular, it is a limit point of $\{y_j\}_{j=1}^{\infty}$.*

Corollary A.9. *The sequence $\{y_j\}_{j=1}^{\infty}$ is convergent (in $\overline{\mathbb{R}}$), with limit $L \in \overline{\mathbb{R}}$, iff its upper and lower limits coincide. In this case, this common value is equal to L: $\lim_{j\to\infty} y_j = L = \underline{\lim}_{j\to\infty}\, y_j = \overline{\lim}_{j\to\infty}\, y_j$.*

Naturally, if $\{y_j\}_{j=1}^{\infty}$ is bounded, then we can replace $\overline{\mathbb{R}}$ by $\mathbb{R}$ in both Theorem A.8 and Corollary A.9. In particular, in Corollary A.9, then $L \in \mathbb{R}$.

Exercise A.10. Prove Theorem A.8 and Corollary A.9.

[*Hint*: For the proof of Corollary A.9, use the results of Exercise A.4.]

Exercise A.11.

(i) Show that if $x_j := (-1)^j$, for each integer $j \geq 1$, then the only limit points of the sequence $\{x_j\}_{j=1}^{\infty}$ are -1 and $+1$. Thus, $\{(-1)^j\}_{j=1}^{\infty}$ is divergent (since it has more than one limit point, see part (i) of Exercise A.4) and

$$\overline{\lim_{j\to\infty}}(-1)^j = +1, \quad \text{while} \ \underline{\lim_{j\to\infty}}(-1)^j = -1. \tag{A.6}$$

(ii) If $x_j := (-1)^j + j^{-a}$, for each $j \geq 1$, with $a > 0$, show that $\{x_j\}_{j=1}^{\infty}$ is also divergent and that the counterpart of (A.6) still holds; namely, $\overline{\lim}_{j\to\infty}\, x_j = +1$, while $\underline{\lim}_{j\to\infty}\, x_j = -1$.

(iii) What are the limit points, as well as the upper and lower limits, of the sequence $\{x_j := \cos(j\pi/2)\}_{j=1}^{\infty}$?

(iv) Can you find a sequence whose set of limit points exactly coincides with $\{j^{-a} : j \geq 1\}$, where $a > 0$ is given? If not, find a sequence with infinitely many limit points, first countably many, and then, a continuum of limit points (such as $[0, 1]$, for example).

Exercise A.12.

(i) Show that $\overline{\lim}_{j\to\infty}(x_j + y_j) \leq \overline{\lim}_{j\to\infty} x_j + \overline{\lim}_{j\to\infty} y_j$, provided that the right-hand side does not involve any "forbidden operation" (namely, $(+\infty) + (-\infty)$ or $(-\infty) + (+\infty)$).

(ii) Prove that if $\lim_{j\to\infty} x_j = L$, with $L \in \mathbb{R}$, then (provided that the right-hand side of the inequality stated in part (i) does not involve any forbidden operation), $\overline{\lim}_{j\to\infty}(x_j + y_j) = L + \overline{\lim}_{j\to\infty} y_j$.

(iii) State and prove an analog of (i) and (ii) for lower (instead of upper) limits.

Exercise A.13.

(i) Show that $-\infty \leq \underline{\lim}_{j\to\infty} y_j \leq \overline{\lim}_{j\to\infty} y_j \leq +\infty$.

(ii) Prove that if $x_j \leq y_j$, for all $j \geq 1$ sufficiently large, then

$$\underline{\lim}_{j\to\infty} x_j \leq \underline{\lim}_{j\to\infty} y_j \quad \text{and} \quad \overline{\lim}_{j\to\infty} x_j \leq \overline{\lim}_{j\to\infty} y_j.$$

(iii) Deduce from parts (i) and (ii) that if $x_j \geq 0$, for all $j \geq 1$ sufficiently large, then $0 \leq \underline{\lim}_{j\to\infty} x_j \leq \overline{\lim}_{j\to\infty} x_j \leq +\infty$. If, in addition, $\{x_j\}_{j=1}^{\infty}$ is bounded away from $+\infty$ (resp., from 0), then show that the last (resp., first) inequality in the latter string of inequalities is strict.

The next problem is a little bit more abstract than the previous ones and may require some more advanced understanding of topology, especially for part (ii).

Exercise A.14.

(i) Show that the set of limit points of a bounded sequence is always closed. Furthermore, show that it is a compact (i.e., in the present case, closed and bounded) subset of $\mathbb{R}$.

(ii) Extend the results of part (i) to the case when the given sequence is unbounded. Accordingly, show that the set of limit points of an unbounded sequence is a compact subset of the extended real line $\overline{\mathbb{R}} = [-\infty, +\infty]$.

[*Hint*: Equip $\overline{\mathbb{R}}$ with the order topology (generated by the natural open intervals of $\overline{\mathbb{R}}$) and show that $\overline{\mathbb{R}}$ is compact. Then, establish that the set of limit points is closed in $\overline{\mathbb{R}}$, much as in part (i), and deduce that it is therefore compact (since the topology of $\overline{\mathbb{R}}$ is Hausdorff).]

A.2. Upper and Lower Limits of Functions

This section will be much briefer than Section A.1 because the definitions and statements (concerning real-valued or even $\overline{\mathbb{R}}$-valued functions) parallel the corresponding ones for sequences given in Section A.1. The key bridge between the case of functions and that of sequences is provided by the following definition—which could have instead become a theorem if we had adopted a different initial definition of the notion of a limit point.

Definition A.15. Let $f : D \subseteq \mathbb{R} \to \mathbb{R}$ and let x_0 be an accumulation point of the domain D of the function f. Then, we say that $L \in \overline{\mathbb{R}} = [-\infty, +\infty]$ is a *limit point* of $f = f(x)$ as $x \to x_0$ if it is the limit (in $\overline{\mathbb{R}}$) of the sequence $\{f(x_j)\}_{j=1}^{\infty}$, for some sequence of points $\{x_j\}_{j=1}^{\infty}$ in $D \setminus \{x_0\}$ such that $\lim_{j\to\infty} x_j = x_0$.

We can then define the upper and lower limits of $f = f(x)$ (or simply of f) as $x \to x_0$ as follows; compare with Theorem A.8.

Definition A.16. Let f and x_0 be as in Definition A.15. Then, the *upper limit* (resp., the *lower limit*) of $f = f(x)$ as $x \to x_0$ is the largest (resp., the smallest) limit point of f in $\overline{\mathbb{R}} = [-\infty, \infty]$. We then denote the upper (resp., lower) limit indifferently by

$$\limsup_{x\to x_0} f(x) \quad \text{or} \quad \overline{\lim_{x\to x_0}}\, f(x) \quad (\text{resp., } \liminf_{x\to x_0} f(x) \text{ or } \underline{\lim_{x\to x_0}}\, f(x)).$$

If f is bounded near x_0, then the upper (resp., lower) limit is a real number.[2]

All of the properties studied in Section A.1 can be easily transferred to functions (instead of sequences). We will not repeat most of them here. Let us simply mention that, as was alluded to earlier, one could define the notion of a limit point of the function f as $x \to x_0$, much as was done for sequences in Definition A.1. More specifically, we could also let

$$\text{(A.7)} \qquad \limsup_{x\to x_0} f(x) := \inf_{\delta>0} (\sup A_f^{(\delta)}) \quad \text{and} \quad \liminf_{x\to x_0} f(x) := \sup_{\delta>0} (\inf A_f^{(\delta)}),$$

where, for each $\delta > 0$, $A_f^{(\delta)} := \{f(x) : |x - x_0| < \delta,\ x \in D \setminus \{x_0\}\}$.

Accordingly, one could alternatively define the notions of upper and lower limits of f much as in Definition A.6, adapted to a continuous (rather than a discrete) variable. Also, the fact that the upper (resp., lower) limit of $f(x)$ as $x \to x_0$ is the largest (resp., smallest) limit point of f would then become a theorem (the exact analog for functions of Theorem A.8 from Section A.1) rather than be used as a primary definition (as in Definition A.16). Finally, we note that just as in the case of sequences, the upper and lower limits of a function always exist (possibly as extended real numbers).

[2]More specifically, for the upper (resp., lower) limit, this is the case if f is bounded from above (resp., from below) in $\mathbb{R}$, near x_0 in D.

The next important result is the precise analog for functions of Corollary A.9 and Exercise A.4 from Section A.1.

Theorem A.17. *The function $f = f(x)$ is convergent (in $\overline{\mathbb{R}}$) to $L \in \overline{\mathbb{R}}$ as $x \to x_0$ iff it has a single limit point (necessarily equal to L) and iff its upper and lower limits as $x \to x_0$ coincide (in $\overline{\mathbb{R}}$). In this case, the common value is equal to L. If, in addition, f is bounded (near x_0), then $L \in \mathbb{R}$ and so the above convergence holds in $\mathbb{R}$.*

We also point out that the properties stated in Exercises A.12 and A.13 have obvious counterparts in the present case of functions. We leave the precise statements and proofs as a simple exercise to the reader.

For further reading on the topic of this appendix, see, for example, the books [**5**], [**81**], [**193**], [**521**], [**523**], and [**571**].

Exercises

Exercise A.18. Prove Theorem A.17.

Exercise A.19. State and prove the counterpart of Exercise A.14 in the case of functions (instead of sequences).

Exercise A.20.

(i) Show that if $f \geq 0$ (near x_0), then

$$0 \leq \underline{\lim}_{x \to x_0} f(x) \, (\leq) \, \overline{\lim}_{x \to x_0} f(x) \leq +\infty.$$

(ii) Assume that $f \geq 0$ (near x_0). Then, show that f is bounded and bounded away from zero (near x_0) iff

$$-\infty < \underline{\lim}_{x \to x_0} f(x) \, (\leq) \, \overline{\lim}_{x \to x_0} f(x) < +\infty.$$

Also, if $f \geq 0$ (near x_0), then f is bounded (near x_0) iff $\overline{\lim}_{x \to x_0} f(x) < +\infty$.

Exercise A.21. Consider the function $f(x) := \sin(1/x)$ from $(0, 1]$ to $\mathbb{R}$. Sketch the graph of f and then determine the set of limit points, as well as the upper and lower limits of f, as $x \to 0^+$.[3]

Exercise A.22. Convince yourself that all of the definitions and results in Section A.2 extend without any difficulty if $f : D \subseteq X \to \mathbb{R}$, where $X = (X, d)$ is a metric space and $x_0 \in X$ is an accumulation point of the domain D of f.

[3]The graph of f is often called the *topologist's sine curve* in the literature.

Appendix B

Carathéodory's Approach to Measure Theory

The purpose of this appendix and of the closely related Chapters 6 and 7 of this book is to provide several important complements of measure theory, pertaining especially to the key notions of σ-algebras, outer measures, and measures, along with their applications to the rigorous construction (in Section B.4) of Lebesgue measure and, more generally, of Lebesgue–Stieltjes measures, as well as of Hausdorff measures in Chapters 6 and 7 (and in the introductory Chapter 5). In particular, we give a complete proof of Carathéodory's Measurability, Restriction, and Extension Theorems stated and used in Chapter 3 (see Theorems 3.19, 3.26, and 3.57, respectively).

Our main immediate goal is to provide (in Section B.1) a proof of the Carathéodory Restriction Theorem (Theorem 3.26), restated below as the second part of Theorem B.3), as well as of its preliminary step, the Carathéodory Measurability Theorem (Theorem 3.19, restated below as the first part of Theorem B.3), establishing that the collection of measurable sets (in the sense of Carathéodory) forms a σ-algebra and that the given outer measure restricted to this σ-algebra becomes a (countably additive) measure.

After having stated (in Section B.2) several related definitions and results concerning the completion of a measure, we then proceed (in Section B.3) to precisely formulate and prove the Carathéodory Extension Theorem (Theorem

B.12, the algebra or elementary family version; and Theorem 3.57, the semialgebra or elementary family version), which will often be used in the applications (for example, in Section B.4 and in Chapter 6) in conjunction with Theorem 3.48 (from Chapter 3) explaining how to construct outer measures out of suitable set functions.

We also lay the foundations for another important result (Theorem 6.14 in Section 6.2) stating that the Borel subsets of a metric space are always measurable (in the sense of Carathéodory) provided that the underlying outer measure is a *metric* outer measure. It will follow that for a metric outer measure μ^*, the σ-algebra of measurable sets (in the sense of Carathéodory, i.e., with respect to μ^*) contains the σ-algebra of Borel sets and hence all of the compact sets; see Corollaries 6.15 and 6.16. Consequently, one deduces that all of the classical fractals (including all self-similar sets—and, more generally, all of the attractors of IFSs—in Euclidean space $\mathbb{R}^m$) are also μ^*-measurable.

These results play a key role in measure theory for justifying the construction and establishing the properties of many important measures, including all Hausdorff measures (of which Lebesgue measure on Euclidean space $\mathbb{R}^m$ is a very special case). As is discussed in Chapters 5–7, the fact that Hausdorff outer measure, when restricted to the σ-algebra of measurable sets, and restricted even further to the Borel σ-algebra of the underlying space, gives rise (in each case) to a measure (called Hausdorff measure) is crucial to explaining some of the key properties of Hausdorff dimension.

Because the proofs of the aforementioned results are fairly technical, we have preferred to include them in an appendix rather than in the main text. However, including these precise statements and their detailed proofs seemed important to us in order to provide significantly more complete foundations for fractal geometry than can be usually found in most textbooks on the subject.

Upon a first reading, some readers may feel more comfortable reading the statements of the main results in Sections B.1–B.3 and then (after having read the first part of the introductory Chapter 5) moving on to Chapters 6 and 7, devoted to the applications to the construction and main properties of Hausdorff outer measures and measures associated with various families—including, as a key special case, the standard (or classic) and the spherical Hausdorff outer measures and measures. In Section B.4 as well, upon a first reading, the reader may choose to look at the main results and omit the proofs, in order to obtain an overview of the underlying logical structure and of the intimate relationship with the material discussed in many parts of Sections B.1–B.3, as well as with the material on metric outer measures and their associated Borel measures discussed in Section 6.2.

The seriously interested and involved reader, however, has the intellectual comfort to know that (with very few exceptions) all of the main results and proofs of Carathéodory's approach to measure theory (in Sections B.1–B.3 and, in one notable instance, in Chapter 3), along with the corresponding applications provided in Section B.4 (for Lebesgue–Stieltjes measures) and in Chapters 5–7 (for Hausdorff outer measures and measures), are discussed in detail and that, as a result, the readers can always return to any part of the present appendix and study in depth the statement and the proof of their choice or else (after having glanced at the proof supplied in the text) find and write up their own proof.

Many exercises, of various levels of difficulty, as well as sometimes requiring a certain amount of creativity and ingenuity, are provided along the way, both in this appendix and in its applications to the construction of Hausdorff measures found in Chapters 5–7—in order to help improve or consolidate the reader's understanding and technical abilities.

B.1. Carathéodory's Measurability and Restriction Theorems

Let X be an arbitrary set. In the sequel, in order to simplify the exposition, given an outer measure μ^* on X (see Definition 3.10), we will often say that a subset E of X which is measurable with respect to μ^* (or measurable in the sense of Carathéodory) is *μ^*-measurable.*

Recall from Definition 3.11 that it means that

$$\mu^*(A) = \mu^*(A \cap E) + \mu^*(A \cap E^c), \quad \text{for all } A \subseteq X. \tag{B.1}$$

The next useful result follows from the subadditivity of μ^*.

Proposition B.1 (Criterion for μ^*-Measurability). *Let μ^* be an outer measure on a set X. Then, $E \subseteq X$ is μ^*-measurable iff*

$$\mu^*(A) \geq \mu^*(A \cap E) + \mu^*(A \cap E^c), \quad \text{for all } A \subseteq X \quad (\text{with } \mu^*(A) < \infty). \tag{B.2}$$

Proof. Let $A, E \subseteq X$ be arbitrary. Since

$$A = (A \cap E) \cup (A \cap E^c), \tag{B.3}$$

the (finite) subadditivity of μ^* implies that $\mu^*(A) \leq \mu^*(A \cap E) + \mu^*(A \cap E^c)$. Hence, in light of the definition of μ^*-measurability given in (B.1), E is μ^*-measurable iff the inequality (B.2) holds, as desired.

Also, we note that when $\mu^*(A) = +\infty$, the inequality (B.2) obviously holds; hence, we may restrict our attention to subsets A of X such that $\mu^*(A) < \infty$, as claimed. □

Remark B.2. We caution the reader that even though the union on the right-hand side of (B.3) is a disjoint union, it does *not* follow in general (that is, for arbitrary $A, E \subseteq X$) from the definition of an outer measure that (B.1) holds. Indeed, μ^* need not be finitely (and let alone countably) additive. The fact that when E is μ^*-measurable, μ^* behaves as though it were (finitely) additive whenever an arbitrary set $A \subseteq X$ is "cut out" (or intersected) by E and its complement E^c (in the restricted sense that (B.1) holds) is the great subtlety of the notion of Carathéodory measurability.

As we will see next, this deceptively simple property suffices for μ^* to become finitely (and even countably) additive, when restricted to the class $\mathcal{M} := \mathcal{M}_{\mu^*}$ of μ^*-measurable sets (as well as for $\mathcal{M}$ to be an algebra or even a σ-algebra); see Theorem B.3 just below.

We are now ready to state and prove the first main result of Carathéodory's theory (Theorem 3.19 and Theorem 3.26 in Chapter 3, here stated as a single theorem).[1]

Theorem B.3 (Carathéodory's Measurability and Restriction Theorem). *Let μ^* be an outer measure on a set X. Then,*

$$\mathcal{M} = \mathcal{M}_{\mu^*} := \{\mu^*\text{-measurable sets}\}, \tag{B.4}$$

the collection of subsets of X which are μ^*-measurable, is a σ-algebra on X.

Furthermore, the restriction of μ^* to the measurable space $(X, \mathcal{M})$ is a (countable additive) measure (denoted by $\overline{\mu}$). Moreover, $\overline{\mu}$ is a complete measure, in the sense of Definition B.5 *in Section* B.2 *below.*

Proof. We will prove this result in several steps. First, we will show (in Step 1) that $\mathcal{M}$ is an algebra (in the sense of Definition 3.2). Second, we will prove (in Step 2) that μ^* is finitely additive on $(X, \mathcal{M})$. Third, we will show (in the first part of Step 3) that $\mathcal{M}$ is a σ-algebra (in the sense of Definition 3.12), and finally (in the latter part of Step 3), that μ^*, when restricted to $(X, \mathcal{M})$, is countable additive and hence is a measure (denoted by $\overline{\mu}$), as claimed in the theorem.

Step 1. *$\mathcal{M}$ is an algebra; that is, it is closed under finite unions and complement. (Clearly, $\emptyset \in \mathcal{M}$.)*

The fact that $E \in \mathcal{M} \Rightarrow E^c \in \mathcal{M}$ is obvious since the definition of μ^*-measurability is symmetric in E and E^c; see (B.1). Thus, all we have to show is that $\mathcal{M}$ is closed under finite unions.

Let $E, F \in \mathcal{M}$. Since we want to show that $E \cup F \in \mathcal{M}$, and in light of Proposition B.1, it suffices to show that for any $A \subseteq X$,

$$\mu^*(A \cap (E \cup F)) + \mu^*(A \cap (E \cup F)^c) \leq \mu^*(A). \tag{B.5}$$

Since $E \in \mathcal{M}$ and $F \in \mathcal{M}$, we have, respectively, for all $B, C \subseteq X$,

$$\mu^*(B) = \mu^*(B \cap E) + \mu^*(B \cap E^c) \quad \text{and} \quad \mu^*(C) = \mu^*(C \cap F) + \mu^*(C \cap F^c). \tag{B.6}$$

[1] In fact, Theorem B.3 contains more information than the combination of Theorems 3.19 and 3.26 because its last statement regards the completeness of the measure $\overline{\mu}$ obtained by restriction of the outer measure μ^*.

Given $A \subseteq X$, we have successively

$$\begin{aligned} \mu^*(A) &= \mu^*(A\cap E)+\mu^*(A\cap E^c) \\ &= \mu^*(A\cap E\cap F)+\mu^*(A\cap E\cap F^c) \\ &\quad+\mu^*(A\cap E^c\cap F)+\mu^*(A\cap E^c\cap F^c) =: \alpha+\mu^*(A\cap(E\cup F)^c), \end{aligned} \tag{B.7}$$

where

$$\alpha := \mu^*(A\cap E\cap F)+\mu^*(A\cap E\cap F^c)+\mu^*(A\cap E^c\cap F). \tag{B.8}$$

The first equality in (B.7) simply follows from (B.1) (which is just (B.6) with $B := A$), since $E \in \mathcal{M}$. Furthermore, since $F \in \mathcal{M}$, the second equality in (B.7) follows from two applications of the right-hand side of (B.6), with $C := A \cap E$ and $C := A \cap E^c$, respectively, according to which

$$\mu^*(A\cap E) = \mu^*(A\cap E\cap F)+\mu^*(A\cap E\cap F^c) \quad \text{and} \tag{B.9}$$

$$\mu^*(A\cap E^c) = \mu^*(A\cap E^c\cap F)+\mu^*(A\cap E^c\cap F^c). \tag{B.10}$$

Indeed, we merely have to observe that the sum of the two identities (B.9) and (B.10) yields the second equality of (B.7).

Finally, in light of (B.8), the third (and last) equality of (B.7) follows from the fact that $(E \cup F)^c = E^c \cap F^c$.

Next, drawing a simple Venn diagram easily shows that[2]

$$E\cup F = (E\cap F)\cup(E\cap F^c)\cup(E^c\cap F) \tag{B.11}$$

and hence,

$$A\cap(E\cup F) = (A\cap E\cap F)\cup(A\cap E\cap F^c)\cup(A\cap E^c\cap F). \tag{B.12}$$

It follows from (B.12), the (finite) subadditivity of μ^*, and then from (B.8) that $\mu^*(A \cap (E \cup F)) \leq \alpha$. Hence, in light of the latter inequality (for the first inequality of (B.13)) and of (B.7) (for the only equality of (B.13)),

$$\mu^*(A\cap(E\cup F))+\mu^*(A\cap(E\cup F)^c) \leq \alpha+\mu^*(A\cap(E\cup F)^c) = \mu^*(A). \tag{B.13}$$

We conclude that the inequality (B.5) holds and thus that $E\cup F \in \mathcal{M}$, as desired. This completes Step 1 since clearly (in light of (B.1)), $\emptyset \in \mathcal{M}$ and so, $\mathcal{M}$ is not empty (as required in the definition of an algebra of sets).

Step 2. *Let $\overline{\mu}$ be the restriction of μ^* to the algebra $\mathcal{M}$. Then, $\overline{\mu}$ is finitely additive.*

This claim follows relatively easily from Step 1. Indeed, we already know from Step 1 that if $E, F \in \mathcal{M}$ are disjoint, then $E\cup F \in \mathcal{M}$. Also, it follows from

[2] The unions appearing on the right-hand sides of (B.11) and (B.12) are actually pairwise disjoint unions but we *cannot* use this fact here; see Remark B.2 above.

(B.1) (applied with $A := E \cup F$ and $E := E$ in order to justify the first equality in (B.14) just below) that we have successively

(B.14)
$$\begin{aligned}\mu^*(E \cup F) &= \mu^*((E \cup F) \cap E) + \mu^*((E \cup F) \cap E^c) = \mu^*(E) + \mu^*(F \cap E^c) \\ &= \mu^*(E) + (\mu^*(F \cap E) + \mu^*(F \cap E^c)) = \mu^*(E) + \mu^*(F).\end{aligned}$$

Hence, as desired, we have $\mu^*(E \cup F) = \mu^*(E) + \mu^*(F)$, if $E, F \in \mathcal{M}$ are *disjoint*.

Next, we briefly justify the equalities in (B.14). We have already explained the first equality. The second equality in (B.14) results from the fact that $F \cap E = \emptyset$ and $E \cap E^c = \emptyset$. The third equality in (B.14) simply follows from the fact that $\mu^*(F \cap E) = \mu^*(\emptyset) = 0$. Finally, the fourth (and last) equality in (B.14) is a consequence of the fact that $E \in \mathcal{M}$. Indeed, in light of (B.1) (applied with $A := F$ and $E := E$), we precisely have that $\mu^*(F) = \mu^*(F \cap E) + \mu^*(F \cap E^c)$.

Finally, as is easily verified by induction on the integer $n \geq 1$, it follows from Step 1 that if $E_1, \ldots, E_n \in \mathcal{M}$ are pairwise disjoint, then $\bigcup_{j=1}^{n} E_j \in \mathcal{M}$ and $\mu^*\left(\bigcup_{j=1}^{n} E_j\right) = \sum_{j=1}^{n} \mu^*(E_j)$, which means that $\overline{\mu}$ is finitely additive on $(X, \mathcal{M})$ and concludes the proof of Step 2.

Step 3. *$\mathcal{M}$ is a σ-algebra and $\overline{\mu}$ (the restriction of μ^* to $\mathcal{M}$) is a measure; that is, $\overline{\mu}$ is countably additive.*

Indeed, since $\mathcal{M}$ is an algebra (by Step 1), in order to show that $\mathcal{M}$ is a σ-algebra, it suffices to check that $\mathcal{M}$ is closed under countable *disjoint* unions. (See Proposition B.4 following the proof of this theorem.) Therefore, let $\{E_j\}_{j=1}^{\infty}$ be a sequence of *pairwise disjoint* elements of $\mathcal{M}$, and let $D := \bigcup_{j=1}^{\infty} E_j$. We want to show that $D \in \mathcal{M}$ and

$$\overline{\mu}(D) = \sum_{j=1}^{\infty} \overline{\mu}(E_j). \tag{B.15}$$

By Proposition B.1, in order to prove that $D \in \mathcal{M}$, it suffices to check that for all $A \subseteq X$, $\mu^*(A) \geq \mu^*(A \cap D) + \mu^*(A \cap D^c)$.

Now, let $A \subseteq X$ be given and let $D_0 := \emptyset$ and $D_n := \bigcup_{j=1}^{n} E_j$, for every integer $n \geq 1$. (Recall that $D := \bigcup_{j=1}^{\infty} E_j$.) Then, by Step 1 (which states that $\mathcal{M}$ is an algebra), we have that each D_n belongs to $\mathcal{M}$. Furthermore, since (for any $n \geq 1$) $E_n \in \mathcal{M}$, we can apply (B.1) to $E := E_n$ (and to $A \cap D_n$ instead of A) in order to obtain that[3]

$$\mu^*(A \cap D_n) = \mu^*(A \cap D_n \cap E_n) + \mu^*(A \cap D_n \cap E_n^c). \tag{B.16}$$

But we claim that (still for $n \geq 1$)

$$D_n \cap E_n = E_n \quad \text{and} \quad D_n \cap E_n^c = D_{n-1}. \tag{B.17}$$

[3] Here and thereafter, for notational simplicity, we write E_n^c instead of $(E_n)^c$.

Indeed, by the above definition of D_n and since the sets E_j are pairwise disjoint, $D_n \cap E_n = \bigcup_{j=1}^{n}(E_j \cap E_n) = E_n$. Also, $D_n \cap E_n^c = \bigcup_{j=1}^{n}(E_j \cap E_n^c) = \bigcup_{j=1}^{n-1} E_j = D_{n-1}$, since $E_n \cap E_n^c = \emptyset$ and $E_j \subseteq E_n^c$ for all integers $j \neq n$.

In light of (B.17), the identity (B.16) becomes (for every $n \geq 1$)

$$\mu^*(A \cap D_n) = \mu^*(A \cap E_n) + \mu^*(A \cap D_{n-1}). \tag{B.18}$$

Applying the identity (B.18) repeatedly (or, equivalently, reasoning by induction on $n \geq 1$), and taking into account the fact that $\mu^*(A \cap D_0) = \mu^*(\emptyset) = 0$, we deduce that[4]

$$\mu^*(A \cap D_n) = \sum_{j=1}^{n} \mu^*(A \cap E_j). \tag{B.19}$$

Next, since $D_n \in \mathcal{M}$, we can apply (B.1) to $E := D_n$ to obtain that

$$\mu^*(A) = \mu^*(A \cap D_n) + \mu^*(A \cap D_n^c), \text{ for all } A \subseteq X. \tag{B.20}$$

Hence, in light of (B.19) and since $\mu^*(A \cap D_n^c) \geq \mu^*(A \cap D^c)$ (because $D_n \subseteq D$ and so $D_n^c \supseteq D^c$),[5] we deduce from (B.20) and footnote 5 that

$$\mu^*(A) \geq \sum_{j=1}^{n} \mu^*(A \cap E_j) + \mu^*(A \cap D^c), \quad \text{for all } n \geq 1. \tag{B.21}$$

We can now let $n \to \infty$ in (B.21) to deduce that

$$\mu^*(A) \geq \sum_{j=1}^{\infty} \mu^*(A \cap E_j) + \mu^*(A \cap D^c). \tag{B.22}$$

We next use the countable subadditivity (see part (iii) of Definition 3.10) of the outer measure μ^* to obtain that (since $A \cap D = \bigcup_{j=1}^{\infty}(A \cap E_j)$ because $D := \bigcup_{j=1}^{\infty} E_j$) $\sum_{j=1}^{\infty} \mu^*(A \cap E_j) \geq \mu^*(A \cap D)$. Combining the latter inequality and (B.22), we deduce that

$$\begin{aligned} \mu^*(A) &\geq \sum_{j=1}^{\infty} \mu^*(A \cap E_j) + \mu^*(A \cap D^c) \\ &\geq \mu^*(A \cap D) + \mu^*(A \cap D^c) = \mu^*(A). \end{aligned} \tag{B.23}$$

The last equality in (B.23) is obtained as follows: First, note that the inequality $\mu^*(A) \geq \mu^*(A \cap D) + \mu^*(A \cap D^c)$ is exactly what was needed to show that $D \in \mathcal{M}$. This shows that $\mathcal{M}$ is a σ-algebra. Now that we know that $D = \bigcup_{j=1}^{\infty} E_j \in \mathcal{M}$,

[4] Note that this does *not* follow from Step 2 (the finite additivity of $\overline{\mu}$ restricted to $\mathcal{M}$) since $A \subseteq X$ is arbitrary and thus, the finite disjoint family $\{A \cap E_j\}_{j=1}^{n}$, whose union is $A \cap D_n$, does not necessarily consist of elements of $\mathcal{M}$.

[5] We also use here the *monotonicity* of the outer measure μ^* (see part (ii) of Definition 3.10) according to which $A \cap D^c \subseteq A \cap D_n^c$ implies that $\mu^*(A \cap D^c) \leq \mu^*(A \cap D_n^c)$.

the last equality of (B.23), namely, $\mu^*(A) = \mu^*(A \cap D) + \mu^*(A \cap D^c)$, follows at once from (B.1), applied to $E := D \in \mathcal{M}$.

A glance at (B.23) shows that, clearly, *we must only have equalities in* (B.23). Hence, we have $\mu^*(A) = \sum_{j=1}^{\infty} \mu^*(A \cap E_j) + \mu^*(A \cap D^c)$, for every $A \subseteq X$. It now suffices to choose $A := D$ in the latter identity (and remember that $\mu^*(\emptyset) = 0$ and $D \cap E_j = E_j$, for all $j \geq 1$) in order to deduce that $\mu^*(D) = \sum_{j=1}^{\infty} \mu^*(E_j)$,[6] as desired; see (B.15) and the discussion surrounding it. This establishes the countable additivity of μ^* restricted to $\mathcal{M}$ (i.e., of $\overline{\mu}$) and concludes the proof of Step 3 as well as of the main part of Theorem B.3.

Step 4. *The measure $\overline{\mu}$ on the measurable space $(X, \mathcal{M})$ obtained in Step* 3 *is complete, in the sense of Definition* B.5 *below.*

[The reader may omit this part of the proof until he or she has read Definition B.5 (in Section B.2 below) and the text surrounding it.]

Indeed, let $D \subseteq X$ be such that $\mu^*(D) = 0$. Then, for any $A \subseteq X$, with $\mu(A) = 0$, we have (by the finite subadditivity of μ^* for the first inequality of (B.24) and by the monotonicity of μ^* for the last inequality in (B.24); see (iii) and (ii) of Definition 3.10, respectively)

$$0 = \mu^*(A) \leq \mu^*(A \cap D) + \mu^*(A \cap D^c) = \mu^*(A \cap D^c) \leq \mu^*(A) = 0, \tag{B.24}$$

so that $D \in \mathcal{M}$ because (B.24) clearly implies that[7]

$$\mu^*(A) = \mu^*(A \cap D) + \mu^*(A \cap D^c), \text{ for all } A \subseteq X. \tag{B.25}$$

Since (B.24) precisely means that $D \in \mathcal{M}$ (see (B.1)), we conclude that the measure $\overline{\mu}$ is complete, as desired. Indeed, in light of Definition B.5, a moment's reflection shows that since μ is the Carathéodory measure associated with the outer measure μ^*, verifying the completeness of μ amounts to showing that $\mu^*(A) = 0$, with $A \subseteq X \Rightarrow A \in \mathcal{M}$, which is exactly what we have proved.

This completes the proof of Step 4 and also of the entire theorem (Theorem B.3). □

In the course of the proof of Step 3 of Theorem B.3, we have used the following very useful (and simple) result.

Proposition B.4. *Let X be a set. If $\mathcal{A}$ is an algebra (on X) that is closed under countable disjoint unions, then $\mathcal{A}$ is a σ-algebra (on X).*

---

[6] Recall that, by definition, $\overline{\mu}$ denotes the restriction of μ^* to the σ-algebra $\mathcal{M}$. Hence, this statement is precisely equivalent to (B.15).

[7] Note that in (B.25), we have also used the monotonicity and the positivity of μ^* in order to deduce that $\mu^*(A \cap D) = 0$; indeed, $0 \leq \mu^*(A \cap D) \leq \mu^*(A) = 0$.

Proof. Since $\mathcal{A}$ is an algebra, $\emptyset \in \mathcal{A}$. Let $\{A_j\}_{j=1}^{\infty}$ be an arbitrary sequence of elements of $\mathcal{A}$. Construct another sequence $\{B_j\}_{j=1}^{\infty}$ of elements of $\mathcal{A}$, as follows: $B_0 := \emptyset$ and $B_j := A_j \setminus \left(\bigcup_{k=1}^{j-1} A_k\right)$, for each integer $j \geq 1$. Clearly, as we have seen in other proofs in Chapter 3, the B_j's are *pairwise disjoint* and $\bigcup_{k=1}^{j} B_k = \bigcup_{k=1}^{j} A_k$, for each integer $j \geq 1$ and also for $j = \infty$. Since, by hypothesis, $\mathcal{A}$ is closed under countable disjoint unions, we have that $\bigcup_{j=1}^{\infty} B_j \in \mathcal{A}$ and thus, by construction, we deduce that $\bigcup_{j=1}^{\infty} A_j = \bigcup_{j=1}^{\infty} B_j \in \mathcal{A}$, as required. This concludes the proof of the proposition. □

B.2. Completion of a Measure

The following definition is useful in showing that a certain σ-algebra is "as large as possible", relative to a given measure on it.

Definition B.5 (*Complete Measure*). A measure μ on a measurable space $(X, \mathcal{M})$ is said to be *complete* if $A \subseteq X$, with $\mu(A) = 0 \Rightarrow A \in \mathcal{M}$, or, more precisely, if the μ-null sets (i.e., the sets of measure zero with respect to μ) belong to the σ-algebra $\mathcal{M}$.

Rephrased: Any subset of a μ-null set must belong to the σ-algebra $\mathcal{M}$; i.e., $B \subseteq A$, with $\mu(A) = 0$ and $A \in \mathcal{M} \Rightarrow B \in \mathcal{M}$.

It is not difficult to check that if $(X, \mathcal{M}, \mu)$ is a complete measure space (i.e., if the measure μ is complete on $(X, \mathcal{M})$, in the sense of Definition B.5 just above), then $\mathcal{N} := \{\mu\text{-null sets}\}$ is closed under countable unions and contains the empty set $\emptyset$. In general, however, $\mathcal{N}$ is not a σ-subalgebra of $\mathcal{M}$ because, in general, $\mathcal{M}$ is not closed under complements.

The following result (Theorem B.6) shows that even if a measure is not complete, then it has a unique completion extending it to a larger σ-algebra. The relatively easy and routine proof of Theorem B.6 is left as an exercise for the interested reader; see Exercise B.8.

Theorem B.6 (Completion Theorem). *Let ν be a measure on a measurable space $(X, \mathcal{A})$, so that X is a set and $\mathcal{A}$ is a σ-algebra on X. Then, ν admits a unique completion $\mu := \overline{\nu}$.*

This means that μ is a measure on a measurable space $(X, \mathcal{M})$, with $\mathcal{M} \supseteq \mathcal{A}$ (and $\mathcal{M}$ being a σ-algebra over the same set X), and the restriction of μ to $\mathcal{A}$ coincides with ν (so that $\mu = \overline{\nu}$ is an extension of ν to $\mathcal{M}$). (One often writes $\mathcal{M} = \overline{\mathcal{A}}$, the completion of $\mathcal{A}$ relative to the measure ν.) Furthermore, any measure with these properties coincides with $\mu = \overline{\nu}$.

More specifically, $B \in \mathcal{M}$ *iff* $B = A \cup C$, *where* $A \in \mathcal{A}$ *and* C *is a subset of a* μ*-null set (i.e.,* $C \subseteq D$, *with* $D \in \mathcal{A}$ *and* $\mu(D) = 0$, *for some* $D \subseteq X$*). Then,* $\mu(B) := \nu(A)$, *independently of the choice of* C *as above.*

The following result is an immediate consequence of the uniqueness part of Theorem B.6.

Corollary B.7. *If* μ *is a complete measure over a measurable space* $(X, \mathcal{A})$*, then* $\overline{\mu} = \mu$, *in the notation of Theorem* B.6; *i.e.,* μ *coincides with its completion. Also, still in the notation of Theorem* B.6, *the* σ*-algebra* $\mathcal{M}$ *coincides with* $\mathcal{A}$.

Exercise B.8. Give a detailed statement and proof of Theorem B.6 and Corollary B.7.

B.3. Extension Theorems for Algebras and Semialgebras

Next, we provide all of the key steps needed to establish the Carathéodory Extension Theorem (Theorem 3.57), also known as the Carathéodory–Hahn Theorem—but which should, in fact, be attributed to Fréchet [**201**], at least in the classic form given in Theorem B.12 below.

B.3.1. Carathéodory's Extension Theorem for Algebras. We will first focus on the following key result (Theorem B.12), which is Carathéodory's original extension theorem, the algebra version, of which we will give a complete proof. For this purpose, let us adapt the notion of a premeasure to the situation at hand.[8]

Definition B.9 (*Premeasure on an Algebra*). Let $\mathcal{A}$ be an algebra on a set X (in the sense of Definition 3.2). We say that $\mu : \mathcal{A} \to [0, +\infty]$ is a *premeasure on the algebra* $\mathcal{A}$ (or just a *premeasure* on $\mathcal{A}$, in short)[9] if $\mu(\emptyset) = 0$ and μ is countably additive, in the sense that if $\{A_j\}_{j=1}^{\infty}$ is a sequence of pairwise disjoint elements of $\mathcal{A}$ such that $\bigcup_{j=1}^{\infty} A_j \in \mathcal{A}$, then $\mu\left(\bigcup_{j=1}^{\infty} A_j\right) = \sum_{j=1}^{\infty} \mu(A_j)$.

Remark B.10. Note that a premeasure on an algebra $\mathcal{A}$ is a measure (in the usual sense of Definition 3.23) iff $\mathcal{A}$ is a σ-algebra, in which case, the condition that $\bigcup_{j=1}^{\infty} A_j \in \mathcal{A}$ is automatically satisfied.

---

[8]Note that Definition B.9 is stronger than the one given in part (ii) of Definition 3.53 for a general set function on a suitable class of subsets of X (and, in particular, for a semialgebra or, equivalently, an elementary family).

[9]In some textbooks, such as [**193**], μ is called a measure (instead of a premeasure) on the algebra $\mathcal{A}$. In the sequel, we will often refer to it as being a "premeasure on the algebra $\mathcal{A}$" in order to distinguish it from a premeasure on the semialgebra $\mathcal{S}$.

Exercise B.11. Show that a premeasure μ on an algebra $\mathcal{A}$ is finitely additive and monotonic, in the obvious sense.[10]

In the following key theorem, $\mu : \mathcal{A} \to [0, +\infty]$ is a premeasure on an algebra $\mathcal{A}$ (in the sense of Definition B.9), while μ^* is the outer measure associated with μ (via Theorem 3.48). Furthermore, $\overline{\mu}$ is the corresponding Carathéodory measure (in the sense of Definition 3.50), defined as the restriction of the outer measure μ^* to the σ-algebra $\mathcal{M} = \mathcal{M}_{\mu^*}$ of all μ^*-measurable sets; see Theorem 3.19 and Theorem 3.26, the Carathéodory Measurability and Restriction Theorems, stated together in Theorem B.3. Finally, $\mathfrak{M} = \mathfrak{M}_{\mathcal{A}}$ is the σ-algebra (of X) generated by $\mathcal{A}$ (i.e., the smallest σ-algebra containing $\mathcal{A}$).

Theorem B.12 (Carathéodory's Extension Theorem: Algebra Version). *Let $\mathcal{A}$ be an algebra over X, in the sense of Definition* 3.2, *and let μ be a premeasure on the algebra $\mathcal{A}$, in the sense of Definition* B.9. *Then, with the notation and terminology recalled just before the statement of this theorem, we have*

$$\mathcal{A} \subseteq \mathfrak{M} \subseteq \mathcal{M}. \tag{B.26}$$

Furthermore, we have the following key results:

(i) (Existence of an Extension of μ). *The Carathéodory measure $\overline{\mu}$ associated with μ is an extension of μ from the algebra $\mathcal{A}$ to the σ-algebra $\mathcal{M}$; i.e.,*

$$\overline{\mu}(E) = \mu(E), \text{ for all } E \in \mathcal{A}. \tag{B.27}$$

(ii) (Uniqueness of the Extension of μ). *Furthermore, if ν is another measure on $\mathcal{M}$ extending μ from $\mathcal{A}$ to $\mathcal{M}$, then*

$$\nu(E) \leq \overline{\mu}(E), \text{ for all } E \in \mathcal{M}, \tag{B.28}$$

with equality (i.e., $\nu(E) = \overline{\mu}(E)$) whenever $\overline{\mu}(E) < \infty$.

Moreover, if the premeasure μ is σ-finite,[11] *then $\overline{\mu}$ is the unique extension of μ to a measure on $\mathcal{M}$; hence, given any measure ν as stated just above, we must have that $\nu = \overline{\mu}$.*

The detailed proof of Theorem B.12 will be provided just after the proof of the following corollary.

Corollary B.13. *Under the hypotheses of Theorem* B.12 (*and with the notation and terminology introduced just before the statement of that theorem*), *let $\widetilde{\mu}$ denote the restriction of $\overline{\mu}$ (or of μ^*) to the σ-algebra $\mathfrak{M}$. (Note that here, for the strong uniqueness statement just below, we assume, in particular, that μ is σ-finite.) Then, $\widetilde{\mu}$ is a measure on $\mathfrak{M}$ and $\overline{\mu}$ (the Carathéodory measure $\overline{\mu}$ on $\mathcal{M}$*

[10] Note that since $\mathcal{A}$ is an algebra, a finite union of elements of $\mathcal{A}$ is still in $\mathcal{A}$. Hence, the statement about finite additivity is a little simpler than the one about the countable additivity of the measure μ.

[11] Just as in the case of a measure in Definition 3.41, this means that X is the union of a finite or countable collection of elements of $\mathcal{A}$, each of which has finite premeasure: $X = \bigcup_{n\geq 1} A_n$, with $A_n \in \mathcal{A}$ and $\mu(A_n) < \infty$, for all $n \geq 1$.

associated with the premeasure μ on the algebra $\mathcal{A}$) is its unique extension (as a measure) from $\mathfrak{M}$ to $\mathcal{M}$. In fact, $\overline{\mu}$ is the (unique) completion of $\widetilde{\mu}$, in the sense of Definition B.5 *and Theorem* B.6.

Proof. This corollary easily follows from Theorem B.12 combined with Theorem B.6 and the last statement of Theorem B.3 according to which the Carathéodory measure $\overline{\mu}$ is a complete measure on the measurable space $(X, \mathcal{M})$.

Indeed, by Theorem B.6 (the Completion Theorem) and in light of the inclusions (B.26) in Theorem B.12, the measure $\widetilde{\mu}$ on $\mathfrak{M} \subseteq \mathcal{M}$ has a unique completion (that is, there is a unique *complete* measure on X extending $\widetilde{\mu}$ to a possibly larger σ-algebra). However, in light of Theorem B.12 and the last statement of Theorem B.3 (the Measurability and Restriction Theorem), $\overline{\mu}$ is precisely such a measure: It is a complete measure and extends the measure $\widetilde{\mu}$ from the σ-algebra $\mathfrak{M}$ to the σ-algebra $\mathcal{M}$.

Therefore, by the uniqueness of the completion (Theorem B.6) and of the extension μ from $\mathcal{A}$ to $\mathcal{M}$ (according to part (ii) of Theorem B.12), we conclude that $\overline{\mu}$ coincides with the completion of $\widetilde{\mu}$. In summary, $\overline{\mu}$ is (or can be viewed as) the completion of $\widetilde{\mu}$, as claimed. Recall that, in this corollary, we are assuming that μ is σ-finite—and hence that we can apply the full strength of the uniqueness statement in part (ii) of Theorem B.12. □

Proof of Theorem B.12. For clarity, the proof of Theorem B.12 is divided into several steps.

Step 1. *The following inclusions hold, as stated in formula* (B.26) *of Theorem* B.12: $\mathcal{A} \subseteq \mathfrak{M} \subseteq \mathcal{M}$.

Indeed, it clearly suffices to prove the inclusion $\mathcal{A} \subseteq \mathcal{M}$ because, by definition, $\mathfrak{M} = \mathfrak{M}_{\mathcal{A}}$ is the σ-algebra generated by $\mathcal{A}$; that is, $\mathfrak{M}$ is the *smallest* σ-algebra containing $\mathcal{A}$. Since (by the Carathéodory Measurability Theorem, Theorem 3.19 or the first part of Theorem B.3) $\mathcal{M} = \mathcal{M}_{\mu^*}$ is a σ-algebra (the σ-algebra of μ^*-measurable sets) and (provided we have established the above inclusion) $\mathcal{M}$ contains $\mathcal{A}$, we must have that $\mathfrak{M} \subseteq \mathcal{M}$. Therefore, (B.26) follows from the stated inclusions.

We now proceed to show that the inclusion $\mathcal{A} \subseteq \mathcal{M}$ holds. Let $A \in \mathcal{A}$ and let us prove that A is μ^*-measurable. In light of Proposition B.1, it suffices to show that for any set $E \subseteq X$ with $\mu^*(E) < \infty$,[12] and for every fixed $\varepsilon > 0$,

$$\mu^*(E) + \varepsilon \geq \mu^*(E \cap A) + \mu^*(E \cap A^c), \tag{B.29}$$

[12] *Caution*: For obvious reasons, we have interchanged here the roles played by the sets A and E in Proposition B.1—and hence also in the definition of a μ^*-measurable set given in (B.1).

from which we deduce (by letting $\varepsilon \to 0^+$, for example) that

$$\mu^*(E) \geq \mu^*(E \cap A) + \mu^*(E \cap A^c), \tag{B.30}$$

as stated in the inequality (B.2) of Proposition B.1 (with the roles of A and E interchanged).

Now, fix $\varepsilon > 0$ and $E \subseteq X$, with $\mu^*(E) < \infty$. By definition of the outer measure μ^* associated with the premeasure μ on $\mathcal{A}$ (see (3.12), defining $\mu^*(E)$ as an infimum), there exists a sequence $\{A_j\}_{j=1}^{\infty}$ of elements of $\mathcal{A}$ such that $A \subseteq \bigcup_{j=1}^{\infty} A_j$; i.e., $\{A_j\}_{j=1}^{\infty}$ is an $\mathcal{A}$-cover of E, in the sense of Definition 3.46 and $\mu^*(E) + \varepsilon \geq \sum_{j=1}^{\infty} \mu(A_j)$. Since the premeasure μ on the algebra $\mathcal{A}$ is countably additive and thus also finitely additive (see Exercise B.11 above), we have that for all integers $j \geq 1$, $\mu(A_j) = \mu(A_j \cap A) + \mu(A_j \cap A^c)$. Hence, by combining the inequality and the identity obtained in the last two paragraphs, we obtain successively

$$\begin{aligned} \mu^*(E) + \varepsilon \geq \sum_{j=1}^{\infty} \mu(A_j) &= \sum_{j=1}^{\infty} \mu(A_j \cap A) + \sum_{j=1}^{\infty} \mu(A_j \cap A^c) \\ &\geq \mu^*(E \cap A) + \mu^*(E \cap A^c), \end{aligned} \tag{B.31}$$

from which we deduce that (B.29) holds for every $\varepsilon > 0$ and thus that (B.30) also holds, as desired. Hence, by Proposition B.1, we conclude that $A \in \mathcal{M}$.

We close the proof of Step 1 by justifying the last inequality in (B.31): Since, clearly, $\{A_j \cap A\}_{j=1}^{\infty}$ is an $\mathcal{A}$-cover of $E \cap A$ (because $\{A_j\}_{j=1}^{\infty}$ is an $\mathcal{A}$-cover of E) and each $A_j \cap A \in \mathcal{A}$ (since $\mathcal{A}$ is an algebra), we have (by the definition of $\mu^*(E)$ in (3.12)), $\sum_{j=1}^{\infty} \mu(A_j \cap A) \geq \mu^*(E \cap A)$. Similarly, interchanging the roles of A and A^c (and because A^c and so, $A_j \cap A^c$ belongs to $\mathcal{A}$, since $\mathcal{A}$ is an algebra), $\sum_{j=1}^{\infty} \mu(A_j \cap A^c) \geq \mu^*(E \cap A^c)$. Thus, the last inequality of (B.31) follows by adding up the just obtained two inequalities. This concludes the proof of Step 1.

Step 2 (Existence of an Extension). *The Carathéodory measure $\overline{\mu}$ is an extension of μ (from the algebra $\mathcal{A}$ to the σ-algebra $\mathcal{M}$)*; i.e., we have *for every $A \in \mathcal{A}$, $\mu^*(A) = \overline{\mu}(A) = \mu(A)$.*

Indeed, since by construction, $\overline{\mu}$ is the restriction of μ^* to the σ-algebra $\mathcal{M}$ of μ^*-measurable sets, it suffices to show that $\mu^*(A) = \mu(A)$, for all $A \in \mathcal{A}$. Assume that $A \in \mathcal{A}$ is given. First, note that obviously, in light of the definition of $\mu^*(A)$ in (3.12), we have that $\mu^*(A) \leq \mu(A)$, since by letting $A_1 := A$ and $A_j := \emptyset$, for every integer $j \geq 2$, we clearly obtain an $\mathcal{A}$-cover of A (in the sense of Definition 3.46), from which the stated inequality follows.

Next, let $\{A_j\}_{j=1}^{\infty}$ be an arbitrary cover of A, where $A \in \mathcal{A}$ is given. We will show that

$$\mu(A) \leq \sum_{j=1}^{\infty} \mu(A_j), \tag{B.32}$$

from which it follows (in light of the definition of $\mu^*(A)$ as an infimum in (3.12))[13] that $\mu^*(A) \geq \mu(A)$. Combining the previous inequality with its (already proved) reverse inequality, we obtain that $\mu^*(A) = \mu(A)$, as claimed above, and therefore, we conclude the proof of Step 2.

Hence, we must now prove the inequality (B.32). Let $B_1 := A_1$ and $B_j := A_j \setminus (\bigcup_{k=1}^{j-1} A_k)$, for each integer $j \geq 2$. Then, clearly, since $\mathcal{A}$ is an algebra, B_j belongs to $\mathcal{A}$, for all $j \geq 1$. Also, the B_j's are pairwise disjoint and $\bigcup_{j=1}^{\infty} B_j = \bigcup_{j=1}^{\infty} A_j \supseteq A$. Moreover, let $C_j := A \cap B_j$, for each $j \geq 1$. Much as above, and in light of the just-mentioned identity, $\{C_j\}_{j=1}^{\infty}$ is a sequence of *pairwise disjoint* elements of $\mathcal{A}$, but now, $\bigsqcup_{j=1}^{\infty} C_j = \bigcup_{j=1}^{\infty}(A_j \cap A) = A$. Since $\bigsqcup_{j=1}^{\infty} C_j = A \in \mathcal{A}$ and μ is a premeasure on the algebra $\mathcal{A}$, it follows from the countable additivity of μ (see Definition B.9) that $\mu(A) = \sum_{j=1}^{\infty} \mu(C_j) \leq \sum_{j=1}^{\infty} \mu(A_j)$, where the last inequality follows from the monotonicity of μ (see Exercise B.11) and the fact that $C_j \subseteq A_j$ for all $j \geq 1$. This completes the proof of Step 2.

Note that Steps 1 and 2 together establish the first part of Theorem B.12 (including part (i) about the existence of an extension, along with the main statement in part (ii) about the uniqueness of an extension).

Step 3 (Uniqueness of the Extension). *This step consists in proving the uniqueness statement in part* (ii) *of Theorem* B.12.

Indeed, let ν be a measure on the σ-algebra $\mathcal{M}$ which is an extension of μ (i.e., $\nu(A) = \mu(A)$, for all $A \in \mathcal{A}$). Then, we must first show that

$$\nu(E) \leq \overline{\mu}(E) = \mu^*(E), \text{ for all } E \in \mathcal{M}, \tag{B.33}$$

as claimed in (B.28). Observe that the equality in (B.33) follows from the fact that, by definition, $\overline{\mu}$ is the restriction of μ^* to $\mathcal{M}$; see also the last equality concluding the statement of Step 2 above.

In order to prove the inequality in (B.33), we proceed much as at the beginning of the proof of Step 2. Let $E \in \mathcal{M}$. We show that for an arbitrary $\mathcal{A}$-cover $\{A_j\}_{j=1}^{\infty}$ of E, we have

$$\nu(E) \leq \sum_{j=1}^{\infty} \mu(A_j), \tag{B.34}$$

[13] Recall that the infimum of $S \subseteq [0, +\infty]$ is the largest of the lower bounds of S.

from which—by the definition of $\mu^*(E)$ in (3.12) as an infimum or greatest lower bound—we deduce that $\nu(E) \le \mu^*(E)$ and hence that (B.33) holds.

We now proceed to prove the inequality (B.34) for any $\mathcal{A}$-cover of E. Let $\{A_j\}_{j=1}^{\infty}$ be such an $\mathcal{A}$-cover; i.e., $A_j \in \mathcal{A}$ for all $j \ge 1$ and $E \subseteq \bigcup_{j=1}^{\infty} A_j =: A$. Then, since, by hypothesis, ν coincides with μ on $\mathcal{A}$, we have

$$\sum_{j=1}^{\infty} \nu(A_j) = \sum_{j=1}^{\infty} \mu(A_j). \tag{B.35}$$

Furthermore, by the monotonicity and the countable subadditivity of the measure ν on $\mathcal{M}$, we have $\nu(E) \le \nu(A) \le \sum_{j=1}^{\infty} \nu(A_j)$. Combining the latter inequalities with (B.35), we deduce (B.34), as desired. This concludes the proof of (B.33).

In the sequel, we will need the following observation:

$$\nu(A) = \lim_{k\to\infty} \nu\Big(\bigcup_{j=1}^{k} A_j\Big) = \lim_{k\to\infty} \overline{\mu}\Big(\bigcup_{j=1,}^{k} A_j\Big) = \overline{\mu}(A), \tag{B.36}$$

where we have used (twice) the fact that for each $k \ge 1$, $\bigcup_{j=1}^{k} A_j \in \mathcal{A} \subseteq \mathcal{M}$ and part (v) of Proposition 3.29, applied to each of the measures ν and $\overline{\mu}$.

Next, we will use this observation in order to show that if we assume that $(\mu^*(E) =)\overline{\mu}(E) < \infty$, then we have the equality $\nu(E) = \overline{\mu}(E)$, as claimed in the last statement of part (ii) of the theorem. Indeed, in this case, given any $\varepsilon > 0$, we can find an $\mathcal{A}$-cover $\{A_j\}_{j=1}^{\infty}$ of E such that (still with $A := \bigcup_{j=1}^{\infty} A_j$)

$$\overline{\mu}(A) \le \sum_{j=1}^{\infty} \overline{\mu}(A_j) \le \mu^*(E) + \varepsilon = \overline{\mu}(E) + \varepsilon.$$

Hence (since $\overline{\mu}(E) < \infty$ and $E \subseteq A$; see part (iii) of Proposition 3.29), we deduce that $\overline{\mu}(A \setminus E) = \overline{\mu}(A) - \overline{\mu}(E) \le \varepsilon$. Thus, by (B.36), used to justify the two equalities,

$$\begin{aligned} \overline{\mu}(E) &\le \overline{\mu}(A) = \nu(A) = \nu(E) + \nu(A \setminus E) \\ &\le \nu(E) + \overline{\mu}(A \setminus E) \le \nu(E) + \varepsilon, \end{aligned}$$

which (since $\varepsilon > 0$ is arbitrary) implies that $\overline{\mu}(E) \le \nu(E)$. Since we have already proved that $\nu(E) \le \overline{\mu}(E)$ (see (B.33)), we conclude that $\overline{\mu}(E) = \nu(E)$, as claimed.

Finally, let us assume that the premeasure μ is σ-finite; this means that there exists $\{E_j\}_{j=1}^{\infty}$ such that $E_j \in \mathcal{A}$ and $\mu(E_j) < \infty$, for all $j \ge 1$, and $X = \bigcup_{j=1}^{\infty} E_j$. Let $E \in \mathcal{M}$. Then, clearly, $E_j \cap E \in \mathcal{M}$ for all $j \ge 1$ and $E = \bigcup_{j=1}^{\infty}(E_j \cap E)$. Let $F_k := \bigcup_{j=1}^{k}(E_j \cap E)$, for all $k \ge 1$. Then, $F_k \in \mathcal{M}$, $\nu(F_k) < \infty$

(since $F_k \subseteq E_k$ and so $\nu(F_k) \leq \nu(E_k) = \mu(E_k) < \infty$)[14] and $F_k \nearrow E$, as $k \to \infty$, by part (v) of Proposition 3.29 (applied to each of the measures ν and $\overline{\mu}$). Hence, it follows that

$$\text{(B.37)} \qquad \nu(F_k) \nearrow \nu(E) \quad \text{and} \quad \overline{\mu}(F_k) \nearrow \overline{\mu}(E), \quad \text{as } k \to \infty,$$

and, by the previous case studied above, that $\nu(F_k) = \overline{\mu}(F_k)$, for all $k \geq 1$.

By passing to the limit as $k \to \infty$ in this last equality, we deduce from (B.37) that $\nu(E) = \overline{\mu}(E)$, as desired. This concludes the proof of Step 3 as well as the proof of the entire Theorem B.12. □

B.3.2. Carathéodory's Extension Theorem for Semialgebras. Next, we give a full proof of Theorem 3.57—the Carathéodory Extension Theorem for Semialgebras, also known incorrectly as the Carathéodory–Hahn Extension Theorem, since it is really due to Fréchet in [**201**]. For the clarity of the exposition of the proof, we will prove separately (in Lemma B.15) the countable additivity of μ, which will show that μ (to be denoted by $\mu_{\mathcal{A}}$) is a premeasure on the algebra $\mathcal{A}$, in the sense of Definition B.9 and will complete the proof of Theorem 3.57. We point out that the proof of Lemma B.15 will rely, in particular, on the finite additivity of $\mu_{\mathcal{A}}$, which will be stated and established in Lemma B.14 and its proof.

Proof of Theorem 3.57. Let $\mathcal{S}$ be a semialgebra (or elementary family) on X. Then, under the hypotheses of Theorem 3.57, one shows that the premeasure μ on the semialgebra $\mathcal{S}$, in the sense of parts (i) and (ii) of Definition 3.53, admits a unique extension to a premeasure (still denoted by μ, for simplicity) on the algebra $\mathcal{A}$, in the sense of Definition B.9, where $\mathcal{A}$ is the σ-algebra generated by the elementary family $\mathcal{S}$.

One then applies Theorem B.12 to the aforementioned premeasure μ defined on the algebra $\mathcal{A}$ and thereby obtains all of the conclusions of Theorem 3.57.

Therefore, it suffices to show that μ extends (uniquely) to a premeasure, in the sense of Definition B.9, on the algebra $\mathcal{A}$. For this purpose, recall that, by Proposition 3.6—according to which, since $\mathcal{A}$ is the algebra generated by the elementary family $\mathcal{S}$, each element $A \in \mathcal{A}$ can be written (not necessarily uniquely) as a finite *disjoint union* of elements of $\mathcal{S}$—we have

$$\text{(B.38)} \qquad A = \bigsqcup_{j=1}^{m} C_j, \text{ with } C_j \in \mathcal{S} \text{ and for each } j = 1, \dots, m.$$

Since the extension of μ from $\mathcal{S}$ to $\mathcal{A}$ (if it exists) is a premeasure (still denoted by μ), it must be finitely additive (see part (ii) of Definition 3.53), and hence,

[14] Indeed, $\nu(E_k) = \mu(E_k)$ since $E_k \in \mathcal{A}$ and ν (being an extension of μ) coincides with μ on $\mathcal{A}$.

we must have that (in light of (B.38))

$$\mu(A) = \sum_{j=1}^{m} \mu(C_j) \tag{B.39}$$

or, equivalently, and with the obvious notation for the premeasure $\mu = \mu_{\mathcal{S}}$ on the semialgebra $\mathcal{S}$ and for the set function $\mu = \mu_{\mathcal{A}}$ on the algebra $\mathcal{A}$,

$$\mu_{\mathcal{A}}(A) = \sum_{j=1}^{m} \mu_{\mathcal{S}}(C_j). \tag{B.40}$$

This already shows that the extension $\mu = \mu_{\mathcal{A}}$ (of $\mu = \mu_{\mathcal{S}}$) from $\mathcal{S}$ to $\mathcal{A}$ is unique, provided it exists. Moreover, this will also show that (with the above notation) $\mu = \mu_{\mathcal{A}}$ is indeed an extension of $\mu = \mu_{\mathcal{S}}$ from $\mathcal{S}$ to $\mathcal{A}$—provided we can prove, as will be done below, that the right-hand side of (B.39) (or, equivalently, of (B.40)) is independent of the choice of the decomposition of A as a finite disjoint union of elements of $\mathcal{S}$.[15]

Next, we show that the extension μ is well defined by (B.39) (or, equivalently, by (B.40)), in spite of the lack of uniqueness of the decomposition of A in (B.38). Let $A = \bigsqcup_{k=1}^{n} D_k$ be another finite disjoint decomposition of A, with $D_k \in \mathcal{S}$, for each $k = 1, \ldots, n$. Then, we claim that

$$\sum_{j=1}^{m} \mu(C_j) = \sum_{k=1}^{n} \mu(D_k), \tag{B.41}$$

which will show that μ is indeed well defined by (B.39), independently of the decomposition of A as a finite disjoint union of elements of $\mathcal{S}$.

To establish (B.41), we note that in light of (B.38), for each fixed $k \in \{1, \ldots, n\}$, $\{C_j \cap D_k\}_{j=1}^{m}$ is a finite disjoint decomposition of D_k into elements of $\mathcal{S}$. Hence, by the finite additivity of the premeasure $\mu = \mu_{\mathcal{S}}$ on the semialgebra $\mathcal{S}$ (see part (ii) of Definition 3.53), we have

$$\mu(D_k) = \sum_{j=1}^{m} \mu(C_j \cap D_k). \tag{B.42}$$

Then, after summing (B.42) over $k = 1, \ldots, n$, we obtain the following identity:

$$\sum_{k=1}^{n} \mu(D_k) = \sum_{k=1}^{n} \sum_{j=1}^{m} \mu(C_j \cap D_k). \tag{B.43}$$

[15] In fact, if we assume that $A \in \mathcal{S}$, then the choice of the obvious decomposition $\{A\}$ of A yields, in light of (B.39) or (B.40), that $\mu_{\mathcal{A}}(A) = \mu_{\mathcal{S}}(A)$, as required.

By symmetry (that is, by exchanging the roles of the C_j's and D_k's), we obtain that (still with $\mu := \mu_{\mathcal{S}}$)

$$\sum_{j=1}^{m} \mu(C_j) = \sum_{j=1}^{m} \sum_{k=1}^{n} \mu(D_k \cap C_j). \tag{B.44}$$

Since, obviously, the right-hand sides of (B.43) and (B.44) are equal, it follows from (B.43) and (B.44) that the equality (B.41) holds, as claimed. This shows that the extension $\mu = \mu_{\mathcal{A}}$ is well defined.

All that remains to be shown is that the well-defined (and uniquely defined) extension $\mu := \mu_{\mathcal{A}}$ from $\mathcal{S}$ to $\mathcal{A}$ is indeed a premeasure on the algebra $\mathcal{A}$. For the details of this last step of the proof of Theorem 3.57, we refer the interested reader to Lemmas B.15 and B.14 below, as well as to the discussion preceding the statement of Lemma B.15. □

In order to complete the proof of Theorem 3.57, we need to prove the following two lemmas (Lemma B.14 and Lemma B.15), which establish, respectively, the finite additivity (Lemma B.14) and the countable additivity (Lemma B.15) of $\mu = \mu_{\mathcal{A}}$ on $\mathcal{A}$,[16] which shows that $\mu_{\mathcal{A}}$ is a premeasure (in the sense of Definition B.9) on the algebra $\mathcal{A}$.

As it turns out, only the countable additivity (Lemma B.15) remained to be proved in the proof of Theorem 3.57. However, in the course of the proof of the countable additivity (i.e., of Lemma B.15), we will make repeated use of the finite additivity (i.e., of Lemma B.14). Therefore, we must state and prove both lemmas.

For clarity, both in the statement and the proof of each of these two lemmas, we will distinguish between $\mu = \mu_{\mathcal{S}}$, which is the original premeasure on the semialgebra (or elementary family) $\mathcal{S}$, and $\mu = \mu_{\mathcal{A}}$, which is the (unique) extension of $\mu_{\mathcal{S}}$ from $\mathcal{S}$ to the algebra $\mathcal{A}$.

Recall the key fact that the latter extension $\mu_{\mathcal{A}}$ exists, is unique, and is well defined (by (B.40) and (B.38), independently of the decomposition of $A \in \mathcal{A}$ as a finite disjoint union of elements of $\mathcal{S}$) was established in the above proof of Theorem 3.57 and relied in an essential manner on the finite additivity of the premeasure $\mu_{\mathcal{S}}$ on the semialgebra $\mathcal{S}$; see part (ii) of Definition 3.53. Finally, we point out that in the proof of Lemma B.15, we will also use in an essential manner the countable subadditivity of $\mu_{\mathcal{S}}$, which is the second main property of a premeasure on the semialgebra $\mathcal{S}$; see again part (ii) of Definition 3.53.

Lemma B.14 (Finite Additivity of $\mu = \mu_{\mathcal{A}}$). *In the proof of Theorem 3.57 given above, the set function $\mu = \mu_{\mathcal{A}}$ (defined as the extension of the premeasure $\mu = \mu_{\mathcal{S}}$*

[16] Naturally, from a logical point of view (and since $\mu_{\mathcal{A}}(\emptyset) = \mu_{\mathcal{S}}(\emptyset) = 0$), the finite additivity of $\mu_{\mathcal{A}}$ is an immediate consequence of its countable additivity. Nevertheless, the finite additivity of $\mu_{\mathcal{A}}$ (Lemma B.14) is required in order to establish Lemma B.15.

from $\mathcal{S}$ *to* $\mathcal{A}$ *by means of* (B.40) *and* (B.38)) *is finitely additive; i.e., if* $\{A_k\}_{k=1}^l$ *is a finite collection of pairwise disjoint elements of* $\mathcal{A}$,[17] *then* $\mu_{\mathcal{A}}\left(\bigcup_{k=1}^l A_k\right) = \sum_{k=1}^l \mu_{\mathcal{A}}(A_k)$.

Proof. Let $\{A_k\}_{k=1}^l$ be given as in the statement of the lemma. Since each $A_k \in \mathcal{A}$, for $k = 1, \dots, l$, we can write (by Proposition 3.6 and since $\mathcal{A}$ is the algebra generated by $\mathcal{S}$) A_k as a finite disjoint union of elements of $\mathcal{S}$:

$$A_k = \bigsqcup_{j=1}^{n_k} C_{j,k}, \text{ with } n_k \in \mathbb{N} \text{ and } C_{j,k} \in \mathcal{S}, \text{ for all } j = 1, \dots, n_k. \tag{B.45}$$

Let $A := \bigsqcup_{k=1}^l A_k$. Thus,

$$\mu_{\mathcal{A}}(A_k) = \sum_{j=1}^{n_k} \mu_{\mathcal{S}}(C_{j,k}) \tag{B.46}$$

(by the definition (B.40) of $\mu_{\mathcal{A}}(A_k)$), and by summing (B.46) for $k = 1, \dots, l$, we obtain that

$$\sum_{k=1}^l \mu_{\mathcal{A}}(A_k) = \sum_{k=1}^l \sum_{j=1}^{n_k} \mu_{\mathcal{S}}(C_{j,k}). \tag{B.47}$$

On the other hand, since $A \in \mathcal{A}$ and A is the *finite* and *disjoint* union of the $C_{j,k}$'s (for $k = 1, \dots, l$ and $j = 1, \dots, n_k$),[18] with each $C_{j,k} \in \mathcal{S}$, we have (according to the definition (B.40) of $\mu_{\mathcal{A}}(A)$ and since finite sums of numbers can be rearranged in any order) $\mu_{\mathcal{A}}(A) = \sum_{k=1}^l \sum_{j=1}^{n_k} \mu_{\mathcal{S}}(C_{j,k})$.[19] Since the right-hand sides of (B.47) and of the latter identity are identical, we deduce that $\mu_{\mathcal{A}}(A) = \sum_{k=1}^l \mu_{\mathcal{A}}(A_k)$, as desired. This completes the proof of the lemma. □

We can next state (and prove) the main lemma completing the proof of Theorem 3.57 given above. As was already mentioned, its proof makes use, in particular, of Lemma B.14 (the finite additivity of $\mu_{\mathcal{A}}$). We continue to use the notation adopted for $\mu = \mu_{\mathcal{S}}$ and $\mu = \mu_{\mathcal{A}}$ in the statement and the proof of Lemma B.14.

---

[17] Note that it is not necessary to require that $\bigcup_{k=1}^l A_k$ belongs to $\mathcal{A}$. Indeed, since $\mathcal{A}$ is an algebra, this is automatically true.

[18] It is easy to check that the $C_{j,k}$'s are pairwise disjoint, as follows: If $(j,k) \neq (j',k')$, then either $j \neq j'$ or $k \neq k'$. In the latter case, we use the fact that A_k and $A_{k'}$ are disjoint, while in the former case (and assuming that $k = k'$, which is clearly the only subcase not covered by the latter case), we appeal to the disjointness of the decomposition of A_k in (B.45) in order to conclude that $C_{j,k} \cap C_{j',k} = \emptyset$.

[19] Without loss of generality, we may assume that each of the $\mu_{\mathcal{A}}(A_k)$'s, and hence also each of the $\mu_{\mathcal{S}}(C_{j,k})$'s, is finite. Otherwise, it is easy to check that $\mu_{\mathcal{A}}(A) = \sum_{k=1}^l \mu_{\mathcal{A}}(A_k) = +\infty$ and hence, the claimed identity trivially holds.

Lemma B.15 (Countable Additivity of $\mu = \mu_{\mathcal{A}}$). *The set function $\mu = \mu_{\mathcal{A}}$ on the algebra $\mathcal{A}$ is countably additive, in the sense of Definition* B.9; *i.e., if $\{A_j\}_{j=1}^{\infty}$ is a countable collection of pairwise disjoint elements of $\mathcal{A}$ such that, in addition, $A := \bigcup_{j=1}^{\infty} A_j$ belongs to $\mathcal{A}$, then*

$$\mu_{\mathcal{A}}(A) = \mu_{\mathcal{A}}\left(\bigcup_{j=1}^{\infty} A_j\right) = \sum_{j=1}^{\infty} \mu_{\mathcal{A}}(A_j). \tag{B.48}$$

Proof. Let $\{A_j\}_{j=1}^{\infty}$ be given as in the statement of the lemma and let $A := \bigsqcup_{j=1}^{\infty} A_j \in \mathcal{A}$.

First, we use the *finite additivity* of the set function $\mu_{\mathcal{A}}$ established in Lemma B.14 to deduce that for each fixed $J \in \mathbb{N}$ and since $\{A_j\}_{j=1}^{J}$ is a *finite disjoint* family of elements of the algebra $\mathcal{A}$, we have

$$\mu_{\mathcal{A}}\left(\bigsqcup_{j=1}^{J} A_j\right) = \sum_{j=1}^{J} \mu_{\mathcal{A}}(A_j). \tag{B.49}$$

Note that since $\bigcup_{j=1}^{\infty} A_j \subseteq A$, it also follows from the monotonicity of $\mu_{\mathcal{A}}$ that[20] $\mu_{\mathcal{A}}\left(\bigcup_{j=1}^{J} A_j\right) \leq \mu_{\mathcal{A}}(A)$. Combining (B.49) and the latter inequality, we obtain that for each $J \in \mathbb{N}$, $\sum_{j=1}^{J} \mu_{\mathcal{A}}(A_j) \leq \mu_{\mathcal{A}}(A)$. It now suffices to let $J \to \infty$ (or to take the supremum over $J \in \mathbb{N}$) in the latter inequality to obtain the following inequality (in $[0, +\infty]$):[21]

$$\sum_{j=1}^{\infty} \mu_{\mathcal{A}}(A_j) \leq \mu_{\mathcal{A}}(A). \tag{B.50}$$

We will next proceed to prove the reverse inequality, namely, $\mu_{\mathcal{A}}(A) \leq \sum_{j=1}^{\infty} \mu_{\mathcal{A}}(A_j)$, which (combined with (B.50)) will imply that the equalities in (B.48) hold, thereby completing the proof of the lemma.

Since each A_j belongs to $\mathcal{A}$, for $j \in \mathbb{N}$, we can write it (by Proposition 3.6 and because $\mathcal{A}$ is the algebra generated by the semialgebra or "elementary family" $\mathcal{S}$) as a finite disjoint union of elements of $\mathcal{S}$, denoted by $\{A_{j,k}\}_{k=1}^{n_j}$, for some $n_j \in \mathbb{N}$:

$$A_j = \bigsqcup_{k=1}^{n_j} C_{j,k}, \text{ for all } j \in \mathbb{N}. \tag{B.51}$$

---

[20] Clearly, the finite additivity of $\mu_{\mathcal{A}}$ (on the algebra $\mathcal{A}$) implies its monotonicity; see Exercise B.11 above.

[21] If $\mu_{\mathcal{A}}(A) < \infty$, then (by (B.50)) $\sum_{j\in\mathbb{N}} \mu_{\mathcal{A}}(A_j) < \infty$ and, in particular, $\mu_{\mathcal{A}}(A_j) < \infty$ for every $j \in \mathbb{N}$. We will not need to make this assumption, however, and will continue instead to work with possibly infinite sums (in $[0, +\infty]$).

According to the definition of $\mu_{\mathcal{A}}(A_j)$ given in (B.40) (or else by Lemma B.14), we then have

$$\mu_{\mathcal{A}}(A_j) = \sum_{k=1}^{n_j} \mu_{\mathcal{S}}(C_{j,k}) = \sum_{k=1}^{n_j} \mu_{\mathcal{A}}(C_{j,k}), \text{ for all } j \geq 1. \tag{B.52}$$

For notational simplicity and in order to simplify some of the arguments, it is useful here to extend the finite disjoint family $\{C_{j,k}\}_{k=1}^{n_j}$ to a countable infinite disjoint family of elements of $\mathcal{S}$, as follows. For each fixed $j \in \mathbb{N}$, let $C_{j,k} := \emptyset$ for all integers $k > n_j$, so that (B.51) and (B.52) become, respectively, $A_j = \bigsqcup_{k\in\mathbb{N}} C_{j,k}$, for all $j \geq 1$, and (since $\mu_{\mathcal{A}}(\emptyset) = 0$) $\mu_{\mathcal{A}}(A_j) = \sum_{k\in\mathbb{N}} \mu_{\mathcal{A}}(C_{j,k})$, for all $j \geq 1$.

We further decompose A and each A_j for $j \in \mathbb{N}$. Indeed, since (by hypothesis) A belongs to $\mathcal{A}$ and in light of Proposition 3.6, we can also write A as the disjoint union of a finite family $\{D_q\}_{q=1}^{l}$ of elements of $\mathcal{S}$:

$$A = \bigsqcup_{q=1}^{l} D_q, \tag{B.53}$$

and so (by Lemma B.14),

$$\mu_{\mathcal{A}}(A) = \sum_{q=1}^{l} \mu_{\mathcal{A}}(D_q). \tag{B.54}$$

Again, it is convenient for the present purpose to extend $\{D_q\}_{q=1}^{l}$ to a countable disjoint family of elements of $\mathcal{S}$, $\{D_q\}_{q=1}^{\infty}$, simply by letting $D_q := \emptyset$, for all $q > l$. Accordingly, (B.53) and (B.54) become, respectively,

$$A = \bigsqcup_{q=1}^{\infty} D_q \quad \text{and (since } \mu_{\mathcal{A}}(\emptyset) = 0) \quad \mu_{\mathcal{A}}(A) = \sum_{q=1}^{\infty} \mu_{\mathcal{A}}(D_q). \tag{B.55}$$

We can now proceed to providing the crux of the argument. Recall that we want to prove that $\mu_{\mathcal{A}}(A) \leq \sum_{j=1}^{\infty} \mu_{\mathcal{A}}(A_j)$. In light of (B.53) and since for each $j \in \mathbb{N}$, $A_j \subseteq A$ (so that $A_j = A_j \cap A$), we have

$$A_j = \bigsqcup_{q=1}^{l} (A_j \cap D_q), \text{ for all } j \in \mathbb{N}. \tag{B.56}$$

Similarly, it follows from (B.51) that we have the following *finite disjoint union* of elements of $\mathcal{S}$: $A_j \cap D_q = \bigsqcup_{k=1}^{n_j} (C_{j,k} \cap D_q)$, for each $j \in \mathbb{N}$ and $q \in \{1, \ldots, l\}$. Hence, for all $j \in \mathbb{N}$, we have the following *finite disjoint union* of elements of $\mathcal{S}$:[22] $A_j = \bigsqcup_{q=1}^{l} \bigsqcup_{k=1}^{n_j} (C_{j,k} \cap D_q)$. It then follows either from Lemma B.14 or from the definition of $\mu_{\mathcal{A}}(A_j)$ in (B.39) or (B.40) that

$$\mu_{\mathcal{A}}(A_j) = \sum_{q=1}^{l} \sum_{k=1}^{n_j} \mu_{\mathcal{S}}(C_{j,k} \cap D_q), \text{ for each } j \in \mathbb{N}. \tag{B.57}$$

Consequently, we can sum (over $j \in \mathbb{N}$) the equalities in (B.57) to obtain that $\sum_{j\in\mathbb{N}} \mu_{\mathcal{A}}(A_j) = \sum_{j\in\mathbb{N}} \sum_{q=1}^{l} \sum_{k=1}^{n_j} \mu_{\mathcal{S}}(C_{j,k} \cap D_q)$. Hence, using the aforementioned extensions of the finite families $\{D_q\}_{q=1}^{l}$ and $\{C_{j,k}\}_{k=1}^{n_j}$ to countable families $\{D_q\}_{q\in\mathbb{N}}$ and $\{C_{j,k}\}_{k\in\mathbb{N}}$, respectively, we can rewrite the latter identity as follows (since $\mu_{\mathcal{S}}(\emptyset) = 0$):[23]

$$\begin{aligned} \sum_{j\in\mathbb{N}} \mu_{\mathcal{A}}(A_j) &= \sum_{j\in\mathbb{N}} \sum_{q\in\mathbb{N}} \sum_{k\in\mathbb{N}} \mu_{\mathcal{S}}(C_{j,k} \cap D_q) \\ &= \sum_{q\in\mathbb{N}} \sum_{(j,k)\in\mathbb{N}^2} \mu_{\mathcal{S}}(C_{j,k} \cap D_q) = \sum_{(j,k,q)\in\mathbb{N}^3} \mu_{\mathcal{S}}(C_{j,k} \cap D_q). \end{aligned} \tag{B.58}$$

Next, we will show that $\mu_{\mathcal{A}}(A)$ is bounded from above by either of the last two sums on the right-hand side of (B.58), thereby completing the proof of the lemma. Let us fix $q \in \mathbb{N}$. We claim that[24]

$$D_q = \bigcup_{(j,k)\in\mathbb{N}^2} (C_{j,k} \cap D_q). \tag{B.59}$$

Indeed, $A_j = \bigcup_{k\in\mathbb{N}} C_{j,k}$ and $A = \bigcup_{j\in\mathbb{N}} A_j = \bigcup_{q\in\mathbb{N}} D_q$, so that (since $D_q \subseteq A$) we have for each $q \in \mathbb{N}$,

$$D_q = A \cap D_q = \bigcup_{j\in\mathbb{N}} (A_j \cap D_q) = \bigcup_{(j,k)\in\mathbb{N}^2} (C_{j,k} \cap D_q).$$

Now, since $D_q \in \mathcal{S}$ and is given as a countable union of elements of $\mathcal{S}$, as in (B.59), we deduce from the *countable subadditivity of the premeasure* $\mu_{\mathcal{S}}$ on the semialgebra $\mathcal{S}$ (see part (ii) of Definition 3.53), along with the fact that $\mu_{\mathcal{A}}$

[22]Note that $C_{j,k} \cap D_q \in \mathcal{S}$ since $C_{j,k} \in \mathcal{S}$ and $D_q \in \mathcal{S}$; indeed, $\mathcal{S}$ is a semialgebra.

[23]The equalities in (B.58), as well as in the other potentially infinite sums written in this proof, hold in $[0, +\infty]$. Furthermore, in the last two equalities of (B.58) we have used the discrete version of the Fubini (or, rather, of the Fubini–Tonelli) Theorem, establishing the equalities of infinite double sums—or, more generally, multiple sums—of nonnegative extended real numbers.

[24]The union in (B.59) is really a disjoint union but this fact is unimportant for this part of the argument.

coincides with $\mu_{\mathcal{S}}$ on $\mathcal{S}$, that for each $q \in \mathbb{N}$,

$$\mu_{\mathcal{A}}(D_q) = \mu_{\mathcal{S}}(D_q) \leq \sum_{(j,k)\in\mathbb{N}^2} \mu_{\mathcal{S}}(C_{j,k} \cap D_q). \tag{B.60}$$

Upon summing over all $q \in \mathbb{N}$ the equalities and inequalities in (B.60), and using the value of $\mu_{\mathcal{A}}(A)$ given by the right-hand side of (B.55), we obtain that

$$\begin{aligned} \mu_{\mathcal{A}}(A) = \sum_{q\in\mathbb{N}} \mu_{\mathcal{A}}(D_q) &\leq \sum_{q\in\mathbb{N}} \sum_{(j,k)\in\mathbb{N}^2} \mu_{\mathcal{S}}(C_{j,k} \cap D_q) \\ &= \sum_{(j,k,q)\in\mathbb{N}^3} \mu_{\mathcal{S}}(C_{j,k} \cap D_q). \end{aligned} \tag{B.61}$$

Finally, according to the identity (B.58), the right-hand side of (B.61) is precisely equal to $\sum_{j=1}^{\infty} \mu_{\mathcal{A}}(A_j)$. We thus conclude from (B.61) that (B.50) holds. Since we have already established the reverse inequality stated in (B.50), towards the beginning of the proof, we deduce that $\mu_{\mathcal{A}}(A) = \sum_{j=1}^{\infty} \mu_{\mathcal{A}}(A_j)$; i.e., (B.48) holds and the proof of the lemma (as well as of Theorem 3.57) is now complete. □

B.4. Application to Lebesgue–Stieltjes and Cantor Measures

We now consider several very important applications of Carathéodory's Measurability, Restriction, and Extension Theorems studied and proved in Sections B.1 and B.3. (We will also use the completeness results stated in Section B.2.) In this section, we will first focus on the construction of Lebesgue measure on $\mathbb{R}$ (Example B.16) or on $\mathbb{R}^m$ (Exercise B.18), respectively, as well as on Lebesgue–Stieltjes measures associated with monotonic functions (Example B.20). We will then specialize the latter discussion to a very important example from the point of view of fractal geometry and classical analysis (as well as for technological, physical, and other scientific applications), namely, the Cantor–Lebesgue measure (or Cantor measure, in short). It is the Lebesgue–Stieltjes measure associated with the Cantor (or Cantor–Lebesgue) function whose graph is the famous Devil's staircase. See, especially, Example B.25 and Exercise B.26 where the Cantor measure is discussed as coinciding with the (normalized) Hausdorff measure on the classic ternary Cantor set.

In Chapter 6, we discuss another key application of Carathéodory's approach to measure theory, this time to the construction of the "δ-approximate" Hausdorff measures on metric spaces. The results obtained in Chapter 6 then play a key role in Chapter 7, where the detailed construction of Hausdorff measures (obtained as limits of Hausdorff δ-approximate measures, as $\delta \to 0^+$) is provided, and many of the main properties of these measures are established.

Example B.16 (*Lebesgue Measure on* $\mathbb{R}$). Let $\mathcal{S}$ be the elementary family (i.e., semialgebra) of $X := \mathbb{R}$,

$$\mathcal{S} = \{(a,b), [a,b], [a,b), (a,b] \,:\, a \le b, -\infty \le a \le b \le +\infty\}, \tag{B.62}$$

much as in (3.1) of Example 3.7.[25]

Then, $\mu : \mathcal{S} \to [0, +\infty]$ is defined by $\mu((a,b]) := +\infty$ if $a = -\infty$ or $b = +\infty$,[26] $\mu(\emptyset) = 0$, and otherwise (i.e., if $(a,b]$ is bounded), by

$$\mu((a,b]) := b - a, \tag{B.63}$$

the length of the bounded interval $(a,b]$. (In the sequel, we let $(+\infty)-(-\infty) := +\infty$.) Similarly, we let

$$\mu([a,b]) = \mu([a,b)) = \mu((a,b]) := b - a \,(= \mu((a,b))).$$

Then, one checks that μ is a premeasure on $(\mathbb{R}, \mathcal{S})$, where $\mathcal{S}$ is the above semialgebra $\mathcal{S}$ (in the sense of part (ii) of Definition 3.53) on $\mathbb{R}$. Alternatively, one can define by (B.39) the extension of μ from $\mathcal{S}$ to $\mathcal{A}$ (the algebra generated by $\mathcal{S}$) and check that μ thus extended is a premeasure on the algebra $\mathcal{A}$, in the sense of Definition B.9.

Also, it is not difficult to check that the premeasure μ is σ-finite on $\mathcal{S}$ (or on $\mathcal{A}$). Indeed, $\mathbb{R} = \bigcup_{n=1}^{\infty}(-n, n]$, with $\mu((-n,n]) = 2n < \infty$, for each integer $n \ge 1$. Therefore, it follows from either Theorem 3.57 or Theorem B.12 that the premeasure μ on $\mathcal{S}$ (or on $\mathcal{A}$) has a unique measure extension $\overline{\mu}$ to the σ-algebra $\mathcal{M}$ of Lebesgue measurable sets (that is, the σ-algebra of μ^*-measurable sets, where μ^* is the outer measure associated to μ via Theorem 3.48 and is also called the *Lebesgue outer measure* on $\mathbb{R}$).

This unique extension $\overline{\mu}$ is the Carathéodory measure associated to μ and is called the *Lebesgue measure* on $\mathbb{R}$ (or the 1-*dimensional Lebesgue measure*); it is denoted by $\mathcal{L}^1$.

Next, we consider $\mathfrak{M}$, the σ-algebra generated by $\mathcal{A}$. Then, it is not difficult to check that $\mathfrak{M} = \mathcal{B}(\mathbb{R})$, the Borel σ-algebra of $\mathbb{R}$. (In fact, it suffices to check that $\mathfrak{M}$ is generated by the open subsets of $\mathbb{R}$, which are the finite or countable disjoint unions of open intervals, by Theorem 2.43.) Let $\widetilde{\mu}$ be the restriction of $\overline{\mu} = \mathcal{L}^1$ to $\mathfrak{M} = \mathcal{B}(\mathbb{R})$, as in Corollary 6.16 of Theorem 6.14 in Section 6.2. Then, it follows from Corollary B.13 that Lebesgue measure $\overline{\mu} = \mathcal{L}^1$ (defined on $\mathcal{M} =: \mathcal{L}(\mathbb{R})$, the σ-algebra of Lebesgue measurable sets) is the (unique) completion of $\widetilde{\mu}$ and, therefore, the unique extension from the σ-algebra $\mathfrak{M} = \mathcal{B}(\mathbb{R})$ of Borel subsets of $\mathbb{R}$ to the σ-algebra $\mathcal{M} = \mathcal{L}(\mathbb{R})$ of Lebesgue measurable subsets of

[25] In (B.62) and throughout Example B.16, we adopt the same convention as in (3.1) of Example 3.7 and the associated footnote 1, on page 75. Namely, every unbounded interval I appearing on the right-hand side of (B.62), and containing at least one of its endpoints (provided it is $\pm\infty$), must be replaced by the corresponding interval $I \cap \mathbb{R} \subseteq \mathbb{R}$.

[26] More precisely, when $b = +\infty$, one should replace $(a,b]$ by $(a,b] \cap \mathbb{R} = (a, +\infty)$ in (B.62), so that $\mathcal{S}$ is defined on $X := \mathbb{R}$.

$\mathbb{R}$. In particular, Lebesgue measure $\mathcal{L}^1$ is a complete measure, a technical fact which is rather useful in the rigorous development of Lebesgue's integration theory.

Exercise B.17. Verify all of the statements in Example B.16 by appealing, in particular, to the results established in Sections B.1–B.3 (as well as to one key result provided in Section 6.2).

Exercise B.18 (*Lebesgue Measure on* $\mathbb{R}^m$).

(i) Extend Example B.16 to $\mathbb{R}^m$, for any integer $m \geq 1$. This shows, in particular, that Lebesgue measure $\mathcal{L}^m$ on $\mathbb{R}^m$ exists and is the unique extension to $\mathcal{M} =: \mathcal{L}(\mathbb{R}^m)$ (the σ-algebra of Lebesgue measurable sets) of the volume function (viewed as a premeasure on the semialgebra $\mathcal{S}$). Here, the semialgebra $\mathcal{S}$ is the natural m-dimensional analog of its 1-dimensional counterpart, given by (B.62), and is defined in terms of half-closed, half-open rectangles. Furthermore, the volume function μ, defined on the semialgebra $\mathcal{S}$ is the exact m-dimensional counterpart of the length function given by (B.63) in Example B.16.

(ii) Also, show that $\mathcal{L}^m$ is the unique measure completion of $\widetilde{\mu}$, the restriction of $\overline{\mu} = \mathcal{L}^m$ to $\mathfrak{M} = \mathcal{B}(\mathbb{R}^m)$, the σ-algebra of Borel subsets of $\mathbb{R}^m$.

Remark B.19. There exist Lebesgue subsets of $\mathbb{R}^m$ that are not Borel measurable. This can only be shown by using the Axiom of Choice; see, e.g., [**193**, **521**, **523**]. Hence, at least within the standard axiomatic set theory used in essentially all of mathematics, we have

$$\mathfrak{M} = \mathcal{B}(\mathbb{R}^m) \subsetneq \mathcal{L}(\mathbb{R}^m) = \mathcal{M}. \tag{B.64}$$

In practice, all classic fractals encountered in the mathematical and scientific literature—for example, the Cantor set, the Cantor curve or "Devil's staircase", the snowflake curve, the Sierpinski gasket and carpet, all self-similar sets (Section 4.3) and self-affine sets (Section 4.4), as well as, more generally, the attractors of arbitrary iterated function systems (see Sections 4.2–4.5), Julia sets and the Mandelbrot set, strange attractors (see, e.g., [**33**, **172**, **394**, **483**])—are Borel subsets of $\mathbb{R}^m$; in fact, all of the aforementioned examples are compact (and hence closed and bounded) subsets of $\mathbb{R}^m$. There are also interesting fractals or fractal-like subsets of $\mathbb{R}^m$ which are not compact, such as the unbounded Sierpinski gasket (see, e.g., [**221**, **529**]) and even more interestingly, Penrose tilings, aperiodic tilings, and mathematical quasicrystals; see, e.g., [**14**, **448**, **544**], [**377**], [**378**], along with [**326**, especially Appendix F] and the many relevant references therein.

Actually, the higher-dimensional theory of complex dimensions enables one to define the fractality of bounded and unbounded sets as the presence of nonreal complex dimensions; see the book [**356**] and the papers [**329**, **355**, **357**–**365**, **498**, **499**], [**106**–**114**], along with Chapter 9 (especially Section 9.3) of the present book. (For the earlier $m = 1$ case, see the books [**370, 374, 375**], along with Section 8.2 of the present book.) One basic fact of the theory is that an arbitrary bounded subset A of $\mathbb{R}^m$ is "fractal" (in the above sense) iff its closure (which is necessarily compact) is "fractal". Note that A itself need not be Lebesgue measurable, and let alone Borel measurable. Consequently, even if A is, for example, a bounded Lebesgue subset of $\mathbb{R}^m$ which is not Borel measurable, it could be fractal in that sense (and necessarily, its closure $\overline{A}$, which is compact, would be fractal as well). It is an open problem whether there are any such fractal sets of geometric, analytic, or number-theoretic interest, but if we assume the Axiom of (Countable) Choice, it is natural to conjecture that it would indeed be the case.

Example B.20 (*Lebesgue–Stieltjes Measures on* $\mathbb{R}$). Let $g : \mathbb{R} \to \mathbb{R}$ be a nondecreasing function which is right continuous, and let $\mathcal{S}$ be the elementary family (or semialgebra) on $X := \mathbb{R}$ considered in Example B.16; see (B.62). Furthermore, let $\mu = \mu_g : \mathcal{S} \to [0, +\infty]$ be defined by[27]

$$\mu((a,b]) = g(b^+) - g(a^+) = g(b) - g(a), \tag{B.65}$$

since g is right continuous (and so, $g(b^+) = g(b)$ and $g(a^+) = g(a)$), for any interval $(a, b] \in \mathcal{S}$ with $-\infty \le a \le b \le +\infty$ (see footnote 26) and with the conventions according to which $g(\pm\infty) := \lim_{x\to\pm\infty} g(x)$, and, as before, $+\infty - (-\infty) := +\infty$.[28]

Similarly, we have that $\mu([a,b)) = g(b^-) - g(a^-)$,

$$\mu((a,b)) = g(b^-) - g(a^+) = g(b^-) - g(a), \text{ and}$$
$$\mu([a,b]) = g(b^+) - g(a^-) = g(b) - g(a^-),$$

where the latter identity yields (for $a = b$) $\mu(\{a\}) = g(a) - g(a^-)$, the jump of g at a.[29] Here, as before, $g(b^+)$ denotes the right-hand limit of g at b and $g(a^-)$ stands for the left-hand limit of g at a.[30]

One then proceeds much as in Example B.16 to construct the *Lebesgue–Stieltjes measure* $\overline{\mu}$ defined on the σ-algebra $\mathcal{M} = \mathcal{M}_{\mu^*}$ of μ^*-measurable (or Lebesgue–Stieltjes measurable) sets. That is, one uses either Theorem 3.57 to extend the premeasure μ on $\mathcal{S}$ to all of $\mathcal{M}$ or else one first extends μ by (B.39) (and (B.38)) to a premeasure (still denoted by μ) on the algebra generated by

[27] Once again, we adopt here the same implicit convention as in (3.1) of Example 3.7 and the associated footnote 1, on page 75.

[28] Clearly, the limits defining $g(\pm\infty)$ exist in $\overline{\mathbb{R}} = [-\infty, +\infty]$ since g is nondecreasing; also, $g(-\infty) = \inf g$ and $g(+\infty) = \sup g$. Furthermore, $g(\mp\infty) \in \overline{\mathbb{R}} = [-\infty, +\infty]$, with by construction, $g(-\infty) \le g(+\infty)$ and with all possible allowable cases ($g(-\infty) \in \mathbb{R}$, $g(+\infty) \in \mathbb{R}$, $g(-\infty) = -\infty$, $g(+\infty) = +\infty$). For example, we may have $g(-\infty) \in \mathbb{R}$ and $g(+\infty) = +\infty$ or $g(\mp\infty) = \mp\infty$ (as is the case for the standard Lebesgue measure on $\mathbb{R}$ discussed in Example B.16).

[29] Indeed, for example, by (B.65), $\mu([a,b]) = \mu([a,b)) + \mu(\{b\}) = (g(b^-) - g(a^-)) + (g(b^+) - g(b^-)) = g(b^+) - g(a^-) = g(b) - g(a^-)$, since g is right continuous at b.

[30] We let $g(b^+) := \lim_{x\to b, x>b} g(x)$ and $g(a^-) := \lim_{x\to a, x<a} g(x)$.

$\mathcal{S}$ and then uses Theorem B.12 to construct $\overline{\mu} = \overline{\mu}_g$. Furthermore, one uses Theorem 6.14 along with Corollaries 6.15 and 6.16 in Section 6.2 in order to obtain the Borel measure $\widetilde{\mu} = \widetilde{\mu}_g$, the restriction of $\overline{\mu}$ to the Borel σ-algebra $\mathfrak{M} = \mathcal{B}(\mathbb{R})$ (which is the σ-algebra generated both by $\mathcal{S}$ and by the algebra $\mathcal{A}$, indifferently).

Moreover, one verifies that the premeasure on $\mathcal{S}$ is σ-finite, from which we conclude (still in light of either Theorem 3.57 or Theorem B.12) that $\overline{\mu}$ is the unique extension of μ to $\mathcal{M}$. Moreover, it follows from Corollary B.13 that $\overline{\mu}$ is the completion of $\widetilde{\mu}$; in particular, it is the unique measure extension of $\widetilde{\mu}$ from $\mathfrak{M} = \mathcal{B}(\mathbb{R})$ to $\mathcal{M}$.

Since, in light of (B.65) (and the text following it), and by the uniqueness of the extension, the measure $\overline{\mu}$ (and hence also $\widetilde{\mu}$) remains unchanged when one adds a real constant to the function g, we may assume without loss of generality and provided g is bounded from below (i.e., $g(-\infty) > -\infty$) that $g(-\infty) = 0$. (Indeed, it then suffices to replace g by $g - g(-\infty)$, which is still a nondecreasing and right continuous function.)

Finally, we note that if $g(x) := x$, for all $x \in \mathbb{R}$, then $g(\pm\infty) = \pm\infty$ and the Lebesgue–Stieltjes measure associated to g is the standard Lebesgue measure on $\mathbb{R}$ (the construction of which is provided in Example B.16). Then, with the obvious notation, $\overline{\mu}_g = \mathcal{L}^1$, $\mathcal{M}_g = \mathcal{L}(\mathbb{R})$, and so, $\widetilde{\mu}_g$ is the Borel measure obtained by restricting $\mathcal{L}^1$ to $\mathcal{B}(\mathbb{R})$.

Exercise B.21. Show that $\mu = \mu_g$ defined by (B.65) (and the text following it) is a premeasure on the semialgebra $\mathcal{S}$—or, alternatively, that μ (defined by (B.65) and the subsequent text) is a premeasure on the algebra $\mathcal{A}$ generated by $\mathcal{S}$. In addition, show that the premeasure μ (and hence also $\overline{\mu}$ and $\widetilde{\mu} = \widetilde{\mu}_g$) is σ-finite.

Exercise B.22. Let $\overline{\mu}_g$ and $\widetilde{\mu}_g$ be the measures constructed in Example B.20. Show directly that $\overline{\mu}_g$ (or, equivalently, $\widetilde{\mu}_g$) is a *continuous* (or *diffuse*) *measure*[31] iff the function g is continuous (on $\mathbb{R}$).

[*Hint*: $(a, b - \frac{1}{n}] \nearrow (a, b]$, so that (by part (v) of Proposition 3.29) we have that $\mu((a, b - \frac{1}{n}]) \to \mu((a, b])$, as $n \to \infty$.]

Definition B.23 (*Baire Measure*)**.** A measure defined on the Borel σ-algebra $\mathcal{B}(\mathbb{R})$ is called a *Baire measure* (on $\mathbb{R}$) if it is finite for all bounded sets of $\mathbb{R}$ (or, equivalently, for all bounded intervals $(a, b]$, with $-\infty < a \leq b < +\infty$).

Clearly, in Example B.20, $\widetilde{\mu} = \widetilde{\mu}_g$ is a Baire measure on $\mathbb{R}$. We can now obtain, in some sense, a converse of the result (and construction) obtained in Example B.20.

[31]This means that $\overline{\mu}_g(\{x\}) = \widetilde{\mu}_g(\{x\}) = 0$, for all $x \in \mathbb{R}$, and similarly for $\widetilde{\mu}_g$. Note that since $\{x\}$ is a closed and, hence, Borel subset of $\mathbb{R}$, we have $\overline{\mu}_g(\{x\}) = \widetilde{\mu}_g(\{x\})$, for all $x \in \mathbb{R}$.

Given a Baire measure ν on $\mathbb{R}$ (as in Definition B.23), define the function $g = g_\nu : \mathbb{R} \to [0, +\infty)$ by $g(x) := \nu((-\infty, x])$, for all $x \in \mathbb{R}$. The map g_ν is called the *cumulative distribution function* of ν. Clearly, g is nondecreasing and right continuous; also, $g(-\infty) = 0$. Thus, $g(x) \in \mathbb{R}$, for all $x \in \mathbb{R}$. Furthermore, for $-\infty \leq a \leq b \leq +\infty$, we have $\nu((a, b]) = g(b) - g(a)$. Also, for every $x \in \mathbb{R}$, $g(x) - g(x^-) = \nu(\{x\})$, from which it follows that g is continuous iff ν is continuous (i.e., $\nu(\{x\}) = 0$, for all $x \in \mathbb{R}$); this is essentially the content of Exercise B.22.

We prove, for example, that $g(-\infty) = 0$. Observe that $(-\infty, -n] \searrow \emptyset$, as $n \to \infty$, and that $g(-1) = \nu((-\infty. -1]) < \infty$; so that (by part (vi) of Proposition 3.29), $g(-n) = \nu((-\infty, -n]) \searrow 0$, as $n \to \infty$, and hence, $g(-\infty) = \inf g = 0$.

Similarly, in order to show that g is right continuous, we note that for any given $x \in \mathbb{R}$, $(-\infty, x + (1/n)] \searrow (-\infty, x]$, as $n \to \infty$; hence (since $g(x + 1) = \nu((-\infty, x + 1]) < \infty$), it again follows from part (vi) of Proposition 3.29 that

$$g\left(x + \frac{1}{n}\right) = \nu\left(\left(-\infty, x + \frac{1}{n}\right]\right) \searrow g(x) = \nu((-\infty, x]), \text{ as } n \to \infty.$$

Therefore, $g(x^+) = g(x)$, as desired. (Indeed, since g is nondecreasing, we have that $g(x^+) = \inf_{y>x} g(y) = \lim_{n\to\infty} g(x + (1/n))$.)

By combining the above discussion and the construction provided in Example B.20, we obtain the results stated in parts (i)–(ii) of the following exercise (which is a little harder than the other exercises in this appendix).

Exercise B.24.

(i) Show that there is a bijective correspondence between the Baire measures ν on $\mathbb{R}$ (in the sense of Definition B.23) and the functions $g : \mathbb{R} \to [0, +\infty)$ which are nondecreasing and right continuous as well as satisfy $g(-\infty) = 0$. In short, we can denote this correspondence in the following manner: $g_\nu := g \leftrightarrow \nu =: \nu_g$.

(ii) Show that in the above correspondence stated in part (i), the Baire measure ν is bounded (i.e., $\nu(\mathbb{R}) < \infty$ if $g(-\infty) = 0$ and hence, $g \geq 0$) iff the function g is bounded (i.e., $g(+\infty) < \infty$). Hence, there is also a bijective correspondence between bounded Baire measures and bounded functions g of the type described in (i).

The following example plays an important role in fractal geometry, as well as in classical analysis.

Example B.25 (*Cantor–Lebesgue Measure*). Let g be the Cantor (or Cantor–Lebesgue) function, as discussed in various places in this book, including in the introduction (Chapter 1) and in Section 4.4, as well as in Chapter 9 (especially in Section 9.4). Recall that $g : [0, 1] \to [0, 1]$ is nondecreasing and continuous, $g(0) = 0$, $g(1) = 1$, and g is constant on every open deleted interval in the construction of the classic ternary Cantor set C.[32] Furthermore, by definition, the graph of g is the celebrated "Devil's staircase". Then, with the notation of Example B.20, we can associate to g the Lebesgue–Stieltjes measure $\overline{\mu}_g$.

Now, as in Example B.20, let $\nu := \widetilde{\mu}_g$ be the Borel measure—i.e., the restriction of the Lebesgue–Stieltjes measure $\overline{\mu}_g$ to the Borel σ-algebra $\mathcal{B}(\mathbb{R})$ or else to $\mathcal{B}([0,1])$ since $\text{supp}(\overline{\mu}_g)$, the support of $\overline{\mu}_g$, is clearly contained in $[0,1]$. Then (see Definition 3.42), $\text{supp}(\nu) = \text{supp}(\overline{\mu}_g) = C$, the classic Cantor set. Furthermore, since g is continuous, $\nu = \nu_g$ is a continuous measure; see Exercise B.22. The bounded Baire measure $\nu = \widetilde{\mu}_g$ is a very interesting measure. As stated just above, its support is the ternary Cantor set C, a set of Lebesgue measure zero (since $\mathcal{L}^1(C) = 0$). For this reason, it is referred to as a *singular* continuous measure.

Next, consider the natural (but *normalized*) Hausdorff measure on the Cantor set C, $\mathcal{H}_C = \mathcal{H}^{\beta}_{C,norm}$ where (necessarily) $\beta := \log 2/\log 3 = \log_3 2$, the (Hausdorff or Minkowski) dimension of the Cantor set. See Chapter 5, especially Sections 5.1 and 5.2, along with Chapter 7. In the sequel, we always normalize $\mathcal{H} = \mathcal{H}_C$ so that it becomes a probability measure: $\mathcal{H}_C([0,1]) = 1$. Since we also have $\nu([0,1]) = g(1) - g(0) = 1$ (because $\nu = \nu_g$ and $g(1) = 1$ while $g(0) = 0$), it is tempting to compare ν and $\mathcal{H}_C$. We claim that (with the above notation)

$$\nu_g = \widetilde{\mu}_g = \mathcal{H}_C, \tag{B.66}$$

which is an interesting (but relatively little known) fact in fractal geometry and, more broadly, in analysis; see Exercise B.26 just below.

Exercises

Exercise B.26 (*Cantor Measure and Hausdorff Measure*). Let $\nu_g = \widetilde{\mu}_g$ and $\mathcal{H}_C$ be defined as in Example B.25 just above.

(i) Show directly that $\text{supp}(\nu_g) = \text{supp}(\mathcal{H}_C) = C$, as claimed in Example B.25; see Definition 3.42. Furthermore, prove that $\nu_g = \mathcal{H}_C$, as claimed in (B.66). Also, give a *geometric interpretation* of this identity; see part (ii).

(ii) (*Three Possible Approaches*). Justify each (or at least one) of the following three possible, but not quite unrelated, methods that can be used for proving the identity (B.66), which is the crux of the matter:

(a) (*Brute Force*). Show directly that $\nu_g((a,b]) = \mathcal{H}_C((a,b])$, for every subinterval $(a,b]$ of $[0,1]$. Since the collection of all such intervals generates the Borel σ-algebra $\mathcal{B}([0,1])$, conclude that $\nu_g = \mathcal{H}_C$; i.e., (B.66) holds.

[32] Alternatively, one can view g as a function $g : \mathbb{R} \to [0,1]$ by letting $g(x) := 0$, for all $x < 0$, and $g(x) := 1$, for all $x > 1$. This definition better fits our present framework. We leave the small necessary adjustments to the reader.

(b) (*Self-Similarity of the Measures*). Show that ν_g and $\mathcal{H}_C$, viewed as Borel probability measures, are *self-similar measures* (in the sense of Section 4.8), with the same scaling ratios 1/3 and the same probability weights 1/2. More specifically, they satisfy the following *self-similar relation:*[33]

$$\gamma(\varphi) = \frac{1}{2}\gamma(\varphi \circ S_1) + \frac{1}{2}\gamma(\varphi \circ S_2), \tag{B.67}$$

for every continuous function φ on $[0, 1]$. Here, S_1 and S_2 (viewed as self-maps of $[0, 1]$) are the similarities defining the ternary Cantor set C as a *self-similar set* (in the sense of Section 4.3 and, especially, of Example 4.23) with scaling ratios equal to 1/3 (so that $C = S_1(C) \sqcup S_2(C)$):

$$S_1(x) = \frac{1}{3}x \quad \text{and} \quad S_2(x) = \frac{1}{3}x + \frac{2}{3}, \text{ for all } x \in [0, 1]. \tag{B.68}$$

Since, according to the results of Section 4.8 (see Theorem 4.82), the fixed point equation (B.67) has a *unique* solution (as a Borel probability measure on $\mathbb{R}$ or, equivalently, on $[0, 1]$), conclude that $\nu_g = \mathcal{H}_C$, as desired.

(c) (*Self-Affinity of the Graphs*). Since, clearly, $\mathcal{H}_C$ is a bounded and continuous Borel measure on $\mathbb{R}$ (or, equivalently, on $[0, 1]$, by part (i)), it follows from both parts (i) and (ii) of Exercise B.24 that (with the notation adopted in Exercises B.22 and B.24) $\mathcal{H}_C = \nu_h$, for some *unique* bounded, nondecreasing and continuous function $h : [0, 1] \to [0, 1]$ such that $h(0) = 0$ and $h(1) = 1$.

Then, show that—like the graph of the Cantor function g, the Devil's staircase $\mathbf{\Gamma}_{DS}$—the graph of h is an *inhomogeneous self-affine set* (in the sense of Example 4.93) with affinity constants 1/3 and 1/2, along the horizontal and vertical directions, respectively. Conclude that $h = g$ and hence, $\nu_h = \nu_g = \mathcal{H}_C$, as desired.

[33] Here, γ is a Borel probability measure on $[0, 1]$ and we use the notation $\gamma(\varphi) := \int_{[0,1]} \varphi \, d\gamma$, for every continuous function φ on $[0, 1]$; see Section 3.6.

Acknowledgments

The research of the first author, Michel L. Lapidus, has been partially supported by the US National Science Foundation under the research grants DMS-8703138, DMS-8904389, DMS-9207098, DMS-9623002, DMS-0070497, DMS-0707524, and DMS-1107750, as well as by the Burton Jones Endowed Chair Fund while he was the holder of the F. Burton Jones Endowed Chair in Pure Mathematics at the University of California, Riverside, between July 2015 and June 2021. Support of the first author by the Institut des Hautes Scientifiques (IHES) in Bures-sur Yvette (near Paris, France) during many stays at the institute (and during which parts of this book were written) is gratefully acknowledged.

The research of the second author, Goran Radunović, was supported by the Croatian Science Foundation under the projects IP-2014-09-2285 and UIP-2017-05-1020.

The first author (MLL) has taught courses based on various parts of the notes eventually giving rise to this book (and to its sequel, [**119**]), on many occasions and at different levels, either Honors Undergraduate Seminars, Upper-Division Undergraduate courses, or Topics Graduate courses, all of which he offered and created at the University of California, Riverside (as well as, in the early 1990s, at Yale University) over many years, and at many research conferences and graduate summer schools throughout the world. Three of his later Ph.D. students (Erin P. J. Pearse, Robert G. Niemeyer, and Hung Lu) attended his undergraduate courses or seminars on fractal geometry. All of his Ph.D. students at the University of California, Riverside, took various parts of his graduate courses and seminars on this subject.

These include Cheryl A. Griffith, Christina Q. He, Erin P. J. Pearse, John A. Rock, Hung (Tim) Lu, Britta K. Daudert, Scot P. Childress, Vicente Alvarez, Jonathan J. Sarhad, Hafedh Herichi, Nishu Lal, Robert G. Niemeyer, Michael A. Maroun, John Quinn, Goran Radunović (University of Zagreb, Croatia), Timothy L. Cobler, Roby A. Scott, Sean Watson, Andrea Arauza Rivera, Frank Klostler, Edward K. Voskanian, Alexander M. Henderson, Adam Yassine, Therese-Marie Landry, Adam D. Richardson, Matthew Overduin, Will Hoffer, Shanna Dobson, Marco Munoz, Russell Ryan, and Gregory P. Leathrum, among others.

The first author is very grateful to many of them for sharing with him their notes of his lectures as well as for their comments, feedback, and useful suggestions. He is especially grateful to Robert G. Niemeyer and John A. Rock, who were, at different times, his collaborators on this project—but, unfortunately, had to leave it, due to other professional or personal commitments. We want to acknowledge here their significant initial contributions.

The first author would also like to thank Professor Claire David for her helpful feedback about Section 9.7.

The project for this book began a long time ago (around 2003) when the current (and then) Publisher of the American Mathematic Society (AMS), Dr. Sergei Gelfand, a man of exceptional energy and talent, asked the first author to write a textbook on fractal geometry for one of the AMS book series.[1] It took quite a while for the project to take shape but, fortunately, we are now at the end of the journey, at least for this book. We wish to thank Sergei Gelfand for his very valuable comments, criticism, and feedback throughout this journey, as well as for his patience and gentle nudging.

The authors are very grateful to Arlene O'Sean, of the American Mathematical Society, for her outstanding and very careful copy editing of the book.

The first author would also like to thank his two recent Administrative Assistants, Yu-Tzu Tsai (from 2011 to 2022) and Janice Carter (since August 2022), for not just having typed large parts of this book, but also for their feedback, guidance, and friendship during sometimes difficult periods. It is not an exaggeration to state that this book could not have been completed without their absolute devotion and professionalism.

[1] Initially, he had in mind the *Student Mathematical Library* (SML) but, as the size of the book grew, it started exceeding the capacities of SML and we eventually agreed that *Graduate Studies in Mathematics* (GSM) was the appropriate series for this textbook.

Bibliography

[1] L. V. Ahlfors, *Complex analysis: An introduction to the theory of analytic functions of one complex variable*, 3rd ed., International Series in Pure and Applied Mathematics, McGraw-Hill Book Co., New York, 1978. MR510197

[2] D. S. Alexander, *A History of Complex Dynamics: From Schröder to Fatou and Julia*, Aspects of Mathematics, vol. 24, Vieweg & Teubner Verlag, Wiesbaden, 2013. (e-Book Archives: Springer Fachmedien, Wiesbaden, 1994.)

[3] D. S. Alexander, F. Iavernaro, and A. Rosa, *Early days in complex dynamics: A history of complex dynamics in one variable during 1906–1942*, History of Mathematics, vol. 38, American Mathematical Society, Providence, RI; London Mathematical Society, London, 2012, DOI 10.1090/hmath/038. MR2857586

[4] P. S. Alexandrov and H. Hopf, *Topologie I* ["Topology I"], Springer, Berlin, 1935 [French].

[5] C. D. Aliprantis and O. Burkinshaw, *Principles of real analysis*, 2nd ed., Academic Press, Inc., Boston, MA, 1990. MR1043168

[6] P. C. Allaart and K. Kawamura, *The Takagi function: a survey*, Real Anal. Exchange **37** (2011/12), no. 1, 1–54. MR3016850

[7] P. Alonso Ruiz, J. P. Chen, L. G. Rogers, R. S. Strichartz, and A. Teplyaev (eds.), *Analysis, probability and mathematical physics on fractals*, Fractals and Dynamics in Mathematics, Science, and the Arts: Theory and Applications, vol. 5, World Scientific Publishing Co. Pte. Ltd., Hackensack, NJ, 2020, DOI 10.1142/11696. MR4472244

[8] R. D. Anderson, *A characterization of the universal curve and a proof of its homogeneity*, Ann. of Math. (2) **67** (1958), 313–324, DOI 10.2307/1970007. MR96180

[9] R. D. Anderson, *One-dimensional continuous curves and a homogeneity theorem*, Ann. of Math. (2) **68** (1958), 1–16, DOI 10.2307/1970040. MR96181

[10] M. Arbeiter and N. Patzschke, *Random self-similar multifractals*, Math. Nachr. **181** (1996), 5–42, DOI 10.1002/mana.3211810102. MR1409071

[11] A. Arneodo, E. Bacry, and J.-F. Muzy, *The thermodynamics of fractals revisited with wavelets*, Physica A **213** (1995), 232–275.

[12] P. Assouad, *Étude d'une dimension métrique liée à la possibilité de plongements dans* $\mathbf{R}^n$ (French, with English summary), C. R. Acad. Sci. Paris Sér. A-B **288** (1979), no. 15, A731–A734. MR532401

[13] A. Avila, J. Kahn, M. Lyubich, and W. Shen, *Local connectivity of Julia sets of unicritical polynomials*, Ann. of Math (2) **178** (2009), 783–797.

[14] M. Baake and U. Grimm, *Aperiodic order. Vol. 1, A mathematical invitation*, with a foreword by Roger Penrose, Encyclopedia of Mathematics and its Applications, vol. 149, Cambridge University Press, Cambridge, 2013, DOI 10.1017/CBO9781139025256. MR3136260

[15] M. Bader, *Space-filling curves: An introduction with applications in scientific computing*, Texts in Computational Science and Engineering, vol. 9, Springer, Heidelberg, 2013, DOI 10.1007/978-3-642-31046-1. MR2985717

[16] A. Baker, *Transcendental number theory*, Cambridge University Press, London-New York, 1975. MR422171

[17] S. Banach, *Sur les opérations dans les ensembles abstraits et leur application aux équations intégrales* (French), Fund. Math. **3** (1922), 133–181, DOI 10.4064/fm-3-1-133-181. MR3949898

[18] S. Banach, Über die Baire'sche Kategorie gewisser Funktionenmengen ["On Baire's category of certain functions"], *Studia Math.* **3** (1931), 174–179 [German].

[19] C. Bandt, K. J. Falconer, and M. Zähle (eds.), *Fractal Geometry and Stochastics V*, Proc. Fifth Internat. Conf. (Tabarz, Germany, March 2014), Progress in Probability, vol. 10, Birkhäuser, Springer International Publishing Switzerland, Cham and Basel, 2015.[1]

[20] C. Bandt, J. Flachsmeyer, and H. Haase (eds.), *Topology, measures, and fractals*, papers from the Conference "Topology and Measure VI" held in Warnemünde, August 1991, Mathematical Research, vol. 66, Akademie-Verlag, Berlin, 1992. MR1226271

[21] K. Barański, *On the complexification of the Weierstrass non-differentiable function*, Ann. Acad. Sci. Fenn. Math. **27** (2002), no. 2, 325–339. MR1921311

[22] K. Barański, *On some lacunary power series*, Michigan Math. J. **54** (2006), no. 1, 65–79, DOI 10.1307/mmj/1144437438. MR2214788

[23] K. Barański, *Hausdorff dimension of the limit sets of some planar geometric constructions*, Adv. Math. **210** (2007), no. 1, 215–245, DOI 10.1016/j.aim.2006.06.005. MR2298824

[24] K. Barański, *On the dimension of graphs of Weierstrass-type functions with rapidly growing frequencies*, Nonlinearity **25** (2012), no. 1, 193–209, DOI 10.1088/0951-7715/25/1/193. MR2864382

[25] K. Barański, *Dimension of the graphs of the Weierstrass-type functions*, Fractal geometry and stochastics V, Progr. Probab., vol. 70, Birkhäuser/Springer, Cham, 2015, pp. 77–91, DOI 10.1007/978-3-319-18660-3_5. MR3558151

[26] K. Barański, B. Bárány, and J. Romanowska, *On the dimension of the graph of the classical Weierstrass function*, Adv. Math. **265** (2014), 32–59, DOI 10.1016/j.aim.2014.07.033. MR3255455

[27] M. T. Barlow, Diffusions on fractals, in: *Lectures on Probability Theory and Statistics* (P. Bernard, ed.), Ecole d'Eté de Probabilités de Saint-Flour XXV–1995, Lecture Notes in Mathematics, vol. 1690, Springer-Verlag, 1998, pp. 1–121.

[28] M. F. Barnsley, *Fractals everywhere*, 2nd ed., revised with the assistance of and with a foreword by Hawley Rising, III, Academic Press Professional, Boston, MA, 1993. MR1231795

[29] M. F. Barnsley, *Superfractals*, Cambridge University Press, Cambridge, 2006, DOI 10.1017/CBO9781107590168. MR2254477

[30] M. F. Barnsley and S. Demko, *Iterated function systems and the global construction of fractals*, Proc. Roy. Soc. London Ser. A **399** (1985), no. 1817, 243–275. MR799111

[31] M. F. Barnsley, R. L. Devaney, B. B. Mandelbrot, H.-O. Peitgen, D. Saupe, and R. F. Voss, *The science of fractal images*, with contributions by Yuval Fisher and Michael McGuire, Springer-Verlag, New York, 1988, DOI 10.1007/978-1-4612-3784-6. MR952853

[32] J. Barrow-Green, *Poincaré and the three body problem*, History of Mathematics, vol. 11, American Mathematical Society, Providence, RI; London Mathematical Society, London, 1997, DOI 10.1090/hmath/011. MR1415387

[33] A. F. Beardon, *Iteration of rational functions: Complex analytic dynamical systems*, Graduate Texts in Mathematics, vol. 132, Springer-Verlag, New York, 1991, DOI 10.1007/978-1-4612-4422-6. MR1128089

[1] This volume is a part of a series of five (or six) volumes with different sets of editors, but the same title and published by Birkhäuser in the same book series.

[34] T. Bedford, *Crinkly Curves, Markov Partitions and Dimensions*, Ph.D. Thesis, Mathematics, University of Warwick, U.K., 1984.

[35] T. Bedford, M. Keane, and C. Series (eds.), *Ergodic theory, symbolic dynamics, and hyperbolic spaces*, papers from the Workshop on Hyperbolic Geometry and Ergodic Theory held in Trieste, April 17–28, 1989, Oxford Science Publications, The Clarendon Press, Oxford University Press, New York, 1991. MR1130170

[36] J. Bell, *On the Theory of Hausdorff Measures in Metric Spaces*, Ph.D. Thesis, Mathematics, University College, London, England, Great Britain, 1994.

[37] A. S. Belov, *On the Salem and Zygmund problem with respect to the smoothness of an analytic function that generates a Peano curve* (Russian), Mat. Sb. **181** (1990), no. 8, 1048–1060, DOI 10.1070/SM1991v070n02ABEH001384; English transl., Math. USSR-Sb. **70** (1991), no. 2, 485–497. MR1076142

[38] M. Benedicks and L. Carleson, *The dynamics of the Hénon map*, Ann. of Math. (2) **133** (1991), no. 1, 73–169, DOI 10.2307/2944326. MR1087346

[39] M. Berger and B. Gostiaux, *Differential geometry: manifolds, curves, and surfaces*, translated from the French by Silvio Levy, Graduate Texts in Mathematics, vol. 115, Springer-Verlag, New York, 1988. MR917479

[40] M. V. Berry, *Distribution of modes in fractal resonators*, Structural stability in physics (Proc. Internat. Symposia Appl. Catastrophe Theory and Topological Concepts in Phys., Inst. Inform. Sci., Univ. Tübingen, Tübingen, 1978), Springer Ser. Synergetics, vol. 4, Springer, Berlin, 1979, pp. 51–53, DOI 10.1007/978-3-642-67363-4_7. MR556688

[41] M. V. Berry, *Some geometric aspects of wave motion: wavefront dislocations, diffraction catastrophes, diffractals*, Geometry of the Laplace operator (Proc. Sympos. Pure Math., Univ. Hawaii, Honolulu, Hawaii, 1979), Proc. Sympos. Pure Math., XXXVI, Amer. Math. Soc., Providence, RI, 1980, pp. 13–28. MR573427

[42] M. V. Berry and Z. V. Lewis, *On the Weierstrass-Mandelbrot fractal function*, Proc. Roy. Soc. London Ser. A **370** (1980), no. 1743, 459–484, DOI 10.1098/rspa.1980.0044. MR570931

[43] A. S. Besicovitch, *An investigation of continuous functions in connection with the question of differentiability*, Mat. Sbornik, **31** (1924), 529–556 [Russian].

[44] A. S. Besicovitch, *On the sum of digits of real numbers represented in the dyadic system*, Math. Ann. **110** (1935), no. 1, 321–330, DOI 10.1007/BF01448030. MR1512941

[45] A. S. Besicovitch, *A general form of the covering principle and relative differentiation of additive functions. II*, Proc. Cambridge Philos. Soc. **42** (1946), 1–10, DOI 10.1017/s0305004100022660. MR14414

[46] A. S. Besicovitch and S. J. Taylor, *On the complementary intervals of a linear closed set of zero Lebesgue measure*, J. London Math. Soc. **29** (1954), 449–459, DOI 10.1112/jlms/s1-29.4.449. MR64849

[47] A. S. Besicovitch and H. D. Ursell, *Sets of fractional dimensions (V): On dimensional numbers of some continuous curves*, J. London Math. Soc. No. 1, **1** (1937), 18–25.

[48] M. Bestvina, *Characterizing k-dimensional universal Menger compacta*, Bull. Amer. Math. Soc. (N.S.) **11** (1984), no. 2, 369–370, DOI 10.1090/S0273-0979-1984-15313-8. MR752801

[49] G. Birkhoff and G.-C. Rota, *Ordinary differential equations*, 3rd ed., John Wiley & Sons, New York-Chichester-Brisbane, 1978. MR507190

[50] C. J. Bishop and Y. Peres, *Fractals in probability and analysis*, Cambridge Studies in Advanced Mathematics, vol. 162, Cambridge University Press, Cambridge, 2017, DOI 10.1017/9781316460238. MR3616046

[51] P. Blanchard, *Complex analytic dynamics on the Riemann sphere*, Bull. Amer. Math. Soc. (N.S.) **11** (1984), no. 1, 85–141, DOI 10.1090/S0273-0979-1984-15240-6. MR741725

[52] H. Bohr, *Almost Periodic Functions* [English translation of the original edition], Dover Publications, Inc., Mineola, 2018. (Reprint of the 1947 Chelsea edition.)

[53] A. Bonifant, M. Lyubich, and S. Sutherland (eds.), *Frontiers in complex dynamics: In celebration of John Milnor's 80th birthday*, Princeton Mathematical Series, vol. 51, Princeton University Press, Princeton, NJ, 2014. MR3289442

[54] M. Bonk, *Quasiconformal geometry of fractals*, International Congress of Mathematicians. Vol. II, Eur. Math. Soc., Zürich, 2006, pp. 1349–1373. MR2275649

[55] K. Borsuk, *Remark on the Cartesian product of two l-dimensional spaces* (English, with Russian summary), Bull. Acad. Polon. Sci. Sér. Sci. Math. Astronom. Phys. **23** (1975), no. 9, 971–973. MR394636

[56] G. Bouligand, *Ensembles impropres et nombre dimensionnel* [*"Improper sets and dimensional number"*], Bull. Sci. Math. (2) **52** (1928), 320–344 and 361–376 [French].

[57] D. M. Bressoud, *A radical approach to Lebesgue's theory of integration*, MAA Textbooks, Cambridge University Press, Cambridge, 2008. MR2380238

[58] R. Brooks and J. P. Matelski, *The dynamics of 2-generator subgroups of* PSL(2, **C**), Riemann surfaces and related topics: Proceedings of the 1978 Stony Brook Conference (State Univ. New York, Stony Brook, N.Y., 1978), Ann. of Math. Stud., No. 97, Princeton Univ. Press, Princeton, NJ, 1981, pp. 65–71. MR624805

[59] J. Brossard and R. Carmona, *Can one hear the dimension of a fractal?*, Comm. Math. Phys. **104** (1986), no. 1, 103–122. MR834484

[60] L. E. J. Brouwer, *On the structure of perfect sets of points*, KNAW, Proceedings **12** (1909–1910), 785–794.

[61] G. Brown, G. Michon, and J. Peyrière, *On the multifractal analysis of measures*, J. Statist. Phys. **66** (1992), no. 3-4, 775–790, DOI 10.1007/BF01055700. MR1151978

[62] A. Bunde and S. Havlin (eds.), *Fractals and disordered systems*, Springer-Verlag, Berlin, 1991, DOI 10.1007/978-3-642-51435-7. MR1130617

[63] A. M. Caetano, *On the search for the asymptotic behaviour of the eigenvalues of the Dirichlet Laplacian for bounded irregular domains: Some aspects of fractals in mathematics and physics (Toulouse, 1993)*, Internat. J. Appl. Sci. Comput. **2** (1995), no. 2, 261–287. MR1369783

[64] G. Calcagni, *Complex dimensions and their observability*, Phys. Rev. D **96** (2017), no. 4, 046001, 7, DOI 10.1103/physrevd.96.046001. MR3852956

[65] L. Calvet and A. J. Fisher, *Multifractal Volatility: Theory, Forecasting, and Pricing*, Academic Press Advanced Finance, Academic Press, San Diego (Elsevier, Amsterdam), 2008.

[66] J. W. Cannon, *A positional characterization of the* $(n-1)$*-dimensional Sierpiński curve in* $S^n (n \neq 4)$, Fund. Math. **79** (1973), no. 2, 107–112, DOI 10.4064/fm-79-2-107-112. MR319203

[67] H. Cantor, *Ueber eine Eigenschaft des Inbegriffs aller reellen algebraischen Zahlen* (German), J. Reine Angew. Math. **77** (1874), 258–262, DOI 10.1515/crll.1874.77.258. MR1579605

[68] G. Cantor, *Ein Beitrag zur Mannigfaltigkeitslehre*, J. Reine Angew. Math. **84** (1878), 242–258, DOI 10.1515/crelle-1878-18788413. MR1581650

[69] G. Cantor, *Ueber unendliche, lineare Punktmannichfaltigkeiten* (German), Math. Ann. **21** (1883), no. 4, 545–591, DOI 10.1007/BF01446819. MR1510215

[70] G. Cantor, *De la puissance des ensembles parfaits de points: Extrait d'une lettre adressée à l'éditeur* (French), Acta Math. **4** (1884), no. 1, 381–392, DOI 10.1007/BF02418423. MR1554642

[71] G. Cantor, Ueber eine elementare Frage der Mannigfaltigkeitslehre, ["On an elementary question in the theory of manifolds"], *Jahresbericht der Deutschen Mathematiker-Vereinigung*, **1** (1891), 75–78. (English translation availble in: W. B. Ewald (ed.), *From Immanuel Kant to David Hilbert: A Source Book in the Foundations of Mathematics*, vol. 2, Oxford Univ. Press, Oxford, 1996, pp. 920–922.)

[72] D. Carfi, M. L. Lapidus, E. P. J. Pearse, and M. van Frankenhuijsen, *Fractal Geometry and Dynamical Systems in Pure and Applied Mathematics*, Parts I & II, Contemporary Mathematics, vols. 600 and 601, Amer. Math. Soc., Providence, R. I., 2013. (I: *Fractals in Pure Mathematics*, vol. 600. II: *Fractals in Applied Mathematics*, vol. 601.)

[73] L. Carleson and T. W. Gamelin, *Complex dynamics*, Universitext: Tracts in Mathematics, Springer-Verlag, New York, 1993, DOI 10.1007/978-1-4612-4364-9. MR1230383

[74] R. Cawley and R. D. Mauldin, *Multifractal decompositions of Moran fractals*, Adv. Math. **92** (1992), no. 2, 196–236, DOI 10.1016/0001-8708(92)90064-R. MR1155465

[75] A. Cayley, *The Newton–Fourier imaginary problem*, Amer. J. Math. **2** (1879), 92.

[76] C. Cellérier, *Note sur les principes fondamentaux de l'analyse* [*"Note on the fundamental principles of analysis"*], Bull. Sci. Math. Ser. 2 **14** (1890), 142–160 [French].

[77] D. Cerveau (ed.), *Complex Dynamics and Geometry*, paperback edition, SMF/AMS Texts and Monographs, vol. 10, Amer. Math. Soc., Providence, R. I., 2003.

[78] E. Cesàro, *Remarques sur la courbe de von Koch* [*"Remarks on the von Koch curve"*], Atti della Reale Accademia delle Scienze Fisiche e Matematiche di Napoli **XII** (1905), 1–12 [French]. (See also [**79**, vol. II, pp. 464–479].)

[79] E. Cesàro, *Opere scelte. Vol. I: Algebra, serie, teoria dei numeri. Parte prima* (Italian), Edizioni Cremonese, Rome, 1964. A cura dell'Unione Matematica Italiana; redattore principale, Carlo Miranda. MR260554

[80] D. R. Chalice, *A characterization of the Cantor function*, Amer. Math. Monthly **98** (1991), no. 3, 255–258, DOI 10.2307/2325032. MR1093958

[81] G. Choquet, *Topology*, translated from the French by Amiel Feinstein, Pure and Applied Mathematics, Vol. XIX, Academic Press, New York-London, 1966. MR193605

[82] G. Choquet, *Outils topologiques et métriques de l'analyse mathématique* (French), Cours rédigé par Claude Mayer, Centre de Documentation Universitaire, Paris, 1969. MR262426

[83] G. Choquet, *Lectures on analysis. Vol. I: Integration and topological vector spaces*, edited by J. Marsden, T. Lance, and S. Gelbart, W. A. Benjamin, Inc., New York-Amsterdam, 1969. MR250011

[84] E. Christensen, C. Ivan, and M. L. Lapidus, *Dirac operators and spectral triples for some fractal sets built on curves*, Adv. Math. **217** (2008), no. 1, 42–78, DOI 10.1016/j.aim.2007.06.009. MR2357322

[85] Z. Ciesielski and S. J. Taylor, *First passage times and sojourn times for Brownian motion in space and the exact Hausdorff measure of the sample path*, Trans. Amer. Math. Soc. **103** (1962), 434–450, DOI 10.2307/1993838. MR143257

[86] F. Cipriani, D. Guido, T. Isola, and J.-L. Sauvageot, *Spectral triples for the Sierpinski gasket*, J. Funct. Anal. **266** (2014), no. 8, 4809–4869, DOI 10.1016/j.jfa.2014.02.013. MR3177323

[87] T. Cobler and M. L. Lapidus, *Towards a fractal cohomology: spectra of Polya-Hilbert operators, regularized determinants and Riemann zeros*, Exploring the Riemann zeta function, Springer, Cham, 2017, pp. 35–65. MR3700037

[88] T. Cobler and M. L. Lapidus, Zeta functions and Weierstrass' factorization theorem via regularized determinants and infinitesimal shifts on weighted Bergman spaces, (tentative title), in preparation, 2024.

[89] N. Cohen, Fractal antenna and fractal resonator primer, in: *Benoît Mandelbrot: A Life in Many Dimensions*, The Mandelbrot Memorial Volume (M. Frame and N. Cohen, eds.), Fractals and Dynamics in Mathematics, Science, and the Arts: Theory and Applications, vol. 1, World Scientific Publishing, Singapore and London, 2015, pp. 207-228.

[90] P. J. Cohen, *Set Theory and the Continuum Hypothesis*, Dover, Publications, Inc., Mineola, New York, 2008. (Reprinted from the original 1966 edition, W. A. Benjamin, Inc.)

[91] D. L. Cohn, *Measure theory*, Birkhäuser, Boston, MA, 1980. MR578344

[92] A. Connes, *Noncommutative geometry*, Academic Press, Inc., San Diego, CA, 1994. MR1303779

[93] J. B. Conway, *Functions of one complex variable*, 2nd ed., Graduate Texts in Mathematics, vol. 11, Springer-Verlag, New York-Berlin, 1978. MR503901

[94] W. A. Coppel, *An interesting Cantor set*, Amer. Math. Monthly **90** (1983), no. 7, 456–460, DOI 10.2307/2975723. MR711644

[95] A. Crannell, S. May, and L. Hilbert, *Shifts of finite type and Fibonacci Harps*, Appl. Math. Lett. **20** (2007), no. 2, 138–141, DOI 10.1016/j.aml.2006.03.007. MR2283900

[96] P. Cvitanović, *Circle maps: irrationally winding*, From number theory to physics (Les Houches, 1989), Springer, Berlin, 1992, pp. 631–658. MR1221112

[97] P. Cvitanović (ed.), *Universality in Chaos*, second edition (of the 1989 edition), Taylor and Francis, New York, 2017.

[98] G. Dafni, R. J. McCann, and A. Stancu (eds.), *Analysis and geometry of metric measure spaces*, Lecture notes from the 50th Séminaire de Mathématiques Supérieures (SMS) held at the Université de Montréal, Montréal, June 27–July 8, 2011, CRM Proceedings & Lecture Notes, vol. 56, American Mathematical Society, Providence, RI, 2013, DOI 10.1090/crmp/056. MR3075414

[99] G. Darboux, *Mémoire sur les fonctions discontinues* (French), Ann. Sci. École Norm. Sup. (2) **4** (1875), 57–112. MR1508624

[100] G. Darboux, *Addition au mémoire sur les fonctions discontinues* (French), Ann. Sci. École Norm. Sup. (2) **8** (1879), 195–202. MR1508677

[101] R. B. Darst, *Some Cantor sets and Cantor functions*, Math. Mag. **45** (1972), 2–7, DOI 10.2307/2688371. MR301153

[102] R. Darst, *The Hausdorff dimension of the nondifferentiability set of the Cantor function is* $[\ln(2)/\ln(3)]^2$, Proc. Amer. Math. Soc. **119** (1993), no. 1, 105–108, DOI 10.2307/2159830. MR1143222

[103] C. David, *Bypassing dynamical systems: a simple way to get the box-counting dimension of the graph of the Weierstrass function* (English, with English and Ukrainian summaries), Proc. Int. Geom. Cent. **11** (2018), no. 2, 53–68, DOI 10.15673/tmgc.v11i2.1028. MR3855920

[104] C. David, *On fractal properties of Weierstrass-type functions* (English, with English and Ukrainian summaries), Proc. Int. Geom. Cent. **12** (2019), no. 2, 43–61, DOI 10.15673/tmgc.v12i2.1485. MR4033034

[105] C. David, Wandering across the Weierstrass function, while revisiting its properties (survey article), in: *Analysis of PDE-Morningside Lecture Notes* (J.-Y. Chemin, P. Zhange and F.-H. Lin, eds.), **6**, 2021.

[106] C. David and M. L. Lapidus, Weierstrass fractal drums, I: A glimpse of complex dimensions, preprint, Sorbonne Université, Paris, France, 2022. (Also: e-print, URL:`https://hal.sorbonne-universite.fr/hal-03642326`, 2022; version 2, 2023.)

[107] C. David and M. L. Lapidus, Weierstrass fractal drums, II: Towards a fractal cohomology, *Math. Zeitschrift* **308** (2024) 35, 56 pp. URL:https://doi.org/10.1007/s00209-024-03547-z. (Also: e-print, URL:https://hal.archives-ouvertes.fr/hal-03758820/document, 2022; version 4, 2024.)

[108] C. David and M. L. Lapidus, Fractal complex dimensions and cohomology of the Weierstrass curve, in: *From Classical Analysis to Analysis on Fractals: A Tribute to Robert Strichartz* (P. Alonso Ruiz, M. Hinz, K. A. Okoudjou, L. G. Rogers and A. Teplyaev, eds.), Applied and Numerical Harmonic Analysis, vol. 2, Birkhäuser, Boston, in press, 2024. (Also: e-print, URL:https://hal.sorbonne-universite.fr/hal-03797595, 2022; version 2, 2023.)

[109] C. David and M. L. Lapidus, Iterated fractal drums—some new perspectives: Polyhedral measures, atomic decompositions and Morse theory in: *Fractal Geometry in Pure and Applied Mathematics* (Proc. Special Session, Joint Internat. Meeting of the AMS-SMF-EMS, Grenoble, France, July 2022), (H. Herichi, M. R. Lancia, A. Rozanova-Pierrat, T.-B. Landry, S. Winter, eds.), *Contemporary Mathematics*, Amer. Math. Soc., Providence, R. I., 2025, in press. (Also: e-print, URL:https://hal.sorbonne-universite.fr/hal-03946104v2, 2023.)

[110] C. David and M. L. Lapidus, Polyhedral neighborhoods vs. tubular neighborhoods: New insights for fractal zeta functions and complex dimensions, preprint, Sorbonne Université, Paris, France, 2023. (Also: e-print, `http://hal.science/hal-04153049`, version 2, 2023.)

[111] C. David and M. L. Lapidus, Understanding fractality: Polyhedral approach to the Koch curve and its complex dimensions, Asymptotic Analysis, in press, 2024. (Also: e-print, `http://hal.science/hal-04348346`, 2023.)

[112] C. David and M. L. Lapidus, New insights for fractal zeta functions: Polyhedral neighborhoods vs. tubular neighborhoods, preprint, 2023.

[113] C. David and M. L. Lapidus, Unveiling fractal cohomology: A Weierstrass approach, preprint, 2023.

[114] C. David and M. L. Lapidus, From Weierstrass to Riemann: The Frobenius pass—When fractal flows come into play with zeta functions, in: *Essays in Analytic Number Theory—in Honor of Helmut Maier's* 70th *Birthday* (J. Friedlander, C. Pomerance and M. Th. Rassias, eds.), Springer Nature, Cham, in press, 2024. (Also: e-print, URL: `https://hal.sorbonne-universite.fr/hal-04614665`.

[115] C. David and M. L. Lapidus, *Iterated Fractal Drums, Complex Dimensions and Fractal Cohomology—A Bridge Between the Discrete and the Continuous: The Key Examples of the Weierstrass and the Koch Snowflake Curves* (tentative title), research monograph, 2024, research monograph in preparation.

[116] C. David and M. L. Lapidus, *Advanced Fractal Geometry: Iterated Fractal Drums, Fractal Curves and Fractal Cohomology, with Applications*, 2024, textbook in preparation.

[117] C. David and M. L. Lapidus, *Towards a Fractal Calculus for the 21st Century*, 2024, textbook in preparation.

[118] C. David, M. L. Lapidus and M. Overduin, Fractional Taylor series: Conditions of existence, computation, and explicit formulas (tentative title), in preparation, 2024.

[119] C. David, M. L. Lapidus and G. Radunović, *A Continuation of Fractal Geometry: Complex Dimensions and Vibrations of Fractal Drums, with Applications*, 2024, textbook in preparation.

[120] C. David and G. Lebeau, *h-Laplacians on singular sets*, J. Fractal Geom. **10** (2023), no. 1-2, 61–108, DOI 10.4171/jfg/126. MR4627102

[121] G. David and S. Semmes, *Fractured fractals and broken dreams: Self-similar geometry through metric and measure*, Oxford Lecture Series in Mathematics and its Applications, vol. 7, The Clarendon Press, Oxford University Press, New York, 1997. MR1616732

[122] E. B. Davies, *Two-dimensional Riemannian manifolds with fractal boundaries*, J. London Math. Soc. (2) **49** (1994), no. 2, 343–356, DOI 10.1112/jlms/49.2.343. MR1260117

[123] B. V. Dekster, *The Jung theorem for spherical and hyperbolic spaces*, Acta Math. Hungar. **67** (1995), no. 4, 315–331, DOI 10.1007/BF01874495. MR1315815

[124] C.-J. de la Vallée Poussin, Recherches analytiques sur la théorie des nombres. Première partie: La fonction $\zeta(s)$ de Riemann et les nombres premiers en général ["Analytic investigations in number theory. First part: The Riemann zeta function $\zeta(s)$ and prime numbers in general"], *Ann. Soc. Sci. Bruxelles Sér. I* **20** (1896), 183–256 [French].

[125] C.-J. de la Vallée Poussin, Sur la fonction $\zeta(s)$ de Riemann et le nombre des nombres premiers inférieurs à une limite donnée ["On the Riemann zeta function $\zeta(s)$ and the number of prime numbers less than a given quantity"], *Mém. Couronnés et Autres Mém. Publ. Acad. Roy. Sci., des Lettres, Beaux-Arts Belg.* **59** (1899–1900) [French].

[126] B. Demír, A. Deníz, Ş. Koçak, and A. E. Üreyen, *Tube formulas for graph-directed fractals*, Fractals **18** (2010), no. 3, 349–361, DOI 10.1142/S0218348X10004919. MR2683935

[127] A. Deniz, Ş. Koçak, Y. Özdemir, and A. E. Üreyen, *Tube formula for self-similar fractals with non-Steiner-like generators*, Proceedings of the Gökova Geometry-Topology Conference 2012, Int. Press, Somerville, MA, 2013, pp. 123–145. MR3203360

[128] A. Deniz, Ş. Koçak, Y. Özdemir, and A. E. Üreyen, *Tube volumes via functional equations*, J. Geom. **106** (2015), no. 1, 153–162, DOI 10.1007/s00022-014-0241-3. MR3320885

[129] A. Deniz, M. Ş. Koçak, Y. Özdemir, A. Ratiu, and A. E. Üreyen, *On the Minkowski measurability of self-similar fractals in* $\mathbb{R}^d$, Turkish J. Math. **37** (2013), no. 5, 830–846, DOI 10.3906/mat-1103-20. MR3105496

[130] L. Denzer, B. Grünbaum, and V. Klee, Helly's theorem and its relatives, in: *Convexity*, Proceedings of Symposia in Pure Mathematics, Amer. Math. Soc., Providence, R. I., pp. 101–190.

[131] G. Derfel, P. J. Grabner, and F. Vogl, *The zeta function of the Laplacian on certain fractals*, Trans. Amer. Math. Soc. **360** (2008), no. 2, 881–897, DOI 10.1090/S0002-9947-07-04240-7. MR2346475

[132] G. de Rham, *Sur une courbe plane* (French), J. Math. Pures Appl. (9) **35** (1956), 25–42. MR90628

[133] G. de Rham, *Sur quelques courbes definies par des equations fonctionnelles* (French), Univ. e Politec. Torino Rend. Sem. Mat. **16** (1956/57), 101–113. MR95227

[134] G. de Rham, *Sur un exemple de fonction continue sans dérivée* (French), Enseign. Math. (2) **3** (1957), 71–72. MR85314

[135] R. de Santiago, *The Recursive Structure of Lattice Strings*, Master's Thesis, Mathematics, California State Polytechnic University, Pomona, United States, 2012.

[136] R. de Santiago, M. L. Lapidus, S. A. Roby and J. A. Rock, Multifractal analysis via scaling zeta functions and recursive structure of lattice strings, in: *Fractal Geometry and Dynamical Systems in Pure and Applied Mathematics I: Fractals in Pure Mathematics* (D. Carfi, M. L. Lapidus, E. P. J. Pearse and M. van Frankenhuijsen, eds.), Contemporary Mathematics, vol. 600, Amer. Math. Soc., Providence, R. I., 2013, 205–238. dx.doi.org/10.1090/conm/600/11930. (Also: E-print, arXiv:1207.6680v3 [math-ph], 2013; IHES preprint, IHES/M/12/19, 2013.)

[137] K. Dettmers, R. Giza, R. Morales, J. A. Rock, and C. Knox, *A survey of complex dimensions, measurability, and the lattice/nonlattice dichotomy*, Discrete Contin. Dyn. Syst. Ser. S **10** (2017), no. 2, 213–240, DOI 10.3934/dcdss.2017011. MR3600644

[138] R. L. Devaney, *An introduction to chaotic dynamical systems*, reprint of the second (1989) edition, Studies in Nonlinearity, Westview Press, Boulder, CO, 2003. MR1979140

[139] R. L. Devaney and L. Keen (eds.), *Chaos and fractals: The mathematics behind the computer graphics*, Lecture notes prepared for the American Mathematical Society Short Course held in Providence, Rhode Island, August 6–7, 1988; AMS Short Course Lecture Notes, Proceedings of Symposia in Applied Mathematics, vol. 39, American Mathematical Society, Providence, RI, 1989, DOI 10.1090/psapm/039. MR1010232

[140] U. Dini, *Su alcune funzioni che in tutto un intervallo non hanno mai derivata ["On a function having no derivative on an entire interval"]*, Annali di Matematica **8** (1877), 122–137 [Italian].

[141] U. Dini, *Fondamenti per la teorica delle funzione di variabili reali* ["Fundamental Principles of the Theory of Functions of a Real Variable"], Tipografia T. Nistri e. C., Pisa, 1878 [Italian].

[142] A. Douady, *Systèmes dynamiques holomorphes* (French), Bourbaki seminar, Vol. 1982/83, Astérisque, vol. 105, Soc. Math. France, Paris, 1983, pp. 39–63. MR728980

[143] A. Douady and J. H. Hubbard, *On the dynamics of polynomial-like mappings*, Ann. Sci. École Norm. Sup. (4) **18** (1985), no. 2, 287–343. MR816367

[144] A. Douady and J. H. Hubbard, *Exploring the Mandelbrot set. The Orsay Notes*, Univ. Paris-Sud, Orsay, France, 2009.

[145] O. Dovgoshey, O. Martio, V. Ryazanov, and M. Vuorinen, *The Cantor function*, Expo. Math. **24** (2006), no. 1, 1–37, DOI 10.1016/j.exmath.2005.05.002. MR2195181

[146] P. G. Drazin, *Fractals*, in: L. F. Richardson, *The Collected Works of Lewis Fry Richardson*, vol. 1, Cambridge Univ. Press, Cambridge, 1993, pp. 45–46.

[147] E. Dubon and J. M. Sepulcre, *On the complex dimensions of nonlattice fractal strings in connection with Dirichlet polynomials*, Exp. Math. **23** (2014), no. 1, 13–24, DOI 10.1080/10586458.2013.853630. MR3177454

[148] P. Dugnac, *Richard Dedekind et les fondements des mathématiques* ["Richard Dedekind and the foundations of mathematics"], No. 24, Vrin, 1976.

[149] B. Duplantier, *Conformal fractal geometry & boundary quantum gravity*, Fractal geometry and applications: a jubilee of Benoît Mandelbrot, Part 2, Proc. Sympos. Pure Math., vol. 72, Amer. Math. Soc., Providence, RI, 2004, pp. 365–482, DOI 10.1090/pspum/072.2/2112128. MR2112128

[150] B. Duplantier and S. Sheffield, *Liouville quantum gravity and KPZ*, Invent. Math. **185** (2011), no. 2, 333–393, DOI 10.1007/s00222-010-0308-1. MR2819163

[151] B. Duplantier, R. Rhodes, S. Sheffield, and V. Vargas, *Log-correlated Gaussian fields: an overview*, Geometry, analysis and probability, Progr. Math., vol. 310, Birkhäuser/Springer, Cham, 2017, pp. 191–216, DOI 10.1007/978-3-319-49638-2_9. MR3821928

[152] G. A. Edgar (ed.), *Classics on Fractals*, Studies in Nonlinearity, Addison-Wesley, Reading, 1993.

[153] G. Edgar, *Measure, topology, and fractal geometry*, 2nd ed., Undergraduate Texts in Mathematics, Springer, New York, 2008, DOI 10.1007/978-0-387-74749-1. MR2356043

[154] G. A. Edgar, *Fractals and universal spaces in dimension theory [book review of MR2460244]*, Bull. Amer. Math. Soc. (N.S.) **47** (2010), no. 1, 163–170, DOI 10.1090/S0273-0979-09-01263-4. MR2566450

[155] G. A. Edgar and R. D. Mauldin, *Multifractal decompositions of digraph recursive fractals*, Proc. London Math. Soc. (3) **65** (1992), no. 3, 604–628, DOI 10.1112/plms/s3-65.3.604. MR1182103

[156] H. M. Edwards, *Riemann's zeta function*, reprint of the 1974 original [Academic Press, New York; MR0466039 (57 #5922)], Dover Publications, Inc., Mineola, NY, 2001. MR1854455

[157] H. G. Eggleston, *The fractional dimension of a set defined by decimal properties*, Quart. J. Math. Oxford Ser. **20** (1949), 31–36, DOI 10.1093/qmath/os-20.1.31. MR31026

[158] T. Eichinger and S. Winter, *Regularly varying functions, generalized contents, and the spectrum of fractal strings*, Horizons of fractal geometry and complex dimensions, Contemp. Math., vol. 731, Amer. Math. Soc., Providence, RI, 2019, pp. 63–94, DOI 10.1090/conm/731/14673. MR3989816

[159] A. Einstein, *Investigations on the theory of the Brownian movement*, edited with notes by R. Fürth, translated by A. D. Cowper, Dover Publications, Inc., New York, 1956. MR77443

[160] K. E. Ellis, M. L. Lapidus, M. C. Mackenzie and J. A. Rock, Partition zeta functions, multifractal spectra, and tapestries of complex dimensions, in: *Benoit Mandelbrot: A Life in Many Dimensions*, The Mandelbrot Memorial Volume (M. Frame and N. Cohen, eds.), Fractals and Dynamics in Mathematics, Science, and the Arts: Theory and Applications, vol. 1, World Scientific Publishing, Singapore and London, 2015, pp. 267–312. (Also: e-print, `arXiv:1007.1467v2 [math-ph]`, 2011; IHES preprint, IHES/M/12/15, 2012.)

[161] R. Engelking, *Dimension theory*, translated from the Polish and revised by the author, North-Holland Mathematical Library, vol. 19, North-Holland Publishing Co., Amsterdam-Oxford-New York; PWN—Polish Scientific Publishers, Warsaw, 1978. MR482697

[162] D. Essouabri, *Singularité des séries de Dirichlet associées à des polynômes de plusieurs variables et applications en théorie analytique des nombres* (French, with English summary), Ann. Inst. Fourier (Grenoble) **47** (1997), no. 2, 429–483, DOI 10.5802/aif.1570. MR1450422

[163] D. Essouabri, *Zeta functions associated to Pascal's triangle mod p*, Japan. J. Math. (N.S.) **31** (2005), no. 1, 157–174, DOI 10.4099/math1924.31.157. MR2153729

[164] D. Essouabri and B. Lichtin, *Zeta functions of discrete self-similar sets*, Adv. Math. **232** (2013), 142–187, DOI 10.1016/j.aim.2012.09.011. MR2989980

[165] D. Essouabri and B. Lichtin, *k-point configurations of discrete self-similar sets*, Fractal geometry and dynamical systems in pure and applied mathematics. I. Fractals in pure mathematics, Contemp. Math., vol. 600, Amer. Math. Soc., Providence, RI, 2013, pp. 21–50, DOI 10.1090/conm/600/11947. MR3203398

[166] D. Essouabri and B. Lichtin, *Thick self similar sets are asymptotically generic*, Ann. Acad. Sci. Fenn. Math. **43** (2018), no. 2, 835–857, DOI 10.5186/aasfm.2018.4349. MR3839839

[167] L. C. Evans and R. F. Gariepy, *Measure theory and fine properties of functions*, revised edition, Textbooks in Mathematics, CRC Press, Boca Raton, FL, 2015. MR3409135

[168] W. D. Evans and D. J. Harris, *Fractals, trees and the Neumann Laplacian*, Math. Ann. **296** (1993), no. 3, 493–527, DOI 10.1007/BF01445117. MR1225988

[169] C. J. G. Evertsz and B. B. Mandelbrot, Multifractal Measures, Appendix B of [**483**], pp. 921–953.

[170] C. J. G. Evertsz, H.-O. Peitgen and R. F. Voss (eds.), *Fractal Geometry and Analysis: The Mandelbrot Fetschrift*, World Scientific Publishing, Singapore and London, 1996. (Reprinted in the special issue of the journal *Fractals* in honor of Benoît B. Mandelbrot's 70th birthday: *Fractals* No. 4, **3** (1995).)

[171] K. J. Falconer, *The geometry of fractal sets*, Cambridge Tracts in Mathematics, vol. 85, Cambridge University Press, Cambridge, 1986. MR867284

[172] K. Falconer, *Fractal geometry: Mathematical foundations and applications*, 3rd ed., John Wiley & Sons, Ltd., Chichester, 2014. MR3236784

[173] K. J. Falconer, *The Hausdorff dimension of self-affine fractals*, Math. Proc. Cambridge Philos. Soc. **103** (1988), no. 2, 339–350, DOI 10.1017/S0305004100064926. MR923687

[174] K. J. Falconer, *On the Minkowski measurability of fractals*, Proc. Amer. Math. Soc. **123** (1995), no. 4, 1115–1124, DOI 10.2307/2160708. MR1224615

[175] K. Falconer, *Techniques in fractal geometry*, John Wiley & Sons, Ltd., Chichester, 1997. MR1449135

[176] K. Falconer, *Fractals*, Very Short Introductions, Oxford University Press, Oxford, 2013, DOI 10.1093/actrade/9780199675982.001.0001. MR3154938

[177] F. Family and T. Vicsek (eds.), *Dynamics of fractal surfaces*, World Scientific Publishing Co., Inc., River Edge, NJ, 1991, DOI 10.1142/1452. MR1171480

[178] P. Fatou, *Sur les équations fonctionnelles, ["On functional equations"], I–III*, Bull. Soc. Math. France; I: **47** (1919), 161–271. II: **48** (1920), 33–94. III: **48** (1920), 208–314 [French].

[179] P. Fatou, *Sur l'itération des fonctions transcendantes Entières* (French), Acta Math. **47** (1926), no. 4, 337–370, DOI 10.1007/BF02559517. MR1555220

[180] J. Feder, *Fractals*, with a foreword by Benoit B. Mandelbrot, Physics of Solids and Liquids, Plenum Press, New York, 1988, DOI 10.1007/978-1-4899-2124-6. MR949210

[181] H. Federer, *Curvature measures*, Trans. Amer. Math. Soc. **93** (1959), 418–491, DOI 10.2307/1993504. MR110078

[182] H. Federer, *Geometric measure theory*, Die Grundlehren der mathematischen Wissenschaften, Band 153, Springer-Verlag New York, Inc., New York, 1969. MR257325

[183] H. Federer, *Colloquium lectures on geometric measure theory*, Bull. Amer. Math. Soc. **84** (1978), no. 3, 291–338, DOI 10.1090/S0002-9904-1978-14462-0. MR467473

[184] V. V. Fedorchuk, *The Fundamentals of Dimension Theory*, North Holland, New York, 1978.

[185] M. J. Feigenbaum, *Universal behavior in nonlinear systems: Order in chaos (Los Alamos, N.M., 1982)*, Phys. D **7** (1983), no. 1-3, 16–39, DOI 10.1016/0167-2789(83)90112-4. MR719045

[186] W. Feller, *An introduction to probability theory and its applications. Vol. II*, John Wiley & Sons, Inc., New York-London-Sydney, 1966. MR210154

[187] D.-J. Feng and K.-S. Lau, *Multifractal formalism for self-similar measures with weak separation condition* (English, with English and French summaries), J. Math. Pures Appl. (9) **92** (2009), no. 4, 407–428, DOI 10.1016/j.matpur.2009.05.009. MR2569186

[188] J. F. Fleron, *A note on the history of the Cantor set and Cantor function*, Math. Mag. **67** (1994), no. 2, 136–140, DOI 10.2307/2690689. MR1272828

[189] P. M. Fitzpatrick and H. L. Royden, *Real Analysis*, 4th ed., Pearson, 2010.

[190] J. Fleckinger and M. L. Lapidus, *Remainder estimates for the asymptotics of elliptic eigenvalue problems with indefinite weights*, Arch. Rational Mech. Anal. **98** (1987), no. 4, 329–356, DOI 10.1007/BF00276913. MR872751

[191] J. Fleckinger-Pellé and D. G. Vassiliev, *An example of a two-term asymptotics for the "counting function" of a fractal drum*, Trans. Amer. Math. Soc. **337** (1993), no. 1, 99–116, DOI 10.2307/2154311. MR1176086

[192] M. Fleischmann, D. J. Tildesley, and R. C. Ball (eds.), *Fractals in the natural sciences*, reprint of Proc. Roy. Soc. London Ser. A **423** (1989), no. 1864 [MR1059261 (90m:00020)], Princeton University Press, Princeton, NJ, 1990, DOI 10.1515/9781400861040. MR1124853

[193] G. B. Folland, *Real analysis: Modern techniques and their applications*, 2nd ed., A Wiley-Interscience Publication, Pure and Applied Mathematics (New York), John Wiley & Sons, Inc., New York, 1999. MR1681462

[194] M. L. Frame and N. Cohen (eds.), *Benoît Mandelbrot: A Life in Many Dimensions*, The Mandelbrot Memorial Volume, Fractals and Dynamics in Mathematics, Science, and the Arts: Theory and Applications, vol. 1, World Scientific Publishing, Singapore and London, 2015.

[195] M. L. Frame and B. B. Mandelbrot (eds.), *Fractals, Graphics and Mathematics*, MAA Notes. vol. 58, Mathematical Association of America, Washington D. C., 2002.

[196] M. Frantz, *Lacunarity, Minkowski content, and self-similar sets in* $\mathbb{R}$, Fractal geometry and applications: a jubilee of Benoît Mandelbrot. Part 1, Proc. Sympos. Pure Math., vol. 72, Amer. Math. Soc., Providence, RI, 2004, pp. 77–91, DOI 10.1090/pspum/072.1/2112101. MR2112101

[197] J. M. Fraser, *Inhomogeneous self-similar sets and box dimensions*, Studia Math. **213** (2012), no. 2, 133–156, DOI 10.4064/sm213-2-2. MR3024316

[198] J. M. Fraser, *Inhomogeneous self-affine carpets*, Indiana Univ. Math. J. **65** (2016), no. 5, 1547–1566, DOI 10.1512/iumj.2016.65.5903. MR3571439

[199] J. M. Fraser, *Assouad dimension and fractal geometry*, Cambridge Tracts in Mathematics, vol. 222, Cambridge University Press, Cambridge, 2021, DOI 10.1017/9781108778459. MR4411274

[200] J. M. Fraser, *Fractal geometry of Bedford-McMullen carpets*, Thermodynamic formalism, Lecture Notes in Math., vol. 2290, Springer, Cham, 2021, pp. 495–516, DOI 10.1007/978-3-030-74863-0_15. MR4436835

[201] M. Fréchet, Des familles et fonctions additives d'ensembles abstraits [On Families and Additives Functions of Abstracts Sets], *Fund. Math.* **5** (1924), 206–251 [French].

[202] U. Freiberg and S. Kombrink, *Minkowski content and local Minkowski content for a class of self-conformal sets*, Geom. Dedicata **159** (2012), 307–325, DOI 10.1007/s10711-011-9661-5. MR2944534

[203] U. Frisch and G. Parisi, Fully developed turbulence and intermittency, in: Proc. Internat. Summer School Phys. Enrico Fermi, North-Holland, Amsterdam, 1985, 84–88.

[204] J. Fröhlich (ed.), *Scaling and self-similarity in physics: Renormalization in statistical mechanics and dynamics*, Lectures presented at the seminar held at the Institut des Hautes Études Scientifiques, Bures-sur-Yvette, 1981/1982, Progress in Physics, vol. 7, Birkhäuser Boston, Inc., Boston, MA, 1983, DOI 10.1007/978-1-4899-6762-6. MR733467

[205] O. A. Frostman, *Potentiel d'équilibre et capacité des ensembles avec quelques applications à la théorie des fonctions ["Equilibrium potential and capacity of sets, with several applications to the theory of functions"]*, Meddel Lund Univ. Math. Sem. **3** (1935), 1–118. (See also: Ph.D. Dissertation, Mathematics, Lund University, Sweden, 1935.)

[206] Y. Fujita, N. Hamamuki, A. Siconolfi, and N. Yamaguchi, *A class of nowhere differentiable functions satisfying some concavity-type estimate*, Acta Math. Hungar. **160** (2020), no. 2, 343–359, DOI 10.1007/s10474-019-01007-3. MR4075406

[207] M. Fukushima and T. Shima, *On a spectral analysis for the Sierpiński gasket*, Potential Anal. **1** (1992), no. 1, 1–35, DOI 10.1007/BF00249784. MR1245223

[208] H. Furstenberg, *Ergodic theory and fractal geometry*, CBMS Regional Conference Series in Mathematics, vol. 120, American Mathematical Society, Providence, RI, 2014, DOI 10.1090/cbms/120. MR3235463

[209] D. Gatzouras, *Lacunarity of self-similar and stochastically self-similar sets*, Trans. Amer. Math. Soc. **352** (2000), no. 5, 1953–1983, DOI 10.1090/S0002-9947-99-02539-8. MR1694290

[210] J. Gerling, *Untersuchungen zur Theorie von Weyl–Berry–Lapidus* ["Investigations in the Weyl–Berry–Lapidus Theory"], Graduate Thesis (diplomarbeit), Physics, Universität Osnabrück, Germany, 1992 [German].

[211] J. Gerling and H.-J. Schmidt, *Self-similar drums and generalized Weierstrass functions: Fractals and disorder (Hamburg, 1992)*, Phys. A **191** (1992), no. 1-4, 536–539, DOI 10.1016/0378-4371(92)90578-E. MR1199397

[212] J. Gerling and H.-J. Schmidt, *Three-term asymptotics of the spectrum of self-similar fractal drums*, J. Math. Sci. Univ. Tokyo **6** (1999), no. 1, 101–126. MR1683321

[213] J. Gerver, *The differentiability of the Riemann function at certain rational multiples of π*, Proc. Nat. Acad. Sci. U.S.A. **62** (1969), 668–670, DOI 10.1073/pnas.62.3.668. MR268336

[214] J. Gerver, *More on the differentiability of the Riemann function*, Amer. J. Math. **93** (1971), 33–41, DOI 10.2307/2373445. MR276422

[215] J. Gleick, *Chaos: Making a new science*, Penguin Books, New York, 1987. MR1010647

[216] E. A. Gorin and B. N. Kukushkin, *Integrals associated with the Cantor staircase* (Russian, with Russian summary), Algebra i Analiz **15** (2003), no. 3, 188–220, DOI 10.1090/S1061-0022-04-00817-9; English transl., St. Petersburg Math. J. **15** (2004), no. 3, 449–468. MR2052168

[217] J.-F. Gouyet, *Physics and fractal structures*, with a foreword by Benoît Mandelbrot, translated from the French by David Corfield, Masson, Paris; Springer-Verlag, New York, 1996. MR1390993

[218] S. Graf, R. D. Mauldin, and S. C. Williams, *The exact Hausdorff dimension in random recursive constructions*, Mem. Amer. Math. Soc. **71** (1988), no. 381, x+121, DOI 10.1090/memo/0381. MR920961

[219] J. Granholm, *Remarkable Curves in the Euclidean Plane*, Thesis (Mathematics), Linköping University, Sweden, 2014.

[220] A. Gray, *Tubes*, 2nd ed., with a preface by Vicente Miquel, Progress in Mathematics, vol. 221, Birkhäuser Verlag, Basel, 2004, DOI 10.1007/978-3-0348-7966-8. MR2024928

[221] D. Guido, T. Isola, and M. L. Lapidus, *A trace on fractal graphs and the Ihara zeta function*, Trans. Amer. Math. Soc. **361** (2009), no. 6, 3041–3070, DOI 10.1090/S0002-9947-08-04702-8. MR2485417

[222] J. Hadamard, *Étude sur les propriétés des fonctions entières et en particulier d'une fonction considérée par Riemann* [*"Study of the properties of entire functions and, in particular, of a function considered by Riemann"*], J. Math. Pures Appl. (4) **9** (1893), 171–215 [French]. (Reprinted in [**224**, pp. 103–147].)

[223] J. Hadamard, *Sur la distribution des zéros de la fonction $\zeta(s)$ et ses conséquences arithmétiques* (French), Bull. Soc. Math. France **24** (1896), 199–220. MR1504264

[224] J. Hadamard, *Oeuvres de Jacques Hadamard* ["Collected Works of Jacques Hadamard"], Tome I, Editions du Centre National de la Recherche Scientifique, Paris, 1968 [French].

[225] B. M. Hambly, *Brownian motion on a random recursive Sierpinski gasket*, Ann. Probab. **25** (1997), no. 3, 1059–1102, DOI 10.1214/aop/1024404506. MR1457612

[226] B. M. Hambly, *On the asymptotics of the eigenvalue counting function for random recursive Sierpinski gaskets*, Probab. Theory Related Fields **117** (2000), no. 2, 221–247, DOI 10.1007/s004400050005. MR1771662

[227] B. M. Hambly and M. L. Lapidus, *Random fractal strings: their zeta functions, complex dimensions and spectral asymptotics*, Trans. Amer. Math. Soc. **358** (2006), no. 1, 285–314, DOI 10.1090/S0002-9947-05-03646-9. MR2171234

[228] G. H. Hardy, *Weierstrass's non-differentiable function*, Trans. Amer. Math. Soc. **17** (1916), no. 3, 301–325, DOI 10.2307/1989005. MR1501044

[229] G. H. Hardy and J. E. Littlewood, *Some problems of diophantine approximation*, Acta Math. **37** (1914), no. 1, 193–239, DOI 10.1007/BF02401834. MR1555099

[230] A. Harrison, *Fractals in Chemistry*, Oxford Chemistry Primers, Oxford Science Publications, Oxford Univ. Press, Oxford and New York, 1995.

[231] M. Hata, *On some properties of set-dynamical systems*, Proc. Japan Acad. Ser. A Math. Sci. **61** (1985), no. 4, 99–102. MR796477

[232] M. Hata, *Singularities of the Weierstrass type functions*, J. Analyse Math. **51** (1988), 62–90, DOI 10.1007/BF02791119. MR963150

[233] M. Hata and M. Yamaguti, *The Takagi function and its generalization*, Japan J. Appl. Math. **1** (1984), no. 1, 183–199, DOI 10.1007/BF03167867. MR839313

[234] F. Hausdorff, *Dimension und äußeres Maß* (German), Math. Ann. **79** (1918), no. 1-2, 157–179, DOI 10.1007/BF01457179. MR1511917

[235] C. Q. He and M. L. Lapidus, *Generalized Minkowski content, spectrum of fractal drums, fractal strings and the Riemann zeta-function*, Mem. Amer. Math. Soc. **127** (1997), no. 608, x+97, DOI 10.1090/memo/0608. MR1376743

[236] A. Heck and J. M. Perdang (eds.), *Applying fractals in astronomy*, Lecture Notes in Physics. New Series m: Monographs, vol. 3, Springer-Verlag, Berlin, 1991, DOI 10.1007/978-3-540-47582-8. MR1180331

[237] M. Hénon, *A two-dimensional mapping with a strange attractor*, Comm. Math. Phys. **50** (1976), no. 1, 69–77. MR422932

[238] H. Herichi and M. L. Lapidus, *Quantized number theory, fractal strings and the Riemann hypothesis—from spectral operators to phase transitions and universality*, Fractals and Dynamics in Mathematics, Science, and the Arts: Theory and Applications, vol. 4, research monograph, World Scientific Publishing Co. Pte. Ltd., Hackensack, NJ, 2021. MR4365930

[239] H. Herichi and M. L. Lapidus, Riemann zeros and phase transitions via the spectral operator on fractal strings, *J. Phys. A: Math. Theor.* **45** (2012), 374005, 23pp. (Special issue in honor of Prof. Stewart Dowker's 75th birthday.)(Also: e-print, `arXiv:1203.4828v2 [math-ph]`, 2012; IHES preprint, IHES/M/12/09, 2012.)

[240] H. Herichi and M. L. Lapidus, *Fractal complex dimensions, Riemann hypothesis and invertibility of the spectral operator*, Fractal geometry and dynamical systems in pure and applied mathematics. I. Fractals in pure mathematics, Contemp. Math., vol. 600, Amer. Math. Soc., Providence, RI, 2013, pp. 51–89, DOI 10.1090/conm/600/11948. MR3203399

[241] H. Herichi and M. L. Lapidus, *Truncated infinitesimal shifts, spectral operators and quantized universality of the Riemann zeta function* (English, with English and French summaries), Ann. Fac. Sci. Toulouse Math. (6) **23** (2014), no. 3, 621–664, DOI 10.5802/afst.1419. MR3266708

[242] H. Herichi and M. L. Lapidus, Spectral operators, infinitesimal shifts and a quantization of Dirichlet convolution and Möbius inversion (tentative title), in preparation, 2024.

[243] C. Hermite and T. J. Stieltjes, *Correspondance d'Hermite et Stieltjes*, tome II, Gauthier-Villars, Paris, 1905 [French]. (See the letter of C. Hermite to T. J. Stieltjes, letter 374, written in 1893.)

[244] M. Hervé and J. Dixmier (eds.), *Oeuvres de Gaston Julia* ["Collected Works of Gaston Julia"], Gauthier-Villars, Paris, 1968–1970 [French]. (See, especially, vol. I, pp. 121–319.)

[245] Y. Heurteaux, *Weierstrass functions with random phases*, Trans. Amer. Math. Soc. **355** (2003), no. 8, 3065–3077, DOI 10.1090/S0002-9947-03-03221-5. MR1974675

[246] Y. Heurteaux, *Weierstrass functions in Zygmund's class*, Proc. Amer. Math. Soc. **133** (2005), no. 9, 2711–2720, DOI 10.1090/S0002-9939-05-07857-3. MR2146218

[247] D. Hilbert, *Ueber die stetige Abbildung einer Line auf ein Flächenstück* (German), Math. Ann. **38** (1891), no. 3, 459–460, DOI 10.1007/BF01199431. MR1510683

[248] T. H. Hildebrandt, *Questions, Discussions, and Notes: A Simple Continuous Function with a Finite Derivative at No Point*, Amer. Math. Monthly **40** (1933), no. 9, 547–548, DOI 10.2307/2301999. MR1522922

[249] E. Hille and J. D. Tamarkin, *Remarks on a Known Example of a Monotone Continuous Function*, Amer. Math. Monthly **36** (1929), no. 5, 255–264, DOI 10.2307/2298506. MR1521732

[250] A. M. Hinz, S. Klavžar, and C. Petr, *The tower of Hanoi—myths and maths*, second edition of [MR3026271], with a foreword by Ian Stewart, Birkhäuser/Springer, Cham, 2018, DOI 10.1007/978-3-319-73779-9. MR3791459

[251] A. Hinkis, *Proofs of the Cantor-Bernstein theorem: A mathematical excursion*, Science Networks. Historical Studies, vol. 45, Birkhäuser/Springer, Heidelberg, 2013, DOI 10.1007/978-3-0348-0224-6. MR3026479

[252] K. Hockett and P. Holmes, *Bifurcation to badly ordered orbits in one-parameter families of circle maps, or angels fallen from the devil's staircase*, Proc. Amer. Math. Soc. **102** (1988), no. 4, 1031–1051, DOI 10.2307/2047355. MR934888

[253] M. Holschneider and P. Tchamitchian, *Pointwise analysis of Riemann's "nondifferentiable" function*, Invent. Math. **105** (1991), no. 1, 157–175, DOI 10.1007/BF01232261. MR1109624

[254] T. Y. Hu and K.-S. Lau, *Fractal dimensions and singularities of the Weierstrass type functions*, Trans. Amer. Math. Soc. **335** (1993), no. 2, 649–665, DOI 10.2307/2154398. MR1076614

[255] J. H. Hubbard, *Local connectivity of Julia sets and bifurcation loci: three theorems of J.-C. Yoccoz*, Topological methods in modern mathematics (Stony Brook, NY, 1991), Publish or Perish, Houston, TX, 1993, pp. 467–511. MR1215974

[256] D. Hug, G. Last, and W. Weil, *A local Steiner-type formula for general closed sets and applications*, Math. Z. **246** (2004), no. 1-2, 237–272, DOI 10.1007/s00209-003-0597-9. MR2031455

[257] B. R. Hunt, *The prevalence of continuous nowhere differentiable functions*, Proc. Amer. Math. Soc. **122** (1994), no. 3, 711–717, DOI 10.2307/2160745. MR1260170

[258] B. R. Hunt, *The Hausdorff dimension of graphs of Weierstrass functions*, Proc. Amer. Math. Soc. **126** (1998), no. 3, 791–800, DOI 10.1090/S0002-9939-98-04387-1. MR1452806

[259] B. R. Hunt and J. A. Yorke, *Smooth dynamics on Weierstrass nowhere differentiable curves*, Trans. Amer. Math. Soc. **325** (1991), no. 1, 141–154, DOI 10.2307/2001663. MR1040043

[260] W. Hurewicz and H. Wallman, *Dimension Theory*, Princeton Mathematical Series, vol. 4, Princeton Univ. Press, Princeton, 1968. (Originally published in 1941.)

[261] J. E. Hutchinson, *Fractals and self-similarity*, Indiana Univ. Math. J. **30** (1981), no. 5, 713–747, DOI 10.1512/iumj.1981.30.30055. MR625600

[262] A. E. Ingham, *The distribution of prime numbers*, reprint of the 1932 original, with a foreword by R. C. Vaughan, Cambridge Mathematical Library, Cambridge University Press, Cambridge, 1990. MR1074573

[263] I. Ivanšić and U. Milutinović, *A universal separable metric space based on the triangular Sierpiński curve: In memory of T. Benny Rushing*, Topology Appl. **120** (2002), no. 1-2, 237–271, DOI 10.1016/S0166-8641(01)00017-7. MR1895494

[264] A. Ivić, *The Riemann zeta-function: The theory of the Riemann zeta-function with applications*, A Wiley-Interscience Publication, John Wiley & Sons, Inc., New York, 1985. MR792089

[265] S. Jaffard, *The spectrum of singularities of Riemann's function*, Rev. Mat. Iberoamericana **12** (1996), no. 2, 441–460, DOI 10.4171/RMI/203. MR1402673

[266] S. Jaffard, *Multifractal formalism for functions*, SIAM J. Math. Anal. **28** (1997), 994–998.

[267] S. Jaffard, *Oscillation spaces: properties and applications to fractal and multifractral functions*, J. Math. Phys. **39** (1998), no. 8, 4129–4141, DOI 10.1063/1.532488. MR1633152

[268] S. Jaffard, *The multifractal nature of Lévy processes*, Probab. Theory Related Fields **114** (1999), no. 2, 207–227, DOI 10.1007/s004400050224. MR1701520

[269] S. Jaffard, *Wavelet techniques in multifractal analysis*, Fractal geometry and applications: a jubilee of Benoît Mandelbrot, Part 2, Proc. Sympos. Pure Math., vol. 72, Amer. Math. Soc., Providence, RI, 2004, pp. 91–151, DOI 10.1090/pspum/072.2/2112122. MR2112122

[270] S. Jaffard and Y. Meyer, *Wavelet methods for pointwise regularity and local oscillations of functions*, Mem. Amer. Math. Soc. **123** (1996), no. 587, x+110, DOI 10.1090/memo/0587. MR1342019

[271] M. Jarnicki and P. Pflug, *Continuous nowhere differentiable functions: The monsters of analysis*, Springer Monographs in Mathematics, Springer, Cham, 2015, DOI 10.1007/978-3-319-12670-8. MR3444902

[272] M. Jašek, *Funkce Bolzanova* [*"Bolzano's function"*], Časopis pro Pěstovani Matenatiky a Fyziky ["Journal for the Cultivation of Mathematics and Physics"], No. 4, **51** (1922), 248–264 [Czech].

[273] T. J. Jech, *The axiom of choice*, Studies in Logic and the Foundations of Mathematics, Vol. 75, North-Holland Publishing Co., Amsterdam-London; American Elsevier Publishing Co., Inc., New York, 1973. MR396271

[274] G. W. Johnson and M. L. Lapidus, *The Feynman Integral and Feynman's Operational Calculus*, Oxford Mathematical Monographs, Oxford Science Publications, Oxford Univ. Press, Oxford and New York, 2000. (Corrected printing and paperback edition, 2002.)

[275] P. W. Jones, *Quasiconformal mappings and extendability of functions in Sobolev spaces*, Acta Math. **147** (1981), no. 1-2, 71–88, DOI 10.1007/BF02392869. MR631089

[276] C. Jordan, *Remarques sur les intégrales définines* [*"Remarks on definite integrals"*], J. Math. Pure Appl. **8** (1892), 69–99.

[277] G. Julia, *Mémoire sur l'itération des fonctions rationnelles* [*"Memoir on the iterational of rational functions"*], J. Math. Pure Appl. Tome 1, **8** (1918), 47-245. [French]. (See also [**244**, vol. I], pp. 121–319.)

[278] H. Jung, *Ueber die kleinste Kugel, die eine räumliche Figur einschliesst* (German), J. Reine Angew. Math. **123** (1901), 241–257, DOI 10.1515/crll.1901.123.241. MR1580570

[279] H. W. E. Jung, *Über den kleinsten Kreis, der eine ebene Figur einschließt* (German), J. Reine Angew. Math. **137** (1910), 310–313, DOI 10.1515/crll.1910.137.310. MR1580792

[280] M. Kac, *Can one hear the shape of a drum?*, Amer. Math. Monthly **73** (1966), no. 4, 1–23, DOI 10.2307/2313748. MR201237

[281] J.-P. Kahane and R. Salem, *Ensembles parfaits et séries trigonométriques* (French), Actualités Scientifiques et Industrielles [Current Scientific and Industrial Topics], No. 1301, Hermann, Paris, 1963. MR160065

[282] J.-P. Kahane, M. Weiss, and G. Weiss, *On lacunary power series*, Ark. Mat. **5** (1963), 1–26, DOI 10.1007/BF02591111. MR190303

[283] L. V. Kantorovich, *Mathematical methods of organizing and planning production*, Management Sci. **6** (1959/60), 366–422, DOI 10.1287/mnsc.6.4.366. MR129016

[284] L. Kantorovitch, *A new method of solving of some classes of extremal problems*, C. R. (Doklady) Acad. Sci. URSS (N.S.) **28** (1940), 211–214. MR3456

[285] L. V. Kantorovič and G. Š. Rubinšteĭn, *On a space of completely additive functions* (Russian, with English summary), Vestnik Leningrad. Univ. **13** (1958), no. 7, 52–59. MR102006

[286] J. L. Kaplan, J. Mallet-Paret, and J. A. Yorke, *The Lyapunov dimension of a nowhere differentiable attracting torus*, Ergodic Theory Dynam. Systems **4** (1984), no. 2, 261–281, DOI 10.1017/S0143385700002431. MR766105

[287] Y. Katznelson, *An introduction to harmonic analysis*, 3rd ed., Cambridge Mathematical Library, Cambridge University Press, Cambridge, 2004, DOI 10.1017/CBO9781139165372. MR2039503

[288] A. S. Kechris, *Classical descriptive set theory*, Graduate Texts in Mathematics, vol. 156, Springer-Verlag, New York, 1995, DOI 10.1007/978-1-4612-4190-4. MR1321597

[289] G. Keller, *A simpler proof for the dimension of the graph of the classical Weierstrass function* (English, with English and French summaries), Ann. Inst. Henri Poincaré Probab. Stat. **53** (2017), no. 1, 169–181, DOI 10.1214/15-AIHP711. MR3606737

[290] M. Kesseböhmer and S. Kombrink, *Fractal curvature measures and Minkowski content for self-conformal subsets of the real line*, Adv. Math. **230** (2012), no. 4-6, 2474–2512, DOI 10.1016/j.aim.2012.04.023. MR2927378

[291] A. Ya. Khinchin, *Continued fractions*, translated from the third (1961) Russian edition, with a preface by B. V. Gnedenko, reprint of the 1964 translation, Dover Publications, Inc., Mineola, NY, 1997. MR1451873

[292] J. Kigami, *Harmonic calculus on p.c.f. self-similar sets*, Trans. Amer. Math. Soc. **335** (1993), no. 2, 721–755, DOI 10.2307/2154402. MR1076617

[293] J. Kigami, *Analysis on fractals*, Cambridge Tracts in Mathematics, vol. 143, Cambridge University Press, Cambridge, 2001, DOI 10.1017/CBO9780511470943. MR1840042

[294] J. Kigami and M. L. Lapidus, *Weyl's problem for the spectral distribution of Laplacians on p.c.f. self-similar fractals*, Comm. Math. Phys. **158** (1993), no. 1, 93–125. MR1243717

[295] J. Kigami and M. L. Lapidus, *Self-similarity of volume measures for Laplacians on p.c.f. self-similar fractals*, Comm. Math. Phys. **217** (2001), no. 1, 165–180, DOI 10.1007/s002200000326. MR1815029

[296] D. A. Klain and G.-C. Rota, *Introduction to geometric probability*, Lezioni Lincee. [Lincei Lectures], Cambridge University Press, Cambridge, 1997. MR1608265

[297] M. Klimeš, P. Mardešić, G. Radunović, and M. Resman, *Reading analytic invariants of parabolic diffeomorphisms from their orbits*, Ann. Scuola Norm-Sci, 2025, DOI: 10.2422/2036-2145.202208_022, 25 pp., in press.

[298] M. Kline, *Mathematical thought from ancient to modern times*, Oxford University Press, New York, 1972. MR472307

[299] M. Kneser, *Einige Bemerkungen über das Minkowskische Flächenmass* (German), Arch. Math. (Basel) **6** (1955), 382–390, DOI 10.1007/BF01900510. MR73679

[300] K. Knopp, *Ein einfaches Verfahren zur Bildüng stetiger nirgends differenzierbarer Funktionen* (German), Math. Z. **2** (1918), no. 1-2, 1–26, DOI 10.1007/BF01212897. MR1544308

[301] K. Knopp, *Theory of Functions, Parts I and II*, Dover books on Mathematics [English translation of *Funktionentheorie, Parts I and II*, Walter de Gruyter, Berlin, 1928], Dover Publications, Inc., Mineola, 1996.

[302] Ş. Koçak and A. V. Ratiu, *Inner tube formulas for polytopes*, Proc. Amer. Math. Soc. **140** (2012), no. 3, 999–1010, DOI 10.1090/S0002-9939-2011-11307-8. MR2869084

[303] S. Kombrink, *Fractal Curvature Measures and Minkowski Content for Limit Sets of Conformal Function Systems*, Ph.D. Thesis, Mathematics, Universität Bremen, Germany, 2011.

[304] S. Kombrink, *A survey on Minkowski measurability of self-similar and self-conformal fractals in* $\mathbb{R}^d$, Fractal geometry and dynamical systems in pure and applied mathematics. I. Fractals in pure mathematics, Contemp. Math., vol. 600, Amer. Math. Soc., Providence, RI, 2013, pp. 135–159, DOI 10.1090/conm/600/11931. MR3203401

[305] S. Kombrink, E. P. J. Pearse, and S. Winter, *Lattice-type self-similar sets with pluriphase generators fail to be Minkowski measurable*, Math. Z. **283** (2016), no. 3-4, 1049–1070, DOI 10.1007/s00209-016-1633-x. MR3519995

[306] H.-J. Kowalsky, *Einbettung metrischer Räume* (German), Arch. Math. (Basel) **8** (1957), 336–339, DOI 10.1007/BF01900142. MR91451

[307] G. Kowalewski, *Über Bolzanos Nichtdifferenzierbare Stetige Funktion* (German), Acta Math. **44** (1923), no. 1, 315–319, DOI 10.1007/BF02403926. MR1555188

[308] T. Kumagai, *Anomalous random walks and diffusions: from fractals to random media*, Proceedings of the International Congress of Mathematicians—Seoul 2014. Vol. IV, Kyung Moon Sa, Seoul, 2014, pp. 75–94. MR3727603

[309] C. Kuratowski, *Topologie. Vol. I* (French), 4ème éd, Monografie Matematyczne [Mathematical Monographs], Tom 20, Państwowe Wydawnictwo Naukowe, Warsaw, 1958. MR90795

[310] J. C. Lagarias, *The Takagi function and its properties*, Functions in number theory and their probabilistic aspects, RIMS Kôkyûroku Bessatsu, B34, Res. Inst. Math. Sci. (RIMS), Kyoto, 2012, pp. 153–189. MR3014845

[311] N. Lal and M. L. Lapidus, *Hyperfunctions and spectral zeta functions of Laplacians on self-similar fractals*, J. Phys. A **45** (2012), no. 36, 365205, 14, DOI 10.1088/1751-8113/45/36/365205. MR2967908

[312] N. Lal and M. L. Lapidus, *The decimation method for Laplacians on fractals: spectra and complex dynamics*, Fractal geometry and dynamical systems in pure and applied mathematics. II. Fractals in applied mathematics, Contemp. Math., vol. 601, Amer. Math. Soc., Providence, RI, 2013, pp. 227–249, DOI 10.1090/conm/601/11959. MR3203865

[313] S. P. Lalley, *The packing and covering functions of some self-similar fractals*, Indiana Univ. Math. J. **37** (1988), no. 3, 699–710, DOI 10.1512/iumj.1988.37.37034. MR962930

[314] S. P. Lalley, *Renewal theorems in symbolic dynamics, with applications to geodesic flows, non-Euclidean tessellations and their fractal limits*, Acta Math. **163** (1989), no. 1-2, 1–55, DOI 10.1007/BF02392732. MR1007619

[315] S. P. Lalley, *Probabilistic methods in certain counting problems of ergodic theory*, Ergodic theory, symbolic dynamics, and hyperbolic spaces (Trieste, 1989), Oxford Sci. Publ., Oxford Univ. Press, New York, 1991, pp. 223–258. MR1130178

[316] S. P. Lalley and D. Gatzouras, *Hausdorff and box dimensions of certain self-affine fractals*, Indiana Univ. Math. J. **41** (1992), no. 2, 533–568, DOI 10.1512/iumj.1992.41.41031. MR1183358

[317] T.-M. Landry, M. L. Lapidus, and F. Latrémolière, *Metric approximations of spectral triples on the Sierpiński gasket and other fractal curves*, Adv. Math. **385** (2021), Paper No. 107771, 43, DOI 10.1016/j.aim.2021.107771. MR4256136

[318] M. L. Lapidus, *Fractal drum, inverse spectral problems for elliptic operators and a partial resolution of the Weyl-Berry conjecture*, Trans. Amer. Math. Soc. **325** (1991), no. 2, 465–529, DOI 10.2307/2001638. MR994168

[319] M. L. Lapidus, *Spectral and fractal geometry: from the Weyl-Berry conjecture for the vibrations of fractal drums to the Riemann zeta-function*, Differential equations and mathematical physics (Birmingham, AL, 1990), Math. Sci. Engrg., vol. 186, Academic Press, Boston, MA, 1992, pp. 151–181, DOI 10.1016/S0076-5392(08)63379-2. MR1126694

[320] M. L. Lapidus, *Vibrations of fractal drums, the Riemann hypothesis, waves in fractal media and the Weyl-Berry conjecture*, Ordinary and partial differential equations, Vol. IV (Dundee, 1992), Pitman Res. Notes Math. Ser., vol. 289, Longman Sci. Tech., Harlow, 1993, pp. 126–209. MR1234502

[321] M. L. Lapidus, Fractal geometry and applications—An introduction to this volume, in: *Fractal Geometry and Applications: A Jubilee of Benoît Mandelbrot* (M. L. Lapidus and M. van Frankenhuijsen, eds.), Proceedings of Symposia in Pure Mathematics, vol. 72 (Part 1: *Analysis, Number Theory and Dynamical Systems.* Part 2: *Multifractals, Probability and Statistical Mechanics, and Applications*), Amer. Math. Soc., Providence, R. I., 2004, Part 1, pp. 1–26.

[322] M. L. Lapidus, *Fractals and vibrations: can you hear the shape of a fractal drum?*, Symposium in Honor of Benoit Mandelbrot (Curaçao, 1995), Fractals **3** (1995), no. 4, 725–736, DOI 10.1142/S0218348X95000643. MR1410291

[323] M. L. Lapidus, *Analysis on fractals, Laplacians on self-similar sets, noncommutative geometry and spectral dimensions*, Topol. Methods Nonlinear Anal. **4** (1994), no. 1, 137–195, DOI 10.12775/TMNA.1994.025. MR1321811

[324] M. L. Lapidus, *Towards a noncommutative fractal geometry? Laplacians and volume measures on fractals*, Harmonic analysis and nonlinear differential equations (Riverside, CA, 1995), Contemp. Math., vol. 208, Amer. Math. Soc., Providence, RI, 1997, pp. 211–252, DOI 10.1090/conm/208/02742. MR1467009

[325] M. L. Lapidus, The Sierpinski gasket and carpet, Kluwer Encyclopedia of Mathematics (M. Hazewinkel, ed.), suppl. vol. III, Kluwer Academic Publisher, Dordrecht & Amsterdam, 2002, pp. 364–368.

[326] M. L. Lapidus, *In search of the Riemann zeros: Strings, fractal membranes and noncommutative spacetimes*, American Mathematical Society, Providence, RI, 2008, DOI 10.1090/mbk/051. MR2375028

[327] M. L. Lapidus, *Towards quantized number theory: spectral operators and an asymmetric criterion for the Riemann hypothesis*, Philos. Trans. Roy. Soc. A **373** (2015), no. 2047, 20140240, 24, DOI 10.1098/rsta.2014.0240. MR3367510

[328] M. L. Lapidus, *The sound of fractal strings and the Riemann hypothesis*, Analytic number theory, Springer, Cham, 2015, pp. 201–252. MR3467400

[329] M. L. Lapidus, *An overview of complex fractal dimensions: from fractal strings to fractal drums, and back*, Horizons of fractal geometry and complex dimensions, Contemp. Math., vol. 731, Amer. Math. Soc., Providence, RI, 2019, pp. 143–265, DOI 10.1090/conm/731/14677. MR3989820

[330] M. L. Lapidus, *From Complex Fractal Dimensions and Quantized Number Theory To Fractal Cohomology: A Tale of Oscillations, Unreality and Fractality*, Fractals and Dynamics in Mathematics, Science, and the Arts: Theory and Applications, World Scientific Publishing, Singapore and London, 2025, to appear.

[331] M. L. Lapidus, *A Panorama of Number Theory: From Euler and Riemann to Weil and Beyond*, series of books in preparation (for Springer/Birkhäuser), 2024.

[332] M. L. Lapidus, J. Lévy-Véhel, and J. A. Rock, *Fractal strings and multifractal zeta functions*, Lett. Math. Phys. **88** (2009), no. 1-3, 101–129, DOI 10.1007/s11005-009-0302-y. MR2512142

[333] M. L. Lapidus and H. Lũ, *Nonarchimedean Cantor set and string*, J. Fixed Point Theory Appl. **3** (2008), no. 1, 181–190, DOI 10.1007/s11784-008-0062-9. MR2402916

[334] M. L. Lapidus and H. Lũ', *Self-similar p-adic fractal strings and their complex dimensions*, p-Adic Numbers Ultrametric Anal. Appl. **1** (2009), no. 2, 167–180, DOI 10.1134/S2070046609020083. MR2566062

[335] M. L. Lapidus and L. Hùng, *The geometry of p-adic fractal strings: a comparative survey*, Advances in non-Archimedean analysis, Contemp. Math., vol. 551, Amer. Math. Soc., Providence, RI, 2011, pp. 163–206, DOI 10.1090/conm/551/10893. MR2882397

[336] M. L. Lapidus, L. Hùng, and M. van Frankenhuijsen, *Minkowski measurability and exact fractal tube formulas for p-adic self-similar strings*, Fractal geometry and dynamical systems in pure and applied mathematics. I. Fractals in pure mathematics, Contemp. Math., vol. 600, Amer. Math. Soc., Providence, RI, 2013, pp. 161–184, DOI 10.1090/conm/600/11949. MR3203402

[337] M. L. Lapidus, H. Lu, and M. van Frankenhuijsen, *Minkowski dimension and explicit tube formulas for p-adic fractal strings*, Fractal and Fractional No. 4, **2** (2018), 26th paper, 1–30. (DOI: 10.3390/fractalfract2040026.) (Also: e-print, `arXiv:1603.09409v2 [math-ph]`, 2018.)

[338] M. L. Lapidus, L. Hùng, and M. van Frankenhuijsen, *p-adic fractal strings of arbitrary rational dimensions and Cantor strings*, p-Adic Numbers Ultrametric Anal. Appl. **13** (2021), no. 3, 215–230, DOI 10.1134/S2070046621030043. MR4308348

[339] M. L. Lapidus and H. Maier, *Hypothèse de Riemann, cordes fractales vibrantes et conjecture de Weyl-Berry modifiée* (French, with English summary), C. R. Acad. Sci. Paris Sér. I Math. **313** (1991), no. 1, 19–24. MR1115940

[340] M. L. Lapidus and H. Maier, *The Riemann hypothesis and inverse spectral problems for fractal strings*, J. London Math. Soc. (2) **52** (1995), no. 1, 15–34, DOI 10.1112/jlms/52.1.15. MR1345711

[341] M. L. Lapidus, R. L. Miller, and R. G. Niemeyer, *Nontrivial paths and periodic orbits of the T-fractal billiard table*, Nonlinearity **29** (2016), no. 7, 2145–2172, DOI 10.1088/0951-7715/29/7/2145. MR3521643

[342] M. L. Lapidus, J. W. Neuberger, R. J. Renka, and C. A. Griffith, *Snowflake harmonics and computer graphics: numerical computation of spectra on fractal drums*, Internat. J. Bifur. Chaos Appl. Sci. Engrg. **6** (1996), no. 7, 1185–1210, DOI 10.1142/S0218127496000680. MR1412217

[343] M. L. Lapidus and R. G. Niemeyer, *Towards the Koch snowflake fractal billiard: computer experiments and mathematical conjectures*, Gems in experimental mathematics, Contemp. Math., vol. 517, Amer. Math. Soc., Providence, RI, 2010, pp. 231–263, DOI 10.1090/conm/517/10144. MR2731085

[344] M. L. Lapidus and R. G. Niemeyer, *Sequences of compatible periodic hybrid orbits of prefractal Koch snowflake billiards*, Discrete Contin. Dyn. Syst. **33** (2013), no. 8, 3719–3740, DOI 10.3934/dcds.2013.33.3719. MR3021377

[345] M. L. Lapidus and R. G. Niemeyer, *The current state of fractal billiards*, Fractal geometry and dynamical systems in pure and applied mathematics. II. Fractals in applied mathematics, Contemp. Math., vol. 601, Amer. Math. Soc., Providence, RI, 2013, pp. 251–288, DOI 10.1090/conm/601/11956. MR3203866

[346] M. L. Lapidus and M. M. H. Pang, *Eigenfunctions of the Koch snowflake domain*, Comm. Math. Phys. **172** (1995), no. 2, 359–376. MR1350412

[347] M. L. Lapidus and E. P. J. Pearse, *A tube formula for the Koch snowflake curve, with applications to complex dimensions*, J. London Math. Soc. (2) **74** (2006), no. 2, 397–414, DOI 10.1112/S0024610706022988. MR2269586

[348] M. L. Lapidus and E. P. J. Pearse, *Tube formulas for self-similar fractals*, Analysis on graphs and its applications, Proc. Sympos. Pure Math., vol. 77, Amer. Math. Soc., Providence, RI, 2008, pp. 211–230, DOI 10.1090/pspum/077/2459871. MR2459871

[349] M. L. Lapidus and E. P. J Pearse, *Tube Formulas and complex dimensions of self-similar tilings*, Acta Applicandae Mathematicae No. 1, **112** (2010), 91–137. (Also: e-print, `arXiv:math.DS/0605527v5`, 2010; Springer Open Access: DOI: 10.1007/S10440-010-9562-x.)

[350] M. L. Lapidus, E. P. J. Pearse, and S. Winter, *Pointwise tube formulas for fractal sprays and self-similar tilings with arbitrary generators*, Adv. Math. **227** (2011), no. 4, 1349–1398, DOI 10.1016/j.aim.2011.03.004. MR2799798

[351] M. L. Lapidus, E. P. J. Pearse, and S. Winter, *Minkowski measurability results for self-similar tilings and fractals with monophase generators*, Fractal geometry and dynamical systems in pure and applied mathematics. I. Fractals in pure mathematics, Contemp. Math., vol. 600, Amer. Math. Soc., Providence, RI, 2013, pp. 185–203, DOI 10.1090/conm/600/11951. MR3203403

[352] M. L. Lapidus and C. Pomerance, *Fonction zêta de Riemann et conjecture de Weyl-Berry pour les tambours fractals* (French, with English summary), C. R. Acad. Sci. Paris Sér. I Math. **310** (1990), no. 6, 343–348. MR1046509

[353] M. L. Lapidus and C. Pomerance, *The Riemann zeta-function and the one-dimensional Weyl-Berry conjecture for fractal drums*, Proc. London Math. Soc. (3) **66** (1993), no. 1, 41–69, DOI 10.1112/plms/s3-66.1.41. MR1189091

[354] M. L. Lapidus and C. Pomerance, *Counterexamples to the modified Weyl-Berry conjecture on fractal drums*, Math. Proc. Cambridge Philos. Soc. **119** (1996), no. 1, 167–178, DOI 10.1017/S0305004100074053. MR1356166

[355] M. L. Lapidus and G. Radunović, Fractal zeta functions and logarithmic gauge Minkowski measurability, in preparation, 2024.

[356] M. L. Lapidus, G. Radunović, and D. Žubrinić, *Fractal Zeta Functions and Fractal Drums: Higher-Dimensional Theory of Complex Dimensions*, Springer Monographs in Mathematics, Springer, New York, 2017. (ISBN: 978-3-319-44706-3.)

[357] M. L. Lapidus, G. Radunović, and D. Žubrinić, Distance and tube zeta functions of fractals and arbitrary compact sets, *Adv. in Math.* **307** (2017), 1215–1267. (dx.doi.org/10.1016/j.aim2016.11.034). (Also: e-print, `arXiv:1506.03525v3 [math-ph]`, 2016; e-print, Institut des Hautes Scientfiques, M/15/15, 2015.)

[358] M. L. Lapidus, G. Radunović, and D. Žubrinić, *Complex dimensions of fractals and meromorphic extensions of fractal zeta functions*, J. Math. Anal. Appl. **453** (2017), no. 1, 458–484, DOI 10.1016/j.jmaa.2017.03.059. MR3641785

[359] M. L. Lapidus, G. Radunović, and D. Žubrinić, *Zeta functions and complex dimensions of relative fractal drums: theory, examples and applications*, Dissertationes Math. **526** (2017), 105, DOI 10.4064/dm757-4-2017. MR3764036

[360] M. L. Lapidus, G. Radunović, and D. Žubrinić, *Fractal tube formulas and a Minkowski measurability criterion for compact subsets of Euclidean spaces*, Discrete Contin. Dyn. Syst. Ser. S **12** (2019), no. 1, 105–117, DOI 10.3934/dcdss.2019007. MR3836595

[361] M. L. Lapidus, G. Radunović, and D. Žubrinić, *Fractal tube formulas for compact sets and relative fractal drums: oscillations, complex dimensions and fractality*, J. Fractal Geom. **5** (2018), no. 1, 1–119, DOI 10.4171/JFG/57. MR3760302

[362] M. L. Lapidus, G. Radunović, and D. Žubrinić, *Minkowski measurability criteria for compact sets and relative fractal drums in Euclidean spaces*, Analysis, probability and mathematical physics on fractals, Fractals Dyn. Math. Sci. Arts Theory Appl., vol. 5, World Sci. Publ., Hackensack, NJ, 2020, pp. 21–98. MR4472246

[363] M. L. Lapidus, G. Radunović, and D. Žubrinić, *Fractal zeta functions and complex dimensions of relative fractal drums*, J. Fixed Point Theory Appl. **15** (2014), no. 2, 321–378, DOI 10.1007/s11784-014-0207-y. MR3298004

[364] M. L. Lapidus, G. Radunović, and D. Žubrinić, *Fractal zeta functions and complex dimensions: a general higher-dimensional theory*, Fractal geometry and stochastics V, Progr. Probab., vol. 70, Birkhäuser/Springer, Cham, 2015, pp. 229–257, DOI 10.1007/978-3-319-18660-3_13. MR3558159

[365] M. L. Lapidus, G. Radunović, and D. Žubrinić, *Essential singularities of fractal zeta functions*, Pure Appl. Funct. Anal. **5** (2020), no. 5, 1073–1094. MR4167443

[366] M. L. Lapidus and J. A. Rock, *Towards zeta functions and complex dimensions of multifractals*, Complex Var. Elliptic Equ. **54** (2009), no. 6, 545–559, DOI 10.1080/17476930802326758. MR2537255

[367] M. L. Lapidus, J. A. Rock, and D. Žubrinić, *Box-counting fractal strings, zeta functions, and equivalent forms of Minkowski dimension*, Fractal geometry and dynamical systems in pure and applied mathematics. I. Fractals in pure mathematics, Contemp. Math., vol. 600, Amer. Math. Soc., Providence, RI, 2013, pp. 239–271, DOI 10.1090/conm/600/11929. MR3203405

[368] M. Lapidus and J. Sarhad, *Dirac operators and geodesic metric on the harmonic Sierpinski gasket and other fractal sets*, J. Noncommut. Geom. **8** (2014), no. 4, 947–985, DOI 10.4171/JNCG/174. MR3310935

[369] M. L. Lapidus and M. van Frankenhuysen, *Complex dimensions of fractal strings and oscillatory phenomena in fractal geometry and arithmetic*, Spectral problems in geometry and arithmetic (Iowa City, IA, 1997), Contemp. Math., vol. 237, Amer. Math. Soc., Providence, RI, 1999, pp. 87–105, DOI 10.1090/conm/237/1710790. MR1710790

[370] M. L. Lapidus and M. van Frankenhuysen, *Fractal geometry and number theory: Complex dimensions of fractal strings and zeros of zeta functions*, Birkhäuser Boston, Inc., Boston, MA, 2000, DOI 10.1007/978-1-4612-5314-3. MR1726744

[371] M. L. Lapidus and M. van Frankenhuysen, *Complex dimensions of self-similar fractal strings and Diophantine approximation*, Experiment. Math. **12** (2003), no. 1, 41–69. MR2002673

[372] M. L. Lapidus and M. van Frankenhuijsen (eds.), *Fractal Geometry and Applications: A Jubilee of Benoît Mandelbrot*, Proceedings of Symposia in Pure Mathematics, vol. 72 (Part 1: *Analysis, Number Theory and Dynamical Systems.* Part 2: *Multifractals, Probability and Statistical Mechanics, and Applications*), Amer. Math. Soc., Providence, R. I., 2004, Part 1, pp. 1–26.

[373] M. L. Lapidus and M. van Frankenhuijsen, *Fractality, self-similarity and complex dimensions*, Fractal geometry and applications: a jubilee of Benoît Mandelbrot. Part 1, Proc. Sympos. Pure Math., vol. 72, Amer. Math. Soc., Providence, RI, 2004, pp. 349–372, DOI 10.1090/pspum/072.1/2112111. MR2112111

[374] M. L. Lapidus and M. van Frankenhuijsen, *Fractal geometry, complex dimensions and zeta functions: Geometry and spectra of fractal strings*, Springer Monographs in Mathematics, Springer, New York, 2006, DOI 10.1007/978-0-387-35208-4. MR2245559

[375] M. L. Lapidus and M. van Frankenhuijsen, *Fractal geometry, complex dimensions and zeta functions: Geometry and spectra of fractal strings*, 2nd ed., Springer Monographs in Mathematics, Springer, New York, 2013, DOI 10.1007/978-1-4614-2176-4. MR2977849

[376] M. L. Lapidus, M. van Frankenhuijsen and E. K. Voskanian, Quasiperiodic patterns of the complex dimensions of nonlattice self-similar strings, via the LLL algorithm, *Mathematics* No. 6, **9** (2021), 591 (35 pages). (Special Issue on "Fractal and Computational Geometry".) (Also: e-print, `arXiv:2009.03493v2 [math.NT]`, 2020.)

[377] M. L. Lapidus, M. van Frankenhuijsen and E. K. Voskanian, *Diffraction measures and patterns of the complex dimensions of self-similar fractal strings. I. The lattice case*, Houston J. Math. No. 4, **4** (2023), 833-859. (Also: e-print, arXiv:2305.09050v1 [math-ph], 2023)

[378] M. L. Lapidus, M. van Frankenhuijsen and E. K. Voskanian, Diffraction measures and patterns of the complex dimensions of self-similar fractal strings. II. The Nonlattice Case (tentative title), in preparation, 2024.

[379] M. L. Lapidus and S. Watson, Fractal zeta functions and complex dimensions of Ahlfors metric measure spaces, preprint, 2024.

[380] K.-S. Lau and S.-M. Ngai, *L^q-spectrum of the Bernoulli convolution associated with the golden ratio*, Studia Math. **131** (1998), no. 3, 225–251. MR1644468

[381] H. Lauwerier, *Fractals: Endlessly repeated geometrical figures*, translated from the Dutch by Sophia Gill-Hoffstädt, Princeton Science Library, Princeton University Press, Princeton, NJ, 1991. MR1113486

[382] G. F. Lawler, *Conformally invariant processes in the plane*, Mathematical Surveys and Monographs, vol. 114, American Mathematical Society, Providence, RI, 2005, DOI 10.1090/surv/114. MR2129588

[383] G. F. Lawler, O. Schramm, and W. Werner, *The dimension of the planar Brownian frontier is* 4/3, Math. Res. Lett. **8** (2001), no. 4, 401–411, DOI 10.4310/MRL.2001.v8.n4.a1. MR1849257

[384] G. F. Lawler, O. Schramm and W. Werner, Values of Brownian intersection exponents, I-III. I: Half-plane exponents, *Acta. Math.* **187** (2001), 237–273; II: Plane exponents, *Acta. Math.* **187** (2001), 275–308; and III: Two-sided exponents, *Ann. Inst. Henri Poincaré, Probab. et Stat.* No. 1, **38** (2002), 109–123.

[385] G. F. Lawler, O. Schramm, and W. Werner, *Conformal invariance of planar loop-erased random walks and uniform spanning trees*, Ann. Probab. **32** (2004), no. 1B, 939–995, DOI 10.1214/aop/1079021469. MR2044671

[386] J.-F. Le Gall, *Random geometry on the sphere*, Proceedings of the International Congress of Mathematicians—Seoul 2014. Vol. 1, Kyung Moon Sa, Seoul, 2014, pp. 421–442. MR3728478

[387] J.-F. Le Gall, *Brownian geometry*, Jpn. J. Math. **14** (2019), no. 2, 135–174, DOI 10.1007/s11537-019-1821-7. MR4007665

[388] J.-F. Le Gall, *The volume measure of the Brownian sphere is a Hausdorff measure*, Electron. J. Probab. **27** (2022), Paper No. 113, 28, DOI 10.1214/22-ejp837. MR4474537

[389] H. Lebesgue, Leçons sur l'intégration et la recherche des fonctions primitives ["Lessons on Integration and the Search for Anti-Derivative Functions"], Gauthier–Villars, Paris, 1904.

[390] F. Ledrappier, *On the dimension of some graphs*, Symbolic dynamics and its applications (New Haven, CT, 1991), Contemp. Math., vol. 135, Amer. Math. Soc., Providence, RI, 1992, pp. 285–293, DOI 10.1090/conm/135/1185095. MR1185095

[391] F. Ledrappier and L.-S. Young, *The metric entropy of diffeomorphisms. II. Relations between entropy, exponents and dimension*, Ann. of Math. (2) **122** (1985), no. 3, 540–574, DOI 10.2307/1971329. MR819557

[392] S. Lefschetz, *On compact spaces*, Ann. of Math. (2) **32** (1931), no. 3, 521–538, DOI 10.2307/1968249. MR1503014

[393] T. Lei, *Similarity between the Mandelbrot set and Julia sets*, Comm. Math. Phys. **134** (1990), no. 3, 587–617. MR1086745

[394] T. Lei (ed.), *The Mandelbrot set, theme and variations*, London Mathematical Society Lecture Note Series, vol. 274, Cambridge University Press, Cambridge, 2000. MR1765080

[395] A. Le Méhauté, *Les géométries fractales: L'espace-temps brisé* ("Fractal Geometries: Broken Space-Time"), Editions Hermès, Paris, 1990.

[396] A. Le Méhauté, *Fractal geometries: Theory and applications*, translated from the 1990 French edition by Jack Howlett, CRC Press, Boca Raton, FL, 1991. MR1153251

[397] M. Levitin and D. Vassiliev, *Spectral asymptotics, renewal theorem, and the Berry conjecture for a class of fractals*, Proc. London Math. Soc. (3) **72** (1996), no. 1, 188–214, DOI 10.1112/plms/s3-72.1.188. MR1357092

[398] J. Lévy Véhel and F. Mendivil, *Multifractal and higher-dimensional zeta functions*, Nonlinearity **24** (2011), no. 1, 259–276, DOI 10.1088/0951-7715/24/1/013. MR2746147

[399] J. Lévy Véhel and C. Tricot, *On various multifractal spectra*, Fractal geometry and stochastics III, Progr. Probab., vol. 57, Birkhäuser, Basel, 2004, pp. 23–42. MR2087130

[400] J. Lévy Véhel and R. Vojak, *Multifractal analysis of Choquet capacities*, Adv. in Appl. Math. **20** (1998), no. 1, 1–43, DOI 10.1006/aama.1996.0517. MR1488230

[401] W. Li and F. M. Dekking, *Hausdorff dimension of subsets of Moran fractals with prescribed group frequency of their codings*, Nonlinearity **16** (2003), no. 1, 187–199, DOI 10.1088/0951-7715/16/1/312. MR1950783

[402] W. Li, L. Olsen, and Z. Wen, *Hausdorff and packing dimensions of subsets of Moran fractals with prescribed mixed group frequency of their codings*, Aequationes Math. **77** (2009), no. 1-2, 171–185, DOI 10.1007/s00010-008-2947-5. MR2495725

[403] E. Lindelöf, *Sur l'application de la méthode des approximations successives aux équations différentielles ordinaires du premier ordre* [*"On an application of the method of successive approximations to first order ordinary differential equations"*], C. R. Acad. Sci. Paris **118** (1894), 454–457. (In this paper, the author discusses generalizations of an earlier work by Charles–Émile Picard.)

[404] S. L. Lipscomb, *Imbedding One-dimensional Metric Spaces*, ProQuest LLC, Ann Arbor, MI, 1973. Thesis (Ph.D.)–University of Virginia. MR2623058

[405] S. L. Lipscomb, *A universal one-dimensional metric space*, TOPO 72—general topology and its applications (Proc. Second Pittsburgh Internat. Conf., Pittsburgh, Pa., 1972; dedicated to the memory of Johannes H. de Groot), Lecture Notes in Math., Vol. 378, Springer, Berlin-New York, 1974, pp. 248–257. MR358738

[406] S. L. Lipscomb, *On imbedding finite-dimensional metric spaces*, Trans. Amer. Math. Soc. **211** (1975), 143–160, DOI 10.2307/1997225. MR380751

[407] S. L. Lipscomb, *An imbedding theorem for metric spaces*, Proc. Amer. Math. Soc. **55** (1976), no. 1, 165–169, DOI 10.2307/2041865. MR391034

[408] S. L. Lipscomb, *Fractals and universal spaces in dimension theory*, Springer Monographs in Mathematics, Springer, New York, 2009, DOI 10.1007/978-0-387-85494-6. MR2460244

[409] S. L. Lipscomb, *The quest for universal spaces in dimension theory*, Notices Amer. Math. Soc. **56** (2009), no. 11, 1418–1424. MR2590725

[410] E. N. Lorenz, Predictability: Does the flap of a butterfly's wings in Brazil set off a tornado in Texas? *Amer. Assoc. Advancement of Science*, 139th Meeting, AAAS Section on Environmental Sciences: New Approaches to Global Weather, Extended Abstract, 12-29-1972, 1972. (Also: *Bull. Amer. Meteorological Soc.* No. 5, **50** (1969), 345-352.)

[411] E. N. Lorenz, *The essence of chaos*, The Jessie and John Danz Lectures, University of Washington Press, Seattle, WA, 1993, DOI 10.4324/9780203214589. MR1285591

[412] M. Lyubich, Dynamics of quadratic polynomials, I and II. I: Combinatorics and geometry of the Yoccoz puzzle, *Acta Math.* **178** (1997), 185–247. II: Rigidity and local connectivity, *Acta Math.* **178** (1997), 247–297. Feigenbaum-Coullet-Tresser universality and Milnor's hairiness conjecture, *Ann. of Math.* (2) **149** (1999), 319-420.

[413] J. M. Mackay, *Assouad dimension of self-affine carpets*, Conform. Geom. Dyn. **15** (2011), 177–187, DOI 10.1090/S1088-4173-2011-00232-3. MR2846307

[414] L. Malozemov and A. Teplyaev, *Self-similarity, operators and dynamics*, Math. Phys. Anal. Geom. **6** (2003), no. 3, 201–218, DOI 10.1023/A:1024931603110. MR1997913

[415] B. B. Mandelbrot, *How long is the coast of Britain? Statistical self-similarity and fractional dimension*, Science **156** (1967), 636–638.[2]

[416] B. B. Mandelbrot, *Intermittent turbulence in self-similar cascades: Divergence of high moments and dimension of the carrier*, J. Fluid. Mech. **62** (1974), 331–358.

[417] B. B. Mandelbrot, *Fractal aspects of the iteration $z \to \lambda z(1-z)$ for complex λ and z*, Annuals of New York Acad. Sci. **357** (1980), 249–259.

[418] B. B. Mandelbrot, *Fractals: Form, Chance and Dimension*, W. H. Freeman, New York, 1977 [French].

[419] B. B. Mandelbrot, *The Fractal Geometry of Nature*, revised and enlarged edition (of the 1977 edition, [**418**]), W. H. Freeman and Company, New York, 1983.

[420] B. B. Mandelbrot, *Les Objets Fractals* ["Fractal Objects"], second and revised edition (of the 1975 edition), Nouvelle Bibliothèque Scientifique, Flammarion, Paris, 1984 [French].

[421] B. B. Mandelbrot, *Measures of fractal lacunarity: Minkowski content and alternatives*, Fractal geometry and stochastics (Finsterbergen, 1994), Progr. Probab., vol. 37, Birkhäuser, Basel, 1995, pp. 15–42. MR1391969

[422] B. B. Mandelbrot, *Fractals and scaling in finance: Discontinuity, concentration, risk*, Selected Works of Benoit B. Mandelbrot, Selecta Volume E, with a foreword by R. E. Gomory, Springer-Verlag, New York, 1997, DOI 10.1007/978-1-4757-2763-0. MR1475217

[423] B. B. Mandelbrot, *Multifractals and $1/f$ noise: Wild self-affinity in physics (1963–1976)*, Selected Works of Benoit B. Mandelbrot, with contributions by J. M. Berger, J.-P. Kahane, and J. Peyrière, Selecta Volume N, Springer-Verlag, New York, 1999, DOI 10.1007/978-1-4612-2150-0. MR1713511

[424] B. B. Mandelbrot, *Gaussian self-affinity and fractals: Globality, the earth, $1/f$ noise, and R/S*, Selected Works of Benoit B. Mandelbrot, Selecta (old or new) Volume H, includes contributions by F. J. Damerau, M. Frame, K. McCamy, J. W. Van Ness, J. R. Wallis, and others, Springer-Verlag, New York, 2002. MR1878884

[425] B. B. Mandelbrot, *Fractals and chaos: The Mandelbrot set and beyond*, Selected Works of Benoit B. Mandelbrot, Selecta Volume C, with a foreword by P. W. Jones and texts co-authored by C. J. G. Evertsz and M. C. Gutzwiller, Springer-Verlag, New York, 2004, DOI 10.1007/978-1-4757-4017-2. MR2029233

[426] B. B. Mandelbrot and R. L. Hudson, *The (mis)behavior of markets: A fractal view of risk, ruin, and reward*, Basic Books, New York, 2004. MR2100767

[427] Y. I. Manin, *The notion of dimension in geometry and algebra*, Bull. Amer. Math. Soc. (N.S.) **43** (2006), no. 2, 139–161, DOI 10.1090/S0273-0979-06-01081-0. MR2216108

[428] P. Mardešić, G. Radunović, and M. Resman, *Fractal zeta functions of orbits of parabolic diffeomorphisms*, Anal. Math. Phys. **12** (2022), no. 5, Paper No. 114, 70, DOI 10.1007/s13324-022-00724-3. MR4474935

[429] O. Martio and M. Vuorinen, *Whitney cubes, p-capacity, and Minkowski content*, Exposition. Math. **5** (1987), no. 1, 17–40. MR880256

[430] P. Mattila, *Geometry of sets and measures in Euclidean spaces: Fractals and rectifiability*, Cambridge Studies in Advanced Mathematics, vol. 44, Cambridge University Press, Cambridge, 1995, DOI 10.1017/CBO9780511623813. MR1333890

[431] R. D. Mauldin and S. C. Williams, *On the Hausdorff dimension of some graphs*, Trans. Amer. Math. Soc. **298** (1986), no. 2, 793–803, DOI 10.2307/2000650. MR860394

[432] V. G. Maz'ja, *Sobolev spaces*, translated from the Russian by T. O. Shaposhnikova, Springer Series in Soviet Mathematics, Springer-Verlag, Berlin, 1985, DOI 10.1007/978-3-662-09922-3. MR817985

[433] S. Mazurkiewicz, *Sur les fonctions non-dérivables* [*"On non-differentiable functions"*], Studia Math. No. 3, **3** (1931), 92–94.

[434] C. McMullen, *The Hausdorff dimension of general Sierpiński carpets*, Nagoya Math. J. **96** (1984), 1–9, DOI 10.1017/S0027763000021085. MR771063

[2]See also the relevant parts of Mandelbrot's early books connected to fractals (especially, [**418**], [**419**] and [**420**]), where the Koch curve and the Koch snowflake play a prominent role.

[435] C. McMullen, *Iteration on Teichmüller space*, Invent. Math. **99** (1990), no. 2, 425–454, DOI 10.1007/BF01234427. MR1031909

[436] C. T. McMullen, *Frontiers in complex dynamics*, Bull. Amer. Math. Soc. (N.S.) **31** (1994), no. 2, 155–172, DOI 10.1090/S0273-0979-1994-00519-1. MR1260523

[437] C. T. McMullen, *Complex dynamics and renormalization*, Annals of Mathematics Studies, vol. 135, Princeton University Press, Princeton, NJ, 1994. MR1312365

[438] C. T. McMullen, *Kleinian groups and John domains*, Topology **37** (1998), no. 3, 485–496, DOI 10.1016/S0040-9383(97)00039-6. MR1604874

[439] K. Menger, *Allgemeine Räume und Cartesische Räume* [*"General spaces and Cartesian spaces"*], Proc. Akad. Wetensch. Amst. **29** (1926), 476–482 [German].

[440] K. Menger, *Dimensiontheorie* ["Dimension Theory"], B. G. Teubner Publishers, Leipzig, 1928 [German].

[441] G. Métivier, *Étude asymtotique des valeurs propres et de la fonction spectrale de problèmes aux limites* ["Study of the Asymptotic Distribution of the Eigenvalues and of the Spectral Function for Boundary Value Problems"], Thèse de Doctorat d'État Ès Sciences Mathématiques, Mathématiques, Université de Nice, France, 1976.

[442] G. Métivier, *Valeurs propres de problèmes aux limites elliptiques irrégulières* (French), Bull. Soc. Math. France Suppl. Mém. **51-52** (1977), 125–219. MR473578

[443] R. Miculescu and A. Mihail, *Lipscomb's space ω^A is the attractor of an infinite IFS containing affine transformations of $l^2(A)$*, Proc. Amer. Math. Soc. **136** (2008), no. 2, 587–592, DOI 10.1090/S0002-9939-07-08981-2. MR2358499

[444] J. Milnor, *Dynamics in one complex variable*, 3rd ed., Annals of Mathematics Studies, vol. 160, Princeton University Press, Princeton, NJ, 2006. MR2193309

[445] H. Minkowski, Theorie der konvexen Körper, insbesondere Begründung ihres Oberflächenbegriffs ["Theory of convex bodies"], in: *Gesammelte Abhandlungen von Hermann Minkowski* (part II, Chapter XXV), Chelsea, New York, 1967, pp. 131–229 [German]. (Originally reprinted in: Gesamm. Abh., vol. II, Leipzig, 1911.)

[446] S. Molchanov and B. Vainberg, *On spectral asymptotics for domains with fractal boundaries*, Comm. Math. Phys. **183** (1997), no. 1, 85–117, DOI 10.1007/BF02509797. MR1461952

[447] P. Montel, *Leçons sur les familles normales* ["A Course on Normal Families"], Ganthier-Villars, Paris, 1927.

[448] R. V. Moody, Model sets: A survey (survey article), in: *From Quasicrystals to More Complex Systems* (F. Axel and J.-P. Gazeau, eds.), Les Editions de Physique, Springer-Verlag, Berlin, 2000, pp. 145–166.

[449] G. Mora and J. M. Sepulcre, *Privileged regions in critical strips of non-lattice Dirichlet polynomials*, Complex Anal. Oper. Theory **7** (2013), no. 4, 1417–1426, DOI 10.1007/s11785-012-0248-4. MR3079865

[450] G. Mora, J. M. Sepulcre, and T. Vidal, *On the existence of exponential polynomials with prefixed gaps*, Bull. Lond. Math. Soc. **45** (2013), no. 6, 1148–1162, DOI 10.1112/blms/bdt043. MR3138483

[451] G. Mora, J. M. Sepulcre, and T. Vidal, *On the existence of fractal strings whose set of dimensions of fractality is not perfect*, Rev. R. Acad. Cienc. Exactas Fís. Nat. Ser. A Mat. RACSAM **109** (2015), no. 1, 11–14, DOI 10.1007/s13398-014-0164-8. MR3315698

[452] P. A. P. Moran, *Additive functions of intervals and Hausdorff measure*, Proc. Cambridge Philos. Soc. **42** (1946), 15–23, DOI 10.1017/s0305004100022684. MR14397

[453] K. Morita, *A condition for the metrizability of topological spaces and for n-dimensionality*, Sci. Rep. Tokyo Kyoiku Daigaku Sect. A **5** (1955), 33–36. MR71754

[454] P. Mörters and Y. Peres, *Brownian motion*, with an appendix by Oded Schramm and Wendelin Werner, Cambridge Series in Statistical and Probabilistic Mathematics, vol. 30, Cambridge University Press, Cambridge, 2010, DOI 10.1017/CBO9780511750489. MR2604525

[455] D. Mumford, C. Series, and D. Wright, *Indra's pearls: The vision of Felix Klein*, Cambridge University Press, New York, 2002, DOI 10.1017/CBO9781107050051.024. MR1913879

[456] J. R. Munkres, *Topology*, Second edition of [MR0464128], Prentice Hall, Inc., Upper Saddle River, NJ, 2000. MR3728284

[457] J.-i. Nagata, *On a necessary and sufficient condition of metrizability*, J. Inst. Polytech. Osaka City Univ. Ser. A **1** (1950), 93–100. MR43448

[458] J.-i. Nagata, *A remark on general imbedding theorems in dimension theory*, Proc. Japan Acad. **39** (1963), 197–199. MR164319

[459] J.-i. Nagata, *Modern dimension theory*, Revised edition, Sigma Series in Pure Mathematics, vol. 2, Heldermann Verlag, Berlin, 1983. MR715431

[460] T. Nakayama and K. Yakubo (eds.), *Fractal Concepts in Condensed Matter Physics*, Physics and Astronomy Online Library, Springer-Verlag, Berlin and Heidelberg, 2003.

[461] E. Netto, *Beitrag zur Mannigfaltigkeitslehre* (German), J. Reine Angew. Math. **86** (1879), 263–268, DOI 10.1515/crll.1879.86.263. MR1579773

[462] S. Nicolay and L. Simons, Building Cantor's bijection, `arXiv:1409.1755v1 [math.HO]` 2014.

[463] R. G. Niemeyer, E. P. J. Pearse, J. A. Rock and T. Samuel (eds.), *Horizons of Fractal Geometry and Complex Dimensions*, Contemporary Mathematics, vol. 731, Amer. Math. Soc., Providence, R. I., 2019.

[464] G. Nöbeling, *Hausdorffsche und mengen theoretische Dimension* ["*Hausdorff and set-theoretic dimension*"], Ergebnisse math. Kolloquium Wien **3** (1931), 24–25 [German].

[465] G. Nöbeling, *Über eine n-dimensionale Universalmenge im* R^{2n+1} (German), Math. Ann. **104** (1931), no. 1, 71–80, DOI 10.1007/BF01457921. MR1512650

[466] E. Noether and J. Cavaillès (eds.), *Briefwechsel Cantor–Dedekind* ["Correspondence between Cantor and Dedekind"], Hermann, Paris, 1937.

[467] T. F. Nonnenmacher, G. A. Losa, and E. R. Weibel (eds.), *Fractals in Biology and Medicine*, Birkhäuser, Basel, 1994.[3]

[468] L. Nottale, *Scale relativity and fractal space-time: A new approach to unifying relativity and quantum mechanics*, Imperial College Press, London, 2011, DOI 10.1142/p752. MR3236401

[469] L. Olsen, *Multifractal tubes: multifractal zeta-functions, multifractal Steiner formulas and explicit formulas*, Fractal geometry and dynamical systems in pure and applied mathematics. I. Fractals in pure mathematics, Contemp. Math., vol. 600, Amer. Math. Soc., Providence, RI, 2013, pp. 291–326, DOI 10.1090/conm/600/11920. MR3203407

[470] L. Olsen, *Multifractal tubes*, Further developments in fractals and related fields, Trends Math., Birkhäuser/Springer, New York, 2013, pp. 161–191, DOI 10.1007/978-0-8176-8400-6_9. MR3185037

[471] L. Olsen, *A multifractal formalism*, Adv. Math. **116** (1995), no. 1, 82–196, DOI 10.1006/aima.1995.1066. MR1361481

[472] L. Olsen, *Multifractal geometry*, Fractal geometry and stochastics, II (Greifswald/Koserow, 1998), Progr. Probab., vol. 46, Birkhäuser, Basel, 2000, pp. 3–37. MR1785619

[473] L. Olsen, *A lower bound for the symbolic multifractal spectrum of a self-similar multifractal with arbitrary overlaps*, Math. Nachr. **282** (2009), no. 10, 1461–1477, DOI 10.1002/mana.200610179. MR2571707

[474] W. Ott and J. A. Yorke, *Prevalence*, Bull. Amer. Math. Soc. (N.S.) **42** (2005), no. 3, 263–290, DOI 10.1090/S0273-0979-05-01060-8. MR2149086

[475] T. N. Palmer, *The Primacy of Doubt: From Quantum Physics to Climae Change. How the Science of Uncertainty Can Help Us Understand Our Chaotic World*, Basic Books, New York, 2022.

[476] G. Parisi and U. Frisch, Fully developed turbulence and intermittency inturbulence, and predictability in geophysical fluid dynamics and climate dynamics, in: *International School of "Enrico Fermi", Course 88* (M. Ghil, ed.), North-Holland, Amsterdam, 1985, pp. 84–88.

[477] S. J. Patterson, *An introduction to the theory of the Riemann zeta-function*, Cambridge Studies in Advanced Mathematics, vol. 14, Cambridge University Press, Cambridge, 1988, DOI 10.1017/CBO9780511623707. MR933558

[3]At least three additional volumes—with the same title as the present one, but with different sets of editors—were published in this series and by the same publisher.

[478] G. Peano, *Applicazioni geometriche del calcolo infinitesimale* ["Geometric Applications of the Infinitesimal Calculus"], Fratelli Bocca, Torino, 1887 [Italian].

[479] G. Peano, *Sur une courbe, qui remplit toute une aire plane* (French), Math. Ann. **36** (1890), no. 1, 157–160, DOI 10.1007/BF01199438. MR1510617

[480] E. P. J. Pearse, *Canonical self-affine tilings by iterated function systems*, Indiana Univ. Math. J. **56** (2007), no. 6, 3151–3169, DOI 10.1512/iumj.2007.56.3220. MR2375713

[481] E. P. J. Pearse and S. Winter, *Geometry of canonical self-similar tilings*, Rocky Mountain J. Math. **42** (2012), no. 4, 1327–1357, DOI 10.1216/RMJ-2012-42-4-1327. MR2981047

[482] J. Peinke, J. Parisi, O. E. Rössler, and R. Stoop, *Encounter with chaos: Self-organized hierarchical complexity in semiconductor experiments*, Springer-Verlag, Berlin, 1992, DOI 10.1007/978-3-642-77625-0. MR1201820

[483] H.-O. Peitgen, H. Jürgens, and D. Saupe, *Chaos and fractals: New frontiers of science*, 2nd ed., with a foreword by Mitchell J. FeigenbaumSpringer-Verlag, New York, 2004, DOI 10.1007/b97624. MR2031217

[484] H.-O. Peitgen and P. H. Richter, *The beauty of fractals: Images of complex dynamical systems*, Springer-Verlag, Berlin, 1986, DOI 10.1007/978-3-642-61717-1. MR852695

[485] Y. Peres, *The self-affine carpets of McMullen and Bedford have infinite Hausdorff measure*, Math. Proc. Cambridge Philos. Soc. **116** (1994), no. 3, 513–526, DOI 10.1017/S0305004100072789. MR1291757

[486] Y. Peres and B. Solomyak, *Absolute continuity of Bernoulli convolutions, a simple proof*, Math. Res. Lett. **3** (1996), no. 2, 231–239, DOI 10.4310/MRL.1996.v3.n2.a8. MR1386842

[487] Y. Peres and B. Solomyak, *Self-similar measures and intersections of Cantor sets*, Trans. Amer. Math. Soc. **350** (1998), no. 10, 4065–4087, DOI 10.1090/S0002-9947-98-02292-2. MR1491873

[488] J. Perrin, *Les Atomes* ["The Atoms"], new edition (of the 1913 original edition), with a foreword by Pierre-Gilles de Gennes, Flammarion, Paris, 1991.

[489] J. C. Perry, *Lipscomb's universal space is the attractor of an infinite iterated function system*, Proc. Amer. Math. Soc. **124** (1996), no. 8, 2479–2489, DOI 10.1090/S0002-9939-96-03554-X. MR1346984

[490] J. Perry and S. Lipscomb, *The generalization of Sierpiński's triangle that lives in 4-space*, Houston J. Math. **29** (2003), no. 3, 691–710. MR1998159

[491] Y. B. Pesin, *Dimension theory in dynamical systems: Contemporary views and applications*, Chicago Lectures in Mathematics, University of Chicago Press, Chicago, IL, 1997, DOI 10.7208/chicago/9780226662237.001.0001. MR1489237

[492] Y. Pesin and V. Climenhaga, *Lectures on fractal geometry and dynamical systems*, Student Mathematical Library, vol. 52, American Mathematical Society, Providence, RI, 2009, DOI 10.1090/stml/052. MR2560337

[493] I. Peterson, *The mathematical tourist: Snapshots of modern mathematics*, W. H. Freeman and Company, New York, 1988. MR1053161

[494] I. Peterson, *The jungles of randomness: A mathematical safari*, John Wiley & Sons, Inc., New York, 1998. MR1472979

[495] Ch. Pommerenke, *Boundary behaviour of conformal maps*, Grundlehren der mathematischen Wissenschaften [Fundamental Principles of Mathematical Sciences], vol. 299, Springer-Verlag, Berlin, 1992, DOI 10.1007/978-3-662-02770-7. MR1217706

[496] A. G. Postnikov, *Tauberian theory and its applications*, Proc. Steklov Inst. Math. **2** (1980), v+138. A translation of Trudy Mat. Inst. Steklov. **144** (1979). MR603991

[497] F. Przytycki and M. Urbański, *On the Hausdorff dimension of some fractal sets*, Studia Math. **93** (1989), no. 2, 155–186, DOI 10.4064/sm-93-2-155-186. MR1002918

[498] G. Radunović, *Fractal Analysis of Unbounded Sets in Euclidean Spaces and Lapidus Zeta Functions*, Ph. D. Thesis, University of Zagreb, Zagreb, Croatia, 2015.

[499] G. Radunović, *Fractality and Lapidus zeta functions at infinity*, Math. Commun. **21** (2016), no. 2, 141–162. MR3517485

[500] R. Rammal, *Spectrum of harmonic excitations on fractals*, J. Physique **45** (1984), no. 2, 191–206, DOI 10.1051/jphys:01984004502019100. MR737523

[501] R. Rammal and G. Toulouse, *Random walks on fractal structures and percolation cluster*, J. Physique Lettres **44** (1983), L13–L22.

[502] J. Rataj and S. Winter, *Characterization of Minkowski measurability in terms of surface area*, J. Math. Anal. Appl. **400** (2013), no. 1, 120–132, DOI 10.1016/j.jmaa.2012.10.059. MR3003969

[503] J. Rataj and M. Zähle, *Curvature measures of singular sets*, Springer Monographs in Mathematics, Springer, Cham, 2019, DOI 10.1007/978-3-030-18183-3. MR3932153

[504] M. Rees, *One hundred years of complex dynamics*, Proc. A. **472** (2016), no. 2185, 20150453, 16, DOI 10.1098/rspa.2015.0453. MR3464803

[505] M. Resman, *Invariance of the normalized Minkowski content with respect to the ambient space*, Chaos Solitons Fractals **57** (2013), 123–128, DOI 10.1016/j.chaos.2013.10.001. MR3128607

[506] D. Revuz and M. Yor, *Continuous martingales and Brownian motion*, 3rd ed., Grundlehren der mathematischen Wissenschaften [Fundamental Principles of Mathematical Sciences], vol. 293, Springer-Verlag, Berlin, 1999, DOI 10.1007/978-3-662-06400-9. MR1725357

[507] F. Rezakhanlou, *The packing measure of the graphs and level sets of certain continuous functions*, Math. Proc. Cambridge Philos. Soc. **104** (1988), no. 2, 347–360, DOI 10.1017/S0305004100065518. MR948919

[508] A. D. Richardson, *The Complex Dimensions of Space-Filling Curves*, ProQuest LLC, Ann Arbor, MI, 2022. Thesis (Ph.D.)–University of California, Riverside. MR4478850

[509] L. F. Richardson, *Weather prediction by numerical process*, 2nd ed. of the 1922 ed., with a foreword by Peter Lynch, Cambridge Mathematical Library, Cambridge University Press, Cambridge, 2007, DOI 10.1017/CBO9780511618291. MR2358797

[510] L. F. Richardson, *Statistics of Deadly Quarrels*, Pacific Grove, Boxwood Press, 1960.

[511] L. F. Richardson, The problem of contiguity: An appendix to "Statistics of Deadly Quarrels" ([**510**]), in: *Yearbook of the Society for the Advancement of General Systems*, Society for General Systems Research, No. 139, **6** (1961), pp. 139–181. (Reprinted in [**512**], vol. 2, pp. 579–627.)

[512] L. F. Richardson, *The Collected Papers of Lewis Fry Richardson*, vol. 1: *Meteorology and Numerical Analysis*; vol. 2: *Quantitative Psychology and Studies of Conflicts* (O. M. Ashford *et al.*, eds.), Cambridge Univ. Press, Cambridge, 1993.

[513] B. Riemann, Über die Dartsellburkeit einer Funktion durch eine trigonometrische Reihe ["On the Representability of a Function by a Trigonometric Series"], Habilitationsschrift, 1854 [German]. (Published in 1868 in [**514**].)

[514] B. Riemann, Über die Dartsellburkeit einer Funktion durch eine trigonometrische Reihe ["On the Representability of a Function by a Trigonometric Series"], *Abh. der. Ges. der Wiss. zu Gött.* **13** (1868), 87–132 [German]. (See also: *G. F. B. Riemann*: *Werke* (Collected Works), pp. 227–264.)

[515] B. Riemann, Ueber die Anzahl der Primzahlen unter einer gegebenen Grösse ["On the number of primes below a given magnitude"], *Monatsb. der Berliner Akad.* 1858/60, pp. 671–680 [German]. (English translation in [**156**, Appendix, pp. 229–305].)

[516] J. A. Rock, *Zeta functions, complex dimensions of fractal strings and multifractal analysis of mass distributions*, ProQuest LLC, Ann Arbor, MI, 2007. Thesis (Ph.D.)–University of California, Riverside. MR2710390

[517] R. T. Rockafellar, *Convex analysis*, reprint of the 1970 original, Princeton Paperbacks, Princeton Landmarks in Mathematics, Princeton University Press, Princeton, NJ, 1997. MR1451876

[518] C. A. Rogers, *Hausdorff measures*, reprint of the 1970 original, with a foreword by K. J. Falconer, Cambridge Mathematical Library, Cambridge University Press, Cambridge, 1998. MR1692618

[519] O. E. Rössler, *An equation for continuous chaos*, Phys. Lett. Ser. A. **57** (1976), 397–398.

[520] P. Rotondo, L. G. Molinari, P. Ratti, and M. Gherardi, *Devil's staircase phase diagram of the fractional quantum Hall effect in the thin-torus limit*, Phys. Rev. Lett. No. 25, **116** (2016). (DOI:10.1103/Phys. RevLett. 116.256803.)

[521] H. L. Royden, *Real analysis*, 3rd ed., Macmillan Publishing Company, New York, 1988. MR1013117

[522] W. Rudin, *Principles of mathematical analysis*, 3rd ed., International Series in Pure and Applied Mathematics, McGraw-Hill Book Co., New York-Auckland-Düsseldorf, 1976. MR385023

[523] W. Rudin, *Real and complex analysis*, 3rd ed., McGraw-Hill Book Co., New York, 1987. MR924157

[524] W. Rudin, *Functional analysis*, 2nd ed., International Series in Pure and Applied Mathematics, McGraw-Hill, Inc., New York, 1991. MR1157815

[525] D. Ruelle, Strange attractors, *Math. Intelligencer* **2** (1980), 126–137. [Reprinted in: *Universality in Chaos* (P. Cvitanovič, ed.), second edition (of the 1989 edition), Taylor and Francis, New York, 2017, pp. 37–48.]

[526] D. Ruelle, *Chaotic evolution and strange attractors: The statistical analysis of time series for deterministic nonlinear systems*, notes prepared and with a foreword by Stefano Isola, Lezioni Lincee. [Lincei Lectures], Cambridge University Press, Cambridge, 1989, DOI 10.1017/CBO9780511608773. MR1017644

[527] D. Ruelle and F. Takens, *On the nature of turbulence*, Comm. Math. Phys. **20** (1971), 167–192. MR284067

[528] C. Sabot, *Integrated density of states of self-similar Sturm-Liouville operators and holomorphic dynamics in higher dimension* (English, with English and French summaries), Ann. Inst. H. Poincaré Probab. Statist. **37** (2001), no. 3, 275–311, DOI 10.1016/S0246-0203(00)01068-2. MR1831985

[529] C. Sabot, *Spectral properties of self-similar lattices and iteration of rational maps* (English, with English and French summaries), Mém. Soc. Math. Fr. (N.S.) **92** (2003), vi+104, DOI 10.24033/msmf.405. MR1976877

[530] C. Sabot, *Spectral analysis of a self-similar Sturm-Liouville operator*, Indiana Univ. Math. J. **54** (2005), no. 3, 645–668, DOI 10.1512/iumj.2005.54.2490. MR2151229

[531] H. Sagan, *Approximating Polygons for Lebesgue's and Schoenberg's Space Filling Curves*, Amer. Math. Monthly **93** (1986), no. 5, 361–368, DOI 10.2307/2323593. MR1540861

[532] H. Sagan, *Space-filling curves*, Universitext, Springer-Verlag, New York, 1994, DOI 10.1007/978-1-4612-0871-6. MR1299533

[533] R. Salem and A. Zygmund, *Lacunary power series and Peano curves*, Duke Math. J. **12** (1945), 569–578. MR15154

[534] B. Sapoval, *Universalité et fractales* ["Universality and Fractals"], Champ Sciences, Flammarion, Paris, 2001 [French].

[535] C. Sargent, Box Counting Zeta Functions of Self-Similar Sets, *Master's Thesis*, California State Polytechnic University, Pomona, 2014.

[536] A. Schief, *Separation properties for self-similar sets*, Proc. Amer. Math. Soc. **122** (1994), no. 1, 111–115, DOI 10.2307/2160849. MR1191872

[537] D. Schleicher (ed.), *Complex dynamics: Families and friends*, A K Peters, Ltd., Wellesley, MA, 2009, DOI 10.1201/b10617. MR1500163

[538] R. Schmalz, *Out of the mouths of mathematicians: A quotation book for philomaths*, MAA Spectrum, Mathematical Association of America, Washington, DC, 1993. MR1254437

[539] W. M. Schmidt, *Diophantine approximation*, Lecture Notes in Mathematics, vol. 785, Springer, Berlin, 1980. MR568710

[540] R. Schneider, *Convex Bodies: The Brunn–Minkowski Theory*, Encyclopedia of Mathematics and its Applications, vol. 44, Cambridge Univ. Press, Cambridge, 2003. (Reprinted from the 1993 edition.)

[541] I. J. Schoenberg, *On the Peano curve of Lebesgue*, Bull. Amer. Math. Soc. **44** (1938), no. 8, 519, DOI 10.1090/S0002-9904-1938-06792-4. MR1563786

[542] M. Schroeder, *Fractals, chaos, power laws: Minutes from an infinite paradise*, W. H. Freeman and Company, New York, 1991. MR1098021

[543] S. Semmes, *Some novel types of fractal geometry*, Oxford Mathematical Monographs, The Clarendon Press, Oxford University Press, New York, 2001. MR1815356

[544] M. Senechal, *Quasicrystals and geometry*, Cambridge University Press, Cambridge, 1995. MR1340198

[545] J.-P. Serre, *A course in arithmetic*, translated from the French, Graduate Texts in Mathematics, No. 7, Springer-Verlag, New York-Heidelberg, 1973. MR344216

[546] W. Shen, *Hausdorff dimension of the graphs of the classical Weierstrass functions*, Math. Z. **289** (2018), no. 1-2, 223–266, DOI 10.1007/s00209-017-1949-1. MR3803788

[547] M. Shishikura, *The Hausdorff dimension of the boundary of the Mandelbrot set and Julia sets*, Ann. of Math. (2) **147** (1998), no. 2, 225–267, DOI 10.2307/121009. MR1626737

[548] C. L. Siegel, *Iteration of analytic functions*, Ann. of Math. (2) **43** (1942), 607–612, DOI 10.2307/1968952. MR7044

[549] C. L. Siegel and J. K. Moser, *Lectures on celestial mechanics*, translated from the German by C. I. Kalme, reprint of the 1971 translation, Classics in Mathematics, Springer-Verlag, Berlin, 1995. MR1345153

[550] W. Sierpinski, *Sur une nouvelle courbe continue qui remplit toute une aire plane* [*"On a new type of continuous curve which fills up a planar area"*], Bull. Int. Acad. Sci. Cracovie Ser. A (1912), 462–478 [French].

[551] W. Sierpinski, *Sur une courbe dont tout point est un point de ramification* [*"On a curve each point of which is a ramification point"*], Comptes Rerdus Acad. Sci. Paris **160** (1915), 302–305 [French].

[552] W. Sierpinski, *On a Cantorian curve which contains a bijective and continuous image of any given curve*, Mat. Sbornik **30** (1916), 267–287 [Russian].

[553] W. Sierpinski, *Un exemple élémentaire d'une fonction croissante qui a presque partout une dérivée nulle* [*"An elementary example of a nondecreasing function which admits a zero derivative almost everywhere"*], Giornale Mat. Battaglini (3) **6** (1916), 314–334.

[554] W. Sierpinski, *On a curve every point of which is a ramification point*, Prace Mat. Fiz **27** (1917), 77–86 [Polish].

[555] W. Sierpinski, *Oeuvres choisies* ["Collected (Selected) Works"], vols. 1 & 2, Pańswowe Wydwnicto Naukowe, Warsaw, 1975 [French].

[556] B. Simon, *Basic complex analysis*, A Comprehensive Course in Analysis, Part 2A, American Mathematical Society, Providence, RI, 2015, DOI 10.1090/simon/002.1. MR3443339

[557] L. Simon, *Lectures on geometric measure theory*, Proceedings of the Centre for Mathematical Analysis, Australian National University, vol. 3, Australian National University, Centre for Mathematical Analysis, Canberra, 1983. MR756417

[558] S. Smirnov, *Critical percolation in the plane: conformal invariance, Cardy's formula, scaling limits* (English, with English and French summaries), C. R. Acad. Sci. Paris Sér. I Math. **333** (2001), no. 3, 239–244, DOI 10.1016/S0764-4442(01)01991-7. MR1851632

[559] S. Smirnov and W. Werner, *Critical exponents for two-dimensional percolation*, Math. Res. Lett. **8** (2001), no. 5-6, 729–744, DOI 10.4310/MRL.2001.v8.n6.a4. MR1879816

[560] H. J. S. Smith, *On the Integration of Discontinuous Functions*, Proc. Lond. Math. Soc. **6** (1874/75), 140–153, DOI 10.1112/plms/s1-6.1.140. MR1576736

[561] S. M. Srivastava, *A Course on Borel Sets*, second edition, Springer, New York, 2010.

[562] L. L. Stachó, *On the volume function of parallel sets*, Acta Sci. Math. (Szeged) **38** (1976), no. 3-4, 365–374. MR442202

[563] M. A. Štan′ko, *Solution of Menger's problem in the class of compacta* (Russian), Dokl. Akad. Nauk SSSR **201** (1971), 1299–1302. MR293607

[564] E. M. Stein, *Singular integrals and differentiability properties of functions*, Princeton Mathematical Series, No. 30, Princeton University Press, Princeton, NJ, 1970. MR290095

[565] E. M. Stein and R. Shakarchi, *Fourier analysis: An introduction*, Princeton Lectures in Analysis, vol. 1, Princeton University Press, Princeton, NJ, 2003. MR1970295

[566] J. Steiner, Über parallele Flächen, *Monatsb. preuss. Akad. Wiss.*, Berlin, 1840, pp. 114–118 [German]. (Reprinted in: *Gesamm. Werke*, vol. II, pp. 173–176.)

[567] S. Sternberg, *Dynamical Systems*, revised edition, Dover Publications, Inc., Mineola, 2013.

[568] I. Stewart, *Does God play dice? The mathematics of chaos*, Basil Blackwell, Oxford, 1989. MR997093

[569] I. Stewart, *Four encounters with Sierpiński's gasket*, Math. Intelligencer **17** (1995), no. 1, 52–64, DOI 10.1007/BF03024718. MR1319993

[570] I. Stewart and A. C. Clarke (eds.), *The Colours of Infinity: The Beauty and Power of Fractals*, Clear Books, Bath, 2004.

[571] R. S. Strichartz, *The Way of Analysis*, Jones and Bartlet Publishers, Boston, 1995.

[572] R. S. Strichartz, *Fourier asymptotics of fractal measures*, J. Funct. Anal. **89** (1990), no. 1, 154–187, DOI 10.1016/0022-1236(90)90009-A. MR1040961

[573] R. S. Strichartz, Self-similar measures and their Fourier transforms, I, *Indiana Univ. Math. J.* **39** (1990), 797–817; II, *Trans. Amer. Math. Soc.* **336** (1993), 335–361; III, *Indiana Univ. Math. J.* **42** (1993), 367–411.

[574] R. S. Strichartz, *Differential equations on fractals: A tutorial*, Princeton University Press, Princeton, NJ, 2006. MR2246975

[575] R. S. Strichartz, A. Taylor, and T. Zhang, *Densities of self-similar measures on the line*, Experiment. Math. **4** (1995), no. 2, 101–128. MR1377413

[576] D. Sullivan, *The density at infinity of a discrete group of hyperbolic motions*, Inst. Hautes Études Sci. Publ. Math. **50** (1979), 171–202. MR556586

[577] D. Sullivan, *Disjoint spheres, approximation by imaginary quadratic numbers, and the logarithm law for geodesics*, Acta Math. **149** (1982), no. 3-4, 215–237, DOI 10.1007/BF02392354. MR688349

[578] D. Sullivan, *Entropy, Hausdorff measures old and new, and limit sets of geometrically finite Kleinian groups*, Acta Math. **153** (1984), no. 3-4, 259–277, DOI 10.1007/BF02392379. MR766265

[579] D. Sullivan, *Quasiconformal homeomorphisms and dynamics. I. Solution of the Fatou-Julia problem on wandering domains*, Ann. of Math. (2) **122** (1985), no. 3, 401–418, DOI 10.2307/1971308. MR819553

[580] D. Sullivan, *Bounds, quadratic differentials, and renormalization conjectures*, American Mathematical Society centennial publications, Vol. II (Providence, RI, 1988), Amer. Math. Soc., Providence, RI, 1992, pp. 417–466. MR1184622

[581] E. Szpilrajn (Marczewski), La dimension et la mesure ["Dimension and measure"], *Fund. Math.* **28** (1937), 81–89.

[582] T. Takagi, A simple example of a continuous function without derivatives, *Phys.-Math. Soc. Japan* **1** (1903), 176–177. (See also: [**583**, pp. 4–5.].)

[583] T. Takagi, *The Collected Papers of Teiji Takagi* (S. Kuroda, ed.), Iwanami, 1973.

[584] S. J. Taylor, *The exact Hausdorff measure of the sample path for planar Brownian motion*, Proc. Cambridge Philos. Soc. **60** (1964), 253–258, DOI 10.1017/s0305004100037713. MR164380

[585] S. J. Taylor and C. Tricot, *Packing measure, and its evaluation for a Brownian path*, Trans. Amer. Math. Soc. **288** (1985), no. 2, 679–699, DOI 10.2307/1999958. MR776398

[586] A. Teplyaev, *Spectral zeta function of symmetric fractals*, Fractal geometry and stochastics III, Progr. Probab., vol. 57, Birkhäuser, Basel, 2004, pp. 245–262. MR2087144

[587] A. Teplyaev, *Spectral zeta functions of fractals and the complex dynamics of polynomials*, Trans. Amer. Math. Soc. **359** (2007), no. 9, 4339–4358, DOI 10.1090/S0002-9947-07-04150-5. MR2309188

[588] E. C. Titchmarsh, *The theory of the Riemann zeta-function*, 2nd ed., edited and with a preface by D. R. Heath-Brown, The Clarendon Press, Oxford University Press, New York, 1986. MR882550

[589] G. Tomkowicz and S. Wagon, *The Banach-Tarski paradox*, 2nd ed., with a foreword by Jan Mycielski, Encyclopedia of Mathematics and its Applications, vol. 163, Cambridge University Press, New York, 2016. MR3616119

[590] C. Tricot Jr., *Two definitions of fractional dimension*, Math. Proc. Cambridge Philos. Soc. **91** (1982), no. 1, 57–74, DOI 10.1017/S0305004100059119. MR633256

[591] C. Tricot, *Mesures et Dimensions* ["Measures and Dimensions"], Thèse de Doctorat d'État Ès Sciences Mathématiques, Université Paris-Sud, Orsay, France, 1983 [French].

[592] C. Tricot, *Dimensions aux bords d'un ouvert* (French, with English summary), Ann. Sci. Math. Québec **11** (1987), no. 1, 205–235. MR912170

[593] C. Tricot, *Curves and fractal dimension*, with a foreword by Michel Mendès France, translated from the 1993 French original, Springer-Verlag, New York, 1995, DOI 10.1007/978-1-4612-4170-6. MR1302173 All references to this book are to the original French edition.

[594] C. Tricot, *General Hausdorff functions, and the notion of one-sided measure and dimension*, Ark. Mat. **48** (2010), no. 1, 149–176, DOI 10.1007/s11512-008-0087-8. MR2594591

[595] M. Tsujii, *Fat solenoidal attractors*, Nonlinearity **14** (2001), no. 5, 1011–1027, DOI 10.1088/0951-7715/14/5/306. MR1862809

[596] J. Uher, Symmetric continuity implies continuity, *Real Anal. Exchange* **13** (1987/88), 36–38.

[597] P. Urysohn, *Zum Metrisationsproblem* (German), Math. Ann. **94** (1925), no. 1, 309–315, DOI 10.1007/BF01208661. MR1512260

[598] M. van den Berg and P. B. Gilkey, *A comparison estimate for the heat equation with an application to the heat content of the S-adic von Koch snowflake*, Bull. London Math. Soc. **30** (1998), no. 4, 404–412, DOI 10.1112/S0024609398004469. MR1620833

[599] B. L. van der Waerden, *Ein einfaches Beispiel einer nicht-differenzierbaren stetigen Funktion* (German), Math. Z. **32** (1930), no. 1, 474–475, DOI 10.1007/BF01194647. MR1545179

[600] M. van Frankenhuijsen, *Arithmetic progressions of zeros of the Riemann zeta function*, J. Number Theory **115** (2005), no. 2, 360–370, DOI 10.1016/j.jnt.2005.01.002. MR2180508

[601] M. van Frankenhuijsen, *Riemann zeros in arithmetic progression*, Fractal geometry and dynamical systems in pure and applied mathematics. I. Fractals in pure mathematics, Contemp. Math., vol. 600, Amer. Math. Soc., Providence, RI, 2013, pp. 365–380, DOI 10.1090/conm/600/11921. MR3203410

[602] G. Vitali, Sulle funzioni integrali ["On the integrable functions"], *Atti. Acad. Sci. Torino Cl. Sci. Fis. Mat. Naturi* **40** (1905), 1021–1034 [Italian].

[603] H. von Koch, Sur une courbe continue sans tangente, obtenue par une construction géométrique élémentaire ["On a continuous curve without tangent, obtained by an elementary geometric construction"], *Arkiv für Matematik, Astrononi och* Fysik, **1** (1904), 681–704 [French].

[604] M. Waldschmidt, P. Moussa, J. M. Luck, and C. Itzykson (eds.), *From number theory to physics*, papers from the Meeting on Number Theory and Physics held in Les Houches, March 7–16, 1989, Springer-Verlag, Berlin, 1992, DOI 10.1007/978-3-662-02838-4. MR1221099

[605] L. N. Vasershtein, *Markov processes over denumerable products of spaces describing large system of automata* (Russian), Problemy Peredači Informacii **5** (1969), no. 3, 64–72, DOI 10.1016/s0016-0032(33)90010-1; English transl., Problems Inform. Transmission **5** (1969), no. 3, 47–52. MR314115

[606] S. R. Watson, *Fractal Zeta Functions: To Ahlfors Spaces and Beyond*, ProQuest LLC, Ann Arbor, MI, 2017. Thesis (Ph.D.)–University of California, Riverside. MR3755432

[607] K. Weierstrass, *Monatsberichte der Königlich Preussischen Akademie der Wissenschaften zu Berlin* ["Monthly Reports of the Royal Prussian Academy of Science"], 1872, p. 560 [German].[4]

[608] K. Weierstrass, Briefe von K. Weierstrass an Paul du Bois-Reymond ["Letter from K. Weierstrass to Paul du Bois-Reymond"], 1873 [German].

[609] K. Weierstrass, Über continuirliche Funktionen eines reellen Arguments die für keinen Werth des letzeren einen bestimmten Differentialquotienen besitzen ["On continuous functions of a real argument which possess a definite derivative for no value of the argument"], in: *Königlich Preussischen Akademie der Wissenschaften* ["Monthly Reports of the Royal Prussian Academy of Science"], in: *Mathematische Werke von Karl Weierstrass* ["Mathematical Works of Karl Weierstrass"], vol. II, Mayer & Mueller, 1895, Berlin, pp. 71-74 [German]. (See also: *Journal für die reine und angewandte Mathematik* ("*Crelles Journal*") **79** (1875), 29–31 [German].)

[610] K. Weierstrass, *Abhandlungen aus der Functionenlehre* ["Treatise on the Theory of Functions"], Julius Springer, Berlin, 1886 [German].

[611] W. Werner, *Random planar curves and Schramm-Loewner evolutions*, Lectures on probability theory and statistics, Lecture Notes in Math., vol. 1840, Springer, Berlin, 2004, pp. 107–195, DOI 10.1007/978-3-540-39982-7_2. MR2079672

[4]On page 560 of this reference, there is a short statement (without an abstract or a full paper) according to which, on July 18th of 1872, "Mr. Weierstrass read [his paper] on continuous functions without definite (i.e., well-defined and finite) derivatives".

[612] H. Weyl, *On the Volume of Tubes*, Amer. J. Math. **61** (1939), no. 2, 461–472, DOI 10.2307/2371513. MR1507388

[613] H. Weyl, *Hermann Weyl: Gesammelte Abhandlungen* ["Hermann Weyl: Collected Works"], vols. I and III, Springer-Verlag, Berlin and New York, 1968 [German].

[614] G. T. Whyburn, *Topological characterization of the Sierpiński curve*, Fund. Math. **45** (1958), 320–324, DOI 10.4064/fm-45-1-320-324. MR99638

[615] N. Wiener, *Differential-space*, J. Math. and Phys. **2** (1923), 131–174. MR4250033

[616] N. Wiener, *The Average value of a Functional*, Proc. London Math. Soc. (2) **22** (1924), 454–467, DOI 10.1112/plms/s2-22.1.454. MR1575720

[617] N. Wiener, The quadratic variation of a function and its Fourier coefficients, *J. Mass. Inst. Tech.* **3** (1924), 73–94.

[618] N. Wiener, *Collected works. Vol. I*, Mathematical philosophy and foundations; potential theory; Brownian movement, Wiener integrals, ergodic and chaos theories, turbulence and statistical mechanics, with commentaries, edited by P. Masani, Mathematicians of Our Time, vol. 10, MIT Press, Cambridge, Mass.-London, 1976. MR532698

[619] S. Winter, Curvature measures and fractals, *Dissertationes Math.* (*Rozprawy Mat.*) **453** (2008), 1–66.

[620] S. Winter and M. Zähle, *Fractal curvature measures of self-similar sets*, Adv. Geom. **13** (2013), no. 2, 229–244, DOI 10.1515/advgeom-2012-0026. MR3038703

[621] Y. Xiao, *Random fractals and Markov processes*, Fractal geometry and applications: a jubilee of Benoît Mandelbrot, Part 2, Proc. Sympos. Pure Math., vol. 72, Amer. Math. Soc., Providence, RI, 2004, pp. 261–338, DOI 10.1090/pspum/072.2/2112126. MR2112126

[622] W. Yahao and W. Min, *Dimensions of modified Besicovitch subsets of Moran fractal*, Chaos Solitons Fractals **42** (2009), no. 5, 2779–2785, DOI 10.1016/j.chaos.2009.03.188. MR2559996

[623] J.-C. Yoccoz, *Théorème de Siegel, nombres de Bruno et polynômes quadratiques: Petits diviseurs en dimension* 1 (French), Astérisque **231** (1995), 3–88. MR1367353

[624] M. Yor, *Marc Yor: La passion du mouvement brownien* ["Marc Yor: The Passion of Brownian Motion"], Gazette des mathématiciens et MATAPLI, Numéro spécial 2015, SMAI and Société Mathématique de France (SMF), Paris, 2015. (Coordinator of the special issue: B. Helffer.)

[625] M. Zähle, *Integral and current representation of Federer's curvature measures*, Arch. Math. (Basel) **46** (1986), no. 6, 557–567, DOI 10.1007/BF01195026. MR849863

[626] M. Zähle, *Curvatures and currents for unions of sets with positive reach*, Geom. Dedicata **23** (1987), no. 2, 155–171, DOI 10.1007/BF00181273. MR892398

[627] M. Zähle, *Approximation and characterization of generalised Lipschitz-Killing curvatures*, Ann. Global Anal. Geom. **8** (1990), no. 3, 249–260, DOI 10.1007/BF00127938. MR1089237

[628] M. Zähle, *Lipschitz-Killing curvatures of self-similar random fractals*, Trans. Amer. Math. Soc. **363** (2011), no. 5, 2663–2684, DOI 10.1090/S0002-9947-2010-05198-0. MR2763731

[629] M. Zähle, *Curvature measures of fractal sets*, Fractal geometry and dynamical systems in pure and applied mathematics. I. Fractals in pure mathematics, Contemp. Math., vol. 600, Amer. Math. Soc., Providence, RI, 2013, pp. 381–399, DOI 10.1090/conm/600/11953. MR3203411

[630] E. Zermelo (ed.), *Gezammelte Abhandlungen Mathematischen und Philosophischen Inhalts* ["Mathematical and Philosphophical Collected Works"], Springer, New York, 1980 [German].

[631] D. Žubrinić, *Minkowski content and singular integrals*, Chaos Solitons Fractals **17** (2003), no. 1, 169–177, DOI 10.1016/S0960-0779(02)00441-1. MR1960771

[632] D. Žubrinić, *Analysis of Minkowski contents of fractal sets and applications*, Real Anal. Exchange **31** (2005/06), no. 2, 315–354. MR2265777

[633] A. Zygmund, *Trigonometric series. 2nd ed. Vols. I, II*, Cambridge University Press, New York, 1959. MR107776

Index of Symbols

[1]The boldface page numbers correspond to the definitions of the symbols.

Author Index

Subject Index

Selected Published Titles in This Series

247 **Michel L. Lapidus and Goran Radunović,** An Invitation to Fractal Geometry, 2024
245 **Bennett Chow and Yutze Chow,** Lectures on Differential Geometry, 2024
244 **John M. Lee,** Introduction to Complex Manifolds, 2024
243 **J. M. Landsberg,** Quantum Computation and Quantum Information, 2024
242 **Jayadev S. Athreya and Howard Masur,** Translation Surfaces, 2024
241 **Thorsten Theobald,** Real Algebraic Geometry and Optimization, 2024
240 **Nam Q. Le,** Analysis of Monge–Ampère Equations, 2024
239 **K. Cieliebak, Y. Eliashberg, and N. Mishachev,** Introduction to the h-Principle, Second Edition, 2024
238 **Julio González-Díaz, Ignacio García-Jurado, and M. Gloria Fiestras-Janeiro,** An Introductory Course on Mathematical Game Theory and Applications, Second Edition, 2023
237 **Michael Levitin, Dan Mangoubi, and Iosif Polterovich,** Topics in Spectral Geometry, 2023
236 **Stephanie Alexander, Vitali Kapovitch, and Anton Petrunin,** Alexandrov Geometry, 2024
235 **Bennett Chow,** Ricci Solitons in Low Dimensions, 2023
234 **Andrea Ferretti,** Homological Methods in Commutative Algebra, 2023
233 **Andrea Ferretti,** Commutative Algebra, 2023
232 **Harry Dym,** Linear Algebra in Action, Third Edition, 2023
231 **Luís Barreira and Yakov Pesin,** Introduction to Smooth Ergodic Theory, Second Edition, 2023
230 **Barbara Kaltenbacher and William Rundell,** Inverse Problems for Fractional Partial Differential Equations, 2023
229 **Giovanni Leoni,** A First Course in Fractional Sobolev Spaces, 2023
228 **Henk Bruin,** Topological and Ergodic Theory of Symbolic Dynamics, 2022
227 **William M. Goldman,** Geometric Structures on Manifolds, 2022
226 **Milivoje Lukić,** A First Course in Spectral Theory, 2022
225 **Jacob Bedrossian and Vlad Vicol,** The Mathematical Analysis of the Incompressible Euler and Navier-Stokes Equations, 2022
224 **Ben Krause,** Discrete Analogues in Harmonic Analysis, 2022
223 **Volodymyr Nekrashevych,** Groups and Topological Dynamics, 2022
222 **Michael Artin,** Algebraic Geometry, 2022
221 **David Damanik and Jake Fillman,** One-Dimensional Ergodic Schrödinger Operators, 2022
220 **Isaac Goldbring,** Ultrafilters Throughout Mathematics, 2022
219 **Michael Joswig,** Essentials of Tropical Combinatorics, 2021
218 **Riccardo Benedetti,** Lectures on Differential Topology, 2021
217 **Marius Crainic, Rui Loja Fernandes, and Ioan Mărcuţ,** Lectures on Poisson Geometry, 2021
216 **Brian Osserman,** A Concise Introduction to Algebraic Varieties, 2021
215 **Tai-Ping Liu,** Shock Waves, 2021
214 **Ioannis Karatzas and Constantinos Kardaras,** Portfolio Theory and Arbitrage, 2021
213 **Hung Vinh Tran,** Hamilton–Jacobi Equations, 2021
212 **Marcelo Viana and José M. Espinar,** Differential Equations, 2021
211 **Mateusz Michałek and Bernd Sturmfels,** Invitation to Nonlinear Algebra, 2021

For a complete list of titles in this series, visit the
AMS Bookstore at **www.ams.org/bookstore/gsmseries/**.